# MODERN MATHEMATICS
## FOR ELEMENTARY EDUCATORS

**TWELFTH EDITION**

**Ruric E. Wheeler** ■ **Ed R. Wheeler**

KENDALL/HUNT PUBLISHING COMPANY
4050 Westmark Drive  Dubuque, Iowa 52002

*Dedicated to*

*Joyce and Claire,*
*our beloved wives and best friends*

**Book Team**

Chairman and Chief Executive Officer   Mark C. Falb
Senior Vice President, College Division   Thomas W. Gantz
Director of National Book Program   Paul B. Carty
Editorial Development Manager   Georgia Botsford
Developmental Editor   Angela Willenbring
Vice President, Production and Manufacturing   Alfred C. Grisanti
Assistant Vice President, Production Services   Christine E. O'Brien
Project Coordinator   Angela Shaffer
Permissions Editor   Renae Heacock
Designer   Jenifer Chapman
Managing Editor, College Field   David Tart
Associate Managing Editor, College Field   Greg DeRosa
Acquisitions Editor, College Field   Adam Cooper

# CONTENTS

* These sections may be omitted without loss of continuity or comprehension.

\* These sections may be omitted without loss of continuity or comprehension.

* These sections may be omitted without loss of continuity or comprehension.

# PREFACE

How can authors revise a text that has been used successfully through eleven editions? The answer is very simple: the authors consult with the faculty and students who have used the most recent edition and ask, "How can this text be improved?" Upon the advice of our users, both faculty and students, we are pleased to announce we are giving added attention to the *Principles and Standards of School Mathematics,* published by the National Council of Teachers of Mathematics. We reference the 10 standards of *Principles and Standards* by listing at the beginning of each section of the book the three or four standards that will be illustrated in that section. More importantly, we adopt the spirit of the *Principles and Standards* in our presentation of material and in the exercises with which we challenge students.

In this edition of *Modern Mathematics for Elementary Educators,* we have included greater use of the calculator. The calculator has always been acknowledged as a critical tool in our text, but in this edition we have integrated careful explanations of calculator use into the text to enhance learning opportunities for our students. Fraction arithmetic on a calculator provides a valuable teaching aid for the classroom teacher. Thus, for maximum effectiveness, we recommend calculators that can do fraction arithmetic and have special menus for such topics as probability and statistics. Our generic explanations should point the way for use of all calculators, but should be especially helpful for the inexpensive Texas Instrument TI 34 II.

You will note many new ideas in this revision. For example, in Chapter 1 we have given more emphasis to one of the most powerful problem solving strategies, *create an algebraic model.* Changes have been made in Chapter 2 to improve clarity and pacing. The revision in Chapter 4 improves the presentation of material connecting arithmetic operations with the applied problems for which they provide a model. In addition to other changes throughout the book, the presentations in the last two chapters more closely reflect the material that is being included in syllabi.

The following statement is found on page 15 of the *Principles and Standards.* **Mathematical thinking and reasoning skills, including making conjectures and developing sound deductive arguments, are important because they serve as a basis for developing new insights and promoting further study.** One way this is accomplished is by using exploratory exercises designed to develop critical and mathematical thinking capability. In our book, we call these **PCR Excursions** to emphasize that they are built around problem solving—**P**, communicating mathematics—**C**, and reasoning—**R**. Each Excursion is built around a set of carefully constructed questions, designed to lead the student to discover ideas and patterns of thought, and to use inductive and deductive reasoning. In the optimal use of these Excursions, the teacher functions as a coach or tutor, supplementing the questions in the Excursions with questions of his or her own. We have classroom tested these Excursions three ways: in a cooperative learning format; as laboratory assignments; and as the stimulus for classroom discussions. We have been pleasantly surprised to find them effective in engaging students and improving mathematical thinking in all three settings.

What must be characteristic of a textbook that has been used by over 1000 institutions to teach over one million students? It must be a text from which it is enjoyable to learn and from which it is enjoyable to teach. In this edition, we retained and improved those features that distinguish this book from others in the field. Unique features of *Modern Mathematics for Elementary Educators* include:

- Topics presented in a flexible format that provides the teacher with maximum freedom in designing his or her own course. Several suggested course outlines are found in the *Instructor's Manual.*

- Emphasis on problem solving (as evidenced by indexing problem solving strategies by page number at the beginning of each chapter).

- Historical sketches to create student interest in the mathematics topic being studied.

- Careful explanation of the need to extend number systems from whole numbers to real numbers.

- A calculator explanation of the arithmetic of fractions.

- Simulation using a table of random digits to illustrate a valuable tool in the study of probability.

- Content and exercises to help the beginning college student learn to think critically and mathematically. Research indicates that students remember few of the facts that we so elegantly teach them, but that if they learn to reason and to solve problems, they will have acquired tools that will serve them a lifetime.

## ANCILLARY MATERIALS

In this edition, the *Instructor's Manual* has been upgraded to be of maximum benefit to teachers. In addition to containing the solutions to all exercises in the book, this manual contains outlines of possible courses, solutions for the Just for Fun problems, instructions for the use of PCR Excursions (including answers), outlines for additional PCR Excursions, suggestions relative to when to emphasize standards, as well as many other valuable teaching ideas.

A test bank is also available.

A 350-page ancillary text, *Mathematics Activities for Teaching* by Jane Barnard and Ed R. Wheeler continues to be available to supplement this text. It contains a rich variety of activities (keyed to the chapters in *Modern Mathematics*) for teaching mathematics concepts to students, kindergarten through middle grades. In addition, it includes "Teaching Notes" that discuss the significance of the activities in the growth and preparation of the young student.

# ACKNOWLEDGMENTS

We appreciate those who made suggestions on the revision for the twelfth edition: Darla Ottaman, Elizabethtown Community College; C.B. Carey, Spring Arbor University; Rhonda Dillow, Shawnee College; Hassan Saffari, Prestonsburg Community College; Steve Shaft, Sauk Valley Community College; Lance Skidmore, Mount St. Mary's College; Michele Starkey, Mount St. Mary's College; C. Varnais, University of Wyoming; and Ralph Willis, Western Carolina University.

One of the nicest things about publishing a successful book is the many letters, e-mail messages, and suggestions received from people throughout the United States and Canada. We express our appreciation to the following people (especially the reviewers) who gave suggestions on previous editions: Marilee Adams, Reza Aklaghi, Bernadette Antkoviak, Thomas Arbutiski, Thomas Autry, D. Becker, Evelyn Bell, Michael Berry, Richard W. Billstein, Wayne Bishop, Barbara Boe, James Boone, Nancy Budner, Catherine Cant, Judy Cantey, Boyd L. Cardon, Charles Carey, Imogene Carrol, Barbara W. Carson, V.C. Cateforis, Gregory Chamblee, Mark Christie, James Gerry Church, Darrell Coates, Ted Coe, Steve Cohen, Mary Louise Collette, Robert Colling, Linda Collins, Thomas Collins, Sister Mary Colman, M. Cox, Keith Craswell, Arnold J. Daerksen, Arthur Daniel, Julie Dewan, R.L. Dieterback, Robert Dolan, Arlene Dowshen, Irving Drooyan, Arnold Durra, Barbara H. Dunn, Fred Ettline, Richard Evans, Daniel J. Ewy, Pierre Fabinski, Hilton G. Falahee, Gayle Farmer, W. Ferry, Mila Fgratti, Iris B. Fitter, David Foreman, John R. Formsma, Grace Foster, Betty Freemal, Richard Fritsche, Elizabeth Frye, Robert Fullerton, Charles Funkhouser, Barbara Gale, David Gardner, Stan Garrelson, Thomas Gibney, Judith Gibbs, Margret Fitting Gifford, Bonnie Gillespie, Robert Godfrey, Reuben Goering, Eunice Goldberg, Louise Goode, Terry Goodman, Janice M. Green, Jerry Grossman, Charles E. Hampton, Francis T. Hamrick, Stanley Hartzler, Elizabeth Harvey, M.L. Helms, Alice Hemingway, George Henderson, Randall Hicks, Mary Hito, Irving Hollingshead, Alan Holz, Larry Hopkins, Freddie Howard, Dwight House, Janice Huang, John Huber, Janet Huddleston, Mary Hudson, E. Duane Huechteman, J.D. Humbern, Lyndal Hutcherson, Dennis Huthnane, Scott Inch, Debra Ann Jagielski, Bob Johnson, Don Johnson, Larry Johnson, Marilyn Johnson, John P. Jones, Pat Jones, Vijay Joshi, Mark Juliani, M. Bonnie Kelterborn, Judy Kidd, James Kilpatrick, M.A. Kirkpatrick, Margaret Kothmann, M. Landers, W.I. Layton, Marie Long, F.J. Lorenzen, Joyne McCannoughby, Daisy McCoy, Thomas A. McCready, Ken McIntyre, Walter McQueen, Lynn Mach, Richard Marshall, Judy Massey, Ann B. Megaw, Marilyn Mercer, William Mitchell, Winston Moody, C. Naples, Deborah Narang, Jenny Neff, Edward Nichols, Ken Ohm, Linda Oldaker, Paul Orsen, Darla Ottman, Allison Owen, Cada Parrish, Deborah Paschal, John Peterson, Gerald Petrella, Jerry W. Phillips, R.B. Phillips, Pamela Pierce, K. Pilger, Robert Poe, Kenneth Pope, Margaret

# CRITICAL THINKING AND PROBLEM SOLVING

# 1

Three sailors and a monkey are shipwrecked on a desert island, where coconuts are the only available food. The men collect coconuts but decide to wait until morning to divide the supply. However, once Bob and Carl are asleep, Mark arises, divides the coconuts into three equal shares, gives the remaining coconut to the monkey, hides his share, and restacks the others. Later, Bob sneaks out of bed, divides the pile of coconuts into three equal shares, has one left over, gives it to the monkey, hides his pile, and restacks the remaining coconuts. Toward morning, Carl, too, arises, divides the pile into three equal shares, has one coconut left over, gives it to the monkey, hides his pile, and restacks the remaining coconuts. In the morning, the sailors meet and divide the pile of coconuts. Again, one coconut is left over for the monkey. What is the least possible number of coconuts in the original pile? (See the solution on p. 39.)

Problem solving lies at the very heart of the creative activity of mathematicians, physicians, urban planners, and corporate managers—indeed, all persons who seek to organize information to achieve a desired goal. Yet, this sounds somewhat curious to the ears of a typical college student who believes that problem solving is what you do when you solve Exercise 5 in Chapter 3 by using Example 4 to guide you. Obviously we are dealing with two different understandings of problem solving.

The National Council of Teachers of Mathematics (NCTM), in its publication, *Principles and Standards for School Mathematics* (2000), states on page 56: "People who reason and think analytically tend to note patterns, structure, or regularity in both real-world situations and symbolic objects; they ask if those patterns are accidental or if they occur for a reason; then they conjecture and prove." Mathematical reasoning and proof often begin with problem solving.

Equally important are skills in critical thinking. In Section 1 of this chapter we will examine the relationship between problem solving and critical thinking. In the remaining sections we will review and practice many of the important mental habits and strategies that have characterized the work of master problem solvers over the years.

## SECTION 1 CRITICAL THINKING

**PROBLEM**

What is the largest number of regions into which a circle can be divided by connecting two points on a circle with a line segment? What is the largest number of regions into which the circle can be divided by connecting three points on the circle with line segments? Study the pictures below, count the number of regions, and verify the answers shown in the table.

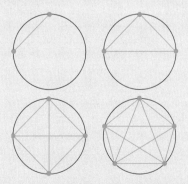

| Number of Points | 2 | 3 | 4 | 5 | 6 |
|---|---|---|---|---|---|
| Number of Regions | 2 | 4 | 8 | 16 | |

Now, on the basis of the pattern you have observed, what is the largest number of regions into which you can divide the interior of a circle by connecting six points on the circumference?

**OVERVIEW**

We illustrate two important concepts with the preceding problem. First, after observing the pattern for a small number of cases, you made an educated guess (a conjecture) about what the pattern should be. By forming such a conjecture, you practiced what is called **inductive reasoning.** (See p. 10.) With inductive reasoning we make conjectures from a small number of cases about what the pattern should be in all cases. Second, we see that with inductive reasoning we never establish the complete certainty of a conjecture. For example, count the number of regions formed in the circle when six points are connected by segments. In the table did you conjecture that 32 regions would be formed? If you count carefully, you will see that only 31 regions were formed. Using deductive reasoning we are able to determine that the conjecture is not true in all cases.

**GOALS**

Problem Solving, Reasoning and Proof are listed as important standards in *Principles and Standards for School Mathematics.* We use each of these in critical thinking.

One pundit has mused that education is what remains when you have forgotten everything you know. Whether this statement is completely valid or not, it points to the fact that your education is much more than the collection of facts you master during your years in school. Much more important than those facts is the set of attitudes and intellectual skills you acquire along with those facts. Chief among those intellectual skills is the facility for critical thinking.

As one would expect from academicians, there is some disagreement about what constitutes critical thinking. Philosophers who write about the topic give one answer, while psychologists give another. Philosophers are most interested in the exercise of logic and reason as tools for moving from knowledge to principle. Hence they stress **deductive reasoning** as the core of critical thinking. By contrast, psychologists are interested in the thinking process and how this process should improve with education. They regard the acquisition of thinking skills as a developmental issue and emphasize experimentation and research as the appropriate vehicles to achieve this development. Experiences in **inductive reasoning** are the heart of their understanding of critical thinking.

Fortunately, a well-designed mathematics course is an ideal laboratory for building critical thinking skills whether you take the position of the philosopher or the psychologist. In a well-designed mathematics course, you will build skills of inductive reasoning and problem solving, learn and practice the basic concepts of logic, and use these concepts in mathematical reasoning.

In an effort to ensure that all who study from this book acquire understanding and not just a handful of facts, we have paid careful attention to three standards in the *Principles and Standards for School Mathematics* published by the National Council of Teachers of Mathematics (NCTM).

1. P: Problem Solving
2. C: Communication
3. R: Reasoning and Proof

Each of these standards or goals is an important element in using mathematics to develop critical thinking skills. One effective way to make progress on these goals is to gather in groups and solve substantial mathematical problems. For this edition of this text, we developed a special set of exercises, the PCR Excursions. You and a group of your classmates are encouraged to find time to work on all of these PCR Excursions. As you begin these problems, discuss carefully such questions as "What's really being asked in this problem?" and "How can we get started on this problem?" Write out your thinking, especially your conjectures. As you near a solution, articulate carefully to one another what you have actually learned. Look back at your solution to see if, as a group, you can find another way to see the solution more clearly. Our experiences with our own students convince us that the PCR Excursions are one of the best vehicles we have developed to involve our students in their **own** learning process.

# SECTION 2 CRITICAL THINKING AND INDUCTIVE REASONING

| **PROBLEM** | Leslie is to be paid 1 penny for her first day of work, 2 pennies for her second day, 4 pennies for her third day, 8 pennies for her fourth day, and so on. How much will she receive for her fifteenth day of work? |
| --- | --- |
| **OVERVIEW** | We solve this problem by observing patterns. What does the word pattern mean to you? Simply defined, a pattern is any kind of regularity recognized by the mind. You will learn to recognize many different kinds of patterns in this section. |
| **GOALS** | Use the Algebra Standard: • understand patterns, relations, and functions; • represent and analyze mathematical situations and structures using algebraic symbols; • use mathematical models to represent and understand quantitative relationships; • analyze change in various contexts.<br>Illustrate the Reasoning and Proof Standard, page 23.*<br>Illustrate the Problem Solving Standard, page 15.*<br>Illustrate the Communication Standard, page 78.*<br><br>* The complete statement of the standard is given on this page of the book. |

## 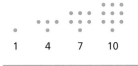 WORKING WITH PATTERNS

1    4    7    10

FIGURE 1-1

Alan H. Schoenfeld of the University of California, Berkeley, was describing a type of critical thinking in mathematics when he wrote, "Mathematics profoundly done is an act of sense-making. It helps people make sense of patterns of things in the real world or of symbolic abstractions."

We start by studying patterns at a very elementary level. The first example is one you might encounter in an elementary-school textbook.

| **EXAMPLE 1** | Suppose a dot printer is programmed to print the series of designs shown in Figure 1-1. What number will the next pattern represent? |
| --- | --- |
| *SOLUTION* | Did you add another row of three dots to the design? The answer is 13. Now let's try a harder pattern. |

**EXAMPLE 2**

Consider the 3 × 3 array of drawings in Figure 1-2. Study the changes that occur from left to right on the first line; then study the second line. Find the patterns that will allow you to draw the correct figure in the lower right-hand corner.

*SOLUTION*

The stems and cross bars seem to cycle from right to left and back. Because each of the three possible stems appears in rows 1 and 2, we draw the stem that is missing in row 3, ⊥ . Similarly, the cross bar that is missing is chosen, ⊥̲. At the end of the cross bar are two circles, • and ○. Can you determine which circle belongs on the left and which belongs on the right?

Patterns occur in many forms, such as the sequence of alphabet letters in Example 3.

**EXAMPLE 3**

Suppose you are asked to identify the next two letters in the following list:

$$B, A, E, D, H, G, K, J, \underline{\hspace{1cm}} , \underline{\hspace{1cm}}$$

Would your answer be one of the following pairs: *P, O*; *S, R*; *N, M*; or *N, O*?

*SOLUTION*

FIGURE 1-2

To emphasize the meaning of each part of our analysis, let's describe what we discover as we work through the problem.

First note that *A*, which follows *B* in the series, is the letter preceding *B* in the alphabet. This is one possible pattern. Do any other pairs of letters repeat this pattern? How about *E, D*; *H, G*; and *K, J*? The two letters we are trying to discover must be such that the second letter precedes the first letter in the alphabet. To determine which two letters to use with this pattern we may need some additional information.

Because the letters evidently go in pairs, look at the first letter of each pair. Study the relationship between *B* and *E*. What do you notice? There are two letters in the alphabet between *B* and *E*. Is this the beginning of a pattern? Yes, there are also two letters between *E* and *H*, and two between *H* and *K*. If there are two letters between *K* and our first unknown letter, the letter will be *N*, and the next letter will be *M*. Thus, from our list of possible answers, *N, M* is correct. (NOTE: In this type of problem, the answer is not necessarily unique, as will be discussed later.)

**EXAMPLE 4**

In an unfamiliar language, *zer mon* means "brown dog," *mil mon miko* means "little brown bird," and *kon miko* means "song bird." What is the word for "little" in this language?

*SOLUTION*

Let's work systematically to discover the pattern used. "Bird" is in both "little brown bird" and "song bird." The common word in *mil mon miko* and *kon miko* is *miko*, so *miko* means "bird." "Brown" is in both "brown dog" and "little brown bird." *Mon* is common to *zer mon* and *mil mon miko*. Therefore, *mon* means "brown." The only word left for "little" in *mil mon miko* is *mil*; so *mil* means "little."

# GEOMETRIC PATTERNS

Among the mathematical patterns studied by the Pythagoreans, a group of ancient Greek mathematicians, was the pattern for a set of numbers called **triangular numbers.** The numbers of blocks in the successive triangles in Figure 1-3 are triangular numbers.

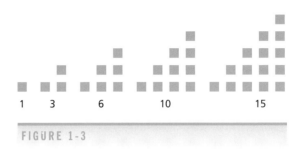

FIGURE 1-3

The preceding discussion serves as an introduction to geometric patterns.

**EXAMPLE 5**

Determine the geometric pattern that governs the triangular numbers, and see if you can write down the next two triangular numbers after 15.

*SOLUTION*

One way to look at this pattern is to observe that each successive triangle is obtained from the preceding one by drawing a new column on the right. In the fourth triangle, a column of four blocks is added; in the fifth, a column of five blocks. What are the next two triangular numbers? Did you get 21 and 28? Draw these.

Another way to look at this pattern is to think of adding a new block at the bottom of each column and then adding a block on the left of the new bottom row. In many cases there is more than one way to describe the same pattern.

Now that you are thinking geometrically, consider the next example.

**EXAMPLE 6**

The top four diagrams in Figure 1-4 form a series that changes according to some rule. First try to discover the rule, and then choose from among the alternatives the diagram that should come next.

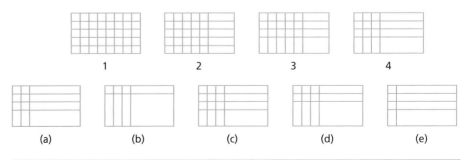

FIGURE 1-4

*SOLUTION*   Look at the four rectangles from left to right, and notice that lines have been removed in each step. Two vertical lines are removed from the first rectangle to obtain the second rectangle. What is removed from the second to obtain the third rectangle? (One horizontal line.) What is removed from the third to obtain the fourth rectangle? (Two vertical lines.) One possible pattern could be to alternate removing two vertical lines and then one horizontal line. Did you get answer (d)?

## NUMBER SEQUENCES

Both philosophers and psychologists would lay claim to the next examples, which introduce number sequences. In addition to observing patterns in the examples, we also make conjectures and sometimes generalize results. These strategies of making guesses and generalizing are easily illustrated by using number sequences.

<table>
<tr><td rowspan="2" style="writing-mode: vertical-rl">DEFINITION</td><td>**Number Sequence**</td><td>A *number sequence* is a collection of numbers arranged in order so that there is a first term, a second term, a third term, and so on. Sequences are arranged from left to right, and the numbers are separated by commas.</td></tr>
</table>

A sequence should contain at least three terms if it is to reveal a pattern.

**EXAMPLE 7**   What pattern can be detected for the following sequence?

$$5, 8, 11, 14, \underline{\quad}, \underline{\quad}, \underline{\quad}, \ldots$$

Find the next three terms.

*SOLUTION*   Each term of the sequence is 3 more than the preceding term. A reasonable continuation of the pattern would be 17, 20, and 23.

   Strictly speaking, there is not a unique answer to this type of problem. For example, you might guess that the next three terms of the sequence

$$10, 13, 18, 25, \underline{\quad}, \underline{\quad}, \underline{\quad}, \ldots$$

are 34, 45, and 58. The reasoning used to obtain these answers involves increasing each term of the sequence by consecutive odd numbers $(3, 5, 7, \ldots)$ as seen in (a).

|   | (a) | (b) |
|---|-----|-----|
|   | $10 + \underline{3} = 13$ | $10 + 3 = 13$ |
|   | $13 + \underline{5} = 18$ | $13 + 5 = 18$ |
|   | $18 + \underline{7} = 25$ | $18 + 7 = 25$ |
|   | $25 + \underline{9} = 34$ | $25 + 11 = 36$ |
|   | $34 + \underline{11} = 45$ | $36 + 13 = 49$ |
|   | $45 + \underline{13} = 58$ | $49 + 17 = 66$ |

However, the sequence 10, 13, 18, 25, . . . may be part of an otherwise entirely different pattern. Another person might reason as seen in (b): This pattern involves increasing each number by consecutive prime numbers (a special kind of number that we will study in Chapter 5). This example illustrates that patterns obtained inductively are not unique or necessarily the pattern that the author of the example intended.

Consider the sequence

$$3, 7, 11, 15, 19, \ldots$$

Notice that we use an ellipsis (denoted by three dots) both to indicate that terms are omitted and (in this case) to signify that the sequence continues in the same manner indefinitely. Notice also that

$$3 + 4 = 7$$
$$7 + 4 = 11$$
$$11 + 4 = 15$$
$$15 + 4 = 19$$

Thus this sequence can be obtained by adding the constant 4 to each term to obtain the next term. Such a sequence is called an **arithmetic sequence,** and 4 is called the **common difference;** these terms are illustrated in Figure 1-5.

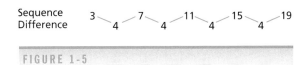

FIGURE 1-5

We can also reason as follows, where the first term is $a = 3$ and the common difference $d = 4$:

$$7 = 3 + 1 \cdot 4 \qquad \text{or} \qquad a + 1 \cdot d$$
$$11 = 3 + 2 \cdot 4 \qquad \text{or} \qquad a + 2 \cdot d$$
$$15 = 3 + 3 \cdot 4 \qquad \text{or} \qquad a + 3 \cdot d$$
$$19 = 3 + 4 \cdot 4 \qquad \text{or} \qquad a + 4 \cdot d$$

Now let's consider an arithmetic sequence whose first term is $a$ and whose common difference is $d$. If the first term is $a$, the second term is $a + d$; the third term is $(a + d) + d$ or $a + 2d$; and the fourth term is $(a + 2d) + d = a + 3d$.

$$a, a + d, a + 2d, a + 3d, \ldots$$

**PRACTICE PROBLEM**

Find a pattern and make a conjecture concerning the next three terms of the sequence 11, 18, 25, 32, ____ , ____ , ____ . . .

*ANSWER*

This is an arithmetic sequence with common difference 7. The next three terms are 39, 46, 53.

**EXAMPLE 8**

Find a pattern and guess the next three terms of the sequence 2, 6, 18, 54, ____ , ____ , ____ . . .

*SOLUTION* | You may have discovered that you can obtain any term by multiplying the preceding term by 3. Thus, the next three terms are 162, 486, and 1458.

The sequence in Example 8 is called a **geometric sequence.** In general, a geometric sequence is obtained by multiplying each term by a constant called the **common ratio.** If the first term is $a$ and the common ratio is $r$, then the second term is $ar^1$; the third term is $(ar)r = ar^2$; and the fourth term is $(ar^2)r = ar^3$. The sequence is thus

$$a, ar, ar^2, ar^3, \ldots$$

**EXAMPLE 9**

Find the next three terms for each of the following patterns.

(a) 1, 4, 1, 7, 1, 10, 1, 13, 1, . . .

(b) 1, 4, 9, 16, 25, . . .

*SOLUTION*

$1 = 1^2$

$4 = 2^2$

$9 = 3^2$

$16 = 4^2$

$25 = 5^2$

In (a), you probably first noted that every other term of the sequence is 1. The alternate terms increase by 3. The sequence is 1, 4, 1, 4 + 3, 1, 7 + 3, 1, 10 + 3, 1, . . . . Therefore, the next three terms are 13 + 3, 1, and 16 + 3; or 16, 1, and 19. In (b), the terms of the sequence are the squares ($x^2 = x \cdot x$ and $3^2 = 3 \cdot 3$) of consecutive counting numbers as seen in the margin.

Therefore, the next three terms are $6^2$, $7^2$, and $8^2$; or 36, 49, and 64. This sequence can also be obtained by adding successive odd numbers to the preceding terms: $1 + \underline{3} = 4$; $4 + \underline{5} = 9$; $9 + \underline{7} = 16$; $16 + \underline{9} = 25$; $25 + \underline{11} = 36$; $36 + \underline{13} = 49$; $49 + \underline{15} = 64$; and so on.

To describe a sequence, we often try to find a pattern that relates the number of a term to the term itself. Let's return to Example 7, where each term was obtained by adding 3 to the preceding term:

$$5, 8, 11, 14, 17, \ldots$$

The second term of the sequence is $8 = 5 + 3$. The third term is $11 = 8 + 3 = (5 + 3) + 3 = 5 + 2(3)$. Term 4 is $14 = 5 + 3(3)$, and the fifth term is $5 + 4(3)$. Can you use this pattern to write the sixth term? What is the connection between the number of the term and the number multiplied by 3 in the term?

| Number of Term | 1 | 2 | 3 | 4 | . . . | $n$ |
|---|---|---|---|---|---|---|
| Term | 5 | 8 | 11 | 14 | . . . | |
| Computation Used | 5 | 5 + 3 | 5 + 2(3) | 5 + 3(3) | . . . | 5 + (n − 1)3 |

With a little algebra, we can see that $5 + (n - 1)3$ is also $2 + 3n$, so that $2 + 3n$ would be a correct answer also.

**EXAMPLE 10**

A sequence is built on the pattern $2n + 1$. Find the terms of the sequence that correspond to $n = 1, 2, 3, 4,$ and 5.

*SOLUTION* | For $n = 1$, $2(1) + 1 = 3$; for $n = 2$, $2(2) + 1 = 5$; for $n = 3$, $2(3) + 1 = 7$; for $n = 4$, $2(4) + 1 = 9$; for $n = 5$, $2(5) + 1 = 11$. Therefore, the terms of the sequence are 3, 5, 7, 9, 11, . . . .

**EXAMPLE 11**

Find an expression for the $n$th term of the arithmetic sequence

$$a, \quad a+d, \quad a+2d, \quad a+3d, \quad \ldots$$

*SOLUTION*

We can describe the sequence of terms as follows:

| Number of Term (n) | 1 | 2 | 3 | 4 ... | n |
|---|---|---|---|---|---|
| Term | a | a + d | a + 2d | a + 3d | ... |

$$
\begin{aligned}
n = 2 & \qquad a + (2-1)d = a + d \\
n = 3 & \qquad a + (3-1)d = a + 2d \\
n = 4 & \qquad a + (4-1)d = a + 3d
\end{aligned}
$$

Thus the $n$th term seems to be $a + (n-1)d$. See if this yields the first term when $n = 1$.

**PRACTICE PROBLEM**

Write a possible $n$th term for the sequence

$$1 \cdot 2^2, \quad 2 \cdot 2^3, \quad 3 \cdot 2^4, \quad 4 \cdot 2^5, \quad 5 \cdot 2^6, \quad \ldots$$

*ANSWER*

A possible $n$th term is $n \cdot 2^{n+1}$.

Now we return to the problem that appeared in the introduction of this section.

**EXAMPLE 12**

Leslie has taken a job in which she is to be paid 1 penny for the first day, 2 pennies for the second day, 4 pennies for the third day, 8 pennies for the fourth day, and so on. How much will she receive for the fifteenth day?

*SOLUTION*

First we tabulate the information we are given:

| Number of Term | 1 | 2 | 3 | 4 | ... | 15 |
|---|---|---|---|---|---|---|
| Term | 1 | $1 \cdot 2$ | $1 \cdot 2^2$ | $1 \cdot 2^3$ | ... | $1 \cdot 2^{14}$ |

This is a geometric sequence and the fifteenth term is $1 \cdot 2^{14} = 16{,}384$ pennies, or \$163.84.

## INDUCTIVE REASONING

The type of reasoning used in this section to predict missing terms of a sequence, to formulate conjectures, and to generalize the $n$th term of a sequence is called *inductive reasoning*.

DEFINITION

**Inductive Reasoning**   Inductive reasoning is the process by which one arrives at a general conclusion from limited observations.

Many times, a scientist makes observations, discovers regularities, and formulates conclusions. In science this is called *experimental research;* in mathematics, we say that the scientist is reasoning inductively. Similarly, business people, lawyers, and, in fact, all of us use inductive reasoning regularly. You should note that inductive reasoning in itself does not prove that a unique general pattern exists. This contrasts with **deductive reasoning,** in which one proceeds carefully from definitions and previously established facts to determine whether a pattern is true in all possible cases.

In this book we emphasize an inductive, intuitive approach for understanding mathematics; that is, we will often use examples to illustrate the reasonableness of a procedure or formula, rather than giving a careful, deductive argument that the procedure or formula is always true. However, be assured that in each case a deductive proof is available. Also be assured that from time to time we will show you examples of deductive thinking so that you may build these skills also.

## Just for Fun

Five students are sitting in a circle. Two pairs of students have the same color hair. Those with the same color hair are not sitting next to each other. Celeste is on the right side of Teresa and on the left side of Jane. Jane has the same color hair as Alice. Ruth is the fifth student. Identify the positions of the students in the circle.

# EXERCISE SET 1

*R* 1. Which word is different from the collection of other words? Why?

   (a) run       (b) ride       (c) walk
   (d) jog       (e) hike

2. Find a pattern and then complete the blanks.

   (a) $\triangle, \square, \square, \triangle, \square, \square,$ _____ , _____ , . . .
   (b) $\triangle, \square, \Circle,$ _____ , _____ , . . .
   (c) $a, b, a, b, b, a, b, b,$ _____ , _____ , . . .
   (d) $x, x + 1, x + 2, x + 3, x + 4,$ _____ , _____ , . . .
   (e) $A, C, E, G, I, K,$ _____ , _____ , . . .
   (f) $1, 2, 3, 5, 6, 7, 9, 10, 11,$ _____ , _____ , . . .

3. Each sequence of letters has a pattern. Find a pattern and write the next three letters.

   (a) $A, Y, X, B, W, V, C, U, T,$ _____ , _____ , _____
   (b) $B, D, B, B, F, B, B, B, H, B, B,$ _____ , _____ , _____
   (c) $M, N, L, P, Q, O, S, T, R,$ _____ , _____ , _____

4. Find a pattern and then complete the blanks.

   (a) $2, 4, 8, 16,$ _____ , _____ , _____
   (b) $1, 3, 9, 27,$ _____ , _____ , _____

5. Find a pattern and write the next two sets of six letters.

       *EFGHIJ*    *JEFGHI*    *IJEFGH*

6. Select the number that should appear next in each of the following sequences.

   (a) $1, 4, 9, 16,$ _____     (b) $2, 5, 10, 17,$ _____

7. In an unfamiliar language, *lobo strino* means "fat dog," *tropo lobo wasca* means "very fat boy," and *tropo bludo* means "very smart." How do you say "smart boy"?

8. In an unfamiliar language, *enic lod nam* means "nice old man," *enic moor* means "nice room," and *elittl nam* means "little man." How do you say "old room"?

9. What are the next three square numbers?

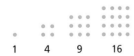

    1      4      9      16

10. A pentagonal number is the sum of a triangular number and a square number. What are the next three pentagonal numbers?

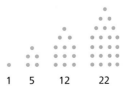

    1     5    12    22

11. Look for a pattern in the first three products, and then find the fourth product.

| (a) | 1 | 11 | 111 | 1111 |
|---|---|---|---|---|
| | · 1 | · 11 | · 111 | · 1111 |
| | 1 | 121 | 12321 | ? |
| (b) | 9 | 99 | 999 | 9999 |
| | · 9 | · 9 | · 9 | · 9 |
| | 81 | 891 | 8991 | ? |

12. Using the expression for the $n$th term of a sequence, find the first five terms of each sequence. Begin with $n = 1$.

   (a) $n^2 + 1$       (b) $n^2 + n$
   (c) $4n - 1$       (d) $3n^2 + 2$

13. Verbally describe a pattern for each sequence.

   (a) $7, 12, 17, 22, . . .$     (b) $5, 8, 11, 14, 17, . . .$
   (c) $3, 9, 27, 81, 243, . . .$   (d) $1, 8, 27, 64, 125, . . .$

14. Determine a pattern and then find the next three terms for each sequence.

   (a) $2, 6, 18, 54, 162, . . .$  (b) $1, 4, 7, 10, . . .$
   (c) $3, 5, 9, 15, 23, . . .$   (d) $1, 3, 1, 8, 1, 13, 1, . . .$
   (e) $4, 1, 5, 2, 6, . . .$     (f) $7, 10, 6, 9, 5, . . .$

15. Complete the blanks so that the following are arithmetic sequences.

   (a) $2, 7, 12,$ _____ , _____ , . . .
   (b) $5,$ _____ , $11, 14, . . .$

16. Complete the blanks so that the following are geometric sequences.

   (a) $1, 3, 9, 27,$ _____ , _____
   (b) $4,$ _____ , $36, 108,$ _____ , _____

17. Determine a possible pattern and sketch the next geometric figure.

   (a)

   (b)

   (c)

   (d)

18. Find an expression for the $n$th term of the following geometric sequence.
$$a, ar, ar^2, ar^3, \ldots$$

$T$ 19. Study the given array and then find the following numbers without actually computing the squares.
$$2^2 - 1^2 = 3$$
$$3^2 - 2^2 = 5$$
$$4^2 - 3^2 = 7$$
$$5^2 - 4^2 = 9$$
   (a) $6^2 - 5^2$    (b) $7^2 - 6^2$
   (c) $8^2 - 7^2$    (d) $9^2 - 8^2$

20. To get a bill passed, the governor called three legislators on Monday. She asked each of them to call three other legislators on Tuesday, and each of these was to call three other legislators on Wednesday. If this process continued for a week, with the last calls being made on Sunday, and if no legislator was called twice, how many legislators were called?

21. The sum of the first $n$ terms of an arithmetic sequence is given by $S = (a + l)n/2$ where $l = a + (n - 1)d$. Use these two formulas to find the sum of the first eight terms of the sequences in Exercise 15.

22. One famous number sequence was first studied by and named for Leonardo de Pisa, also called Fibonacci. Determine the pattern and the next three terms of the Fibonacci sequence:
$$1, 1, 2, 3, 5, 8, 13, 21, 34, \ldots$$

23. At the beginning of the year, I owned two rabbits, a male and a female. At the end of the first month and each month thereafter, this original pair gave birth to a new pair, male and female. Furthermore, each new pair, in the second month after their birth, gave birth to another pair and continued to do so in subsequent months. Thus, at the end of the first month, I had two pairs; at the end of the second month, I had three pairs; and at the end of the third month, I had five pairs. How many did I have after 4 months? 5 months? 6 months? Compare your results with the sequence in Exercise 22.

24. In biology, one learns that a male bee has only one parent, his mother. A female bee has both father and mother. Complete the picture of the family tree of a male bee, where ♂ represents male and ♀ represents female. Go back a couple of generations. Do you notice any relationship between the Fibonacci numbers of Exercise 22 and

   (a) the number of ancestors in each generation?
   (b) the number of male ancestors in each generation?
   (c) the number of female ancestors in each generation?

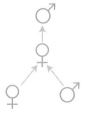

$C$ 25. Consider a sequence that begins
$$1, 4, 9, \underline{\hspace{1cm}}, \underline{\hspace{1cm}}, \ldots$$

   (a) Describe a rule that would yield 16 as the fourth term. What would be the fifth term?
   (b) Describe a rule that would yield 22 as the fourth term. What would be the fifth term?

26. While playing basketball, John notices that with daily practice he can increase the number of free throws he makes before missing. On the first day, he makes only 1; on the second day, 5; on the third day, 14; and on the fourth day, 30. If this pattern continues, how many free throws can he expect to make on the fifth day?

27. Find an expression for the $n$th term of each part of Exercise 13, where $n$ represents a counting number.

28. Find the sum of the first four terms of the Fibonacci sequence (see Exercise 22); the first five terms; the first six terms. Describe a pattern.

## LABORATORY ACTIVITY
•  •  •  •  •  •  •  •  •  •  •  •  •  •  •  •  •  •  •  •  •  •  •  •  •  •

29. Using your calculator,
   (a) Place 3 in the calculator; then multiply by 3 five times. This is denoted by $3^6$. What is the value of $3^6$?
   (b) Key in 3 ⌐∧⌐ 6 ⌐=⌐. Did you get the same value as in (a)? Thus 3 ⌐∧⌐ 6 is another way of computing $3^6$. (Some calculators use ⌐$y^x$⌐.)
   (c) Find the sum of the first six terms of the geometric sequence 10, 100, 1000, 10,000, . . . .
   (d) Find the sum of the first eight terms of the geometric sequence 4, 16, 64, 256, . . . .
   (e) Find the sum of the first seven terms of 4, 12, 36, 108, . . . .

30. (a) Find the value of the expression $n^2 - n + 41$ for each value of $n$ ($n = 1, 2, 3, \ldots, 10$) in order to complete the table below. For $n = 1$ the value is $1^2 - 1 + 41 = 41$; for $n = 2$, $2^2 - 2 + 41 = 43$.

| $n$ | 1 | 2 | 3 | 4 | 5 | 6 | 7 | 8 | 9 | 10 |
|---|---|---|---|---|---|---|---|---|---|---|
| Value | 41 | 43 | | | | | | | | |
| Numbers that Divide the Value | 1, 41 | 1, 43 | | | | | | | | |

(b) For each value of the expression that results from the evaluation, use your calculator to find all of the counting numbers that divide it evenly. (For example, 3 divides 12 evenly, getting an answer of 4.) By testing possible divisors of 41 with your calculator, you will find that only 1 and 41 evenly divide 41. Likewise, only 1 and 43 evenly divide 43.

(c) Summarize what you observe about the counting numbers that evenly divide the values obtained in the table.

(d) What conjecture might you make as a result of this analysis?

(e) Evaluate the expression when $n = 41$. What are the counting numbers that evenly divide the resulting value?

(f) Compare (d) and (e). What does this say about inductive reasoning?

*The following problem requires some knowledge of the elementary operations of high school algebra.*

## ∴ PCR Excursion ∴

31. Consider the sum

$$S = 4 + 12 + 36 + 108 + 324 + 972$$

A. (a) The numbers of this sum comprise what kind of a sequence?

(b) What is the common ratio?

(c) Let us explore a pattern for computing this sum without adding the individual terms. Multiply $S$ and the terms of the sum equal to $S$ by the common ratio of the sequence:

$$S = 4 + 12 + 36 + 108 + 324 + 972$$

$$3S = 12 + 36 + 108 + 324 + 972 + 2916$$

Subtract each side of the "$S =$" equation from the "$3S =$" equation. On the left side you get $3S - S$. What do you get on the right side?

(d) In high school you learned to write

$$3 \cdot S - S = 3 \cdot S - 1 \cdot S = (3 - 1) \cdot S$$

Thus

$$(3 - 1) \cdot S = 2916 - 4$$

Divide both sides by $(3 - 1)$, getting $S =$?

B. (a) Let us revisit this exploration using the sum $S$ written in terms of powers of 3.

$$S = 4 + 4 \cdot 3^1 + 4 \cdot 3^2 + 4 \cdot 3^3 + 4 \cdot 3^4 + 4 \cdot 3^5$$

Follow the steps in (c) above to find an expression for $(3 - 1) \cdot S$.

(b) Follow the steps in (d) above to find a value for $S$. Did you get

$$S = \frac{4(3^6 - 1)}{3 - 1}?$$

C. Consider the sum

$$S = a + ar + ar^2 + ar^3 + ar^4 + \ldots + ar^{n-1}$$

Note that

$$rS = ar + ar^2 + ar^3 + ar^4 + \ldots + ar^{n-1} + ar^n$$

Follow the reasoning you used in sections A and B to compute the sum $S$.

D. The sum of the terms of the geometric sequence (the first $n$ terms)

$$S = a + ar^1 + ar^2 + ar^3 + \ldots + ar^{n-1}$$

is

$$S = \frac{a(r^n - 1)}{r - 1}$$

Is this what you learned in C? If so, congratulations! You have used deductive reasoning to develop a formula for the sum of $n$ terms of a geometric sequence. Now use this formula to find the sums in Exercise 29 (c), (d), and (e) of this section.

# SECTION 3 INTRODUCTION TO PROBLEM SOLVING

| | |
|---|---|
| **PROBLEM** | Can you use a pencil to connect these dots with four straight line segments without lifting the pencil from the paper? |

• • •

• • •

• • •

| | |
|---|---|
| **OVERVIEW** | In this section, we investigate ways that will help you answer the preceding question. As we discuss techniques for problem solving, bear in mind that problem solving is much more than a matter of memorizing techniques. It is a process whereby you apply your knowledge and skills to understand and to satisfy the demands of unfamiliar situations. |

| | |
|---|---|
| **GOALS** | Use the Problem Solving Standard: • build new mathematical knowledge through problem solving; • solve problems that arise in mathematics and in other contexts; • apply and adapt a variety of appropriate strategies to solve problems; • monitor and reflect on the process of mathematical problem solving. Illustrate the Algebra Standard, page 4.* Illustrate the Representation Standard, page 50.* Use the Communication Standard, page 78.* |

\* The complete statement of the standard is given on this page of the book.

In the 1950s, a very successful research mathematician named George Polya began writing a series of insightful articles and books on problem solving. The four steps that Polya used to describe the mathematical problem-solving process follow.

| | |
|---|---|
| **Steps in Problem Solving** | 1. Understand the problem.    2. Devise a plan.<br>3. Carry out the plan.    4. Look back. |

Students are often eager to rush through the first step, understand the problem. Yet time spent obtaining a careful understanding of the problem will often make finding a solution much easier. In fact, in some instances, once the problem is correctly understood, the solution is immediate. The following suggestions will help you understand many problems.

| **Understand the Problem** | 1. Identify what you are trying to find (the unknown). 2. Summarize the information that is available to you (the data). | 3. Strip the problem of irrelevant details. 4. Do not impose conditions that do not exist. |
|---|---|---|

As we begin each problem, we should write a short description of what we are looking for in the problem, the unknown. Then we must focus on the information that is available to help us in our efforts, the data. Asking a series of questions will sometimes help us dislodge information hidden in the problem. What questions would be useful in digesting the information in this problem?

**EXAMPLE 1**

On their way back to the university, Joy, Beth, and Dill took turns driving. Joy drove 50 miles more than Beth, and Beth drove twice as far as Dill. Dill only drove 10 miles. How many miles is the trip back to the university?

*SOLUTION*

**UNDERSTANDING THE PROBLEM**    What are we trying to find? (The distance back to the university.) Do you know how far anyone drove? (Yes, Dill drove 10 miles.) How many more miles did Joy drive than Beth? (50 miles.) What is the relationship between the number of miles driven by Beth and the number driven by Dill? (Beth drove twice as far as Dill.)

Many times problems are difficult to solve because they contain too much information, some of it irrelevant. This is particularly true of "real life" mathematical problems as opposed to problems that have been neatly packaged for a textbook.

**EXAMPLE 2**

Dan buys only blue and brown socks. He keeps all his socks in one drawer, and in that drawer he has eight blue socks and six brown socks. If he reaches into the drawer without looking, what is the smallest number of socks he must take out to be sure of getting two of the same color?

*SOLUTION*

We are looking for how many socks he must take from the drawer to get a matched pair. Students often mistakenly concentrate on the numbers of blue socks and brown socks in the drawer when trying to solve this problem. These numbers are irrelevant. To solve this problem, we need to analyze only what happens as Dan draws socks one by one from the drawer. Suppose Dan has drawn two socks; they are either the same color or different colors. If they are the same color, he has a pair in two draws. If they are different colors, the next sock he draws must match one of them. Thus, at most three draws are necessary to get a pair of socks that match.

We make many problems in mathematics and in life more difficult because we impose conditions on the solution that are not present in the problem. Consider the example with which we began this section.

EXAMPLE 3

Without lifting your pencil from the paper, connect the dots in Figure 1-6 with four straight line segments.

*SOLUTION*

Many fine mathematicians have puzzled for several minutes over this problem. Perhaps because of our experience connecting dot-to-dot puzzles as children, many of us feel constrained to keep the pencil lines within the array of dots. Is this a requirement of the problem? As soon as we realize that we can wander outside the box of dots, the solution in Figure 1-7 becomes obvious. It is important not to impose conditions that are not required by the problem.

FIGURE 1-6

## DEVISE A PLAN

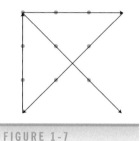

FIGURE 1-7

In the preceding examples, as soon as we understood the problem, we had a solution. This, of course, is not usually the case. After we have spent time understanding the problem, Polya asks us to devise a plan and then to carry out the plan. Often a considerable amount of creativity is required to formulate a plan. However, a number of plans or strategies have been used successfully in many different problem-solving situations. Because of their wide applicability, we will study them with the intention that they will give us insight as we devise our own plans. We will examine four such strategies in this section and six more in the next.

| Strategies | 1. Make a chart or a table. | 3. Guess, test, and revise. |
| --- | --- | --- |
| | 2. Draw a picture. | 4. Form an algebraic model. |

## MAKE A CHART OR A TABLE

This first strategy sometimes clarifies the situation described in a problem.

EXAMPLE 4

The Astros are in first place in a six-team Grapefruit League, and the Braves are in fifth place. The Dodgers are midway between the two. If the Reds are ahead of the Braves and the Giants are immediately behind the Dodgers, what place are the Padres in?

*SOLUTION*

By writing down the six positions and placing the available information in the table, we immediately see the answer. We begin by placing the Astros in first place and the Braves in fifth place. (See Table 1-1.) For the Dodgers to be midway between these two, they must be in third place. The Giants are immediately behind the Dodgers, so they must be in fourth place. Since the Reds are ahead of the Braves, the Reds must be in second place. The Padres must be in sixth place, the only remaining open slot.

**TABLE 1-1**

| First Place | Second Place | Third Place | Fourth Place | Fifth Place | Sixth Place |
|---|---|---|---|---|---|
| Astros | | Dodgers | | Braves | |

# DRAW A PICTURE

The strategy of drawing a picture often makes solving a problem easier. Whether or not drawing a picture leads immediately to a solution, it is helpful for understanding the problem. We will use this strategy throughout this text, drawing arrow diagrams, one-dimensional graphs, two-dimensional graphs, histograms, frequency polygons, circle charts, and many other pictures. In each case, our goal will be understanding the problem and sometimes actually solving the problem.

**EXAMPLE 5**

A beetle is at the bottom of a bottle that is 6 inches deep and it wants out. Each day, the beetle climbs up 1 inch, and each night it slides back $\frac{1}{2}$ inch. How long will it take the beetle to climb out of the bottle?

*SOLUTION*

To understand this problem, examine the sketch of the bottle in Figure 1-8. On the first day, the beetle reaches the 1-in. plateau but slides back to $\frac{1}{2}$ in. at night. On the second, it reaches $1\frac{1}{2}$ in., but at night it slides back to the 1-in. mark. This process continues until the 11th day. When the beetle reaches the top on the 11th day, it crawls out, rather than sliding back.

Before we illustrate the strategy "guess, test, and revise," we will discuss Polya's last step, **look back.**

11th day — 6"
10th day
9th day — 5"
8th day
7th day — 4"
6th day
5th day — 3"
4th day
3rd day — 2"
2nd day
1st day — 1"

**FIGURE 1-8**

## LOOK BACK

Polya's last step, look back, is very important. Once we have found the solution, we may be tempted to skip this fourth step. Yet in this final step we are forced to examine the thinking that produced the solution. Here are four components of a successful review of a solution:

**Steps in Looking Back**

1. Interpret the results with an English sentence.
2. Check the results.
3. Ask if there is another way to solve the problem.
4. Ask if there are other problems that can be solved by using the techniques used in this problem.

# ⠠⠶ Guess, Test, and Revise

No other single strategy applies to as many problems as does guess, test, and revise. Although this strategy may be time-consuming and cumbersome to complete and may yield only an approximate answer, we will see that it provides a solid starting point that inspires more efficient plans.

**EXAMPLE 6**

In the barnyard there are pigs and chickens. They have 55 heads and 148 legs. How many of the animals are pigs and how many are chickens?

*SOLUTION*

**UNDERSTAND THE PROBLEM**   Like all good problem solvers, we will first carefully identify the unknowns. It is often useful to give names to those unknowns. Let $P$ = number of pigs and $C$ = number of chickens. Next, we will carefully summarize the information given to us as well as the information that we bring to the problem from our personal knowledge base:

> 55 heads: chickens and pigs have 1 head each.
> 148 legs: chickens have 2 legs and pigs have 4 legs.

**DEVISE A PLAN**   We will use guess, test and revise. We will guess values for the number of chickens ($C$) and the number of pigs ($P$), test to determine if we are correct, and, if not, we will revise our guess to make a more informed guess.

**CARRY OUT THE PLAN**
*First Guess*: $C = 10$, $P = 10$
Test: We know that there must be 55 heads. We can quickly see that for 10 chickens and 10 pigs

$$10 + 10 \neq 55$$

Clearly we must choose $C$ and $P$ to be larger so that the number of heads is 55.
*Revised Second Guess*: $C = 25$, $P = 30$
Test: Checking heads: $25 + 30 = 55$
  Checking legs: $2(25) + 4(30) = 170 \neq 148$
Since we have too many legs, we must revise our guess to have fewer pigs with four legs each.
*Revised Third Guess*: $C = 35$, $P = 20$
Test: Checking heads: $35 + 20 = 55$
  Checking legs: $2(35) + 4(20) = 150$
We are now very close. We realize that replacing one pig with a chicken will reduce the number of legs by 2 while keeping the heads at 55.
*Revised Fourth Guess*: $C = 36$, $P = 19$
Test: Checking heads: $36 + 19 = 55$
  Checking legs: $2(36) + 4(19) = 148$

**LOOK BACK**
1. Interpret the solution with an English sentence: There are 36 chickens and 19 pigs in the barnyard.
2. Check the results: Because 36 chickens and 19 pigs have $36 + 19 = 55$ heads and $2(36) + 4(19) = 148$ legs, the solution is correct.

3. Is there another way to solve this problem? Look back at the test for each of the guesses. Do you see a pattern? For each guess of $C$ chickens and $P$ pigs we tested to determine whether the following two statements were true.

Number of heads: Is $C + P = 55$?

Number of legs: Is $2C + 4P = 148$?

# Just for Fun················

1. A farmer has 17 calves. All but 9 die. How many does he have left?
2. Is it legal for a man to marry his widow's sister?
3. There was a blind beggar who had a brother, but this brother had no brother. What was the relationship between the two?
4. If Tom's father is Dick's son, how are Tom and Dick related?
5. A rectangular house is built so that every wall has a window opening on the south. A bear is seen from one of the windows. What color is the bear?
6. Five apples are in a basket. How can you divide them among five girls so that each gets an apple, but one apple remains in the basket?
7. Two United States coins total 30¢, but one of the coins is not a nickel. Can you explain?
8. Ms. A.: How much will 1 cost?
   Salesman: 30¢.
   Ms. A.: How much will 15 cost?
   Salesman: 60¢.
   Ms. A.: I need 615.
   Salesman: That will be 90¢.
   Explain.

# EXERCISE SET 2

*In Exercises 1-22, (a) identify the unknown, (b) summarize the information (What questions would be helpful?), (c) solve the problem, and (d) interpret your solution with a sentence. After each problem is completed, determine which problem-solving hints and strategies were helpful.*

*R*  1. The Browns have six daughters. Each daughter has two brothers. How many children are there in the Brown family?

2. Two ducks before a duck, two ducks behind a duck, and a duck in the middle are how many ducks?

3. Jodi has 8 coins totaling 57 cents. What are the coins if she does not have a half dollar?

4. Players on the football team are trying to gain weight. Tim, Jim, George, and Rick check their weights. Tim weighs twice as much as Jim. Jim weighs 50 pounds less than George. Rick weighs 10 pounds more than George. If Rick weighs 200 pounds, how much does Tim weigh?

5. Four boxes together weigh 60 pounds. Each box is twice as heavy as the next box, and the lightest box weighs 20 pounds less than the two middle ones together. What are the weights of the boxes?

6. Ron buys old comic books for 80¢ each, old buttons for 4¢ each, and old toys for 60¢ each. He sells each of them for 5¢ profit. How much does he make if he buys and sells four comic books, four buttons, and two toys?

7. Lou drives a small car, getting 28 miles per gallon in town and 34 miles per gallon on the highway. Her fuel tank holds 15 gallons. Regular gasoline sells for $1.30 a gallon. What is the cost of a full tank of gasoline?

8. A train 1 mile long moving at a rate of 1 mile per hour enters a 1 mile long tunnel. How long does it take the train to pass completely through the tunnel?

9. In the exercises listed, which facts could be classified as irrelevant?

    (a) Exercise 6          (b) Exercise 7

10. In the barnyard is an assortment of chickens and pigs. Counting heads, one finds 13; counting legs, one finds 46. How many pigs and how many chickens are there?

11. In a parking lot there are motorcycles and cars. There are 45 motors and 146 tires. How many motorcycles and how many cars are in the parking lot?

12. John had 24 coins in his pocket. When he emptied his pocket, he found that his pocket contained nickels and dimes, and that the contents totaled $1.50. How many nickels and how many dimes did he have in his pocket?

*T* 13. Write an exercise to be solved by using the guess, test, and revise strategy.

14. There are ten posts in Brooke's backyard. To decorate for a party, she strings crepe paper between posts. How many streamers will she need?

15. A container holds a mixture of 37 peanuts, 14 cashews, and 25 pecans. How many nuts must you take out to be sure of getting at least two of one kind? At least three of one kind?

16. Twelve clothespins are placed on a line at 8-foot intervals. How far is the first from the last?

17. In how many ways can one make change for a quarter using an unlimited number of pennies, nickels, and dimes?

18. A frog sits at the bottom of a 15-foot-deep well. Each day he climbs up 3 feet, and each night he slides back 2 feet. How long will the frog take to get out of the well?

19. How long will it take to cut a 10-foot log into five 2-foot lengths, allowing 2 minutes per cut?

20. In a machine, a gum ball costs 5¢. There are five colors of gum balls in the machine. How much money do you need to be certain you get three gum balls of the same color?

21. Newlywed Sonya makes toast in a pan that holds only two slices. After browning one side of a slice, she turns it over. Each side takes 30 seconds. How can she brown both sides of three slices in the shortest period of time?

22. You have eight sticks: four of them are exactly half the length of the other four. Enclose three equal squares with them, making sure that there is no overlapping.

*C* 23. One container holds 2 quarts of water, and another holds 1 quart of red grape juice. One cup of water is transferred to the container of grape juice and mixed thoroughly. Then one cup of the mixture is transferred back to the container of water. Is there more grape juice in the water or more water in the grape juice?

24. Each hour, an airplane leaves New York for London. The trip across takes 6 hours. At the same times, airplanes leave London for New York. How many planes will a given plane meet on the trip from New York to London?

25. Form four triangles by arranging six matchsticks so that they touch only at their endpoints.

## REVIEW EXERCISES

• • • • • • • • • • • • • • • • • • • • • • • • • • • • • • •

26. Consider the following pattern.

(a) Is the 31st square shaded?
(b) Is the 501st square shaded?

27. Find the next three terms of each sequence.
   (a) $A, Z, C, X, E, V, G, \ldots$
   (b) $C, B, A, G, F, E, K, J, I, \ldots$
   (c) $1, 5, 9, 13, \ldots$
   (d) $3, 6, 9, 18, 21, 42, 45, \ldots$

### ∵ PCR Excursion ∵

28. A clerk at Yute's store claims to be able to weigh all counting-number weights up to 40 pounds on a two-pan balance scale by using only 4 weights. Using guess, test, and revise and a great deal of reasoning, show how you would find the 4 weights so that you can weigh objects up to 40 pounds.

   (a) Observe that a 1-pound weight and a 3-pound weight will enable you to weigh all weights up to 4 pounds.

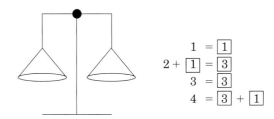

$$1 = \boxed{1}$$
$$2 + \boxed{1} = \boxed{3}$$
$$3 = \boxed{3}$$
$$4 = \boxed{3} + \boxed{1}$$

   (b) Determine a third weight that will enable you to weigh up to 13 pounds.
   (c) Determine a fourth weight that will enable you to weigh up to 40 pounds.
   (d) Show how you would weigh all weights up to 40 pounds.

SECTION

# 4  STRATEGIES FOR PROBLEM SOLVING

**PROBLEM**

Fifteen people are in a room, and each shakes hands once and only once with everyone else. How many handshakes are there?

**OVERVIEW**

Below you will find Polya's four-step description of the mathematical problem-solving process juxtaposed with the four steps that some psychologists use to describe problem solving in a more general context.

Psychologists
1. Preparation
2. Incubation
3. Inspiration
4. Verification

Polya
1. Understand the problem
2. Make a plan
3. Carry out the plan
4. Look back

Both lists suggest that successful problem solving will require significant investments of time and patience. Polya's list tends to be more active. Polya's encouragement to "make a plan" and "carry it out" describes what to do while waiting for "inspiration."

| **GOALS** | Use the Reasoning and Proof Standard: • recognize reasoning and proof as fundamental aspects of mathematics; • make and investigate mathematical conjectures; • develop and evaluate mathematical arguments and proofs; • select and use various types of reasoning and methods of proof.<br>Illustrate the Problem Solving Standard, page 15.<br>Illustrate the Algebra Standard, page 4.<br>Illustrate the Communication Standard, page 78. |
|---|---|

Polya's second major step in problem solving is "Make a plan." The strategies discussed next, alone or in concert, provide the rough outline of plans that will solve a great many problems and use the thinking skills we have discussed in the previous three sections. The first four strategies were discussed in Section 3.

| **Strategies for Making a Plan** | 1. Make a chart or table.<br>2. Draw a picture or diagram.<br>3. Guess, test, and revise.<br>4. Form an algebraic model.<br>5. Look for a pattern. | 6. Try a simpler version of the problem.<br>7. Work backward.<br>8. Restate the problem.<br>9. Eliminate impossible situations.<br>10. Use reasoning. |
|---|---|---|

## LOOK FOR A PATTERN; TRY A SIMPLER VERSION

One of the key techniques in problem solving is to look for a pattern. Further, if you examine a simpler version of the problem, you may find it is easier to detect a pattern if one exists. This dual strategy of looking for a pattern in a simple version of the problem provides a potent weapon in solving problems similar to the following.

**EXAMPLE 1**

How many people can be seated at ten square, 4-person tables that are lined up end to end?

*SOLUTION*

First look at simpler versions of the problem. As we proceed, we will confirm our thinking with pictures and summarize what we learn in a chart (Figure 1-9).

| **Arrangement and Number of Tables** | | **Number of Seats** |
|---|---|---|
|  | 1 table | 2 on sides + 2 on ends = 4 |
| | 2 tables | 2 · 2 on sides + 2 on ends = 6 |
| | 3 tables | 2 · 3 on sides + 2 on ends = 8 |
| | 4 tables | 2 · 4 on sides + 2 on ends = 10 |
| | $n$ tables | $2n + 2$ |

FIGURE 1-9

Thus for 10 tables ($n = 10$),

$$2(10) + 2 = 22$$

Twenty-two people can be seated.

**EXAMPLE 2**

Fifteen people are in a room, and each shakes hands once and only once with everyone else. How many handshakes occur?

*SOLUTION*

We consider simple versions of this problem to see if a pattern emerges.

Two people:         $A \leftrightarrow B$
                    (1 handshake)
Three people:       $A \leftrightarrow B$      $B \leftrightarrow C$
                    $A \leftrightarrow C$
                    $2 + 1 = 3$ handshakes
Four people:        $A \leftrightarrow B$      $B \leftrightarrow C$      $C \leftrightarrow D$
                    $A \leftrightarrow C$      $B \leftrightarrow D$
                    $A \leftrightarrow D$
                    $3 + 2 + 1 = 6$ handshakes
Five people:        $A \leftrightarrow B$      $B \leftrightarrow C$      $C \leftrightarrow D$      $D \leftrightarrow E$
                    $A \leftrightarrow C$      $B \leftrightarrow D$      $C \leftrightarrow E$
                    $A \leftrightarrow D$      $B \leftrightarrow E$
                    $A \leftrightarrow E$
                    $4 + 3 + 2 + 1 = 10$ handshakes

From the pattern, we infer that 15 people shake hands this many times:

$$14 + 13 + \cdots + 3 + 2 + 1 = 105$$

(See Exercise 33 for some ideas on how to compute this sum.)

# WORK BACKWARD

Some problems do not ask for a numerical answer, but rather ask how one can achieve a specified goal. In these circumstances, it is often useful to work backward from the desired goal.

**EXAMPLE 3**

How can you measure 1 liter of water from a faucet if you have only a 4-liter container and a 7-liter container?

*SOLUTION*

Our goal is to have 1 liter of water in either the 4-liter or 7-liter container. Let us imagine that we have achieved the goal (see Figure 1-10a) and then think about steps that could have preceded it.

We would have 1 liter left in the 4-liter container if we had filled the 4-liter container and poured off exactly 3 liters. How could we manage to pour off exactly 3 liters? If the 7-liter container contained exactly 4 liters, it would be easy to accomplish this task. See Figure 1-10(b).

Now, how could we get exactly 4 liters into the 7-liter container? Clearly, we could fill the 4-liter container and pour it into the 7-liter container (Figure 1-10c).

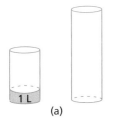

1 L

(a)

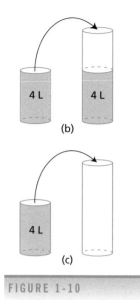

(b)

(c)

In summary, to obtain 1 liter, we (a) fill the 4-liter container, (b) pour it into the 7-liter container, (c) fill the 4-liter container, and (d) pour as much as possible into the 7-liter container. This sequence of steps leaves 1 liter in the 4-liter container.

As you have noticed, we do not always discuss all four of Polya's steps on each problem, and you will probably not do so either as you write up your solutions. However, Polya's steps should serve as a mental outline as you proceed through your problems. Further, as you begin a problem, it is always good practice to identify the unknowns on your paper and to summarize the available information, preferably in a picture or table. After the problem is solved, you should always interpret the result with an English sentence. Finally, mentally review the solution both to check results and to look for generalizations and other problems that your work suggests. As we discuss the next two examples, we will give you guidance by referring more explicitly to Polya's steps.

## RESTATE THE PROBLEM

It is always good practice to restate the problem in your own words, turning and testing the meanings to gain a better understanding. On some occasions, you will restate the problem so clearly that the solution to the problem becomes almost obvious.

**EXAMPLE 4**

To select the best two-person debate team in the United States, 117 teams are selected, with the understanding that when a team loses a debate, the team is dropped from the tournament. How many debates must be completed to select a winner?

*SOLUTION*

**UNDERSTAND THE PROBLEM**   What is unknown? How many debates will it take to determine the winner? What are the data? There are 117 teams, and it is a single-elimination tournament. In other words, if a team loses one debate, it is out of the tournament. If Team A competes against Team B and Team A wins, Team B exits from the tournament and Team A competes again.

**DEVISE A PLAN**   Let us try to reword this problem so that the answer is clearer.

**CARRY OUT THE PLAN**   The unknown is "How many debates until a champion is determined?" Rephrased:

How many teams must lose in order to choose a champion? 116
How many teams lose per debate? 1
How many debates until a champion is determined? 116

**LOOK BACK**   The championship team is determined after 116 debates. Can we verify this answer another way? Suppose we had used the twin strategies of looking for a pattern in a simpler case. (See Figure 1-11).

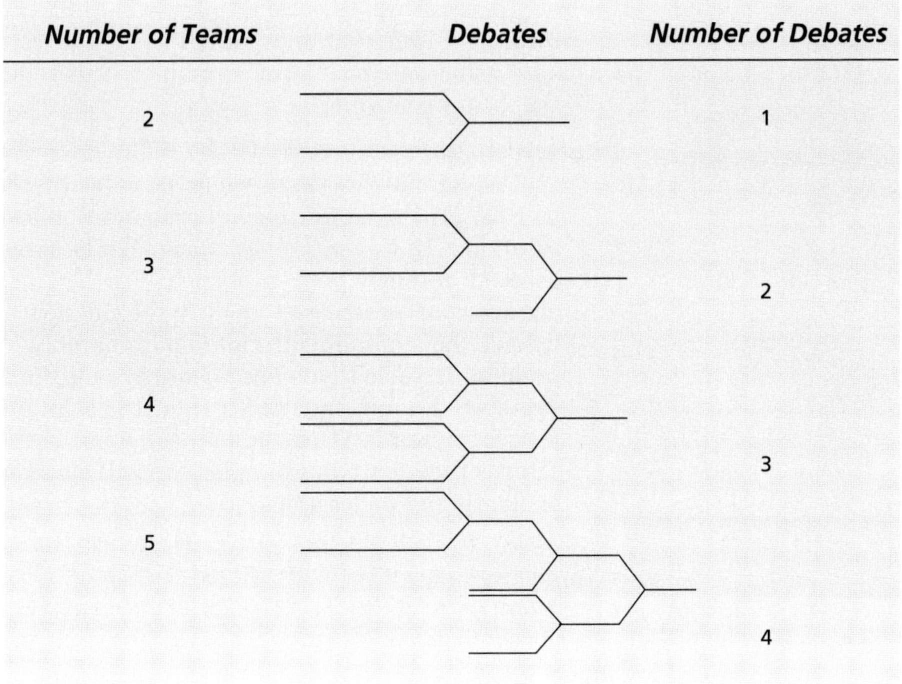

| Number of Teams | Debates | Number of Debates |
|---|---|---|
| 2 | | 1 |
| 3 | | 2 |
| 4 | | 3 |
| 5 | | 4 |

**FIGURE 1-11**

We observe that the required number of debates is 1 fewer than the number of teams competing. Indeed, 117 teams will require 116 debates to determine a winner.

# Use Reasoning; Eliminate Impossible Situations

Example 5 involves the combined strategies of elimination and reasoning, as well as the familiar strategy of making a table. The table furnishes us with an organized method for eliminating incorrect situations. Selecting the clues that lead to elimination is very important.

**EXAMPLE 5**

Five students—Leon, Sarah, Russo, Sue, and Sharon—participate in a debate tournament in which each team must have at least one affirmative debater and one negative debater. Two of the students attend Harp College, and three attend Sloan. Three are affirmative debaters, and two are negative. Leon and Sue attend the same college. Russo and Sharon attend different schools. Sarah and Russo are on the same side. Sue and Sharon are on opposite sides. A negative debater from Harp College was selected outstanding debater. Who was that person?

*SOLUTION*

**UNDERSTAND THE PROBLEM**   Two of the five students attend Harp, and three attend Sloan. We are given facts about Leon and Sue, Russo and Sharon, Sarah and Russo, and Sue and Sharon. We also know that a negative debater from Harp College was named outstanding debater. We are to find out which of the five debaters received this honor.

**DEVISE A PLAN**  We will construct a table (Table 1-2) and reason from given facts to answer "yes" or "no" for each student for each of the categories affirmative, negative, Harp College, and Sloan College.

| Category | Leon | Sarah | Russo | Sue | Sharon |
|---|---|---|---|---|---|
| Affirmative | No | Yes | Yes | | |
| Negative | Yes | No | No | | |
| Harp College | No | Yes | | No | |
| Sloan College | Yes | No | | Yes | |

**TABLE 1-2**

**CARRY OUT THE PLAN**  Leon and Sue attend the same college. Russo and Sharon attend different colleges. What does this tell you? (Leon and Sue, along with either Russo or Sharon attend Sloan.) Sarah must attend Harp. Why? Sarah and Russo are affirmative debaters along with either Sue or Sharon. Leon is a negative debater. Why? The other negative debater is either Sue or Sharon. Sue is not from Harp College. Sharon can be from Harp College. Therefore, the outstanding debater is Sharon.

If you do not yet feel like an expert problem solver, do not be overly worried. We will revisit these same ideas throughout the text and use many more strategies in the days ahead. What you should understand now is that no single strategy is a magic wand that you can use for all problems, but that all strategies provide ideas that you can use in devising your own plans to solve problems.

# Just for Fun

What is the least number of pitches that a pitcher can throw in the course of pitching a full-length baseball game?

# EXERCISE SET 3

*R* 1. If it takes 12 minutes to cut a log into 3 pieces, how long would it take to cut the log into 4 pieces? (HINT: Draw a picture.)

2. Charles and Gary earn the same amount of money, although one works 5 days more than the other. If Charles earns $56 a day and Gary earns $96 a day, how many days does each work? (HINT: Try guess, test, and revise on this one.)

3. Without adding all 30 numbers, find the sum of the first 30 odd counting numbers. (HINT: Look at simpler cases, and look for a pattern.)

4. Jack goes to the well to get some water. He has a 3-liter pail and a 5-liter pail, and he wishes to return with exactly 1 liter of water. How can he do this? (HINT: Work backward!)

5. Jack's 5-liter pail in Exercise 4 springs a leak, and he must carry a 3-liter pail and a 7-liter pail to the well. How can he return with 1 liter of water in this case?

*In each of Exercises 6 through 22, record the strategies that you try as you look for a solution, even if you are not successful.*

6. Forty-four players enter a single-elimination tennis tournament on the Fourth of July. How many matches must be played to determine a winner?

7. Each member of the Wheeler family gives a gift at Christmas to all other members of the family. Find the total number of gifts given by the ten members of the family.

8. Arrange the first 6 counting numbers in the 6 circles so that the sum of each side (3 circles) of the triangle totals (a) 9 and (b) 10. (c) Are there other possible common sums?

*T* 9. Eighty-six baseball teams participate in a double-elimination tournament. (A team continues to compete until it has lost two games.) How many games must be played to determine a champion (a) if the champion does not lose a game? (b) if the champion loses one game?

10. How can one use a 3 minute timer and a 5 minute timer to measure 7 minutes?

11. Draw a straight line across the face of a clock so that the sums of the numbers on each side of the line are the same.

12. What is the largest sum of money—all in coins, with no silver dollars—that I could have in my pocket without being able to give change for a dollar, half dollar, quarter, dime, or nickel?

13. You have $1 in change. You have at least one of each coin less than a half dollar. You do not have a half dollar. What is the smallest number of coins you can have?

14. Jodi traded dollar bills for dimes and quarters and received the same number of each. What is the smallest number of dollars she could have had?

15. How many cubes are there in this figure? (Assume that all hidden cubes are contained in the figure.)

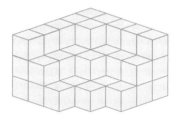

16. Twenty-five coins total $1. If one of the coins is a quarter, what are the other coins?

17. Can you determine which of ten dimes is counterfeit (lighter than others) with only three weighings on a balance scale? How?

18. These 4 steps are made of cubes. How many cubes are needed to make 20 steps?

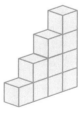

19. How can you obtain 6 gal. of cider from a tank if you have only a 4-gal. container and a 9-gal. container?

*C* 20. If a bird and a half lays an egg and a half in a day and a half, how many eggs will 33 birds lay in 11 days?

21. If 3 cats catch 3 rats in 100 minutes, how many cats will catch 100 rats in 100 minutes?

22. If 3 hens lay 3 eggs in 3 days, how many eggs will 300 hens lay in 300 days?

23. Arno, Terry, Aaron, Mike, and Chris are starters on the Windsor Forest basketball team. Two shoot with their left hands, and three shoot with

their right hands. Two are over 6 ft. tall, and three are under 6 ft. Arno and Aaron shoot with the same hand; Mike and Chris use different hands to shoot. Terry and Chris are in the same height range, but Aaron and Mike are in different height ranges. The player who plays center is over 6 ft. tall and is left-handed. Who is he? (HINT: Try using reasoning to eliminate impossible cases on this one. Completing the following table might be helpful.)

|  | Arno | Terry | Aaron | Mike | Chris |
|---|---|---|---|---|---|
| **Left Hand** |  |  |  |  |  |
| **Right Hand** |  |  |  |  |  |
| **Taller than 6 ft.** |  |  |  |  |  |
| **Shorter than 6 ft.** |  |  |  |  |  |

24. Kanisha and Byron are playing a game called NIM. Each of them has a large pile of pennies, and the game pile starts empty. In turn, they move 1, 2, or 3 pennies from their pile to the game pile. The player who brings the total in the game pile to 24 wins the game. Byron plays first. Devise a strategy so that Kanisha can be sure of winning. (HINT: Since the goal is clearly stated, try working backward to find the best strategy for Kanisha to use. If there are 21, 22, or 23 pennies on Byron's last move, he can win. Thus Kanisha should arrange for Byron to have a 20 on his last move; clearly she can win if he adds 1, 2, or 3 to 20. Similarly, she should arrange for Byron to play with a game pile of 16 on his next to last move. Describe the rest of the strategy.)

25. Three soccer teams—the Cats, the Dogs, and the Ants—play a three-game tournament. We have the following information: Each team played two games. The Cats won two games; the Dogs tied one game. The Dogs scored a total of two goals; the Ants scored three goals. One goal was scored against the Cats, four against the Dogs, and seven against the Ants. Find the scores of all three games.

26. It takes 8 days for a truck to cross the desert. The truck can carry gasoline for only 5 days. What is the smallest number of trucks to start the trip in order for one to get across and the others to return?

27. Cut a 3-by-8 rectangle into two pieces with one cut, and form a 2-by-12 rectangle with the pieces.

28. Labels were incorrectly placed on each of the three boxes. Explain how it is possible to determine the correct labels by removing just one ball from one box.

| 2 red balls | 1 red ball 1 black ball | 2 black balls |
|---|---|---|

29. A standard 8-by-8 checkerboard is shown below. Is it possible to cover this checkerboard with dominos (one domino covers two blocks) so that the two corners labeled with * are left uncovered? Explain. (HINT: Experiment with a miniature 2-by-2 checkerboard, a 3-by-3 checkerboard, and a 4-by-4 checkerboard until you understand the pattern.)

30. Thirty points are marked on a circle, and straight line segments are drawn to connect each point with all the other points. How many line segments are drawn?

## REVIEW EXERCISES

31. (a) Conjecture a pattern for this sequence, and complete the next three terms: 1, 2, 4, _____ , _____ , _____ .
    (b) Conjecture a different pattern for this same sequence and fill in the next three terms.

32. Jane has 20 coins consisting of nickels and dimes, totaling $1.20. How many of each does she have?

#### ∴ PCR Excursion ∴

33. The sequence of numbers 1, 3, 5, 7, 9, . . . is called the *sequence of odd counting numbers*. The sums 1, $1 + 3$, and $1 + 3 + 5$ are the first three partial sums of the odd counting numbers and are represented geometrically below:

| 1 | 1 + 3 | 1 + 3 + 5 |
|---|---|---|

A. (a) Represent the next 3 of these partial sums geometrically.
   (b) Because the dots in $1 + 3$ form a square with 2 dots on a side,

$$1 + 3 = 2^2$$

Because the dots in $1 + 3 + 5$ form a square with 3 dots on a side,

$$1 + 3 + 5 = \underline{\hspace{1cm}}$$

Describe the next 3 of these partial sums in a similar way.

(c) Use the pattern you observed in (b) to find the sum of the first 8 odd counting numbers; the first 100 odd counting numbers.

(d) Write the sum of odd counting numbers that is equal to $n^2$.

B. Look at the following rectangular arrays of dots:

2 by 3     3 by 4     4 by 5

(i)     (ii)     (iii)

(a) Observe that because rectangle (i) has 2 rows and 3 columns, it has $2 \cdot 3$ dots. Alternatively, looking at the dots on diagonal lines, we see that it has $1 + 2 + 2 + 1$ dots. We can conclude from this that

$$1 + 2 + 2 + 1 = 2 \cdot 3$$

Because rectangle (ii) has 3 rows and 4 columns, it has $3 \cdot 4$ dots. Alternatively, the dots on diagonal lines indicate that it has $1 +$

$2 + 3 + 3 + 2 + 1$ dots. What can we conclude from rectangle (ii)? What similar relationship can we find in rectangle (iii)?

(b) Draw a rectangle of dots with 5 rows and 6 columns. What can we learn? How about a rectangle with 6 rows and 7 columns?

(c) Using the pattern you observed above, what fact can you learn about this sum?

$$1 + 2 + 3 + 4 + 5 + 6 + 7 + 7 + 6 + \cdots + 1$$

(d) Write a sum equal to $9 \cdot 10$.

(e) Write a sum equal to $n \cdot (n + 1)$.

(f) Because $1 + 2 + 3 + 3 + 2 + 1$ is equal to both $2(1 + 2 + 3)$ and $4 \cdot 3$, we can conclude that $2(1 + 2 + 3) = 4 \cdot 3$. Dividing by 2 yields

$$1 + 2 + 3 = \frac{4 \cdot 3}{2}$$

Find a similar pattern for the sums $1 + 2 + 3 + 4$ and $1 + 2 + 3 + 4 + 5$.

(g) Use the reasoning of step (f) to determine a formula for the sum $1 + 2 + 3 + 4 + \cdots + n$.

# SECTION 5 FORMING ALGEBRAIC MODELS

**PROBLEM**

An engineer is paid $24 per hour and his helper is paid $16 per hour. Together they worked 38 hours on a job and made a total of $744. How many hours did the engineer work?

**OVERVIEW**

In Section 3 we observed that the strategy of guess, test, and revise often reveals patterns that can be summarized by algebraic equations. When we form an equation or inequality that represents the problem, we have formed an algebraic model for the problem. In this section, we will use guess, test, and revise to lead us to algebraic equations. You may wish to solve these equations from your knowledge of algebra, or you can guess, test, and revise to find soutions.

**GOALS**

Illustrate the Algebra Standard, page 4.
Illustrate the Representation Standard, page 50
Illustrate the Problem Solving Standard, page 15.
Illustrate the Communication Standard, page 78.

The ability to translate some verbal information into precise algebraic form requires the use of one or more variables.

**EXAMPLE 1**

A delivery man has 4 boxes weighing a total of 60 pounds. Box B weighs twice as much as Box A, Box C weighs twice as much as Box B and Box D weighs twice as much as Box C. Express these relationships mathematically.

SOLUTION

To summarize this information mathematically, we need to be able to represent the weight of Box A without knowing its numerical value. To do so we use a variable like $x$ or $y$ or $\square$. Let $\square$ be the weight of the lightest box, Box A. Then

Weight of Box B $= 2 \cdot \square$
Weight of Box C $= 2(2 \cdot \square) = 4 \cdot \square$
Weight of Box D $= 2(4 \cdot \square) = 8 \cdot \square$
Total weight: $\square + 2 \cdot \square + 4 \cdot \square + 8 \cdot \square = 60$

**EXAMPLE 2**

Translate the following phrases into mathematical language by using the variable $x$ in place of the unspecified number.
(a) Six added to the product of nine times a certain number
(b) Eight less than the product of three times a number

SOLUTION

(a) $9 \cdot x + 6$        (b) $3 \cdot x - 8$

In the preceding example, $9 \cdot x$ can also be written as $9x$, and $3 \cdot x$ as $3x$.

**EXAMPLE 3**

In each of the following expressions, replace the variable $x$ with the number 3 and evaluate.
(a) $8x + 4$        (b) $2(x + 1) - x$

SOLUTION

(a) $8 \cdot 3 + 4 = 28$    (b) $2(3 + 1) - 3 = 5$

A very important step in solving many problems involves representing the unknown values by variables and then writing the equation that summarizes mathematically the information given in the problem. When you have used variables and have written an equation, *you have designed an algebraic model* for the problem. $x + 4 = 7$ is an equation that may represent information in a problem. It is sometimes called an *open sentence*. When the variables in an equation or open sentence are replaced by numerical values, the equation can then be classified as true or false. Replacing $x$ by 2, $2 + 4 = 7$ (false); but replacing $x$ by 3, $3 + 4 = 7$ (true).

**EXAMPLE 4**

Translate the following expressions into mathematical sentences.
(a) A number added to 15 is equal to 23.
(b) Two times some number plus 7 equals 15.

SOLUTION

(a) $x + 15 = 23$      (b) $2\square + 7 = 15$

**EXAMPLE 5**

Using the equations of the preceding example, replace $x$ by 8 in (a) and $\square$ by 3 in (b); then classify the mathematical sentences as true or false.

SOLUTION

(a) $8 + 15 = 23$  (true)      (b) $2 \cdot 3 + 7 = 15$  (false)

If a replacement turns an equation into a true sentence, then the replacement is called a **solution** of an equation. The possible replacements for a variable of an equation make up the **domain** of the variable.

**EXAMPLE 6**

*SOLUTION*

Is there a solution for $x + 3 = 8$ on the domain {1, 2, 3, 4, 5, 6}?

(HINT: Use guess and test) Yes, $x$ replaced by 5 gives a true statement: $5 + 3 = 8$. So 5 is a solution of $x + 3 = 8$.

In this section most equations will have solutions in the domain of the counting numbers {1, 2, 3, . . .}. In Chapter 4 we will solve equations on the domain of whole numbers, and in subsequent chapters we will solve equations on the domain of integers, rational numbers, and real numbers.

However, in this section we will not focus on the techniques of solving equations. Rather we will focus on the problem solving strategy of using a variable to create equations (algebraic models) that represent the problems. Specifically, we will use "guess and test" to help us see the patterns that determine the equations.

Once an algebraic model for a problem is created, you can solve the problem merely by solving the equation using standard techniques from algebra, or you can solve the algebraic equations of this section by wise use of guess, test, and revise.

**UNDERSTAND THE PROBLEM** The important first step is understanding the problem. In Section 3 we discussed several aids for doing this. It is especially important to identify the unknown and name it with a variable.

**WRITE THE EQUATION** In some cases, you will see clearly how to translate the information in the problem into an algebraic equation. On other occasions, you will be stumped. On those occasions we recommend that you use the "guess, test, and revise" strategy to help you discover the algebraic model.

**CARRY OUT THE PLAN; LOOK BACK** Because the plan we are using in this section involves representing the problem with an algebraic equation, to carry out the plan, we must solve the equation. In the "look back" step, at the very least we must write a sentence representing what we learn from the solution of the equation, and we must check the solution against the requirements of the problem.

**EXAMPLE 7**

*SOLUTION*

A rectangle is 6 feet longer than it is wide. If the perimeter is 28 feet, how long is the rectangle? Find an algebraic model for this problem.

**Unknown:** Let $x$ = the length of the rectangle in feet.
**Data:** The length is 6 feet more than the width. The perimeter of the rectangle is 28 feet. We know (or we look up) the fact that perimeter is the sum of the lengths of the four sides of the rectangle.

**WRITE THE EQUATION** To help us see more clearly the relationships in the problem, we guess a value for the unknown. (See Figure 1-12(a).)

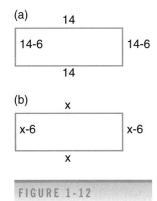

(a)

(b)

FIGURE 1-12

| | Length | Width | Perimeter |
|---|---|---|---|
| **Guess** | 14 | $(14 - 6)$ | $14 + 14 + 8 + 8 \overset{?}{=} 28$ |

Examining our work, we see that this guess fails because the sum is 44 feet and not 28 feet. However, we also see more clearly how to write our equation. (See Figure 1-12(b).)

| | Length | Width | Perimeter |
|---|---|---|---|
| **Using a Variable** | $x$ | $x - 6$ | $x + x + x - 6 + x - 6 = 28$ |

The algebraic model for this problem is the equation:

$$x + x + x - 6 + x - 6 = 28 \qquad \text{or} \qquad 4x - 12 = 28$$

**EXAMPLE 8**

Complete the solution for Example 7 by solving the equation that represents the information in the problem and interpreting the results.

*SOLUTION*

We will illustrate two solutions to this problem. We will show how to find the solution of the equation $4x - 12 = 28$ by judicious use of guess, test and revise. We will also review the algebraic solution of the problem.

*Solution 1:*

| **Guess** | **Substitute in $4x - 12 = 28$** |
|---|---|
| $x = 6$ | $4(6) - 12 = 12 \neq 28$ (need larger guess) |
| $x = 8$ | $4(8) - 12 = 20 \neq 28$ (need larger guess) |
| $x = 10$ | $4(10) - 12 = 28$ |
| The solution is $x = 10$ | |

*Solution 2:* To solve the equation algebraically, we wish to isolate $x$ on one side of the equation. We will accomplish this by adding 12 to both sides of the equation and then dividing both sides of the equation by 4.

$$4x - 12 = 28$$
$$4x - 12 + 12 = 28 + 12$$
$$4x = 40$$
$$4x/4 = 40/4$$
$$x = 10$$

**LOOK BACK**  We first state the result in a sentence reflecting what has been learned about the problem: The length of the rectangle is 10 feet. If the length is 10 feet, the width is $10 - 6 = 4$ feet. Thus the perimeter is $10 + 10 + 4 + 4 = 28$.

**EXAMPLE 9**

An engineer is paid $24 per hour and his helper is paid $16 per hour. Together they worked 38 hours on a job and made a total of $744. How many hours did the engineer work?

*SOLUTION*

**Unknown:** Let $h$ = the number of hours that the engineer worked on the job.

**Data:** A total of 38 hours are worked.

A total of $744 was earned.

The engineer earned $24 per hour; the helper, $16 per hour.

Total wages = (dollars per hour) (number of hours)

**WRITE THE EQUATION** To help us understand the relationships in this problem, we will guess that $h = 10$ hours and carefully test this guess.

| Guess 1 | Engineer's Hours | Helper's Hours | Total Wages |
|---|---|---|---|
| $h = 10$ | 10 | (38 − 10) | $24(10) + 16(38 - 10) \overset{?}{=} 744$ |

Since the total wages in this case are $688, we must guess more hours for the engineer.

| Guess 2 | Engineer's Hours | Helper's Hours | Total Wages |
|---|---|---|---|
| $h = 15$ | 15 | (38 − 15) | $24(15) + 16(38 - 15) \overset{?}{=} 744$ |

Again, the total wages in this case are $728 so we do not yet have a solution. However, we use the variable $h$ to express the pattern we see in the computations above. Then we get the algebraic model for this problem.

| Engineer's Hours | Helper's Hours | Total Wages |
|---|---|---|
| $h$ | (38 − h) | $24h + 16(38 - h) = 744$ |

**CARRY OUT THE PLAN** We could proceed to find this solution by guess, test and revise. Alternatively, we can solve the problem algebraically. Since $16(38 - h) = 608 - 16h$, the equation can be solved as follows. Remember that we wish to isolate $h$ on one side of the equation.

$$24h + 16(38 - h) = 744$$
$$24h + 608 - 16h = 744$$
$$24h - 16h + 608 - 608 = 744 - 608$$
$$8h = 136$$
$$h = 17$$

**Look back**   The engineer worked 17 hours. Therefore the helper worked $(38 - 17) = 21$ hours. We can quickly check that

$$24(17) + 16(21) = 744$$

**EXAMPLE 10**

A car averages 32 miles per gallon on interstate highways and averages 22 miles per gallon on city streets. On a recent trip the car traveled 600 miles and used 20 gallons of gasoline. How many of the gallons of gasoline were used while driving on interstate highways?

*SOLUTION*

**Unknown:**   Let $g$ = the number gallons used while driving on interstate highways.

**Data:**   The total distance travelled on the trip was 600 miles. Twenty gallons of gasoline were used.

The car produced 32 miles per gallon on interstates and 22 miles per gallon on city streets.

Total distance = (miles per gallon) (gallons).

**Write the equation**   To help see the relationships present in this problem, we guess that the number of gallons used on interstate highways $(g) = 8$. To test this guess we note that the number of gallons used on city streets must be $(20 - 8)$.

| Guess | Miles on Interstates | Miles on City Streets | Total Miles |
|---|---|---|---|
| $g = 10$ | $32(8)$ | $22(20 - 8)$ | $32(8) + 22(20 - 8) \overset{?}{=} 600$ |

Although our guess was wrong since $32(8) + 22(20 - 8)$ is not equal to 600, we can see the equation that must be solved.

$$32g + 22(20 - g) = 600$$

**Carry out the plan**   To complete the algebraic solution we follow the steps below:

$$32g + 22(20 - g) = 600$$
$$32g + 440 - 22g = 600$$
$$10g + 440 = 600$$
$$10g + 440 - 440 = 600 - 440$$
$$10g = 160$$
$$g = 16$$

**Look back**   Sixteen gallons were used in driving on interstate highways and $(20 - 16) = 4$ gallons were used in driving on city streets. Clearly

$$32(16) + 22(4) = 600$$

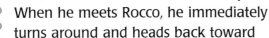

Ricco and Rocco live in towns that are 32 miles apart. At 8:00 on Saturday morning, they start bicycling toward one another. Each travels at a constant rate of 8 miles per hour. Ricco's faithful dog, Fido, runs ahead at a rate of 20 mph. When he meets Rocco, he immediately turns around and heads back toward Ricco. He continues to race between the two boys until they meet midway between their homes. How far does Fido run?

## EXERCISE SET 4

*R* 1. Translate the following verbal phrases into mathematical expressions or sentences using variables.

(a) Five added to the product of four times a certain number.

(b) Seven less than the product of two times a number.

(c) The quotient of four times a number and six.

(d) The difference of five times a number and seven (or less seven).

2. Evaluate each of the following mathematical expressions at $x = 4$.

(a) $8x + 2$      (b) $2x - 5$
(c) $4(x + 1) + 1$      (d) $2(x - 3) + 2$

*For each of Exercises 3–8, name the unknown and use guess and test to find an algebraic model (an equation) that represents the information in the problem.*

3. A rectangle is 3 feet longer than it is wide. If the perimeter is 26 feet, how long is the rectangle?

4. A short order cook earns $8.00 an hour and her assistant earns $5.00 an hour. During one long night they worked a total of 18 hours and earned a total of $129. How many hours did the short order cook work?

5. A car averages 28 miles per gallon on interstate highways but only averages 20 miles per gallon on city streets. On a recent trip the car traveled 784 miles and used 30 gallons of gasoline. How many gallons of gasoline were used driving on interstate highways?

6. If 6 times a number is increased by 10 the result is 8 times the number. What is the number?

7. Marcellus has $3.90 (390 cents) worth of change in his pockets consisting exclusively of nickels and quarters. If he has a total of 30 coins, how many quarters does he have?

8. Kisha is 12 years old and her brother Albert is 2 years old. In how many years will Kisha be twice as old as Albert?

*T* 9. Solve the following equations. If you use a procedure other than guess, test, and revise, check each solution by making certain that when the equation is evaluated at the solution, the resulting sentence is true.

(a) $2x + 4 = 12$      (b) $3x - 5 = 10$
(c) $5x = 50$      (d) $x + 3 = 12$
(e) $3x - 6 = 9$      (f) $5x + 8 = 13$

10. For each of the following exercises, solve the equation that you previously created and interpret the solution with an English sentence.

(a) Exercise 3      (b) Exercise 4
(c) Exercise 5      (d) Exercise 6
(e) Exercise 7      (f) Exercise 8

*Create an algebraic model for each of the following problems, solve the problem, and interpret your solution with an English sentence.*

11. A delivery man has four boxes weighing a total of 60 pounds. Box B weighs twice as much as Box A, Box C weighs twice as much as Box B and Box D weighs twice as much as Box C. How much does Box A weigh?

*C* 12. Ricco and Rocco live in towns that are 40 miles apart. At 8:00 A.M. on Saturday morning, they start toward each other on their bicycles. Ricco travels 8 miles per hour and Rocco travels 12 miles per hour. How long until they meet each other? HINT: Distance = (rate) (time).

13. A swimming pool is empty. At 1:00 P.M. a hose is turned on and begins to deliver 5 gallons per minute into the swimming pool. Thirty minutes later a second hose is turned on and pours 7 gallop is per minute into the pool. At what time will there be 1710 gallons of water in the pool?

14. Theater tickets for children cost $2.00 while tickets for adults cost $4.00. If the sale of 512 tickets resulted in revenue of $1026, how many adult tickets were sold?

15. Ferrante rents a compact car for $25 a day plus 20 cents per mile. If he rents the car for one day, how far can he go for $50?

16. A walker starts down a path at a rate of 80 meters per minute. Ten minutes later a second walker starts down the path at a rate of 100 meters per minute. How long until the second walker catches the first?

17. Savannah and Atlanta are 280 miles apart. Grandpa leaves Atlanta at 1:00 P.M. and drives toward Savannah at a rate of 40 miles per hour. Grandson leaves Savannah at 2:00 P.M. and travels toward Atlanta at a rate of 80 miles per hour. When will they meet?

18. A brick mason earns $18 per hour for his first 40 hours of work each week and earns time and one half for overtime (hours in excess of 40 hours each week). One week he earned $1098. How much overtime did he work?

## REVIEW EXERCISES

19. Find the pattern.

$$1 = 1$$
$$3 + 5 = 8$$
$$7 + 9 + 11 = 27$$
$$13 + 15 + 17 + 19 = 64$$
$$21 + 23 + 25 + 27 + 29 = 125$$

Predict the sum of $31 + 33 + 35 + 37 + 39 + 41$.

20. Three mothers—Mary, Eve, and Jean—have a total of 14 children. Mary has 2 boys and Eve has the same number of girls. Eve has no more children than Mary, who has 3 children. Jean has 6 more girls than boys and the same number of boys as Mary has girls. How many girls each do Jean and Mary have?

## LABORATORY ACTIVITY

21. *(Tower of Hanoi)* Draw a rectangle and divide it into three equal sections to create a playing board. Place coins of different denominations (such as a quarter, a nickel, and a penny) as in the figure.

The object is to move the coins one at a time from Section I to Section III with as few moves as possible, subject to the following rules.

(i)   When two or more coins are in the same section, the coins must be in a stack, with the greatest value on the bottom and with coins of decreasing value stacked up to the top.

(ii)   A coin may not be moved into a section if it must be placed on a coin of smaller value.

(a) Can you move two coins to Section III in $2^2 - 1$ moves?

(b) Can you move the three coins in $2^3 - 1$ moves?

(c) Find an expression for the number of moves necessary to move four coins.

(d) How many moves are necessary for $n$ coins?

### ∴ PCR Excursion ∴

22. A. Follow the instructions on each of the following "tricks" by choosing at least two counting numbers for each of (i), (ii), (iii), and (iv).

(i) Think of a counting number.
   Add 4 to this number.
   Multiply your answer by 2.
   Subtract 6 from your answer.
   Divide by 2.
   Subtract the number with which you started.
   Your answer is 1.

(ii) Choose a counting number.
    Add the next consecutive counting number.
    Add 3.
    Divide by 2.
    Subtract 1.
    Subtract the original number.
    Your answer is 1.

(iii) Choose a counting number.
    Double it.
    Add 12.
    Add your original number.
    Divide by 3.
    Add 3.
    Subtract the original number.
    Your result is 7.

(iv) Choose a counting number.
    Add 6.
    Double the result.
    Subtract 8.
    Divide by 2.
    Subtract the original number.
    Your result is 2.

B. Suppose in (i) of part A we let the specified number be denoted by a box and the quantity we add to that number by circles.

Think of a number: □
Add 4: □○○○○
Multiply by 2: { □○○○○
(same as doubling)   □○○○○
Subtract 6: { □○
(take away 6)   □○
Divide by 2: □○
(same as taking half)
Subtract the original number: ○

Because the box represents any number and a circle represents 1, we have used deductive reasoning to show that the result in (i) will always be 1.

(a) Use similar reasoning to show (ii) is always true.
(b) Do the same for (iii).
(c) Do the same for (iv).

C. (a) Now in (i) let the number you think of be represented by $x$.

Show that the result is always 1.

(b) Use $x$ to prove the result in (ii).
(c) Use $x$ to prove the result in (iii).
(d) Use $x$ to prove the result in (iv).

D. Now write a number trick of your own and verify that it always works.

# SOLUTION TO INTRODUCTORY PROBLEM

**UNDERSTANDING THE PROBLEM.** First, let's restate the problem. Three times the following procedure is executed:

(a) The remaining pile is divided into three equal piles with one extra coconut.

(b) One of the three piles is hidden, while the other two equal piles are combined and left.

Finally, the remnant after these three incursions is divided into three equal piles, with one coconut remaining.

**DEVISING A PLAN.** Work the problem backward. Notice that the sum of the last three piles plus 1 must be an even number because it is the sum of two equal previous piles. This situation must be true for the total number of coconuts at all times except at the beginning. This fact allows us to rule out possibilities as we look at various cases.

**CARRYING OUT THE PLAN.** Let $n$ be the number of coconuts received by each sailor at the end. Now $n$ cannot be even, because $n + n + n + 1$ would be odd, which is a contradiction. If $n = 1$, then

$1 + 1 + 1 + 1 = 4$ (Carl has divided into piles of 2.)

$2 + 2 + 2 + 1 = 7$ (odd)

which is a contradiction. If $n = 3$, then

$3 + 3 + 3 + 1 = 10$ (Carl has divided into piles of 5.)

$5 + 5 + 5 + 1 = 16$ (Bob has divided into piles of 8.)

$8 + 8 + 8 + 1 = 25$ (odd)

which is a contradiction. If $n = 5$,

$5 + 5 + 5 + 1 = 16$ (Carl has divided into piles of 8.)

$8 + 8 + 8 + 1 = 25$ (odd)

which is a contradiction. If $n = 7$,

$7 + 7 + 7 + 1 = 22$ (Carl has divided into piles of 11.)

$11 + 11 + 11 + 1 = 34$ (Bob has divided into piles of 17.)

$17 + 17 + 17 + 1 = 52$ (Mark has divided into piles of 26.)

$26 + 26 + 26 + 1 = 79$

The least possible number of coconuts in the pile is 79.

**LOOKING BACK.** Mark divides the pile into 3 stacks of 26, hides his 26, and gives the extra coconut to the monkey. This leaves 52 coconuts in the pile. Bob divides the pile into 3 stacks of 17, hides his 17, and gives the extra coconut to the monkey. This leaves 34 coconuts in the pile. Carl divides the pile into stacks of 11, hides his 11, and gives the extra to the monkey. This leaves 22 coconuts in the pile. In the morning they divide the pile, giving each sailor 7 coconuts and the monkey 1.

# SUMMARY AND REVIEW

## NUMBER SEQUENCES

1. An arithmetic sequence is characterized by a common difference between terms.

2. In a geometric sequence, a term divided by the preceding term gives a common ratio.

## PROBLEM SOLVING

*Polya's four steps:*

1. Understand the problem.
2. Devise a plan.
3. Carry out the plan.
4. Look back.

*Understanding a problem:*

1. Identify what you are trying to find (the unknowns).

2. Summarize the information that is available to you (the data).
3. Don't impose conditions that do not exist.
4. Strip the problem of irrelevant details.

*Strategies for problem solving:*

1. Make a chart or a table.
2. Draw a picture or a diagram.
3. Guess, test, and revise.
4. Look for a pattern.
5. Try a simpler version of the problem.
6. Restate the problem.
7. Eliminate impossible situations.
8. Use reasoning.
9. Form an algebraic model.
10. Work backward.

# CHAPTER TEST

1. Study two rows at a time of the accompanying triangle to determine patterns of regularity. Note that the sum of two consecutive numbers in a row gives the number between them in the next row. In the fourth row, for example, the sum of 1 and 3 equals 4 (circled) in the fifth row, and the sum of 3 and 3 is 6 (circled). Complete the sixth and seventh rows of the array.

$$
\begin{array}{ccccccccc}
 &  &  &  & 1 &  &  &  & \\
 &  &  & 1 &  & 1 &  &  & \\
 &  & 1 &  & 2 &  & 1 &  & \\
 & 1 &  & 3 &  & 3 &  & 1 & \\
1 & & \textcircled{4} & & \textcircled{6} & & 4 & & 1
\end{array}
$$

2. A farmer was asked by a passing stranger how many chickens and how many goats he had in his farmyard. The farmer answered that his animals had 90 legs and 62 eyes. How many of each animal did the farmer have?

3. A small boy determines to increase his physical stamina. Each morning he does pushups, increasing the number completed by 6 each day. After 5 days, he has done a total of 100 pushups. How many did he do each day?

4. A developer wants to place lots around each of four sides of a square block. If each of the four sides can be divided into seven lots (including corner lots), how many lots are available?

5. One of a group of nine dimes is counterfeit and lighter than the others. Explain how you could identify the counterfeit dime with only two weighings on a balance scale.

6. What is the 36th term of the arithmetic sequence 17, 20, 23, 26, . . . ?

7. Determine the next three entries in each sequence.
   (a) 2, 7, 12, 17, . . .
   (b) 1, 5, 1, 1, 5, 1, 1, 1, 5, . . .
   (c) A, C, D, E, G, H, I, K, L, M, . . .

8. Three times Ralph's weight added to 54 kg is equal to 300 kg. How much does Ralph weigh?

9. Can you determine a pattern?

$$1 + 2 = 3$$
$$4 + 5 + 6 = 7 + 8$$
$$9 + 10 + 11 + 12 = 13 + 14 + 15$$

Find the next two lines.

10. How can one measure 2 quarts from a faucet with a 6-quart container and a 10-quart container?

11. If 2 carpenters can build one deck in 1 day, how many days will it take 4 carpenters to build 8 decks?

12. Yolanda leaves Atlanta early in the morning for a very important presentation in Houston. She usually drives at a speed of 60 miles per hour. One hour later, Rolf discovers that Yolanda has forgotten her file needed for this presentation. He immediately starts after her at a rate of 75 miles per hour. How many hours will it take for Rolf to catch Yolanda?

13. The expression for the $n$th term of a sequence is $3n - 1$. Write the first five terms of this sequence.

14. Consider this sequence: 2, 6, 18, 54, . . .
    (a) Find the next three terms.
    (b) Find the tenth term.
    (c) Write an expression for the nth term.

15. Admission to a certain movie was $6 for adults and $3 for children. If 300 tickets were sold for a total of $1320, how many adult tickets were sold? How many children's tickets were sold?

# THE LANGUAGE OF LOGIC

**2**

**MALL PARKING - ORANGE LEVEL**

Elevators

Dale, Sigmund, and Bob each own a car. One of them owns a Saturn, one owns a Honda, and one owns a Plymouth. One car is white, one car is blue, and one car is gray.

Dale does not own the Saturn, and his car is not blue.

If the Honda does not belong to Dale, then it is white.

If the blue car is either the Saturn or belongs to Sigmund, then the Plymouth is white.

If the Honda is either gray or white, then Sigmund does not own the Saturn.

Determine the color and owner of each car.

Mathematics contributes many things to our culture, but at the most fundamental level, mathematics is a language. It is the language of choice when one must speak or write with great precision. In this chapter and the next, we will examine two important components of this language. In this chapter we will introduce the vocabulary and concepts of logic, and in the next chapter we will introduce the vocabulary and concepts of sets.

In Chapter 1 we quite often used inductive reasoning to help us solve problems; that is, we looked at several examples and succeeded in detecting the important pattern that led to a solution. On one or two occasions in Chapter 1, we probed more deeply and used deductive reasoning to discover why the pattern worked and why it was true in all possible cases. The formal language of deductive reasoning is the language of logic. In this chapter we shall examine the basic ideas of this language of logic. In so doing, we will become more precise users of language on a day-to-day basis and obtain a glimpse of how mathematicians regularly use deductive reasoning to establish the validity of conjectures.

Problem-solving strategies emphasized in this chapter include the following: ▪ Draw a picture. (56, 68, 71, 72) ▪ Use reasoning. (used throughout the chapter) ▪ Make a chart or table. (used throughout the chapter) ▪ Work backward. (47, 58, 63) ▪ Use a counterexample. (70)

# SECTION 1 — AN INTRODUCTION TO LOGIC

| PROBLEM | "Uncle Fred will bring turkey on Thanksgiving Day or he will bring cole slaw." The preceding statement can be true in three different ways. Can you enumerate them? The answer to this question will become more obvious as you study this section. |
|---|---|
| OVERVIEW | To this point in this text, you have improved your critical thinking skills by studying problem-solving strategies and by practicing inductive reasoning. However, philosophers who write about developing critical thinking skills identify logic as an essential step in the process. |
| GOALS | Illustrate the Reasoning and Proof Standard, page 23.* <br> Illustrate the Problem Solving Standard, page 15.* <br> Illustrate the Communication Standard, page 78.* <br> Illustrate the Algebra Standard, page 4.* <br><br> *The complete statement of the standard is given on this page of the book. |

## CONNECTIVES

At the heart of the language of mathematics lies the language associated with elementary logic. Elementary logic provides the terminology and structure that are used in defining mathematical terms and creating mathematical argument. Logic also provides insight in the design of computer programs and electronic circuits. In this chapter we take the first steps toward mastering the fundamentals of logics.

Sentences that can be identified as either *true* or *false* but not both, are called **statements** or **propositions.**

**EXAMPLE 1**

Determine which of the following sentences is a statement. If a sentence is a statement, determine whether it is true or false.

(a) Atlanta is the capital of Georgia.
(b) $2 + 3 = 7$.
(c) John Smith was the first president of the United States.
(d) For all integers, $n^2 + n = n(n + 1)$.
(e) Keep your eyes on the road.
(f) What time is it?
(g) Oh, what a beautiful sunset!
(h) $x + 2 = 6$.
(i) She is captain of the soccer team.
(j) I am lying to you.

*SOLUTION*

Sentences (a), (b), (c), and (d) are statements. Statements (a) and (d) are true, denoted by T; statements (b) and (c) are false, denoted by F. Sentence (e) is a command, sentence (f) a question, and sentence (g) an explanation. They are not statements because they cannot be assigned a truth value. Sentences (h) and (i) are not statements because we do not know the identity of "$x$" in (h) and the identity of "she" in (i). Sentence (j) is not a statement because it is self-contradictory, sometimes called a paradox.

Sentences like (h) and ( i ) that become statements when an unknown is assigned a specific value are called **open sentences** or **propositional functions.** The sentence "$x + 1 = 5$" is an open sentence. If $x = 10$, the resulting statement is false. If $x = 4$, the statement is true.

Our primary concern in this text will be sentences that are statements or open sentences. We will represent such sentences using lower-case letters such as *p, q,* and *r.* When a statement is true we will denote the truth value by T, and when a statement is false by F.

**EXAMPLE 2**

Determine the truth value of statements *p* and *q.*

> *p:* President Lincoln was born in Texas.
> *q:* $2 + 3 = 5$.

*SOLUTION*

Statement *p* is false and hence has truth value F. Statement *q* is true with truth value T.

In ordinary speech and writing we often build more complex sentences from simple ones. For example, we take two sentences and build a new sentence using **connectives** such as *and* and *or.*

1. The birth rate in Freedom County has increased, or there are more young married couples in the county.
2. The number of majors in sociology at Graduatum College has increased, and the number of majors in physics has decreased.

| DEFINITION | **Compound Statement** | A statement is a **compound statement** if it consists of simple statements joined by connectives; otherwise, the statement is a **simple statement.** |

Let's look now at the meaning of the terms *opposite* and *negation*. The opposite of white is commonly accepted to be black, but the negation of white is anything not white.

| DEFINITION | **Negation of a Statement** | The *negation of statement p* is the statement "It is not true that $p$" (denoted by $\sim p$). If $p$ is true, then $\sim p$ is false; if $p$ is false, then $\sim p$ is true. |

The negation of "New York is the capital of the United States" is "It is not true that New York is the capital of the United States." The use of the grammatical form "It is not true that $p$" sometimes results in awkward sentence structures. For many statements, the negation may be more simply expressed by negating the predicate. For instance, the negation of the preceding statement could be written as "New York is not the capital of the United States." By definition, the negation of a statement is false when the statement is true, and true when the statement is false.

Below, the definitions of conjunction and disjunction are given in terms of statements; however, the same definitions hold for sentences, too.

#  TRUTH TABLES

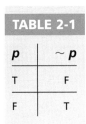

**TABLE 2-1**

| $p$ | $\sim p$ |
|-----|----------|
| T   | F        |
| F   | T        |

Table 2-1 summarizes the two possibilities for $p$ and $\sim p$. Such a table, called a **truth table,** emphasizes the fact that $p$ and $\sim p$ cannot be simultaneously true or simultaneously false.

Consider the statement "Fred ordered hamburgers and Leslie selected hot dogs." We accept this statement as true, provided that on the day in question Fred actually requested hamburgers and Leslie actually helped herself to hot dogs. By contrast, we consider the statement false if Fred did not order hamburgers, regardless of Leslie's preference. Likewise, the statement is false if Leslie did not pick hot dogs, even if Fred did order hamburgers.

| DEFINITION | **Conjunction** | Given any statements $p$ and $q$, the statement "$p$ and $q$" (denoted by $p \wedge q$) is called a *conjunction*. The *conjunction* "$p$ and $q$" is true if both $p$ and $q$ are true; otherwise, "$p$ and $q$" is false. |

Because $p$ is either true or false and $q$ is either true or false, there are four possibilities for the combined truth values of $p$ and $q$. Table 2-2 illustrates these possibilities. The third column of the table identifies the truth value of $p \wedge q$ for each of the four possibilities. When $p$ is true and $q$ is true, then $p \wedge q$ is true. When $p$ is true and $q$ is false, then $p \wedge q$ is false. When $p$ is false and $q$ is true, then $p \wedge q$ is false. Finally, when $p$ is false and $q$ is false, then $p \wedge q$ is false.

The words *but* and *nevertheless* are sometimes used in place of the word *and*. Thus the statement "Takira is 16, but she does not yet drive" is a conjunction.

**TABLE 2-2**

| $p$ | $q$ | $p \wedge q$ | Examples |
|---|---|---|---|
| T | T | T | 2 + 3 = 5 and 1 is less than 2. |
| T | F | F | 2 + 3 = 5 and 4 is less than 3. |
| F | T | F | December has 34 days and Christmas is on December 25. |
| F | F | F | December has 32 days and January has exactly 22 days. |

If you had a true–false question involving "or"—for instance, "The Democrats have control of the House of Representatives or the Republicans control the Senate"—would the answer be true or false?

**DEFINITION**

**Disjunction**

Given any statements $p$ and $q$, the statement "$p$ or $q$" (denoted by $p \vee q$) is called a *disjunction*. The *disjunction* "$p$ or $q$" is true if $p$ is true, if $q$ is true, or if both $p$ and $q$ are true.

The *or* as used in this definition is called an *inclusive or*. This means that statements using the connective *or* are false only when both parts are false. Table 2-3 illustrates the truth values for the statement "$p$ or $q$" for all possible combinations of truth values of $p$ and of $q$.

**TABLE 2-3**

| $p$ | $q$ | $p \vee q$ | Examples |
|---|---|---|---|
| T | T | T | Sugar is sweet or December has 31 days. |
| T | F | T | Sugar is sweet or the earth is flat. |
| F | T | T | 2 + 3 = 7 or 2 is less than 7. |
| F | F | F | 2 + 3 = 7 or 2 is greater than 5. |

**EXAMPLE 3**

Suppose we have the following two statements:

$p$: I exercise daily.
$q$: I am fat.

Translate the following statements into words.
(a) $p \vee q$        (b) $p \wedge q$      (c) $\sim p$
(d) $\sim (p \vee q)$        (e) $(\sim p) \wedge q$

*SOLUTION*

(a) I exercise daily or I am fat.
(b) I exercise daily and I am fat.
(c) I do not exercise daily.
(d) It is not the case that I exercise daily or that I am fat.
(e) I do not exercise daily and I am fat.

**EXAMPLE 4**

Let $p$ be false (F), and let $q$ be true (T). Find the truth values of the following.
(a) $p \wedge q$        (b) $p \vee q$

*SOLUTION*

(a) $p \wedge q$ has truth value F $\wedge$ T, which is F. Thus, the statement is false for this case.
(b) $p \vee q$ has truth value F $\vee$ T, which is T. Thus, the statement is true in this case.

**PRACTICE PROBLEM**

Let $p$ be T, and let $q$ be F. Find the truth values of the following.
(a) $\sim (p \wedge q)$        (b) $(\sim p) \wedge q$

*ANSWER*

(a) The statement $\sim (p \wedge q)$ is true.
(b) The statement $(\sim p) \wedge q$ is false.

Notice that $\sim (p \wedge q)$ and $(\sim p) \wedge q$ have different truth values for the same truth values of $p$ and $q$.

**EXAMPLE 5**

Construct a truth table for $\sim (p \wedge \sim q)$.

*SOLUTION*

The required truth table appears in Table 2-4.

| TABLE 2-4 | | | | |
|---|---|---|---|---|
| $p$ | $q$ | $\sim q$ | $p \wedge \sim q$ | $\sim (p \wedge \sim q)$ |
| T | T | F | F | T |
| T | F | T | T | F |
| F | T | F | F | T |
| F | F | T | F | T |

When working with truth tables, you must be careful to list all possible truth values. In statements involving $p$, $q$, and $r$, for example, $p$ can be T or F, $q$ can be T or F, and $r$ can be T or F. Thus, eight combinations are needed to cover all possibilities.

**EXAMPLE 6**

Make a truth table for $p \wedge (q \vee r)$.

*SOLUTION*

The required truth table appears in Table 2-5.

**TABLE 2-5**

| $p$ | $q$ | $r$ | $q \vee r$ | $p \wedge (q \vee r)$ |
|---|---|---|---|---|
| T | T | T | T | T |
| T | T | F | T | T |
| T | F | T | T | T |
| T | F | F | F | F |
| F | T | T | T | F |
| F | T | F | T | F |
| F | F | T | T | F |
| F | F | F | F | F |

# Just for Fun

Three students—Joan, James, and Edward—agree to be part of an experiment in reasoning. First, the students are shown five ribbons—three blue and two red. The students are told that a ribbon will be placed on each of their backs and that each is to try to determine the color of the ribbon. Joan can see the ribbons on James and Edward. James can see the ribbon on Edward, but Edward cannot see the other ribbons. When asked the color of the ribbon on her back, Joan says that she does not know. Similarly, James cannot reason the answer. However, when Edward is asked the question, he is able to give a correct answer. Why?

# EXERCISE SET 1

*R* 1. Determine which of the following are statements, and classify each statement as either true or false.
   (a) President George Washington was born in Alabama.
   (b) Good morning.  (c) $5 + 4 = 9$
   (d) $x + 5 = 8$  (e) Close the door.
   (f) $2 \cdot 3 = 7$  (g) $3x = 6$
   (h) Help stop inflation.

2. If $p$ is T and $q$ is F, find the following truth values. The form $\sim(\ )$ represents the negation of everything inside the parentheses.
   (a) $(\sim p) \wedge q$  (b) $(\sim p) \vee (\sim q)$
   (c) $\sim(p \vee q)$  (d) $\sim(p \wedge q)$
   (e) $\sim[(\sim p) \wedge q]$  (f) $\sim(p \vee \sim q)$

3. Find the truth values of the statements in Exercise 2 if $p$ is T and $q$ is T.

4. Find the truth values of the statements in Exercise 2 if $p$ is F and $q$ is F.

5. In each of the parts that follow, construct two statements—one using *or* and the other using *and*—and find the truth values of each.
   (a) November has 30 days. Thanksgiving is always on November 25.
   (b) The smallest counting number is 2; 10 is not a multiple of 5.
   (c) $2 + 3 = 4 + 1$  $8 \cdot 6 = 4 \cdot 12$
   (d) Triangles are not squares; 3 is smaller than 5.

6. Translate the following statements into symbolic form, using $A, B, C, D, \wedge, \vee$, and $\sim$, where $A, B, C$, and $D$ denote the following statements:
   > $A$: It is snowing.
   > $B$: The roofs are white.
   > $C$: The streets are not slick.
   > $D$: The trees are beautiful.
   (a) It is snowing and the trees are beautiful.
   (b) The trees are beautiful or it is snowing.
   (c) The streets are not slick and the roofs are not white.
   (d) The trees are not beautiful and it is not snowing.

7. Using the statements of Exercise 6, translate the following symbolic statements into English sentences: (If statements are grouped by parentheses, set off by commas.)
   (a) $A \wedge \sim B$  (b) $(\sim B) \vee (\sim C)$
   (c) $A \wedge (B \vee C)$  (d) $(A \vee \sim C) \wedge D$
   (e) $\sim(A \wedge \sim D)$  (f) $\sim(A \wedge \sim C)$

8. If "It is snowing" and "The trees are beautiful" are true and if "The roofs are white" and "The streets are slick" are false, classify the following statements as true or as false.
   (a) It is snowing and the trees are beautiful.
   (b) The trees are beautiful or it is snowing.
   (c) It is not snowing and the roofs are white.
   (d) It is not snowing or the streets are slick.

9. Make up statements where $p$ is true and $q$ is false to illustrate the following:
   (a) $\sim$ T is F.  (b) $\sim$ F is T.
   (c) T $\wedge$ F is F.  (d) T $\vee$ F is T.

10. Classify the following compound statements as true or false:
    (a) $2 + 1 = 3$  or  $3 + 5 = 9$
    (b) $3 + 1 = 4$  and  $4 + 3 = 6$
    (c) $2 + 3 = 6$  and  $4 + 5 = 9$

*T* 11. Construct truth tables to show the truth values for the following:
    (a) $p \vee \sim q$  (b) $p \wedge \sim p$

12. The Law of Double Negation says that $\sim(\sim p)$ can be replaced by $p$ in every logical expression. Complete the truth table below and use the results to explain the Law of Double Negation.

| $p$ | $\sim p$ | $\sim(\sim p)$ |
|-----|----------|----------------|
| T   |          |                |
| F   |          |                |

13. In constructing a truth table, how many different combinations of T's and F's do you need for statements involving
    (a) $p$ only?  (b) $p$ and $q$?
    (c) $p, q$, and $r$?  (d) $p, q, r$, and $s$?

14. Let $p$ represent "The stock market is bullish." Let $q$ represent "The Dow average is increasing." Finally, let $r$ represent "The price of utilities is decreasing." Translate each of the following into words. When two statements are grouped by parentheses in the logical expression, group them with a comma in the English sentence.
    (a) $(\sim p) \wedge q$  (b) $(\sim p) \vee \sim r$
    (c) $(p \vee r) \wedge q$  (d) $(\sim r \vee \sim p) \wedge q$

15. Translate the following sentences into symbols:
    (a) Either I invest my money in stocks, or I put my money in a savings account.
    (b) I do not invest in stocks or deposit my money in a savings account, but I buy gold.

16. Let $p$ represent the statement "The price of automobiles will rise," and let $q$ represent the statement "The inflation rate will increase." Translate each of the following sentences into symbols:

(a) The price of automobiles will rise, and the inflation rate will increase.

(b) The price of automobiles will not rise, but the inflation rate will increase.

17. Construct truth tables to show the truth values of the following:

(a) $(p \wedge q) \vee q$     (b) $(p \vee q) \wedge p$

 18. Find the truth value of each of the following when $q$ is true and $p$ and $r$ are both false.

(a) $(p \vee q) \wedge \sim(\sim p \vee r)$     (b) $(p \wedge \sim q) \vee (r \wedge \sim q)$

(c) $\sim(r \wedge p) \wedge \sim(q \vee r)$     (d) $(r \wedge p) \vee (\sim q \wedge r)$

19. Find the truth values for the expressions in Exercise 18 if $q$ and $r$ are false and $p$ is true.

20. Construct truth tables to show the truth values for the following.

(a) $(p \vee \sim q) \wedge r$     (b) $p \wedge (q \vee r)$

(c) $(p \vee \sim p) \wedge q$     (d) $\sim(p \vee q) \wedge r$

---

### ∴ PCR Excursion ∴

21. A. Without looking ahead, experiment with some fast ways to compute the sum

$$1 + 2 + 3 + \cdots + 100$$

Explain your solutions to others in your group.

B. When the famous German mathematician Carl Friedrich Gauss was presented with this problem as a child, his teacher expected him to spend the next hour solving it. Instead, he handed in a solution immediately. Here is his strategy:

Write the sum two times: once in ascending order, once in descending order.

$$1 + \quad 2 + \quad 3 + \quad 4 + \quad 5 + \cdots + \quad 99 + 100$$
$$100 + \quad 99 + \quad 98 + \quad 97 + \quad 96 + \cdots + \quad 2 + \quad 1$$
$$101 \quad 101 \quad 101 \quad 101 \quad 101 \qquad 101 \quad 101$$

By adding down he discovered that he had 100 summands of 101. Realizing that he had added each number twice, he computed the sum as

$$\frac{100(101)}{2} = 5050$$

Use Gauss's strategy to compute the sum

$$1 + 2 + 3 + \cdots + 50$$

C. Use Gauss's strategy to add the sums:

(a) $1 + 2 + 3 + \cdots + 30$

(b) $1 + 3 + 5 + \cdots + 65$

(c) $3 + 7 + 11 + \cdots + 43$

D. Can you use Gauss's strategy to find a formula for the sum $1 + 2 + 3 + 4 + 5 + \cdots + n$? Congratulations! You have developed a formula for finding the sum of the first $n$ counting numbers. This formula works for any counting number $n$.

---

## SECTION 2  CONDITIONALS AND EQUIVALENT STATEMENTS

**PROBLEM**

Aaron is quite excited. After sitting through a tongue-lashing from his father concerning his failing grades, Aaron overhears his father tell his mother, "If Aaron makes an A in senior English, then I will buy him a Jeep Cherokee." Aaron is suddenly quite interested in the truth value of this statement. When is this conditional true, and when is it false?

**OVERVIEW**

The answer to the preceding question will be obvious after you study this section.

Statements of the form "If $p$, then $q$" are called **conditionals** and are among the most important kinds of statements we use both in the language of mathematics and in daily discourse. Let $p$ represent "Aaron makes an A in senior English," and let $q$ represent "His father will buy him a Jeep Cherokee." "If Aaron makes an A in senior English, then his father will buy him a Jeep Cherokee" is denoted by "If $p$, then $q$." $p$ is called the **hypothesis** (or antecedent) and $q$ the **conclusion** (or consequent).

In order to decide when to classify a conditional as true and when to classify it as false, we shall think of "If $p$, then $q$" as a promise. Aaron has overheard the promise "If Aaron makes an A in senior English, then his father will buy him a Jeep Cherokee." In this situation there are four possibilities, as laid out in Table 2-6.

**TABLE 2-6**

| p: Aaron Makes A | q: Gets Cherokee | Promise Kept |
|:---:|:---:|:---:|
| Yes | Yes | Yes |
| Yes | No | No |
| No | Yes | Yes |
| No | No | Yes |

The only case in which the promise is broken occurs when Aaron makes an A, but his father does not buy the car. In the last two cases, the promise is not broken because Aaron does not make an A; hence, whether the father does or does not buy the car does not affect the promise. This example suggests the following definition of a conditional.

**DEFINITION**

**Conditional**

A **conditional** has the form "If $p$, then $q$" where $p$ and $q$ are statements or open sentences. It is denoted by $p \rightarrow q$. Suppose $p$ and $q$ are statements. The conditional $p \rightarrow q$ is false when $p$ is true and $q$ is false, and is true otherwise.

The truth values of conditional statements are illustrated in Table 2-7.

| | | | |
|---|---|---|---|
| **TABLE 2-7** | | | |
| *p* | *q* | *p* → *q* | **Examples** |
| T | T | T | If there is severe water pollution in Lake Erie, then fish are dying. |
| T | F | F | If 2 + 3 = 5, then 2 + 4 = 7. |
| F | T | T | If 4 + 1 = 7, then 2 · 3 = 6. |
| F | F | T | If 2 + 3 = 7, then 2 + 4 = 8. |

In ordinary language we use conditionals to describe circumstances in which there is a relationship between the hypothesis and the conclusion. In the conditional "If it rains, then I will not water the lawn," there is a clear connection between the hypothesis, "It rains," and the conclusion, "I will not water the lawn." In logic, an "if-then" statement does not necessarily imply relationship between the hypothesis and conclusion. Because in logic we are concerned with the form of statements rather than with the truth of the subject matter, we permit conditionals like "If 3 + 4 = 10, then Tony Blair is president of the United States." Moreover, this conditional is true since the hypothesis is false. (Note the difference between a conditional that is true and a conditional with a true conclusion). In logic any two statements can be connected with the conditional connective, and the result can then be classified as true or false.

**EXAMPLE 1**

(a) Determine whether statements *p* and *q* are true or false.

   *p:* Baseball is played with a racket.
   *q:* Football is played with a bat.

(b) Determine the truth value of the conditional "If baseball is played with a racket, then football is played with a bat."

*SOLUTION*

(a) Neither statement is true.

(b) The conditional is true because a conditional with a false hypothesis and a false conclusion is true.

Conditionals are used so often both in ordinary language and in mathematics that they occur in many different forms. The conditional "If *p*, then *q*" can also be written as:

| | |
|---|---|
| *q* if *p*. | When *p* then *q*. |
| *p* implies *q*. | *p* only if *q*. |

**EXAMPLE 2**

Rewrite the following conditional in four additional forms.

   If it snows, then the streets are slick.

*SOLUTION*

(a) The streets are slick if it snows.
(b) Snowy weather implies the streets are slick.
(c) When it snows, then the streets are slick.
(d) It snows only if the streets are slick.

# RELATED CONDITIONALS

Closely related to a conditional are its *converse,* its *inverse,* and its *contrapositive.*

| DEFINITION | **Converse of a Conditional** | The *converse* of the conditional "If $p$, then $q$" is the conditional "If $q$, then $p$"; that is, the converse of $p \rightarrow q$ is $q \rightarrow p$. |
|---|---|---|

"If it is cold, then it is snowing" is the converse of "If it is snowing, then it is cold." This example shows that the truth of a conditional in no way ensures the truth of its converse. "If it is snowing, then it is cold" is a true conditional, but the fact that it is cold does not imply that it is snowing. Assuming that the truth of a conditional ensures the truth of its converse is a common fallacy, often subtly used in political and advertising campaigns. For instance, "If a person is a communist, that person is a socialist" is a true conditional. Yet its converse has incorrectly been asserted at times to be true: "If a person is a socialist, that person is a communist."

Similarly, the truth of a conditional certainly does not require that its converse be false. "If $2 + 2 + 2 = 6$, then $3 \cdot 2 = 6$" and its converse, "If $3 \cdot 2 = 6$, then $2 + 2 + 2 = 6$," are both true.

| DEFINITION | **Inverse of a Conditional** | The *inverse* of a given conditional is the conditional that results when $p$ and $q$ are replaced by their negations; that is, the inverse of $p \rightarrow q$ is $(\sim p) \rightarrow (\sim q)$. |
|---|---|---|

The inverse of "If a polygon is a rectangle, then it is a parallelogram" is "If a polygon is not a rectangle, then it is not a parallelogram."

Although the truth of a conditional does not ensure the truth of the converse or inverse of the conditional, much advertising assumes that you believe that it does.

**EXAMPLE 3**

"If you are a great basketball player, then you wear expensive Nika basketball shoes." Certainly a quick check of shoes on the feet of the major basketball stars will convince you that this is a true statement. Advertisers hope that you will assume that the converse is true, "If you wear expensive Nika basketball shoes, you will be a great basketball player."

| DEFINITION | **Contrapositive of a Conditional** | The *contrapositive* of the conditional "If $p$, then $q$" is "If $\sim q$, then $\sim p$"; that is, the contrapositive of $p \rightarrow q$ is $(\sim q) \rightarrow (\sim p)$. |
|---|---|---|

The contrapositive of "If it is snowing, then it is cold" is "If it is not cold, then it is not snowing." Common sense dictates that these two statements say the same thing. The fact that the truth of the contrapositive ensures the truth of the associated conditional will be verified later in this section.

To summarize the definitions of *converse, inverse,* and *contrapositive,* consider the following example:

**EXAMPLE 4**

(a) *Conditional:* If it rains, then I buy a new umbrella.
(b) *Converse:* If I buy a new umbrella, then it is raining.
(c) *Inverse:* If it does not rain, then I do not buy a new umbrella.
(d) *Contrapositive:* If I do not buy a new umbrella, then it is not raining.

**PRACTICE PROBLEM**

Make up examples that satisfy (a) T → T, (b) T → F, (c) F → T, and (d) F → F.

*ANSWER*

One possible set of answers follows.
(a) If $6 = 2 + 4$, then $6 + 5 = (2 + 4) + 5$.
(b) If $2 \cdot 4 = 8$, then $2 \cdot 5 = 8 + 1$.
(c) If $2 + 4 = 7$, then $3 + 5 = 8$.
(d) If $2 + 4 = 7$, then $3 \cdot 4 = 7 + 3$.

 # LOGICALLY EQUIVALENT STATEMENTS

When two statements $r$ and $s$ have the same truth values in every possible situation, we say that $r$ is logically *equivalent* to $s$.

<table>
<tr><td>DEFINITION</td><td>Equivalent</td><td>In logic, two statements $r$ and $s$ are called *equivalent* statements if they have identical truth tables.</td></tr>
</table>

The observation that two statements are logically equivalent is very important. It means that in both mathematical discourse and in ordinary conversation the two statements can be used interchangeably without changing the meaning of the communication. If statements that are not logically equivalent are substituted for one another, however, there is a risk that the meaning will be changed.

**EXAMPLE 5**

Show that a statement and its contrapositive are equivalent.

*SOLUTION*

In Table 2-8 we record all possible truth values for both $p \rightarrow q$ and $\sim q \rightarrow \sim p$. Observe that the truth values for $p \rightarrow q$ and $\sim q \rightarrow \sim p$ are the same.

**TABLE 2-8**

| p | q | ~p | ~q | p → q | ~q → ~p |
|---|---|----|----|-------|---------|
| T | T | F | F | T | T |
| T | F | F | T | F | F |
| F | T | T | F | T | T |
| F | F | T | T | T | T |

↑ _____ same _____ ↑

| Equivalent Conditionals | A conditional and its contrapositive are equivalent. |
|---|---|

Because a statement and its contrapositive are equivalent, the statement "If you wash your hair with Shed shampoo, your dandruff disappears" can be used interchangeably with the statement "If your dandruff does not disappear, then you do not wash your hair with Shed shampoo."

**EXAMPLE 6**   Using a truth table, show that the converse of a conditional is not equivalent to the conditional.

*SOLUTION*   Table 2-9 shows that the truth values for $p \rightarrow q$ and $q \rightarrow p$ are not the same for some values of $p$ and of $q$. Therefore, a conditional and its converse are not equivalent.

**TABLE 2-9**

| p | q | p → q | q → p |
|---|---|-------|-------|
| T | T | T | T |
| T | F | F | T |
| F | T | T | F |
| F | F | T | T |

# OTHER EQUIVALENT STATEMENTS

In Exercise 13 you will use truth tables to verify that the following statements are equivalent.

| De Morgan's Laws | For any statements $p$ and $q$, the following pairs of statements are equivalent:<br>$\sim(p \vee q)$ is equivalent to $(\sim p) \wedge (\sim q)$<br>$\sim(p \wedge q)$ is equivalent to $(\sim p) \vee (\sim q)$ |
|---|---|

**EXAMPLE 7**

Find the negation of "I will eat a hamburger or I will not eat a bowl of soup."

*SOLUTION*

Let $p$ be "I will eat a hamburger," and let $q$ be "I will not eat a bowl of soup." Then $\sim p$ is "I will not eat a hamburger," and $\sim q$ is "I will eat a bowl of soup." We want the negation of $p \vee q$, which is $\sim(p \vee q)$. But $\sim(p \vee q)$ is equivalent to $(\sim p) \wedge (\sim q)$. Thus, the negation is "I will not eat a hamburger, and I will eat a bowl of soup."

---

**DEFINITION**

**The Negation of a Conditional**

The negation of the conditional $p \rightarrow q$ is equivalent to $p \wedge \sim q$; that is $\sim(p \rightarrow q)$ is equivalent to $p \wedge \sim q$.

---

In Exercise 12 we will verify that these conditionals are equivalent using truth tables. However, we can intuitively understand this definition by observing that if Manuel promises, "If it rains, then I will bring a umbrella," the one circumstance in which his date Maria would protest would be in the event that "It rained and Manuel did not bring an umbrella." (The equivalence of $\sim(p \rightarrow q)$ and $p \wedge \sim q$ is the basis of a method of mathematical proof called Proof by Contradiction.) In Exercise 11 we will also observe that $p \vee q$ and $\sim p \rightarrow q$ are logically equivalent.

**EXAMPLE 8**

Write the negation of this conditional in two ways: If $n^2$ is even, then $n$ is even.

*SOLUTION*

The standard way to form the negation of a statement is to precede it by the words "It is not true that. . . ." Hence, an awkward form of the negation of this conditional is "It is not true that if $n^2$ is even then $n$ is even." Since $\sim(p \rightarrow q)$ is equivalent to $p \wedge \sim q$, another form of the negation is

$n^2$ is even and $n$ is not even.

$n^2$ is even and $n$ is odd.

To this point we have used negations, conditionals, *and,* and *or* to create complex statements from simpler components. Yet another way to connect two statements or open sentences involves the use of the phrase *if and only if.* The statement "A triangle is equilateral if and only if it is equiangular" is an example of a biconditional.

---

**DEFINITION**

**Biconditional**

The **biconditional** formed from $p$ and $q$ is "$p$ if and only if $q$" (denoted by $p \leftrightarrow q$). Suppose $p$ and $q$ are statements. The biconditional is true when $p$ and $q$ have the same truth values. It is false otherwise.

In the third column of Table 2-10 are found the truth values for the biconditional $p \leftrightarrow q$. Compare Column 3 and Column 6 in this table. Observe that the truth values in these two columns are the same, which proves that the biconditional $p \leftrightarrow q$ is equivalent to $(p \rightarrow q) \wedge (q \rightarrow p)$. This fact has important consequences in mathematics. To prove a theorem of the form $p \leftrightarrow q$ we prove two separate theorems, $p \rightarrow q$ and $q \rightarrow p$.

**TABLE 2-10**

| $p$ | $q$ | $p \leftrightarrow q$ | $p \rightarrow q$ | $q \rightarrow p$ | $(p \rightarrow q) \wedge (q \rightarrow p)$ |
|-----|-----|-----------------------|-------------------|-------------------|----------------------------------------------|
| T | T | T | T | T | T |
| T | F | F | F | T | F |
| F | T | F | T | F | F |
| F | F | T | T | T | T |

# Just for Fun

Three couples on a hike had to cross a river in a small boat with a maximum capacity of two. Because the men were extremely jealous, no woman could be left with a man unless her date was present. How did they manage to cross the river?

# EXERCISE SET 2

*R* 1. Let *A* represent "It is snowing"; *B*, "The roofs are white"; *C*, "The streets are slick"; and *D*, "The trees are beautiful." Write the following in symbolic notation.

   (a) If it is snowing, then the trees are beautiful.
   (b) If it is not snowing, then the roofs are not white.
   (c) If the streets are not slick, then it is not snowing.
   (d) If the streets are slick, then the trees are not beautiful.

2. In Exercise 1, assume that *A* is true, that *B* and *C* are both false, and that *D* is true. Classify the conditional statements as either true or false.

3. State the converse, the inverse, and the contrapositive of each of the following conditionals.

   (a) If a triangle is a right triangle, then one angle has a measure of 90°.
   (b) If a number is a prime, then it is odd.
   (c) If two lines are parallel, then alternate interior angles are equal.
   (d) If Joyce is smiling, then she is happy.

4. Write each of the following statements in "if-then" form.

   (a) Triangles are not squares.
   (b) Birds of a feather flock together.
   (c) Honest politicians do not accept bribes.

5. Write the converse and contrapositive of each part of Exercise 4.

6. Using the notation of Exercise 1, translate the following into sentences.

   (a) $\sim A \to B$      (b) $\sim C \to \sim B$
   (c) $(\sim B \wedge \sim C) \to A$    (d) $(A \vee B) \to \sim C$
   (e) $\sim(A \to \sim B)$     (f) $(\sim C \vee \sim A) \to \sim B$

7. Using the truth values of Exercise 2, classify the statements in Exercise 6 as either true or false.

8. Use the following statements to form a conditional, its converse, its inverse, and its contrapositive:

   (a) I teach third grade. I do not teach in high school.
   (b) I own a car. I have a driver's license.

9. Commercials and advertisements are often based on the assumption that naive audiences will accept that the converse, inverse, contrapositive, and original statement are all true. Write the converse, inverse, and contrapositive of the following statements. Do they have the same truth values?

   (a) If you brush your teeth with White-as-Snow, then you have fewer cavities.
   (b) If you like this book, then you love mathematics.
   (c) If you want to be strong, then you eat Barlies for breakfast.
   (d) If you use Wave, then your clothes are bright and colorful.

10. Write the following in "if-then" notation:

   (a) Carelessness leads to accidents.
   (b) Whenever I see June, my heart throbs.
   (c) A triangle is isosceles if two sides are congruent.
   (d) I will be happy if I pass.

*T* 11. (a) Use truth tables to show that $p \vee q$ and $\sim p \to q$ are logically equivalent.
   (b) Use the result from (a) to rewrite the statement "Either I pass English or I go to summer school."
   (c) Use the result from (a) to rewrite "If I do not sleep tonight, then I will skip the luncheon."

12. (a) Show that $p \wedge \sim q$ is equivalent to $\sim(p \to q)$.
   (b) Use the result from (a) to rewrite "Nakisha is a music major, but she does not sing well."
   (c) Use the result from (a) to write the negation of the conditional "If my dog has fleas, then it scratches often."

13. Use truth tables to verify De Morgan's laws.

   (a) $\sim(p \vee q)$ is equivalent to $\sim p \wedge \sim q$.
   (b) $\sim(p \wedge q)$ is equivalent to $\sim p \vee \sim q$.
   (c) Use (a) to write the negation of "I will take history or I will take physics."
   (d) Use (b) to write the negation of "I am on the baseball team and I am on the basketball team."

14. Using truth tables, determine whether the following pairs of statements are equivalent.

   (a) $\sim p \vee q; p \to q$
   (b) $\sim(p \wedge q); \sim p \wedge \sim q$
   (c) $\sim(p \vee q); \sim p \vee \sim q$
   (d) $p \to q; \sim p \to \sim q$

15. Write the following as conditionals or as the negations of conditionals:

   (a) There will be inflation, or interest rates will increase.
   (b) Charles sings, and Jane dances.
   (c) I will carry my umbrella, and it will not rain.
   (d) The moon is made of cheese, or flight to Mars is impossible.

16. Use De Morgan's laws in writing the contrapositives of the following statements:

    (a) If a pair of lines do not intersect, they are either parallel or skew.

    (b) If $p$ is true and $q$ is true, then the conjunction $p \wedge q$ is true.

17. Write each of the following statements in "if-then" form.

    (a) Triangles are not squares.

    (b) Birds of a feather flock together.

    (c) Honest politicians do not accept bribes.

18. A *contradiction* is a compound statement that is false for all possible truth values of the simple statements that compose it. A contradiction is said to be *logically false*. Use truth tables to show that the following statements are contradictions:

    (a) $p \wedge \sim p$.      (b) $(p \rightarrow q) \wedge (p \wedge \sim q)$

19. A *tautology* is a compound statement that is true for all possible values of the simple statements that compose it. A tautology is said to be *logically true*. Use truth tables to determine which of the following statements are tautologies.

    (a) $p \rightarrow p$      (b) $[p \wedge (p \rightarrow q)] \rightarrow \sim q$
    (c) $(p \rightarrow q) \rightarrow p$      (d) $[(\sim p \vee q) \wedge p] \rightarrow q$
    (e) $(p \rightarrow q) \rightarrow (p \vee q)$      (f) $(\sim p \vee q) \rightarrow (p \rightarrow q)$

20. Using a truth table, determine whether $r \vee (s \wedge t)$ and $(r \vee s) \wedge (r \vee t)$ are logically equivalent.

21. Write in "if-then" form:

    (a) Joe plays quarterback, or his team loses.

    (b) 3 divides 12, or 3 is not a factor.

    (c) You study regularly, or you fail this course.

22. In this section we saw that the biconditional "$p$ if and only if $q$" is equivalent to the statement "If $p$, then $q$ and if $q$, then $p$." Use this fact to rewrite the following statements.

    (a) Triangle A is congruent to Triangle B if and only if Triangle B is congruent to Triangle A.

    (b) If a dancer is graceful, then she is athletic, and if a dancer is athletic then she is graceful.

## REVIEW EXERCISES

• • • • • • • • • • • • • • • • • • • • • • • • • • • • • • • • • •

23. Let $p$ represent "Logic is easy," let $q$ represent "Algebra is hard," and let $r$ represent "Arithmetic is useful." Write each statement in words.

    (a) $r \vee (p \wedge q)$      (b) $\sim(p \vee q)$
    (c) $\sim r \wedge \sim q$      (d) $p \wedge (q \vee \sim r)$

---

∵ **PCR Excursions** ∵

24. Work in pairs. Complete each part independently, then compare and discuss solutions.

    A. Write each of these statements in the form "if $p$ then $q$."
       (a) Hai can vote if he is 18.
       (b) A muddy path implies I will wear boots.
       (c) When I finish studying, then I will go to bed.
       (d) Number $x$ is a whole number only if $x$ is an integer.

    B. Write each statement in A in the form if $\sim q$ then $\sim p$. Discuss the equivalence of the statements in A and B.

    C. One of De Morgan's Laws states for any statements $p$ and $q$, $\sim (p \vee q)$ is equivalent to $\sim p \wedge \sim q$. Find examples of statements $p$ and $q$ so this law is easily understood.

    D. Do the same as in C for $\sim (p \wedge q)$ and $\sim p \vee \sim q$.

25. A game consists of first arranging $n$ coins in a horizontal row. You and your opponent alternate picking up from 1 to $r$ coins at a time. The object of the game is to force your opponent to pick up the last coin. Thus, you want to determine a strategy to win the game. First, determine how many coins you should leave on your last move to force your opponent to take the last one. Next, determine how many coins you should leave on your next-to-last play in order to be certain that you will be able to leave the desired number on the last play; that is, you will solve this problem by using the problem-solving strategies of working backward and using reasoning.

    A. In the $r = 2$ game, on each play, the participant may remove 1 or 2 coins.

       (a) Suppose that the game starts with 9 coins in a row.
          i. Explain why you wish to leave 1 coin on your last move.
          ii. Explain why you wish to leave 4 coins on your next-to-last move.
          iii. Explain why you wish to leave 7 coins after your first move.
          iv. Describe carefully how you can always win the 9-coin game if you go first.

       (b) Can you always win if the game starts with 12 coins and you go first? Explain.

       (c) Suppose the game starts with 10 coins. Can you always win if you go first? Can you always win if you go second? Explain.

(d) Think about games that start with 11 coins, 12 coins, 13 coins, ..., 20 coins. Which of these games can always be won if you go first? Which of these games can always be won if you go second? Explain.

B. In the $r = 3$ game, on each play, the participant may remove 1, 2, or 3 coins.

   (a) Suppose the game begins with 10 coins. Can you devise a strategy to always win if you go first? Explain.

(b) Suppose the game begins with 13 coins. Can you always win if you go first? Can you always win if you go second? Explain.

(c) Describe the strategy that will be required to win the game with 19 coins if you go first.

(d) Examine the games that begin with 10, 11, 12, ..., 20 coins and determine which games can always be won if you go first and which games can always be won if you go second. Explain.

# MAKING USE OF DEDUCTIVE LOGIC

| | |
|---|---|
| **PROBLEM** | If Daniel studies, then he will make an A.<br>If Daniel makes an A, then he will make the Dean's List.<br><u>Daniel studies.</u> |
| **OVERVIEW** | Can you imagine a conclusion that follows from the above argument? Was your conclusion, "Daniel makes the Dean's List"? If so, then you are skilled at one of the classic patterns of reasoning. |
| **GOALS** | Illustrate the Reasoning and Proof Standard, page 23.<br>Illustrate the Problem Solving Standard, page 15.<br>Illustrate the Representation Standard, page 50. |

 DEDUCTIVE REASONING

The reasoning you used in the Overview to reach a conclusion that Daniel makes the Dean's List is an example of a deductive argument. In mathematics and in everyday affairs, we often need to deduce a correct conclusion from a given set of statements. In order to discuss the process by which we do this, we will introduce the notion of an **argument.** An argument consists of two parts: a set of two or more statements called **premises** and a single statement called the **conclusion.** An argument is **valid** if the conclusion is true in every circumstance that the conjunction of the premises is true. If, in some case, the conjunction of the premises is true and the conclusion is false, then the argument is **invalid.** Invalid arguments are sometimes called **fallacies.** Note that *the validity of an argument is independent of the truth values of its premises and its conclusion.*

Arguments are often written symbolically in the following manner:

| | |
|---|---|
| If inflation occurs, the price of automobiles increases. | $p \rightarrow q$ |
| Inflation occurs. | $p$ |
| Therefore, the price of automobiles increases. | $\therefore q$ |

The three dots $\therefore$ are read as *therefore*.

When verbal arguments are converted into symbolic form, the same patterns often occur. For example, the pattern in the previous paragraph, $(p \rightarrow q) \wedge p;\ \therefore\ q$, occurs frequently in testing the validity of arguments and is called the **rule of detachment.** In Example 1 we show that this argument is always valid.

**EXAMPLE 1**

Use a truth table to verify the rule of detachment.

*SOLUTION*

Table 2-11 presents the required truth table.

**TABLE 2-11**

| $p$ | $q$ | $p \rightarrow q$ | $(p \rightarrow q) \wedge p$ |
|---|---|---|---|
| T | Ⓣ | T | Ⓣ |
| T | F | F | F |
| F | T | T | F |
| F | F | T | F |

Notice that for one case in which $(p \rightarrow q) \wedge p$ is true, $q$ is also true. Thus the argument is valid.

**Rule of Detachment**

Because in every case that $(p \rightarrow q) \wedge p$ is true, $q$ is true, the argument

$$p \rightarrow q$$
$$p$$
$$\overline{\therefore q}$$

is valid.

**EXAMPLE 2**

Determine the validity of the following argument.

| | |
|---|---|
| If John is a thief, he is a lawbreaker. | $p \rightarrow q$ |
| John is a thief. | $p$ |
| Therefore, John is a lawbreaker. | $\therefore q$ |

*SOLUTION*

This argument is valid because of the rule of detachment. Observe that because we recognize this previously established pattern, we do not have to do a truth table to show that the argument is valid.

Another often-used pattern of argument, which can be stated as a theorem and proved by means of a truth table, is the **chain rule,** or classically, *hypothetical syllogism.*

| | | | |
|---|---|---|---|
| **Chain Rule** | The following argument is valid and is called the *chain rule*. | | |

$$p \to q$$
$$q \to r$$
$$\overline{\therefore\ p \to r}$$

Table 2-12 verifies that this argument is valid. Observe that in each of the four cases in which the conjunction of the premises is true, the conclusion is true.

**TABLE 2-12**

| $p$ | $q$ | $r$ | $p \to q$ | $q \to r$ | $(p \to q) \land (q \to r)$ | $p \to r$ |
|---|---|---|---|---|---|---|
| T | T | T | T | T | Ⓣ | Ⓣ |
| T | T | F | T | F | F | F |
| T | F | T | F | T | F | T |
| T | F | F | F | T | F | F |
| F | T | T | T | T | Ⓣ | Ⓣ |
| F | T | F | T | F | F | T |
| F | F | T | T | T | Ⓣ | Ⓣ |
| F | F | F | T | T | Ⓣ | Ⓣ |

**EXAMPLE 3**

Friends of Joyce know that the two conditionals

If Joyce is smiling, then she is happy.
If Joyce is happy, then she is polite.

are both true statements. So they know, by the chain rule, that "If Joyce is smiling, then she is polite" is a true statement.

An extension of the chain rule gives

$$[(p \to q) \land (q \to r) \land (r \to s) \land \ldots \land (w \to z)] \to (p \to z)$$

**EXAMPLE 4**

Show that the following argument is valid.

| | |
|---|---|
| If a person is a good speaker, she or he is a good teacher. | $p \to q$ |
| If a person is a good teacher, she or he is friendly. | $q \to r$ |
| If a person is friendly, she or he is polite. | $r \to s$ |
| If a person is polite, she or he is well liked. | $s \to t$ |
| Therefore, if a person is a good speaker, she or he is well liked. | $\therefore\ p \to t$ |

*SOLUTION*

This argument is valid by the chain rule.

In the next example, we use our knowledge of equivalent statements to test the validity of the argument.

**EXAMPLE 5**

Show that the following argument is valid:

> If John is telling the truth, then Earl is guilty.
> Earl is not guilty.
> Therefore, John is not telling the truth.

*SOLUTION*

Because a conditional and its contrapositive are equivalent, $p \rightarrow q$ may be replaced by $\sim q \rightarrow \sim p$. Thus, the argument becomes $(\sim q \rightarrow \sim p) \wedge \sim q$; therefore $\sim p$. This argument is valid by the rule of detachment.

In Example 5, we used the fact that a conditional can always be replaced by its contrapositive to arrive at a valid argument. This same reasoning can be used to derive the rule of contraposition from the rule of detachment.

**Rule of Contraposition**

The argument

$$p \rightarrow q$$
$$\underline{\sim q}$$
$$\therefore \sim p$$

is valid and is called the *rule of contraposition*.

Of course, we could also justify the rule of contraposition with truth tables.

**EXAMPLE 6**

Show that the following argument is valid:

> If I go bowling, I will not study.       $p \rightarrow \sim q$
> If I do not study, I will take a nap.   $\sim q \rightarrow r$
> I will not take a nap.                          $\underline{\sim r}$
> Therefore, I will not go bowling.        $\therefore \sim p$

*SOLUTION*

Now $p \rightarrow \sim q$ and $\sim q \rightarrow r$ can be replaced by $p \rightarrow r$ by the chain rule. By the rule of contraposition, $p \rightarrow r$ and $\sim r$ yield the valid conclusion, $\sim p$.

**EXAMPLE 7**

Show that the following argument is not valid.

> If someone is a college graduate, then he or she is intelligent.   $p \rightarrow q$
> Juan is intelligent.                                                                            $q$
> Therefore, Juan is a college graduate.                                       $\therefore p$

*SOLUTION*

This is not a valid argument. If $p$ is false and $q$ is true, the conjunction of the premises is true, but the conclusion is false. Said another way, this fails to be a valid argument because a statement and its converse are not equivalent. This can be shown by the truth table in Table 2-13.

| **TABLE 2-13** | | | | |
|---|---|---|---|---|
| *p* | *q* | *p → q* | *(p → q) ∧ q* | *p* |
| T | T | T | T | T |
| T | F | F | F | T |
| F | T | T | T | F |
| F | F | T | F | F |

Observe that when $p$ is F and $q$ is T, the conjunction of the premises, $(p \rightarrow q) \wedge q$, is T whereas $p$ is F. Thus the argument is not valid.

We need to remember that there is a difference between a valid argument and a true conclusion. **A conclusion drawn from an argument is true if the argument is valid and if all the premises are true.**

**EXAMPLE 8**

Determine the validity of the following argument and the truth of its conclusion.

If a person is a college professor, then he or she is brilliant.
Dr. Wheeler is a college professor.
Therefore, Dr. Wheeler is brilliant.

*SOLUTION*

The argument is valid, but many generations of students have wondered about the truth of the first premise. Therefore, the truth of the conclusion may be contested.

# Just for Fun

A and B play a game in which they alternate selecting any one of the numbers 1, 2, 3, 4, 5, and 6. When a number is selected (and each number may be selected more than once), it is added to the sum of the numbers already selected. To win the game, a player must make the sum 50. If A goes first, there is a way for A to win every game. What is it? (Hint: Work backward.)

# EXERCISE SET 3

*In Exercises 1 through 3, rewrite the argument in symbols, using the symbols that are suggested.*

*R* 1. If you eat your squash (*s*), then you may go out and play (*p*).
You eat your squash.
Therefore, you may go out and play.

2. If you studied Latin (*L*), then Spanish is easy (*S*).
Spanish is not easy.
Therefore, you did not study Latin.

3. If prices increase (*p*), then consumers will complain (*c*).
If consumers complain, then managers will fret (*m*).
Therefore, if prices increase, then managers will fret.

4. (a)–(c) For each of the arguments in Exercises 1–3, identify which of the rules of logic ensures that the argument is valid.

5. Identify the rule or rules of logic used to find the valid conclusion in each of the following.

(a) $p \rightarrow \sim q$
$\underline{q}$
$\therefore \sim p$

(b) $a \rightarrow b$
$b \rightarrow c$
$\underline{a}$
$\therefore c$

(c) If two sides of a triangle are equal, the angles opposite these sides are equal.
Side *BC* equals side *AB* in triangle *ABC*.
Therefore, the angles opposite *BC* and *AB* are equal.

(d) If the movie is not over, then they will buy popcorn.
They do not buy popcorn.
Therefore, the movie is over.

(e) If Kahleel wins, he is happy.
If Kahleel is happy, he treats his sister well.
Therefore, if Kahleel wins, he treats his sister well.

6. Find a valid conclusion to each of the following sets of premises and indicate the rule or rules of logic that you use.

(a) If the sun shines, the flowers grow.
If the flowers grow, the garden flourishes.

(b) If Fred hits a home run, the game is won.
Fred hits a home run.

(c) If Devron is not sick, he will be at the wedding.
Devron is not at the wedding.

(d) If Jermonte studies logic, then mathematics is easy.
If mathematics is easy, then physics is a snap.
Jermonte studies logic.

7. Draw a valid conclusion from each of the following sets of statements; then explain why your conclusion is valid.

(a) If Susan is a freshman, then Susan takes mathematics. Susan is a freshman.

(b) You will fail this test if you do not study. You do not fail the test.

(c) You cry if you are sad. You are sad.

(d) You will receive your allowance if you cut the grass. You do not receive your allowance.

8. (a) Complete this truth table which will show that the rule of contraposition $((p \rightarrow q) \wedge \sim q)$; $\sim p$, is a valid argument.

| *p* | *q* | $\sim q$ | $p \rightarrow q$ | $(p \rightarrow q) \wedge \sim q$ | $\sim p$ |
|-----|-----|----------|-------------------|-----------------------------------|----------|
| T | T | F | | | |
| T | F | T | | | |
| F | T | F | | | |
| F | F | T | | | |

(b) Explain why the truth table you completed shows that this argument is valid.

*T* 9. Arrange the following statements in logical order to prove that Joan teaches seventh grade, given that Joan is tall.

(a) If Joan has red hair, she teaches seventh grade.

(b) If Joan is tall, she wears contacts.

(c) Therefore, if Joan is tall, she teaches seventh grade.

(d) If Joan wears contacts, she has red hair.

10. Margaret said "If I have hamburgers tonight, it will be raining. If I do not have hamburgers, I will have chicken. If I have chicken, I will have tossed salad." Margaret did not have tossed salad. How was the weather?

11. For each of the following arguments, determine whether the argument is valid by using truth tables. Specifically, check whether the conclusion is true for each case in which the conjunction of the premises is true.

(a) $p \rightarrow q$
$\underline{p \rightarrow r}$
$\therefore q \rightarrow r$

(b) $p \vee q$
$\underline{p}$
$\therefore \sim q$

(c) $p \rightarrow \sim q$
$q \vee r$
$\underline{p}$
$\therefore r$

(d) $p \vee \sim q$
$\underline{\sim p \vee r}$
$\therefore q \rightarrow r$

(e) $q \rightarrow \sim p$
$\underline{q \wedge r}$
$\therefore \sim p \rightarrow \sim r$

(f) $q \rightarrow r$
$\sim p \vee q$
$\underline{p}$
$\therefore r$

12. Write each of the following arguments in symbolic form. Then determine if any of the three rules of this section can be used to determine that the arguments are valid. Otherwise, check validity with truth tables.

   (a) If a man is a good speaker, he is a good teacher.
      Mr. Faulkner is a good teacher.
      Therefore, he is a good speaker.

   (b) If I can't go to town, I will go to the shopping center.
      I can go to town.
      Therefore, I will not go to the shopping center.

   (c) When the movie is over, we must go home.
      The movie is over.
      Therefore, we must go home.

   (d) You will fail this course if you do not study enough.
      You are not studying enough.
      Therefore, you will fail this course.

   (e) If $R$, then not $S$.
      If not $S$, then $T$.
      Therefore, if $R$, then $T$.

   (f) If a quadrilateral is a parallelogram, then opposite sides are parallel. Opposite sides of quadrilateral $ABCD$ are parallel.
      Therefore, $ABCD$ is a parallelogram.

   (g) If $q$, then not $p$.
      $q$ and $r$ are true.
      Therefore, if not $p$, then not $r$.

   (h) If $a$'s are $b$'s, then $c$'s are $d$'s.
      $a$'s are not $b$'s.
      Therefore, $c$'s are not $d$'s.

   (i) If $c$'s are $d$'s, then $a$'s are $e$'s.
      $a$'s are not $e$'s.
      Therefore, $c$'s are not $d$'s.

$C$ 13. Determine the validity of the following arguments:

   (a) If the wheel needs greasing, then it squeaks. If it squeaks, then Bob gets the chills. The wheel is not squeaking. Therefore, Bob does not have the chills.

   (b) If you like mathematics, then you will like this course. You do not like mathematics. Therefore, you will not like this course.

*In everyday discourse, we sometimes make errors by violating our understanding of elementary logic (for example, using two statements as equivalent when, in fact, they are not). Other common errors have their source in misinterpretation and misrepresentation of human experience. In Exercises 14 through 17 we describe four such errors. Explain what is wrong with the fallacies in these exercises.*

14. **Fallacy of False Experts.** Tiger Woods addressed a Senate panel on the issue of aid to the country of Zimberia.

15. **Fallacy of the Loaded Question.** (a) Have you stopped beating your wife? (b) You look good in this suit; will it be cash or charge?

16. **Fallacy of Composition.** (a) Your composition is too long; therefore, each sentence in your composition is too long. (b) The choral presentation was too loud; therefore, the tenors were too loud.

17. **Fallacy of False Cause.** I saw a black cat; I lost my pocketbook. Seeing a black cat brings bad luck.

## REVIEW EXERCISES

18. Write the converse and the contrapositive of each statement.

   (a) If it does not rain, we shall play tennis.
   (b) If I study diligently, I shall pass the course.

19. Find the truth value of each of the following when $q$ is false and $p$ and $r$ are both true.

   (a) $(p \vee q) \wedge \sim (p \vee r)$
   (b) $(p \wedge \sim q) \vee (r \wedge \sim q)$
   (c) $\sim (r \wedge p) \wedge \sim (q \vee r)$
   (d) $(r \wedge p) \vee (\sim q \wedge r)$

### ∴ PCR Excursion ∴

20. Three star basketball players truthfully reported their grades to their coach.

   Anders:  If I passed math, then so did Barker.
      I passed English if and only if Carlos did.

   Barker:  If I passed math, then so did Anders.
      Anders did not pass Spanish.

   Carlos:  Either Anders passed Spanish or I did not pass it.
      If Barker did not pass English, then neither did Anders.

   Each of the three players passed at least one subject. Each subject was passed by at least one of the three. Carlos did not pass the same number of subjects as either of the other players. Use the problem-solving strategies of reasoning and making a chart to determine which subjects each of the three players passed.

   (a) Both Anders and Barker made statements about mathematics. In view of their statements, we see that either they both passed mathematics or they both failed mathematics. We have summarized these two cases in charts.

|   | A | B | C |
|---|---|---|---|
| **M** | P | P | P |
| **E** |   |   |   |
| **S** |   |   |   |

CASE I

|   | A | B | C |
|---|---|---|---|
| **M** | F | F | P |
| **E** |   |   |   |
| **S** |   |   |   |

CASE II

In the second case, how do we know that Carlos must have passed mathematics?

(b) Use Anders's statement about English to place more information in the charts. Observe that this will require you to create two possible charts for Case I and two possible charts for Case II.

(c) Use the other information in the problem to eliminate three of the cases and determine the academic success of the three basketball players.

## SECTION 4 — QUANTIFIERS, VENN DIAGRAMS, AND VALID ARGUMENTS

**PROBLEM**

All serious environmentalists support Senate Bill 137. Senator Snodgrass supports Senate Bill 137. Therefore, it is clear that Senator Snograss is a serious environmentalist.

**OVERVIEW**

If something about this reasoning distrubs you, then you are wise. We need to know more about quantified statements (using words like "all" and "some"). Then we will be able to better evaluate Senator Snodgrass' commitment.

**GOALS**

Illustrate the Problem Solving Standard, page 15.
Illustrate the Reasoning and Proof Standard, page 23.
Illustrate the Representation Standard, page 50.

One way to make sentences such as "He had blond hair" and "$x + 2 = 5$" into statements is to use words like *all* and *some*. "All men have blond hair" is clearly false, and "For some $x$, $x + 2 = 5$" is clearly true. Words such as *all*, *some*, and *no* are called **quantifiers** and give information about "how many" in the statements where they occur. Examples of quantified statements include the following:

Some women have red hair.          No professors are bald-headed.
All bananas are yellow.            Some students do not work hard

**DEFINITION**

**Quantified Statement**

A statement that indicates something about "how many" is said to be *quantified*.

# UNIVERSAL QUANTIFIERS

Certain quantified statements are intended to be *universally* true for the factors under consideration. Such statements usually contain the words *all, every,* or *each,* which are called **universal quantifiers.**

> All prime numbers greater than 2 are odd.
> Every automobile pollutes the atmosphere.
> For each counting number $x$, $x + 3 = 3 + x$.
> All men have black hair.

Expressions involving *all* may be expressed as conditionals in the following manner: "All illegal acts are immoral" can be written as "If an act is illegal, then it is immoral." Similarly, the expression "No men use hair curlers" can be written as "If a person is a man, then he does not use hair curlers."

# EXISTENTIAL QUANTIFIERS

Other quantified statements are intended to indicate that there exists at least one case in which the statement is true. Such statements generally involve the **existential quantifiers** *some, there exist,* or *there exists at least one.*

> There exist students who do not work hard.
> There exists at least one student who does not work hard.
> Some men have black hair.

Quantifiers are quite useful in rewriting sentences that cannot be classified as true or false. Consider the assertion "Men are tall." By adding *all* to obtain "All men are tall," we can conclude that the statement is false. However, by adding *some* to obtain "Some men are tall," we can conclude that the statement is true. Similarly, consider the open sentence "$x + 4 = 6$." It cannot be classified as true or false. However, the statement "There exists a counting number $x$ such that $x + 4 = 6$" is true, because $x = 2$ is such a counting number. "For all counting numbers $x$, $x + 4 = 6$" is false, because (for example) there exists the counting number 5 such that $5 + 4$ is not equal to 6. However, the statement "For all $x$, $3 + x = x + 3$" is true.

# VENN DIAGRAMS

We now introduce Venn diagrams, which are sometimes called *Euler circles* in honor of the first mathematician to use them, Leonhard Euler.

Venn diagrams can be used to depict statements involving quantifiers, as seen in the following example.

**EXAMPLE 1**

Draw Venn diagrams for the following statements.
(a) All $D$ are $A$. (For instance, all dogs are animals.)
(b) Some $P$ are $D$. (For instance, some people are Democrats.)
(c) No $C$ is a $D$. (For instance, no cat is a dog.)

*SOLUTION*

(a) Let's consider the dogs as being interior to one circle and the animals as being interior to another circle. Since all dogs are animals, the circle of $D$'s lies entirely within the circle of $A$'s. See Figure 2-1 (a). If a part of the circle of $D$'s were outside the circle of $A$'s, this would imply that some dogs were not animals. Similarly, if the circle of $A$'s were entirely within the circle of $D$'s, this would imply that all animals were dogs.

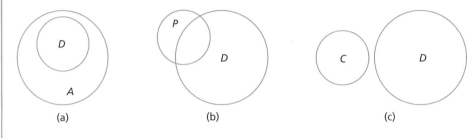

(a)          (b)          (c)

**FIGURE 2-1**

(b) The only thing we know for certain is that at least one $P$ is a $D$. Ordinarily we illustrate this with a diagram like the one shown in Figure 2-1 (b).
(c) The circle representing cats is disjoint from the circle representing dogs, because not one $C$ is a $D$. See Figure 2-1 (c).

**EXAMPLE 2**

(a) Can "Some bullies are students" be validly inferred from "Some students are bullies"?
(b) Can "All animals are people" be validly inferred from "All people are animals"?

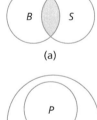

(a)

*SOLUTION*

(a) Yes       (b) No
These answers become clearer when we examine the Venn diagrams in Figure 2-2. In (a), the shaded region represents either "Some $S$ in $B$" or "Some $B$ in $S$." Thus, one can be inferred from the other. In (b), $x$ represents an animal that is not necessarily a person.

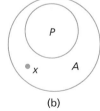

(b)

**FIGURE 2-2**

# NEGATION OF QUANTIFIERS

Sometimes confusion arises when we try to write the negation of a quantified statement. For example, consider "All letters in a box are vowels." How would you write the negation of this statement? Let's look at three possibilities, two of which may erroneously be used as negations:

## Leonhard Euler
### 1707–1783

Undoubtedly the most productive mathematician of all time—both in quantity and quality of work—was Leonhard Euler (rhymes with *boiler*). During his lifetime, he published more than 500 books and research papers. His collected works, including posthumous entries now being published under Swiss sponsorship, will run to about 75 large volumes. Nothing halted Euler's productivity—not the fact that he fathered 13 children or even the fact that overwork cost him his sight in one eye at age 28 and in the other at age 64. Indeed, after becoming blind, his output of publications—with the help of two sons and other collaborators—increased!

Euler was born in Basel, Switzerland, the first of six children in a minister's family. His father taught him mathematics in the cramped quarters of their two-room house. He attended high school in Basel and took mathematics lessons from a college student. At age 13 he entered the University of Basel and received his master's degree at age 16.

To his mathematical studies, Euler added astronomy, medicine, oriental languages, and physics. Although German was his native tongue, he usually wrote in Latin or French.

In 1727, Catherine I of Russia appointed Euler to the faculty of the Academy of Sciences in St. Petersburg. His first appointment was in medicine and physiology, followed by physics. In 1733, at the age of 26, he became the Academy's chief mathematician. From 1741 to 1766, Euler was called by Frederick the Great to be professor of mathematics at the Berlin Academy of Sciences.

It is almost impossible to find any area of mathematics today that has not been influenced deeply by Euler. It can be argued that he was a founder of what we know today as *pure* mathematics. His many contributions helped to formulate and to mold today's curriculum and methods of teaching in algebra, trigonometry, analytic geometry, theory of equations, calculus, number theory, geometry, and probability. In Chapter 11 we will study a formula in three-dimensional geometry developed by Euler. In this chapter we use pictures to illustrate statements that involve quantifiers and to describe relationships between sets. These are called *Venn diagrams* in this text but could also be called *Euler circles* in honor of the man who first used them, Leonhard Euler.

(a) No letters in the box are vowels.
(b) All letters in the box are not vowels.
(c) Some letters in the box are not vowels.

To be a negation, the newly formulated expression must have truth values that are the opposite of the truth values of the original statement in every possible situation. Let's make up some examples and check to see which of the preceding possibilities is the negation. Suppose that in the box we have

Case 1        | a  e  i  o |
Case 2        | a  e  t  s |
Case 3        | r  s  t  b |

Table 2-14 presents a truth table for these cases.

**TABLE 2-14**

| Statement and Possible Negations | Case 1 | Case 2 | Case 3 |
|---|---|---|---|
| All letters are vowels. | T | F | F |
| (a) No letters are vowels. | F | F | T |
| (b) All letters are not vowels. | F | F | T |
| (c) Some letters are not vowels. | F | T | T |

Only the third possible negation (c) has truth values that are the opposite of the truth values of the original statement in all cases. Thus we can use "Some letters in the box are not vowels" as a negation of "All letters in a box are vowels." (a) and (b) are not negations of the original statement, because they yield the same truth value as the original in Case 2.

As the preceding discussion suggests, it is sometimes difficult to state correctly the negations of statements involving quantifiers. In such instances, the following rule is helpful: **To negate a quantified statement that involves a single quantifier, change the quantifier as illustrated below.**

| Statement | Negation |
|---|---|
| All $p$ are $q$. | Some $p$ are not $q$. |
| Some $p$ are $q$. | No $p$ is $q$. |
| Some $p$ are not $q$. | All $p$ are $q$. |
| No $p$ is $q$. | Some $p$ are $q$. |

**EXAMPLE 3**

Write the negation of each statement.
(a) Some women have red hair. (At least one woman has red hair.)
(b) All bananas are yellow.
(c) No professor is bald-headed.
(d) Some students do not work hard. (At least one student does not work hard.)

*SOLUTION*

(a) No woman has red hair.
(b) Some bananas are not yellow.
(c) Some professors are bald-headed. (At least one professor is bald-headed.)
(d) All students work hard.

To show that a universally quantified statement is false, you need only find one case for which the statement is false; that is, you need only identify one **counterexample.** However, in showing that an existentially quantified statement is false, you must show that it is false for all possibilities. Similarly, an existentially quantified statement is true if you can find one case for which it is true, but a universally quantified statement is true only if it is true for all cases.

# ARGUMENTS USING VENN DIAGRAMS

When arguments are given using statements with quantifiers, we sometimes use Venn diagrams to determine whether a conclusion is a valid consequence of the given premises. It is easy to show that reasoning is invalid by using Venn diagrams; however, we must be careful to consider all possibilities when using a Venn diagram to show that reasoning is valid.

**EXAMPLE 4**

Suppose that you are given the following premises:

> All chickens are fowls.
> All leghorns are chickens.

Now consider the following questions based on these premises.
(a) Can you be certain that all leghorns are fowls?
(b) Can you be certain that all fowls are leghorns?

*SOLUTION*

(a) In Figure 2-3, let the interior of the small circle represent leghorns. According to the second premise, all leghorns (*L*) are chickens (*C*), so the circle of leghorns is *within* the circle of chickens. Likewise, the circle of chickens (*C*) is within the square of fowls (*F*). Thus anything within the circle of leghorns (*L*) is automatically within the square of fowls (*F*). Therefore, all leghorns are fowls.

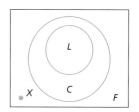

FIGURE 2-3

(b) Figure 2-3 includes a fowl (at *X*) that is not a leghorn. Thus the answer to the second question is *no*.

**EXAMPLE 5**

Use a Venn diagram to determine whether the following argument is valid or invalid.

> Some college professors are absentminded.
> Mr. Smith is a college professor.
> Therefore, Mr. Smith is absentminded.

*SOLUTION*

The premises (the first two sentences) of this argument are illustrated in two ways by Venn diagrams. In Figure 2-4(a) Mr. Smith is a college professor and also is absentminded. However, Figure 2-4(b) demonstrates that Mr. Smith can be a college professor without being absentminded. Thus, the truth of the premises does not require the truth of the conclusion, so Mr. Smith may or may not be absentminded. Because the argument contains a conclusion that is not an inescapable result of the premises, the reasoning is invalid.

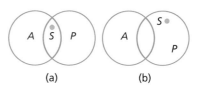

*A:* Absentminded people
*P:* Professors
*S:* Mr. Smith

FIGURE 2-4          (a)          (b)

Let's turn now to the problem posed in the introduction.

**EXAMPLE 6**

Use a Venn diagram to determine whether the following argument is valid or invalid.

> People who are concerned about the environment support the clean air bill.
> Senator Dalton supports the clean air bill.
> Therefore, Senator Dalton is concerned about the environment.

*SOLUTION*

Let $E$ represent people who are concerned about the environment. All of these people support the clean air bill ($C$). But as Figure 2-5 indicates, Senator Dalton ($D$) may support the clean air bill ($C$) without necessarily being inside $E$. Therefore, the argument is invalid.

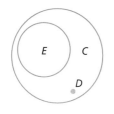

**FIGURE 2-5**

We need to remember that there is a difference between a valid argument and a true conclusion. **A conclusion drawn from an argument is true if the argument is valid and if all the premises are true.**

# Just for Fun

The mobster Big Maf's body was found floating in his swimming pool. When Detective Columbus was called to investigate the murder, he determined a number of facts involving the prime suspects. Either the butler, James, was at home or Big Maf's girlfriend, Blondie B, was out. If the private movie projector was on, Blondie B was at home. But if it was off, Big Maf's wayward child, Goodson, murdered Big Maf. Then again, if Goodson did not commit the murder, the butler did it. However, James was not at home at the time of the murder; a witness verified that he was at the local curb market. Detective Columbus quickly made his arrest. Whom did he arrest? What reasoning did he follow?

# EXERCISE SET 4

*R* 1. Make a Venn diagram showing each of the following:

   (a) All ims are elms.
   (b) Some men are fat.
   (c) No cams are dills.
   (d) All men are persons, and some men are lazy.

2. Make a Venn diagram showing

   (a) the relationship among cats, dogs, and animals
   (b) the relationship among boys, girls, and blond people
   (c) some $x$'s are $y$'s; some $x$'s are $z$'s; no $y$'s are $z$'s
   (d) some $y$'s are $x$'s; no $z$'s are $x$'s; some $z$'s are $y$'s
   (e) all $x$'s are $y$'s; some $y$'s are $z$'s; no $x$'s are $z$'s

3. Assign a value for $x$ so as to make each statement true.

   (a) $x + 4 = 7$      (b) $x - 7 = 4$
   (c) $x^2 = 16$       (d) $x + 3 = 3 + x$

4. If possible, assign a value for $x$ that makes each statement in Exercise 3 false.

5. Use a quantifier to make the statements in Exercise 3 true.

6. Use a quantifier to make the statements in Exercise 3 false.

7. Write the negation of each of the following statements, without using the expression "It is not true that."

   (a) All athletes over 7 ft. tall play basketball.
   (b) Some students work hard at their studies.
   (c) All men use Bob-Bob hair oil.
   (d) Some professors are not intelligent.
   (e) No man weighs more than 500 pounds.
   (f) Some rabbits have white tails.
   (g) Some bunnies are not rabbits.

8. Write the negation of each statement. Do not use "It is not true that."

   (a) There exists a counting number greater than 50.
   (b) Not all counting numbers are greater than 5.
   (c) There exists an $x$ such that $x + 2 = 7$.
   (d) For all $x$, $(x + 3) + 2 = 3 + (2 + x)$.
   (e) For all $x$, $x(x + 2) = x^2 + 2x$.
   (f) For every $a$, $a(a + 1) = a^2 + a$.

9. Write each of these statements as a conditional:

   (a) All frogs are green.
   (b) All cats are intelligent.
   (c) Each college professor is brilliant.
   (d) No people with long legs are athletes.

*For Exercises 10 through 13, assume that the first two statements are correct. Use a Venn diagram to answer the questions.*

10. All cats are elephants. Some elephants are red. Can you be certain that some cats are red?

11. Some cows are birds. All birds have two legs. Can you be certain that some cows have two legs? Can you be certain that all birds are cows? Can you be certain that all cows have two legs?

12. Some clothes are old. No old things are good. Can you be certain that some clothes are good? Can you be certain that some clothes are not good?

13. Some beans are green. All green things are edible. Can you be certain that all beans are edible? Can you be certain that some beans are not edible?

14. In each of the following, two premises are given. Find a conclusion that is a logical result of the premises.

   (a) All college students are clever.
      Diana is a college student.
   (b) All right angles are equal.
      Angles $A$ and $B$ are not equal.
   (c) All college women are beautiful.
      Kay is not beautiful.

15. Write a quantified statement to describe the relationship represented by the Venn diagram.

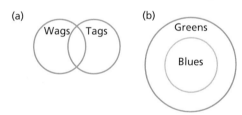

(a) Wags Tags     (b) Greens Blues

*T* 16. Show that the following statements are false by finding a counterexample:

   (a) All counting numbers are even.
   (b) All governors of states are men.

17. Write each of these conditionals as a universally quantified statement:

   (a) If a figure is a square, then it is a rectangle.
   (b) If a number is divisible by 4, then it is divisible by 2.

18. (a–b) Write a negation for each part of Exercise 17 by using the fact that $\sim (p \to q)$ is equivalent to $p \wedge \sim q$.

   (c–d) Write a negation for each part of Exercise 17 by using the rules for forming negations of universally quantified statements.

19. Describe the number $x$ as completely as possible. $I$ represents integers; $R$, rational numbers; $Ir$, irrational numbers; $Re$, real numbers; $Im$, imaginary numbers; and $C$, complex numbers.

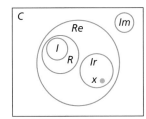

*In Exercises 20 through 27, use Venn diagrams to determine whether the third statement must be correct if it is assumed that the first two statements are true.*

20. All fast runners are athletes. No people with long legs are athletes. Therefore, no people with long legs are fast runners.

21. All spheres are round. All circles are round. Therefore, all circles are spheres.

22. Carelessness always leads to accidents. Mrs. Yeager had an accident. Therefore, Mrs. Yeager was careless.

23. All cats are intelligent animals. Fido is an intelligent animal. Therefore, Fido is a cat.

24. Some ems are red. All ims are ems. Therefore, some ims are red.

25. All athletes are fast runners. Some people with short legs are not fast runners. Therefore, no athletes have short legs.

26. All wise boys are interested in girls. George is interested in girls. Therefore, George is wise.

27. Write a valid conclusion for the following premises: All sales must come from production, and consumption must come from sales.

*In Exercises 28 through 30, use Venn diagrams to select valid conclusions from the given possibilities.*

*C* 28. All $x$'s are $y$'s.
Some $x$'s are $z$'s.
Some $x$'s are $w$'s.

Therefore,
(a) all $w$'s are $y$'s.　(b) all $z$'s are $y$'s.
(c) some $y$'s are $w$'s.　(d) some $x$'s are not $y$'s.

29. All whole numbers are integers.
All integers are rationals.
No irrationals are rational.
All rationals are reals.
All irrationals are reals.
$x$ is an integer.

Therefore,
(a) $x$ is a whole number.
(b) $x$ is a real number.
(c) $x$ is an irrational number.
(d) $x$ is not a rational number.

30. All authors are intelligent.
Some people from New York are authors.
Some people from Philadelphia are not intelligent.

Therefore,
(a) no person from Philadelphia is an author.
(b) some New York people are intelligent.
(c) no authors are from New York.

31. Use Venn diagrams to test whether the conclusion is a necessary result of the given facts.

(a) All students who study hard make A's.
All math students study hard.
Fred is a math student.
Therefore, Fred makes an A.
(b) Some business majors find jobs.
All marketing majors are business majors.
Therefore, some marketing majors find jobs.

32. Place $\varnothing$ (indicating empty) and • (indicating at least one) in each region of the accompanying Venn diagram to indicate each of the following relationships:

(a) All $A$'s are $B$'s.

(b) No $A$'s are $B$'s.

(c) Some $A$'s are $B$'s.

(d) Some $A$'s are not $B$'s.

## REVIEW EXERCISES

*Write the negations of the following conditionals:*

33. If it is raining, I will carry an umbrella.

34. If $x + 3 = 7$, then $x = 4$.

35. If Lucy is late, she will miss the bus.

36. If the Dow–Jones average goes down, then the price of gold will go up.

37. Complete the following truth table.

| $p$ | $q$ | $p \rightarrow q$ | $p \wedge q$ | $p \vee q$ |
|---|---|---|---|---|
|  |  |  |  |  |
|  |  |  |  | F |
| T |  | F |  |  |
|  |  |  |  | T |

# SOLUTION TO INTRODUCTORY PROBLEM

**UNDERSTAND THE PROBLEM.** We must match each car with its owner and color in such a way that each of the following statements in the problem are true.

(1) Dale does not own the Saturn, and his car is not blue.

(2) If the Honda does not belong to Dale, then it is white.

(3) If the blue car is either the Saturn or belongs to Sigmund, then the Plymouth is white.

(4) If the Honda is either gray or white, then Sigmund does not own the Saturn.

**DEVISE A PLAN.** Because the first statement asserts that Dale does not own the Saturn, he must own either the Honda or the Plymouth. We will examine these two cases. We will use reasoning and our knowledge of the truth values of statements involving conditionals and the connectives *and* and *or* to determine which case is correct.

**CARRY OUT THE PLAN.**

Case 1: Dale owns the Plymouth. By (1) the Plymouth is either white or gray. By (2) the Honda is white, so the Plymouth is gray. This means that the Saturn is blue. But then (3) implies that the Plymouth is white. This is a contradiction, so Case I is not possible.

Case 2: Dale owns the Honda. By (1), the Honda is either white or gray. We get no information from (2). From (4) we learn that Sigmund does not own the Saturn, so he owns the Plymouth. Thus, Bob owns the Saturn. Now, from (3) we learn that the Plymouth is white. Thus the Honda is gray and the Saturn is blue.

**LOOK BACK.** The gray Honda belongs to Dale, the blue Saturn belongs to Bob, and the white Plymouth belongs to Sigmund. This solution satisfies all conditions of statements (1) through (4).

# SUMMARY AND REVIEW

1. Understand
   (a) negation, $\sim p$
   (b) conjunction, $p \wedge q$
   (c) disjunction, $p \vee q$
   (d) conditionals, $p \rightarrow q$
   (e) biconditional, $p \leftrightarrow q$

2. Associated with every conditional "if $p$ then $q$" is
   (a) its converse, if $q$ then $p$.
   (b) its contrapositive, if $\sim q$ then $\sim p$.
   (c) its inverse, if $\sim p$ then $\sim q$.

3. Two statements are equivalent if they have identical truth values in a truth table.

4. The contrapositive is logically equivalent to the original proposition.

5. Quantifiers that indicate that quantified statements are always true are called *universal quantifiers*.

6. Quantifiers that indicate that there exists at least one case in which a quantified statement is true are called *existential quantifiers*.

7. *Rule of detachment:* The facts that $p \rightarrow q$ is true and that $p$ is true imply that $q$ is true.

8. *Chain rule:* If the conditionals $p \rightarrow q$ and $q \rightarrow r$ are both true, then $p \rightarrow r$ is true.

9. *Rule of contraposition:* The facts that $p \rightarrow q$ is true and $\sim q$ is true imply that $\sim p$ is true.

10. An argument is valid if the conclusion is true when the conjunction of the premises is true.

# CHAPTER TEST

1. Which of the following are statements?

    (a) James Earl Carter was president of the U.S.
    (b) He is tall and slender.
    (c) How can you do that problem?
    (d) All dogs have fleas.

2. Use Venn diagrams to represent these statements.

    (a) Some Democrats support free trade.
    (b) All athletes are fast.

3. Write the following statements as conditionals.

    (a) All wolves are mammals.
    (b) Tom is late or my watch is fast.

4. Use a truth table to show the truth values of

    (a) $p \wedge (q \vee r)$      (b) $(p \wedge q) \vee (p \wedge r)$

5. Are the statements in (a) and (b) of Problem 4 equivalent?

6. (a) Complete the accompanying truth table.
   (b) Are $p \rightarrow q$ and $q \rightarrow p$ logically equivalent? Explain.

   | $p$ | $q$ | $p \rightarrow q$ | $q \rightarrow p$ |
   |-----|-----|-------------------|-------------------|
   | T | T | | |
   | T | F | | |
   | F | T | | |
   | F | F | | |

7. Write the converse and contrapositive of "If a course is hard, then it is good."

8. Use Venn diagrams to determine whether the third statement is a necessary consequence of the first two.

    (a) Some widgets are fidgets. All fidgets are midgets. Therefore, some widgets are midgets.
    (b) All basketball players are tall. Jamie is not a basketball player. Therefore, Jamie is not tall.

9. Write the negations of the following.

    (a) All birds are black.
    (b) For some $x$, $x + 3 = 5$.

10. Find the truth value of each of the following when $q$ is false, $p$ is true, and $r$ is true.

    (a) $(\sim r \wedge p) \rightarrow q$      (b) $r \rightarrow \sim(p \vee q)$

11. Find a valid conclusion for the following premises:

    All triangles are polygons.
    All polygons are plane figures.

12. Find a conclusion that makes each of the arguments valid, and specify the rule of logic used.

    (a) If it is hot, I will go swimming.
        If I go swimming, I will miss supper.
    (b) If Aaron sleeps late, then he will be fired.
        Aaron was not fired.
    (c) If the budget passes, the stock market will rise.
        The budget passes.
    (d) If the sun shines, we will go on a picnic.
        We do not go on a picnic.

13. (a) Use truth tables to show that the negation of the conditional $p \rightarrow q$ is equivalent to $p \wedge \sim q$.
    (b) Use this fact to write the negation of "If it is hot, I will go to the beach."

# SETS, RELATIONS, AND FUNCTIONS

# 3

In a certain apartment complex live 20 men: 8 are married and own automobiles, 12 own automobiles and have television sets, and 11 are married and have television sets. What is the largest possible number of married men?

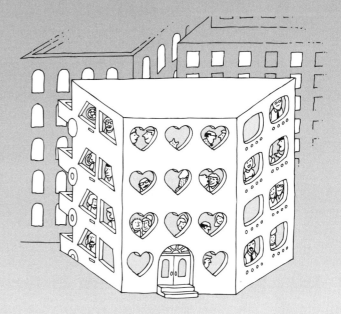

In this chapter we will discuss sets (and the related concepts of relations and functions), an important idea used throughout the language of mathematics. We will use the language of sets at several points in this book, such as when we look carefully at the properties of whole numbers, when we study probability, and when we examine the fundamental concepts of geometry. However, be assured that the language of sets has much broader use than this. The language of sets lives at the heart of all discussions in higher mathematics and has found application in such diverse disciplines as computer science and the social sciences.

Problem-solving strategies emphasized in this chapter include the following: ▪ Look for a pattern. (87–89) ▪ Draw a picture. (throughout the chapter) ▪ Use reasoning. (81, 96, 101, 102) ▪ Try a simpler version of the problem. (87) ▪ Make a chart or table. (throughout the chapter) ▪ Use an equation. (throughout Section 4) ▪ Substitute in a formula. (102, 103, 107, 109) ▪ Make a list of all possibilities. (81, 89, 96)

# SECTION 1 — AN INTRODUCTION TO SETS

**PROBLEM**

Place eight tennis balls in three baskets of different sizes so that there is an odd number of balls in each basket.

  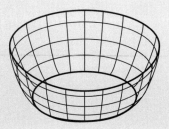

**OVERVIEW**

The preceding is a trick problem involving subsets.

**GOALS**

Use the Communication Standard: ● organize and consolidate mathematical thinking through communication; ● communicate mathematical thinking coherently and clearly to peers, teachers, and others; ● analyze and evaluate the mathematical thinking and strategies of others; ● use the language of mathematics to express mathematical ideas precisely.
Illustrate the Problem Solving Standard, page 15.*
Illustrate the Representation Standard, page 50.*
Illustrate the Reasoning and Proof Standard, page 23.*

*The complete statement of the standard is given on this page of the book.

Although this book gives little emphasis to the formal theory of sets, some familiarity with the notation and language of sets is useful and important. During the latter part of the 19th century, while working with mathematical entities called *infinite series,* Georg Cantor found it helpful to borrow a word from common usage to describe a mathematical idea. The word he borrowed was **set.**

## DESCRIBING SETS

Our purposes will be served if we intuitively describe a set as a collection of objects possessing a property that enables us to determine whether a given object is in the collection. We sometimes say that a set is a *well-defined* collection—meaning that, given an object and a set, we are able to determine whether or not the object is in the set. The individual objects in a set are called **elements** of the set. They are said **to belong to** or **to be members of** or **to be in** the set. The relationship between objects in a set and the set itself is expressed in the form *is an element of* or *is a member of.*

To denote this relationship we use the following notation:

$x \in A$ means $x$ is an **element** of set $A$.

$x \notin A$ means $x$ is **not an element** of set $A$.

Often it is possible to specify a set by listing its members within braces. This method of describing a set is called the **tabulation** method (sometimes called the *roster method*). For example, the set of counting numbers less than 10 can be written as $\{1, 2, 3, 4, 5, 6, 7, 8, 9\}$.

A set remains the same, regardless of the order in which the elements are tabulated. Thus $\{1, 2, 3\}$ is the same set as $\{2, 1, 3\}$, $\{3, 2, 1\}$, $\{3, 1, 2\}$, $\{1, 3, 2\}$, or $\{2, 3, 1\}$. In fact, two sets are said to be **equal** if they contain exactly the same elements. $A = B$, **if and only if $A$ and $B$ have exactly the same elements.** If there is at least one element in either $A$ or $B$ that is not in the other, then $A \neq B$.

| EXAMPLE 1 | Glenda, Mark, and Martia are the only counselors in the admissions office. They constitute the set $A = \{$Glenda, Mark, Martia$\}$. Mark $\in A$; Linda $\notin A$. Can you identify other elements of set $A$? |

Sometimes sets have so many elements that it is tedious, difficult, or even impossible to tabulate them. Sets of this nature may be indicated by a descriptive statement or a rule. The following sets are well specified without a tabulation of members: the counting numbers less than 10, the even numbers less than 1000, the past presidents of the United States, and the football teams in Pennsylvania.

The difficulty of tabulating sets can be minimized by using **set-builder notation,** which encloses within braces a letter or symbol representing an element of the set followed by a qualifying description of the element. For example, let $A$ represent the set of counting numbers less than 10; then

$$A = \{n \mid n \text{ is a counting number less than 10}\}$$

### Georg Cantor
### 1845–1918

An exciting feature that distinguishes mathematics from other disciplines is that mathematics deals with ideas of the infinite. Concepts of infinity plagued many mathematicians until the seminal work of Georg Cantor. We struggle with the notion of infinity because things that are infinite are beyond our realm of normal physical experience; they are therefore *abstractions* that we can deal with only in our minds. Yet Cantor showed that it makes sense to talk about the number of elements in any set, finite or infinite, and, surprisingly, showed that an *infinite* sequence of higher infinities can also be described.

Cantor was born in St. Petersburg, Russia, to Norwegian parents. The family moved to Frankfurt, Germany, when Georg was 11 years old. Resisting his father's encouragement to study engineering, Georg focused on mathematics, philosophy, and physics in his studies at the Universities of Zurich,

Gottingen, and Berlin. He obtained a position at the University of Halle, in Wittenberg, Germany. Although his goal was an appointment to the University of Berlin, he spent most of his life in Wittenberg.

For not attaining his goal to be appointed to the University of Berlin, Cantor blamed a Berlin mathematician named Leopold Kronecker, who did not accept any of Cantor's work on infinite sets. Kronecker openly attacked Cantor's findings, contributing to Cantor's mental collapse in 1884. This collapse was the first of many that occurred throughout the rest of his life. He died in an institution in Wittenberg in 1918, before his results were widely accepted.

In the last decade, a new chapter has been added to Cantor's work. The study of chaos theory and its geometric progeny, fractals, has renewed interest in some of Cantor's most unusual work.

This notation is read "the set of all elements $n$ such that $n$ is a counting number less than 10." Notice that the vertical line is read "such that."

| | |
|---|---|
| **EXAMPLE 2** | Use set-builder notation to denote the set of current United States senators. |
| *SOLUTION* | $\{x \mid x$ is a U.S. senator$\}$. The set is read "the set of all $x$ such that $x$ is a U.S. senator." |

Frequently, three dots (called an **ellipsis**) are used to indicate the omission of terms. The set of even counting numbers less than 100 may be written as $\{2, 4, 6, \ldots, 98\}$. This notation saves time in tabulating elements of large sets, but it can be ambiguous unless the set has been specified completely by another description. For example, $\{2, 4, \ldots, 16\}$ could be $\{2, 4, 8, 16\}$ or it could be $\{2, 4, 6, 8, 10, 12, 14, 16\}$.

An ellipsis is also used to indicate that a sequence of elements continues indefinitely. For example, consider the set of **natural** (or **counting**) **numbers.**

$$\{1, 2, 3, 4, 5, \ldots\}$$

The set of natural numbers is an example of an **infinite set,** described informally as one that contains an unlimited number of elements. In contrast, a **finite set** contains zero elements or a number of elements that can be specified by a natural number.

| | | |
|---|---|---|
| **DEFINITION** | **Empty Set** | A set that contains no elements is called the *empty* or *null set* and is denoted by either $\varnothing$ or { }. |

The relationship between two sets such as $A = \{1, 3, 5, 7\}$ and $B = \{1, 2, 3, 4, 5, 6, 7, 8\}$ is described by the term *subset*.

| | | |
|---|---|---|
| **DEFINITION** | **Subset** | Set $A$ is said to be a *subset* of set $B$, denoted by $A \subseteq B$, if and only if each element of $A$ is an element of $B$. |

**EXAMPLE 3**   If $A = \{x, y\}$ and $B = \{w, x, y, z\}$, then $A \subseteq B$ because each element of $A$ is an element of $B$.

**EXAMPLE 4**   If $P = \{1, 4, 7\}$ and $Q = \{4, 7, 1\}$, then $P \subseteq Q$ because each element of $P$ is an element of $Q$. Moreover, $Q \subseteq P$, and $P \subseteq P$.

To show that $A$ is not a subset of $B$ (denoted by $A \not\subseteq B$), we must find at least one element in $A$ that is not in $B$. Let's use this idea to examine whether $\varnothing$ is a subset of some set $B$. Using our problem-solving techniques, let's list all possibilities:

**1.** Either $\varnothing$ is a subset of $B$      **2.** Or $\varnothing$ is not a subset of $B$

In Possibility 2, if $\varnothing$ is not a subset of $B$, then there must be an element of $\varnothing$ not in $B$. But $\varnothing$ has no elements. Consequently, Possibility 2 cannot be true. The only alternative is for Possibility 1 to be true.

| | |
|---|---|
| **Subset $\varnothing$** | For any set $B$, $\varnothing \subseteq B$. |

Because all dogs are animals, the set of dogs is a subset of the set of animals. Moreover, we call the set of dogs a "proper" subset of the set of animals because some animals are not dogs.

| DEFINITION | **Proper Subset** | Set $A$ is said to be a *proper subset* of set $B$, denoted by $A \subset B$, if and only if each element of $A$ is an element of $B$ and at least one element of $B$ is not an element of $A$; that is, $A \subset B$ if $A \subseteq B$ and $A \neq B$. |
|---|---|---|

 ## OPERATIONS ON SETS

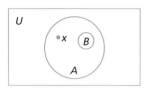

FIGURE 3-1

If a discussion is limited to a fixed set of objects and if all elements to be discussed are contained in this set, then this overall set is called the **universal set,** or simply the **universe.** A very useful device for visualizing and discussing sets is the **Venn diagram.** Circles or other closed curves are used in Venn diagrams (named after the English logician, Robert Venn) to represent sets. The universe can be represented as the region bounded by a rectangle, and the set under consideration as the region bounded by a circle (or some other closed region) within the rectangle. In Figure 3-1, $x \in A$ means that $x$ is some point in the circular region $A$. Also in Figure 3-1, set $B$ is a proper subset of set $A$.

The region outside set $A$ and inside the universe represents the complement of $A$, denoted by $\overline{A}$ (or sometimes by $A'$ or $\sim A$) and read "$A$ bar" or "bar $A$."

| DEFINITION | **Complement of a Set** | The *complement* of set $A$, denoted by $\overline{A}$, is the set of elements in the universe that are not in set $A$. If $A$ is a subset of the universe $U$, then $$\overline{A} = \{x \mid x \in U \text{ and } x \notin A\}.$$ |
|---|---|---|

**EXAMPLE 5**    If $U = \{a, b, c, d\}$ and $A = \{b, c\}$, find $\overline{A}$.

*SOLUTION*    $\overline{A} = \{a, d\}$.    Elements in $U$ but not in $A$

**EXAMPLE 6**    If the universe is all college students and if $A$ is the set of college students who have made all A's, then all college students who have made at least one grade lower than an A is the complement of $A$ (that is, $\overline{A}$) and is represented by the shaded region of Figure 3-2.

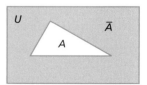

FIGURE 3-2

| **Complement of $A$ Relative to $B$** | The complement of $A$ relative to $B$ is the set of elements in $B$ that are not in $A$. This may be written as $B - A$ or in set-builder notation as $$\{x \mid x \in B \quad \text{and} \quad x \notin A\}$$ |
|---|---|

$B - A$ is sometimes read "the set difference of $B$ and $A$." $B - A$, represented by the shaded region in Figure 3-3, is sometimes called a *relative complement*.

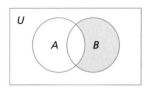

FIGURE 3-3

**EXAMPLE 7**

If $A = \{x, y, z, w\}$ and $B = \{u, v, x, y\}$, then

$$B - A = \{u, v\}$$
$$A - B = \{z, w\}$$
$$B - B = \varnothing$$

Special notations are used for discussing the relationships among members of two or more sets.

---

**DEFINITION**

**Intersection of Sets**

The *intersection* of any two sets $A$ and $B$, denoted by $A \cap B$, is the set of all elements common to both $A$ and $B$. $A \cap B = \{x \mid x \in A$ and $x \in B\}$.

---

**DEFINITION**

**Union of Sets**

The *union* of any two sets $A$ and $B$, denoted by $A \cup B$, is the set of all elements in set $A$ or in set $B$ or in both $A$ and $B$. $A \cup B = \{x \mid x \in A$ or $x \in B\}$.

---

**EXAMPLE 8**

Let $A$ represent a committee consisting of {Rose, Dave, Sue, John, Jack}, and let $B$ represent a second committee consisting of {Sue, Edward, Cecil, John}. The intersection of these two sets, $A \cap B$, is {Sue, John}.

**EXAMPLE 9**

If $A = \{1, 2, 3\}$ and $B = \{6, 7\}$, find $A \cup B$.

*SOLUTION*

$A \cup B = \{1, 2, 3, 6, 7\}$        Elements in either $A$ or $B$

Recall that $A \cup B$ is the set of all elements that belong to $A$ or to $B$ or to both $A$ and $B$. Any elements common to both sets are listed only once in the union. Thus, for

$$A = \{a, b, c, d, e\} \quad \text{and} \quad B = \{c, d, e, f, g\} \text{ we find that}$$
$$A \cup B = \{a, b, c, d, e, f, g\}$$

**EXAMPLE 10**

If $A = \{1, 2\}$, then $A \cup A = \{1, 2\}$.

The shaded regions in Figure 3-4 compare intersection and union under different situations for sets $A$ and $B$. Notice in (a) that $A$ and $B$ overlap or have elements in common. In (b), $A$ is a proper subset of $B$. In (c), $A$ and $B$ have no elements in common, so $A \cap B = \varnothing$; here $A$ and $B$ are said to be **disjoint.**

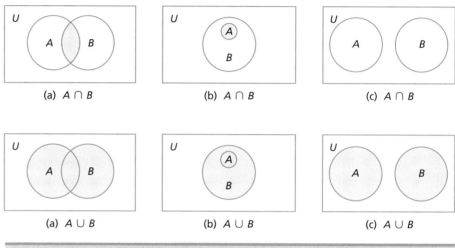

(a) $A \cap B$          (b) $A \cap B$          (c) $A \cap B$

(a) $A \cup B$          (b) $A \cup B$          (c) $A \cup B$

FIGURE 3-4

| DEFINITION | **Disjoint Sets** | Two sets $A$ and $B$ are said to be *disjoint* if and only if $A \cap B = \varnothing$—that is, if the two sets have no elements in common. |
|---|---|---|

**EXAMPLE 11**

If $U = \{x \mid x$ is a counting number less than 10$\}$, $A = \{2, 4, 6\}$, $B = \{1, 2, 3, 4, 5\}$, and $C = \{3, 5, 7\}$, find $A \cap B$, $A \cap C$, $A \cup C$, and $\overline{B}$.

SOLUTION

$A \cap B = \{2, 4\}$          Elements in common
$A \cap C = \varnothing$          No elements in common, disjoint
$A \cup C = \{2, 3, 4, 5, 6, 7\}$          Elements in either $A$ or $C$ or both
$\overline{B} = \{6, 7, 8, 9\}$          Elements in the universe, not in $B$

**PRACTICE PROBLEM**

Find the intersection and union of $A = \{1, 2, 3, \ldots, 100\}$ and $B = \{60, 61, \ldots, 1000\}$.

ANSWER

$A \cap B = \{60, 61, \ldots, 100\}$, $A \cup B = \{1, 2, 3, \ldots, 1000\}$

# PROPERTIES OF SET OPERATIONS

The following properties of the intersection and union of sets can be illustrated by using Venn diagrams.

| | |
|---|---|
| **Properties of Intersection and Union** | For all sets $A$, $B$, and $C$:<br>1. *Commutative properties:* $A \cup B = B \cup A$; $A \cap B = B \cap A$<br>2. *Associative properties:* $A \cup (B \cup C) = (A \cup B) \cup C$;<br>$A \cap (B \cap C) = (A \cap B) \cap C$<br>3. *Identity properties:* $A \cup \varnothing = A$; $A \cap U = A$ |

Notice that the *commutative* properties of intersection and union indicate that order is not important in performing these operations. The *associative* properties of intersection and union indicate that grouping is not important in performing these operations. The *identity* property of union indicates that there is a special set (the null set) with the property that its union with a set is always that same set. The universal set serves the "identity" role for intersection.

Figure 3-5(a) represents the formation of $(A \cap B) \cap C$. Similarly, Figure 3-5(b) represents the formation of $A \cap (B \cap C)$. The double-shaded region representing $(A \cap B) \cap C$ is the same as the region representing $A \cap (B \cap C)$; this illustrates the associative property, $(A \cap B) \cap C = A \cap (B \cap C)$.

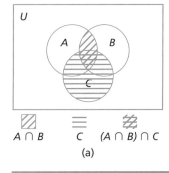

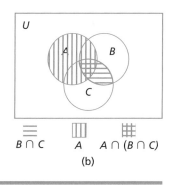

FIGURE 3-5

**EXAMPLE 12**

Let $A = \{x \mid x$ is a state whose name begins with $A\}$ and $B = \{x \mid x$ is a state east of the Mississippi River$\}$. $A \cap B = \{$Alabama$\}$, and $B \cap A = \{$Alabama$\}$; therefore, $A \cap B = B \cap A$, illustrating the commutative property of intersection.

The commutative property of union is demonstrated in the next example.

**EXAMPLE 13**

Consider $A = \{1, 3\}$ and $B = \{3, 5, 7\}$. Then $A \cup B = \{1, 3, 5, 7\}$ and $B \cup A = \{1, 3, 5, 7\}$; thus, $A \cup B = B \cup A$.

**EXAMPLE 14**

If $A = \{1, 2, 3, 4\}$ and $B = \varnothing$, then

$$A \cup B = \{1, 2, 3, 4\} \cup \varnothing = \{1, 2, 3, 4\} = A$$

**EXAMPLE 15**

In Figure 3-6, (a) and (b) represent the formation of $(A \cup B) \cup C$, and (c) and (d) represent $A \cup (B \cup C)$. Notice in (b) and (d) that $(A \cup B) \cup C$ and $A \cup (B \cup C)$ are represented by the same shaded region. This demonstrates that the associative property holds for union.

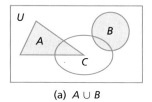  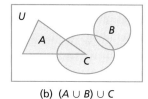  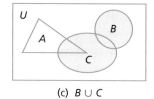  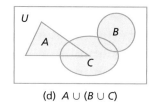

(a) $A \cup B$     (b) $(A \cup B) \cup C$     (c) $B \cup C$     (d) $A \cup (B \cup C)$

FIGURE 3-6

**PRACTICE PROBLEM**

Shade the portion of the diagram in Figure 3-7 that represents $A \cap (B \cap C)$.

*ANSWER*

The answer is given in two steps in Figure 3-8(a) and (b).

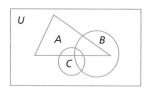  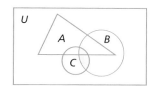  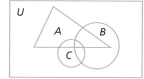

FIGURE 3-7        FIGURE 3-8

**EXAMPLE 16**

The introductory problem asks that we place eight tennis balls in three baskets of different sizes so that there is an odd number of balls in each basket.

*SOLUTION*

The problem did not state that the baskets must be disjoint. In fact, the hint suggested using subsets. So place one basket in another. One solution is given in Figure 3-9.

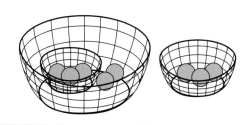

FIGURE 3-9

# Just for Fun

Use problem-solving techniques to discover the number of subsets in a set with $n$ elements. (Hint: Use the strategy of trying simple versions of the problem; then look for a pattern.)

# EXERCISE SET 1

*In this exercise set we shall use the counting numbers* $\{1, 2, 3, \ldots\}$, *the odd counting numbers* $\{1, 3, 5, 7, \ldots\}$, *and the even counting numbers* $\{2, 4, 6, \ldots\}$.

*R* 1. Let $A$ be the set of all counting numbers less than 16. Which of the following statements are true, and which are false? Justify your answer.

(a) $11 \in A$  (b) $81 \in A$
(c) $\{5\} \in A$  (d) $\{1, 2, 3, \ldots, 15\} \in A$
(e) $14 \in A$  (f) $0 \in A$

2. List within braces the members of the following sets.

(a) The counting numbers less than or equal to 16.
(b) The set of even counting numbers.
(c) The set of women presidents of the United States.

3. Express the sets in Exercise 2 in set-builder notation.

4. For each set in the left column, choose the sets from the right column that are subsets of it.

(a) $\{a, b, c, d\}$          (i) $\{\ \}$
(b) $\{o, p, k\}$            (ii) $\{1, 4, 8, 9\}$
(c) Set of letters in the    (iii) $\{o, k\}$
    word *book*             (iv) $\{12\}$
(d) $\{2, 4, 6, 8, 10, 12\}$  (v) $\{b\}$

5. Classify each statement as true or as false.

(a) $\{z, r\} \subseteq \{x, y, z, r\}$
(b) $\{$Brenda, Sharon, Glenda$\} \subseteq \{$Brenda$\}$
(c) $\{7, 2, 6\} \subseteq \{2, 6, 7\}$
(d) $6 \subseteq \{4, 5, 6, 7\}$
(e) $\{p, q, r\} \in \{p, q, r, s\}$

6. Which of the following sets are well defined?

(a) The set of great baseball players
(b) The set of beautiful horses
(c) The set of students in this class
(d) The set of counting numbers smaller than a million

7. Insert the appropriate symbol $\{\in, \notin, \subset$ or $\subseteq\}$ to make the following statements true:

(a) 3 _____ $\{1, 2, 3\}$
(b) $\{2\}$ _____ $\{1, 2, 3\}$
(c) $\{1, 2, 3\}$ _____ $\{1, 2, 3\}$
(d) 0 _____ $\{1, 2, 3\}$

8. Form the union and the intersection of the following pairs of sets:

(a) $R = \{5, 10, 15\}$, $T = \{15, 20\}$
(b) $M = \{1, 2, 3\}$, $N = \{101, 102, 103, 104\}$
(c) $A = \{0, 10, 100, 1000\}$, $B = \{10, 100\}$
(d) $G = \{$odd counting numbers less than 100$\}$, $H = \{$even counting numbers between 1 and 31$\}$
(e) $A = \{x, y, z, t\}$, $B = \{x, y, r, s\}$

9. Using sets $A$ and $B$, describe in words the elements in regions (a), (b), (c), and (d).

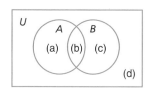

10. If $U = \{a, b, c, d, e, f, g, h\}$, $A = \{b, d, f, h\}$, and $B = \{a, b, e, f, g, h\}$, find the following:

(a) $\overline{A}$        (b) $\overline{B}$        (c) $A \cap B$
(d) $\overline{A \cap B}$  (e) $\overline{A} \cap \overline{B}$  (f) $\overline{A \cup B}$

11. Let $U$ be the set of students at Myschool; let $A$ be the set of female students, $B$ the set of male students, $C$ the set of students who ride the bus, and $D$ the set of members of athletic teams. Denote each of the following symbolically:

    (a) The set of female athletes

    (b) The set of male athletes who ride the bus

    (c) The set of all males who neither ride the bus nor are athletes

    (d) The set of females who either ride the bus or are not athletes

12. Shade the portion of the diagram that illustrates each of the following sets:

    (a) $A \cap B$    (b) $\overline{A} \cap C$    (c) $\overline{A} \cap \overline{B}$

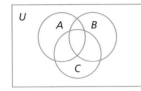

13. Shade the portion of the diagram that will illustrate each of the following sets:

    (a) $A \cup B$    (b) $\overline{A} \cap \overline{C}$
    (c) $A \cap (B \cap C)$    (d) $A \cup (B \cup C)$

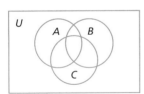

14. (a) Is there any distinction between $\{\emptyset\}$ and $\emptyset$?
    (b) Is $\emptyset \in \{\ \ \}$?    (c) Is $0 \in \{\ \ \}$?
    (d) Is $\emptyset \subseteq \{0\}$?    (e) If $A$ is any set, is $A \subset \emptyset$?
    (f) What is the distinction between 3 and $\{3\}$?

*T* 15. Draw a Venn diagram that illustrates each situation.

    (a) In Ourtown, no elementary teacher teaches in the high school, but some high school teachers teach in the college. Let $U$ be all teachers in Ourtown.

    (b) At Ourtown High, all mathematics teachers have a chalkboard. Some have an overhead projector, and some have a video cassette recorder, but no mathematics teacher has all three. Let $U$ be all teachers at Ourtown High.

16. In the accompanying figure, the sets of elements of closed regions are indicated by (a), (b), . . . , (h). Express the following in terms of these sets:

    (a) $\overline{A}$    (b) $A \cap B$
    (c) $A \cap B \cap C$    (d) $\overline{A} \cup \overline{B}$

    (e) $A \cup B \cup C$    (f) $A \cap \overline{B}$
    (g) $\overline{A \cup B}$    (h) $\overline{A \cup B \cup C}$

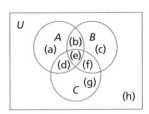

17. (a) List all the subsets of $\{a, b\}$.
    (b) List all the subsets of $\{a, b, c\}$.
    (c) List all proper subsets of $\{a, b, d\}$.

18. Let $A = \{3, 4, 5, 6\}$ and $B = \{5, 6, 7, 8\}$. Find

    (a) $B - A$    (b) $A - B$

*C* 19. In a Venn diagram, a single set partitions the universe into two distinct regions. Two sets partition the universe into a maximum of four regions.

    (a) What is the maximum number of regions into which three sets will partition the universe?
    (b) Four sets?
    (c) Five sets?
    (d) Use problem-solving techniques to conjecture an answer for $n$ sets.

20. Using $A$, $B$, and $C$, describe the elements of the closed regions denoted by (a), (b), (c), (d), (e), (f), (g), and (h).

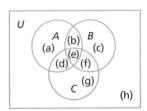

21. Use Venn diagrams to illustrate De Morgan's laws.

    (a) $\overline{A \cap B} = \overline{A} \cup \overline{B}$    (b) $\overline{A \cup B} = \overline{A} \cap \overline{B}$

22. Demonstrate with Venn diagrams the following property, called the *distributive property of union over intersection.*

    $$A \cup (B \cap C) = (A \cup B) \cap (A \cup C)$$

23. Demonstrate with Venn diagrams the following property, called the *distributive property of intersection over union.*

    $$A \cap (B \cup C) = (A \cap B) \cup (A \cap C)$$

24. (a) Illustrate the associative property of union where $U = \{a, b, c, d, e, f, g\}$, $A = \{a, b\}$, $B = \{b, c, d, e\}$, and $C = \{d, e, f\}$.

    (b) Use the sets listed in (a) to illustrate the associative property of intersection.

25. (a) If $x \in A \cap B$, is $x \in A \cup B$? Explain your answer.
    (b) If $x \in A \cup B$, is $x \in A \cap B$? Explain your answer.

26. How many subsets does $\{a, b, c, d\}$ have? Which are proper subsets?

*Set theory can be applied to analyze the power of voting coalitions. A winning coalition consists of any set of voters who can carry a proposal. Answer the questions about voting coalitions in Exercises 27 and 28.*

27. A committee has five members: $A, B, C, D$, and $E$. For a proposal to be passed, it must have at least three votes. List all the possible winning coalitions.

    *Example:* $\{A, B, D\}$ is a winning coalition; $\{B, D\}$ is not a winning coalition.

28. A town council consists of 5 members whose votes are weighted according to the number of citizens in their districts. Member $A$ has 6 votes; member $B$, 5 votes; $C$, 4 votes; $D$ and $E$, 1 vote apiece. Nine or more votes are required to carry an issue.
    (a) List all winning coalitions.

    *Example:* $\{A, B\}$ and $\{A, C, D\}$ are winning coalitions.

    (b) Review the list in (a), and pick all winning coalitions having the property that if any voter were removed from this coalition, the coalition would fail to win.

    *Example:* $\{A, B\}$ has this property, but if $D$ were removed from $\{A, C, D\}$, the resulting coalition $\{A, C\}$ would still win.

    (c) Do members $D$ and $E$ appear in any coalition that you listed in (b)?

    (d) What does this say about the power of $D$ and $E$? (NOTE: Similar arguments have been used in the courts to force modification of certain voting schemes.)

29. Describe in symbols the shaded portion of each of the following Venn diagrams.

(a)    (b)    (c)

---

**∵ PCR Excursion ∵**

30. A. Consider the sequence 1, 8, 16, 24, 32, . . . . Now consider the sequence of partial sums
$$1 = 1$$
$$1 + 8 = 9$$
$$1 + 8 + 16 = 25$$
    (a) Write the next three partial sums.
    (b) Associate with these partial sums the number of dots around closed figures in a square such as

$$1 + 8 = 3^2 \qquad 1 + 8 + 16 = 5^2$$

   Draw the geometric figure for $1 + 8 + 16 + 24$, and find an expression for this sum.
    (c) Draw a $9 \times 9$ square and find the corresponding sum; an $11 \times 11$ square.
    (d) What is the sum of the first $n$ terms of the sequence 1, 8, 16, 24, 32, . . . ?

   B. We can represent the partial sums of the sequence 1, 2, 3, 4, . . . by using blocks.

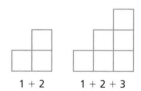

$$1 + 2 \qquad\qquad 1 + 2 + 3$$

    (a) Draw a similar diagram to represent $1 + 2 + 3 + 4$.
    (b) You can use this idea to find the sum of a given number of counting numbers. For example, find the sum of the first 4 counting numbers. To do this, place together, as shown, two diagrams representing $1 + 2 + 3 + 4 = \dfrac{4 \cdot 5}{2}$.

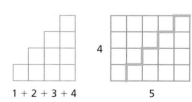

$$1 + 2 + 3 + 4 \qquad\qquad 5$$

   Use this reasoning to find the sum of the first 5 counting numbers.
    (c) Verbally describe the pattern you have discovered.
    (d) Find the sum of the first 100 counting numbers.

# CARTESIAN PRODUCT AND RELATIONS

| | |
|---|---|
| **PROBLEM** | What is the connection between a relationship like "less than" and sets of ordered pairs like {(1, 2), (1, 3), (1, 4), (2, 3), (2, 4), (3, 4)}? |
| **OVERVIEW** | To answer this question and to define a relation, we first need definitions of two terms: *ordered pair* and *Cartesian product.* |
| **GOALS** | Illustrate the Number and Operations Standard. page 119.* <br> Illustrate the Representation Standard, page 50.* <br> Illustrate the Algebra Standard, page 4.* <br> Illustrate the Problem Solving Standard, page 15.* <br><br> * The complete statement of the standard is given on this page of the book. |

## CARTESIAN PRODUCT

Let us now turn our attention to Cartesian product. The union or intersection of two sets produces a set containing some or all of the same elements. The Cartesian product is not expressed in the same form as either set; instead, it consists of ordered pairs of elements.

The concept of an **ordered pair** can be found frequently in real life. Consider, for instance, the following actions: put on socks, put on shoes; fry chicken, eat chicken; start mower, cut grass. Do you perform these actions in the order given? Does the order in which they are performed make a difference? In an ordered pair, the order of the elements is important.

---

**DEFINITION**

**Ordered Pair**

An *ordered pair* is an entity consisting of a first component and a second component. When the pair is denoted by $(x, y)$, $x$ is the first component and $y$ is the second component.

---

If $x$ and $y$ are elements of any two sets (or the same set), they can be used to form an ordered pair denoted by $(x, y)$. This pair is ordered in the sense that $(x, y)$ and $(y, x)$ are not equal unless $x = y$. Equality for ordered pairs is defined as follows.

---

**DEFINITION**

**Equality of Ordered Pairs**

$(a, b) = (c, d)$ if and only if $a = c$ and $b = d$.

---

**EXAMPLE 1**

A traveler can go from Chicago to Miami by auto, plane, train, or bus and from Miami to Nassau by plane or ship. The different ways of traveling from Chicago to Nassau through Miami can be described in terms of the following set of ordered pairs:

{(auto, plane), (auto, ship), (plane, plane), (plane, ship),

(train, plane), (train, ship), (bus, plane), (bus, ship)}

If $A$ represents {auto, plane, train, bus} and $B$ represents {plane, ship}, then the set of ordered pairs representing the different means of travel is denoted by $A \times B$ and is called the *Cartesian product* of sets $A$ and $B$.

---

**DEFINITION**

**Cartesian Product**

The *Cartesian product,* denoted by $A \times B$, of two sets $A$ and $B$ is the set of all ordered pairs $(a, b)$ with the first element chosen from $A$ and the second element chosen from $B$. In set-builder notation,

$$A \times B = \{(a, b) \mid a \in A \quad \text{and} \quad b \in B\}$$

---

**EXAMPLE 2**

If $A = \{r, s, t\}$ and $B = \{w, u, v\}$, find $A \times B$.

*SOLUTION*

$A \times B = \{(r, w), (r, u), (r, v), (s, w), (s, u), (s, v), (t, w), (t, u), (t, v)\}$.

The Cartesian product includes every pair that can be constructed by choosing a left partner from $A$ and a right partner from $B$. $A \times B$ is sometimes called the **cross product** of $A$ and $B$ and is read "$A$ cross $B$." The definition does not specify that the sets used in forming the cross product must be distinct. Examples illustrate that they may be the same or different.

**EXAMPLE 3**

(a) If $A = \{1, 2, 3\}$ and $B = \{4, 5\}$, form $A \times B$.
(b) If $A = \{a, b\}$, form $A \times A$.

*SOLUTION*

(a) $A \times B = \{(1, 4), (1, 5), (2, 4), (2, 5), (3, 4), (3, 5)\}$
(b) By definition, the ordered pairs $(x, y) \in A \times A$ must be such that $x \in A$ and $y \in A$. Thus,

$$A \times A = \{(a, a), (a, b), (b, a), (b, b)\}$$

**PRACTICE PROBLEM**

If $A = \{a, b\}$ and $B = \{c\}$, show that $A \times B \neq B \times A$.

*ANSWER*

$A \times B = \{(a, c), (b, c)\}$, but $B \times A = \{(c, a), (c, b)\}$; thus, $A \times B \neq B \times A$.

 **RELATIONS**

Everyone is familiar with such common nonmathematical relations as "Glenda is engaged to Jerry" and "Leigh is the daughter of Richard." The expressions "is engaged to" and "is the daughter of" are connectives that define relations on the set of all people. From the biblical narrative, the connective "is the father of" gives a relation between Adam and Abel but not between Cain and Abel.

Relations can often be better understood by using a visual representation. In the diagram of a relation, we draw an arrow pointing from $A$ to $B$ if $A$ is related to $B$. Let $R$ represent Richard; $T$, Tom; $S$, Sanders; $L$, Leigh; $G$, Glenda; and $M$, Maxine. From Figure 3-10, can you pick out who is the daughter of whom?

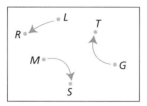

Leigh is the daughter of Richard.
Maxine is the daughter of Sanders.
Glenda is the daughter of Tom.

**FIGURE 3-10**

**EXAMPLE 4**

Each of the following sets of ordered pairs suggests a relation.
(a) (Hawthorne, *The Scarlet Letter*), (Crane, *The Red Badge of Courage*), (Steinbeck, *Of Mice and Men*), (Twain, *Huckleberry Finn*) are included in the relation "$x$ is the author of the book $y$."
(b) {(Harrison Ford, *K-19: The Widow Maker*), (Russell Crowe, *Master and Commander*), (Johnny Depp, *Pirates of the Carribbean*), (Tom Cruise, *The Last Samurai*)} are included in the relation "$x$ is a star in movie $y$."

The following example illustrates intuitively why the mathematical definition of a relation involves the concept of a subset of a Cartesian product.

**EXAMPLE 5**

Consider the Cartesian product of the set $A$ = {Atlanta, New York, Chicago, Buffalo} and the set $B$ = {Georgia, New York}. $A \times B$ is {(Atlanta, Georgia), (New York, Georgia), (Chicago, Georgia), (Buffalo, Georgia), (Atlanta, New York), (New York, New York), (Chicago, New York), (Buffalo, New York)}. If $x$ is an element of $A$ and $y$ is an element of $B$, the relation "$x$ is a city in $y$" would be defined by the set {(Atlanta, Georgia), (New York, New York), (Buffalo, New York)}. Notice that this new set is a subset of $A \times B$. Indeed, all relations on sets can be understood in this way.

**DEFINITION**

**Relation**

A *relation* $R$ from set $A$ to set $B$ is a set of ordered pairs whose first elements are in $A$ and whose second elements are in $B$; that is, $R \subseteq A \times B$.

**EXAMPLE 6**

Let $A$ represent all jockeys, and $B$ represent all horses. Consider $R$, the set of all ordered pairs (rider, horse) for this year's Kentucky Derby. Then $R$ is a subset of $A \times B$; thus, $(c, d) \in R$ means that $c$ is the rider of horse $d$.

As a special case of this definition, a relation $R$ from set $A$ to set $A$ is a subset of $A \times A$. In this case, we say that $R$ is a relation on $A$.

**EXAMPLE 7**

The relation "less than" on the set $A = \{1, 2, 3, 4\}$ is defined by $\{(1, 2), (1, 3), (1, 4), (2, 3), (2, 4), (3, 4)\}$, which is a subset of $A \times A$.

We now classify relations on a set as being *reflexive, symmetric,* and/or *transitive.*

---

**Reflexive Relation**

A relation $R$ on a set $A$ is *reflexive* if and only if $(a, a) \in R$ for every $a \in A$.

---

For instance, the relation "is the same species as" is reflexive on the set of animals. My dog Arthur has the same species as himself. In mathematics, equality is a reflexive relation on the set of counting numbers because $1 = 1$, $2 = 2$, and, in general, $a = a$.

There is a loop at every point of the visual representation of a relation that is reflexive.

If $D$ represents my dog, Arthur; $C$, my cat, Punkin; and $H$, my horse, Trigger; we have the loops shown in Figure 3-11(a) for "is the same species as."

The relation "is the mother of" is not reflexive because one is not the mother of oneself. Although geometric *congruence* is reflexive, *less than* and *greater than* are not.

(a) 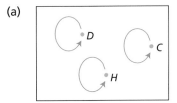   (b)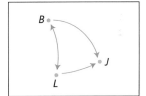

FIGURE 3-11

---

**Symmetric Relation**

A relation $R$ on set $A$ is said to be *symmetric* if and only if $(a, b) \in R$ implies that $(b, a) \in R$.

---

For example, the relation "is the brother of" is a symmetric relation for the set of all males. If Bob is the brother of Louis, then Louis is also the brother of Bob.

In terms of a visual representation of a symmetric relation, every correspondence must have an arrow pointed in both directions. "Is a brother of" is not symmetric on the set of all people, as seen using three children from the same family, $B$ (Bob), $L$ (Louis), and $J$ (Janice). In Figure 3-11(b), the arrow on the connection between Janice and the boys goes in only one direction.

In a symmetric relation, whenever $a$ is related to $b$, $b$ must be related to $a$. Perpendicularity on the set of lines in a plane is symmetric even though it is not reflexive. However, "is the mother of" is not symmetric: If Amy is the mother of Patty, Patty cannot be the mother of Amy. Similarly, "is less than" and "is greater than" are not symmetric.

| **Transitive Relation** | A relation $R$ on set $A$ is *transitive* if and only if $(a, b) \in R$ and $(b, c) \in R$ implies that $(a, c) \in R$. |
| --- | --- |

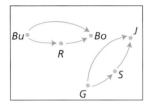

**FIGURE 3-12**

Consider the relation "is taller than," which is transitive. If Burt is taller than Rick and Rick is taller than Bob, then Burt is taller than Bob. This transitive relation is illustrated in Figure 3-12. Interpret Figure 3-12 for Gail, Susan, and Janice.

The transitive property for relations states that, whenever $a$ is related to $b$ and $b$ is related to $c$, then $a$ is related to $c$. Notice that we are not excluding the possibility that $a$ and $c$ are the same. "Is equal to," "is greater than," and "is less than" are good examples of relations for which the transitive property holds. However, a relation like "is the mother of" is not transitive. Elsie is the mother of Amy, and Amy is the mother of Patty, but Elsie is not the mother of Patty. Perpendicularity is also not a transitive relation.

The transitive property of equality can be generalized to involve any finite number of equalities; for instance, if

$$a = b, b = c, c = d, d = e, \text{ and } e = f$$

then
$$a = f$$

| **DEFINITION** | **Equivalence Relation** | A relation on a set $A$ that is reflexive, symmetric, and transitive is called an *equivalence relation*. |
| --- | --- | --- |

"Is the same age as" and "is equal to" are examples of equivalence relations. "Is greater than" is not an equivalence relation. Why not?

**PRACTICE PROBLEM**

Classify "is shorter than" as reflexive, symmetric, or transitive.

*ANSWER*

Transitive only.

**EXAMPLE 8**

Determine whether the relation $R = \{(1, 1), (1, 2), (1, 3), (2, 1), (2, 2), (2, 3), (3, 1), (3, 2), (3, 3)\}$ is an equivalence relation on $A$ where $A = \{1, 2, 3\}$.

*SOLUTION*

The relation is reflexive because all three elements $(1, 1)$, $(2, 2)$, and $(3, 3)$ are members of the relation.

The relation is symmetric because, for each $(a, b) \in R$, $(b, a)$ is also in $R$. For instance, $(1, 2)$ and $(2, 1)$, $(1, 3)$ and $(3, 1)$, and $(2, 3)$ and $(3, 2)$ are in the set.

The relation is transitive. For (1, 2) and (2, 1), the element (1, 1) ∈ R; for (1, 3) and (3, 2), the element (1, 2) ∈ R; for (2, 3) and (3, 1), the element (2, 1) ∈ R. For every $(a, b)$ and $(b, c)$ in the relation, the element $(a, c)$ is in the relation.

Therefore, the relation is an equivalence relation.

An interesting feature of an equivalence relation on $A$ is that the relation partitions or separates the set $A$ into disjoint subsets known as **equivalence classes.** The equivalence relation "is the same sex as" would partition the set {Leo, Mary, William, Sue, Jane} into two equivalence classes {Leo, William} and {Mary, Sue, Jane}. Notice that these two sets are disjoint and that every member of the equivalence class is related to every other member in the class.

**EXAMPLE 9**

Consider the relation "is the same shape as" on the set of geometric objects in Figure 3-13. Is this relation an equivalence relation? If it is an equivalence relation, what are the equivalence classes?

*SOLUTION*

This relation is a reflexive relation because each of the geometric figures has the same shape as itself. If Figure $A$ has the same shape as Figure $B$, then certainly Figure $B$ has the same shape as Figure $A$. Thus the relation is symmetric. Finally, if Figure $A$ has the same shape as Figure $B$ and Figure $B$ has the same shape as Figure $C$, then Figure $A$ has the same shape as Figure $C$. The relation is transitive and is an equivalence relation. The equivalence classes are enclosed by curves in Figure 3-13. Notice that all the circles are in the same class because they have the same shape.

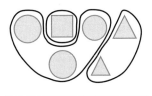

**FIGURE 3-13**

We conclude this section with a discussion of properties of sets needed as a basis for developing the properties of addition and multiplication of whole numbers.

| **Distributive Properties of Cartesian Product over Union** | For all sets $A$, $B$, and $C$:<br>(a)  $C \times (A \cup B) = (C \times A) \cup (C \times B)$<br>(b)  $(A \cup B) \times C = (A \times C) \cup (B \times C)$ |
|---|---|

**EXAMPLE 10**

Consider $A = \{1, 2\}$, $B = \{2, 3\}$, and $C = \{1, 5\}$. Show that $(A \cup B) \times C = (A \times C) \cup (B \times C)$ and $C \times (A \cup B) = (C \times A) \cup (C \times B)$.

*SOLUTION*

$$(A \cup B) \times C = \{1, 2, 3\} \times \{1, 5\}$$
$$= \{(1, 1), (2, 1), (3, 1), (1, 5), (2, 5), (3, 5)\}$$

Now

$$A \times C = \{1, 2\} \times \{1, 5\} = \{(1, 1), (2, 1), (1, 5), (2, 5)\}$$

and

$$B \times C = \{2, 3\} \times \{1, 5\} = \{(2, 1), (3, 1), (2, 5), (3, 5)\}$$

so

$$(A \times C) \cup (B \times C) = \{(1, 1), (2, 1), (3, 1), (1, 5), (2, 5), (3, 5)\}$$
$$= (A \cup B) \times C$$

Similarly,

$$C \times (A \cup B) = \{1, 5\} \times \{1, 2, 3\}$$
$$= \{(1, 1), (1, 2), (1, 3), (5, 1), (5, 2), (5, 3)\}$$
$$C \times A = \{1, 5\} \times \{1, 2\} = \{(1, 1), (1, 2), (5, 1), (5, 2)\}$$

and

$$C \times B = \{1, 5\} \times \{2, 3\} = \{(1, 2), (1, 3), (5, 2), (5, 3)\}$$

Hence,

$$(C \times A) \cup (C \times B) = \{(1, 1), (1, 2), (1, 3), (5, 1), (5, 2), (5, 3)\}$$
$$= C \times (A \cup B)$$

Example 10 illustrates the distributive property of the Cartesian product over union. The Cartesian product fails to be either associative or commutative, as illustrated in Exercises 5 and 19. However, $A \times C$ always has the same number of elements as $C \times A$, and $(A \times B) \times C$ has the same number of elements as $A \times (B \times C)$.

Consider a Cartesian product involving the null set, such as $A \times \varnothing$. $A \times \varnothing$ consists of ordered pairs of elements in which the first element is from $A$ and the second element is from $\varnothing$. Because $\varnothing$ has no elements, it is impossible to form any ordered pairs of elements for $A \times \varnothing$. Thus,

$$A \times \varnothing = \varnothing$$

By similar reasoning, $\varnothing \times A = \varnothing$.

# Just for Fun

You misplaced your football schedule of the Northeast Conference, but you read the picks (of winners) of three sportswriters for this weekend.

A: Reds, Blues, Bulldogs, Tigers
B: Blues, Panthers, Reds, Rockets
C: Rockets, Bulldogs, Jets, Reds

(Notice that no one picked the Angels.) Use this information to construct a schedule for the weekend.

# EXERCISE SET 2

*R*  1. Identify a relation between the elements of the ordered pairs described in each of the following sets:

(a) {(washing machine, Maytag), (book, Kendall/ Hunt), (iron, Sunbeam), (sewing machine, Singer), (tractor, John Deere)}

(b) {(bicycle, 2), (car, 4), (tricycle, 3), (unicycle, 1), (boat, 0), (stagecoach, 4)}

2. Draw a visual representation and then examine each of the following relations on the set of all people to determine whether each is reflexive, symmetric, or transitive.

(a) Is heavier than
(b) Is the uncle of
(c) Lives in the same city as
(d) Runs faster than
(e) Works at the same time as
(f) Is married to
(g) Is a parent of

3. Classify each of the following mathematical relationships on the specified sets as reflexive, symmetric, or transitive. Determine whether each is an equivalence relation.

(a) Is not equal to (counting numbers)
(b) ⊆ (subsets of some universal set)
(c) Is less than (counting numbers)
(d) Has the same area as (triangles)
(e) Is greater than (counting numbers)
(f) ⊂ (subsets of some universal set)
(g) Is the same length as (line segments)
(h) Intersects (lines)
(i) Is perpendicular to (lines)
(j) Is less than or equal to (counting numbers)

4. Determine whether each of the following relations on set $A$ is reflexive, symmetric, or transitive.

(a) {(1, 1), (1, 2), (2, 1), (2, 2)} on $A = \{1, 2\}$
(b) {(3, 3), (3, 4), (4, 3)} on $A = \{3, 4\}$
(c) {(c, c), (c, d), (c, e), (d, d), (d, e), (e, e)} on $A = \{c, d, e\}$

5. If $A = \{a, b, c\}$ and $B = \{r, s, t\}$, tabulate the elements of the indicated Cartesian products.

(a) $A \times B$   (b) $A \times A$
(c) $B \times A$   (d) $B \times B$

6. Let $A = \{a, b, c\}$, $B = \{c, d, e\}$, and $C = \{a, c, x\}$. Tabulate the following Cartesian product sets.

(a) $A \times (B \cap C)$   (b) $A \times (B \cup C)$
(c) $(A \cap B) \times C$   (d) $(A \cup C) \times B$

7. Find $B \times C$ for each pair of sets.

(a) $B = \{3\}$; $C = \{0\}$   (b) $B = \{3\}$; $C = \{\varnothing\}$

8. The Cartesian product $B \times C$ is given by the sets shown. Find a $B$ and a $C$ for each.

(a) {(1, 1), (1, 2), (1, 3), (4, 1), (4, 2), (4, 3)}
(b) {(1, 4), (1, 5), (0, 4), (0, 5)}
(c) {(6, 6), (6, 7), (6, 8)}
(d) {(x, x), (x, y), (x, z), (y, x), (y, y), (y, z), (z, x), (z, y), (z, z)}

9. Three professors at the University of Mystate have student assistants. If $P = \{$Marie, Boyd, Kiefer$\}$ and $S = \{$Janet, Jane, Marion$\}$, find all the possible combinations of professors and assistants by listing the elements of $P \times S$.

*T* 10. (a) Count the number of elements in $A \times B$ and $A \times A$ of Exercise 5.

(b) If $A$ has four elements and $B$ has three elements, how many elements are in $A \times B$? In $B \times B$? In $B \times A$? In $A \times A$?

(c) Let $A$ be a set with $r$ elements, and let $B$ be a set with $s$ elements. How many elements are in $A \times A$? In $A \times B$? In $B \times B$?

11. If $A = \{1, 2, 3, 4\}$ and $B = \{2, 5\}$, tabulate the elements $(x, y)$ of $A \times B$ so that:

(a) $x$ is less than $y$.   (b) $x$ is unequal to $y$.
(c) $x$ is equal to $y$.
(d) $x$ is greater than or equal to $y$.

12. Which of the relations defined on $A$ in Exercise 4 are equivalence relations?

13. If $A = \{a_1, a_2, a_3, a_4\}$ and $B = \{b_1, b_2, b_3\}$, classify the statements below as true or as false.

(a) $(a_1, b_2) \in A \times B$
(b) $\{(a_2, b_1), (a_2, b_2), (a_2, b_3)\} \subset B \times A$
(c) $(b_3, a_2) \in B \times A$
(d) $A \times B = B \times A$
(e) $(b_1, b_2) \in B \times B$

14. Consider the set of names {Asa, Jim, Allen, Ben, John, Betty, Bob, Joe, Jack, Alto, Alisa, Amy, Bill, Jill}. Identify the equivalence classes for the following equivalence relations:

(a) Has the same number of letters in name
(b) Name begins with the same letter

*C* 15. Consider the following visual representation of a relation on the set $\{w, x, y, z\}$.

(a) Find the ordered pairs in the relation.
(b) Determine if the relation is reflexive, symmetric, or transitive.

16. (a) Use a diagram like the one found in Exercise 15 to represent the relation $R$ on the set $\{a, b, c\}$ where $R = \{(a, a), (b, b), (a, b), (b, c), (a, c)\}$.

    (b) Determine if the relation is reflexive, symmetric, or transitive.

17. Give an example of a relation that is reflexive and symmetric but not transitive.

18. Relation $R$ is defined on the set $\{1, 3, 5, 7\}$. Give an $R$ that satisfies the following conditions.

    (a) $R$ is transitive but not reflexive and not symmetric.

    (b) $R$ is transitive, reflexive, and symmetric.

    (c) $R$ is reflexive and transitive but not symmetric.

19. If $A = \{1, 2\}$, $B = \{3, 4, 5\}$, and $C = \{4, 5\}$, tabulate the elements of the following sets.

    (a) $A \times (B \times C)$    (b) $(A \times B) \times C$

20. Given $A = \{1, 2, 3, 4\}$, $B = \{3, 4, 5\}$, and $C = \{4, 5, 6\}$, tabulate the following sets.

    (a) $(A \times A) \cup (B \times B)$    (b) $(C \times A) \cap (C \times B)$
    (c) $(A \times A) \cap (B \times B)$    (d) $(A \times B) \cup (B \times A)$
    (e) $(A \times C) \cap (A \times B)$    (f) $(C \times A) \cap (A \times A)$

## REVIEW EXERCISES

• • • • • • • • • • • • • • • • • • • • • • • • • • • • • •

21. If $A = \{a, b, c\}$, $B = \{b, c, d, e\}$, and $C = \{c, e, g\}$, demonstrate the following properties of sets.

    (a) Commutative property of intersection (use $A$ and $B$)
    (b) Commutative property of union (use $B$ and $C$)
    (c) Associative property of intersection
    (d) Associative property of union

22. Given sets $A = \{1, 2, \ldots, 5\}$ and $B = \{4, 5, \ldots, 10\}$ find each of the following.

    (a) $A \cup B$      (b) $A \cap B$      (c) $A \cap (B \cup A)$

23. In Exercise 22, is $A \subseteq B$?

24. In Exercise 22, find $B - A$.

### ∴ PCR Excursion ∵

25. Let A be the set of all fractions $\frac{a}{b}$ where $a$ and $b$ are taken from $B = \{1, 2, 3, 4, 5, 6\}$, and $a$ is less than $b$. That is, $A$ contains fractions like $\frac{3}{5}$ because both 3 and 5 are in $B$ and 3 is less than 5. $\frac{4}{3}$ is not in $A$ because 4 is not less than 3. Likewise $\frac{5}{7}$ is not in $A$ because 7 is not in $B$.

    (a) List the fifteen elements of $A$.

    (b) Consider on $A$ a relation that for two fractions $\frac{a}{b} \sim \frac{c}{d}$ if $ad = bc$; that is, $\frac{1}{2} \sim \frac{2}{4}$ since $1 \cdot 4 = 2 \cdot 2$. Find other such pairs of fractions.

    (c) Use examples to suggest that this relation is reflexive.

    (d) Use examples to suggest that this relation is symmetric.

    (e) Use examples to suggest that this relation is transitive.

    (f) Do your answers in (c), (d), and (e) suggest that this relation is an equivalence relation?

    (g) Find the equivalence classes determined by this relation.

## SECTION 3 — THE NUMBER OF ELEMENTS IN A SET

**PROBLEM**

A pollster found that, in a group of 100 people, 40 use brand *A*, 25 use both brand *A* and brand *B*, and everyone uses brand *A* or brand *B*. How many people use brand *B*?

**OVERVIEW**

By associating a number with the elements in a set, using Venn diagrams, and utilizing problem-solving techniques, you should be able to solve the preceding problem.

The number of elements in a set provides the basis for counting. Is the number of this page a cardinal number, an ordinal number, or a nominal number? You will learn in this section the distinctions among such numbers.

| GOALS | Determining the number of elements in a set seems very simple. However, in probability theory (Chapter 9), we encounter some challenging problems involving the number of elements in sets.<br><br>Illustrate the Number and Operations Standard, page 119.<br>Illustrate the Problem Solving Standard, page 15.<br>Illustrate the Representation Standard, page 50.<br>Illustrate the Connections Standard, page 147. |
| --- | --- |

The number concept held by mathematicians today is the result of many centuries of study. Leopold Kronecker is supposed to have said, "God created the natural numbers; all else is the work of man." His statement has been interpreted by some as meaning that the human mind is naturally endowed with the power to comprehend the concept of the counting numbers, whereas all other numbers are a result of human inventiveness.

 EQUIVALENT SETS

Before discussing the concept of number, we need to understand one-to-one correspondence and equivalence.

| DEFINITION | **One-to-One Correspondence** | A *one-to-one correspondence* between sets $A$ and $B$ is a pairing of the elements of $A$ and $B$ such that for every element $a$ of $A$ there corresponds exactly one element $b$ of $B$, and for every element $b$ of $B$ there corresponds exactly one element $a$ of $A$. |
| --- | --- | --- |

**EXAMPLE 1**

Consider the sets $A = \{1, 2, 3\}$ and $B = \{$John, Jane, Jill$\}$. One way of placing these two sets in one-to-one correspondence is to use double-headed arrows:

$$
\begin{array}{ccc}
1 & 2 & 3 \\
\updownarrow & \updownarrow & \updownarrow \\
\text{John} & \text{Jane} & \text{Jill}
\end{array}
$$

Of course, these two sets could be placed in one-to-one correspondence in several other ways, as well. For instance, $1 \leftrightarrow$ Jill, $2 \leftrightarrow$ John, and $3 \leftrightarrow$ Jane. Now try to establish a one-to-one correspondence between

$$A = \{1, 2, 3, 4\} \qquad \text{and} \qquad B = \{\text{John, Jane, Jill}\}$$

If we let John correspond to 1, Jane to 2, and Jill to 3, no element is left to correspond to 4. Thus, a one-to-one correspondence cannot be set up between these two sets.

A number of sets can be placed in one-to-one correspondence with a given set. For example, $\{a, b, c, d\}$ can be placed in one-to-one correspondence with

{1, 2, 3, 4}, and so can {□, △, □, ○}, {x, y, z, w}, and {140, 180, 190, 200}. This characteristic leads to the following definition.

| DEFINITION | | |
|---|---|---|
| | **Equivalent Sets** | If a one-to-one correspondence exists between two sets, the sets are said to be *equivalent* (or *matched*). |

**EXAMPLE 2**

If a room contains 70 seats, with one person sitting in each of the seats and no one standing, then the set of seats and the set of people are equivalent.

The term *equivalent* should not be confused with *equality*. If two sets are equal, each element of one set must equal a corresponding element of the other, and conversely. However, sets are equivalent if a one-to-one correspondence exists between elements. Thus, equal sets are always equivalent, but equivalent sets are not always equal.

**EXAMPLE 3**

{a, b, c, d} and {b, c, a, d} are equal sets; however, {a, b, c, d} is equivalent to but not equal to {x, y, z, w}.

The relation "is equivalent to" is a relation on a collection of sets. Further, in the language of Section 2, it is reflexive (every set can be placed in one to one correspondence with itself) and symmetric (a one-to-one correspondence from A to B is also a one-to-one correspondence from B to A). With a little more thought one can see that this relation is also transitive. Thus, the relation "is equivalent to" on a collection of sets is an equivalence relation on that collection. Two sets are equivalent if they belong to the same equivalence class. This is the basis for the idea of a cardinal number discussed in the next paragraph.

# CARDINAL, ORDINAL, AND NOMINAL NUMBERS

Consider four sets: {1, 2, 3}, {*, #, %}, {a, b, c}, and {○, △, □}. What do these sets have in common? The answer is that they are equivalent. Intuitively, another property is probably obvious to you: they have a property of "threeness." {1, 2}, {a, b}, {x, y}, and {○, △} are equivalent and share a property of "twoness." A property that equivalent sets have in common is called their **cardinal number.**

| *n(A)* | The cardinal number of set $A$, denoted by the symbol $n(A)$, represents the number of elements in set $A$. It will be read as **the number of A.** |
|---|---|

This number is the cardinal number of all sets equivalent to $A$. Suppose that there are 30 students in a classroom. If $S$ stands for the set of students, then $n(S) = 30$.

If $A = \{a, b, c, d\}$, then $n(A) = 4$. If $B = \{x, y, z, w\}$, then $n(B) = 4$. $A$ is equivalent to $B$, and $n(A) = n(B)$. However, $A \neq B$. Just because $n(A) = n(B)$, it is not necessarily true that $A = B$. The number of a set is another set property that does not depend in any way on the kind of items represented as elements or on the order in which the elements are listed. The use of counting numbers (also called natural numbers) to tell how many objects are in a set is called *cardinal usage.* "There are 24 hours in a day" uses the counting number 24 in a cardinal sense.

In the phrase "the fifth section of the second chapter of this book," **ordinal numbers** are used. *Fifth* and *second* refer to position or order. The following example illustrates cardinal and ordinal uses of counting numbers.

| **EXAMPLE 4** | The administration building is five stories high (cardinal). The president's office is on the second floor (ordinal). |

Counting numbers may also be used in a nominal sense for naming things. Social security numbers, bank account numbers, and zip codes are **nominal numbers.**

Let's return now to the problem in the introduction of the section.

| **EXAMPLE 5** | A pollster found that in a group of 100 people, 40 use brand $A$, 25 use both brand $A$ and brand $B$, and everyone uses brand $A$ or brand $B$. How many people use brand $B$?

Can you help find the answer? |

*SOLUTION*

To solve this problem, we use Polya's four-step method, outlined in Chapter 1.

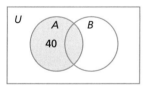

**FIGURE 3-14**

**UNDERSTANDING THE PROBLEM** Let's draw a Venn diagram (Figure 3-14) to visualize the given information. Now $n(A \cup B) = 100$. Why? The region bounded by the circle and designated as $A$ represents the people who use brand $A$; thus, $n(A) = 40$. The region bounded by the circle and designated as $B$ represents the people who use brand $B$. What does the region common to $A$ and $B$ represent? Is $n(A \cap B) = 25$?

**DEVISING A PLAN** We can easily solve the problem by placing numbers in each closed region of Figure 3-15. Since 25 of the 40 people who use brand $A$ also use brand $B$, 15 use brand $A$ only. Now let's see if we can determine how many use brand $B$ only.

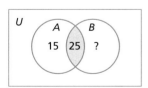

**FIGURE 3-15**

**CARRYING OUT THE PLAN** Because 100 people were surveyed and because everyone contacted uses either brand A or brand B, the total number in the three regions bounded by the circles of Figure 3-15 is $n(A \cup B) = 100$, or

$$15 + 25 + ? = 100$$

Thus, the number using only brand B is 60. Therefore, the total number using brand B is

$$n(B) = 25 + 60 = 85$$

**LOOKING BACK** In Figure 3-15 (replacing ? with 60), does the total in all regions equal 100? (Yes, $15 + 25 + 60 = 100$.) Do 40 people use brand A? (Yes, $15 + 25 = 40$.) Do 25 people use brand A and brand B? (Yes, $n(A \cap B) = 25$.)

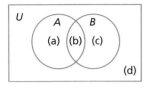

**FIGURE 3-16**

The following discussion demonstrates the theory of the preceding example. We want to find the number of elements that are in either set $A$ or set $B$, denoted by $n(A \cup B)$. The answer would not necessarily be

$$n(A) + n(B)$$

as we discover in Figure 3-16. In this figure, the number of elements in the closed regions of the Venn diagram are denoted by (a), (b), (c), and (d). Thus,

$$n(A \cup B) = (a) + (b) + (c)$$
$$n(A) = (a) + (b)$$
$$n(B) = (b) + (c)$$
$$n(A) + n(B) = (a) + (b) + (b) + (c)$$

So $n(A) + n(B)$ contains (b) more elements than $n(A \cup B)$. Because (b) = $n(A \cap B)$, we have the following theorem:

| **Number of Elements in $A \cup B$** | The number of elements that are in either $A$ or $B$ is <br> $$n(A \cup B) = n(A) + n(B) - n(A \cap B)$$ <br> where $A \cap B$ consists of elements common to $A$ and $B$ |
|---|---|

**EXAMPLE 6**

Out of 1500 freshmen, 13 students failed English, 12 students failed mathematics, and 7 students failed both English and mathematics. How many students failed English or mathematics?

*SOLUTION*

$$n(E) = 13$$
$$n(M) = 12$$
$$n(E \cap M) = 7$$

To find $n(E \cup M)$,

$$n(E \cup M) = n(E) + n(M) - n(E \cap M)$$
$$= 13 + 12 - 7$$
$$= 18$$

**PRACTICE PROBLEM**

Forty members of Alpha Phi are enrolled in English or history. If 30 are enrolled in English and 25 are enrolled in history, how many are enrolled in both English and history?

*ANSWER*

$n(E \cap H) = 15$.

| **Number of Elements in $A \cup B \cup C$** | For any three sets $A$, $B$, and $C$, $$n(A \cup B \cup C) = n(A) + n(B) + n(C) - n(A \cap B) - n(A \cap C)$$ $$- n(B \cap C) + n(A \cap B \cap C)$$ |

**EXAMPLE 7**

Out of 100 freshmen at Lance College, 60 are taking English, 50 are taking history, 30 are taking mathematics, 30 are taking both English and history, 16 are taking both English and mathematics, 10 are taking both history and mathematics, and 6 are taking all three courses. How many of the freshmen are enrolled in English, history, or mathematics?

*SOLUTION*

Let $E$ represent those enrolled in English; $H$, those in history; and $M$, those in mathematics. The example states that

$$n(E) = 60 \qquad n(E \cap H) = 30$$
$$n(H) = 50 \qquad n(E \cap M) = 16$$
$$n(M) = 30 \qquad n(H \cap M) = 10$$
$$n(E \cap H \cap M) = 6$$

We are to find $n(E \cup H \cup M)$, assuming that every freshman is enrolled in one of these courses.

$$n(E \cup H \cup M) = n(E) + n(H) + n(M) - n(E \cap H) - n(E \cap M)$$
$$- n(H \cap M) + n(E \cap H \cap M)$$
$$n(E \cup H \cup M) = 60 + 50 + 30 - 30 - 16 - 10 + 6 = 90$$

# Just for Fun

Jennifer has 16 pieces of candy. Nine contain caramel; ten contain coconut; and six contain both caramel and coconut. Use problem-solving techniques along with sets to find out how many of the candies contain neither caramel nor coconut.

# EXERCISE SET 3

*R* 1. Place a C by the phrases involving cardinal number usage and an O by those involving ordinal number usage.

(a) 16 sheep      (b) Page 9
(c) 82nd Congress      (d) Third grade
(e) Five people      (f) Fifth session
(g) 7 days in a week      (h) 10th round
(i) 12 months in a year      (j) $15
(k) A dozen eggs      (l) Second period

2. Let set $A = \{1, 2, 3, 4, 5, 6\}$. Find all the sets in the following listing that are equivalent to set $A$.

(a) $\{a, b, c, d, e, f\}$      (b) $\{1, 2, 3, 4, 5, 6, \ldots\}$
(c) $\{4, 5, \ldots, 9\}$      (d) $\varnothing$
(e) {Jane, Mary, Sue, Louise}
(f) $\{8, 9, 10, 11, 12, 13\}$

3. Show a one-to-one correspondence between one of the sets $\{1\}$, $\{1, 2\}$, $\{1, 2, 3\}$, $\{1, 2, 3, 4\}$, or $\{1, 2, 3, 4, 5\}$, and the following sets:

(a) $\{b\}$      (b) $\{\varnothing, \lambda\}$
(c) $\{a, b, c, d, e\}$      (d) $\{4, 6, 8\}$
(e) $\{10, 40, 30, 50, 70\}$      (f) $\{2\}$

4. In the accompanying figure, the numbers indicate the cardinal number of the closed region. Find the following:

(a) $n(A)$      (b) $n(B)$
(c) $n(A \cap B)$      (d) $n(A \cup B)$
(e) $n(\overline{A})$      (f) $n(\overline{B})$
(g) $n(U)$      (h) $n(\overline{A} \cap \overline{B})$
(i) $n(\overline{A} \cup \overline{B})$      (j) $n(\overline{A \cap B})$

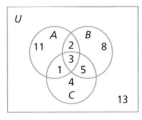

5. Indicate which of the following statements are true and which are false.

(a) Sets $A = \{\bigcirc, \triangle, \diamondsuit, \square, \square\}$, $B = \{1, 6, 8, 3, 0\}$, and $C = $ {Rick, Steve, Allan, Paul, Ed} are equivalent.
(b) Sets $A$ and $B$ in (a) have the same cardinal number because they are equal.
(c) Although sets $A$, $B$, and $C$ in (a) are not equal, they still have the same cardinal number.
(d) The smallest element in the set of natural numbers is 0.
(e) A subset of a finite set may be infinite.

(f) A subset of an infinite set may be finite.
(g) A subset of an infinite set may be infinite.
(h) $n(\{\varnothing\}) = 0$      (i) $n(\varnothing) = 0$
(j) $n(\{\ \ \}) = 1$      (k) $n(\{0, 1, 2\}) = 3$

6. If $A = \{a, b, c, d\}$ and $R = \{r, s, t, v\}$, explain why:

(a) $A$ and $R$ are not equal sets.
(b) $A$ and $R$ are equivalent sets.

7. Define $A = \{1, 2, 3\}$, $B = \{1\}$, $C = \varnothing$, and $D = \{a, b, c, d\}$. Determine whether the statements shown are true or false.

(a) $n(A \times B) = n(B \times A)$      (b) $A \times B = B \times A$
(c) $n(A \times B) = n(A)$      (d) $n(A \times D) = 12$
(e) $A \times C = C \times D$
(f) $n[(A \times B) \times C] = n[A \times (B \times C)]$

*T* 8. For $A = \{2, 3, \ldots, 6\}$ and $B = \{6, 7, \ldots, 10\}$, show that

(a) $n(A \cup B) = n(A) + n(B) - n(A \cap B)$
(b) $n(A \times B) = n(A) \cdot n(B)$
(c) $n(A \times A) = n(A) \cdot n(A)$

9. In Atlanta, 600,000 people read newspaper $A$, 450,000 read newspaper $B$, and 160,000 read both newspapers. How many read either newspaper $A$ or newspaper $B$?

10. Indicate whether the following statements are true or false for all sets $A$ and $B$. If a statement is false, give an example to demonstrate that it is false.

(a) $n(A \cup B) = n(A) + n(B)$
(b) $n(A \cap B) = n(A) - n(B)$
(c) If $n(A) = n(B)$, then $A = B$.
(d) If $A = B$, then $n(A) = n(B)$.
(e) $n(A \times B) = n(A) \cdot n(B)$
(f) $n(A) + n(B) = n(A \cup B) - n(A \cap B)$
(g) $n(\overline{A} \cup \overline{B}) = n(\overline{A \cup B})$
(h) $n(A \cap B) = n(A \cup B) - n(A \cap \overline{B}) - n(\overline{A} \cap B)$

11. Fifty students in Birmingham High were asked to choose among three recreational areas—Six Flags, Disney World, and Opryland—for a class trip. Some students indicated more than one choice. Their choices are shown in the accompanying Venn diagram.

Opryland    Disney World

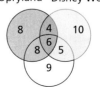

Six Flags

How many students preferred each of the following?

(a) Only Disney World
(b) Both Opryland and Six Flags
(c) Both Opryland and Disney World but not Six Flags
(d) Both Disney World and Six Flags but not Opryland

*C* 12. In your class of 30 students, 20 students indicate that they exercise at least 4 times a week, and 15 indicate that they average 8 hours of sleep a night.

(a) What is the greatest possible number of students who do neither?
(b) What is the greatest possible number who do both?
(c) What is the least possible number who do both?

13. Of 1000 first-year women at a certain university, 300 pledged sororities, 200 live off campus, and 150 pledged sororities and live off campus.

(a) Draw a Venn diagram labeling and counting the four relevant parts of the diagram, where *S* represents sorority pledges and *O* represents living off campus.
(b) How many pledged sororities or live off campus?
(c) How many did not pledge sororities or live off campus?
(d) How many live off campus and did not pledge sororities?
(e) How many pledged sororities and do not live off campus?
(f) How many pledged sororities or do not live off campus?

14. A poll of 100 students was taken at a nonresidential college to find out how they got to campus. The results were as follows: 28 mentioned car pools; 31 used buses; and 42 said that they drove to school alone. In addition, 9 used both car pools and buses, 10 both used a car pool and sometimes drove their own cars, and 6 used buses as well as their own cars. Of the 100 respondents, only 4 used all three methods of arriving on campus.

(a) Draw a Venn diagram, labeling each of the *eight* relevant pieces.
(b) How many students used none of the three methods of transportation?
(c) How many used car pools exclusively to arrive on campus?
(d) How many used buses exclusively to get to campus?

15. A discount store ran a special sale on hair spray, paper cups, and boys' pants. One salesperson reported that 50 people took advantage of the sale. Thirty people purchased hair spray; 25 purchased paper cups; 35 purchased boys' pants; 14 purchased both hair spray and paper cups; 15 purchased both paper cups and boys' pants; 22 purchased both hair spray and boys' pants; and 10 purchased all three. Was the salesperson's report accurate? Explain.

16. How many one-to-one correspondences are there between 2 sets with

(a) 2 members each?    (b) 3 members each?
(c) 4 members each?    (d) *n* members each?

17. A pollster on a college campus interviewed students relative to the athletic games attended. Of the 7900 on campus, she reported that 4500 attended football games, 4850 attended basketball games, and 4750 attended soccer games. Of those who attended games, 3000 attended both football and basketball games, 2750 attended both basketball and soccer, 2250 attended football and soccer, and 1750 attended all three. Was this poll accurate? Why or why not?

18. Draw a Venn diagram so that sets *A*, *B*, and *C* intersect. Place numbers in each small region of the diagram to represent the number of elements in that region if
$$n(A) = 26 \quad n(A \cap B) = 14$$
$$n(B) = 25 \quad n(A \cap C) = 10$$
$$n(C) = 22 \quad n(B \cap C) = 12$$
$$n(A \cap B \cap C) = 3$$

## REVIEW EXERCISES

19. Complete the following.
(a) $A \cap \varnothing = ?$    (b) $A \cup \varnothing = ?$
(c) $A \cap U = ?$    (d) $A \cup U = ?$
(e) $A \cap \overline{A} = ?$    (f) $A \cup \overline{A} = ?$
(g) $U \cap \varnothing = ?$    (h) $U \cup \varnothing = ?$

20. Considering sets $A = \{1, 2, \ldots, 5\}$, $B = \{4, 5, \ldots, 10\}$, and $C = \{6, 7, 8, \ldots, 15\}$, find each of the following.
(a) $A \cap (B \cup C)$    (b) $A \cap (B \cap C)$
(c) $C \cup (B \cup A)$    (d) $B \cup (A \cap C)$

21. Using symbolic notation, describe the set indicated by the shaded part of each diagram.

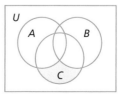

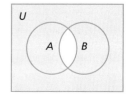

(a)                                    (b)

# FUNCTIONS

**PROBLEM**

One of the two graphs below represents a function; the other one does not. In this chapter you will learn to recognize graphs that represent functions.

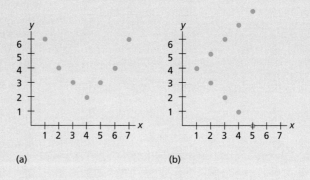

(a)                                         (b)

**OVERVIEW**

A very important concept that is closely related to both sets and relations is the concept of a function from a set *A* to a set *B*. This concept is important because almost any time a physical scientist, a social scientist, or an economist prepares to describe the world he or she is studying, a function is used in that description. In this section we will look at a number of different ways of thinking about this important mathematical object and identifying its important properties.

**GOALS**

Illustrate the Algebra Standard, page 4.
Illustrate the Representation Standard, page 50.
Illustrate the Problem Solving Standard, page 15.
Illustrate the Connections Standard, page 147.

## FUNCTIONS

To investigate the concept of a function, we start with a correspondence between two sets. Figure 3-17 (a), (b), and (c) each displays a correspondence between the elements of set *A* and the elements of set *B*. Later you will learn that two of these represent functions and one does not.

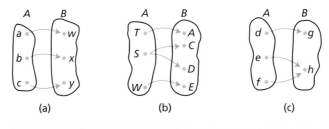

(a)                     (b)                     (c)

FIGURE 3-17

In life, such correspondences occur naturally:

The number of items sold and the revenue received
The amount of money borrowed and the interest paid
Each person and each person's social security number

We are interested in the rule, procedure, or method that gives the revenue from the number of items sold or the interest from money borrowed.

We introduce the concept of a function by considering a rule that sets up a correspondence between set $A$ and set $B$ so that every element of $A$ has as its counterpart a **unique** element of $B$. This way of thinking of a function can be demonstrated by what is called an *input–rule–output* machine. The first set can be considered the input (or domain); then a given rule allows the second set, or some subset of it, to be produced as the output (or range).

**EXAMPLE 1**

As Figure 3-18 illustrates, we insert an input, operate with the rule, and obtain an output. Suppose that the rule is "Add 4 to the input." This rule can expressed as $x + 4$ when the input is $x$. When the input is 3, the rule operates to give $3 + 4$, and the output is 7.

**EXAMPLE 2**

Consider the rule described in Figure 3-19. This rule sets up a correspondence. By examining the correspondence closely, you can see that each element of the input is paired with exactly one output element. Specific examples of this correspondence are given in Figure 3-20.

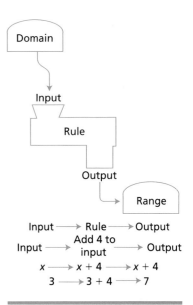

FIGURE 3-18

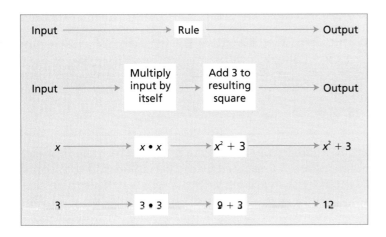

FIGURE 3-19

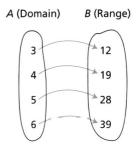

FIGURE 3-20

The preceding discussion suggests that a rule (procedure or method) describes a function if it produces a correspondence between an input set of elements called the **domain** and an output set of elements called the **range,** in such a way that to each element in the domain there corresponds one and only one element in the range.

**EXAMPLE 3**

Which of the diagrams in Figure 3-17 represents functions from $A$ to $B$?

*SOLUTION*

Both (a) and (c) represent functions; 17(b) does not represent a function because the element $S$ is paired with both $C$ and $D$.

From the rule in Example 1, we can obtain the set of ordered pairs

$$\{(1, 5), (2, 6), (3, 7), \ldots\}$$

From the rule in Example 2, we get the set

$$\{(1, 4), (2, 7), (3, 12), \ldots\}$$

These suggest the following definition.

**DEFINITION**

**Function**

A function from $A$ to $B$ is a relation from $A$ to $B$ in which each element of $A$ is paired with one and only one element of $B$. The set of first elements of the ordered pairs make up the domain, and the set of second elements comprise the range.

**EXAMPLE 4**

Consider the following sets of ordered pairs $(x, y)$, where $x$ is an element of the domain $\{2, 3, 4\}$ and $y$ is an element of $\{3, 4, 5, 6\}$:

(a) $S = \{(2, 3), (3, 4), (4, 5)\}$
(b) $S = \{(3, 3), (3, 4)\}$
(c) $S = \{(2, 3), (3, 4), (4, 5), (2, 6)\}$
(d) $S = \{(2, 5), (3, 5), (4, 5)\}$

Which of these sets of ordered pairs are functions? What is the range of each function?

*SOLUTION*

Using the preceding definition, it is easy to verify that (a) and (d) are functions. The set in (b) is not a function because 3 is paired with both 3 and 4, and the set in (c) is not a function, because 2 is paired with both 3 and 6. The range of (a) is $\{3, 4, 5\}$, and the range of (d) is $\{5\}$.

**TABLE 3-1**

| $x$ | 1 | 3 | 4 | 6 |
|---|---|---|---|---|
| $y$ | 4 | 6 | 12 | 18 |

**TABLE 3-2**

| $x$ | 1 | 0 | 1 | 2 |
|---|---|---|---|---|
| $y$ | 2 | 3 | 4 | 5 |

Sometimes a set of ordered pairs is given in a table where the first elements of the ordered pairs are represented by the variable $x$ and the second elements are represented by the variable $y$. (See Table 3-1.) This table represents a function because to each $x$ there corresponds only one $y$. In contrast, Table 3-2 does not represent a function, because the two ordered pairs $(1, 2)$ and $(1, 4)$ have the same first element and different second elements.

**PRACTICE PROBLEM**   Does the set of ordered pairs {(2, 1), (3, 2), (4, 5), (4, 6), (5, 9)} represent a function?

*ANSWER*   No, because (4, 5) and (4, 6) have the same first element.

| $x$ | 1 | 4 | 0 | 2 |
|---|---|---|---|---|
| $y$ | 3 | 3 | 3 | 3 |

## Common Error

Students often feel that the table at left does not represent a function. This table does express a constant function: $y$ is always 3.

# Function Notation

The rule in Example 1 can be written as $y = x + 4$, and the rule in Example 2 as $y = x^2 + 3$. Thus, in both cases, the rule that determines the set of ordered pairs can be expressed in equation form. The equation representing a function assigns to each $x$ in the domain a unique value $y$ in the range.

Often, we write a function as $y = f(x)$, or possibly as $y = F(x)$ or $y = h(x)$ (read "$y$ equals a function of $x$" or "$y$ equals the value of the function at $x$"). This indicates that each $x$ has associated with it a unique $y$. Suppose that we write $y = f(x) = x + 7$. When $x = 2$, we substitute 2 for $x$ to obtain

$$y = f(2) = 2 + 7 = 9$$

Likewise, we can determine that $f(4) = 11$ and $f(6) = 13$. This is another way of saying that ordered pairs (4, 11) and (6, 13) are in the function.

The notation $y = f(x)$ actually states a rule for determining values of $y$ when values are assigned to $x$. For example, if $y = f(x) = x^2$ where $x$ is any real number, then $f(2) = 2^2 = 4$ or $y = 4$. Similarly, if $f(x) = x^2 - 3x + 2$ then $f(3) = 3^2 - 3(3) + 2 = 2$.

**EXAMPLE 5**   Given $f(t) = 6 + 3t$, find $f(1), f(2)$, and $f(0) + f(4)$.

*SOLUTION*
$$f(1) = 6 + 3(1) = 9$$
$$f(2) = 6 + 3(2) = 12$$
$$f(0) = 6 + 3(0) = 6 \quad \text{and} \quad f(4) = 6 + 3(4) = 18$$
$$f(0) + f(4) = 6 + 18 = 24$$

**PRACTICE PROBLEM**   Given $f(t) = t^2 + 1$, find $f(0) \div f(1)$.

*ANSWER*   $\dfrac{f(0)}{f(1)} = \dfrac{1}{2}$

# Graphs of Functions

To this point we have seen that functions can be described by using arrow diagrams (see Figure 3-17 (a) and (c)), as input–rule–output machines (see Figures 3-18 and 3-19), as sets of ordered pairs (see Example 4 (a) and (d)), as tables (see Table 3-1), and by using $f(x)$ notation (see Example 5). A geometric way to describe relations and functions is with a graph.

A set of ordered pairs can be represented geometrically by plotting the points on a coordinate system formed by a horizontal line and a vertical line. If we let the first elements of the ordered pairs represent points on the horizontal line and the second elements represent points on the vertical line, we can plot each of the ordered pairs in the relation.

**EXAMPLE 6**

(a) Plot the graph of the relation {(1, 2), (2, 3), (3, 4), (3, 5), (4, 5)}.
(b) Use the graph to determine if this relation is a function.

*SOLUTION*

(a) The graph of this relation is found in Figure 3-21.
(b) Notice that the vertical line passing through point 3 on the horizontal axis passes through two points on our graph. This indicates that the first element 3 is paired with two different second elements (4 and 5). This is not a geometric representation of a function.

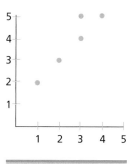

FIGURE 3-21

The observation that we made in Example 6 is true in general.

**Vertical Line Test**

If each vertical line intersects the graph of a relation in at most one point, then the relation represents a function. If any vertical line passes through two or more points of the graph, then the relation is not a function.

**EXAMPLE 7**

Determine if either of the graphs in Figure 3-22 is a graph of a function.

*SOLUTION*

The graph in Figure 3-22(a) represents a function because each vertical line intersects the graph in at most one point. The graph in Figure 3-22(b) does not represent a function because several vertical lines intersect the graph in more than one point.

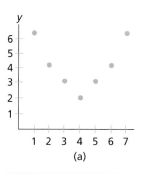

(a)

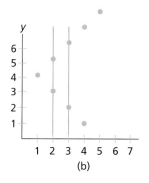

(b)

FIGURE 3-22

**EXAMPLE 8**

So that drivers will not have to make change, a taxicab company charges $3 for the first mile (or fraction thereof) and $1 for each additional mile (or fraction thereof). Express this function (a) in a table, (b) as a set of ordered pairs, (c) as a correspondence with arrows (Figure 3-23), and (d) as a graph (Figure 3-24).

*SOLUTION*

(a)

| Miles or fraction thereof | 1 | 2 | 3 | 4 | 5 . . . |
|---|---|---|---|---|---|
| Charge in dollars | 3 | 4 | 5 | 6 | 7 . . . |

(b) {(1, 3), (2, 4), (3, 5), (4, 6), (5, 7) . . .}

(c)

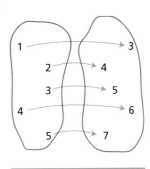

FIGURE 3-23

(d)

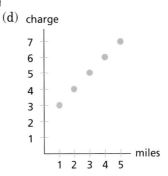

FIGURE 3-24

# Just for Fun

House A faces left and house B faces right. Can you move one line in house A to make it face right?

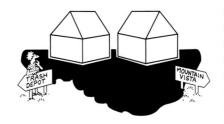

# EXERCISE SET 4

*R* 1. Which of the following sets of ordered pairs are functions?

    (a) {(1, 3), (3, 3), (5, 3)}    (b) {(1, 3), (3, 3), (5, 7)}
    (c) {(1, 3), (3, 5), (5, 1)}    (d) {(1, 1), (3, 3), (5, 5)}
    (e) {(3, 4), (5, 10), (6, 4), (7, 1)}
    (f) {(1, 5), (1, 6), (2, 5), (3, 10)}
    (g) {(3, 7), (7, 3), (8, 3)}    (h) {(4, 6), (5, 6)}
    (i) {(5, 3), (5, 4)}          (j) {(5, 5), (6, 6)}

2. Which of the following tables define functions?

(a)

| x | 2 | 2 |
|---|---|---|
| y | 4 | 1 |

(b)

| x | 1 | 3 |
|---|---|---|
| y | -1 | -1 |

(c)

| x | 1 | 1 |
|---|---|---|
| y | 1 | 2 |

(d)

| x | 0 | 0 |
|---|---|---|
| y | 0 | 1 |

(e)

| x | 1 | 2 |
|---|---|---|
| y | 1 | 1 |

(f)

| x | 0 | 2 | 3 | 4 |
|---|---|---|---|---|
| y | 2 | 4 | 7 | 2 |

(g)

| x | 0 | 1 | 3 | 2 | 1 |
|---|---|---|---|---|---|
| y | 1 | 2 | 4 | 3 | 4 |

(h)

| x | 1 | 2 | 4 | 2 |
|---|---|---|---|---|
| y | 3 | 4 | 6 | 5 |

3. Classify each correspondence as representing or not representing a function.

| Domain | Range |
|--------|-------|
| 0 → | 4 |
| 1 → | 6 |
| 2 → | 8 |
| 3 → | 10 |

(a)

| Domain | Range |
|--------|-------|
| 1 → | 1 |
| 2 → | 4 |
| 3 → | 9 |
| 4 → | 16 |

(b)

| Domain | Range |
|--------|-------|
| 3 → | 4 |
|   | −4 |
| 4 → | 3 |
|   | −3 |

(c)

| Domain | Range |
|--------|-------|
| 1 |   |
| 2 → 4 |   |
| 3 |   |
| 4 |   |

(d)

4. Given the following function rules, find $f(1)$ and $f(5)$.

    (a) $f(x) = x + 5$      (b) $f(x) = 2x + 3$
    (c) $f(x) = 3x + 1$    (d) $f(x) = x^2$
    (e) $f(x) = x^2 + 2x + 5$

5. Consider the following functions given by the rule $y = f(x)$. Compute $f(2)$, $f(3)$, and $f(4)$.

    (a) $f(x) = x - 1$     (b) $f(x) = 2x - 1$
    (c) $f(x) = x^2$       (d) $f(x) = x^2 + 2$

*T* 6. Which of the following graphs represent functions?

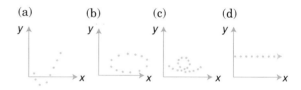

(a)    (b)    (c)    (d)

7. If $y = f(x) = x^3 + 2x$, find $f(2)/f(1)$ and $f(2) \cdot f(4)$.

*C* 8. If $y = f(x) = (x + 2)(x + 1)$, find $f(1)$ and $f(2)$.

9. If $f(x) = 3x + 1$, and the domain is {0, 1, 2, 3}, describe the function with

    (a) Ordered pairs
    (b) Arrows connecting two sets
    (c) A table
    (d) A graph

10. Consider the following functions described with tables. Use your skill at detecting patterns to find a rule of the form $y = f(x)$ that describes how to get $y$ from $x$.

(a)

| x | 1 | 2 | 3 | 4 | 5 |
|---|---|---|---|---|---|
| y | 3 | 4 | 5 | 6 | 7 |

(b)
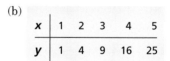

| x | 1 | 2 | 3 | 4 | 5 |
|---|---|---|---|---|---|
| y | 1 | 4 | 9 | 16 | 25 |

(c)
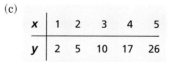

| x | 1 | 2 | 3 | 4 | 5 |
|---|---|---|---|---|---|
| y | 2 | 5 | 10 | 17 | 26 |

(d)
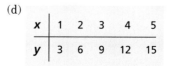

| x | 1 | 2 | 3 | 4 | 5 |
|---|---|---|---|---|---|
| y | 3 | 6 | 9 | 12 | 15 |

11. Use the vertical line test to determine if the following graphs are graphs of functions.

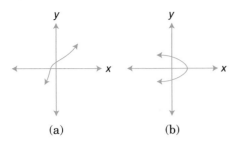

(a)                     (b)

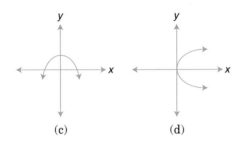

(c)                     (d)

12. A travel agency books a flight to Europe for a group of college students. The fare per student in a 200-passenger airplane is $500 plus $10.00 for each vacant seat.

(a) Write the total revenue as a function of empty seats $x$.
(b) What is the domain of this function?
(c) Calculate the function for 5 empty seats, for 10 empty seats, for 20 empty seats, for 40 empty seats, and for 100 empty seats.

13. A parking lot charges by the hour as follows:
$$f(x) = \$3.00 + \$1.50(x - 1) \qquad \text{for } x \geq 1$$
Find the charge for 1 hour, for 2 hours, for 10 hours, and for 24 hours. Can you state the parking rate in words?

14. The functional relationship between Celsius temperature and Fahrenheit temperature is
$$C(F) = \frac{5(F - 32)}{9}$$
Find $C(0°)$, $C(32°)$, and $C(98.6°)$. (98.6°F is the normal human body temperature.)

15. The equation
$$N(x) = 1000 - 150x \qquad 0 \leq x \leq 6$$
represents the number of bacteria in a culture $x$ hours after an antibacterial treatment has been administered. Find $N(0)$, $N(2)$, $N(4)$, and $N(6)$.

## REVIEW EXERCISES

16. Draw Venn diagrams to verify that $A - (B \cup C) = (A - B) \cap (A - C)$

17. Draw Venn diagrams to verify that $A \cap \overline{B} = \overline{A} \cup B$

18. Write two sets that are equivalent but not equal to $\{a, b, c, d, e\}$.

19. If $A = \{a, b, c, d\}$ and $B = \{c, d, e, f\}$, find each of the following.
(a) $B - A$        (b) $A \cap (B - A)$
(c) $A - (B \cup A)$        (d) $A \cup (B \cap A)$

# SOLUTION TO INTRODUCTORY PROBLEM

In a certain apartment complex live 20 men: 8 are married and own automobiles; 12 own automobiles and have television sets; and 11 are married and have television sets. What is the largest possible number of married men?

**UNDERSTANDING THE PROBLEM** Let $M$ represent married men; $A$, those who own automobiles; and $T$, those who own television sets. We are given

$$n(M \cap A) = 8$$
$$n(M \cap T) = 11$$
$$n(A \cap T) = 12$$
$$n(U) = 20$$

We wish to find the largest possible value for $n(M)$.

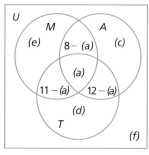

FIGURE 3-25

**DRAWING A PICTURE** The Venn diagram in Figure 3-25 assists in finding a solution.

**DEVISING A PLAN** The strategy employed is to reason backward to a solution. $n(M)$ will be at a maximum when $n(\overline{M})$ is a minimum. $n(\overline{M})$ will be a minimum when (f), (c), (d), and $12 - $ (a) hold the smallest values the conditions of the problem will allow. No conditions are imposed on (f), (c), and (d), so assume that they each are zero. $12 - $ (a) is smallest when (a) is largest, and the largest value (a) can assume is 8. Thus, in order for the total number of men to be 20,

$$\text{(e)} + 8 + 3 + 4 = 20$$
$$\text{(e)} = 5$$

In this case

$$n(M) = 5 + 8 + 3 = 16$$

Therefore, the maximum number of married men is 16.

# SUMMARY AND REVIEW

*Sets*

1. Notation:
   (a) $x \in A$ means $x$ is an element of $A$.
   (b) $\{x \mid x$ has property $Q\}$ is read "the set of all $x$ such that $x$ has property $Q$."
   (c) $\varnothing$ or $\{\ \}$ represents the *empty set*.
   (d) $A \subseteq B$ means that $A$ is a *subset* of $B$.
   (e) $A \subset B$ means that $A$ is a *proper subset* of $B$ ($A \subseteq B$ but $A \neq B$).
   (f) $\overline{A}$ is the *complement* of set $A$.
   (g) $A \cap B$ is the *intersection* of sets $A$ and $B$.
   (h) $A \cup B$ is the *union* of sets $A$ and $B$.
   (i) $A - B$ is the complement of $B$ relative to $A$.
   (j) $A \times B$ is the *Cartesian product* of $A$ and $B$.
   (k) $n(A)$ is the number of elements in $A$.

2. Operations on sets:
   (a) The complement of $A$ is the set of elements in the universe that are not in $A$.
   (b) The union of sets $A$ and $B$ is the set of all elements in $A$ or $B$ or both.
   (c) The intersection of sets $A$ and $B$ is the set of elements common to both $A$ and $B$.
   (d) The difference $A - B$ is the set of all elements of $A$ that are not in $B$.
   (e) The Cartesian product of sets $A$ and $B$ is the set of all ordered pairs formed by taking the first element from $A$ and the second element from $B$.

3. Properties of set operations:

   (a) *Identity properties.* An identity property indicates that there is a special set such that the operation with that set and set A gives set A.

      (i) $A \cup \varnothing = A$

      (ii) $A \cap U = A$

   (b) *Commutative properties.* A commutative property indicates that the order in which the operation is performed does not affect the results.

      (i) $A \cup B = B \cup A$

      (ii) $A \cap B = B \cap A$

   (c) *Associative properties.* An associative property indicates that the manner in which objects are grouped for multiple application of an operation does not affect the results.

      (i) $A \cup (B \cup C) = (A \cup B) \cup C$

      (ii) $A \cap (B \cap C) = (A \cap B) \cap C$

   (d) *Distributive properties.* A distributive property indicates that two operations interact according to a special pattern.

      (i) $A \cup (B \cap C) = (A \cup B) \cap (A \cup C)$ is called the *distributive property of union over intersection.*

      (ii) $A \cap (B \cup C) = (A \cap B) \cup (A \cap C)$ is called the *distributive property of intersection over union.*

      (iii) $A \times (B \cup C) = (A \times B) \cup (A \times C)$ and $(A \cup B) \times C = (A \times C) \cup (B \times C)$ are *distributive properties of the Cartesian product over union.*

   (e) Observe that the operation of the Cartesian product

      (i) does not have an identity set.

      (ii) is not commutative.

      (iii) is not associative.

4. If there exists a one-to-one correspondence between sets A and B, the sets are equivalent.

5. $n(A \cup B) = n(A) + n(B) - n(A \cap B)$

*Numbers*

1. Natural numbers (counting numbers): {1, 2, 3, 4, 5, . . .}

2. Number usage:

   (a) cardinal usage: describes the number of elements in a set.

   (b) ordinal usage: describes position or order.

   (c) nominal usage: names things.

*Relations*

1. A relation from A to B is a subset of $A \times B$.

2. A relation on A is a relation from A to A. A relation R on A is

   (a) reflexive if $(a, a) \in R$ for all $a \in A$.

   (b) symmetric if $(a, b) \in R$ implies that $(b, a) \in R$.

   (c) transitive if $(a, b) \in R$ and $(b, c) \in R$ implies that $(a, c) \in R$.

3. A relation on set A that is reflexive, symmetric, and transitive is called an *equivalence relation.* Equivalence relations partition the set A into disjoint sets called *equivalence classes.*

*Functions*

1. A function from A to B is a relation in which each element of A is paired with exactly one element of B. The set A is called the *domain* of the function; the set of second elements of the ordered pairs is called the *range* of the function.

2. Functions can be described by:

   (a) arrow diagrams

   (b) sets of ordered pairs

   (c) tables

   (d) $y = f(x)$ notation

   (e) graphs

   (f) input-rule–output machines

# CHAPTER TEST

1. Write definitions for the following sets in set-builder notation.

   (a) $A \cap B$      (b) $A \cup B$      (c) $A \times B$

2. Classify the following as true or false.

   (a) $A \times B = B \times A$

   (b) If $A \cap B = A$, then $A = B$.

   (c) $\subset$ is transitive.

   (d) {4, 5, 1} is not a proper subset of {1, 4, 5}.

   (e) $A \cup (B \cap C) = (A \cup B) \cap (A \cup C)$

   (f) $n(A \cup B) = n(A) + n(B)$

   *For (g) through (l), A = {1, 2, 3, 4, 7}.*

   (g) $\varnothing \in A$                    (h) $4 \subset A$

   (i) $n(A) = 7$                      (j) $A \times \varnothing = \varnothing \times A$

   (k) {2, 7} $\subset A$                 (l) $A \subseteq \varnothing$

   (m) In the expression "page 5," 5 is a cardinal number.

   (n) $n(A \cap B) = n(A) - n(B)$

3. If $U = \{1, 3, 5, 7, 11, 13, 17, 19, 23\}$, $A = \{3, 5, 7, 11\}$, $B = \{3, 11, 13, 23\}$, and $C = \{1, 5, 11, 17\}$, find the following.

(a) $C \cap A$     (b) $\overline{B}$     (c) $A \cup C$

(d) $\overline{A \cap B}$     (e) $A \cup (B \cap C)$

4. The shaded portion of the figure represents what set?

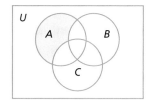

5. Indicate whether the following statements are true or false. If false, explain why they are false.

(a) The transitive property holds true for the relation

$$\{(a, b), (b, c), (c, d), (a, c), (a, d)\}$$

(b) If a relation on set $A$ is reflexive, then it is also symmetric.

(c) The set $\{1, 2, 3, \ldots, 1,000,000\}$ is an infinite set.

(d) The commutative, associative, and identity properties hold for both union and intersection.

(e) The commutative and associative properties hold for the Cartesian product.

(f) If two sets are equivalent, they are equal.

(g) If sets $A$ and $B$ are equal, they are equivalent.

6. If $A = \{3, 2, 7, 1\}$, $B = \{6, 8, 10\}$, and $C = \{3, 6, 5\}$, find $(A \cup B) \cap (B \cup C)$.

7. Classify the following relations as reflexive, symmetric, or transitive. Specify whether or not each relation is an equivalence relation.

(a) is married to

(b) $\subset$

(c) $\{(3, 3), (3, 4), (4, 3), (4, 4)\}$ on $A = \{3, 4\}$

8. The accompanying diagram shows the cardinal number of each closed region. Find the following.

(a) $n(A \cap \overline{B})$     (b) $n(A \cap B \cap C)$

(c) $n(A \cup B)$

(d) $n(A) + n(B) - n(A \cap B)$

(e) What is the relationship between (c) and (d)?

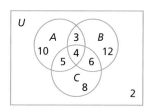

9. (a) Draw a Venn diagram representing $\overline{A \cap B}$.

(b) Draw a Venn diagram representing $\overline{A} \cup \overline{B}$.

(c) What property of sets have you demonstrated?

10. If $f(x) = 3x + 2$, find

(a) $f(2)$     (b) $f(0)$

11. List all the non-empty proper subsets of $\{a, b, c, d\}$.

12. Which of the following sets of ordered pairs are functions?

(a) $\{(1, 2), (2, 7), (3, 5), (1, 4)\}$

(b) $\{(2, 3), (4, 3), (5, 3), (6, 3)\}$

(c) $\{(4, 4), (5, 5), (6, 6)\}$

13. Shade $(A - B) \cap C$.

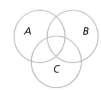

14. $A = \{1, 2\}$, $B = \{4, 5\}$, and $C = \{3, 6\}$. Find $(A \cup B) \times C$.

# WHOLE NUMBERS AND NUMERATION

# 4

E ach letter represents a different digit in these additions. Further, both additions give the same sum. What are X, A, B, G and H?

In this chapter you will be introduced to the first of several number systems we will study: the whole numbers. Addition, multiplication, subtraction, and division will be defined on this system, and the properties of each will be discussed. Make sure you understand the associative and commutative properties of addition and multiplication, the distributive property of multiplication over addition, and the existence of additive and multiplicative identities. We will continue to use these properties as we advance from the whole numbers to the integers, then to the rational numbers, and finally to the real numbers. The number line is introduced to give a visual representation of the operations and to illustrate the concept of order.

Also in this chapter, you will examine several ancient numeration systems. For example, in their numeration system the Egyptians made ink marks on papyri (reed paper), and the Babylonians cut shapes in clay tablets. A short historical account of the development of numeration will help you understand and appreciate our base ten numeration system. The simplicity of the base ten numeration system leads to very efficient algorithms, or computational schemes. Finally, practicing computations in bases other than ten will help you understand the base ten system better.

Problem-solving strategies demonstrated in this chapter include the following:   ▪ Make a chart. (168)   ▪ Guess, test, and revise. (129, 143, 177, 189)   ▪ Try a simpler version of the problem. (158, 159, 190)   ▪ Look at all possibilities. (129, 189)   ▪ Look for a pattern. (154, 159, 177, 190)   ▪ Use reasoning. (154, 168)   ▪ Use properties of numbers. (168, 179, 191)   ▪ Use an equation. (131, 145)   ▪ Use an algorithm. (throughout chapter)

## SECTION 1  WHOLE NUMBER ADDITION, SUBTRACTION, AND ORDER

| PROBLEM | | |
|---|---|---|

Pam adds down.

```
6
3    6 + 3 =  9
4    9 + 4 = 13
2   13 + 2 = 15
15
```

Roy adds up.

```
6    9 + 6 = 15
3    6 + 3 =  9
4    2 + 4 =  6
2
15
```

Sue adds every other term.

```
6
3    6 + 4 = 10
4    3 + 2 =  5      10 + 5 = 15
2
15
```

All get the same answer. Why?

We have been competent arithmeticians for so long that we use the properties of addition without any thought. At our convenience we change the order of addition (3 + 2 = 2 + 3) and regroup addition (2 + 3) + 4 = 2 + (3 + 4), scarcely conscious that we are using powerful properties that have an impact throughout much of mathematics. In this section, you will learn a definition of *addition* in terms of sets, and then you will observe how the properties of addition are developed from the properties of sets.

**GOALS**

Use the Number and Operations Standard: ● understand numbers, ways of representing numbers, relationships among numbers, and number systems; ● understand meanings of operations and how they relate to one another; ● compute fluently and make reasonable estimates.
Illustrate the Representation Standard, page 50.*
Illustrate the Problem Solving Standard, page 15.*
Illustrate the Connections Standard, page 147.*

* The complete statement of the standard is given on this page of the book.

We begin by considering a new set consisting of the natural numbers {1, 2, 3, 4, . . .} and the number 0.

**DEFINITION**

**Whole Numbers**    The set composed of the natural numbers and zero is called the set of *whole numbers.*

Consider the number zero for a moment. The familiar statement "Zero is nothing" is false. **Zero** is defined as the cardinal number of the null set, $n(\varnothing) = 0$, and it answers such questions as "How many elephants are in your class?" The next step is to examine the operations on the set of whole numbers.

Suppose you have |||| Delicious apples and ||| Jonathans. (See Figure 4-1 (a).) How many apples do you have? Here, |||| represents the number of Delicious apples, and ||| gives the number of Jonathan apples. Now place the Delicious apples and the Jonathan apples in a fruit bowl. By combining the abstract number representations of the number of Delicious apples and the number of Jonathans, we see that ||||||| now represents the number of apples. (See Figure 4-1 (b).) This idea of putting together (or taking the union) leads to the definition of addition. Let's look at some examples.

**EXAMPLE 1**    Let set $A = \{a, b, c\}$ and set $B = \{e, f, g, h\}$, with $n(A) = 3$ and $n(B) = 4$. Let's combine or take the union of the two sets. The result is shown in Figure 4-2.

EXAMPLE 2 | Let $A = \{a, b, c\}$ and $B = \{c, d, e, f\}$. Again, let's combine the two sets. The result is shown in Figure 4-3.

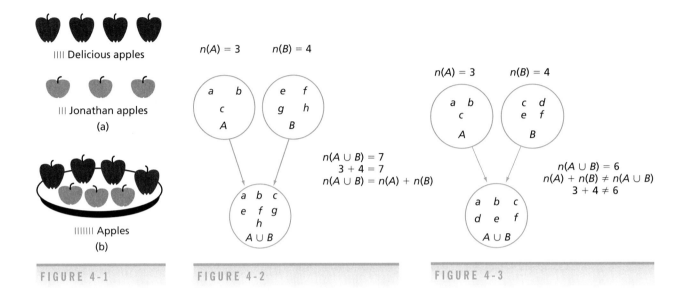

FIGURE 4-1

FIGURE 4-2

FIGURE 4-3

What is the distinction between these two examples? In the first one, the sets are disjoint, whereas in the second the letter $c$ is common to both sets.

<table>
<tr><td>DEFINITION</td><td>Addition of Whole Numbers</td><td>Let $a$ and $b$ represent any two whole numbers, and choose $A$ and $B$ to be disjoint finite sets so that $n(A) = a$ and $n(B) = b$. Then<br><br>$$a + b = n(A \cup B)$$</td></tr>
</table>

Notice that the operation of addition is defined in terms of two numbers, $a$ and $b$; that is, addition is a **binary operation.**

If $A = \{r, s, c, d, e\}$ and $B = \{x, y, z, w\}$, then $n(A) = 5$ and $n(B) = 4$. Notice that $A$ and $B$ are disjoint because $A \cap B = \varnothing$. Next,

$$A \cup B = \{r, s, c, d, e, x, y, z, w\}$$

and $n(A \cup B) = 9$. It follows that

$$n(A \cup B) = n(A) + n(B)$$

or                                  $9 = 5 + 4$

The 5 and 4 are called **addends,** and 9 is called the **sum.**

## ░░ MODELS FOR ADDITION

Now let's look at the representation of whole numbers on a number line to obtain a measurement model of addition. A **number line** is a representation of a geometric line extending endlessly in two opposite directions. The line is marked with two fundamental points, one representing 0 and one representing 1, as in Figure 4-4(a). These points mark the ends of what is called a **line segment** of length 1 (a **unit segment**). Once the length of a unit segment is established, subsequent points are marked to the right of 0 on the number line at the same unit distance apart, as in Figure 4-4(b).

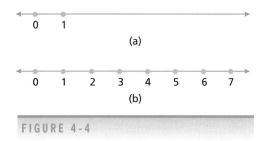

FIGURE 4-4

Thus, to the right of 0, a number line is divided into equal divisions by points, with each point representing a whole number. No matter how large a whole number may be, it can be matched to a point on a number line.

We use parts of lines with arrows on the end, called **arrow diagrams** or **directed line segments,** to represent numbers involved in operations on the whole numbers. For example, let's consider the sum of 2 and 3. Starting at 0 on the number line in Figure 4-5, draw an arrow to the right 2 units long. At the end of the first arrow, draw another arrow to the right 3 units long. Note that the point of the second arrow is at 5. Therefore, $2 + 3 = 5$.

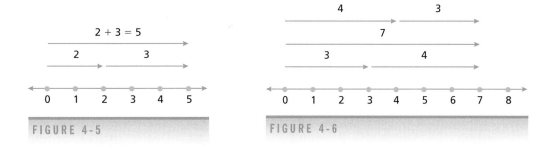

FIGURE 4-5                    FIGURE 4-6

Suppose $A$ and $B$ are disjoint finite sets with $n(A) = a$ and $n(B) = b$, where $a$ and $b$ are whole numbers. It is intuitively evident that $A \cup B$ exists and is a finite set. Thus, from the definition $a + b = n(A \cup B)$, we see that $a + b$ always exists. The fact that we always obtain a whole number when we add two whole numbers is summarized by the closure property.

| **Closure Property for Addition** | For any two whole number $a$ and $b$, there exists a unique whole number $a + b$. |

Notice that closure has two parts: uniqueness and existence. Existence stipulates that an answer exists in the set of whole numbers for the addition of any two whole numbers. Uniqueness stipulates that only one answer exists; that is, when 2 and 7 are added, the unique answer 9 is obtained.

The closure property for addition of whole numbers can be extended to any finite number of whole numbers; thus, for example, $4 + 5 + 0 + 1 + 3 = 13$, a whole number.

We should note that not every set of numbers is closed under the operation of addition. For example, the set of odd numbers is not closed under addition. Why not?

| | |
|---|---|
| **EXAMPLE 3** | Which of the following sets are closed under addition? <br> (a) {0, 1, 2, 3} <br> (b) $\{x \mid x$ is an even whole number} |
| *SOLUTION* | (a) $2 + 3 = 5$, and 5 is not in the set. Therefore, the set is not closed under addition. <br> (b) The sum of any two even whole numbers is an even whole number. For example, $2 + 6 = 8$ (which is even). Therefore, the set is closed under addition. |

By addition, $3 + 4 = 7$ and $4 + 3 = 7$. Are you surprised that the answers are the same? This equality is demonstrated in the measurement model in Figure 4-6. If you add the arrow diagram for 4 to the end of the one for 3, you terminate at the same point (namely 7) where you terminate by adding the arrow diagram for 3 at the end of 4. This property for the addition of whole numbers is called *commutativity*.

| | |
|---|---|
| **Commutative Property of Addition** | For all whole numbers $a$ and $b$, $$a + b = b + a$$ |

Although the commutative property of addition is obvious from the visual representation in Figure 4-6, let's see if this result can be obtained from the definition of addition in terms of the union of sets.

Now let $A$ and $B$ be any two sets such that $a = n(A)$ and $b = n(B)$, with $A \cap B = \varnothing$. Because $A \cup B = B \cup A$, then $A \cup B$ and $B \cup A$ are equivalent, or $n(A \cup B) = n(B \cup A)$. Thus,

$$a + b = n(A \cup B) = n(B \cup A) = b + a$$

When adding 3 or more numbers, we must group the numbers to specify the order in which the operations are to be performed. Suppose that we wish to add 2, 3, and 7. Should we add 2 and 3 and then 7, or should we add the sum of 3 and 7 to 2? Usually we indicate grouping of additions by using parentheses ( ) or brackets [ ]. $(2 + 3) + 7$ means that 2 and 3 are added and then 7 is added to that sum. $2 + (3 + 7)$ means that the sum of 3 and 7 is added to 2. The parentheses serve as a form of mathematical punctuation.

The set theory model shown in Figure 4-7(a) and (b) provides additional understanding of the property that governs grouping in addition, the **associative property of addition.**

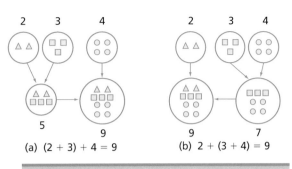

(a) $(2 + 3) + 4 = 9$     (b) $2 + (3 + 4) = 9$

FIGURE 4-7

| Associative Property of Addition | For all whole numbers $a$, $b$, and $c$, $$(a + b) + c = a + (b + c)$$ |
| --- | --- |

**EXAMPLE 4**

$$3 + 4 + 7 = (3 + 4) + 7 = 7 + 7 = 14$$

or

$$3 + 4 + 7 = 3 + (4 + 7) = 3 + 11 = 14$$

## CALCULATOR HINT

You need to use calculators in this course that follow algebraic logic instead of arithmetic logic. Algebraic logic utilizes the following order of computation.

(a) Does computations within parenthesis first.
(b) Calculates exponentials next.
(c) Performs multiplications and divisions from left to right.
(d) Performs additions and subtractions from left to right.

For ease in reading we denote calculator keys by rectangles or squares. ON turns the calculator on. Many calculators have a key often marked 2ⁿᵈ, which combined with another key doubles the number of commands on a keyboard. For example, 2ⁿᵈ ON may turn the calculator off, and 2ⁿᵈ × may give a comma , . Use either ENTER or = to enter something into the calculator or to secure an answer. After each problem one deletes computations and answers with CLEAR .

Perform these operations with your calculator, and see if you get the same answer.

$$2 \boxed{+} \boxed{(} 3 \boxed{+} 7 \boxed{)} \boxed{=} 12$$

or

$$\boxed{(} 2 \boxed{+} 3 \boxed{)} \boxed{+} 7 \boxed{=} 12$$

Because the answers are the same, you have illustrated the associative property of addition.

When 4 numbers are to be added, the numbers may be associated or grouped in several ways. For example,

$$(3 + 2) + (5 + 6) = 5 + 11 = 16$$

$$3 + [(2 + 5) + 6] = 3 + [7 + 6] = 3 + 13 = 16$$

$$[3 + (2 + 5)] + 6 = [3 + 7] + 6 = 10 + 6 = 16$$

That they yield the same number results from a generalization or extension of the associative property of addition.

In simple proofs such as the one in the next example, we often use the transitive property of equality: "If $a = b$ and $b = c$, then $a = c$." In the next example, we show that $(2 + 3) + 5 = 5 + (2 + 3)$ (or $a = b$). Then we show that $5 + (2 + 3) = (5 + 2) + 3$ (or $b = c$). Therefore, $(2 + 3) + 5 = (5 + 2) + 3$ (or $a = c$). Notice also that, because of the closure property of addition, we consider $(2 + 3)$ as one term.

**EXAMPLE 5**

Using the commutative and associative properties for addition, show that $(2 + 3) + 5 = (5 + 2) + 3$.

*SOLUTION*

$(2 + 3) + 5 = 5 + (2 + 3)$    Commutative property of addition

$5 + (2 + 3) = (5 + 2) + 3$    Associative property of addition

Therefore,

$(2 + 3) + 5 = (5 + 2) + 3$    Transitive property of equality

Consider the addition $4 + 0$. We may associate with 4 the set $A = \{a, b, c, d\}$, and with 0 the null set $\varnothing$. (See Figure 4-8.) The union of $A$ and $\varnothing$ is $A$, and $A \cap \varnothing = \varnothing$. Thus, $n(A) + n(\varnothing) = n(A \cup \varnothing) = n(A)$. Hence, $4 + 0 = 4$. We call 0 the *additive identity for addition*.

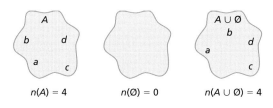

$n(A) = 4$    $n(\varnothing) = 0$    $n(A \cup \varnothing) = 4$

FIGURE 4-8

**Additive Identity**

Zero is called the *additive identity* because, for any whole number $a$,

$$a + 0 = a \quad \text{and} \quad 0 + a = a$$

**EXAMPLE 6**

$$4 + 0 = 4$$

$$0 + 3 = 3$$

$$0 + 0 = 0$$

# SUBTRACTION

We introduce subtraction as the **inverse of addition.** A compact disc costs $25 but John has only $10. How much money does he need to purchase the disc? The problem can be written as $25 − $10 = $□. The answer is $15 because $15 + $10 = $25.

Consider the sum                    $8 + 3 = 11$

Observe that            $11 − 3 = 8$    and    $11 − 8 = 3$

Thus, the addition problem gives rise to two subtraction problems. In general, when $b$ is subtracted from $a$, the result is the number $c$ (called the **difference**). The answer $c$ has the property that, when $b$ is added to $c$, the result is $a$.

**DEFINITION**

**Subtraction**   The difference $a − b$ in the *subtraction* of whole number $b$ from whole number $a$ is equal to whole number $c$ if and only if $c + b = a$.

**EXAMPLE 7**

$16 − 5 = 11$    because    $11 + 5 = 16$

$8 − 3 = 5$    because    $5 + 3 = 8$

$10 − 0 = 10$    because    $10 + 0 = 10$

Figure 4-9 demonstrates the answer to the question, "What whole number added to 3 gives an answer of 8?" As the figure shows, 8 minus 3 gives an answer of 5.

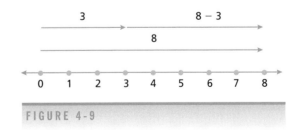

FIGURE 4-9

The difference $a − b$ may be read as "$a$ minus $b$," or "$a$ take away $b$." The number $a$ is called the **minuend,** and $b$ is called the **subtrahend.**

The definition of subtraction does not imply that the whole number $c$ exists for all pairs $a$ and $b$. For instance, subtract 5 from 3 ($3 − 5 = ?$). What whole number added to 5 will give an answer of 3? There is no such number in the system of whole numbers. Thus, the set of whole numbers is *not closed* under the operation of subtraction.

# MODELS OF SUBTRACTION

In order to use the operation of subtraction in real-world problems, we must learn to recognize problems in terms of one or more of the four models for subtraction: the take-away model, the comparison model, the missing addend model, and the set/subset model.

In the **take-away model**, a set is identified and part of it is "taken away." The number of elements that remain in the set is the difference. For example, *Aimee has seven tennis balls and gives Racine two. How many does Aimee have left?* $(7 - 2 = 5)$

In the **comparison model**, two disjoint sets are identified, and the elements in each set are compared by matching them in one-to-one correspondence. The question is asked: How many more elements are in one set than in the other? For example, *Kahlil has six DVDs and Terrell has two DVDs. How many more DVDs does Kahlil have than Terrell?* (or *How many fewer DVDs does Terrell have than Kahlil?*) $(6 - 2 = 4)$

In the **missing addend subtraction model**, the question is asked: How many more are needed to make some total number? For example, *Serina has \$4. She needs \$9 to buy a ticket to the ice skating show. How much more does she need to buy a ticket?* (This can be written symbolically as $4 + q = 9$.) $(9 - 4 = 5)$

In the **set/subset model**, we identify a set. Some of the elements of the set have a specific attribute. The rest of the elements do not have that attribute. The question is asked: How many elements in the set do not have the attribute? For example, *Mr. Price has seven dogs. Five of them are beagles. How many of the dogs are not beagles?* $(7 - 5 = 2)$

The **take-away model** of subtraction is illustrated in the following example.

| EXAMPLE 8 | Five alphabet blocks are in the baby's playpen. The baby shoves 2 blocks out of the playpen and onto the floor. How many blocks remain in the playpen? |
|---|---|
| *SOLUTION* | In this example of the subtraction $5 - 2$, 2 blocks are isolated or are taken away from the 5 blocks, leaving 3. (See Figure 4-10.) |

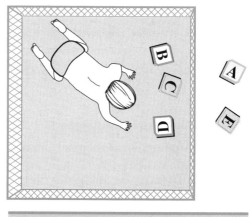

FIGURE 4-10

This approach can be stated as follows: If $a$ and $b$ are any whole numbers and if $A$ and $B$ are sets such that $a = n(A)$, $b = n(B)$, and $B \subseteq A$, then

$$a - b = n(A - B)$$

This concept of subtraction may also be represented on a number line. In the subtraction $5 - 2$, for example, the arrow representing 2, points toward the left and terminates at 3. (See Figure 4-11.)

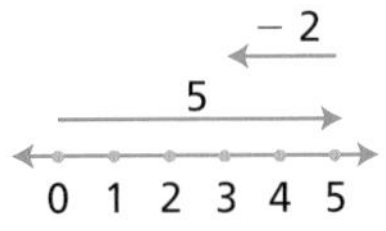

Consider the problem "If Aaron can do 11 push-ups and Mark can do 5 push-ups, how many more push-ups can Aaron do than Mark?" or "How many fewer can Mark do than Aaron?" Neither the missing-addend model nor the take-away model adequately explains this problem. Instead we are comparing two disjoint sets, the set of pushups done by Aaron and the set of pushups done by Mark. We can identify this as a subtraction problem by thinking of it as an example of the comparison model. In Figure 4-12 we have set up a one-to-one correspondence between the two sets. We then find the difference

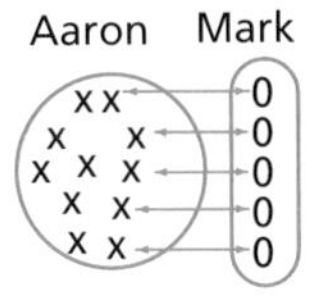

$$11 - 5 = 6$$

Consider the problem, "Twenty children are in Mrs. Barnard's class (C). Eleven are girls (G). How many are boys (B)?" We identify this as subtraction by understanding it as an example of the set/subset model.

$$n(B) = n(C - G) = 20 - 11 = 9$$

## ⬡ ORDER

In studying the basic operations for whole numbers, we have unconsciously been working with order relations. If $a$ and $b$ are different whole numbers, then there is a natural number $n$ such that either $a + n = b$ or $b + n = a$. For instance, 2 is less than 6 because $2 + 4 = 6$; but 6 is greater than 5 because $6 = 5 + 1$.

| DEFINITION | **Less Than or Greater Than** | 1. If $a$ and $b$ are any whole numbers, then $a$ is said to be *less than b*, denoted by $a < b$, if and only if there exists a nonzero whole number $n$ such that $a + n = b$. <br> 2. If $a$ and $b$ are any whole numbers, then $a$ is said to be *greater than b*, written $a > b$, if and only if there exists a nonzero whole number $d$ such that $a = b + d$. |
| --- | --- | --- |

**EXAMPLE 9**

(a) $6 < 14$ because there exists a nonzero whole number 8 such that $6 + 8 = 14$.

(b) $10 > 4$ because there exists a nonzero whole number 6 such that $10 = 4 + 6$.

FIGURE 4-11

Aaron Mark

FIGURE 4-12

If $a$ is greater than $b$, then $b$ is less than $a$; and if $a$ is less than $b$, then $b$ is greater than $a$. Thus the expressions $a < b$ and $b > a$ express the same idea. Sometimes the equality relation is combined with "less than" to give a **less than or equal to** relation. The symbol $a \leq b$ means that either $a < b$ or $a = b$.

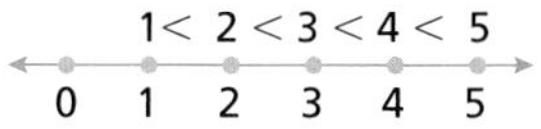

By using a number line, we can easily discuss the relations "greater than" and "less than." When we compare two numbers on a number line, the number on the right is greater than the one on the left, and the one on the left is less than the one on the right. Some obvious relations are $1 < 2$, $2 < 3$, $3 < 4$, and so forth, as seen in Figure 4-13.

$$1 < 2 < 3 < 4 < 5$$

0   1   2   3   4   5

**FIGURE 4-13**

## SOLUTION SETS

In Chapter 2 we learned sentences that can be identified as either true or false are called **statements.** Sentences expressing a relationship of equality are called **equations.** However, many sentences involving variables cannot be classified either as true or as false. These are not called statements but **open sentences** or sentences. For example, $x + 4 = 7$ is both an equation and an example of an open sentence.

**EXAMPLE 10**

Translate the following expressions into mathematical sentences.
(a) A number added to 8 is equal to 23.
(b) Two times some number plus 7 equals 21.

SOLUTION

(a) $x + 8 = 23$      (b) $2\square + 7 = 21$

Variables such as $x$ in (a) and $\square$ in (b) of Example 10 are actually placeholders for replacements. After a replacement is made, the open sentence becomes a statement because it can be classified as true or as false.

**EXAMPLE 11**

Using the open sentences of the preceding example, replace $x$ by 15 in (a) and $\square$ by 3 in (b); then classify the statements as true or false.

SOLUTION

(a) $15 + 8 = 23$     (true)     (b) $2 \cdot 3 + 7 = 21$     (false)

If a replacement turns an open sentence into a true statement, then the replacement is called a **solution** of an equation, and the set of all solutions is called the **solution set.** The possible replacements for a variable of an equation make up the **domain** of the variable. We will work with the set of whole numbers as the domain and use the relationship between addition and subtraction to find the solution. Thus, if

$$x + a = b \qquad \text{then} \qquad x = b - a$$

or if $\qquad x - c = d \qquad \text{then} \qquad x = c + d$

**EXAMPLE 12**

Find the solution set for $x + 7 = 11$, using the set of whole numbers as the domain.

*SOLUTION*

If $x + 7 = 11$, then $x = 11 - 7$. Therefore, $x = 4$ or {4} is the solution set.

**PRACTICE PROBLEM**

Find the solution set for $x - 6 = 4$.

*ANSWER*

$x = 10$ or {10} is the solution set.

We can solve open sentences such as $x + 4 < 6$ on the domain of whole numbers by using the guess, test, and revise strategy. This involves substituting whole numbers for $x$. Thus,

$x = 0$ is a solution, because $0 + 4 < 6$.
$x = 1$ is a solution, because $1 + 4 < 6$.
$x = 2$ is not a solution, because $2 + 4 \not< 6$.

It should be obvious that the sum of 4 and all whole numbers larger than 2 cannot be less than 6. Therefore, the solution set on the domain of whole numbers is {0, 1}.

**PRACTICE PROBLEM**

Find the solution set of $x > 3$ and $x < 5$ on the domain of whole numbers.

*ANSWER*

Because both inequalities must be satisfied, {4} is the solution set.

# Just for Fun

In a magic square, the sum of the numbers in every row, column, and diagonal is the same. Rearrange the numbers shown to make a magic square with a magic sum of 12.

# EXERCISE SET 1

*R* 1. For the given pairs of sets, state whether $n(R) + n(S) = n(R \cup S)$.

(a) $R = \{3, 2, 7, 1\}, S = \{4, 6, 5\}$
(b) $R = \{3, 2, 4, 7\}, S = \{1, 4, 6, 5\}$
(c) $R = \{x \mid x$ is a whole number greater than 1 and less than 6$\}$,
$S = \{x \mid x$ is a whole number greater than 4 and less than 9$\}$
(d) $R = \{x \mid x$ is a whole number greater than 5 and less than 8$\}$,
$S = \{x \mid x$ is a whole number greater than 6 and less than 7$\}$
(e) $R = \varnothing, S = \{\varnothing\}$

2. The following problems illustrate properties of addition described in this section. State which properties are being applied in each example.

(a) $2 + 3 = 3 + 2$
(b) $4 + (2 + 3) = (4 + 2) + 3$
(c) $6 + 0 = 6$
(d) $(a + c) + d = a + (c + d)$
(e) $[2 + (4 + 6)] + 8 = (2 + 4) + (6 + 8)$
(f) $2 + (5 + 7) = 2 + (7 + 5)$
(g) $5 + 0 = 0 + 5$
(h) $(8 + 2) + 3 = 3 + (8 + 2)$
(i) $(4 + 2) + 3 = 3 + (4 + 2)$
(j) $(2 + 5) + 3 = (5 + 2) + 3$

3. Addition can be simplified and often performed mentally by using the associative property of addition.

*Example:*
$27 + 46 = 20 + (7 + 46) = 20 + 53 = 73$

(a) $49 + 58 = 40 + \underline{\hspace{1.5cm}} =$
(b) $56 + 78 = \underline{\hspace{1.5cm}} + (6 + 78) =$
(c) $263 + 85 = 260 + \underline{\hspace{1cm}} =$

4. Frames provide an easy transition to the study of algebra and a clear way of stating problems. Fill in the frames to make the following number statements true.

(a) $3 + (5 + 7) = (5 + \square) + 3$     (b) $5 + 0 = \square$
(c) $7 + \square = 9 + 7$
(d) $5 + (\square + 6) = (6 + 5) + 9$

5. Show the following additions on a number line.

(a) $3 + 5$
(b) $5 + 3$
(c) What do (a) and (b) demonstrate?
(d) Show on a number line that $(2 + 3) + 4 = 2 + (3 + 4)$, illustrating the associative property of addition.

6. Determine what property of addition is illustrated in each of the following.

(a) $a + (b + c) = (a + b) + c$
(b) $a + 0 = a$
(c) $a + c = c + a$
(d) $a + (b + c) = a + (c + b)$

7. Perform the operations shown, if possible; otherwise state that no answer exists for whole numbers.

(a) $6 - 4$          (b) $2 - 7$
(c) $4 - (3 + 2)$    (d) $(8 + 4) - 7$
(e) $6 - (8 - 5)$    (f) $6 - (12 - 3)$

8. Illustrate on the number line the following computations.

(a) $7 - 5 = 2$      (b) $14 - 6 = 8$
(c) $5 + 7 = 12$     (d) $6 + 8 = 14$
(e) $12 - 7 = 5$     (f) $10 - 4 = 6$

9. Using the definition of *less than,* classify the inequalities as either true or false.

(a) $6 < 9$          (b) $8 < 8$
(c) $1 < 1$          (d) $5 < 7$
(e) $7 + 3 \geq 10$  (f) $8 + 9 > 13$
(g) $3 + 3 \leq 6$   (h) $4(6 + 2) \leq 27$

10. For each of the following statements, use the definition of subtraction to write two other statements that can be derived from the given statement.

(a) $5 + 16 = 21$        (b) $(7 + 4) - 3 = 8$
(c) $a + c = f$          (d) $(21 + 5) - 4 = 22$

11. Use a counterexample to prove that, for the operation of subtraction on the whole numbers,

(a) closure does not hold.
(b) commutativity does not hold.
(c) associativity does not hold.

12. Which of the following sets have closure with respect to addition?

(a) $\{1, 2, 3, 4, 5, \ldots\}$      (b) $\{2, 4, 6, 8, 10, \ldots\}$
(c) $\{1, 3, 5, 7, \ldots\}$         (d) $\{0, 1\}$
(e) $\{x \mid x$ is a whole number greater than 15$\}$
(f) $\{x \mid x$ is a natural number less than 10$\}$
(g) $\{0, 3, 6, 9, 12 \ldots\}$

13. Find the solution set on the domain of whole numbers.

(a) $x > 5$ or $x < 7$        (b) $x \geq 5$ and $x \leq 7$
(c) $x > 7$ and $x > 5$       (d) $x > 0$ or $x > 5$
(e) $x + 4 = 8$ or $x + 2 = 4$
(f) $x + 4 = 8$ and $x + 2 = 4$

14. List the elements in the solution sets for the following, using the set of whole numbers as the domain.

   (a) $x + 4 = 9$      (b) $6 + x = 8$
   (c) $8 = 5 + x$      (d) $x + 6 = 8$
   (e) $x + 5 = x + 4$   (f) $3 + x = x + 3$

15. Identify the following problems as subtraction problems by recognizing the problem as involving one or more of the models for subtraction. Explain your answer by identifying the set from which something is "taken away," the two disjoint sets that are being compared, the total that is being sought (in the missing addend model), or the attribute characterizing the subset in the set/subset model.

   (a) Mrs. Ourts has 25 students in her *Transition Mathematics* class and 14 of them are girls. How many boys are in her class?
   (b) Teri wants to collect all 30 of the trading cards of the Atlanta Braves. If she now has 8 cards, how many more does she need to collect?
   (c) Eli has 8 pencils. If he gives 3 to his sister, how many will he have then?
   (d) Mrs. Spero has 11 jars but only 8 lids. How many jars will not have lids?
   (e) If Jerry buys 4 more notebooks, he'll have a dozen. How many notebooks does he have now?
   (f) Duchess gave birth to 8 puppies. Five of them were female. How many male puppies were there?
   (g) Kendle wrote 11 letters to her Space Academy friends. Kate wrote only 3. How many fewer letters did Kate write than Kendle?
   (h) Mr. Snope has a group of seventh graders on his MathCounts team. If he gets 5 eighth graders to join the group, he'll have 12 students on the team. How many seventh graders are on the team?

   *In Exercises 16 through 23, write an equation that serves as a model for or represents the given problem. Then find the solution set to get the answer for the problem.*

*T* 16. John won the long jump by jumping 2 m farther than his nearest competitor. If John jumped 7 m, how far did his nearest competitor jump?

17. Marlyn is 5 years older than Jean, and the sum of their ages is 27. Find Jean's age.

18. The width of a rectangle is 2 cm less than its length. If the perimeter (distance around) is 32 cm, find the length of the rectangle.

19. If Leslie had 4 more tickets to the football game, she would have 9 tickets. How many tickets does she have?

20. Mr. Roe jogged a total of 7200 yards in 3 nights. If each night he increased his distance by 400 yards, how far did he jog on the first night?

21. In a triangle, one side is 5 less than a second side and 10 less than a third side. If the perimeter (distance around) of the triangle is 75, what are the three sides?

22. The length of the smallest side of a triangle is 2 m less than the second side, and the length of the longest side is 4 m more than the second side. Find the length of the three sides of the triangle if the perimeter is 32 m.

23. Let $c$ be the cost of a book to the bookstore. To this cost the bookstore adds a profit of $12 and a tax of $2 to get a selling price of $62. What was the cost of the book?

24. What conditions must we place on the values of $a$, $b$, and $c$ so that the following are defined as whole numbers?

   (a) $a - b$               (b) $(a - b) - c$
   (c) $(a - c) - b$
   (d) When $(a - b) - c$ and $(a - c) - b$ are both defined, are they necessarily equal?

*C* 25. Working from left to right, give explicit reasons for each step in verifying that the following statements are true.

   (a) $8 + (5 + 2) = 2 + (8 + 5)$
   (b) $6 + (9 + 1) = 9 + (6 + 1)$
   (c) $(a + b) + (c + d) = (b + d) + (a + c)$

26. Verify that each of the following is equal to $a + (b + c)$.

   (a) $a + (c + b)$         (b) $(a + b) + c$
   (c) $(b + c) + a$         (d) $(a + c) + b$
   (e) $c + (a + b)$         (f) $(c + b) + a$

27. Demonstrate the (a) associative and (b) commutative properties of addition by separating the set $\{l, m, n, o, p, q, r, s, t, u\}$ into disjoint sets, the union of which is the given set, and by adding the cardinal numbers of each disjoint set.

28. The accompanying table is similar to the ordinary addition table. This table defines a binary operation $\oplus$ for a set $S = \{1, 2, 3\}$. Study the patterns in the table, and answer the following questions.

| $\oplus$ | 1 | 2 | 3 |
|----------|---|---|---|
| **1**    | 2 | 3 | 1 |
| **2**    | 3 | 1 | 2 |
| **3**    | 1 | 2 | 3 |

   (a) Is $S$ closed under the operation $\oplus$
   (b) Is the operation $\oplus$ commutative?

(c) Is the operation associative? (Test several examples.)

(d) Is there an identity element for ⊕ in $S$? If so, name it.

29. Let us invent a new binary operation on the whole numbers. The new operation is defined to be the first number added to twice the second.

    (a) Is this operation commutative?
    (b) Is it associative?
    (c) Is there an identity for this operation?

30. Define a new operation # such that, for any whole numbers $a$ and $b$, $a \# b = 3a + 4b$. Find $a \# b$ for the following values of $a$ and $b$.

    (a) $a = 2, b = 0$      (b) $a = 10, b = 13$
    (c) $a = 6, b = 23$      (d) $a = 2, b = 2$
    (e) $a = 11, b = 1$      (f) $a = 7, b = 17$

**LABORATORY ACTIVITY**

• • • • • • • • • • • • • • • • • • • • • • • • • • • • •

31. Write a word problem that can be solved by subtraction $7 - 4$, using each of the following models.

    (a) Missing-addend model
    (b) Take-away model
    (c) Comparison model
    (d) Set/subset model.

# SECTION 2 WHOLE NUMBER MULTIPLICATION AND DIVISION

**PROBLEM**

The coaches of a squad of 56 football players were authorized to purchase 3 sets of uniforms. An assistant coach reasoned that they would need to purchase

$$3 \cdot 24 \text{ (extra large)} + 3 \cdot 32 \text{ (regular size) uniforms.}$$

However, the head coach submitted a purchase order for $3 \cdot 56$ uniforms. Who was correct?

**OVERVIEW**

Because $72 + 96 = 168$ and $3 \cdot 56 = 168$, both are correct. When we prove that this happens in general, we have established the distributive property of multiplication over addition. However, we must first define *multiplication*. This we will do in terms of repeated addition. An alternative definition will be given in terms of Cartesian product.

**GOALS**

Illustrate the Number and Operations Standard, page 119.*
Illsutrate the Problem Solving Standard, page 15.*
Illustrate the Algebra Standard, page 4.*
Illustrate the Representation Standard, page 50.*

\* The complete statement of the standard is given on this page of the book.

# MULTIPLICATION

Now that the properties of the addition of whole numbers have been defined and developed, we focus on **multiplication,** the operation of finding the product of two numbers. The first approach to multiplication of whole numbers is to consider multiplication as a shorthand for addition. For example,

$$2 + 2 + 2 = 3 \cdot 2$$

By combining three sets of two elements each (as in Figure 4-14), we have a pictorial representation of $3 \cdot 2$.

The next example (sometimes called an *array model*) gives another way to visualize multiplication geometrically.

**EXAMPLE 1**

If a classroom has four rows of desks with six desks in each row, how many desks are in the room?

*SOLUTION*

Designate the desks in each row $a$, $b$, $c$, $d$, $e$, and $f$, and number the rows 1, 2, 3, and 4. (See Figure 4-15.) There are obviously 4 rows, each with 6 desks. Thus the number of desks in the room (using addition) is

$$4 \cdot 6 = 6 + 6 + 6 + 6 = 24$$

There are $4 \cdot 6 = 24$ desks in the room.

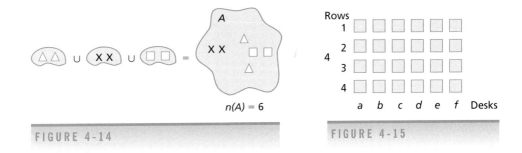

FIGURE 4-14                    FIGURE 4-15

These examples illustrate the following interpretation of multiplication.

| | |
|---|---|
| **Multiplication as Repeated Addition** | If $a$ is a natural number and $b$ is a whole number, then $a \cdot b = b + b + b + \cdots + b$, where $b$ occurs $a$ times. |

**EXAMPLE 2**

$$5 \cdot 4 = 4 + 4 + 4 + 4 + 4 = 20$$

$$2 \cdot 5 = 5 + 5 = 10$$

$$7 \cdot 0 = 0 + 0 + 0 + 0 + 0 + 0 + 0 = 0$$

**EXAMPLE 3** | Demonstrate the multiplication of $3 \cdot 4$ and of $4 \cdot 3$ on a number line as repeated addition. (See Figure 4-16.)

(a)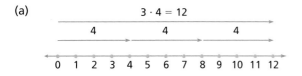

(b)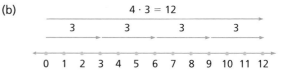

FIGURE 4-16

An alternate definition of multiplication can be given in terms of the Cartesian product. Suppose a football stadium has six lettered exits and four numbered gates. (See Figure 4-17.) If you are inside the park, in how many different ways can you leave through a lettered exit and a numbered gate?

Suppose you list the possibilities as

| | |
|---|---|
| $(A, 1), (A, 2), (A, 3), (A, 4),$ | $(D, 1), (D, 2), (D, 3), (D, 4),$ |
| $(B, 1), (B, 2), (B, 3), (B, 4),$ | $(E, 1), (E, 2), (E, 3), (E, 4),$ |
| $(C, 1), (C, 2), (C, 3), (C, 4),$ | $(F, 1), (F, 2), (F, 3), (F, 4)\}$ |

Does the preceding tabulation look familiar? It is the Cartesian product

$$\{A, B, C, D, E, F\} \times \{1, 2, 3, 4\}$$

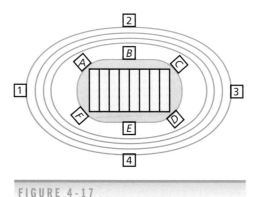

FIGURE 4-17

FIGURE 4-18

The number of ways you can leave the stadium is the cardinal number of the Cartesian product of the set of lettered exits and the set of numbered gates:
$$n(\{A, B, C, D, E, F\} \times \{1, 2, 3, 4\})$$

**EXAMPLE 4** | Set $A = \{a, b, c\}$ has the cardinal number 3, and set $B = \{k, l\}$ has the cardinal number 2. The cardinal number of the set

$$A \times B = \{(a, k), (a, l), (b, k), (b, l), (c, k), (c, l)\}$$

is the number of squares in Figure 4-18 representing the elements of $A \times B$, or 6. Notice that $3 \cdot 2 = n(A) \cdot n(B) = n(A \times B) = 6$.

This discussion leads to the following definition.

| DEFINITION | **Multiplication of Whole Numbers** | Let $a$ and $b$ represent any two whole numbers. Select finite sets $A$ and $B$ such that $n(A) = a$ and $n(B) = b$. Then $$a \cdot b = n(A \times B)$$ |

This definition states that, to multiply two numbers $a$ and $b$, we count the number of terms in the Cartesian product $A \times B$. In an expression indicating multiplication, such as $4 \cdot 3 = 12$, the 3 and 4 are called **factors** and 12 is the **product.** Relative to $ab = c$, we have

$$\underbrace{a \cdot b}_{\text{factors}} = \underset{\text{product}}{c}$$

We also write $a \cdot b$ as $ab$, as $(a)(b)$, as $a(b)$, as $(a)b$, and sometimes as $a \times b$. However, the notation $a \times b$ will not be used in this book because it is easily confused with the Cartesian product.

If $a$ and $b$ are whole numbers, then $A$ and $B$ are finite sets, and it is intuitively evident that $A \times B$ exists and is a finite set. Thus, from the definition $ab = n(A \times B)$, $ab$ always exists. The fact that when you multiply two whole numbers, you always get a whole number is summarized by the following property.

| **Closure Property for Multiplication** | For any two whole numbers $a$ and $b$, there exists a unique whole number $a \cdot b$. |

## MULTIPLICATION MODELS

An **array model** is useful in visualizing the commutative property of multiplication of two whole numbers, as illustrated in the following example.

**EXAMPLE 5**

We learned in Example 1 that a classroom with 4 rows of desks with 6 desks in a row contained 24 desks. Likewise, a classroom with 6 rows of desks with 4 desks per row contains 24 desks. This result is shown in the array model in Figure 4-19.

Sets of blocks (called *flats*) are also effective for demonstrating properties of multiplication. These blocks may be arranged in either of the ways shown in Figure 4-20.

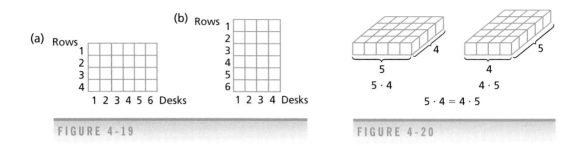

FIGURE 4-19          FIGURE 4-20

The preceding examples illustrate the *commutative property of multiplication*.

| | |
|---|---|
| **Commutative Property of Multiplication** | For all whole numbers $a$ and $b$, $$a \cdot b = b \cdot a$$ |

To prove this property, choose sets $A$ and $B$ so that $n(A) = a$ and $n(B) = b$. Now $A \times B$ is equivalent to $B \times A$, so

$$n(A \times B) = n(B \times A)$$

or $\qquad\qquad a \cdot b = n(A \times B) = n(B \times A) = b \cdot a$

## ASSOCIATIVE PROPERTY

In multiplying $3 \cdot 5 \cdot 4$, notice that the same answer is obtained from different groupings:

$$(3 \cdot 5) \cdot 4 = 15 \cdot 4 = 60$$

$$3 \cdot (5 \cdot 4) = 3 \cdot 20 = 60$$

By the transitive property of equality,

$$(3 \cdot 5) \cdot 4 = 3 \cdot (5 \cdot 4)$$

### CALCULATOR HINT

Compute $12 \cdot (13 \cdot 8)$ and then $(12 \cdot 13) \cdot 8$.

12 ⊠ ⎣ 13 ⊠ 8 ⎦ ⎓ 1248

⎣ 12 ⊠ 13 ⎦ ⊠ 8 ⎓ 1248      Therefore, $12 \cdot (13 \cdot 8) = (12 \cdot 13) \cdot 8$

These two examples illustrate the *associative property of multiplication of whole numbers*.

| **Associative Property of Multiplication** | For all whole numbers $a$, $b$, and $c$, $$(ab)c = a(bc)$$ |
| --- | --- |

Visually the associative property of multiplication is illustrated in Figure 4-21.

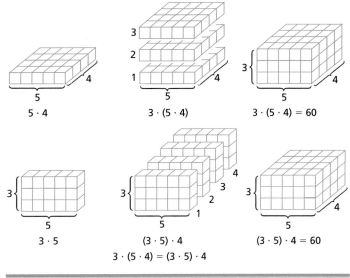

**FIGURE 4-21**

**EXAMPLE 6**

Use the properties of multiplication to show that $b \cdot (4 \cdot 3) = 3 \cdot (4 \cdot b)$.

*SOLUTION*

$$b \cdot (4 \cdot 3) = (4 \cdot 3) \cdot b \quad \text{Commutative property of multiplication}$$

$$(4 \cdot 3) \cdot b = (3 \cdot 4) \cdot b \quad \text{Commutative property of multiplication}$$

$$(3 \cdot 4) \cdot b = 3 \cdot (4 \cdot b) \quad \text{Associative property of multiplication}$$

Therefore,

$$b \cdot (4 \cdot 3) = 3 \cdot (4 \cdot b) \quad \text{Transitive property of equality}$$

The associative property can be used to give meaning to the product of any number of factors. Consider the multiplication of four factors, such as $abcd$. If we use a generalization or repetition of the associative property, we can restate this product as $(ab) \cdot (cd)$, or $a[b(cd)]$, or $[a(bc)]d$, all of which give the same answer.

**EXAMPLE 7**

Compute $3 \cdot 5 \cdot 2 \cdot 4$ in several different ways.

*SOLUTION*

$$(3 \cdot 5) \cdot (2 \cdot 4) = 15 \cdot 8 = 120$$

$$3[5(2 \cdot 4)] = 3[5 \cdot 8] = 3 \cdot 40 = 120$$

$$[3(5 \cdot 2)] \cdot 4 = [3 \cdot 10] \cdot 4 = 30 \cdot 4 = 120$$

EXAMPLE 8 | Consider the multiplication $3 \cdot 1$. If $Q = \{a, b, c\}$ and $R = \{a\}$, then $n(Q) = 3$ and $n(R) = 1$.

$$Q \times R = \{(a, a), (b, a), (c, a)\}$$

By counting, we see that the number of elements in $Q \times R$ is the same as the number of elements in $Q$. Therefore, $n(Q \times R) = n(Q) = 3$. $R \times Q = \{(a, a), (a, b), (a, c)\}$, and $n(R \times Q) = n(Q) = 3$. So

$$3 \cdot 1 = 1 \cdot 3 = 3$$

The preceding discussion illustrates the following.

| Multiplicative Identity for Whole Numbers | 1 is the multiplicative identity because, for any whole number $a$, $$1 \cdot a = a \qquad \text{and} \qquad a \cdot 1 = a$$ |
| --- | --- |

What is the product $a \cdot b$ if $b = 0$? For example, consider $3 \cdot 0$.

$$3 \cdot 0 = 0 + 0 + 0 = 0$$

In general, we would expect $a \cdot 0$ to be 0.

The preceding example and the commutative property of multiplication illustrate the following multiplicative property of zero.

| Multiplicative Property of Zero | If $a$ is any whole number, then $$0 \cdot a = 0 \qquad \text{and} \qquad a \cdot 0 = 0$$ |
| --- | --- |

(a)

(b)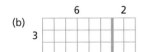

FIGURE 4-22

Sarah spent 2 hours each day studying biology and 4 hours each day in a laboratory. How many hours did she spend on these activities in 3 days? You can obtain the answer in two ways: add the time spent each day and then multiply by the number of days or $2 + 4 = 6$ and $3(2 + 4) = 3 \cdot 6 = 18$, or multiply the time spent on each activity by the number of days and then add. $3 \cdot 2 = 6$ and $3 \cdot 4 = 12$ so the total is $6 + 12 = 18$; that is, $3(2 + 4) = 3 \cdot 2 + 3 \cdot 4$.

To visualize this property, look at the 3-by-8 array in Figure 4-22 (a). This array shows that $3 \cdot 8 = 24$. Because $8 = 6 + 2$, it follows that

$$3 \cdot 8 = 3(6 + 2) = (3 \cdot 6) + (3 \cdot 2) = 18 + 6 = 24$$

as shown in Figure 4-22 (b).

## CALCULATOR HINT

On your calculator find both $17(21 + 5)$ and $17 \cdot 21 + 17 \cdot 5$.

$17 \boxed{\times} \boxed{(} 21 \boxed{+} 5 \boxed{)} \boxed{=} 442$

$17 \boxed{\times} 21 \boxed{+} 17 \boxed{\times} 5 \boxed{=} 442$.    Therefore, $17(21 + 5) = 17 \cdot 21 + 17 \cdot 5$.

These examples illustrate an important property of whole numbers called the *distributive property of multiplication over addition.*

| Distributive Property of Multiplication over Addition | For all whole numbers $a$, $b$, and $c$, $$a(b + c) = ab + ac \quad \text{and} \quad (b + c)a = ba + ca$$ |
| --- | --- |

**EXAMPLE 9**

$$(5 + 4)(8) = 9 \cdot 8 = 72$$
$$(5 + 4)(8) = (5 \cdot 8) + (4 \cdot 8) = 40 + 32 = 72$$

The distributive property is very important in our study of place value later in this chapter.

**EXAMPLE 10**

$$30 + 40 = 3 \cdot 10 + 4 \cdot 10$$
$$= (3 + 4) \cdot 10$$
$$= 7 \cdot 10 = 70$$

The distributive property is useful in the process of multiplying mentally. For instance, $(34)(21)$ could be computed without pencil and paper if considered as $34(20 + 1)$. The distributive property gives

$$(34)(21) = (34)(20) + (34)(1) = 680 + 34 = 714$$

**EXAMPLE 11**

Using the distributive property, mentally compute $63 \cdot 4$.

$$63 \cdot 4 = (60 + 3) \cdot 4$$
$$= 60 \cdot 4 + 3 \cdot 4$$
$$= 240 + 12 = 252$$

The distributive property of multiplication over addition is fundamental to many important manipulations in algebra.

**EXAMPLE 12**

$$3(x + 2y) = 3x + 6y$$
$$(2pq + 3q) = (2p + 3)q$$
$$a[b + c + d] = a[(b + c) + d] = a(b + c) + ad = ab + ac + ad$$

 **DIVISION**

Frequently, you do something and then undo it. You put on your coat, then take it off; open the door, then close it; pull down the window shade, then raise it. Such pairs of opposites are also encountered in mathematics. These opposites are called **inverses**—inverse operations, inverse elements, and inverse functions. Multiplication has an inverse operation: *division.* Just as subtraction "undoes" addition, division "undoes" multiplication.

A basketball squad of 20 boys is divided into 4 teams. How many are on each team? A companion problem might be "How many 5-person teams can you have with a squad of 20 boys?" Both of the answers for $20 \div 4 = \square$ and $20 \div 5 = \square$ are obtained from the fact that $4 \cdot 5 = 20$. In other words, $20 \div 4 = 5$ because $20 = 5 \cdot 4$, and $20 \div 5 = 4$ because $20 = 5 \cdot 4$.

Division of whole numbers is defined in the following manner.

DEFINITION

**Division of Whole Numbers**

If $x$ is a whole number and $y$ is a natural number, then $x$ divided by $y$ is equal to a whole number $z$ if and only if $z \cdot y = x$.

Symbolically, the operation of division may be denoted in any of four ways, where $z$ is the missing factor.

$$x \div y = z, \qquad x/y = z, \qquad \frac{x}{y} = z, \qquad \text{or} \qquad y\overline{)x}^{\,z}$$

In each of these cases, $x$ is called the **dividend;** $y$, the **divisor;** and $z$, the **quotient.**

**EXAMPLE 13**

Use a number line to illustrate $12 \div 3$.

*SOLUTION*

On the number line in Figure 4-23 we are seeking 3 arrow diagrams of equal length whose sum is 12. The diagram shows that $12 \div 3 = 4$.

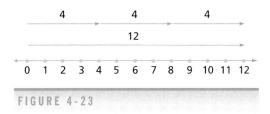

FIGURE 4-23

**EXAMPLE 14**

(a) $12 \div 3 = 4$  because  $4 \cdot 3 = 12$.
(b) $45 \div 9 = 5$  because  $5 \cdot 9 = 45$.
(c) $16 \div 5 = \boxed{?}$  because  $\boxed{?} \cdot 5 = 16$.

There is no whole number that when multiplied by 5 gives 16. Thus, division is not closed on the set of whole numbers. Later in this chapter, we study a concept called a *remainder,* such that

$$16 \div 5 = 3 \qquad \text{with a remainder of 1}$$

### CALCULATOR HINT

Many calculators have a built-in function for finding remainders. For example to find $50 \div 11$ use: 50 $\boxed{\text{INT} \div}$ 11 $\boxed{=}$ to obtain 4 , 6. The 4 is the quotient and the remainder is 6.

# DIVISION MODELS

We understand when to use the operation of division in a problem by recognizing the problem in terms of one or more of the following models:

In the **partition model**, a set is identified as well as the number of equivalent subsets into which we wish to divide it. The question is "how many elements will be in each of the equivalent subsets?" *There are 72 fifth graders at East Broad Elementary. Three buses are available to carry them on a field trip. If each bus is loaded with the same number of students, how many students will be on each bus?*

In the **measurement model**, a set is identified that we wish to divide into equivalent subsets. We also identify the size of each subset. The question is "How many subsets are required." *Twenty jawbreakers are to be placed into bags with five jawbreakers per bag. How many bags are used?* The measurement model can be easily understood using repeated subtraction.

In the **missing factor model**, a total is identified as well as a factor. The question is asked: How many times must the factor be repeated to achieve the total? For example, *Tonya walked three miles in an hour. How long should it take her to walk 12 miles.* (This can be written symbolically as $3 \cdot q = 12$.)

Consider this problem. It can be identified as a division problem using the partition model. Suppose that Joyce has 12 pieces of candy. She wishes to share her candy with her two friends, Bill and June. In Figure 4-24, the set of 12 pieces of candy has been partitioned into 3 equivalent sets; thus each person has 4 pieces of candy, or $12 \div 3 = 4$.

The following example illustrates the measurement or the **repeated subtraction** model of division.

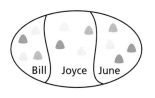

FIGURE 4-24

**EXAMPLE 15**

In a cleanup of the OZT fraternity house, 48 empty cola bottles were collected. These were placed in cases holding 12 bottles each. How many cases were needed to collect the empty bottles?

*SOLUTION*

The freshman pledge of OZT who was assigned the task was not proficient in division, so he computed the number of cases as follows.

| | | | |
|---|---|---|---|
| 48  bottles | 36 | 24 | 12 |
| − 12  bottles in a case | − 12 | − 12 | − 12 |
| 36 | 24 | 12 | 0 |

By counting, he noted that he had used 4 cases for the empty bottles.

Consider this problem. "Sam washed 12 cars in a day. How many days should he expect it to take him to wash 48 cars?" We understand that we are looking for the missing factor D in the equation, $12 \cdot D = 48$. We solve the problem with the division $48 \div 12 = 4$.

## COMMON ERROR

$a \div 0 = a$. This is incorrect. Study the next two paragraphs to understand why.

The number 0 plays a very special role in division. Consider the problem $0 \div 9 = ?$ We need to know what number multiplied by 9 yields 0. The answer is clearly 0. Thus, $0 \div 9 = 0$. We may generalize and say that 0 divided by any whole number other than zero is always 0.

## CALCULATOR HINT

Find $4 \div 0$ on your calculator. $4 \boxed{\div} 0 \boxed{=}$. (The result is Error.) In other words the calculator cannot divide by 0.

$4 \div 0$ as a multiplication problem can be written as $4 = 0(?)$. There is no solution to this problem because 0 multiplied by any number is 0 and not 4. Also, consider $0 \div 0 = ?$ This division can be written as $0 = 0(?)$. Note that any number substituted for the question mark would make this statement true, and thus $0 \div 0$ has no unique answer. We avoid these situations by excluding all division by 0. We say that division by 0 is undefined.

| Division Involving Zero | If $a \neq 0$, then $0 \div a = 0$ because $a \cdot 0 = 0$. If $a \neq 0$, then $a \div 0$ is **undefined** because $a = 0 \cdot d$ is false for any $d$. $0 \div 0$ is undefined because there is no unique quotient. |
|---|---|

Using a combination of subtraction and division, you can solve open sentences of the form $2x + 3 = 9$.

**EXAMPLE 16**

Find the solution set for $2x + 3 = 9$.

*SOLUTION*

By the definition of subtraction, $2x = 9 - 3$, so $2x = 6$. By the definition of division, $x = 6 \div 2$, so $x = 3$.

**PRACTICE PROBLEM**

Find the solution of $3x - 1 = 8$ on the domain of whole numbers.

*ANSWER*

$x = 3$

Sometimes additions and multiplications are contained in the same problem without any notation to indicate the order of performing operations. In such cases we use the following order.

| Order of Operations | 1. First, perform any operations enclosed in parentheses.<br>2. Perform all operations involving exponents.<br>3. Next, perform all multiplications and divisions as they occur from left to right. | 4. Finally, perform additions and subtractions as they occur from left to right. |
|---|---|---|

**EXAMPLE 17**

(a)
$$2 + (3 + 5 \cdot 4) = 2 + (3 + 20)$$
$$= 2 + 23$$
$$= 25$$

(b)
$$5 - 3(4 \cdot 3 - 2 - 3 \cdot 3) = 5 - 3(12 - 2 - 9)$$
$$= 5 - 3(1) = 2$$

## CALCULATOR HINT

Compute the value of $4(7 - 3 \cdot 2) - (2 \cdot 5 - 16 \div 2)$ and then check with a calculator.

$$4 \boxed{\times} \boxed{(} 7 \boxed{-} 3 \boxed{\times} 2 \boxed{)} \boxed{-} \boxed{(} 2 \boxed{\times} 5 \boxed{-} 16 \boxed{\div} 2 \boxed{)} \boxed{=}$$

The answer is 2 in both cases.

**EXAMPLE 18**

A florist is paid \$12 per hour and her helper is paid \$8 per hour. Together they worked 30 hours on a job and made a total of \$320. How many hours did the florist work?

*SOLUTION*

**Unknown:**  Let $h$ = the number of hours that the florist worked on the job.

**Data:**  A total of 30 hours was worked.
A total of \$320 was earned.
The florist earned \$12 per hour; the helper, \$8 per hour.
Total wages = (dollars per hour) (number of hours).

The equation is

$$12h + 8(30 - h) = 320$$
$$12h + 240 - 8h = 320$$
$$4h + 240 = 320$$
$$4h = 80 \qquad \text{Definition of subtraction}$$
$$h = 20 \qquad \text{Definition of division}$$

The florist worked 20 hours. Her helper worked $30 - 20 = 10$ hours.

# Just for Fun

The equation below needs some operation signs inserted to make it a true statement. Can you make the sides balance?

$$6 \quad 6 \quad 6 \quad 6 = 5$$

# EXERCISE SET 2

*R* 1. Classify each of the following statements as either true or false.
   (a) $8 \cdot 0 = 8$      (b) $(16 + 1) + 0 = 17$
   (c) $7 + 0 = 0$      (d) $(ab)(0) = 0$
   (e) $(16 + 1)0 = 17$      (f) $(1 + 1)1 = 1$

2. Determine what property of multiplication is illustrated by each of the following.
   (a) $x \cdot y = y \cdot x$      (b) $1 \cdot x = x$
   (c) $0 \cdot x = 0$
   (d) $x \cdot (y + z) = x \cdot y + x \cdot z$
   (e) $x \cdot (y \cdot z) = (x \cdot y) \cdot z$

3. Using the associative and commutative properties of multiplication, complete the following statements, as illustrated in the example.
   *Example:* $16 \cdot 3 = (8 \cdot 2) \cdot 3 = 8 \cdot (2 \cdot 3) = 8 \cdot 6$
   (a) $18 \cdot 5 = 9 \cdot$ \_\_\_\_\_      (b) $24 \cdot 9 = 8 \cdot$ \_\_\_\_\_
   (c) $8 \cdot 36 =$ \_\_\_\_\_ $\cdot 9$      (d) $35 \cdot 14 =$ \_\_\_\_\_ $\cdot 70$

4. Rename the following numbers, using the distributive property of multiplication over addition.
   (a) $2(3 + 4)$      (b) $5a + 3a$
   (c) $(4 \cdot 2) + (4 \cdot 3)$      (d) $ax + a$
   (e) $4xay + 2xz$      (f) $ax + bx + cx$
   (g) $ax + 5x$      (h) $a(a + 3) + 2(a + 3)$
   (i) $2(x + 4) + 3(x + 4)$

5. Name the property illustrated in each case.
   (a) $(6 + 5) \cdot (4 + 8) = (5 + 6) \cdot (8 + 4)$
   (b) $(5 \cdot 4) \cdot 7 = 5 \cdot (4 \cdot 7)$
   (c) $(6 + 5) \cdot (4 + 8) = (4 + 8) \cdot (6 + 5)$
   (d) $(5 \cdot 4) \cdot 7 = (4 \cdot 5) \cdot 7$
   (e) $4 + 3 = (4 + 3) + 0$
   (f) $(7 \cdot 3) \cdot 2 = 7 \cdot (3 \cdot 2)$
   (g) $1 \cdot (4 + 2) = (2 + 4)$
   (h) $(a)(c) + b = (c)(a) + b$
   (i) $c(a + b) = ca + cb$
   (j) $(a + c)d + (a + c)e = (a + c)(d + e)$
   (k) $5 + (2 + 7) = (2 + 7) + 5$
   (l) $6(5 \cdot 4) = (5 \cdot 4) \cdot 6$
   (m) $6(5 \cdot 8) = 6(8 \cdot 5)$

6. Identify the following problems as division problems by recognizing the problem as involving one or more of the models for division. Explain your answer by identifying the subset being partitioned and the number of equivalent subsets (the partition model), the set being "measured" and the size of the equivalent subsets (the measurement model), or the total that is being sought as well as the factor (in the missing factor model).
   (a) Bottles are packed 24 to a case. How many cases are needed to pack 480 bottles?

   (b) A teacher calculates the test average for Juanita who made 92, 88, and 84 on three tests. What is Juanita's average?
   (c) How many dimes are in a dollar?
   (d) Cassie shares a bag of 36 "Mary Jane's" with 3 of her friends. How many pieces of candy will each person get?
   (e) Two hundred eighty-eight students at DeRenne Middle School are to be divided into nine homerooms. How many students will be in each homeroom?
   (f) Jamal has 24 pencils. He stands at the classroom door and gives 2 pencils to each student who comes into the class. How many students get pencils?
   (g) Eggs are sold by the dozen. How many dozen eggs are in a basket with 72 eggs?

*T* 7. The distributive property of multiplication over addition can be generalized as
$$a(b + c + d + \cdots) = ab + ac + ad + \cdots$$
Compute the following in two ways.
   (a) $5(3 + 7 + 5 + 6)$      (b) $11(5 + 8 + 10 + 4)$

8. Compute where possible.
   (a) $72 \div 9$      (b) $14 \div 0$      (c) $0 \div 16$
   (d) $3(5 + 1) \div 2$      (e) $8(3) \div 4$      (f) $(3 \cdot 2) \div 6$

9. Consider the following subtraction. By counting the number of times 8 must be subtracted from 24 to obtain 0, determine the quotient $24 \div 8$.

| $24$ | | $16$ | | $8$ |
|---|---|---|---|---|
| $-8$ | then | $-8$ | then | $-8$ |
| $16$ | | $8$ | | $0$ |

10. Use the technique shown in Exercise 9 to divide the following.
   (a) $36 \div 9$      (b) $126 \div 21$

11. The distributive property of multiplication over subtraction is stated as
$$a(b - c) = ab - ac$$
Illustrate this property by working the following problems in two ways.
   (a) $6(8 - 5)$      (b) $8(5 - 3)$      (c) $17(10 - 4)$

12. Label each equation as either true or false.
   (a) $(48 \div 12) \div 2 = 48 \div (12 \div 2)$
   (b) $(12 + 6) \div 3 = (12 \div 3) + (6 \div 3)$
   (c) $18 \div (3 + 3) = (18 \div 3) + (18 \div 3)$
   (d) $3 + (9 \cdot 4) = 3 \cdot (9 + 4)$

13. Each of the following statements is false, in general. Find an example in each case that illustrates this fact. Then find a special example for each statement in which the statement is true.

    (a) $x - y = y - x$     (b) $(x - y) - z = x - (y - z)$
    (c) $x - 0 = 0 - x = x$    (d) $(x + y) \div z = x + y \div z$

14. Find the solution set on the domain of whole numbers.

    (a) $2x + 2 = 6$      (b) $3a + 6 = 9$
    (c) $5b + 4 = 4$      (d) $4b + 4 = 12$

*In Exercises 15 through 19, write an equation that serves as a model for or represents the given problem. Then find the solution set to get the answer for the problem.*

15. One fall the Flemings spent \$675 outfitting their 2 children for school. If the clothes for the older child cost twice as much as those for the younger child, how much did they spend on each?

16. The average of five numbers is 100. Four of the numbers are 1, 2, 3, and 4. Find the fifth number.

17. The length of a rectangle is 4 cm more than the width. Find the length and the width if the perimeter is 48 cm.

18. The length of a rectangle is 3 times the width. The difference between the length and the width is 20 units. Find the width of the rectangle. Find the length.

19. Two sides of a triangle have the same length, which is 3 times the third side. If the difference in the longer side and shorter side is 10 units, find the shorter side of the triangle. Find the perimeter.

20. Illustrate each of the following properties with an example.

    (a) Distributive property of division over addition for the set of whole numbers
    (b) Distributive property of division over subtraction for the set of whole numbers

21. (a) For what whole number values of $a$ and $b$ does $a \div b = b \div a$?
    (b) What is the value of $24 \div (4 \div 2)$?
    (c) What is the value of $(24 \div 4) \div 2$?
    (d) What can you conclude from (b) and (c) about the associativity of division for whole numbers?

22. Perform the following operations and then check with a calculator.

    (a) $420 \div 3 - 50 \cdot 2$     (b) $210 \cdot 2 \div 3 - 4$
    (c) Find the remainder for $4329 \div 37$.

23. Determine whether the following sets of numbers are closed with respect to multiplication. Does the set contain an identity element for multiplication?

    (a) {1, 2, 3, 4, 5}     (b) {1}
    (c) The even whole numbers
    (d) The union of {0} and the odd whole numbers
    (e) $\{x \mid x$ is a whole number greater than 10$\}$
    (f) $\{x \mid x$ is a whole number less than 2$\}$
    (g) {1, 4, 7, 10, 13, 16, . . .}     (h) {0, 1}

24. Discuss whether the following statements are correct or incorrect.

    (a) Addition is distributive over addition.
    $$a + (b + c) = (a + b) + (a + c)$$
    (b) Addition is distributive over multiplication.
    $$a + (bc) = (a + b) \cdot (a + c)$$
    (c) Multiplication is distributive over multiplication.
    $$a(b \cdot c) = (ab) \cdot (ac)$$

25. Use the distributive property to write each indicated product as a sum. (NOTE: $x \cdot x = x^2$.)

    (a) $4(x + 3)$         (b) $a(a + 7)$
    (c) $x(x + a)$         (d) $4(a + 3b + 2c)$
    (e) $4y(y + x + 2z)$   (f) $x(y + z + 2)$
    (g) $3x(4 - 2y)$      (h) $5ab(2c + 4d)$

26. Use the distributive property to complete the following calculations.

    (a) $3(12) = 3(10 + 2) =$
    (b) $3(242) = 3(200 + 40 + 2) =$
    (c) $3(132) =$      (d) $2(3,243) =$

27. Use your calculator to verify that multiplication is repeated addition with the following problems.

    (a) $8 \cdot 31$      (b) $7 \cdot 63$

C 28. If the product of two numbers is even and their sum is odd, describe the two numbers.

29. The following illustration depicts the use of grid paper or a peg board to produce a model of multiplication for $3 \cdot 5$.

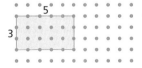

Illustrate in a similar manner multiplication for
    (a) $2 \cdot 3$     (b) $4 \cdot 3$     (c) $2 \cdot 0$

30. Use repeated applications of the distributive property of multiplication over addition to show that $(x + y)^2 = x^2 + 2xy + y^2$. (HINT: $(x + y)^2 = (x + y)(x + y)$.)

31. Use the distributive property to write each indicated product as a sum.

    (a) $(x + 3)(x + 2)$     (b) $(a + 4)(a + 1)$
    (c) $(x + 2y)(x + 3y)$   (d) $(a + 2b)(3a + b)$

32. Use the distributive property to write each sum as a product.

    *Example:* $4x + 12 = 4(x + 3)$

    (a) $8x + 16$        (b) $4ab + b^2 + 3b$
    (c) $6x^2 + 18x$      (d) $ax^2 + bx$
    (e) $3(x + 2) + x(x + 2)$    (f) $a(x + 3) + 6(x + 3)$

33. The accompanying table is similar to an ordinary multiplication table. The binary operation $\odot$ is defined on $S = \{1, 2, 3\}$ by the table.

    (a) Is $S$ closed under the operation $\odot$?
    (b) Is the operation $\odot$ commutative?
    (c) Is the operation $\odot$ associative? (Test several examples.)
    (d) Is there an identity in $S$ for $\odot$? If so, name it.

| $\odot$ | 1 | 2 | 3 |
|---------|---|---|---|
| **1**   | 1 | 2 | 3 |
| **2**   | 2 | 1 | 3 |
| **3**   | 3 | 3 | 3 |

34. Consider the operation given in the accompanying table, and answer the same questions as in Exercise 33. A search for patterns should be helpful.

| * | x | y | z |
|---|---|---|---|
| **x** | x | y | z |
| **y** | y | y | z |
| **z** | z | z | z |

35. Working from left to right, give explicit reasons for each step in verifying that the following statements are true.

    (a) $(2 \cdot 3)4 = (4 \cdot 2)3$
    (b) $(8 + 2)7 = 7(2 + 8)$
    (c) $(ab)c = (ca)b$

## REVIEW EXERCISE

36. What property (or properties) of operations on whole numbers is used in each of the following?

    (a) $(4 + 5) + 7 = 4 + (5 + 7)$
    (b) $(3 + 4) + 5 = 5 + (3 + 4)$

## LABORATORY ACTIVITY

37. Write a word problem that can be answered by the division $12 \div 4$, using the

    (a) missing factor model
    (b) measurement model
    (c) partition model

# SECTION 3 HISTORY OF NUMERATION SYSTEMS

**PROBLEM**

While demolishing an old building last year, a construction crew was surprised to find the numeral

M C M X C I X

on the cornerstone. Why were they surprised?

**OVERVIEW**

In this section, we will examine some ancient numeration systems, including the Roman system. We will also take a close look at our own base ten numeration system and marvel at its simplicity and usefulness. The creation of a numeration system using place value ranks as one of the great accomplishments in human history. It is hard to realize the importance of developing an efficient system of representing numbers.

**GOALS**

Use the Connections Standard: ● recognize and use connections among mathematical ideas; ● understand how mathematical ideas build on one another; ● apply mathematics in contexts outside of mathematics.
Illustrate the Number and Operations Standard, page 119.
Illustrate the Representation Standard, page 50.
Illustrate the Connections Standard, page 15.
Illustrate the Communication Standard, page 78.

One of the earliest systems of numeration was the tally system, which used marks or strokes to represent numbers. For instance, | | | | | | might represent a herd of 6 animals. Even today, tally marks are used to represent numerals like 1, 2, 3, . . . when keeping score in certain games. Thus, 5 and ⅢⅡ are symbols for the same number. As we shall see in this chapter, **numerals**—names for numbers—have changed through the years.

In general, early systems of numeration are characterized as additive systems, multiplicative systems, and place-value systems. Additive systems rely primarily on the addition of numbers represented by symbols; that is, a number represented by a set of symbols is the sum of the numbers represented by the individual symbols.

# EGYPTIAN SYSTEM

The **Egyptian system** is one of the earliest additive systems recorded by history. The Egyptians used hieroglyphics, or picture symbols, along with tally marks to represent numbers. (See Table 4-1).

| TABLE 4-1 | | |
|---|---|---|
| **Egyptian Numerals** | | |
| 1 | | | Stroke |
| 10 | ∩ | Heelbone |
| 100 | ⑨ | Scroll |
| 1000 | ⚸ | Flower |
| 10,000 | ⌠ | Pointed finger |
| 100,000 | ⌇ | Burbot fish |
| 1,000,000 | ⚲ | Astonished person |

**EXAMPLE 1**

(a) 𝟿𝟿∩∩∩||||          = 234
(b) ∩|||||              = 15
(c) ⌈♯♯𝟿𝟿𝟿∩||||||      = 12,316
(d) ♯𝟿𝟿𝟿𝟿𝟿𝟿∩∩∩∩|||| = 1644

Because the Egyptian system is an additive system, symbols could be placed in any order to represent a number.

$$∩∩∩∩||| = 10 + 10 + 10 + 10 + 3 = 43$$

$$∩|∩|∩|∩ = 10 + 1 + 10 + 1 + 10 + 1 + 10 = 43$$

Thus, there was no "place value" in this system (whereas in our system 916 is different from 619).

## Babylonian System

About 3000 B.C., the great **Babylonian** civilization flourished in what is today Iraq. In this period, people wrote numerals on clay tablets with pieces of wood. A vertical wedge ▼ represented 1, and the symbol ❬ represented 10. (See Table 4-2.) Symbols for numbers from 1 to 59 were formed by additive use of these symbols. For example, ❬❬▼▼▼ represented 23. The Babylonian numeration system used the idea of place value, with the places representing different multiples of base 60. The various places are indicated by including a wider space between the characters or by inverting the order of ▼ and ❬ ; that is, if a ▼ appears to the left of a ❬ , it lies in a new place.

### TABLE 4-2

***Babylonian Symbols***

| | |
|---|---|
| ▼ | represents 1 |
| ❬ | represents 10 |

**EXAMPLE 2**

| | | |
|---|---|---|
| ▼  ❬❬▼ | or | $60 + 21 = 81$ |
| ❬▼  ❬▼▼▼ | or | $11(60) + 13 = 673$ |
| ▼▼  ❬❬  ❬❬❬ | or | $2(60)(60) + 20(60) + 30 = 8430$ |
| ▼▼ ❬ | or | $2(60) + 10 = 130$ |

Despite its clever use of place value, the Babylonian numeration system was flawed by its lack of a symbol for zero. This meant that whether ▼ ❬ represented $1(60) + 10$ or $1(60)(60) + 10(60)$ could only be determined by the context in which the numeral was used.

## Roman Numeration System

At the peak of its civilization, around A.D. 100, the Roman Empire needed an elaborate system of numeration for record-keeping and accounting, a situation brought about by the demands of tax collecting and commerce in the vast empire.

The basic Roman numerals and their modern equivalents are shown in Table 4-3.

| TABLE 4-3 | | | | | | |
|---|---|---|---|---|---|---|
| I | V | X | L | C | D | M |
| 1 | 5 | 10 | 50 | 100 | 500 | 1000 |

Essentially, the **Roman system** was an additive system with subtractive and multiplicative features. **If symbols decreased in value from left to right, their values were to be added.** Thus, for example,

$$CX = 100 + 10 = 110 \quad \text{and} \quad MLV = 1000 + 50 + 5 = 1055$$

The position of a letter was important, because XC and CX stood for different numbers, but the Romans did not use place value as we use it today. **If a symbol had a smaller value than the symbol to its right, it was to be subtracted from the symbol to the right.** Hence,

$$XC = 100 - 10 = 90 \quad \text{and} \quad CD = 500 - 100 = 400$$

Ordinarily, not more than 3 identical symbols were used in succession. For example, IV is usually used for 4 instead of IIII. Likewise, no more than 2 symbols are ever involved in the subtractive feature. **The only subtractive symbols allowed are IV, IX, XL, XC, CD, and CM.**

$$LXXVI = L \quad XX \quad VI \quad \text{or} \quad 50 + 20 + 6 = 76$$

$$XLIV = XL \quad IV \quad \text{or} \quad 40 + 4 = 44$$

Writing large numbers in Roman numerals involved using a multiplication feature involving bars above the symbols. For example, $\overline{V}$ indicates 5 multiplied by 1000, or 5000; and $\overline{\overline{V}}$ represents 5,000,000. Thus, a symbol with a bar above it represents the value of the symbol multiplied by 1000; a double bar means multiplication by 1,000,000.

---

**EXAMPLE 3**

Express the following Roman numerals in our numeration system.
(a) $\overline{IV}$DCXLVII     (b) $\overline{\overline{L}}$MDXXI

*SOLUTION*

(a) Consider $\overline{IV}$DCXLVII as $\overline{IV}$ DC XL VII. This is equivalent to

$$4000 + 600 + 40 + 7 = 4647$$

(b) Consider $\overline{\overline{L}}$MDXXI as $\overline{\overline{L}}$ M D XXI. This is equivalent to

$$50,000,000 + 1000 + 500 + 21 = 50,001,521$$

Now we consider the introductory problem to this section.

---

**EXAMPLE 4**

While demolishing an old building last year, a construction crew was surprised to find the numeral

$$M \quad C \quad M \quad X \quad C \quad I \quad X$$

on the cornerstone. Why were they surprised?

*SOLUTION*

M C M X C I X represents 1999. Surely there was a mistake on the cornerstone.

## Mayan System

Another interesting numeration system comes from the early civilizations of northern Central America. Although evidence suggests that the numeration system was developed before the advent of the Mayan Indians, most people still describe this system as the **Mayan numeration system.** At one time it seems to have been based on 5, but in later forms it was based on 20. Instead of using figures to represent a number, the system was based on dots and horizontal lines. Each dot was a unit, and each line represented 5. The largest number recorded in one place with these symbols was 19, three bars and four dots: ≣. Numbers greater than 19 were written in terms of base 20, using a symbol for zero that looked somewhat like a football: ⊗.

In calculating values for the third place in their system, the Mayans used 18(20) instead of the base squared (20²)—perhaps because they thought that the year consisted of approximately 18(20) = 360 days. The Mayans also used 18(20)² instead of (20)³ for the fourth place, and so on; but in spite of this discrepancy, their place-value system of numeration was ingenious. The Mayans wrote their numerals in vertical form, as shown in the next example.

| EXAMPLE 5 |
|---|

(a)  •  
 ⊗  
$$1(20) = 20$$
$$0 = \underline{\phantom{0}0}$$
$$20$$

(b)  ••  
 ••••  
$$2(20) = 40$$
$$4 = \underline{\phantom{0}4}$$
$$44$$

(c)  •  
 ⊗  
 —  
$$1(18)(20) = 360$$
$$0(20) = \phantom{00}0$$
$$5 = \underline{\phantom{00}5}$$
$$365$$

(d)  ≞  
 ≞  
 ⊗  
$$6(18)(20) = 2160$$
$$6(20) = \phantom{0}120$$
$$0 = \underline{\phantom{000}0}$$
$$2280$$

(e)  ••  
 ≞  
 ⊗  
 •••  
$$2(18)(20)^2 = 14{,}400$$
$$6(18)(20) = \phantom{0}2{,}160$$
$$0(20) = \phantom{00,00}0$$
$$8 = \underline{\phantom{00,00}8}$$
$$16{,}568$$

## Hindu–Arabic System

Because the numerals we use today were invented in India by Hindus and brought to Europe by Arabs, our numeration system is called the **Hindu–Arabic system.** It has four important characteristics.

| **Characteristics of the Hindu-Arabic System** | 1. Each symbol represents a cardinal number.<br>2. The position of the symbol in the numeral has a place meaning. | 3. All numbers are constructed from ten basic symbols or digits.<br>4. There is a representation for zero. |
|---|---|---|

Although the Babylonian and Mayan systems are place-value systems, the Hindu–Arabic numeration system is much superior to these. Because the Hindu–Arabic system is based on groups of ten, it is called a **decimal system.** To count in sets of ten, we need ten symbols in our numeration system. These symbols, called **digits,** are written 0, 1, 2, 3, 4, 5, 6, 7, 8, and 9. With these ten digits, we can represent every whole number. (The word *digit* means "a finger or toe," and it is natural to assume that our number system is based on ten digits *because* we have ten fingers.)

As we have seen, the Babylonian system did not make provisions for distinguishing between a base and a base squared. This problem does not exist in the decimal system, because of the existence of the digit 0.

The decimal system uses the idea of **place value** to represent the size of a group in a grouping process. The size of the group represented by a digit depends on the position of the digit in the numeral. The digit also tells how many there are of a particular group. In the numeral 234, the 2 represents two groups of 100 (that is, 200), the 3 represents three groups of 10 (30), and the 4 represents four groups of 1 (4). The idea of place value makes the decimal system very convenient in many ways.

**EXAMPLE 6**

Table 4-4 shows the place value of the digits in 243, 104,679, and 13,580.

**TABLE 4-4**

*Place Value Table*

| | Millions | Hundred thousands | Ten thousands | Thousands | Hundreds | Tens | Ones |
|---|---|---|---|---|---|---|---|
| 243 | | | | | 2 | 4 | 3 |
| 104,679 | | 1 | 0 | 4 | 6 | 7 | 9 |
| 13,580 | | | 1 | 3 | 5 | 8 | 0 |

**EXAMPLE 7**

The digit 4 is found in each of the numerals 654, 456, and 546. Because of its place, however, it is multiplied by different values in each case. In 654, the 4 represents four 1's; in 456, it represents four 100's; and in 546, it represents four 10's.

In a base ten numeral, each successive place to the left represents a group 10 times the size of the preceding group. Beginning at the right and moving toward the left, the first place digit tells how many groups of 1; the second place digit indicates the number of groups of 10; the third place digit

## The Hindu Number System

T H E N & N O W

Today we give little thought to what makes our numeration system so special. Yet, as Eric T. Bell states in *The Development of Mathematics*, the use of 0 as a symbol denoting the absence of units and the use of certain powers of 10 make the system one of the greatest inventions of all time.

$$1\ 2\ 3\ 4$$
$$9\ 6\ 7\ 8\ 9$$
$$5\ 6\ 7\ 8\ 9$$

The Hindu–Arabic system we use today probably evolved over many centuries. The first use of Hindu symbols is found in *Aryabhatiya*, the work of the Indian mathematician Aryabhata (476–550). His phrase "from place to place each is ten times the preceding" suggests that his new system is the ancestor of our modern decimal numeration system. Aryabhata mentions only nine symbols in his book. Thus it seems likely that the introduction of the symbol zero may not have occurred for several years. The first recorded Hindu zero appears in an inscription dated 876 A.D., though many believe that it was in use long before this time.

Much of the work of the early Hindu mathematicians has been lost or destroyed. Yet there is ample evidence that from 400 to 1200, the Hindus developed a mathematical system superior to that of the Greeks, with the possible exception of geometry, although the Hindu mathematicians did some excellent work in this field. Aryabhata, for example, calculated the value of $\pi$ as follows:

Add four to one hundred, multiply by eight, and then add sixty-two thousand; the result is approximately the circumference of a circle of diameter twenty thousand. By this rule the relation of the circumference to diameter is given. In other words,

$$\pi = \frac{\text{circumferences}}{\text{diameter}} \approx \frac{8(100+4)+(62{,}000)}{20{,}000} = 3.1416.$$

In Chapter 12 we will see that this is a very close approximation to $\pi$.

Hindu mathematicians were the first to use negative numbers sometime between 400 and 800. Today negative numbers are an essential part of our real number system, which we will study in the next few chapters.

Why do we call these numerals Hindu–Arabic numerals instead of Hindu numerals? Arabian military conquests and worldwide trade spread them throughout the world. Arabian mathematicians also translated Hindu mathematics for the rest of the world. From one Arabian translator al-Khwarizmi (A.D. 780–850) comes the derivation of the word *algorithm*. A Latin translation of one of his works gives a full account of Hindu numbers. Although al-Khwarizmi made no claim to originality, work with these new numerals became known as that of al-Khwarizmi or algorisimi—and eventually as algorithms.

tells how many groups of 10 times 10; the fourth place digit gives the number of groups of 10 times 10 times 10; and so on. Thus, 2346 is an abbreviation for

$$2(10 \cdot 10 \cdot 10) + 3(10 \cdot 10) + 4(10) + 6(1)$$

or
$$2(1000) + 3(100) + 4(10) + 6$$

which is called the **expanded notation** for 2346.

**PRACTICE PROBLEM**

Write 3,046,050 in words.

*ANSWER*

Three million forty-six thousand fifty.

**EXAMPLE 8**

Write the following numerals in expanded notation.
(a) 7642      (b) 80,002      (c) 36,200

*SOLUTION*

(a) $7642 = 7(1000) + 6(100) + 4(10) + 2$
(b) $80,002 = 8(10,000) + 2$
(c) $36,200 = 3(10,000) + 6(1000) + 2(100)$

Sometimes (b) in Example 8 is written as

$$80,002 = 8(10,000) + 0(1000) + 0(100) + 0(10) + 2(1)$$

and (c) is written as

$$36,200 = 3(10,000) + 6(1000) + 2(100) + 0(10) + 0(1)$$

The numerals 7642, 80,002, and 36,200 are said to be written in base ten notation.

When we work with the metric system, we will discover the value of multiplying and dividing by powers of ten. For example, consider the following multiplication:

$$10^2(400) = 100 \cdot 400 \qquad \text{\footnotesize $10^2$ means $10 \cdot 10$}$$
$$= 40,000$$

More simply, this result can be obtained by annexing the two zeros implied by $10^2$ to 400. In a like manner, we can divide a number by a power of 10. Thus, $2000 \div 10^2$ can be calculated as

$$\frac{2000}{10^2} = \frac{2000}{100} = 20$$

More simply, this result can be obtained by removing two zeros (as again implied by $10^2$) from 2000.

# MODELS FOR REPRESENTING WHOLE NUMBERS

Several excellent models are available for understanding the characteristics of whole numbers. The following are found in many elementary school textbooks.

**MULTIBASE PIECES**   Multibase pieces offer one of the best ways of representing numbers smaller than 10,000. (See Figure 4-25.)

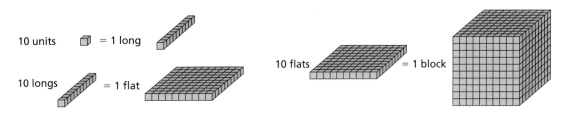

10 units        = 1 long

10 longs        = 1 flat

10 flats        = 1 block

FIGURE 4-25

**EXAMPLE 9**

Represent number 1234 with multibase pieces.

*SOLUTION*

1234 is represented by the pieces shown in Figure 4-26.

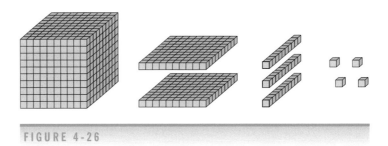

FIGURE 4-26

**BASE TEN ABACUS** In a base ten abacus where $10^5 = 100,000$, $10^4 = 10,000$, $10^3 = 1000$, and $10^2 = 100$, we represent the number 1042 by placing one counter on the $10^3$ column, no counters on the $10^2$ column, four counters on the $10^1$ column, and two counters on the units column. (See Figure 4-27.)

**CHIP ABACUS** In a chip abacus, we represent the number 2134 by placing chips on sections of a cardboard mat representing places for $10^3$, $10^2$, $10^1$, and 1. (See Figure 4-28.)

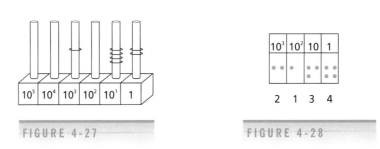

FIGURE 4-27

FIGURE 4-28

# Just for Fun

Consider the symbols that appeared on a recently unearthed Babylonian tablet. Translate these numerals and give your opinion as to what purpose the tablet served.

# EXERCISE SET 3

*R* 1. Write the following numerals in expanded notation.

*Example:* 4321 = 4(1000) + 3(100) + 2(10) + 1

(a) 43,736        (b) 354,555
(c) 14,136        (d) 315,161
(e) 3,111,005     (f) 20,004

2. Write each expanded number in base ten notation.

(a) 3(10,000) + 4(1000) + 7(100) + 8(10) + 6
(b) 4(1000) + 2(10) + 9
(c) 7(100) + 4(10)       (d) 5(10,000) + 6

3. State the place value of the digit 6 in each number.

(a) 6374        (b) 6
(c) 376         (d) 456,987,243

4. In the numeral 34,576, what place value does each of the following represent?

(a) 4      (b) 7      (c) 3      (d) 5

5. In the numeral 12,435, which is greater?

(a) The value represented by 2 or the value represented by 3
(b) The value represented by 1 or the value represented by 4

6. Write the following as Roman numerals.

(a) 76      (b) 101     (c) 189
(d) 44      (e) 148     (f) 948

7. Suppose that you are given the digits 3, 5, and 7.

(a) What is the largest number that can be represented in base ten by these three digits?
(b) What is the smallest number that can be represented in base ten by these three digits?

8. As a basis of comparison for the various systems, complete the accompanying table. If one system does not use a single symbol, use a combination of symbols.

| Hindu–Arabic | 0 | 1 | 5 | 10 | 50 | 100 |
|---|---|---|---|---|---|---|
| Egyptian | | | | | | |
| Roman | | | | | | |
| Babylonian | | | | | | |
| Mayan | | | | | | |

9. Work Exercise 7 for the digits 0, 1, 2, 3, and 4. (0 cannot be used as the first digit in 5-digit numbers.)

10. What numbers are represented by the following Roman symbols?

(a) XXIX            (b) MCMLXXVI
(c) MDCCLXXVI      (d) $\overline{\text{XIXI}}$
(e) $\overline{\overline{\text{X}}}$DCXLIX      (f) $\overline{\overline{\overline{\text{III}}}}$DCXLIV

11. What numbers are represented by these Egyptian symbols?

(a) ||||
(b) ∩∩||
(c) ∩∩∩||||
(d) ⚑∩∩⚑
(e) 𝟡𝟡𝟡∩∩⌒⌒
(f) ⌐⌐⚡⚡⚡

12. Write the Hindu–Arabic symbols that are equivalent to the Babylonian numerals shown.

(a) ▼▼          (b) ◀◀
(c) ◀▼          (d) ◀ ◀ ▼▼
(e) ▼ ◀◀▼ ▼▼    (f) ◀▼ ◀◀ ◀

13. Answer the following questions for a model using multibase pieces.

(a) What is the largest number that can be represented with units only (before exchanging for longs)?
(b) What is the largest number that can be represented with longs only?
(c) What is the largest number that can be represented with longs and units?
(d) What is the largest number that can be represented with flats only?
(e) With longs and flats?
(f) With longs, flats, and units?

14. What numbers are represented by the following Mayan symbols?

(a) •̣      (b) ≐      (c) ••      (d) ≣
(e) ••••    (f) ≡      (g) ••       (h) —
                                •̣
                              •••        ⊛
                                         ⊛

15. Write the following numerals in Egyptian symbols.

(a) 14      (b) 76      (c) 184      (d) 1728

16. Write the numerals in Exercise 15 in Babylonian symbols.

17. Write the numerals in Exercise 15 in Mayan symbols.

18. Display on a calculator the greatest number that has exactly six digits, with no two digits repeated.

*T* 19. Arrange the three numbers in each set from largest to smallest.

(a)

(b) ∩∩∩||||    ▼    ◀◀▼    M X C I I

20. Using the same numeration system, write the numeral for the number that precedes and the number that follows each of the following.

    (a) MLXXXIX

    (b) ◀◀◀◀◀▼▼▼▼▼▼▼▼

    (c) ∩I∩I∩I∩I∩I∩I∩I∩I∩I∩I

21. Find the sum and the difference of each pair of numbers.

    (a) LXIV and XIV

    (b) MCMLXXVI and MDCCLXXVI

    (c) ◀▼ ◀◀▼▼ and ▼▼▼ ◀▼▼

    (d) 𝟗𝟗∩∩III and 𝟗∩∩∩∩II

 22. Kathryn had a dream in which she was selling roses in an international market. She started out with 143 roses. Then her first customer, an Egyptian, bought ∩IIIIII of them. Soon afterward a Babylonian asked to buy ◀◀◀▼▼▼▼▼▼▼▼ roses, and later XIV roses were bought by a Roman. At the end of the day a Mayan asked Kathryn to sell him all the roses that remained. How many roses did he buy? Write the answer in Mayan numerals.

23. You are a farmer in ancient times who produced a crop of 150 bushels of barley, which is 40 more bushels of barley than you produced last year and twice as many bushels as the bushels of corn you produced this year. You hope to increase production of both corn and barley next year by at least 15 bushels each.

    (a) Using Egyptian symbols, record the number of bushels of barley you produced this year and the number you hope to produce next year.

    (b) Using Babylonian symbols, record this year's production of barley and corn.

## REVIEW EXERCISES

24. Find the solution set on the domain of whole numbers.

    (a) $3x + 2 = 8$        (b) $3x + 2 < 8$

    (c) $0 < x - 4$        (d) $2x - 7 = 3$

    (e) $2x + 3 = 7$    or    $x + 3 < 5$

    (f) $2x + 3 = 7$    and    $x + 3 < 5$

25. What property is used in each of the following?

    (a) $3(5 \cdot 10 + 6) = 3(5 \cdot 10) + 3 \cdot 6$

    (b) $xy + yx = xy + xy$

    (c) $b = 1 \cdot b$

26. Show that the following are true. Give the reason for each step.

    (a) $(110)(9) - (9)(6) = 9(110 - 6)$

    (b) $3(a + 0) = (a)(3)$

    (c) $4(23 \cdot 25) = 25(23 \cdot 4)$

    (d) $(108)(6) - 6(108) = 0$

    (e) $c(d \cdot e) = e(d \cdot c)$

    (f) $a + (b + c) = c + (b + a)$

## LABORATORY ACTIVITIES

27. Represent 1253 with multibase pieces.

28. Represent 1253 on a chip abacus.

29. Represent the following numbers with multibase pieces.

    (a) 362        (b) 2495

30. Represent the following numbers on a chip abacus.

    (a) 719        (b) 410023

SECTION
4

# PATTERNS FOR NUMERATION: DECIMAL AND NONDECIMAL BASES

**PROBLEM**

Linda learned that numbers may often be expressed in a form different from base ten numeration. She was surprised when her professor asked her, "Which expressions name the same number 18?"

(a) 3 half dozens        (b) one dozen and six

(c) $33_{\text{five}}$        (d) $2(7) + 4$

**OVERVIEW**

A study of numeration systems with bases other than ten helps provide a more complete understanding of the decimal numeration system. In this section, we introduce different number bases, with particular emphasis on the base seven, base five, and base two numeration systems.

| GOALS | Illustrate the Number and Operations Standard, page 119. |
|---|---|
| | Illustrate the Algebra Standard, page 4. |
| | Illustrate the Representation Standard, page 50. |
| | Illustrate the Problem Solving Standard, page 15. |

# EXPONENTIAL NOTATION

An exponential is a type of abbreviation in mathematics. For instance, $5 \cdot 5 \cdot 5 \cdot 5 \cdot 5 \cdot 5$ is written in exponential form as $5^6$. The symbol $a^2$ means $a \cdot a$; the symbol $a^3$ means $a \cdot a \cdot a$. These ideas lead to the following definition.

**DEFINITION**

**Exponents**

If $n$ is a counting number, then $a^n$ is the product obtained by using $a$ as a factor $n$ times. Symbolically,

$$a^n = a \cdot a \cdot a \cdot a \cdot \ldots \cdot a$$
$$n \text{ factors}$$

and $a^1$ is defined to be $a$.

Each part of the exponential expression $a^n$ has a name. The superscript $n$ is called the **exponent;** the number $a$ is called the **base;** and the complete symbol is read "the $n$th power of $a$" or "$a$ to the $n$th power." A number expressed in the form $a^n$ is said to be written in exponential form.

**EXAMPLE 1**

In the expression $6^3$, 6 is the base and 3 is the exponent; moreover, $6^3$ is read "the third power of 6" or "6 to the third power" or "the cube of 6" or "6 cubed," and it means "$6 \cdot 6 \cdot 6$." Similarly, $5^2 = 5 \cdot 5 = 25$; here, 5 is the base, and $5^2$ is read "the square of 5" or "5 squared."

## CALCULATOR HINT

Although some calculators use $\boxed{y^x}$ for computing powers, we use $\boxed{\wedge}$. Compare $2^5 \cdot 2^3$ with $2^{5+3}$ or $2^8$.

$2 \boxed{\wedge} 5 \boxed{\times} 2 \boxed{\wedge} 3 \boxed{=} 256$, and $2 \boxed{\wedge} 8 \boxed{=} 256$. This calculator example suggests that $a^m \cdot a^n = a^{m+n}$.

Likewise compare $2^5 \cdot 3^5$ and $(2 \cdot 3)^5$.

$2 \boxed{\wedge} 5 \boxed{\times} 3 \boxed{\wedge} 5 \boxed{=} 7776$, and $\boxed{(} 2 \boxed{\times} 3 \boxed{)} \boxed{\wedge} 5 \boxed{=} 7776$, suggesting $a^m b^m = (ab)^m$.

In a like manner compare $(2^4)^3$ and $2^{4 \cdot 3}$. $\boxed{(} 2 \boxed{\wedge} 4 \boxed{)} \boxed{\wedge} 3 \boxed{=} 4096$, and $2 \boxed{\wedge} 12 = 4096$, suggesting $(a^m)^n = a^{mn}$.

We now use two of the strategies of problem solving to find expressions for $a^m \cdot a^n$, $(ab)^m$, and $(a^m)^n$. We shall consider simple examples and then look for a pattern. First write

$$(a) \cdot (a^2) = (a) \cdot (a \cdot a) = a \cdot a \cdot a = a^3 \qquad\qquad = a^{1+2}$$

$$(a) \cdot (a^3) = (a) \cdot (a \cdot a \cdot a) = a \cdot a \cdot a \cdot a = a^4 \qquad\qquad = a^{1+3}$$

$$(a) \cdot (a^4) = (a) \cdot (a \cdot a \cdot a \cdot a) = a \cdot a \cdot a \cdot a \cdot a = a^5 \qquad\qquad = a^{1+4}$$

Thus, it appears that

$$(a) \cdot (a^n) = (a) \cdot \underbrace{(a \cdot a \cdot a \cdot \ldots \cdot a)}_{n \text{ factors}} \qquad\qquad = a^{1+n}$$

Now write

$$(a^2) \cdot (a^4) = (a \cdot a) \cdot (a \cdot a \cdot a \cdot a) = a \cdot a \cdot a \cdot a \cdot a \cdot a = a^6 \qquad = a^{2+4}$$

$$(a^3) \cdot (a^4) = (a \cdot a \cdot a) \cdot (a \cdot a \cdot a \cdot a) = a \cdot a \cdot a \cdot a \cdot a \cdot a \cdot a = a^7 \quad = a^{3+4}$$

Thus, it appears that

$$(a^n) \cdot (a^4) = \underbrace{(a \cdot a \cdot a \cdot \ldots \cdot a)}_{n \text{ factors}} \cdot (a \cdot a \cdot a \cdot a) \qquad\qquad = a^{n+4}$$

More generally, if $m$ and $n$ are natural numbers,

$$a^m \cdot a^n = \underbrace{(a \cdot a \cdot a \cdot \ldots \cdot a)}_{m \text{ factors}} \cdot \underbrace{(a \cdot a \cdot a \cdot \ldots \cdot a)}_{n \text{ factors}} = a^{m+n}$$

It follows that, when multiplying like bases, we add the exponents.
Notice that

$$(2^3)^2 = 2^3 \cdot 2^3 = 2^{3+3} = 2^6 = 2^{3 \cdot 2}$$

Similarly,

$$(a^2)^4 = a^2 \cdot a^2 \cdot a^2 \cdot a^2 = a^{2+2+2+2} = a^8 = a^{2 \cdot 4}$$

In general, $(a^m)^n = a^{m \cdot n}$. It follows that, when raising a power to a power, we multiply the exponents.
Now consider the expression

$$(ab)^2 = (ab) \cdot (ab) = a \cdot a \cdot b \cdot b = a^2 \cdot b^2$$

What properties of multiplication are used? Likewise, consider the expression

$$(ab)^3 = (ab) \cdot (ab) \cdot (ab) = a \cdot a \cdot a \cdot b \cdot b \cdot b = a^3 \cdot b^3$$

What properties of multiplication are used? In general,

$$(ab)^m = \underbrace{ab \cdot ab \cdot ab \cdot \ldots \cdot ab}_{m \text{ factors}} = a^m b^m$$

In later chapters, we shall discuss powers involving exponents other than counting numbers. In the meantime you should know that, for the sake of consistency with these later results, $a^0$ must be defined to be 1. Therefore, **$a^0 = 1$ for all $a \neq 0$.**

**EXAMPLE 2**

(a) $10^0 = 1$

(b) $10^3 \cdot 10^2 = 10^{3+2} = 10^5$

(c) $(10x)^3 = 10^3 \cdot x^3$

(d) $x^4 \cdot x^9 = x^{13}$

The preceding properties of exponents were demonstrated for counting-number exponents. These same properties hold for all exponents used in this book. A summary of these properties follows.

**Exponential Properties**

If $a$ and $b$ are any numbers unequal to 0 and $m$ and $n$ are any exponents, then the following properties hold.

(a) $a^m \cdot a^n = a^{m+n}$

(b) $a^0 = 1$

(c) $a^1 = a$

(d) $(a^m)^n = a^{m \cdot n}$

(e) $(ab)^m = a^m b^m$

## CALCULATOR HINT

Find the value of $2^4(2 \cdot 3)^3$.

$2 \boxed{\wedge} 4 \boxed{\times} \boxed{(} 2 \boxed{\times} 3 \boxed{)} \boxed{\wedge} 3 \boxed{=} 3456$

One way to combine exponential factors is to change the bases so that they are identical.

**EXAMPLE 3**

Simplify $3 \cdot 9 \cdot 27$.

*SOLUTION*

Now $9 = 3^2$ and $27 = 3^3$, so

$$3 \cdot 9 \cdot 27 = 3 \cdot (3^2) \cdot (3^3)$$
$$= 3^{1+2+3} = 3^6$$

**EXAMPLE 4**

Find the value of $n$ to make the following true.

$$3^n \cdot (9) = 81$$

*SOLUTION*

Replace 9 with $3^2$ and 81 with $3^4$.

$$3^n \cdot 3^2 = 3^4$$
$$3^{n+2} = 3^4$$

Therefore, $n + 2 = 4$ or $n = 2$.

## COMMON ERRORS

Have you ever made this mistake?

$$x^n \cdot y^m = (xy)^{n+m} \qquad \text{or} \qquad (xy)^{nm}$$

Both answers are incorrect. You cannot combine the two expressions because both the bases and the powers are different.

## BASE TEN NUMERATION

Because 10 is the base of our numeration system, let us examine successive powers of 10.

$$10^0 = 1$$
$$10^1 = 10$$
$$10^2 = (10)(10) = 100$$
$$10^3 = (10)(10)(10) = 1000$$
$$10^4 = (10)(10)(10)(10) = 10{,}000$$
$$10^5 = (10)(10)(10)(10)(10) = 100{,}000$$

Therefore, 2346 can be written in terms of decreasing powers of 10 as

$$2(10)^3 + 3(10)^2 + 4(10)^1 + 6(10)^0$$

Notice that the digits represent or "count" powers of 10 in a strictly decreasing order:

$$4321 = 4(10)^3 + 3(10)^2 + 2(10)^1 + 1(10)^0$$

$$15 = 1(10)^1 + 5(10)^0$$

$$12{,}000 = 1(10)^4 + 2(10)^3 + 0(10)^2 + 0(10)^1 + 0(10)^0$$

These numbers are said to be written in **expanded form in powers of the base.**

---

**EXAMPLE 5**

$3(10)^4 + 4(10) + 2 \neq 342$, because the terms involving $(10)^3$ and $(10)^2$ are missing, implying that the terms are $0(10)^3$ and $0(10)^2$. Thus,

$$3(10)^4 + 4(10) + 2 = 3(10)^4 + 0(10)^3 + 0(10)^2 + 4(10) + 2$$

$$= 30{,}042$$

## NONDECIMAL BASES

To better understand the base ten system, we now study numeration systems with bases other than base ten. Base ten has ten digits (0, 1, 2, 3, 4, 5, 6, 7, 8, 9). Base seven has only seven (0, 1, 2, 3, 4, 5, 6), and base two has only two (0, 1).

Counting in base seven begins 1, 2, 3, 4, 5, 6, 10, . . . . However, this 10 is not the 10 that we are accustomed to seeing. $10_{\text{seven}}$ represents one group of 7 and no 1's. The next counting number, $11_{\text{seven}}$, represents one 7 and one 1. The counting process continues with 12, 13, 14, 15, 16, and then 20; $20_{\text{seven}}$ represents two 7's and no 1's. If you continue counting in this numeration system, you will soon reach the numeral $66_{\text{seven}}$. What is the next numeral? Be sure you understand why it is $100_{\text{seven}}$.

Base four has only four digits (0, 1, 2, 3). Thus you would count 1, 2, 3, 10, 11, 12, 13, 20, 21, . . . ; $12_{\text{four}}$ is one base and two 1's.

| XXXXXXX | XX |
|---|---|

1(base) + 2 = $12_{seven}$

(a)

| XXXXXXX | |
|---|---|
| XXXXXXX | |
| XXXXXXX | XXXXX |

3(base) + 5 = $35_{seven}$

(b)

FIGURE 4-29

## COMMON ERRORS

(a) One of the most common errors when working with nondecimal bases involves reading "10" as "ten."

(b) Another common mistake is incorrectly reading numerals such as 11, 12, 13, 14, . . . , in bases other than ten. Do not use base ten names such as eleven, twelve, thirteen, and fourteen.

Examine the $x$'s listed in groups of 7 in Figure 4-29. In part (a), the numeral representing the $x$'s could be written $12_{seven}$, representing one group of 7 and two 1's. If the $x$'s corresponded to days of the week, then the $12_{seven}$ would represent 1 week and 2 days. The numeral representing the $x$'s in part (b) could be written as $35_{seven}$, meaning three groups of 7 and five 1's. We write the subscript "seven" to show that the numeral is not the "35" we usually mean in base ten.

Consider now a base seven numeral such as $243_{seven}$ written in expanded notation. The base squared is equivalent to $7^2 = 49$ in base ten.

| ***Place values in base 10*** | $7^2$ | 7 | 1 |
|---|---|---|---|
| ***243**$_{seven}$* | 2 | 4 | 3 |

$$243_{seven} = [2(7)^2 + 4(7) + 3]_{ten}$$

This example provides an easy method for changing $243_{seven}$ to a base ten numeral. Thus,

$$243_{seven} = (98 + 28 + 3)_{ten} = 129_{ten} = 129$$

Multibase pieces illustrate the place value concept of numerals in different bases.

**EXAMPLE 6** Use multibase pieces to represent $2434_{five}$.

*SOLUTION* The required pieces are shown in Figure 4-30.

$2(\text{base})^3 + 4(\text{base})^2 + 3(\text{base}) + 4$

FIGURE 4-30

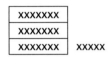

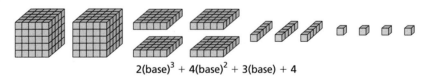

| EXAMPLE 7 | Use multibase pieces to facilitate changing $1143_{five}$ to base ten. |
|---|---|
| *SOLUTION* | The required pieces are shown in Figure 4-31, and followed by the numeric conversion. |

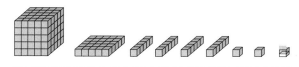

FIGURE 4-31

| Place values in base 10 | $5^3$ | $5^2$ | 5 | 1 |
|---|---|---|---|---|
| *1143*$_{five}$ | 1 | 1 | 4 | 3 |

$$1143_{five} = 1(5)^3 + 1(5)^2 + 4(5) + 3$$
$$= 1(125) + 1(25) + 4(5) + 3(1) = 173$$

# BINARY NUMERALS

Binary numerals are based on groups of two, just as the preceding examples are based on groups of five and seven and just as the decimal system is based on groups of ten. The groups in Figure 4-32 are set up for counting in the binary system. Figure 4-32(a) represents one group of $(2)^2$ elements, one group of 2 elements, and one 1 element, or $111_{two}$. In part (b), there are no groups of two elements, so the symbol becomes $1101_{two}$. The first eleven counting symbols in the base two system are 1, 10, 11, 100, 101, 110, 111, 1000, 1001, 1010, 1011. What is the next symbol?

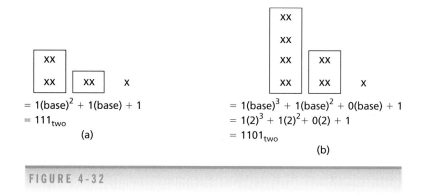

FIGURE 4-32

It is easy to convert a binary numeral to a base ten numeral, as the following examples illustrate.

**EXAMPLE 8**

What decimal numeral is equivalent to $1001_{two}$?

*SOLUTION*

| Place values in base 10 | $2^3$ | $2^2$ | 2 | 1 |
|---|---|---|---|---|
| $1001_{two}$ | 1 | 0 | 0 | 1 |

$$1001_{two} = 1(2)^3 + 0(2)^2 + 0(2) + 1 = 9$$

# CHANGING FROM BASE TEN TO A GIVEN BASE

Suppose that we wish to change 23 in base ten to a base seven numeral. What is the largest power of 7 contained in 23? Is there a $7^2$ contained in 23? No, $7^2 = 49$ is larger than 23. Is there a 7 contained in 23? Yes, $3(7) = 21$ is contained in 23, with a remainder of 2. Hence, $23_{ten} = [3(7) + 2]_{ten} = 32_{seven}$. Do you see that the pattern or rule for changing from base ten to base seven requires division by powers of 7?

Multibase pieces are also helpful in writing a base ten numeral in terms of another base.

**EXAMPLE 9**

Change 532 to a base five numeral.

*SOLUTION*

In base five, a block ⬚ contains 125 units, a flat ⬚ contains 25 units, and a long ⬚ contains 5 units. Therefore,

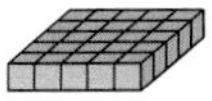

$$\begin{array}{r} 532 \\ \text{Four blocks utilize } 4(125) = 500 \text{ units} \dots - 500 \\ \hline 32 \\ \text{There is one flat in 32} \dots\dots\dots\dots - 25 \\ \hline 7 \\ \text{There is one long in 7} \dots\dots\dots\dots - 5 \\ \hline \text{There are two remaining units} \dots\dots 2 \end{array}$$

It follows that the number in base five is as shown in Figure 4-33.

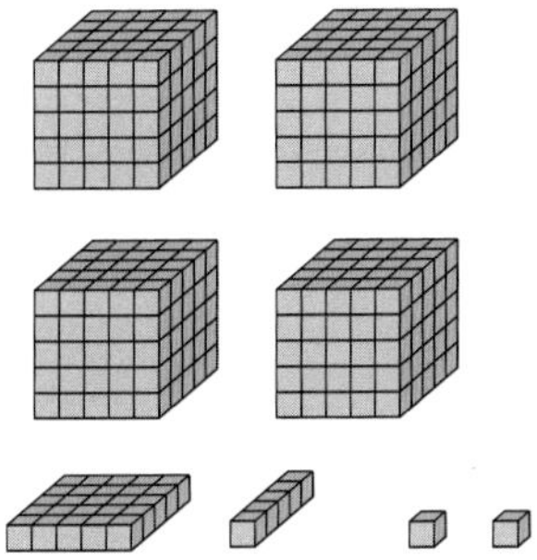

or $4112_{five}$

FIGURE 4-33

**EXAMPLE 10**

Express 59 as a numeral in base two.

*SOLUTION*

Because the numeral is to be expressed in base two, remember that in base ten the powers of 2 are 1, 2, 4, 8, 16, 32, 64, and so on. Because 59 is less than 64, it is first necessary to find the number of 32's in 59.

$$\begin{array}{r} 32 \overline{) 59} \quad \underline{1} \\ 32 \\ \hline 27 \end{array}$$

The remainder is 27. The next step is to determine how many 16's are in 27.

$$
\begin{array}{r|r|l}
32 & 59 & 1 \\
& 32 & \\
16 & 27 & 1 \quad \text{How many 16's are in 27?} \\
& 16 & \\
8 & 11 & 1 \quad \text{How many 8's are in 11?} \\
& 8 & \\
4 & 3 & 0 \quad \text{How many 4's are in 3?} \\
& 0 & \\
2 & 3 & 1 \quad \text{How many 2's are in 3?} \\
& 2 & \\
1 & 1 & 1 \quad \text{How many 1's are in 1?} \\
& 1 & \\
& 0 &
\end{array}
$$

Hence,

$$59_{\text{ten}} = [1(32) + 1(16) + 1(8) + 0(4) + 1(2) + 1(1)]_{\text{ten}}$$

$$= [1 \cdot 2^5 + 1 \cdot 2^4 + 1 \cdot 2^3 + 0 \cdot 2^2 + 1 \cdot 2 + 1]_{\text{ten}}$$

Thus, 59 is expressed in base two as $111011_{\text{two}}$.

**EXAMPLE 11**

(a) Give a visual representation of $234_{\text{five}}$.
(b) Find a decimal numeral equal to $2201_{\text{three}}$.
(c) Change 4695 to a base eight numeral.

*SOLUTION*

(a) The required representation is given in Figure 4-34.
(b) $2201_{\text{three}} = [2 \cdot 3^3 + 2 \cdot 3^2 + 1]_{\text{ten}}$
    $= 54 + 18 + 1$
    $= 73$
(c) We first calculate that $8^2 = 64$, $8^3 = 512$, and $8^4 = 4096$. Therefore,

$$
\begin{array}{r|r|l}
4096 & 4695 & 1 \\
& 4096 & \\
512 & 599 & 1 \\
& 512 & \\
64 & 87 & 1 \\
& 64 & \\
8 & 23 & 2 \\
& 16 & \\
1 & 7 & 7 \\
& 7 & \\
& 0 &
\end{array}
$$

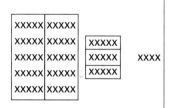

FIGURE 4-34

Hence,

$$4695 = [1 \cdot 8^4 + 1 \cdot 8^3 + 1 \cdot 8^2 + 2 \cdot 8 + 7]_{\text{ten}}$$

$$= 11127_{\text{eight}}$$

**PRACTICE PROBLEM**

Write 231 as a base four numeral.

*ANSWER*

$3213_{\text{four}}$

# ⋮⋮⋮ DUODECIMAL SYSTEM

A system that has received wide attention and has even been suggested as a replacement for the decimal system is the base twelve **duodecimal system.** An argument for having twelve as a base is that twelve has more divisors than ten; the only counting numbers that divide 10 are 1, 2, 5, and 10, whereas the divisors of 12 are 1, 2, 3, 4, 6, and 12. Georges Buffon, a French naturalist, suggested more than 200 years ago that the base twelve system be universally adopted, a suggestion that has been carried into this century. Even today, the Duodecimal Society of America advocates the base twelve system as a replacement for our present system.

| | |
|---|---|
| **EXAMPLE 12** | Rob has 6 gross (a dozen dozen, or 144) of bumper stickers in his room, 3 dozen bumper stickers in his car, and 4 bumper stickers in his briefcase. Rob is to sell all of these for his fraternity. Write the number of bumper stickers to be sold as a base twelve numeral. |
| *SOLUTION* | Six gross is $6(144) = 6(12)^2$ stickers; 3 dozen is $3(12)$. Rob must sell $$6 \cdot 12^2 + 3 \cdot 12 + 4 = 634_{\text{twelve}} \text{ stickers.}$$ |

Figure 4-35 illustrates how we group by dozens. The $x$'s in part (a) are denoted by $13_{\text{twelve}}$ because there is one group of 12 and three 1's. In part (b), there are two groups of 12 and nine 1's, written $29_{\text{twelve}}$.

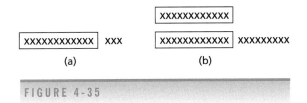

(a)          (b)

FIGURE 4-35

To group and count by twelves, we must introduce two new symbols. T and E are the symbols commonly used for what we usually call ten and eleven. The base twelve numerals are as follows: 1, 2, 3, . . . , 8, 9, T, E, 10, 11, 12, . . . , 19, 1T, 1E, 20, 21, . . . . You recognize that $T_{\text{twelve}} = 10_{\text{ten}}$; $E_{\text{twelve}} = 11_{\text{ten}}$; and $10_{\text{twelve}} = 12_{\text{ten}}$.

| | |
|---|---|
| **EXAMPLE 13** | Change $23E_{\text{twelve}}$ to a base ten numeral. |
| *SOLUTION* | $23E_{\text{twelve}} = [2 \cdot 12^2 + 3 \cdot 12 + 11]_{\text{ten}} = 288 + 36 + 11 = 335$ |

# Just for Fun

Scientists use the binary system in the coded messages they send into space, hoping that some intelligent being might decipher the message. Use the guide below, where each space is labeled either 1 or 0 according to the message below, and darken the spaces marked by 0. What is this message that has been sent into space?

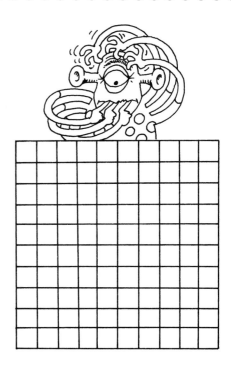

```
0  1  0  1  0  0  1  0  1  0
0  1  0  1  0  1  1  0  1  0
0  0  1  1  0  0  1  1  0  1
0  1  0  1  0  1  1  1  0  1
0  1  0  1  0  0  1  1  0  1
1  1  1  1  1  1  1  1  1  1
0  0  0  1  1  1  1  1  1  1
0  0  0  0  0  0  0  0  0  0
0  0  0  1  1  1  1  0  1  0
1  1  1  1  1  1  1  0  1  0
```

# EXERCISE SET 4

*R* 1. Complete the following table.

| | | |
|---|---|---|
| $10^9$ | 1,000,000,000 | billion |
| _____ | 100,000,000 | hundred million |
| $10^7$ | | ten million |
| $10^6$ | _____ | |
| _____ | 100,000 | _____ |
| $10^4$ | _____ | _____ |
| $10^3$ | _____ | _____ |
| _____ | _____ | hundred |
| _____ | _____ | ten |
| $10^0$ | _____ | _____ |

2. Using exponents, write the following in simpler form.

(a) (2)(2)(2)(2)(2)   (b) (4)(4)(4)(3)(3)
(c) $7 \cdot 7 \cdot 7 \cdot 5 \cdot 5$   (d) $3 \cdot 3 \cdot 3 \cdot 2 \cdot 2 \cdot 2 \cdot 2$
(e) (3)(3)(3)(3)(3)(3)   (f) $x \cdot x \cdot x \cdot y \cdot y$

3. Perform the indicated operations.

(a) $3^4$   (b) $(x^4)(x^7)$   (c) $9^0$   (d) $(2^3)(2^0)$

4. Write the answer in the form $a(10^b)$.

(a) $(3 \cdot 10^4)(4 \cdot 10^3)$   (b) $(7 \cdot 10^4)(8 \cdot 10^6)$

5. Write the following in standard form.

(a) $4 \cdot 10^3 + 5 \cdot 10^2 + 7 \cdot 10^1 + 8 \cdot 10^0$
(b) $9 \cdot 10^4 + 7 \cdot 10^2 + 3 \cdot 10^0$
(c) $8 \cdot 10^3 + 3 \cdot 10^1 + 5 \cdot 10^0$
(d) $3 \cdot 10^4 + 5 \cdot 10^3 + 7 \cdot 10^1$
(e) $4 \cdot 10^2 + 5 \cdot 10^0$
(f) $2 \cdot 10^0$
(g) $3 \cdot 10^2 + 4 \cdot 10^4 + 6 \cdot 10^3 + 7$

6. Write the first 20 counting numbers, using each of the following as a base.

(a) Seven   (b) Two

7. Group 15 dots to find the corresponding numeral in each of the following bases.

   (a) Base ten
   (b) Base seven
   (c) Base two

8. For each of the following, write in the indicated base the preceding counting number and the next consecutive counting number.

   (a) $16_{seven}$     (b) $666_{seven}$
   (c) $111_{two}$     (d) $1011_{two}$

9. Write each of the given numerals as a base ten numeral.

   (a) $601_{seven}$     (b) $660_{seven}$
   (c) $1110_{two}$     (d) $101101_{two}$

10. Change each base ten numeral to a base seven numeral.

    (a) 9     (b) 35     (c) 55
    (d) 285     (e) 1000     (f) 5280

11. Repeat Exercise 10 by changing each numeral to a base two numeral.

12. How many different digits are needed for base five? Base eight? Base sixteen?

13. Find the missing digit for each of the following.

    (a) $1 \underline{\hspace{1em}} 011_{two} = 19_{ten}$     (b) $\underline{\hspace{1em}} 34_{seven} = 221_{ten}$
    (c) $3 \underline{\hspace{1em}} 4_{seven} = 186_{ten}$     (d) $11 \underline{\hspace{1em}} 11_{two} = 27_{ten}$

14. Write the first 50 counting numbers in the duodecimal system.

15. Work Exercise 14 for base five.

16. Write in the indicated base the counting numbers that precede and follow each of the following.

    (a) $EEE_{twelve}$     (b) $607_{eight}$     (c) $222_{three}$

17. Write each of the following as a base ten numeral.

    (a) $157_{nine}$     (b) $E6_{twelve}$     (c) $504_{six}$
    (d) $2010_{four}$     (e) $430_{eight}$     (f) $T0E2_{twelve}$

18. Work Exercise 10 by changing each expression to base twelve.

19. Work Exercise 10 by changing each expression to base five.

T 20. Choose the largest number from each of the following lists.

    (a) $5_{six}$, $11_{four}$, $101_{three}$
    (b) $122_{three}$, $112_{four}$, $76_{eight}$
    (c) $ET_{twelve}$, $101_{eight}$, $11110_{two}$
    (d) $325_{twelve}$, $523_{six}$, $10122_{three}$

21. One way to combine exponential factors is to change the bases so that they are identical. Simplify the following problems by obtaining the smallest common base, such as 2 in (a). Check your answer with a calculator by finding the value before and after simplifying.

    (a) $2^3 \cdot 8^4$     (b) $5^5 \cdot 125$     (c) $16^2 \cdot 2^2$
    (d) $3^4 \cdot 27^2$     (e) $16^6 \cdot 4$     (f) $9^2 \cdot 27^1$

22. Find a value for $n$ to make each of the following a true statement.

    (a) $2^n = 16$     (b) $n^2 = 64$
    (c) $3^4 = 9^n$     (d) $(4^2)^n = 2^8$
    (e) $(a^2)(a^3)(a^n) = a^8$     (f) $(a^6)^n = a^6$
    (g) $n^n = 1$     (h) $n^n = 2n$
    (i) $2^n = n^2$

C 23. Use a calculator in solving these problems.

    (a) Order the following exponential numbers from smallest to largest (increasing order): $10^2$, $7^3$, $4^0$, $2^{10}$, $3^7$, $9^4$, $4^9$.
    (b) Order the following exponential numbers from largest to smallest (decreasing order): $2^5$, $10^2$, $3^4$, $9^1$, $136^0$, $5^4$, $8^2$.
    (c) $3^2 + 4^2 = 5^2$ and $10^2 + 11^2 + 12^2 = 13^2 + 14^2$. Extend this pattern with four terms on the left and three terms on the right.
    (d) By trial and error fill in the blank $(\underline{\hspace{2em}})^4 = 279{,}841$.
    (e) Find the ones digit for $15^9$; $52^8$; $43^6$ without a calculator. Then check with a calculator.

24. Find the base indicated by the letter $b$.

    (a) $67_{ten} = 61_b$     (b) $12_{ten} = 1100_b$
    (c) $234_{ten} = 176_b$

25. A bookstore ordered 8 gross, 7 dozen, and 5 erasers. Express the number of erasers as a base ten numeral.

26. Hubert decided to fill out an application for employment as a freight-car loader in base six. He listed his age as 25, his height as 145 inches, and his weight as 302 pounds. The supervisor, glancing over the application, thought Hubert a giant and decided to hire him. What were Hubert's actual (base ten) statistics?

27. (a) Write the largest 3-digit number found in each of base four, base nine, and base twelve.
    (b) Write the smallest 4-digit number in each of these bases.
    (c) What base ten numbers equal these 6 numbers?

28. Change the following numerals to the base indicated.

    (a) $231_{four}$ to base twelve
    (b) $27TE_{twelve}$ to base six

29. Is it true that $(a^x)(a^y) = (a^y)(a^x)$, where $a$, $x$, and $y$ are counting numbers? Why?

30. Rewrite each number so that each coefficient of a power of 10 is less than 10. (In $4(10)^2$, 4 is the coefficient of $10^2$.) Give reasons for each step.

    (a) $5(10)^2 + 13(10)$
    (b) $7(10)^3 + 25(10)^2 + 5(10) + 17$

(c) $16(10) + 17$    (d) $1(10)^3 + 39(10) + 9$

31. Can you identify even and odd numbers merely by looking at the units digit of a given number when it is expressed in base two? In base three? In base four? In base five?

32. Write the Biblical phrase "three score and ten" as an expression in base twenty, using the symbol T for ten. Do the same for "four score and seven" from the Gettysburg Address. What numbers do these expressions represent in base ten?

33. You have two quarters, four nickels, and two pennies. Use base five to write the number indicating your financial wealth.

34. Mike has 4 gross, 10 dozen, and eleven T-shirts. Write this as a base twelve numeral and find its base ten value.

## REVIEW EXERCISES
••••••••••••••••••••••••••••••••••

35. Express 109 in

    (a) Egyptian numerals.  (b) Babylonian numerals.
    (c) Mayan numerals.     (d) Roman numerals.

36. A time machine took an adventurer back to visit many ancient civilizations. The adventurer found that her work in elementary mathematics helped her understand these civilizations. What answers did she find to the following problems? (Write the answers in base ten.)

    (a) CXLIV + DCLXVI
    (b) ▼▼▼▼▼ ‹ + ‹ ▼▼▼▼
    (c) ⚡ ୨୲୨୨ ∩୲∩∩ − ୨୲୨∩୨୲୨୲୨୨୲

## LABORATORY ACTIVITIES
••••••••••••••••••••••••••••••••••

37. Use multibase pieces to represent the following numbers.

    (a) $101_{two}$      (b) $5342_{seven}$

38. Use multibase pieces to assist in changing the following numbers to base ten.

    (a) $1010_{two}$     (b) $3123_{four}$

39. Use multibase pieces to illustrate changing the following to the given base.

    (a) 14 to base two      (b) 498 to base five

---

### ∴ PCR Excursion ∴

40. Magicians often use tricks based on mathematics to make you think they have unnatural abilities. Consider the following trick problem. Select a number between 1 and 15. Tell on which card or cards—C, A, R, and/or D—the number appears. The magician can tell you the number that you selected. For example, suppose you select a number and tell the magician it is on cards C and D. Immediately, the magician tells you your number is 9. How does the magic work? You will discover the trick as you answer the following questions.

| C | A | R | D |
|---|---|---|---|
| 1 | 2 | 4 | 8 |
| 3 | 3 | 5 | 9 |
| 5 | 6 | 6 | 10 |
| 7 | 7 | 7 | 11 |
| 9 | 10 | 12 | 12 |
| 11 | 11 | 13 | 13 |
| 13 | 14 | 14 | 14 |
| 15 | 15 | 15 | 15 |

A. Write each of the numbers 1 through 15 in binary notation.

B. (a) List separately the binary representations of the numbers on card A and describe verbally a common characteristic of all these representations. (HINT: Are 1's or 0's repeated in a given position of the representations?)
   (b) Repeat the instructions in (a) for card C.
   (c) Card R.      (d) Card D.

C. (a) If a number is on card A only, what is the number?
   (b) What is the number if the number is on card R only?  (c) Card C only?  (d) Card D only?

D. (a) For numbers listed on two cards, use the characteristics observed in B to determine the number selected. For example, suppose you select a number that is only on cards C and A. If it is on card C, it will have a last digit of 1. If it is on card A, the next to the last digit will be 1. So the number is $1 + 1(2) = 3$. What is the number if it is only on C and R?
   (b) Determine the number if it is only on cards C and D.
   (c) A and R.      (d) A and D.      (e) R and D.

E. (a) Suppose the number is on three cards—say, C, A, and D. What is the number?
   (b) Determine the number if it is only on C, R, and D.  (c) On A, R, and D.  (d) On C, A, and R.

F. What would the number be if it were on all four cards?

G. If you were the magician, explain verbally how you would do this trick.

# ADDITION AND SUBTRACTION ALGORITHMS

**PROBLEM**

Use problem-solving techniques to place each of the digits 0 to 9 in one of the blocks to form an addition problem. A guess has been made about placement of two of the digits to start you on your way and to reduce the number of different possible answers.

**OVERVIEW**

The preceding problem makes use of what we call an *algorithm for addition*. Whenever we perform arithmetic operations with single-digit numerals, we write the answer from memory. But to handle operations involving larger numbers, we need to have a pattern or procedure to follow. Such a pattern or procedure is called an **algorithm**. We study several patterns of computation (or algorithms) in the next two sections. The algorithms we use today are refined versions of ones that have been passed down from generation to generation. Many good patterns for performing the various operations exist.

**GOALS**

Illustrate the Algebra Standard, page 4.
Illustrate the Number and Operations Standard, page 119.
Illustrate the Problem Solving Standard, page 15.
Illustrate the Reasoning and Proof Standard, page 23.
Illustrate the Communication Standard, page 78.

## ALGORITHMS FOR ADDITION AND SUBTRACTION

We shall use expanded notation, exponents in place-value charts, and the traditional algorithms to illustrate addition and subtraction.

**EXAMPLE 1**

Perform the addition $421 + 176$.

*SOLUTION*

**Expanded form**
$$400 + 20 + 1$$
$$+\ 100 + 70 + 6$$
$$500 + 90 + 7 = 597$$

**Place-value chart**

| | $10^2$ | $10$ | $1$ |
|---|---|---|---|
| | 4 | 2 | 1 |
| Add | 1 | 7 | 6 |
| Answer | 5 | 9 | 7 |

**Traditional algorithm**
$$421$$
$$+\ 176$$
$$597$$

**EXAMPLE 2**

Show the addition of 48 and 73, using expanded form, a place-value chart, and the traditional algorithm.

*SOLUTION*

**Expanded form**

$$
\begin{array}{r}
40 + 8 \quad 48 \\
\underline{70 + 3} \quad \underline{+\ 73} \\
110 + 11 \quad 11 \\
\underline{\qquad 110} \\
121
\end{array}
$$

**Place-value chart**

| 10² | 10 | 1 |
|---|---|---|
|  | 4 | 8 |
|  | 7 | 3 |
|  | 11 | 11 |
|  | 1 | 1 |
|  | 12 | 1 |
| 1 | 2 |  |
| 1 | 2 | 1 |

Add

**11 = 1(10) + 1**
Add

**12 = 1(10) + 2**

Answer

**Traditional algorithm**

$$
\begin{array}{r}
{}^{1}\phantom{0} \\
48 \\
\underline{+\ 73} \\
121
\end{array}
$$

## MENTAL ADDITION

If you visualize the numerals in expanded form, you can perform additions by using mental mathematics. For example,

| *Traditional algorithm* | *Mental mathematics* | |
|---|---|---|
| $\begin{array}{r} {}^{1}\phantom{0} \\ 56 \\ +\ 35 \\ \hline 91 \end{array}$ | $50 + 30 = 80$ | Add the tens |
| | $6 + 5 = \underline{11}$ | Add the units |
| | $91$ | Add the sum |

You can also use the mental arithmetic to break up numerals on the basis of your knowledge of their expanded form. Here, first add the tens term (30) in one number to the entire other number, and then add the units term (5) to that:

$$
\begin{array}{ccccc}
56 & & 56 & & 56 \\
\underline{+35} & \text{or} & \underline{+30} & \text{or} & \underline{+30} \\
& & +5 & & 86 + 5 = 91
\end{array}
$$

In **trading off,** you use parts of one number to turn another number into a multiple of 10. Of course, if you add to one number, you must compensate by subtracting from another. For example,

$$
\begin{array}{ll}
48 & 48 + 2 = 50 \quad \text{Add 2 to make a multiple of 10.} \\
\underline{+36} & 36 - 2 = \underline{34} \quad \text{Subtract 2 to compensate.} \\
& \qquad\quad 84 \quad \text{Add the two numbers.}
\end{array}
$$

In using **compatible numbers,** you select part of one number to add to another to make the answer a multiple of 10 or 100 or the like. For example,

$$35$$
$$+87$$

$35 + 65 = 100$    Add 65 to make 100.

$100 + 22 = 122$    Add the $87 - 65 = 22$ that remains.

## COMPUTATIONAL ESTIMATION

When you use a calculator to obtain a sum, a good way to estimate the answer is to add rounded numbers.

Several rounding techniques are built into different computers and hand calculators. We briefly discuss two of them. The first simply **truncates** or chops off all digits beyond the number to be displayed by the machine.

By contrast, many machines round off to the nearest digit. This procedure is more precisely described by the following rules.

| Steps in Rounding Numbers to the Nearest Digit | 1. Drop all digits to the right of the place to which you are rounding. <br> 2. If the first (leftmost) digit to be dropped is 5, 6, 7, 8, or 9, | increase the preceding digit by one. <br> 3. If the first (leftmost) digit to be dropped is 0, 1, 2, 3, or 4, leave the preceding digit as is. |
|---|---|---|

**EXAMPLE 3**

*SOLUTION*

Add the following set of numbers with a calculator, and then check the addition by rounding to the nearest hundred.

| Calculator computation | Rounded to nearest hundred |
|---|---|
| 246 | 200 |
| 142 | 100 |
| 374 | 400 |
| 562 | 600 |
| 1324 | 1300 |

The estimate of the answer is 1300. The actual answer therefore seems reasonable.

## REGROUPING FOR SUBTRACTION

To perform some subtractions, we may have to rename or regroup in the opposite direction.

**EXAMPLE 4**

Perform the subtraction $843 - 267$, using expanded form, a place-value chart, and the traditional algorithm.

*SOLUTION*

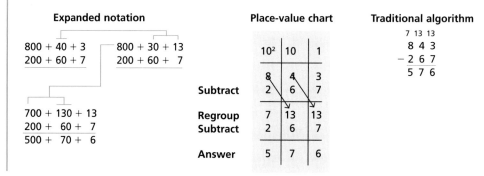

The chip abacus (Figure 4-36) can be used to illustrate the preceding example of subtraction.

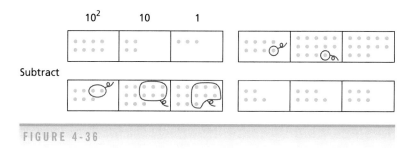

FIGURE 4-36

# MENTAL SUBTRACTIONS AND ESTIMATIONS

Many subtractions can be handled mentally by subtracting the appropriate multiple of 100, then subtracting the multiple of 10, and finally subtracting the number of units.

**EXAMPLE 5**

Find $86 - 42$ mentally.

*SOLUTION*

$$\begin{array}{r} 86 \\ -42 \end{array}$$   $86 - 40 = 46$   Subtract the tens
$46 - 2 = 44$   Subtract the units from the first line

Notice that both 40 and 2 have been subtracted.

**EXAMPLE 6**

Estimate an answer for $684 - 421$ by rounding to the nearest hundred.

*SOLUTION*

$$\begin{array}{r} 684 \\ -421 \end{array}$$    rounds to    $$\begin{array}{r} 700 \\ -400 \\ \hline 300 \end{array}$$

Our estimate of the answer is 300. The actual answer, 263, thus seems reasonable.

# OTHER ALGORITHMS

The algorithms we have presented so far are commonly used and fairly efficient. However, there are many other excellent algorithms; a brief discussion of four of these follows. A technique called **subtraction by equal additions** achieved great popularity in European schools during the 15th and 16th centuries. This algorithm is based on the fact that the difference between two numbers is unchanged if the same amount is added to both numbers. In the subtraction

$$\begin{array}{r} 25 \\ -\phantom{0}8 \\ \hline \end{array}$$

you can add 10 in the units digit of 25 and 1 in the tens digit of 8 to obtain

$$\begin{array}{r} \overset{1}{\phantom{0}} \\ 25 \\ -18 \\ \hline 17 \end{array}$$

| EXAMPLE 7 | Use subtraction by equal additions to calculate $\begin{array}{r} 327 \\ -148 \\ \hline \end{array}$ |

SOLUTION

| Step 1 | Step 2 | Step 3 |
|--------|--------|--------|
| $\overset{1}{\phantom{0}}$ | $\overset{1\,1}{\phantom{00}}$ | $\overset{1\,1}{\phantom{00}}$ |
| $327$ | $327$ | $327$ |
| $-\,158$ | $-\,258$ | $-\,258$ |
| $\phantom{00}9$ | $\phantom{0}79$ | $179$ |

The **scratch method of addition** is very useful for adding several numbers.

| EXAMPLE 8 | Add $67 + 56 + 48$, using the scratch method of addition. |

SOLUTION

$$\begin{array}{r} \overset{2}{\phantom{0}} \\ 67 \\ {}_3\,\,\cancel{56}\,\,{}_3 \\ 48\,\,{}_1 \\ \hline 171 \end{array}$$

Start with $7 + 6 = 13$. Scratch out the 6 and replace it with a 3. Then $3 + 8 = 11$. Scratch out the 8 and replace it with a 1. The number of scratches in the first column is 2, so add this number to the tens column. There is one scratch in the second column, so add 1 to the third column.

The **scratch method of subtraction** begins on the left.

> **EXAMPLE 9**

*SOLUTION*

Use the scratch method of subtraction to calculate $73 - 47$.

$$
\begin{array}{r}
\overset{13}{7\cancel{3}} \\
-47 \\
\hline
\cancel{3} \\
26
\end{array}
$$

First subtract 4 from 7, getting 3. Then scratch through the 3, changing it to 2, and use the 1 to change the 3 in 73 to 13; then subtract 7 from 13.

The **Austrian method of subtraction** consists of the steps shown in Example 10.

> **EXAMPLE 10**

*SOLUTION*

Calculate $764 - 348$.

$$(700 + 60 + 4) - (300 + 40 + 8) =$$
$$(700 + 60 + 10 + 4) - (300 + 40 + 10 + 8) = \quad \text{Add 10 in each set of parentheses}$$
$$(700 + 60 + 14) - (300 + 50 + 8) = \quad \text{Group so 8 will subtract from 14}$$
$$(700 - 300) + (60 - 50) + (14 - 8) = \quad \text{Group by 100's, 10's, and 1's}$$
$$400 + 10 + 6 = 416$$

**Subtraction by taking complements** consists of adding the same sum to both numbers. This sum is selected so that the number left to subtract is a multiple of 10.

> **EXAMPLE 11**

Calculate $1734 - 468$, using the method of subtraction by taking complements.

$$
\begin{array}{r}
1734 \\
-468 \\
\hline
\end{array}
\qquad
\begin{array}{r}
1734 \\
+532 \\
-532 \\
-468 \\
\hline
\end{array}
\qquad
\begin{array}{r}
2266 \\
-1000 \\
\hline
1266
\end{array}
\qquad \text{\small 532 + 468 = 1000}
$$

# ⠐⠆ Addition, Nondecimal Bases

In base seven, $2 + 1 = 3$, $2 + 2 = 4$, $2 + 4 = 6$ and $3 + 3 = 6$ because all sums are less than the base. However, $6 + 1 = 10_{\text{seven}}$, $6 + 2 = 11_{\text{seven}}$ (one base and one 1), and $6 + 3 = 12_{\text{seven}}$ (one base and two 1's).

> **EXAMPLE 12**

Use multibase pieces to show the addition

$$5_{\text{seven}} + 6_{\text{seven}}$$

SOLUTION    The required representation is given in Figure 4-37.

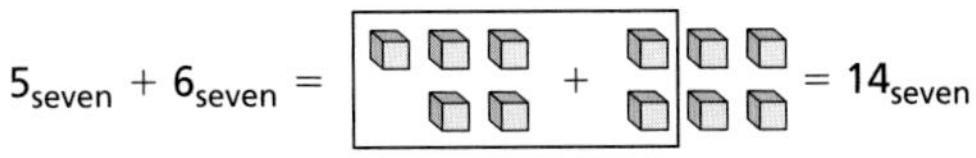

$$5_{seven} + 6_{seven} = \boxed{\text{⬡⬡⬡ ⬡⬡ + ⬡⬡⬡ ⬡⬡⬡}} = 14_{seven}$$

FIGURE 4-37

To perform operations in any base, you need to have appropriate addition tables. So that you won't have to memorize tables in base two and base seven, appropriate tables are provided in this section. Verify the additions in Tables 4-5 and 4-6.

| TABLE 4-5 | | |
|---|---|---|
| **Addition Table, Base Two** | | |
| + | 0 | 1 |
| 0 | 0 | 1 |
| 1 | 1 | 10 |

| TABLE 4-6 | | | | | | |
|---|---|---|---|---|---|---|
| **Addition Table, Base Seven** | | | | | | |
| + | 0 | 1 | 2 | 3 | 4 | 5 | 6 |
| 0 | 0 | 1 | 2 | 3 | 4 | 5 | 6 |
| 1 | 1 | 2 | 3 | 4 | 5 | 6 | 10 |
| 2 | 2 | 3 | 4 | 5 | 6 | 10 | 11 |
| 3 | 3 | 4 | 5 | 6 | 10 | 11 | 12 |
| 4 | 4 | 5 | 6 | 10 | 11 | 12 | 13 |
| 5 | 5 | 6 | 10 | 11 | 12 | 13 | 14 |
| 6 | 6 | 10 | 11 | 12 | 13 | 14 | 15 |

EXAMPLE 13    Find $5_{seven} + 6_{seven}$.

SOLUTION    Looking at the row and the column designated by arrows in Table 4-6 we find $5_{seven} + 6_{seven} = 14_{seven}$.

It is easy to see that the algorithm for the addition of numbers in base ten is applicable to the addition of numbers expressed in other bases.

EXAMPLE 14    Compute $26_{seven} + 34_{seven}$.

SOLUTION    In the following computation, use Table 4-6 to confirm that $6 + 4 = 13$ in base seven.

$$\begin{array}{r} {}^{1}\phantom{0} \\ 26_{seven} \\ + \ 34_{seven} \\ \hline 63_{seven} \end{array}$$

EXAMPLE 15    Use Table 4-5 to find $111_{two} + 101_{two}$.

SOLUTION

$$\left[ \begin{array}{r} {}^{1\,1}\phantom{0} \\ 111 \\ + \ 101 \\ \hline 1100 \end{array} \right]_{two}$$

Some people are more successful in working this type of problem when they "think" base ten (that is, when they perform all operations in base ten) and write the answers in terms of the given base. With this approach, you do not need an addition table.

**EXAMPLE 16**

Illustrate the addition $224_{\text{five}} + 344_{\text{five}}$, using a place-value chart and then the traditional algorithm.

*SOLUTION*

**Place-value charts**

Traditional algorithm

| $5^3$ | $5^2$ | 5 | 1 |
|---|---|---|---|
|  | 2 | 2 | 4 |
| + | 3 | 4 | 4 |
|  |  |  | 13 |

$4 + 4 = 8$
$8 = 1(5) + 3$

or

| $5^3$ | $5^2$ | 5 | 1 |
|---|---|---|---|
|  |  | 1 |  |
|  | 2 | 2 | 4 |
| + | 3 | 4 | 4 |
|  |  | 12 | 3 |

$1 + 2 + 4 = 7$
$7 = 1(5) + 2$

or

| $5^3$ | $5^2$ | 5 | 1 |
|---|---|---|---|
|  | 1 | 1 |  |
|  | 2 | 2 | 4 |
| + | 3 | 4 | 4 |
| 1 | 1 | 2 | 3 |

$1 + 2 + 3 = 6$
$6 = 1(5) + 1$

$$\begin{array}{r} {\scriptstyle 1\,1} \\ 224 \\ +\ 344 \\ \hline 1123 \end{array}_{\text{five}}$$

## ⠿ SUBTRACTION, NONDECIMAL BASES

**EXAMPLE 17**

What is $11_{\text{seven}} - 2_{\text{seven}}$?

*SOLUTION*

Looking in Table 4-6, we find that the answer is $6_{\text{seven}}$. (HINT: What number must be added to $2_{\text{seven}}$ to obtain $11_{\text{seven}}$?)

**EXAMPLE 18°**

Compute $624_{\text{seven}} - 246_{\text{seven}}$.

*SOLUTION*

$$\begin{array}{r} 624_{\text{seven}} \\ -\ 246_{\text{seven}} \end{array} \quad \text{or} \quad \begin{array}{r} {\scriptstyle 14} \\ 6\cancel{1}4_{\text{seven}} \\ -\ 246_{\text{seven}} \\ \hline 5_{\text{seven}} \end{array} \quad \begin{array}{r} {\scriptstyle 11\,14} \\ 5\cancel{1}\cancel{4}_{\text{seven}} \\ -\ 246_{\text{seven}} \\ \hline 345_{\text{seven}} \end{array}$$

We now work the same problem by performing the operations in base ten.

**Place-value charts**

| $7^2$ | 7 | 1 |
|---|---|---|
| 6 | 2 | 4 |
| − 2 | 4 | 6 |

or

| $7^2$ | 7 | 1 |
|---|---|---|
|  |  | 11 |
| 6 | 1 | $\cancel{4}$ |
| − 2 | 4 | 6 |
|  |  | 5 |

or

| $7^2$ | 7 | 1 |
|---|---|---|
|  | 8 | 11 |
| 5 | $\cancel{1}$ | $\cancel{4}$ |
| − 2 | 4 | 6 |
| 3 | 4 | 5 |

$7 + 4 = 11$
$11 - 6 = 5_{\text{ten}}$

$7 + 1 = 8$
$8 - 4 = 4_{\text{ten}}$

**Traditional algorithm**

$$
\begin{array}{r}
624 \\
-\ 246 \\
\end{array}
\quad \text{or} \quad
\begin{array}{r}
\overset{1\ 11}{6\cancel{2}4} \\
-\ 246 \\
\hline
5 \\
\end{array}
\quad \text{or} \quad
\begin{array}{r}
\overset{5\ 8\ 11}{6\cancel{2}\cancel{4}} \\
-\ 246 \\
\hline
345 \\
\end{array}
$$

Operations performed in base ten

The answer is $345_{\text{seven}}$.

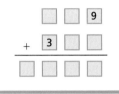

FIGURE 4-38

We now return to the problem at the beginning of this section and use the strategy of guess, test, and revise to obtain a solution. Place each of the digits 0 to 9 in one of the blocks to form an addition problem. The 3 and 9 have already been placed in position.

Let us first consider the leftmost digit of the answer. Because this block will be filled as the consequence of a regrouping process, it must be 1. (The biggest number we can ever regroup in 2-number addition results from $9 + 9 = 18$). To yield the 1 we just placed, the number to be added to 3 must be 6, 7, or 8. Let's start with 8. Because 1 has been used, the second blank in the answer is 2 $(8 + 3 + 1 = 12)$. Therefore, the sum of the two numbers in the second column must be at least 10. Trials of 5 and 6 and 6 and 7 lead to complications. Then we try 4 and 5; $(4 + 5) + 1 = 10$. This leaves 7 and 6 for the last two blocks. One answer is

$$
\begin{array}{r}
859 \\
+\ 347 \\
\hline
1206 \\
\end{array}
$$

Can you find another answer?

# Just for Fun

Jack discovered, in climbing his giant beanstalk, that the giant had a numeration system based on "fee, fie, foe, fum." When the giant counted his golden eggs, Jack heard him count "fee, fie, foe, fum, fot, feefot, fiefot, foefot, fumfot, fotfot, feefotfot, . . . ." Jack believes that the giant has 20 eggs. Can you guess the names of the other 9 numbers that the giant used to finish the counting?

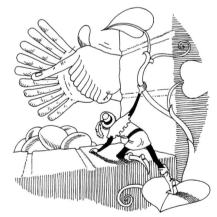

# EXERCISE SET 5

*R* 1. Find the missing digits in each addition or subtraction.

(a)
$$\begin{array}{r} \text{---} \\ + \ 967 \\ \hline 1441 \end{array}$$

(b)
$$\begin{array}{r} 743 \\ + \ \text{---} \\ \hline 1321 \end{array}$$

2. Complete the blanks in the following calculations.

(a)
$$\begin{array}{r} 768 \\ + \ 574 \end{array} \quad \text{or} \quad \begin{array}{l} 700 + \_\_ + 8 \\ \underline{500 + 70 + 4} \\ \_\_ + \_\_ + \_\_ = \_\_ \end{array}$$

(b)
$$\begin{array}{r} 865 \\ - \ 378 \end{array} \quad \text{or} \quad \begin{array}{l} 800 + \_\_ + \_\_ \\ \underline{- \ (300 + 70 + 8)} \end{array} \quad \text{or}$$

$$\begin{array}{l} 800 + \_\_ + \_\_ \\ \underline{- \ (300 + 70 + 8)} \end{array} \quad \text{or} \quad \begin{array}{l} \_\_ + \_\_ + 15 \\ \underline{- \ (300 + 70 + 8)} \\ \_\_ + \_\_ + \_\_ = \_\_ \end{array}$$

3. Find an estimate of the sum

$$\begin{array}{r} 4862 \\ 6375 \\ 8601 \\ 9493 \\ + \ 2140 \end{array}$$

by rounding to thousands. Then add the numbers and round the answer to thousands. Did you expect the answers to be the same?

4. Perform the following additions mentally.

(a)
$$\begin{array}{r} 146 \\ + \ 234 \end{array}$$

(b)
$$\begin{array}{r} 289 \\ + \ 147 \end{array}$$

5. Perform the following subtractions mentally

(a)
$$\begin{array}{r} 46 \\ - \ 23 \end{array}$$

(b)
$$\begin{array}{r} 844 \\ - \ 568 \end{array}$$

6. Place digits 1, 2, 3, 4, 5, 6, 7, and 8 in the boxes

$$\begin{array}{r} \square\ \square\ \square\ \square \\ + \ \square\ \square\ \square\ \square \end{array}$$

to obtain

(a) the greatest sum.  (b) the least sum.

7. In Exercise 6, place the digits to obtain

(a) the greatest difference.
(b) the least whole number difference.

8. Perform the following computations using expanded notation vertical format, a place-value chart, and the traditional algorithm.

(a) $46 + 75$  (b) $48 + 74$
(c) $60 - 37$  (d) $136 - 29$

9. Perform the following subtractions.

(a) $1100_{two} - 11_{two}$  (b) $100101_{two} - 10011_{two}$
(c) $1111_{seven} - 555_{seven}$  (d) $404_{seven} - 65_{seven}$

10. Perform the following additions.

(a) $304_{seven} + 366_{seven}$  (b) $1111_{two} + 101_{two}$
(c) $1011_{two} + 1001_{two}$  (d) $562_{seven} + 453_{seven}$

11. Our time system is a base 60 system. Add and subtract the following pairs of times.

(a) 6 hours 40 minutes 31 seconds
    3 hours 51 minutes 37 seconds
(b) 8 hours
    6 hours 10 minutes 15 seconds

12. Make an addition table for the following bases.

(a) Three  (b) Twelve  (c) Five

*T* 13. Perform the following operations.

(a) $35_{twelve} + 15_{twelve}$  (b) $2304_{five} + 121_{five}$
(c) $12_{twelve} - 9_{twelve}$  (d) $121_{three} + 22_{three}$
(e) $8T2_{twelve} + 26E_{twelve}$  (f) $321_{five} - 143_{five}$
(g) $503_{twelve} - 2TE_{twelve}$  (h) $714_{twelve} - ET_{twelve}$
(i) $320_{five} - 43_{five}$  (j) $3E8_{twelve} + 3TT_{twelve}$

14. Show the following computations, using expanded notation, and then draw lines to show as clearly as possible the connection between this algorithm and our traditional algorithm.

*Example:*

$$\begin{array}{r} 26 \\ + \ 46 \end{array} \qquad \begin{array}{r} 20 + 6 \\ + \ 40 + 6 \\ \hline 60 + 12 \end{array} \longrightarrow \begin{array}{r|r} 26 & \overset{1}{26} \\ + \ 46 & + \ 46 \\ \hline 12 & 72 \\ 60 & \\ \hline 72 & \end{array}$$

(a)
$$\begin{array}{r} 34 \\ + \ 27 \end{array}$$

(b)
$$\begin{array}{r} 63 \\ - \ 36 \end{array}$$

(c)
$$\begin{array}{r} 167 \\ + \ 245 \end{array}$$

(d)
$$\begin{array}{r} 378 \\ - \ 196 \end{array}$$

15. Use Austrian subtraction to subtract

(a)
$$\begin{array}{r} 5706 \\ - \ 3407 \end{array}$$

(b)
$$\begin{array}{r} 329 \\ - \ 146 \end{array}$$

(c)
$$\begin{array}{r} 1634 \\ - \ 985 \end{array}$$

16. Compute the answers in Exercise 15 by using complements.

17. Work Exercise 15 by using the scratch method.

18. Perform the following additions and subtractions in three ways. (i) Write each number in expanded notation. (ii) Write each number in a place-value chart. (iii) Use the traditional algorithm.

(a)
$$\begin{array}{r} 364 \\ + \ 423 \end{array}$$

(b)
$$\begin{array}{r} 476 \\ - \ 324 \end{array}$$

(c)
$$\begin{array}{r} 426 \\ - \ 14 \end{array}$$

(d)
$$\begin{array}{r} 1758 \\ + \ 241 \end{array}$$

(e)
$$\begin{array}{r} 1894 \\ - \ 562 \end{array}$$

(f)
$$\begin{array}{r} 8547 \\ - \ 2413 \end{array}$$

19. Determine the error in each calculation, and then correct each mistake.

    (a) $\begin{array}{r} 11010010_{\text{two}} \\ -\ \ \ 11111_{\text{two}} \\ \hline 10998899_{\text{two}} \end{array}$

    (b) $\begin{array}{r} 2132_{\text{nine}} \\ -\ 2004_{\text{nine}} \\ \hline 128_{\text{nine}} \end{array}$

    (c) $\begin{array}{r} 47E_{\text{twelve}} \\ +\ 145_{\text{twelve}} \\ \hline 606_{\text{twelve}} \end{array}$

    (d) $\begin{array}{r} 435_{\text{seven}} \\ 313_{\text{seven}} \\ +\ \ 111_{\text{seven}} \\ \hline 1060_{\text{seven}} \end{array}$

20. (a) If $243 + 461 = 724$, what base is being used?

    (b) If $576 + 288 = 842$, what base is being used?

    (c) If $4203 + 434 = 10142$, what base is being used?

    (d) If $4203 + 434 = 5041$, what base is being used?

    (e) If $4203 + 434 = 4640$, what base is being used?

21. For what possible bases is the following addition correct?
    $$\begin{array}{r} 4203 \\ 434 \\ \hline 4637 \end{array}$$

$C$ 22. Perform the following operations. Leave the answer in terms of the base of the numeral on the left.

    (a) $563_{\text{seven}} - ET_{\text{twelve}}$

    (b) $11011_{\text{two}} + 323_{\text{four}} + TE_{\text{twelve}}$

    (c) $10011_{\text{two}} + 423_{\text{five}}$   (d) $E4T_{\text{twelve}} - 11111_{\text{two}}$

## REVIEW EXERCISES

23. Determine the next counting number in the original base for each of the following, and write the answer as a base ten numeral.

    (a) $10110_{\text{two}}$   (b) $46_{\text{seven}}$   (c) $466_{\text{seven}}$

24. Write the following numbers in base seven and in base two.

    (a) 47   (b) 106   (c) 421   (d) 621

25. Write the following as base ten numerals.

    (a) $543_{\text{seven}}$   (b) $1101_{\text{two}}$

26. 333 is a 3-digit numeral. If you were permitted to change one of its digits from 3 to 4, which digit should be changed to

    (a) alter the size of the number the least?

    (b) alter the size of the number the most?

27. How much smaller will the number 476 be if

    (a) 3 is subtracted from the second digit?

    (b) 4 is subtracted from the units digit?

    (c) 1 is subtracted from the hundreds digit?

28. Find the value of $k$ that makes the following statements true.

    (a) $x^3 \cdot x^4 \cdot x^k = x^{10}$   (b) $(x^2)^k = x^4$

    (c) $3^{12} = 9^k$   (d) $2^k = 32$

    (e) $5^k \cdot 125 = 5^7$   (f) $3^k \cdot 27 = 3^8$

## LABORATORY ACTIVITIES
• • • • • • • • • • • • • • • • • • • • • • • • • • • • • • • •

29. Illustrate the addition $254 + 167$, using

    (a) Multibase pieces   (b) Chip abacus

30. Illustrate the addition $246 + 385$, using

    (a) Chip abacus   (b) Multibase pieces

31. Illustrate the subtraction $542 - 264$, using

    (a) Multibase pieces   (b) Chip abacus

### ∴ PCR Excursion ∴

32. Form numbers that will sum to 191 by using *all* of the digits 1 through 7 *only once,* where no number in the sum has more than 2 digits. This problem could be worked by guess, test, and revise, but instead let's work the problem with systematic reasoning. Because we have only 7 digits, the following are the only cases we need to consider:

    1. 7 1-digit numbers
    2. 1 2-digit number and 5 1-digit numbers
    3. 2 2-digit numbers and 3 1-digit numbers
    4. 3 2-digit numbers and a 1-digit number

Clearly write your answers to the following questions.

    (a) Reason why you can eliminate possibility 1—namely, adding the 7 1-digit numbers.

    (b) Reason why you can eliminate possibility 2.

    (c) Why are there only 2 cases under possibility 3?

        (i)  The sum of the units column is 11.

        (ii) The sum of the units column is 21.

    (d) With the sum of the units column equal to 11, reason that the sum of the tens column cannot be 19.

    (e) With the sum of the units column equal to 21, reason that the sum of the tens column cannot be 19.

    (f) Repeat the reasoning of possibility 3 for possibility 4. State the reasons we can eliminate possibility 4.

    (g) Is it possible to arrange 7 digits, 1 through 7, so that the sum is 191 if the largest number is a 2-digit number?

    (h) Work this problem where the sum is 181. Do you get a different answer?

# MULTIPLICATION AND DIVISION ALGORITHMS

**PROBLEM**

Consider this algorithm for multiplying (256)(324):

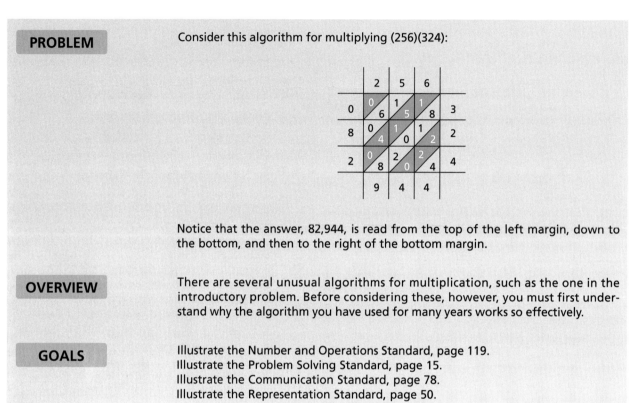

Notice that the answer, 82,944, is read from the top of the left margin, down to the bottom, and then to the right of the bottom margin.

**OVERVIEW**

There are several unusual algorithms for multiplication, such as the one in the introductory problem. Before considering these, however, you must first understand why the algorithm you have used for many years works so effectively.

**GOALS**

Illustrate the Number and Operations Standard, page 119.
Illustrate the Problem Solving Standard, page 15.
Illustrate the Communication Standard, page 78.
Illustrate the Representation Standard, page 50.

Several models will help you to understand our traditional algorithm for multiplication. Suppose that you want to multiply $3 \cdot 16$. This multiplication can be written as $3(10 + 6)$; and a visual representation of it is shown in Figure 4-39.

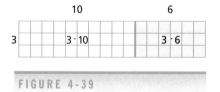

FIGURE 4-39

Now we apply the distributive property of multiplication over addition to obtain an answer for $3 \cdot 16$ (called a *horizontal format*). Locate each part of the sum $(3 \cdot 10 + 3 \cdot 6)$ in Figure 4-39.

**EXAMPLE 1**

Multiply $3 \cdot 16$ by using a horizontal format.

SOLUTION

$$3 \cdot 16 = 3(10 + 6) \qquad \text{Definition of addition}$$
$$= 3 \cdot 10 + 3 \cdot 6 \qquad \text{Distribution property of multiplication over addition}$$
$$= 30 + 18 \qquad \text{Multiplication}$$
$$= 48 \qquad \text{Addition}$$

**EXAMPLE 2**

Compute $3 \cdot 16$, using a place-value chart (repeated addition) and a vertical format (partial products), and connect this work with the traditional algorithm for multiplication.

SOLUTION

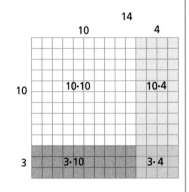

|  | Place-value chart | | Partial products | | Traditional algorithm |
|---|---|---|---|---|---|
|  | 10 | 1 | 16 | | 1 |
| Repeated Addition | 1 | 6 | · 3 | | 16 |
|  | 1 | 6 | 18 | (3 · 6) | 3 |
|  | 1 | 6 | 30 | (3 · 10) | 48 |
|  |  |  | 48 | | |
| Regroup | 3 | 18 | | | |
|  | 1 | 8 | | | |
| Answer | 4 | 8 | | | |

FIGURE 4-40

We now show the multiplication of $13 \cdot 14$ as four partial products. (See Figure 4-40). This represents two applications of the distributive property of multiplication over addition, as demonstrated in the next example.

**EXAMPLE 3**

Show $13 \cdot 14$ in horizontal format.

SOLUTION

$$13 \cdot 14 = 13(10 + 4)$$
$$= 13 \cdot 10 + 13 \cdot 4$$
$$= (10 + 3) \cdot 10 + (10 + 3) \cdot 4$$
$$= 10 \cdot 10 + 3 \cdot 10 + 10 \cdot 4 + 3 \cdot 4 = 182$$

**EXAMPLE 4**

Perform the multiplication $47 \cdot 53$ in horizontal and vertical formats and in intermediate and traditional algorithms.

SOLUTION

In this example the distributive property of multiplication over addition must be applied twice.

**Horizontal format**

$$47 \cdot 53 = 47(50 + 3)$$
$$= 47 \cdot 50 + 47 \cdot 3 \qquad \text{(Why?)}$$
$$= (40 + 7) \cdot 50 + (40 + 7) \cdot 3$$
$$= 40 \cdot 50 + 7 \cdot 50 + 40 \cdot 3 + 7 \cdot 3 \qquad \text{(Why?)}$$
$$= 2000 + 350 + 120 + 21$$
$$= 2491$$

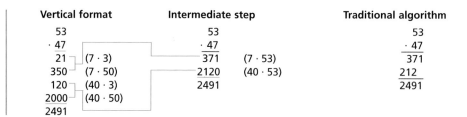

## MENTAL MULTIPLICATION

You can use expanded notation to do mental arithmetic. For example,

$$
\begin{array}{ll}
43 & 40 \cdot 5 = 200 \\
\underline{\cdot 5} & 3 \cdot 5 = \underline{\phantom{0}15} \\
& \phantom{0}215 \quad \text{Add the two products}
\end{array}
$$

Sometimes you can use doubling and halving to find the answer mentally. For example,

$$
\begin{array}{ll}
46 & 23 \quad \text{Take half of 46} \\
\underline{\cdot 5} & \underline{\cdot 10} \quad \text{Double the 5} \\
& 230
\end{array}
$$

Sometimes, when working with several numbers, you can regroup for mental multiplication. For example,

$$8 \cdot 2 \cdot 15 \cdot 5 = (8 \cdot 15) \cdot (2 \cdot 5)$$
$$= 120 \cdot 10 = 1200$$

## ESTIMATION IN MULTIPLICATION

When performing multiplications, especially with a calculator, rounding and using powers of 10 enable you to estimate the answer quickly.

**EXAMPLE 5**

Estimate $4326 \cdot 724{,}832$, using only the first rounded digit in each number.

*SOLUTION*

$$4326 \approx 4000 = 4 \cdot 10^3$$
$$724{,}832 \approx 700{,}000 = 7 \cdot 10^5$$
$$4326 \cdot 724{,}832 \approx (4 \cdot 10^3)(7 \cdot 10^5)$$
$$\approx 28 \cdot 10^8 \approx 3{,}000{,}000{,}000$$

The exact answer is $3{,}135{,}623{,}232$.

### COMMON ERROR

$$
\begin{array}{llll}
\text{What is the common error?} & 26 & & 42 \\
& \underline{\cdot 31} & & \underline{\cdot 34} \\
& 26 & & 168 \\
& \underline{78} & & \underline{126} \\
& 104 & & 294
\end{array}
$$

## ADDITIONAL MULTIPLICATION ALGORITHMS

| Halving | Doubling |
|---------|----------|
| 37      | 168      |
| ~~18~~  | ~~336~~  |
| 9       | 672      |
| ~~4~~   | ~~1344~~ |
| ~~2~~   | ~~2688~~ |
| 1       | 5376     |
|         | 6216     |

37 · 168 = 6216

**FIGURE 4-41**

There are several other interesting algorithms for multiplication. For brevity, we consider only two: Russian peasant multiplication and lattice multiplication. Russian peasant multiplication, used in medieval Europe, involves repeatedly doubling one of the numbers to be multiplied, and halving the other. Any time a remainder occurs in the halving process, the remainder is dropped. All lines in which the entry in the halving column is an even number are then crossed out, and the entries in the doubling column are added to get the answer.

**EXAMPLE 6**

Multiply 37 · 168, using Russian peasant multiplication.

*SOLUTION*

See Figure 4-41.

The lattice method is explained in the next example.

**EXAMPLE 7**

Multiply 476 · 75 by the lattice method of multiplication.

*SOLUTION*

The problem is to find the product of 476 and 75. Because the 2 numbers to be multiplied are represented by numerals with 2 and 3 digits, respectively, a rectangle containing 6 small squares of equal size is drawn. (See Figure 4-42.) The numerals 4, 7, and 6 are written across the top of the rectangle and 7 and 5 are written along the right side from top to bottom. Diagonals (from lower left to upper right) are drawn in each of the squares to form a *lattice* design. Products of pairs of digits taken from the top and the right side of the rectangle are then entered in the squares. The tens digit of the product is written above the diagonal and the units digit is written below the diagonal. Thus, in the first square, 6 · 7 = 42, so 4 is above the diagonal and 2 is below it. Now add the elements between adjacent diagonals, beginning at the lower right corner. There is only one element below the first diagonal. It is a 0, so write a 0 in the space at the bottom of this diagonal. The sum of the elements between the next two diagonals is 2 + 3 + 5, and the answer 10 is recorded as follows. When a sum is more than 9, record the units digit as before and regroup the tens digit to be added to the elements between the next two adjacent diagonals. When all diagonal elements have been totaled, the answer is read down the left side and along the bottom. The answer in this case is 35,700.

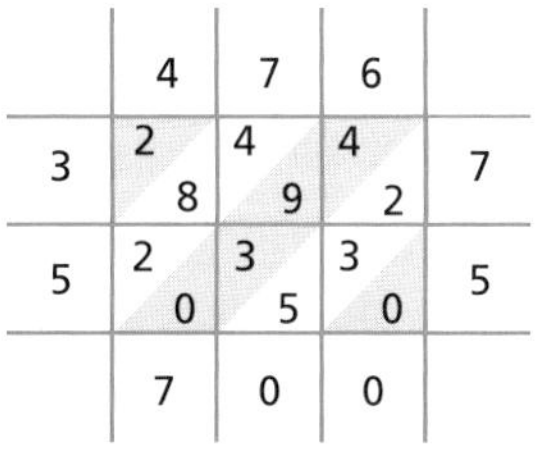

**FIGURE 4-42**

## ALGORITHMS FOR DIVISION

In Section 2 of this chapter, division was defined as the inverse of multiplication. If $a$ and $b$ $(b \neq 0)$ are any two whole numbers, then $a \div b$ is some whole number $c$ (if it exists) such that $a = bc$. Clearly, division of this type is not possible for every pair of whole numbers. For example, it is not possible to compute $8 \div 5$ in the system of whole numbers. It is possible, however, in every division problem involving whole numbers (where the divisor is unequal to zero), to find an answer consisting of a quotient and a remainder that are both whole numbers.

| **Division Algorithm** | If $a$ and $b$ $(b \neq 0)$ are whole numbers, then there exist unique whole numbers $q$ and $r$ such that $a = bq + r$, where $0 \leq r < b$. |
|---|---|

This theorem states that, for any two whole numbers $a$ and $b$ $(b \neq 0)$, $a$ can be divided by $b$ to obtain two numbers, $q$ (the quotient) and $r$ (the remainder), such that $a = bq + r$, where $0 \leq r < b$. If we call $b$ the *divisor* and $a$ the *dividend,* then

$$\textbf{dividend} = \textbf{divisor} \cdot \textbf{quotient} + \textbf{remainder}$$

**EXAMPLE 8**

If $a = 47$ and $b = 7$, write $a$ in the form $bq + r$, where $0 \leq r < b$.

*SOLUTION*

We compute $47 = 7(6) + 5$; therefore, $q = 6$ and $r = 5$, and (as required) $0 \leq 5 < 7$.

**PRACTICE PROBLEM**

If $a = 54$ and $b = 9$, write $a$ in the form $bq + r$, where $0 \leq r < b$.

*ANSWER*

$54 = 9(6) + 0$; therefore, $q = 6$ and $r = 0$.

## Calculator Hint

Consider the division $a \div b$. To find the quotient $q$ and the remainder $r$ guaranteed by the division algorithm, use the following to find $350 \div 21$:

$$350 \boxed{\text{INT}\div} 21 \boxed{=} \text{ gives } 16 \quad 14.$$

$$q = 16 \text{ and } r = 14.$$

Many procedures are known for performing the operation of division. Generally, they are expressed in such a way that the properties of whole numbers relative to the operations involved are completely obscured. Most people use a process of guessing, multiplying, subtracting, and then guessing again. If an initial guess proves wrong, it is replaced by a new guess. This process is called **long division.** The division algorithm states that the answer exists, so after a sufficient number of guesses a person should always find the correct answer.

However, long division becomes more understandable when viewed as repeated subtraction. For instance, the division $16 \div 5$ asks the question

"How many 5's are contained in 16?" We find the answer by repeated subtractions:

$$
\begin{array}{r}
16 \\
-\phantom{0}5 \\
\hline
11 \\
-\phantom{0}5 \\
\hline
6 \\
-\phantom{0}5 \\
\hline
1
\end{array}
$$

Thus, there are three 5's in 16, with a remainder of 1. This format (sometimes called **scaffolding**) is used in the next 2 examples.

**EXAMPLE 9**

Find $20 \div 6$ by repeated subtraction, using a scaffold.

*SOLUTION*

Notice that in (b), instead of subtracting 6 three times, we subtract 18 or $(3 \cdot 6)$ once.

(a)
$$
\begin{array}{r|l}
6\phantom{|}\overline{\phantom{0}20} & \\
\phantom{6|}\underline{\phantom{0}6} & 1\ (6) \\
\phantom{6|}14 & \\
\phantom{6|}\underline{\phantom{0}6} & 1\ (6) \\
\phantom{6|}\phantom{0}8 & \\
\phantom{6|}\underline{\phantom{0}6} & 1\ (6) \\
\phantom{6|}\phantom{0}2 & 3
\end{array}
$$

(b)
$$
\begin{array}{r|l}
6\phantom{|}\overline{\phantom{0}20} & \\
\phantom{6|}\underline{18} & 3\ (6\text{'s}) \\
\phantom{6|}\phantom{0}2 &
\end{array}
$$

The answer is $20 \div 6 = 3$ with a remainder of 2.

Observe that the "guessing" portion of long division really amounts to asking the question, "How large a multiple of the divisor can I subtract and have a remainder that is a whole number?"

**EXAMPLE 10**

Compute $7410 \div 23$, using a scaffold.

*SOLUTION*

In (a), the largest multiple of the form "(power of 10) times 23" that we can subtract from 7410 is (100)(23). After (100)(23) is subtracted 3 times, (10)(23) can be subtracted twice. Finally (1)(23) is subtracted twice, leaving a remainder of 4. In (b), instead of subtracting 100(23) 3 times, we subtract 300(23). Likewise, instead of subtracting 10(23) twice, we subtract 20(23).

(a)
$$
\begin{array}{r|l}
23\phantom{|}\overline{7410} & \\
\underline{2300} & 100\ (23\text{'s}) \\
5110 & \\
\underline{2300} & 100\ (23\text{'s}) \\
2810 & \\
\underline{2300} & 100\ (23\text{'s}) \\
\phantom{0}510 & \\
\underline{\phantom{0}230} & 10\ (23\text{'s}) \\
\phantom{0}280 & \\
\underline{\phantom{0}230} & 10\ (23\text{'s}) \\
\phantom{00}50 & \\
\underline{\phantom{00}23} & 1\ (23) \\
\phantom{00}27 & \\
\underline{\phantom{00}23} & 1\ (23) \\
\phantom{000}4 & 322
\end{array}
$$

(b)
$$
\begin{array}{r|l}
23\phantom{|}\overline{7410} & \\
\underline{6900} & 300(23\text{'s}) \\
\phantom{0}510 & \\
\underline{\phantom{0}460} & 20(23\text{'s}) \\
\phantom{00}50 & \\
\underline{\phantom{00}46} & 2(23\text{'s}) \\
\phantom{000}4 & 322
\end{array}
$$

How many 23's in 7410?

How many 23's in 510?

How many 23's in 50?

The answer is $7410 \div 23 = 322$ with a remainder of 4. As a check, we can compute $23 \cdot 322 + 4 = 7410$.

Notice that, in part (b) of Example 10, we subtracted $(300)(23)$ instead of subtracting $(100)(23)$ 3 times and we subtracted $(20)(23)$ instead of subtracting $(10)(23)$ twice. The shortcut in (b) of Example 10 will be employed in future uses of the scaffold format in this text.

An intermediate algorithm helps bridge the gap between the scaffold format for division and the format of our traditional algorithm.

**EXAMPLE 11**

Compute $8134 \div 38$, using the scaffold format, an intermediate format, and the traditional algorithm.

*SOLUTION*

| Scaffold format | Intermediate algorithm | Traditional algorithm |
|---|---|---|
| | 4 | |
| | 10 | |
| | 200 | 214 |
| 38 ⟌ 8134 | 38 ⟌ 8134 | 38 ⟌ 8134 |
| 7600    200(38) | 7600    How many 38's in 8134? | 76 |
| 534 | 534    How many 38's in 534? | 53 |
| 380    10(38) | 380 | 38 |
| 154 | 154    How many 38's in 154? | 154 |
| 152    4(38) | 152 | 152 |
| 2    214 | 2 | 2 |

Therefore, $8134 \div 38 = 214$ with a remainder of 2. As a check, we compute $38 \cdot 214 + 2 = 8134$.

The thinking involved in mental division makes use of the basic algorithms of division we have just studied.

# Mental Division and Estimation

Sometimes dividends are such that you can arrange them as sums (or differences) that you can divide mentally. Here are two examples:

$$6 \, ⟌ \, 462 \quad \text{is} \quad \begin{array}{r} 70 + 7 = 77 \\ 6 \, ⟌ \, \overline{420 + 42} \end{array}$$

$$3 \, ⟌ \, 201 \quad \text{is} \quad \begin{array}{r} 70 - 3 = 67 \\ 3 \, ⟌ \, \overline{210 - 9} \end{array}$$

**EXAMPLE 12**

Estimate $2{,}954{,}424 \div 2356$, using only the first rounded digit of each numeral.

*SOLUTION*

$$2{,}954{,}424 \approx 3{,}000{,}000 = 3(10^6) = 30(10^5)$$
$$2356 \approx 2000 \qquad = 2(10^3)$$
$$\frac{2{,}954{,}424}{2356} \approx \frac{30(10^5)}{2(10^3)} = 15(10^2) = 1500$$

The exact answer is 1254.

## COMMON ERROR

What is the error?

$$
\begin{array}{r}
15 \\
8\,\overline{)\,48} \\
\underline{8} \\
40 \\
\underline{40}
\end{array}
$$

# MULTIPLICATION, NONDECIMAL BASES

To perform multiplications and divisions in base two and base seven we need Tables 4-7 and 4-8. First let's verify some entries in Table 4-8. In the case of $4 \cdot 2$, we have $4 \cdot 2 = (2 + 2 + 2) + 2$. Here, $2 + 2 + 2 = 6$ and 2 more is one base and one 1, or $11_{\text{seven}}$. Similarly, for $3 \cdot 5$ we have

$$
\left[
\begin{aligned}
3 \cdot 5 &= 5 + 5 + 5 \\
&= 13 + 5 \\
&= 21
\end{aligned}
\right]_{\text{seven}}
$$

**TABLE 4-7**

*Multiplication Table, Base Two*

| · | 0 | 1 |
|---|---|---|
| **0** | 0 | 0 |
| **1** | 0 | 1 |

**TABLE 4-8**

*Multiplication Table, Base Seven*

| · | 0 | 1 | 2 | 3 | 4 | 5 | 6 |
|---|---|---|---|---|---|---|---|
| **0** | 0 | 0 | 0 | 0 | 0 | 0 | 0 |
| **1** | 0 | 1 | 2 | 3 | 4 | 5 | 6 |
| **2** | 0 | 2 | 4 | 6 | 11 | 13 | 15 |
| **3** | 0 | 3 | 6 | 12 | 15 | 21 | 24 |
| **4** | 0 | 4 | 11 | 15 | 22 | 26 | 33 |
| **5** | 0 | 5 | 13 | 21 | 26 | 34 | 42 |
| **6** | 0 | 6 | 15 | 24 | 33 | 42 | 51 |

**EXAMPLE 13**

Multiply 3 and 5 in base seven.

*SOLUTION*

From Table 4-8, in the row with label 3 and in the column with label 5, we find 21. Thus,

$$[3 \cdot 5 = 21]_{\text{seven}}$$

**EXAMPLE 14**

Find the product of $216_{\text{seven}}$ and $14_{\text{seven}}$.

*SOLUTION*

$$
\left[
\begin{array}{r}
{\scriptstyle 1\,3} \\
216 \\
\cdot\ \ 14 \\
\hline
1203 \\
216\ \ \\
\hline
3363
\end{array}
\right]_{\text{seven}}
$$

$4 \cdot 6 = 33$ in Table 4-8

$4 \cdot 1 + 3 = 10$

$4 \cdot 2 + 1 = 11 + 1 = 12$

Instead of using Table 4-8, we now work the same problem by performing operations in base ten and converting to base seven.

$$\begin{bmatrix} & & \overset{1(7)^2}{} & \overset{3(7)}{} \\ & 2(7)^2 & + 1(7) & + 6 \\ & & 1(7) & + 4 \\ \hline 1(7)^3 & + 2(7)^2 & + 0(7) & + 3 \\ 2(7)^3 & + 1(7)^2 & + 6(7) & \\ \hline 3(7)^3 & + 3(7)^2 & + 6(7) & + 3 \end{bmatrix}_{\text{ten}} \quad \begin{bmatrix} 4 \cdot 6 = 24 = 3(7) + 3 \\ 4 \cdot 1(7) + 3(7) = 7(7) = 1(7)^2 + 0(7) \\ 4 \cdot 2(7)^2 + 1(7)^2 = 9(7)^2 = [1(7) + 2](7)^2 \\ = 1(7)^3 + 2(7)^2 \end{bmatrix}_{\text{ten}}$$

The answer is $3363_{\text{seven}}$.

Now using the traditional algorithm, but thinking in base 10, we have

$$\begin{bmatrix} & \overset{1\;3}{} & \\ & 216 & \\ & 14 & \\ \hline & 1203 & \\ & 216 & \\ \hline & 3363 & \end{bmatrix}_{\text{seven}} \quad \begin{bmatrix} 4 \cdot 6 = 24 = 3(7) + 3 \\ 4 \cdot 1 + 3 = 7 = 1(7) + 0 \\ 4 \cdot 2 + 1 = 9 = 1(7) + 2 \\ \text{first line } 1203 \\ 1 \cdot 6 = 6, \; 1 \cdot 1 = 1, \text{ and } 1 \cdot 2 = 2 \\ \text{second line } 216 \end{bmatrix}_{\text{ten}}$$

**EXAMPLE 15**

Multiply 1011 and 101 in base two.

*SOLUTION*

$$\begin{bmatrix} 1011 \\ 101 \\ \hline 1011 \\ 1011 \\ \hline 110111 \end{bmatrix}_{\text{two}} \qquad \text{Use Table 4-5 for addition.}$$

# DIVISION, NONDECIMAL BASES

Division will be illustrated with the scaffold format.

**EXAMPLE 16**

Divide $11010_{\text{two}}$ by $1011_{\text{two}}$.

*SOLUTION*

$$1011 \overline{\smash{)}\begin{array}{r} 11010 \\ 10110 \\ \hline 100 \end{array}} \qquad 10(1011)$$

The answer is $10_{\text{two}}$ with a remainder of $100_{\text{two}}$.

**EXAMPLE 17**

Divide $1662_{\text{seven}}$ by $24_{\text{seven}}$.

*SOLUTION*

First use Table 4-8 to make a table of the products of possible quotients (0, 1, 2, 3, 4, 5, and 6) and the divisor.

| Possible quotient | 0 | 1 | 2 | 3 | 4 | 5 | 6 |
|---|---|---|---|---|---|---|---|
| Quotient times divisor (24) | 0 | 24 | 51 | 105 | 132 | 156 | 213 |

$_{\text{seven}}$

Notice that $5(24) = 156$ is just less than 166.

$$
\begin{array}{r}
52 \\
24 \,\overline{\smash{\big)}\, 1662} \\
\underline{156} \\
102 \\
\underline{51} \\
21 \\
\end{array}
$$
seven

The answer is $52_{\text{seven}}$ with a remainder of 21. As a check, we compute

$$
\left[
\begin{aligned}
52 \cdot 24 + 21 &= 1641 + 21 \\
&= 1662
\end{aligned}
\right]_{\text{seven}}
$$

**EXAMPLE 18**

Divide $34211_{\text{five}}$ by $23_{\text{five}}$.

*SOLUTION*

First make a multiplication table for possible quotients: 0, 1, 2, 3, and 4.

Then perform the division:

$$
\begin{array}{r}
1222 \\
23 \,\overline{\smash{\big)}\, 34211} \\
\underline{23} \\
112 \\
\underline{101} \\
111 \\
\underline{101} \\
101 \\
\underline{101} \\
\end{array}
$$
five

| Possible quotient | 0 | 1 | 2 | 3 | 4 | |
|---|---|---|---|---|---|---|
| Quotient times divisor | 0 | 23 | 101 | 124 | 202 | five |

The answer is exactly $1222_{\text{five}}$. As a check, we compute

$$[23 \cdot 1222 + 0 = 34211]_{\text{five}}$$

# Just for Fun

In the computations below, a given letter
stands for the same digit no matter where it
occurs. Can you unravel the puzzle and find
the digit represented by each letter?

$$
\begin{array}{l}
\phantom{2X5}\overline{\phantom{)}\;X\;V\;U} \\
2X5\,\overline{\smash{\big)}\,U\,7\,X\,9\,Z} \\
\phantom{2X5)}\underline{Y\,X\,Z} \\
\phantom{2X5)}\underline{X\,Z\,T\,9} \\
\phantom{2X5)}\underline{X\,5\,W\,5} \\
\phantom{2X5)XZ}T\,4\,Z \\
\phantom{2X5)XZ}\underline{T\,4\,Z} \\
\end{array}
\qquad
\begin{array}{r}
P\;Q\;R \\
\cdot\,Q\;Q \\
\hline
S\;R\;Q\;Q \\
S\;R\;Q\;Q \\
\hline
P\;M\;P\;R\;Q \\
\end{array}
$$

# EXERCISE SET 6

*R* 1. Draw rectangles as in Figure 4-39 for each product to illustrate the distributive property of multiplication over addition.

   (a) $6 \cdot 28$     (b) $8 \cdot 34$     (c) $3(26)$

2. Using Exercise 1, perform the following computations in the four ways demonstrated in the examples: (i) horizontal format, (ii) place-value chart, (iii) vertical format (partial products), and (iv) traditional algorithm.

   (a) $6 \cdot 28$     (b) $8 \cdot 34$     (c) $3(26)$

3. For each part of Exercise 2, draw lines connecting equivalent parts of (iii) and (iv).

4. For the following pairs of numbers, let $a$ be the first number of the pair and let $b$ be the second number. Find whole numbers $q$ and $r$ for each pair such that $a = bq + r$, where $0 \le r < b$. Then check your answer with a calculator.

   (a) 72, 11     (b) 16, 9     (c) 11, 18

   (d) 106, 13    (e) 51, 14    (f) 25, 39

   (g) 54, 9      (h) 176, 21

5. Perform the following divisions in the three ways demonstrated in Example 11 in this section: (i) scaffold format, (ii) intermediate algorithm, and (iii) traditional algorithm.

   (a) $166 \div 6$     (b) $324 \div 7$     (c) $1425 \div 8$

6. Draw lines connecting the equivalent parts of (i) and (ii) for each part of Exercise 5.

7. Fill in the missing numerals.

   (a)        4  7
             · 8  4
          1  _  8
       3  _  6
       3  _  _  8

   (b)         _  2
          21 ) 6  7  8
               6  _
               _     8
               4  2

*T* 8. Follow the directions (i), (iii) and (iv) of Exercise 2 or (i), (ii), or (iii) of Exercise 5 and perform the following computations.

   (a) $26 \cdot 32$     (b) $84 \cdot 76$

   (c) $74 \cdot 92$     (d) $1075 \div 27$

9. Draw lines connecting the equivalent parts of (iii) and (iv) for multiplication and of (i) and (ii) for division in Exercise 8.

10. Perform the following multiplications mentally, using expanded notation.

   (a) $8 \cdot 63$     (b) $4 \cdot 72$

11. Perform the following multiplications mentally, using doubling and halving.

   (a) $5 \cdot 168$     (b) $5 \cdot 276$

12. Perform the following divisions mentally.

   (a) $249 \div 3$     (b) $234 \div 3$

*Estimate each product or quotient as in the examples in this chapter. Then determine whether a calculator answer makes sense relative to your estimate.*

13. $8316 \cdot 7536$     14. $9126 \cdot 904$

15. $85{,}968 \div 27$     16. $14{,}632 \div 31$

17. Perform the following multiplications.

   (a) $(1101_{two})(11_{two})$     (b) $(366_{seven})(34_{seven})$
   (c) $(454_{seven})(205_{seven})$     (d) $(1101_{two})(101_{two})$

18. Make a multiplication table for the given bases.

   (a) Three     (b) Twelve     (c) Five

19. Perform the indicated operations.

   (a) $(44_{five})(4_{five})$     (b) $(22_{twelve})(5_{twelve})$
   (c) $(323_{five})(43_{five})$     (d) $(40E_{twelve})(3T_{twelve})$
   (e) $(2012_{three})(21_{three})$     (f) $(2222_{three})(22_{three})$

20. For each of the bases given, name the largest and the smallest number represented by a 3-digit numeral of the form $abc$, where $a \ne 0$. Find the difference between each of these pairs of numbers.

   (a) Five     (b) Eleven     (c) Three     (d) Six

21. Perform the following divisions.

   (a) $11101_{two} \div 11_{two}$     (b) $460_{seven} \div 23_{seven}$
   (c) $1666_{seven} \div 102_{seven}$

22. Estimate the answer mentally.

   (a) $7648 \div 18$     (b) $92{,}456 \div 28$

*C* 23. A multiplication is given by

   If all different symbols represent different digits, find the digits represented by △, □, ⊙,    , and ◇.

24. Multiply 15,873 by 7, 14, and 21. From the pattern, guess the answer for 49(15,873). Check your answer.

25. Multiply $11 \cdot 11$, $111 \cdot 111$, and $1111 \cdot 1111$. Now guess the answer for $11111 \cdot 11111$. Check your guess.

26. Perform the following divisions in three ways: (i) scaffold format, (ii) intermediate algorithm, and (iii) traditional algorithm.

   (a) $1728 \div 32$     (b) $2301 \div 47$

   (c) $23{,}410 \div 37$     (d) $2435 \div 867$

27. Compute, using the lattice method of multiplication.

   (a) 34 · 176      (b) 56 · 742

28. Multiply, using Russian peasant multiplication.

   (a) 14 · 36   (b) 54 · 17

 29. Perform the following division by repeated subtraction.

$$2473 \div 23$$

   Can you check your answer by using the calculator to perform the operation another way?

 30. Perform the following divisions.

   (a) 16,048 ÷ 21      (b) 16,044 ÷ 21

   Explain how you can tell if the division is exact or if there is a remainder. Find the remainder.

## REVIEW EXERCISES

31. Determine the property illustrated by the following.

   (a) $1(x + 2) = x + 2$      (b) $(x + 1)x = x(x + 1)$
   (c) $0(x + 3) = 0$      (d) $x(yz) = (xy)z$
   (e) $x(y + z) = xy + xz$

32. (i) Use expanded notation in vertical format, (ii) show the computation with a place-value chart, and (iii) use the traditional algorithm for addition (or subtraction) in each problem.

   (a)　436　　(b)　524　　(c)　476　　(d)　584
   　　+ 243　　　+ 361　　　− 243　　　− 361

---

33. The Acme Potato Chip Company received 6 freight cars that were to be filled with 100-lb. bags of potatoes. It was learned that the automatic bag-weighing machine malfunctioned for a while and that some of the freight cars were full of 90-lb. bags of potatoes. Devise a plan for taking a few bags from each freight car so that in one weighing you will discover which freight cars have the 90-lb. bags.

A. *Outline of a plan.* Label or name the 6 freight cars: one unit, base two, base two squared, base two cubed, base two to the fourth power, and base two to the fifth power. Then load a truck by taking from each freight car the number of bags of potatoes as given in the name of the freight car. How many pounds of potatoes would be on the truck if the bag-weighing machine had worked?

B. *Testing the plan.* (a) Suppose the weight of all bags is 6290 lb. How many 90-lb bags do you have, and from what freight car did they come?
   (b) Answer this question if the total weight of all bags is 6280 lb.
   (c) Answer this question if the total weight of all bags is 6220 lb.
   (d) 6140 lb.      (e) 5980 lb.

C. *Using the plan.* (a) Think of other possible weights of potatoes that could be on the truck. In each case try to determine how many 90-lb. bags are on the truck and from which freight cars they came. For example, suppose the potatoes weighed 6060 lb. How many 90-lb. bags were there and from which freight car did they come?
   (b) 5940 lb.      (c) 6130 lb.      (d) 5740 lb.

---

# SOLUTION TO INTRODUCTORY PROBLEM

UNDERSTANDING THE PROBLEM Each letter represents a different digit in these additions. Further, both additions give the same sum. What are X, A, B, G, and H?

$$\begin{aligned}
XXX &= 888 \\
AAA &= 555 \\
\underline{BBB} &= \underline{666} \\
CDEF &= 2109
\end{aligned}$$

DEVISING A PLAN By guess, test, and revise and by your knowledge of numeration systems, you should obtain the following answer.

$$\begin{aligned}
XXX &= 888 \\
GGG &= 444 \\
\underline{HHH} &= \underline{777} \\
CDEF &= 2109
\end{aligned}$$

# SUMMARY AND REVIEW

## WHOLE NUMBERS

1. The numbers $\{0, 1, 2, 3, 4, 5, \ldots\}$ are called *whole numbers*.

2. If $A$ and $B$ are disjoint sets with $n(A) = a$ and $n(B) = b$, then $a + b$ is defined by $n(A \cup B)$.

3. Properties of addition of whole numbers:

   (a) *Closure property of addition:* If $a$ and $b$ are whole numbers, there is a unique whole number $c$ with $a + b = c$.

   (b) *Commutative property of addition:* $a + b = b + a$

   (c) *Associative property of addition:* $(a + b) + c = a + (b + c)$

   (d) *Identity for addition:* 0, since $a + 0 = 0 + a = a$

4. $a < b$ if there is a nonzero whole number $n$ with $a + n = b$.

5. For whole numbers $a$ and $b$, subtraction is defined by $a - b = c$ if there is a whole number $c$ with $a = b + c$. Thus, subtraction is the *inverse* of addition.

6. $a \cdot b$ is defined to be $b$ added $a$ times.

7. Properties of multiplication of whole numbers.

   (a) *Closure property of multiplication:* If $a$ and $b$ are whole numbers, there is a unique whole number $c$ with $a \cdot b = c$.

   (b) *Commutative property of multiplication:* $a \cdot b = b \cdot a$

   (c) *Associative property of multiplication:* $(a \cdot b) \cdot c = a \cdot (b \cdot c)$

   (d) *Distributive property of multiplication over addition:* $a \cdot (b + c) = (a \cdot b) + (a \cdot c)$

   (e) *Identity for multiplication:* 1 because $a \cdot 1 = a$

   (f) *Zero property of multiplication:* $a \cdot 0 = 0$

8. If $a$ and $b$ are whole numbers, $b \neq 0$, division is defined by $a \div b = c$ if $b \cdot c = a$. Division is the inverse operation of multiplication. Division by 0 is undefined.

## EXPONENTIAL NOTATION

1. For whole numbers $n$, $a^n$ is defined by

   (a) $a^n = \underbrace{a \cdot a \cdot a \cdot \ldots \cdot a}_{n \text{ factors}}$ if $n$ *is* a counting number

   (b) $a^0 = 1$

2. Operations involving exponents

   (a) $a^x \cdot a^y = a^{x+y}$      (b) $(a^x)^y = a^{xy}$

   (c) $(ab)^x = a^x \cdot b^x$

3. In the symbol $b^a$, $b$ is called the *base* and $a$ is called the *exponent;* $b^a$ is called the $a$th power of $b$.

## NUMERATION SYSTEMS

1. If the numerals are written as sums of products of powers of $b$, we call the system a base $b$ numeration system.

   (a) The binary numeration system is a base two system.

   (b) The decimal numeration system is a base ten system.

2. Historical numeration systems

   (a) The Egyptian system was an additive numeration system with neither a concept of place value nor a zero. The Egyptian numerals with their modern equivalents are (1, |), (10, ∩), (100, 𝟡), (1000, ⚡), (10,000, ⌒), (100,000, ☡), and (1,000,000, ✻).

   (b) The Babylonian system was an additive system that included a base sixty place-value feature but had no zero. The symbols used were ◄ for ten and ▼ for one. The numbers 1 to 59 were formed additively, and the place-value system was used to form larger numbers.

   (c) The Roman system was an additive system with one peculiar place-value principle: If a symbol were preceded by a symbol of lesser value, the lesser value was subtracted. The Roman numerals with their modern equivalents are (1, I), (5, V), (10, X), (50, L), (100, C), (500, D), and (1000, M). A bar over a numeral multiplied the value of the numeral by 1000.

   (d) The Mayan numeration system was a place-value system with a base of 20. A dot represented 1, a line represented 5, and ⊛ represented zero. The numerals were written in vertical form. One eccentricity of this base twenty system is the fact that the third place represented $(18)(20)$ instead of $(20)^2$, and the fourth place represented $(18)(20)^2$ instead of $(20)^3$.

3. (a) When a base ten number is written as a sum of multiples of 1, 10, 100, 1000, 10,000, $\ldots$, we say that the number is written in expanded notation.

   (b) When a base $b$ number is written as a sum of multiples of powers of $b$, such as 1, $b^1$, $b^2$, $b^3$, $b^4$, $\ldots$, we say that the number is written in expanded notation in terms of its base.

4. Make sure you understand addition, multiplication, subtraction, and division algorithms.

# CHAPTER TEST

1. Write as a base ten number.
   (a) ⚡️🍩🍩∩||     (b) ▼◀ ◀▼▼

2. Rewrite each expression as indicated.
   (a) $4^3$ as a power of 2     (b) $2^5 \cdot 2^7$
   (c) $x^5 \cdot y^5$ as a power of $xy$
   (d) $9^2 \cdot 3^5$ as a power of 3

3. (a) Write 2,004,006 in expanded notation, using powers of 10.
   (b) What is the meaning of the 7 in 463,742?
   (c) Write in the usual form: $2(10)^4 + 3(10)^2 + 4(10)$.

4. Add   24
       + 37
   (a) in expanded notation vertical format.
   (b) with a place-value chart.
   (c) with the traditional algorithm.

5. Draw a rectangle to show the double application of the distributive property of multiplication over addition in the multiplication of $16 \cdot 24$.

6. Multiply 24 using a vertical format (with partial
       $\cdot$ 16
   products) and then using the traditional algorithm. Draw lines to show equivalent parts.

7. Determine the next counting number in base two after $101011_{two}$.

8. (a) Convert 276 to a binary numeral.
   (b) Convert 276 to a base seven numeral.

9. Apply the division algorithm to $27 \div 5$, finding $q$ and $r$.

10. Determine the properties of whole numbers used in the following equalities.
    (a) $(3 + 7) + 9 = 3 + (7 + 9)$
    (b) $3(x + y) = 3x + 3y$     (c) $x = 1 \cdot x$
    (d) $xz + yz = (x + y)z$

11. Select the computations that equal 0.
    (a) $5 - 5$             (b) $0 \div 4$
    (c) $4 \div (3 - 3)$     (d) $4(2 - 2)$
    (e) $(5 - 5) \div 2$     (f) $(2 \cdot 3) \div 0$
    (g) $(7 - 7) \div (4 - 4)$

12. Select the true statements from the following list.
    (a) If $n + 6 = 11$, then $n \geq 5$.
    (b) If $n + 89 = 91$, then $n < 2$.
    (c) If $n + 13 = 12$, then $n$ is not a whole number.
    (d) If $n + 16 = 14$, then $1 < n < 2$.

(e) If $0 - 6 = n$, then $n$ is a whole number.
(f) If $0 \div 6 = n$, then $n$ is a whole number.
(g) If $a$ is a natural number, then $(a - 0)/a = 1$.
(h) If $a$ is a natural number, then $(a - a)/a = 1$.

13. Find the solution set of $2x < 9$ on the set of whole numbers.

14. Find $20 \div 4$ by repeated subtraction. Show your work.

15. John runs 6 miles less 2 times the distance that Lisa runs. Together they run 4 miles. Set up an equation to represent this information.

16. Solve the equation in Problem 15.

17. Use the following examples to show why subtraction and division are inverse operations for addition and multiplication.
    (a) $12 - 8 = 4$ because    (b) $15 \div 3 = 5$ because

18. Demonstrate the following operations on a number line.
    (a) $4 + 5$     (b) $3 \cdot 4$     (c) $8 - 6$

19. Compute
    (a) $110111_{two} + 1011_{two}$     (b) $465_{seven} - 244_{seven}$
    (c) $111_{two} \cdot 1011_{two}$     (d) $254_{seven} - 6_{seven}$

20. Write the following as base ten numerals.
    (a) $406_{seven}$     (b) $10001_{two}$

21. Perform the subtraction $2436 - 568$ in the following three ways.
    (a) Write each number in expanded notation (vertical format).
    (b) Show the subtraction with a place-value chart.
    (c) Use the traditional algorithm.

22. Perform the division $2435 \div 56$ in the following 3 ways.
    (a) Scaffold format
    (b) An intermediate algorithm
    (c) The traditional algorithm
    (d) Draw lines to show equivalent parts of (a) and (b).

23. Perform the indicated operations.
    (a) $(455_{six})(24_{six})$     (b) $(58E_{twelve})(E_{twelve})$

24. Name the base used in the following calculations.
    (a) $452 - 263 = 156$     (b) $(604)(35) = 31406$

25. Working from left to right, show that $(6 + 2)5 = 2 \cdot 5 + 5 \cdot 6$ and give reasons for each step.

# THE SYSTEM OF INTEGERS AND ELEMENTARY NUMBER THEORY

# 5

A trainer in an athletic department was asked to arrange the towels in the locker room in stacks of equal size. When he separated the towels into sets of four, one was left over. When he tried stacks of five, one was left over. The same was true for stacks of six. However, he was successful in arranging the towels in stacks of seven each. What is the smallest possible number of towels in the locker room?

The great mathematician Carl Friedrich Gauss (1777–1855) said, "Mathematics is the queen of the sciences, and arithmetic is the queen of mathematics." By arithmetic, Gauss meant the subject of this chapter—number theory.

"Number theory provides rich opportunities for explorations that are interesting, enjoyable, and useful." Many topics of this chapter on number theory are found in the elementary-school mathematics curriculum: divisibility tests, factors, primes, multiples, greatest common divisors (or factors), clock arithmetic, and others.

The usefulness of number theory applies not only to whole numbers but also to a system of numbers which we call integers. When studying subtraction in Chapter 4, we found a deficiency in the set of whole numbers. The set of whole numbers is not closed under subtraction because there is no solution for operations like 5 − 7. Consequently, we need to extend the system of whole numbers so that subtraction will always have meaning.

After studying integers and the foundations of number theory, you will be presented with problems such as these: If Thanksgiving is on November 26, on what day of the week does Christmas fall? If July 4 is on Sunday, what will be the date of Labor Day? Problems of this type are easily solved with modular systems or clock arithmetic.

Problem-solving strategies illustrated in this chapter include the following: ▪ Guess, test, and revise. (209, 215, 228, 249) ▪ Use reasoning. (207, 209, 219, 228, 230, 242) ▪ Solve an equivalent problem. (251) ▪ Restate the problem. (219) ▪ Use a formula. (207) ▪ Use a variable. (207, 208, 216, 228) ▪ Use an equation. (208, 216) ▪ Use number characteristics. (throughout the chapter)

# THE SYSTEM OF INTEGERS

**PROBLEM**

The temperature in Chicago at noon was 0°C. Before bedtime, the temperature had dropped by 9°. What was the new temperature?

**OVERVIEW**

The everyday need for negative numbers is readily apparent. Such numbers provide us with a convenient method of distinguishing between 10°C above zero and 10°C below zero, 1 mile above sea level and 1 mile below sea level, a gain of 6 yards and a loss of 6 yards on the football field, and "I owe $50" and "I have $50." Negative numbers, part of the system of integers, are important in scientific study, business transactions, computer utilization, and so on.

The whole numbers we have studied in previous chapters are a subset of the integers and are called *nonnegative integers*.

**GOALS**

Illustrate the Number and Operations Standard, page 119.*
Illustrate the Representation Standard, page 50.*
Illustrate the Algebra Standard, page 4.*
Illustrate the Problem Solving Standard, page 15.*
Illustrate the Reasoning and Proof Standard, page 23.*

* The complete statement of the standard is given on this page of the book.

# ADDITIVE INVERSES

We begin our discussion of integers by finding additive inverses. Is it possible, for example, to find numbers $x$, $y$, and $z$ such that $1 + x = 0$, $2 + y = 0$, or $5 + z = 0$? If by *numbers* we mean the set of whole numbers 0, 1, 2, . . . , the answer is no. However, we can invent some new numbers that will satisfy such requirements. If you earn \$10 and then spend the \$10, your net holding is \$0. If the money spent is symbolized as ⁻\$10, we can say that

$$\$10 + {}^-\$10 = \$0$$

Similarly, the flood waters go up 6 feet and then go back down six feet. What is the effect of this? The water level is the same as it was before the flood, or $6 + {}^-6 = 0$. Let ⁻1 be a number such that $1 + {}^-1 = 0$; let ⁻2 be a number such that $2 + {}^-2 = 0$; and let ⁻5 be a number such that $5 + {}^-5 = 0$. In general, ⁻$n$ is a number such that $n + {}^-n = 0$.

| | |
|---|---|
| **D E F I N I T I O N** | **Additive Inverse**   We form a new set of numbers in which each number $n$ has a unique *additive inverse* ⁻$n$. For each such pair of numbers $n$ and ⁻$n$, $$n + {}^-n = {}^-n + n = 0$$ |

Thus, ⁻1 is the only number that, when added to 1, gives 0; ⁻20 is the only number that, when added to 20, gives 0; and in general ⁻$n$ is the only number that, when added to $n$, gives 0. Furthermore, if ⁻$n$ is the additive inverse of $n$, then $n$ is the additive inverse of ⁻$n$. For example, 6 is the additive inverse of ⁻6 because

$$6 + {}^-6 = 0$$

From the preceding definition, however, ⁻(⁻6) is an additive inverse of ⁻6, because

$${}^-({}^-6) + {}^-6 = 0$$

But because the additive inverse of ⁻6 must be unique, we stipulate that ⁻(⁻6) = 6. **Note: In general ⁻(⁻n) = n. Although ⁻3 can be read as "negative 3," ⁻x should not be read as "negative x" but as "the additive inverse of x" or "the opposite of x."** If x happens to be negative—for example, x = ⁻5—then

$$^-(^-5) = 5$$

which is positive.

What is the additive inverse of 0? It is a number ⁻0 such that 0 + ⁻0 = 0. But 0 itself has the property that 0 + 0 = 0. In order that the additive inverse be unique, we therefore insist that ⁻0 be the same as 0.

## CALCULATOR HINT

(a) The additive inverse of a number is obtained on a calculator by either a change of sign key ⎡+/−⎤ or a minus key ⎡(−)⎤. For example, 9 ⎡+/−⎤ = ⁻9, or ⎡(−)⎤ 9 ⎡=⎤ ⁻9

(b) On your calculator find ⁻(37 − 243) and ⁻(47 − 23 − 24).

⎡(−)⎤ ⎡(⎤ 37 ⎡−⎤ 243 ⎡)⎤ ⎡=⎤ 206

⎡(−)⎤ ⎡(⎤ 47 ⎡−⎤ 23 ⎡−⎤ 24 ⎡)⎤ ⎡=⎤ 0

---

**DEFINITION**

**Set of Integers**

The *set of integers I* is the set

$$I = W \cup {}^-N$$

where ⁻N represents the set of additive inverses of the natural numbers. Thus,

$$I = \{\ldots\ ^-3,\ ^-2,\ ^-1,\ 0,\ 1,\ 2,\ 3,\ \ldots\}$$

---

The natural numbers, as part of the integers, are often called **positive integers.** The set of positive integers is sometimes written as { +1, +2, + 3, . . . } or as {⁺1, ⁺2, ⁺3, . . . }, but in this book the plus signs are omitted. The positive integers and 0 are called the **nonnegative integers.** The set

$$\{\ldots,\ ^-4,\ ^-3,\ ^-2,\ ^-1\}$$

is called the set of **negative integers.**

Negative numbers are needed when a situation has two opposite directions. Examples are bank deposits and withdrawals, the ups and downs of temperature, profits and losses on the stock market, and gains and losses in football. In fact, we will need to learn how to add, subtract, multiply, and divide using such numbers.

## MODELS FOR INTEGERS

Many different models are available to assist you in understanding integers and operations with integers. Long ago, before the notion of a negative number was formally developed, Chinese traders used a system of colored rods to keep records of their transactions. One color denoted credits and another color denoted debits. We borrow this Chinese scheme to provide a concrete representation of integers.

Obtain a set of different-colored blocks or chips, and use one color for debits (or negative numbers) and another color for credits (or positive numbers). In the following diagrams, the colored blocks represent debits and the uncolored blocks represent credits. Notice that □ and ▪, □ □ and ▪ ▪, and □ □ □ and ▪ ▪ ▪ represent additive inverses.

□ ▪ represents $1 + {}^-1 = 0$, ▪ ▪ represents $2 + {}^-2 = 0$, and ▪ ▪ ▪ represents $3 + {}^-3 = 0$. ▪ ▪ ▪ or ▪ ▪ ▪ represents ${}^-1$, since □ □ and ▪ ▪ are additive inverses.

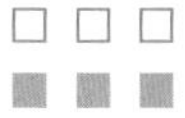

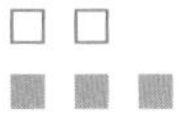

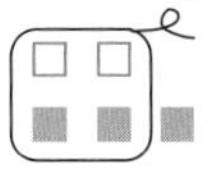

Similarly, ▪ ▪ ▪ or □ □ □ □ □ represents 2. Of course, the representation of integers by a combination of blocks is not unique. For example, ${}^-3$ may be represented by

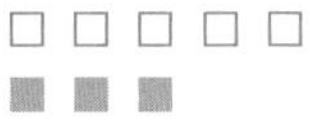

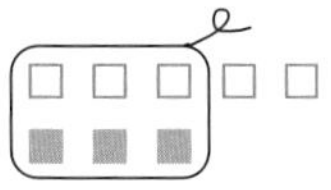

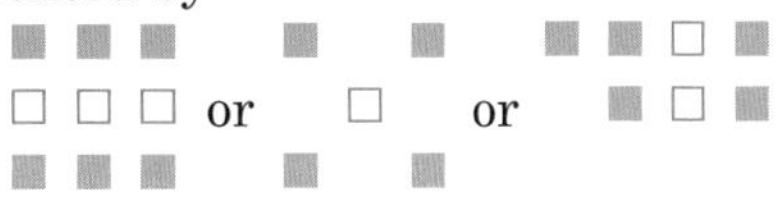

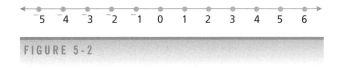

Likewise, integers may be represented by positive and negative electrical charges, where ⊕ represents a positive charge and ⊖ a negative charge. Thus, $\left(\frac{+}{-}\right)$ represents 0, and $\left(\frac{+}{--}\right)$ or $\left(\frac{++}{---}\right)$ represents ${}^-1$. (This model is very similar to the Chinese block model.)

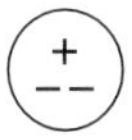

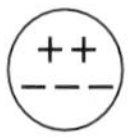

Possibly the most familiar model involves a thermometer. (See Figure 5-1.) Actually, the scale of a thermometer is a vertical number line, and a number line is an excellent model for integers.

A number line can be constructed for the integers just as one was constructed for the whole numbers. Mark an arbitrarily chosen point as 0, called the **origin.** Then measure equal segments to the right to determine points labeled 1, 2, 3, 4, . . . and to the left to determine points labeled ${}^-1$, ${}^-2, {}^-3, {}^-4, . . . ,$ as illustrated in Figure 5-2. By this process, you set up a one-to-one correspondence between a subset of points on the line and the integers. The positive integers extend to the right of zero, and the negative integers extend to the left of zero. Pairs of additive inverses, such as ${}^-5$ and 5, are represented by points at equal distances from zero.

FIGURE 5-1

FIGURE 5-2

Although the formal definition of *less than* is given in the next section, notice for now that ${}^-5$ lies to the left of ${}^-4$ on the number line in Figure 5-2 and ${}^-4$ lies to the left of ${}^-3$. In Chapter 4 when we observed that 2 was to the left of 3 on a number line, we said $2 < 3$. Therefore, intuitively from Figure 5-2, we can say

$$\ldots\, ^-5 < {}^-4 < {}^-3 < {}^-2 < {}^-1 < 0$$

whereas

$$0 < 1 < 2 < 3 < 4 < 5 \ldots$$

# ADDITION OF INTEGERS

By considering the different-colored blocks or chips as elements of sets, we can use a set model to illustrate the addition of integers. For example, one way to represent $^-5 + {}^-3$ is by

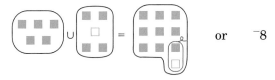

or $\qquad ^-5 + {}^-3 = {}^-8$

Another (and equally valid) way to represent this operation is by

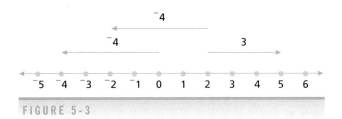

or $\qquad ^-8$

Integers may also be represented by arrow diagrams, as in Figure 5-3. The integer 3 may be represented by an arrow diagram that is directed toward the right and is of the same length as the line segment from 0 to 3 on the number line; for example, 3 can be represented by the line segment from 2 to 5 on the number line. Thus an arrow diagram may be *translated* or moved along the number line, as shown in Figure 5-3. In the same figure, the integer $^-4$ is represented by each of two arrow diagrams of length 4; notice that both arrow diagrams are directed toward the left.

FIGURE 5-3

In Chapter 4, we illustrated the addition of whole numbers by using arrow diagrams. We illustrate this procedure again for the following problem and then extend the concept for a similar problem. On the first down the Howard College Bulldogs threw for a pass gaining 5 yards. On the second down, the fullback ran for 3 yards. How many yards did the team gain on the two downs? In Figure 5-4(a) an arrow diagram shows the 5 yard gain and a second arrow on the end of the first one shows the 3 yard gain. The two gains amount to $5 + 3 = 8$ yards.

On the next set of downs, the right halfback tried an around the end play losing 5 yards. Then the fullback tried to go through the middle of the line and lost 3 yards. Note in Figure 5-4(b) the arrow diagram representing the 5 yards lost and on the end of this an arrow diagram representing the 3 yards lost. In this diagram it is easy to see that on the two downs the team lost 8 yards or a $^-8$ in the diagram. It is obvious that $^-5 + {}^-3 = {}^-8$.

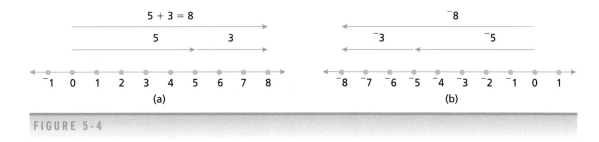

FIGURE 5-4

From our set model and from the number line model, do you agree with Part 2 of the following definition?

<table>
<tr><td>DEFINITION</td><td>**Addition of Integers**</td><td>Let $A$ and $B$ be disjoint sets such that $n(A) = a$ and $n(B) = b$. Then<br>1. $a + b = n(A \cup B)$<br>2. $^-a + {}^-b = {}^-(a + b)$<br>3. $a + {}^-b = {}^-b + a = (a - b)$    if $a > b$<br>4. $a + {}^-b = {}^-b + a = 0$    if $a = b$<br>5. $a + {}^-b = {}^-b + a = {}^-(b - a)$    if $a < b$</td></tr>
</table>

**EXAMPLE 1**

Find the following sums.
(a) $^-4 + {}^-3$        (b) $^-2 + {}^-7$

SOLUTION

(a) $^-7$            (b) $^-9$

**EXAMPLE 2**

Use a set-theory model to illustrate Part 3 of the previous definition for $6 + {}^-2$.

SOLUTION

or      $6 + {}^-2 = 4$

**EXAMPLE 3**

Again, on one set of downs the Bulldogs gained 6 yards and then on the next play lost 2 yards. How many yards did they lose or gain? Or what is $6 + {}^-2$?

SOLUTION

The gain and loss are represented by arrow diagrams in Figure 5-5. It is easy to see a four yard gain, or $6 + {}^-2 = 4$.

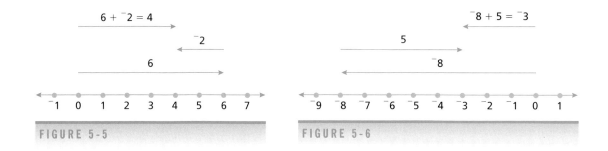

FIGURE 5-5                          FIGURE 5-6

**EXAMPLE 4**    From the pattern established in the preceding examples or from the definition of integer addition, find the following.
(a) $5 + {}^-3$        (b) $6 + {}^-2$        (c) $7 + {}^-2$

*SOLUTION*    (a) 2            (b) 4            (c) 5

**EXAMPLE 5**    Use a set-theory model to illustrate Part 5 of the previous definition for
${}^-8 + 5$.

*SOLUTION*

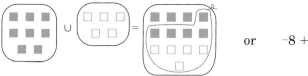          or      $-8 + 5 = -3$

**EXAMPLE 6**    Jill always checked her window thermometer to determine what clothes to
wear. Overnight the temperature dropped by 8 degrees. By noon the next
day the temperature had increased by five degrees. What was the result of
these two changes? On the number line this problem becomes: Find the sum
of ${}^-8 + 5$.

*SOLUTION*    In Figure 5-6 we see the temperature had decreased by 3 degrees.

**EXAMPLE 7**    From the pattern established in the preceding examples or from the definition of integer addition, find the following.
(a) ${}^-7 + 2$        (b) $4 + {}^-9$        (c) ${}^-5 + 3$

*SOLUTION*    (a) ${}^-5$            (b) ${}^-5$            (c) ${}^-2$

In the previous examples, we motivated the definition of integer addition using number lines and chips. However, it is important to observe that
the definition given is the only definition that will ensure that the properties of addition on the whole numbers are preserved as we extend the whole
numbers to the integers.

Recall that the system of whole numbers consists of the set $W = \{0, 1, 2, 3, \ldots\}$ and that the operations $+$ and $\cdot$ satisfy several properties: closure, associative, commutative, and distributive properties; and the existence of additive and multiplicative identities. Extending the whole numbers to the system of integers is done in such a way that the two operations $+$ and $\cdot$ retain these properties but add one more: the existence of a unique additive inverse for each element. The operations of $+$ and $\cdot$ are defined so that the operations involving whole numbers within the system of integers are the same as those occurring entirely within the system of whole numbers.

**System of Integers**

The *system of integers* consists of the set $I$ of integers $\{ \ldots, \ ^-4, \ ^-3, \ ^-2, \ ^-1, 0, 1, 2, 3, 4, \ldots \}$ and two binary operations, addition, $+$, and multiplication, $\cdot$, with the following properties for any integers $a$, $b$, and $c$.

**Closure Properties**
1. $a + b$ is a unique integer.       2. $ab$ is a unique integer.

**Commutative Properties**
3. $a + b = b + a$       4. $ab = ba$

**Associative Properties**
5. $(a + b) + c = a + (b + c)$       6. $(ab)c = a(bc)$

**Distributive Properties of Multiplication over Addition**
7. $a(b + c) = ab + ac$       8. $(b + c)a = ba + ca$

**Identity Elements**
9. There is a unique integer 0 such that
$$0 + a = a + 0 = a$$
10. There is a unique integer 1 such that
$$1 \cdot a = a \cdot 1 = a$$

**Additive Inverses**
11. For each $a$ in $I$, there exists a unique additive inverse $^-a$ such that
$$a + {}^-a = {}^-a + a = 0$$

(This is the property that distinguishes the integers from the whole numbers.)

Let us now investigate the addition of two integers, using the properties listed in the preceding box. We consider first the case where both integers are negative.

**EXAMPLE 8**    Use the preceding properties to show that $^-5 + {}^-3 = {}^-8$.

*SOLUTION*

$$(^-5 + {}^-3) + (5 + 3) = (^-5 + {}^-3) + (3 + 5) \qquad \text{Commutative property of addition}$$
$$= {}^-5 + [^-3 + (3 + 5)] \qquad \text{Associative property of addition}$$
$$= {}^-5 + [(^-3 + 3) + 5] \qquad \text{Associative property of addition}$$
$$= {}^-5 + [0 + 5] \qquad \text{Additive inverse}$$
$$= {}^-5 + 5 \qquad \text{Additive identity}$$
$$= 0 \qquad \text{Additive inverse}$$

Therefore,

$$(^-5 + {}^-3) + (5 + 3) = 0 \qquad \text{Transitive property of equality}$$

Thus the number $^-5 + {}^-3$ is the additive inverse of $5 + 3$. But, by definition, the unique additive inverse of $5 + 3$ is $^-(5 + 3) = {}^-8$. Hence,

$$^-5 + {}^-3 = {}^-8$$

The algebraic reasonableness of other parts of the definition of integer addition will be considered in the exercise set for this section.

 **ABSOLUTE VALUE**

Many books state the definition of integer addition in terms of absolute value.

---

**DEFINITION**

**Absolute Value**    The *absolute value* of an integer $n$ is the distance of $n$ from the origin 0, where distance is always nonnegative. The absolute value of an integer $n$ is written as $|n|$. Thus

$$|n| = n \qquad \text{if } n > 0$$
$$|n| = 0 \qquad \text{if } n = 0$$
$$|n| = {}^-n \qquad \text{(the additive inverse of } n) \text{ if } n < 0$$

---

$$|^-6| = 6; \qquad |^-3| = 3; \qquad |^-4| = 4; \qquad |0| = 0$$

**EXAMPLE 9**    Clearly, the absolute value of an integer is always a nonnegative integer.

We can now state in words the rule for the addition of integers.

---

**DEFINITION**

**Addition of Signed Numbers**    1. To add two *signed numbers* that have the same sign, add their absolute values and affix their common sign.
2. To add two signed numbers that have unlike signs, find the difference in their absolute values, and affix to the answer the sign of the number whose absolute value is larger.

---

**EXAMPLE 10**

Verify the following examples.
$$2 + (^-5) = ^-(5 - 2) = ^-3$$
$$7 + (^-4) = 7 - 4 = 3$$
$$^-3 + (^-5) = ^-(3 + 5) = ^-8$$
$$^-7 + 9 = 9 + (^-7) = 9 - 7 = 2$$
$$^-9 + 5 = 5 + (^-9) = ^-(9 - 5) = ^-4$$
$$^-16 + (^-17) = ^-(16 + 17) = ^-33$$

**PRACTICE PROBLEM**

Find $(^-5 + 3) + ^-4$.

*ANSWER*

$^-6$

## CALCULATOR HINT

To add negative numbers on a calculator, use a $\boxed{+/-}$ key or a minus key, $\boxed{(-)}$. With the $\boxed{(-)}$ key you add $4 + ^-7$ and $(^-5 + 3) + ^-4$ as follows:

$4 \boxed{+} \boxed{(-)} 7 \boxed{=}$    Did you get an answer of $^-3$?

$\boxed{(}\boxed{(-)} 5 \boxed{+} 3 \boxed{)} \boxed{+} \boxed{(-)} 4 \boxed{=}$    Did you get $^-6$?

# SUBTRACTION OF INTEGERS

For whole numbers $x$, $y$, and $z$, $x - y = z$ if and only if $x = y + z$. Subtraction for integers will be defined consistently with the definition of subtraction for whole numbers.

**DEFINITION**

**Subtraction of Integers**

If $x$ and $y$ are integers, the **difference** in the *subtraction* of $y$ from $x$ (denoted by $x - y$) is the integer $z$ if and only if $x = y + z$.

**EXAMPLE 11**

$$6 - 4 = 2 \quad \text{because} \quad 6 = 4 + 2$$
$$^-2 - 5 = ^-7 \quad \text{because} \quad ^-2 = 5 + ^-7$$

Now $^-4 - 2 = ^-6$, because $^-4 = 2 + ^-6$. Likewise, $^-4 + ^-2 = ^-6$. So $^-4 - 2 = ^-4 + ^-2$. This example illustrates that subtraction is equivalent to addition of the additive inverse of the number being subtracted.

**Subtraction Property**

If $x$ and $y$ are integers, then there always exists a unique integer $z$ such that $x - y = z$. This integer $z$ can be written as $z = x + ^-y$. That is,

$$x - y = x + ^-y$$

## COMMON ERROR

$$3 - (^-2) \neq 1; \quad 3 - (^-2) = 3 + 2 = 5$$

**To subtract in the set of integers, change the sign of the number being subtracted and then proceed as in addition.** This process enables you to reduce subtractions involving positive and negative integers to calculations involving addition only.

**EXAMPLE 12**

(a) $8 - 6 = 8 + {}^-6 = 2$      ⁻6 is the additive inverse of 6.
(b) $^-12 - {}^-10 = {}^-12 + 10 = {}^-2$      10 is the additive inverse of ⁻10.

## CALCULATOR HINT

To find $^-12 - {}^-10$ using your calculator, press the following sequence of keys:

$\boxed{(-)}$ 12 $\boxed{-}$ $\boxed{(-)}$ 10 $\boxed{=}$      Did you get ⁻2, as in the preceding example?

**EXAMPLE 13**

With a set model, use $^-2 - 3$ to show subtraction as the addition of the additive inverse.

*SOLUTION*

⁻2    +    ⁻3    =    ⁻5

(□ □)  ∪  (□ □ / □)  =  (□ □ □ / □ □ □)

From the preceding property of subtraction, we can consider subtraction on a number line as reversing the direction of the arrow of the number being subtracted.

**EXAMPLE 14**

Find $^-3 - 4$.

*SOLUTION*

The arrow for 4, which is being subtracted from $^-3$, points to the right. Change the direction of the arrow to point toward the left, representing $^-4$, and add. The answer, as illustrated in Figure 5-7 is $^-7$.

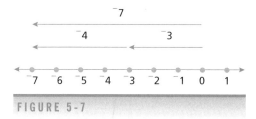

FIGURE 5-7

**EXAMPLE 15**

Find the solution set of $x + {}^-4 = {}^-6$ on the domain of integers.

*SOLUTION*

If $x + {}^-4$ equals ${}^-6$, then, by the definition of subtraction, $x = {}^-6 - {}^-4 = {}^-6 + 4 = {}^-2$. This solution is represented by the black dot in Figure 5-8.

$$ \overset{\text{-3 -2 -1 0 1 2 3}}{\longleftrightarrow} $$

**FIGURE 5-8**

Because the operation of subtracting 5 is equivalent to adding ${}^-5$, most books use the same notation for the sign of a negative number and for the operation of subtraction. That is, ${}^-5$ is indicated by $-5$, the same notation used in subtracting 5 from 8 ($8 - 5$). Starting in Chapter 6, this book will use the standard notation.

## COMMON ERROR

Subtraction for integers is commutative. (No!)

$$ {}^-2 - {}^-3 \neq {}^-3 - {}^-2 $$
$$ 1 \neq {}^-1 $$

# Just for Fun

A boy and a girl ran a 100-meter race. The girl crossed the finish line when the boy had gone 95 meters, so she won the race by 5 meters. When they raced a second time, the girl wanted to make the contest more even, so she started at −5 meters. If the two ran at the same constant speed as before, who won the race? (If you think it was a tie, you had better start using your problem-solving techniques.)

# EXERCISE SET 1

*R*1. What are the additive inverses of the following integers?

    (a) 5      (b) ⁻3

    (c) 0      (d) ⁻8

    (e) $a$      (f) ⁻$a$

2. Draw arrow diagrams on a number line to represent the following integers. Start each arrow diagram at 0.

    (a) ⁻2     (b) ⁻4     (c) 6

3. Draw arrow diagrams starting at ⁻3 for each integer in Exercise 2.

4. From the following list, select the pairs that are additive inverses.

    (a) ⁻2, 2        (b) ⁻3, (⁻3)

    (c) 7, ⁻8        (d) (2 · 3), ⁻(3 + 3)

    (e) (2 + 3), ⁻(2 + 3)   (f) 0, 0

    (g) 2, 2        (h) 1, ⁻1

5. Perform the indicated operations. Then check (a)–(n) with your calculator.

    (a) 2 − 5       (b) 3 − ⁻1

    (c) ⁻6 − 2      (d) 4 − 13

    (e) ⁻10 + ⁻3    (f) 6 − ⁻2

    (g) ⁻8 + ⁻3     (h) 13 − ⁻6

    (i) ⁻9 − 7      (j) 500 − ⁻5

    (k) (6 − 4) − ⁻2   (l) ⁻7 − (6 − 4)

    (m) (⁻4 + 8) + ⁻6  (n) (6 + ⁻6) + ⁻2

    (o) ($a$ + $b$) + ⁻$b$   (p) ($x$ + $y$) + ⁻($x$ + $y$)

6. Find the following sums by using the number line.

    (a) ⁻6 + ⁻5    (b) 6 + ⁻3    (c) ⁻5 + ⁻3

    (d) 8 + ⁻1     (e) 2 + ⁻6    (f) 0 + ⁻7

7. For each of the properties shown, state whether the property holds for subtraction of integers. If a property does not hold, disprove it by giving a counterexample. If a property is valid, illustrate it with an example.

    (a) associative    (b) commutative

8. If the domain is the set of integers, find the solution set for each equation and graph the solution set on a number line.

    (a) $x$ + ⁻3 = 2     (b) $x$ − ⁻5 = 4

    (c) $x$ + ⁻5 = ⁻2    (d) $x$ − ⁻6 = ⁻3

9. Compute the following without a calculator. Then check with a calculator.

    (a) ⁻3 + 4 + 7 + ⁻6 + ⁻8

    (b) ⁻6 − 7 + 8 − ⁻6 + ⁻5

*For Exercises 10 through 14, write an arithmetic expression that provides a model for the situation and perform the calculations.*

10. Suppose that you have $89 in the bank. If you write one check for $79 and then deposit $69, how much money do you have in the bank? How much do you have in the bank if you write a second check for $100?

11. The temperature was recorded last night as ⁻16°C. At noon, the temperature had increased by 8°. What was the temperature at noon?

12. Determine whether Fred or Ed has more points. Fred has earned 10 points, lost 6, lost 2, earned 5, lost 7, lost 3, and earned 8. Ed has earned 6 points, earned 2, earned 4, lost 7, and lost 5.

13. Dennis has a checking account. At the first of the month, he deposited $448, adding to a previous balance of $121. He later wrote six checks for $135, $225, $12, $20, $139, and $80 and then made another deposit of $250. If the service charge for the overdrawn account was $20, what balance did Dennis have at the end of the month?

14. The University of Mystate football team, the Wild Geese, made the following plays in their first game. In each case, tell whether the team made the 10 yards for a first down, and give the number of yards they made over the necessary 10 or the number of yards they lacked to make the first down.

    (a) The Geese received the ball on the 4-yard line and gained 15 yards.

    (b) The Geese completed a pass for 8 yards and then were pushed back 4 yards. After one futile attempt to move the ball, they ran around the left end for 5 yards.

    (c) The Geese marched down the field 7 yards. Then the quarterback fell down while trying to complete a pass and lost 3 yards. They made 4 yards on each of the next 2 plays.

    (d) The Geese received a 10-yard penalty on the first play for being offsides. They completed a 7-yard pass and a 9-yard pass and then twice tested the middle for 3 yards on each try.

15. Express these changes as positive or negative integers.

    (a) The change in elevation from an airplane 10,000 feet above sea level to a submarine 500 feet below sea level

(b) The change in position of the ball if a football team first loses 24 yards and then gains 8 yards

(c) The change in elevation from a desert 100 feet below sea level to the top of a mountain 3400 feet above sea level

16. Classify as either true or false.

(a) Every integer is a whole number.

(b) The set of integers is the same as the set of additive inverses of whole numbers.

(c) Every counting number is an integer.

(d) Every integer has an additive inverse.

(e) The intersection of the positive integers and the negative integers is the null set.

(f) The union of the negative integers and the counting numbers is the set of integers.

17. (a) With a signed number, describe the position of a vein of coal 150 meters under ground level.

(b) Aristotle was born in 384 B.C., and Euclid was born in 365 B.C. Who was born first?

(c) Which temperature is higher, $^-22°C$ or $^-12°C$?

(d) Explain each of the above with a diagram.

18. Evaluate.

(a) $\left|^-7\right|$   (b) $\left|0\right|$   (c) $\left|12\right|$   (d) $\left|^-6\right| + \left|^-5\right|$
(e) $\left|2\right| + \left|^-3\right|$   (f) $\left|^-8\right| - \left|^-6\right|$

19. Find the additive inverse of each of the following.

(a) $^-(21 + 5)$   (b) $^-(a - b)$   (c) $^-c$
(d) $\left|x\right|$   (e) $^-\left|5\right|$   (f) $\left|x - y\right|$

20. Perform the following operations with your calculator.

(a) $^-411 + 648$   (b) $^-782 + {}^-647$
(c) $^-286 - 450$   (d) $239 - 684$

21. (a) What is the smallest nonnegative integer?

(b) What is the smallest positive integer?

(c) What is the largest positive integer?

(d) What is the smallest negative integer?

(e) What is the largest negative integer?

(f) Is 0 different from $^-0$?

22. Find the following sums.

(a) $^-2x + 3x + {}^-7x - 4x$

(b) $^-4a - 6a - (^-2a) - 5a$

(c) $^-10y^2 + (^-3y^2) - 4y^2 - (^-6y^2)$

(d) $^-8k^2 - (^-3k^2) - 2k^2 + (^-4k^2)$

23. If $W$ stands for the whole numbers, $I$ for the integers, $^-I$ for the negative integers, and $^+I$ for the positive integers, find the following.

(a) $^-I \cap {}^+I$   (b) $^-I \cap W$   (c) $^-I \cup {}^+I$
(d) $^-I \cup W$   (e) $^+I \cap W$   (f) $W \cup {}^+I$

24. Use a number line to illustrate the following:

(a) $^-5 - {}^-3$   (b) $^-6 - {}^-2$   (c) $^-4 - 3$

25. Use a number line to illustrate each of the following additions.

(a) $^-2 + {}^-4$   (b) $^-3 + 2$

26. Prove that $^-(^-x) = x$.

27. Give the reasons in the following illustration of why $6 + {}^-2 = 6 - 2$.

$$6 + {}^-2 = (4 + 2) + {}^-2 \qquad \text{Why?}$$
$$= 4 + (2 + {}^-2) \qquad \text{Why?}$$
$$= 4 + 0 \qquad \text{Why?}$$
$$= 4$$

But $6 - 2 = 4$. Thus,

$$6 + {}^-2 = 6 - 2 \qquad \text{Why?}$$

28. Give the reason for each step.

$$^-8 + 5 = (^-5 + {}^-3) + 5 \qquad \text{Why?}$$
$$= (^-3 + {}^-5) + 5 \qquad \text{Why?}$$
$$= {}^-3 + (^-5 + 5) \qquad \text{Why?}$$
$$= {}^-3 + 0 \qquad \text{Why?}$$
$$= {}^-3$$

But $^-(8 - 5) = {}^-3$. Thus,

$$^-8 + 5 = {}^-(8 - 5) \qquad \text{Why?}$$

29. Working from left to right, identify the property or properties that make the following true.

(a) $(8 + 4) + {}^-2 = (8 + {}^-2) + 4$

(b) $(6 + {}^-6) + 0 = (0 + 6) + {}^-6$

(c) $(^-4 + 6) + {}^-2 = (^-2 + 6) + {}^-4$

## LABORATORY ACTIVITY

30. Use colored blocks to illustrate each part of Exercise 25.

### ❖ PCR Excursion ❖

31. Use only once the integers $^-4$, $^-3$, $^-2$, $^-1$, 0, 1, 2, 3, 4 to form a $3 \times 3$ additive magic square. (The sums of the integers in every row, column, and diagonal are equal.)

(a) What is the sum of all integers given? What does this tell you?

(b) Now form your $3 \times 3$ magic square.

(c) We wish to use the integers $^-6$, $^-5$, $^-4$, $^-3$, $^-2$, $^-1$, 0, 1, 2 to form a $3 \times 3$ magic square. Again, what does the sum of the given integers indicate to you?

(d) Form your $3 \times 3$ magic square.

(e) We wish to use $^-6$, $^-5$, $^-4$, $^-3$, $^-2$, $^-1$, 0, 1, 2, 3, 4, 5, 6, 7, 8, 9 to form a $4 \times 4$ magic square. What does the sum of the given integers indicate to you?

(f) Form your $4 \times 4$ magic square.

# INTEGER MULTIPLICATION AND DIVISION

**PROBLEM**

The football team at Mystate U had some difficulty in its last game. On each of four sets of downs in the first quarter, the team lost 6 yards. How many yards did the team lose during the first quarter?

**OVERVIEW**

In this section, we study the operations of multiplication and division for integers. Again, the extension from whole numbers to integers is made in such a way that the properties of multiplication and division for whole numbers hold for integers.

**GOALS**

Illustrate the Number and Operations Standard, page 119.*
Illustrate the Representation Standard, page 50.*
Illustrate the Algebra Standard, page 4.*
Illustrate the Connections Standard, page 147.*

\* The complete statement of the standard is given on this page of the book.

## MULTIPLICATION OF INTEGERS

Read again the problem presented above. Of course, this example can be considered as a multiplication problem: $4 \cdot {}^-6 = ?$ To solve such problems, we define multiplication for the integers so that the closure, commutative, associative, and identity properties of multiplication, the multiplicative property of 0, and the distributive property of multiplication over addition, as discussed for the whole numbers, hold for the set of integers.

**EXAMPLE 1**

The football team at Mystate U had some difficulty in the last game. On each of four downs at the end of the first quarter the team lost 6 yards. How many yards did the team lose on these four downs?

*SOLUTION*

Our problem can be replaced by the question "How do we define $4 \cdot {}^-6$ in order to be consistent with the idea of multiplication of whole numbers?" In whole numbers we defined $4 \cdot 6$ to be 6 added four times. Similarly, we define $4 \cdot {}^-6$ to be

$$^-6 + {}^-6 + {}^-6 + {}^-6 = {}^-24$$

Thus, the football team lost 24 yards in the first quarter.

In interpreting multiplication as repeated addition, notice that $4({}^-6) = {}^-24$ on the number line in Figure 5-9.

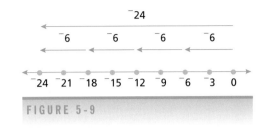

**FIGURE 5-9**

$3 \cdot 4 = 12$
$2 \cdot 4 = 8$
$1 \cdot 4 = 4$
$0 \cdot 4 = 0$
$^-1 \cdot 4 = \underline{\hspace{1cm}}$
$^-2 \cdot 4 = \underline{\hspace{1cm}}$
$^-3 \cdot 4 = \underline{\hspace{1cm}}$
$^-4 \cdot 4 = \underline{\hspace{1cm}}$
$^-5 \cdot 4 = \underline{\hspace{1cm}}$
$^-6 \cdot 4 = ^-24$

Since multiplication is commutative, it follows that

$$^-6 \cdot 4 = 4 \cdot ^-6 = ^-24$$

This idea of multiplication is emphasized by the pattern in the margin.

Evidently, the answer is decreased by 4 each time the multiplier is decreased by 1. Can you complete the blanks?

**EXAMPLE 2**

How should we define $^-3 \cdot ^-4$, if the properties of multiplication of whole numbers are to hold for integers?

*SOLUTION*

Consider $^-3 \cdot ^-4 + ^-3 \cdot 4 = ^-3(^-4 + 4) = ^-3(0) = 0$. Thus, $^-3 \cdot ^-4$ is the additive inverse of $^-3 \cdot 4$, and $^-3 \cdot 4 = ^-12$. But the unique additive inverse of $^-12$ is 12, so $^-3 \cdot ^-4 = 12$.

$3 \cdot ^-4 = ^-12$
$2 \cdot ^-4 = ^-8$
$1 \cdot ^-4 = ^-4$
$0 \cdot ^-4 = 0$
$^-1 \cdot ^-4 = \underline{\hspace{1cm}}$
$^-2 \cdot ^-4 = \underline{\hspace{1cm}}$
$^-3 \cdot ^-4 = \underline{\hspace{1cm}}$

The preceding product can also be inferred from the pattern in the margin.

Notice that, as the multipliers decrease by 1, the answers increase by 4. Complete the blanks, and see if your answers agree with the preceding example.

Such examples illustrate the following definition of the multiplication of integers.

**DEFINITION**

**Multiplication of Integers**

The **product of two integers** is defined by the following cases, where $a$ and $b$ are whole numbers.

1. $a \cdot b = n(A \times B)$, where $a = n(A)$ and $b = n(B)$ for sets $A$ and $B$.
2. $^-a \cdot ^-b = ab$
3. $^-a \cdot b = b \cdot ^-a = ^-(ab)$

We can restate this definition in two parts as follows:

1. **The product of two numbers with unlike signs is negative.**

2. **The product of two numbers with like signs is positive.**

**EXAMPLE 3**

$$5(^-3) = {}^-15$$
$$(^-7)(6) = {}^-42$$
$$(^-6)(^-17) = 6 \cdot 17 = 102$$
$$(^-2)(^-5)(^-3) = (^-2)[(^-5)(^-3)]$$
$$= (^-2) \cdot 15$$
$$= {}^-30$$

## COMMON ERROR

$(^-3)^4$ is not the same as $^-3^4$.   $(^-3)^4 = (^-3)(^-3)(^-3)(^-3) = 81$
$^-3^4 = {}^-(3)(3)(3)(3) = {}^-81$

## CALCULATOR HINT

To find $^-211(^-362 - 48)$, press the following sequence of keys:

$\boxed{(-)}$ 211 $\boxed{\times}$ $\boxed{(}$ $\boxed{(-)}$ 362 $\boxed{-}$ 48 $\boxed{)}$ $\boxed{=}$   Did you get 86,510?

## DIVISION OF INTEGERS

Multiplication has an inverse operation, division. For instance, because $(2)(8) = 16$, then 16 divided by 8 equals 2, and 16 divided by 2 equals 8. Division for integers is defined in the following manner.

| DEFINITION | Division of Integers | If $x$ and $y$ are integers, with $y \neq 0$, then $x$ divided by $y$ is equal to an integer $z$ if and only if $x = y \cdot z$. |
|---|---|---|

To determine whether the answer in a particular division is positive or negative, we refer to the definition of multiplication. Because $x \div y = z$ if and only if $x = y \cdot z$, the sign of $z$ will be fixed so that the product equals $x$. We may summarize as follows.

| Sign of a Quotient | The quotient (if it exists) of two positive or two negative integers is a positive integer; the quotient (if it exists) of a negative and a positive integer, in either order, is a negative integer. |
|---|---|

**EXAMPLE 4**

$$^-12 \div 3 = {}^-4 \qquad \text{because} \qquad 3 \cdot {}^-4 = {}^-12$$
$$45 \div {}^-5 = {}^-9 \qquad \text{because} \qquad {}^-5 \cdot {}^-9 = 45$$
$$^-24 \div {}^-6 = 4 \qquad \text{because} \qquad {}^-6 \cdot 4 = {}^-24$$
$$^-16 \div 5 = {}?$$

Division of integers is not closed, because $^-16 \div 5$ has no meaning in the set of integers. There is no integer that, when multiplied by 5, yields $^-16$. The number system will be extended in Chapter 6 to overcome this deficiency.

## CALCULATOR HINT

Division can be handled in two ways on a calculator. Consider $^-33 \div {}^-11$.

[(−)] 33 [÷] [(−)] 11 [=]  or  [(−)] 11 [x⁻¹] [×] [(−)] 33 [=]   The [x⁻¹] divides 1 by $^-11$.

The answer is 3 in both cases.

Remember that zero plays a very special role in division.

**Properties of Zero**

Let $a$ be any integer. If $a \neq 0$, then $0 \div a = 0$, because $a \cdot 0 = 0$. If $a \neq 0$, then $a \div 0$ is undefined, because $a = 0 \cdot d$ is false for any $d$. $0 \div 0$ is undefined, because there is no unique quotient.

**EXAMPLE 5**

Solve $^-2x + 3 = 9$, using the facts that addition and subtraction are inverse operations and multiplication and division are inverse operations.

*SOLUTION*

$$^-2x + 3 = 9$$
$$^-2x = 9 - 3 \qquad \text{If } a + b = c, \text{ then } a = c - b.$$
$$= 6$$
$$x = 6 \div {}^-2 \qquad \text{If } ab = c, \text{ then } a = c \div b.$$
$$= {}^-3$$

To check this result, we can compute

$$^-2({}^-3) + 3 = 9$$
$$6 + 3 = 9$$
$$9 = 9$$

## ⬠ ORDERING INTEGERS

In the previous section we noticed that, if one integer is less than another, it is located to the left of the second number on a number line. For example, $^-1$ is less than 2, and it lies to the left of 2 on the number line. If you add the positive integer 3 to $^-1$, you get an answer of 2:

$$^-1 + 3 = 2$$

Similarly, $^-3 < ^-1$ and $^-3 + 2 = ^-1$. (Again, the difference, 2, is a positive integer.) This suggests a definition for *less than* for the set of integers very similar to that for the set of whole numbers.

**DEFINITION**

**Less Than**

For integers $a$ and $b$, $a$ is said to be less than $b$, denoted by $a < b$, if and only if there exists a positive integer $c$ such that $a + c = b$.

Of course, when we state that $a$ is less than $b$, we imply that $b$ is greater than $a$.

**EXAMPLE 6**

Show that $^-3 < 2$ and $^-6 < ^-4$, by finding in each case a positive integer $c$ as described in the preceding definition.

SOLUTION

$^-3 < 2$     because     $^-3 + 5 = 2$     5 is the positive integer.

$^-6 < ^-4$     because     $^-6 + 2 = ^-4$     2 is the positive integer.

**Transitive Property of Less Than**

Let $a$, $b$, and $c$ be any integers. If $a < b$ and $b < c$, then

$$a < c$$

For example, $^-3 < ^-1$ and $^-1 < 2$; therefore, $^-3 < 2$.

**EXAMPLE 7**

Find the solution set of $x + ^-4 < ^-3$ on the domain of integers, and graph it on a number line.

SOLUTION

Using the strategy of guess, test, and revise, we discover the set of solutions.

Try $x = 2$:     $2 + ^-4 = ^-2 \nless ^-3$     No solution

Try $x = 1$:     $1 + ^-4 = ^-3 \nless ^-3$     No solution

Try $x = 0$:     $0 + ^-4 = ^-4 < ^-3$     Solution

Try $x = -1$:     $^-1 + ^-4 = ^-5 < ^-3$     Solution

The sum of $^-4$ and any integer larger than 2 is larger than $^-2$. Because this sum is not less than $^-3$, such an integer is not a solution.

The sum of $^-4$ and any integer less than $^-1$ is less than $^-5$. Because this sum is less than $^-3$, such an integer is a solution. It follows that all integers less than 1 satisfy the given inequality. (See Figure 5-10.)

$^-3 \quad ^-2 \quad ^-1 \quad 0 \quad 1 \quad 2 \quad 3$

FIGURE 5-10

# Just for Fun

Insert a pair of parentheses to make the following a true statement.

$$1 - 2 + 3 - 4 + 5 - 6 = {}^{-}1$$

# EXERCISE SET 2

**R** 1. Write the following products and quotients as integers.

(a) $5({}^{-}9)$   (b) ${}^{-}7({}^{-}4)$   (c) $0 \div {}^{-}9$
(d) ${}^{-}24 \div {}^{-}3$   (e) ${}^{-}7(46)$   (f) ${}^{-}5(42)$
(g) ${}^{-}6({}^{-}5)$   (h) ${}^{-}4(3)$   (i) $(3 \cdot {}^{-}2) \div {}^{-}3$
(j) $8({}^{-}3) \div {}^{-}4$   (k) ${}^{-}9({}^{-}5 \cdot {}^{-}3)$   (l) ${}^{-}3({}^{-}2 \cdot 4)$

2. Perform the following computations without a calculator. Then check your answer with a calculator on all except (k).

(a) ${}^{-}3({}^{-}4 \cdot 6)$       (b) $({}^{-}2 \cdot {}^{-}5) \cdot 7$
(c) $({}^{-}6 \cdot {}^{-}4) \cdot (8 \cdot {}^{-}2)$       (d) $({}^{-}6 \cdot {}^{-}4) \cdot (0 \cdot 3)$
(e) $(23 - 5) \div {}^{-}6$       (f) $({}^{-}6 + {}^{-}18) \div {}^{-}4$
(g) $({}^{-}27 + {}^{-}5) \div {}^{-}4$       (h) ${}^{-}3(5 + {}^{-}1) \div 6$
(i) $({}^{-}4 + {}^{-}8) \cdot (5 + {}^{-}7)$       (j) $({}^{-}6 + {}^{-}1)({}^{-}3 + {}^{-}2)$
(k) $({}^{-}x \cdot {}^{-}y) \cdot {}^{-}z$       (l) $({}^{-}4 + 4)(3 + {}^{-}6)$

3. Compute each part in two different ways, using the distributive property.

(a) ${}^{-}1(3 + {}^{-}5)$       (b) $(3 + {}^{-}2)(4)$

4. Generalize the distributive property for integers, and expand each of the following into the sum of three terms.

(a) ${}^{-}5(a + b + c)$       (b) $3(2x + y + z)$
(c) ${}^{-}y(3 + {}^{-}2x + 4z)$       (d) ${}^{-}x(2 + 3z + y)$

5. Under what conditions will the product of two factors be

(a) greater than 0?       (b) equal to 0?
(c) less than 0?

6. (a) For what integral values of $a$ and $b$ does $a \div b = b \div a$?

(b) What is the value of ${}^{-}24 \div (4 \div {}^{-}2)$?
(c) What is the value of $({}^{-}24 \div 4) \div {}^{-}2$?
(d) What can you conclude from parts (b) and (c) about the associativity of division for integers?

7. Solve on the domain of integers.

(a) ${}^{-}2x + 6 = 4$       (b) ${}^{-}3x - 6 = {}^{-}9$
(c) ${}^{-}2x - 3 = 5$       (d) ${}^{-}3x + 4 = {}^{-}5$

*In Exercises 8 through 18, first find an equation that serves as a model to represent each problem. Then solve the equation.*

8. Signs are placed at regular intervals along a mountain road to indicate altitude. At Greenbriar Lodge, the sign indicates an elevation of 876 meters. But 17 kilometers down the road, another sign indicates that the elevation is 672 meters. What is the average elevation change per kilometer traveled from the lodge?

9. Find 3 consecutive integers whose sum is ${}^{-}75$.

10. On a particular Monday the Dow-Jones Industrial average closed at 10,000. On Wednesday it closed down twice as many points as it gained Tuesday. On Thursday it lost 10 points. On Friday it gained 50 points and closed at 9960. How much did the market lose on Wednesday?

11. The Dalley Doughnut Shop began operations $12,000 in debt. It has been losing about $900 a month and is now in debt $16,500. How long has the shop been in operation?

12. A plane flying at an altitude of 10,000 meters descends at a rate of 800 meters per minute for 10 minutes. What is the altitude of the plane relative to a mountain of height 2000 meters?

13. In northeastern Alaska, the average temperature has fallen by 4°C per day for a week. If the temperature is now $^-45$°C, what was the temperature 5 days ago?

14. (a) In Exercise 13, what was the temperature 2 days ago?
    (b) If the pattern continues, what will be the temperature 2 days from now?

15. The temperature now at the North Pole is $^-40$°C. The temperature is expected to drop by 6° per day for the next few days. Write an open sentence for the temperature $d$ days from now.

16. In Exercise 15, what will be the temperature 4 days from now?

17. The highest temperature during the day in Atlanta is 94°F. The high temperature is expected to decrease by 2°F per day for several days. When will the high temperature be 88°F?

18. In Dome, Alaska, the temperature has been falling 3° per day for the last 10 days. If the temperature is now 30° below 0, what was the temperature a week ago? If the pattern continues, what will be the temperature 1 week from today?

$T$ 19. Write the next term of the following sequences:
    (a) $^-5, ^-11, ^-17, ^-23,$ _____
    (b) $8, 5, 2, ^-1,$ _____
    (c) $1, ^-3, 9, ^-27,$ _____
    (d) $^-2, 8, ^-32, 128,$ _____
    (e) Do you remember the names of these sequences?

20. Simplify the following expressions.
    (a) $\dfrac{(^-x)^2}{(^-x)^5}$
    (b) $(^-x)^2 \cdot (x)^2$
    (c) $\dfrac{(^-x)^3}{x^2} \cdot \dfrac{(^-x)^2}{x}$
    (d) $\dfrac{x^2}{(^-x)^3} \cdot \dfrac{(^-x)^2}{x}$

21. Show the following multiplications on a number line.
    (a) $2(^-5)$   (b) $3(^-2)$

22. Let
$$W = \{0, 1, 2, 3, \ldots\}$$
$$M = \{0, ^-1, ^-2, ^-3, \ldots\}$$
$$I = \{\ldots, ^-3, ^-2, ^-1, 0, 1, 2, 3, \ldots\}$$
Which sets are closed under
    (a) addition?         (b) subtraction?
    (c) multiplication?   (d) division?

23. For integers $x$ and $y$, under what condition will $x = ^-y$?

24. Complete the following table for integers by answering yes or no when the operation is
    (a) addition           (b) subtraction
    (c) multiplication     (d) division

| Property | Operation | Yes or No |
|---|---|---|
| Closure | | |
| Commutativity | | |
| Associativity | | |
| Identity | | |
| Inverse | | |
| Distributive over addition | | |
| Distributive over subtraction | | |

25. Identify whether the following numbers are positive or negative.
    (a) $(^-5)^2$                    (b) $^-(^-5)^3$
    (c) $^-(^-5)^4$                  (d) $(^-5)^5$
    (e) $(^-5)^n$ if $n$ is even     (f) $(^-5)^n$ if $n$ is odd

26. Evaluate the following expressions, and check your answers with a calculator.
    (a) $(^-72 - ^-64) \div (^-56 + 60)$
    (b) $(84 + ^-92) \div (^-76 - ^-84)$
    (c) $(^-144 \div 12) \div (^-3 \cdot 2)$
    (d) $^-144 \div [12 \div ^-3] \cdot 2$

27. (a) Find the largest (if it exists)
    (i)   whole number      (ii) integer
    (iii) negative integer
    (b) Find the smallest (if it exists)
    (i)   whole number       (ii) integer
    (iii) negative integer   (iv) positive integer

28.
$$a(b - c) = a(b + ^-c) \quad \text{Why?}$$
$$= ab + ^-ac \quad \text{Why?}$$
$$= ab - ac \quad \text{Why?}$$
$$a(b - c) = ab - ac \quad \text{Why?}$$

This verifies the _____ property of multiplication over subtraction.

29. Use Exercise 28 to simplify the following.

    (a) $2(3 - 5)$      (b) $^-3(4 - 8)$

30. (a) Tom's five test grades in his mathematics class are 90, 74, 46, 67, and 53. What is his average grade?

    (b) Subtract the average grade from each of his five test grades. Add the five numbers obtained. Are you surprised at the answer?

31. $(a + b) \cdot (a - b) = (a + b) \cdot (a + {}^-b)$
    $\qquad = (a + b) \cdot a + (a + b) \cdot {}^-b$
    $\qquad = a^2 + b \cdot a + a \cdot {}^-b + b \cdot {}^-b$
    $\qquad = a^2 + ba - ba - b^2$
    $\qquad = a^2 - b^2$

    Use this to simplify the following expressions.

    (a) $(x - 2)(x + 2)$      (b) $(3x + 5)(3x - 5)$
    (c) $(2x - y)(2x + y)$    (d) $(3x + 4y)(3x - 4y)$

32. Use Exercise 31 to write each of the following as a product.

    (a) $x^2 - 9$       (b) $4y^2 - 25$
    (c) $9x^2 - y^2$    (d) $4x^2 - 9y^2$

33. Solve each of the following on the domain of integers.

    (a) $2x < {}^-6$      (b) $3x < {}^-9$
    (c) $2x + 4 > 2$      (d) $3x + 5 > 8$

34. For the domain of integers, graph the solution set on a number line.

    (a) $x < 4$ and $x > {}^-3$     (b) $x < 2$ and $x < {}^-3$
    (c) $x < 2$ or $x < {}^-3$      (d) $x > {}^-1$ or $x < 5$

## REVIEW EXERCISES

35. Perform the indicated calculations without a calculator. In (a) through (e) check your answer with a calculator.

    (a) $36 - ({}^-7) - 18$
    (b) $^-12 - 15 + ({}^-7)^2 - ({}^-26)$
    (c) $50 - 64 - 13 + 10$       (d) $0 - [2 - ({}^-3)^2]$
    (e) $^-6 - ({}^-6)^2$          (f) $^-x + (2x + {}^-3x)$

36. Show the following operations on a number line.

    (a) $^-2 + 3$   (b) $^-2 + {}^-3$   (c) $4 + {}^-2$
    (d) $3 + {}^-3$   (e) $^-5 + 2$   (f) $4 + {}^-3$

---

# SECTION 3  DIVISIBILITY, PRIMES, COMPOSITES, AND FACTORIZATION

| | |
|---|---|
| **PROBLEM** | Can a teacher arrange 61 students as a rectangle (that is, in rows with an equal number of students in each)? |
| **OVERVIEW** | You will be able to answer the preceding problem when you learn what is meant by a prime number and how to factor a positive integer as a product of primes. Prime numbers form a subset of positive integers. |
| **GOALS** | Illustrate the Algebra Standard, page 4.<br>Illustrate the Number and Operations Standard, page 119.<br>Illustrate the Representation Standard, page 50.<br>illustrate the Problem Solving Standard, page 15.<br>Illustrate the Connections Standard, page 147.<br>Illustrate the Communication Standard, page 78. |

# DIVISIBILITY

Let us begin by defining the expression "$a$ divides $b$."

| DEFINITION | **Divides** | If $a$ and $b$ are any integers, then $a$ is said to *divide* $b$ (denoted by $a \mid b$) if and only if there exists an integer $c$ such that $b = ac$. |

**EXAMPLE 1**

The symbolic expression $a \mid b$ is read as "$a$ divides $b$."

Using divisibility concepts, give three ways of expressing $24 = 8 \cdot 3$.

SOLUTION

> 8 divides 24
> 24 is divisible by 8
> $8 \mid 24$

If $a$ divides $b$, then it can be said, equivalently, that

> $a$ is a **divisor** of $b$
> $a$ is a **factor** of $b$
> $b$ is a **multiple** of $a$

**EXAMPLE 2**

$2 \mid 16$ (read as "2 divides 16") because $2 \cdot 8 = 16$. Thus, 2 is a factor of 16, and 16 is a multiple of 2.

**EXAMPLE 3**

Classify the following as either true or false, where $a$ is a positive integer.
(a) $1 \mid a$    (b) $a \mid a$    (c) $3 \mid 42$    (d) $4 \mid 0$

SOLUTION

(a) $1 \mid a$ is true because $a = 1 \cdot a$.
(b) $a \mid a$ is true because $a = a \cdot 1$.
(c) $3 \mid 42$ is true because $42 = 3 \cdot 14$.
(d) $4 \mid 0$ is true because $0 = 4 \cdot 0$.

FIGURE 5-11

**Be careful to note the distinction between the statement $a \mid b$ ($a$ divides $b$) and the symbol $a/b$, an alternate notation for the operation $a \div b$.** But notice that $a \mid b$ does mean that $b/a$ is an integer. We use $a \nmid b$ to indicate that $a$ does not divide $b$; for example, $3 \nmid 8$ because there is no integer $x$ such that $8 = 3x$.

A stack of chips provides a concrete model to illustrate divisibility.

**EXAMPLE 4**

Arranging 12 chips in three equal stacks, as shown in Figure 5-11, illustrates that $3 \mid 12$. If $3 \nmid 12$, you would not be able to arrange 12 chips in 3 stacks with the same number in each stack.

**PRACTICE PROBLEM**

Why does $5 \mid {}^{-}30$? Identify a factor and a multiple in this relationship.

*ANSWER*

$5 \mid {}^{-}30$ because $5({}^{-}6) = {}^{-}30$. Thus, 5 is a factor and ${}^{-}30$ is a multiple.

## PROPERTIES OF DIVISIBILITY

| Properties of Divisibility | Let $x$, $y$, and $z$ be integers, with $x \neq 0$.<br>(a)  $1 \mid y$ and $x \mid x$.<br>(b)  *Transitive property:* If $x \mid y$ and $y \mid z$, then $x \mid z$.<br>(c)  If $x \mid y$ and $x \mid z$, then $x \mid (y + z)$ and $x \mid (y - z)$.<br>(d)  If $x \mid y$ and $x \mid (y + z)$ or $x \mid (y - z)$, then $x \mid z$.<br>(e)  If $x \mid y$ or $x \mid z$, then $x \mid yz$.<br>(f)  If $x \mid y$ and $x \nmid z$, then $x \nmid (y + z)$ and $x \nmid (y - z)$. |
|---|---|

**EXAMPLE 5**

(a) $1 \mid 7$ and $7 \mid 7$ illustrate Property (a).
(b) $3 \mid 6$ and $6 \mid 24$; therefore, $3 \mid 24$. This example illustrates the transitive property of divisibility given in Property (b).
(c) $7 \mid 49$ and $7 \mid 84$; hence, $7 \mid 133$ because $133 = 49 + 84$. $6 \mid 30$ and $6 \mid 18$, so $6 \mid 12$, because $12 = 30 - 18$. These examples illustrate the property of divisibility in Property (c).
(d) $6 \mid 30$ and $6 \mid 42$. But $30 = 42 + {}^{-}12$, so $6 \mid (42 + {}^{-}12)$. Therefore, by Property (d), $6 \mid {}^{-}12$.
(e) Because $3 \mid 18$, then $3 \mid (18 \cdot 20)$ by Property (e), even though 3 does not divide 20.
(f) $3 \mid 12$ and $3 \nmid 5$; therefore, $3 \nmid (12 + 5)$ or $3 \nmid 17$, by Property (f).

Now let's use problem-solving strategies to verify some of these properties.

**EXAMPLE 6**

Show that, if $x \mid y$ and $x \mid z$, then $x \mid (y + z)$. See Property (c).

*SOLUTION*

*What information is given?* We are given two facts, $x \mid y$ and $x \mid z$. *Can we restate these data in another way?* Yes. Because $x \mid y$, the definition of *divides* asserts that there exists an integer $k$ for which $y = kx$. Similarly, there is an integer $l$ for which $z = lx$. *What is unknown? What are we trying to show?* We are trying to show that $x \mid (y + z)$. Look at the definition of *divides* to see what this means. We need to find some integer $m$ such that $y + z = mx$. *What is our plan?* We shall try to find the connection between the information we know and the desired conclusion.

$$y + z = kx + lx$$

$$y + z = (k + l)x \qquad \text{Why?}$$

Because $k + l$ is an integer, it can be the $m$ desired. We thus have $y + z = mx$. The definition of *divisibility* tells us that $x \mid (y + z)$.

Can this be turned around? Is the converse true? If $x \mid (y + z)$, must $x \mid y$ and $x \mid z$?

**EXAMPLE 7**

If $x \mid (y + z)$, $x$ may or may not divide either $y$ or $z$. For instance, $3 \mid 24$; $24 = 11 + 13$, so $3 \mid (11 + 13)$, but $3 \nmid 11$ and $3 \nmid 13$.

Property (c) can be generalized to include the sum of any finite number $n$ of integers. If $a \mid b_1, a \mid b_2, a \mid b_3, \ldots$, and $a \mid b_n$ then $a \mid (b_1 + b_2 + b_3 + \cdots + b_n)$

Similarly, Property (d) may be generalized to include the sum of any finite number $n$ of integers. If $a \mid b_1, a \mid b_2, \ldots, a \mid b_{n-1}$ and if $a \mid (b_1 + b_2 + \cdots + b_{n-1} + b_n)$, then $a \mid b_n$

## COMMON ERROR

If $d \nmid a$ and $d \nmid b$, then $d \nmid (a + b)$. Show this is incorrect with an example.

**EXAMPLE 8**

A room contains 348 boxes. Can these be arranged in (a) 3 rows? (b) 4 rows? (c) 5 rows? (d) 6 rows? (e) 7 rows? (f) 8 rows?

*SOLUTION*

(a) Yes, $3 \mid 348$    (b) Yes, $4 \mid 348$    (c) No, $5 \nmid 348$
(d) Yes, $6 \mid 348$    (e) No, $7 \nmid 348$    (f) No, $8 \nmid 348$

The process of determining mentally whether a number is divisible by another number often becomes a matter of inspection. The following properties indicate some tests that may be used to investigate divisibility by several small positive integers.

**Divisibility Criteria**

An integer $x$ is **divisible**
1. by 2 (or 5) if and only if the number named by the last digit of $x$ is divisible by 2 (or 5)
2. by 3 (or 9) if and only if the number named by the sum of its digits is divisible by 3 (or 9)
3. by 4 if and only if the last two digits of $x$ represent a number divisible by 4; by 8 if and only if the last three digits of $x$ represent a number divisible by 8
4. by 6 if and only if $x$ is divisible by both 2 and 3
5. by 7 if and only if it satisfies the following property: Subtract the double of the right-hand digit from the positive number represented by the remaining digits. If the difference is divisible by 7, then the original number is divisible by 7.
6. by 10 if and only if $x$ ends in 0
7. by 11 if and only if the difference in the sum of the digits in the odd-numbered digit places and the sum of the digits in the even-numbered digit places is divisible by 11

**EXAMPLE 9**

(a) 756 is divisible by 3 because $7 + 5 + 6 = 18$, and $3 \mid 18$.
(b) $4 \mid 536$ because $4 \mid 36$.
(c) 678,342,570 and 417,235 are both divisible by 5, but 736 is not divisible by 5.
(d) 6294 is divisible by 6 because 4 is divisible by 2, and $6 + 2 + 9 + 4 = 21$, and $3 \mid 21$.
(e) 362,789,576 is divisible by 8 because $8 \mid 576$.
(f) 31,383 is divisible by 11 because $(3 + 3 + 3) - (8 + 1) = 9 - 9 = 0$, and 0 is divisible by 11.
(g) To determine whether $7 \mid 6055$ (and in general to investigate divisibility by 7), repeat the operation described in Criterion 5 until the difference is small enough to make the divisibility (or not) by 7 obvious. Thus,

$$605 - 2(5) = 595$$

$$59 - 2(5) = 49$$

and 7 divides 49. It follows that $7 \mid 6055$.

## CALCULATOR HINT

The process of determining whether one number is divisible by another is very easy to determine on a calculator. $7 \nmid 27$ because $27 \div 7 = 3.2857....$ Whenever the quotient contains a decimal, divisibility does not hold because the decimal part of the answer represents the remainder in the division algorithm. One can also quickly determine if the remainder is zero by using $\boxed{\text{INT} \div}$.

$$27 \boxed{\text{INT} \div} 7 \boxed{=} \text{ has a remainder of 6}$$

$$2187 \boxed{\text{INT} \div} 9 \boxed{=} \text{ has a remainder of 0.}$$

## COMMON ERRORS

(a) Every even number is divisible by 4.
(b) If a number is divisible by 3, it is divisible by 9.

The divisibility criteria for small positive integers are a result of the properties of divisibility listed on p. 220. To understand the criterion for divisibility by 2, consider whether $2 \mid 54,236$. Recall that 54,236 can be written as

$$54,236 = 5(10,000) + 4(1000) + 2(100) + 3(10) + 6$$

Now $2 \mid 10,000$, so $2 \mid 5(10,000)$ by Property (e). Likewise, $2 \mid 4(1000)$, $2 \mid 2(100)$, and $2 \mid 3(10)$. Now if $2 \mid 6$ (and it does), then by the generalization of Property (c), $2 \mid 54,236$. So we can tell if a number is divisible by 2 just by looking to see whether 2 divides its last digit.

## COMMON ERROR

A student states that a positive integer is divisible by 24 because it is divisible by 4 and by 6. Show with an example that this is incorrect. For $a \mid c$ and $b \mid c$ to imply that $ab \mid c$, what must be true?

The following theorem is very helpful in testing divisibility by products of small positive integers.

**Divisibility by a Product**

Suppose that $a$ and $b$ are positive integers and that the only positive integer that divides both $a$ and $b$ is the number 1. An integer is divisible by the product of $a$ and $b$ if and only if it is divisible by both $a$ and $b$.

**EXAMPLE 10**

Show that $18 \mid 10{,}134$ using divisibility criteria.

*SOLUTION*

$$2 \mid 10{,}134 \qquad 2 \mid 4$$

$$9 \mid 10{,}134 \qquad 9 \mid (1 + 1 + 3 + 4) \text{ or } 9 \mid 9$$

Because there is no positive integer other than 1 that divides both 2 and 9, it follows that $18 \mid 10{,}134$.

## PRIMES

We introduce primes by a procedure called the **Sieve of Eratosthenes.**

Make an array of the first 100 positive integers. (See Table 5-1.) Cross out the 1. Then cross out all the numbers greater than 2 that are divisible by 2 (every even number except 2). Next, cross out all numbers divisible by 3, but do not cross out 3. Notice that 4, the next consecutive number, has already been crossed out. Cross out all numbers divisible by 5 except 5 itself. Continue this process for every natural number up to 100 that has not been crossed out. When this task is completed, circle the numbers not crossed out; the array will resemble Table 5-1.

The numbers circled in this array are the prime numbers less than or equal to 100. Notice that none of these circled numbers is a multiple of any positive integer that precedes it (other than 1).

**TABLE 5-1**

| | | | | | | | | | |
|---|---|---|---|---|---|---|---|---|---|
| ~~1~~ | 2 | 3 | ~~4~~ | 5 | ~~6~~ | 7 | ~~8~~ | ~~9~~ | ~~10~~ |
| 11 | ~~12~~ | 13 | ~~14~~ | ~~15~~ | ~~16~~ | 17 | ~~18~~ | 19 | ~~20~~ |
| ~~21~~ | ~~22~~ | 23 | ~~24~~ | ~~25~~ | ~~26~~ | ~~27~~ | ~~28~~ | 29 | ~~30~~ |
| 31 | ~~32~~ | ~~33~~ | ~~34~~ | ~~35~~ | ~~36~~ | 37 | ~~38~~ | ~~39~~ | ~~40~~ |
| 41 | ~~42~~ | 43 | ~~44~~ | ~~45~~ | ~~46~~ | 47 | ~~48~~ | ~~49~~ | ~~50~~ |
| ~~51~~ | ~~52~~ | 53 | ~~54~~ | ~~55~~ | ~~56~~ | ~~57~~ | ~~58~~ | 59 | ~~60~~ |
| 61 | ~~62~~ | ~~63~~ | ~~64~~ | ~~65~~ | ~~66~~ | 67 | ~~68~~ | ~~69~~ | ~~70~~ |
| 71 | ~~72~~ | 73 | ~~74~~ | ~~75~~ | ~~76~~ | ~~77~~ | ~~78~~ | 79 | ~~80~~ |
| ~~81~~ | ~~82~~ | 83 | ~~84~~ | ~~85~~ | ~~86~~ | ~~87~~ | ~~88~~ | 89 | ~~90~~ |
| ~~91~~ | ~~92~~ | ~~93~~ | ~~94~~ | ~~95~~ | ~~96~~ | 97 | ~~98~~ | ~~99~~ | ~~100~~ |

<table>
<tr><td rowspan="2">DEFINITION</td><td>**Prime Number**</td><td>A positive integer $p$ greater than 1 is *prime* if and only if the only positive integer divisors of $p$ are 1 and $p$.</td></tr>
</table>

**Notice that the smallest prime is 2 and that 2 is the only even prime.** The first 10 primes are

$$\{2, 3, 5, 7, 11, 13, 17, 19, 23, 29\}$$

How many more primes exist? Is there a largest prime? Can you find a prime greater than 100? Can you find a prime greater than 100,000? About 300 B.C., Euclid proved that there are infinitely many primes.

To test whether a number $N$ is prime, we could divide it by all primes less than $N$. For instance, to determine whether 101 is prime, we could test all primes less than 101. To reduce this work, however, let's return to the Sieve of Eratosthenes and make an observation as we cross out numbers.

| Prime | Observation |
|-------|-------------|
| 2 | The first number (not crossed out) that 2 divides is $4 = 2^2$. |
| 3 | The first number (not crossed out) that 3 divides is $9 = 3^2$. |
| 5 | The first number (not crossed out) that 5 divides is $25 = 5^2$. |
| 7 | The first number (not crossed out) that 7 divides is $49 = 7^2$. |

Therefore, to test whether a number such as 101 is a prime, we would test its divisibility by all primes up to the first prime whose square was larger than 101. Such a prime would be 11, because $(11)^2 = 121$. Any prime greater than or equal to 11 that divided 101 would give a quotient of less than 11, and we would already have located that quotient in our testing procedure. Therefore, there would be no point in testing primes larger than 11.

| **Finding a Prime** | When testing to see whether a number is *prime,* we need to try as divisors only primes whose squares are less than or equal to the number being tested. |
|---|---|

## COMPOSITE NUMBERS

In the Sieve of Eratosthenes (Table 5-1), the numbers other than 1 that are crossed out are composite numbers.

| DEFINITION | **Composite Number** | A positive integer is *composite* if it has a positive integer divisor other than itself and 1. |
| --- | --- | --- |

The first ten composite positive integers are

$$\{4, 6, 8, 9, 10, 12, 14, 15, 16, 18\}$$

Once again, there are many more composite numbers. **Notice that 1 is neither prime nor composite. 1 is called a *unit*.** The key difference between a prime number and a composite number is that the former has only two positive divisors—itself and 1—and the latter has more than two divisors.

When a composite number is written as a product of all its prime divisors, we say that we have a **prime factorization.** Two methods are commonly employed to find all the prime factors of a composite number. The first method consists of repeated division starting with the smallest prime (2) and dividing by it as long as it divides the last quotient; then going to the next larger prime; and continuing until all prime factors have been obtained. The second method involves factoring the number into any two easily recognized factors and then factoring the factors.

**EXAMPLE 11**

What are the prime factors of 72? Use both methods.

*SOLUTION*

Using repeated division, starting with the smallest prime, we have

| Regular form | Compact form |
| --- | --- |
| $72 \div 2 = 36$ | $2\overline{)72}$ |
| $36 \div 2 = 18$ | $2\overline{)36}$ |
| $18 \div 2 = 9$ | $2\overline{)18}$ |
| $9 \div 3 = 3$ | $3\overline{)9}$ |
| $3 \div 3 = 1$ | $3\overline{)3}$ |
| | $1$ |

$$72 = 2 \cdot 2 \cdot 2 \cdot 3 \cdot 3 = 2^3 \cdot 3^2$$

Using factoring of easily recognized factors, we get

$$72 = (12)(6) = (4 \cdot 3)(3 \cdot 2) = (2 \cdot 2 \cdot 3) \cdot (3 \cdot 2)$$

$$= 2 \cdot 2 \cdot 2 \cdot 3 \cdot 3$$

Some elementary textbooks employ **factor trees** to illustrate the procedure for factoring a number into its prime factors. (See Figure 5-12). Can you draw still another factor tree for 210?

NOTE: The problem of factoring very large composite numbers is currently an important area of study in research mathematics.

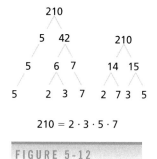

$$210 = 2 \cdot 3 \cdot 5 \cdot 7$$

FIGURE 5-12

# FUNDAMENTAL THEOREM OF ARITHMETIC

Interestingly, the factorization of a composite number into prime factors is unique, except for the order of the factors.

| | |
|---|---|
| **Fundamental Theorem of Arithmetic** | Every composite positive integer can be factored uniquely (except for the order of the factors) into a product of primes. |

Recall that 1 was not defined to be a prime. If it had been, this theorem would not be true. For example,

$$6 = 2 \cdot 3 \quad \text{and} \quad 6 = 1 \cdot 2 \cdot 3$$

so the factorization into primes would not be unique.

The Fundamental Theorem of Arithmetic asserts that, if $x$ is any composite number, it can be written as $x = p_1 p_2 \ldots p_n$, where each $p_i$ is a prime and where the primes are unique except for the order in which they are written. The factors of 70 (2, 5, and 7) are unique, but 70 may be written as $2 \cdot 7 \cdot 5$, as $5 \cdot 7 \cdot 2$, and so on. **It is possible for some (or all) of the $p$'s to be equal; that is, a factor may be used more than once in the Fundamental Theorem.** Thus, for example, $24 = 2 \cdot 2 \cdot 2 \cdot 3$.

**EXAMPLE 12**

To illustrate the uniqueness of the factors in the prime factorization of a composite number, we factor 360 in two ways, as shown in Figure 5-13. In both cases, $360 = 2 \cdot 2 \cdot 2 \cdot 3 \cdot 3 \cdot 5$.

The Fundamental Theorem of Arithmetic holds for any composite number in the system of positive integers. Let us look at a number system for which the Fundamental Theorem does not hold.

**EXAMPLE 13**

Consider the system of numbers of the form $\{3x + 1 \mid x = 0, 1, 2, 3, \ldots\}$. This set of numbers can be written as $\{1, 4, 7, 10, 13, 16, \ldots, 25, \ldots\}$. In this system 4, 10, and 25 are all primes because they have no factors or divisors greater than 1 in the system. Notice also the two prime factorizations of 100: $100 = 10 \cdot 10$ and $100 = 4 \cdot 25$. Because 100 can be expressed as a product of primes in two ways, the Fundamental Theorem of Arithmetic is not true for this system.

# NUMBER-THEORY PROBLEMS

Many people believe that all mathematical problems have been solved—that mathematics today is simply a matter of learning what has already been done. However, the opposite is true. More mathematics has been developed in the last century than in all recorded history before that date. The more mathematics is developed, the more problems arise. Each day there are numerous new problems to be solved. Because it contains so many easily stated and famous problems that have defied solution through the years, number theory is fascinating to many people. In the remainder of this section, we shall present some of these unsolved problems.

Let us consider the sum of all the **proper divisors** of a positive integer. (The proper divisors of a positive integer are all its positive divisors except the number itself.) The sum of the proper divisors of a positive integer is sometimes equal to, sometimes less than, and sometimes greater than the positive integer. This observation leads to the following definitions.

---

**DEFINITION**

**Perfect, Deficient, Abundant**

If the sum of the proper divisors of a positive integer is equal to the positive integer itself, the positive integer is called a *perfect* number; if this sum is less than the positive integer, it is called a *deficient* number; if it is greater than the positive integer, it is called an *abundant* number.

---

**EXAMPLE 14**

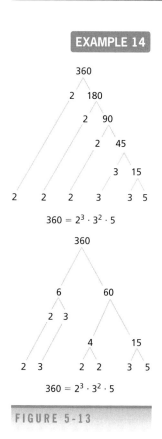

FIGURE 5-13

From Table 5-2, it is obvious that 6 is a perfect number; 4, 8, 9, 10, 14, and 15 are deficient numbers; and 12 is an abundant number.

**TABLE 5-2**

| Number | Sum of Proper Divisors |
|:---:|:---:|
| 4 | $1 + 2 = 3$ |
| 6 | $1 + 2 + 3 = 6$ |
| 8 | $1 + 2 + 4 = 7$ |
| 9 | $1 + 3 = 4$ |
| 10 | $1 + 2 + 5 = 8$ |
| 12 | $1 + 2 + 3 + 4 + 6 = 16$ |
| 14 | $1 + 2 + 7 = 10$ |
| 15 | $1 + 3 + 5 = 9$ |

Perfect numbers seem to be quite rare. The next one after 6 is 28 ($1 + 2 + 4 + 7 + 14 = 28$); the next to appear is 496; the fourth is 8128; and the fifth is 33,550,336. Clearly, finding perfect numbers is a rather difficult task. How many perfect numbers are there? No one knows. Perhaps there are infinitely many, but can you prove it? Are there any odd perfect numbers? All the perfect numbers that have been found thus far are even numbers.

It has been proved (by Euclid) that, if $2^n - 1$ is a prime, then $N = 2^{n-1}(2^n - 1)$ is perfect, and every known perfect number is of this form.

**EXAMPLE 15**

(a) $2^2 - 1 = 3$ is a prime, so $N = 2(3) = 6$ is perfect.

(b) $2^3 - 1 = 7$ is a prime, so $N = 2^2(2^3 - 1) = 28$ is perfect.

(c) $2^5 - 1 = 31$ is a prime, so $N = 2^4(2^5 - 1) = 496$ is perfect:

$$1 + 2 + 4 + 8 + 16 + 31 + 62 + 124 + 248 = 496$$

(d) $2^7 - 1 = 127$ is a prime, so $N = 2^6(2^7 - 1) = 8128$ is perfect.

## T H E N & N O W

### Emmy Noether
### 1882–1935

Amalie Emmy Noether was born in Erlangen, Germany in March 1882. Known as Emmy, she grew up to become a world-famous mathematician.

As a young girl, Emmy attended a local public school for girls, where she was taught feminine manners and household arts, along with academic subjects. At 18, she passed the exams qualifying her to teach at girls' schools. But she was determined to study at a university, despite the fact—and this was usual in Europe at the time—that women were not allowed to attend as registered students, but only as auditors.

After the laws had changed to allow women full student status, Noether, along with only 3 other women, was enrolled in 1904 at the University of Erlangen. She concentrated on mathematics and completed her doctoral dissertation in 1907, graduating at age 23.

For the next 4 years, Noether lectured and did mathematical research at Erlangen, without any pay or academic rank. In 1915, after gaining recognition for her published mathematical research, she was invited to join the new Mathematical Institute at Göttingen, but still with no pay or rank. Her studies at Göttingen led her to many original mathematical theorems that she published in 43 research papers throughout her life. In her publications she essentially established the field of mathematics called *modern algebra*. Her experience of World War I and its violent aftermath in Germany persuaded her to become a pacifist. Rising Naziism and the fact that Noether was Jewish brought a close to life as she knew it in Göttingen. In 1933, she left Germany for the United States and a professorship at Bryn Mawr College, a Quaker school for women in suburban Philadelphia. She was well received in the United States and often invited to lecture at colleges and other academic institutions. Albert Einstein in 1935 called her "the most significant creative mathematical genius thus far produced since the higher education of women began."

Today the mathematics community remembers her many achievements, not the least of which is her work on the Fundamental Theorem of Arithmetic. This theorem was known even before Euclid [365(?) B.C.–275(?) B.C.], but it was Emmy Noether who deepened our understanding of it. As you know, this theorem asserts that every positive integer can be factored in only one way as a product of prime numbers. Without this property of numbers, it would make no sense for elementary-school teachers to ask children to write a fraction in lowest terms, a problem you will encounter in the next chapter. Also without this property, the ideas of greatest common divisor and least common multiple could not be developed the way that you will find them in this chapter.

Some additional unsolved problems involve amicable numbers. From Greek to medieval times, amicable numbers were believed to have a mystical relationship to human friendship.

| DEFINITION | | |
|---|---|---|
| **Amicable** | | Two positive integers are said to be *amicable* if each is the sum of the proper divisors of the other. |

One example of an amicable number pair is 220 and 284. The proper divisors of 220 are 1, 2, 4, 5, 10, 11, 20, 22, 44, 55, and 110, whose sum is 284, and the sum of the proper divisors of 284 is $1 + 2 + 4 + 71 + 142 = 220$.

The fact that 1184 and 1210 are amicable was discovered by a 16-year-old boy. 17,296 and 18,416 are also amicable. Perhaps you can discover a pair. How many amicable pairs are there? Are there infinitely many? No one knows.

# Just for Fun

Professor Abstract's car license plate consists of a 3-digit number. He challenges the class to find this number from the following clues. If you add 1 to the number, it is divisible by 7; if you add 4, it is divisible by 8; and if you add 7, it is divisible by 9. What is the number?

# EXERCISE SET 3

*R* 1. List three distinct positive integers that are divisors of each of the following.

   (a) 35    (b) 216    (c) 1001    (d) 299

2. Classify as true or false. If a statement is false, explain why.

   (a) 39 is a multiple of 9.
   (b) 11 is a factor of 111.    (c) 4 is a multiple of 8.
   (d) 9 is a factor of 12.    (e) 8 is divisible by 24.
   (f) If every digit of a number is divisible by 3, the number is divisible by 3.
   (g) 12 divides 6.
   (h) $5 \mid 25$ and $5 \nmid 4$; therefore, $5 \nmid 25 \cdot 4$.
   (i) If $a \mid b$, then $a \mid b^2$ $(a \neq 0)$.
   (j) $5 \nmid 11$, $5 \nmid 4$; therefore, $5 \nmid (11 + 4)$.

3. Test each for divisibility by 4 and 8.

   (a) 8116    (b) 9320    (c) 26,428

4. Which of the following numbers are divisible by 11?

   (a) 704    (b) 7051    (c) 81,928

5. Which of the following numbers are divisible by 7?

   (a) 6468    (b) 1848    (c) 8162

6. Which of the following numbers are divisible by 5? By 10?

   (a) 7280    (b) 625    (c) 89,001

7. Factor into prime factors.

   (a) 144    (b) 592  (c) 612  (d) 162

8. Which of the following are primes?

   (a) 149    (b) 89    (c) 87
   (d) 43    (e) 737    (f) 411
   (g) 91    (h) 1003

9. Find the sum of the prime factors of each composite number.

   (a) 24    (b) 33    (c) 57
   (d) 2000    (e) 108    (f) 299

10. Express each number as the product of its prime factors.

   (a) 54    (b) 36    (c) 120
   (d) 51    (e) 141    (f) 76
   (g) 144    (h) 178    (i) 256

11. (a) How many primes are less than 50? Less than 100? Less than 150?
   (b) How many composites are less than 50? Less than 100? Less than 150?

12. Use a factor tree to find the prime factorization of each of the following.

    (a) 1512        (b) 810        (c) 1836

13. Use the compact form of the repeated division technique to write the numbers in Exercise 12 as a product of primes.

14. What is the largest prime you need to use

    (a) in checking whether 689 is a prime?
    (b) in testing 7001?

*T* 15. Classify the following numbers as perfect (P), deficient (D), or abundant (A).

    (a) 12        (b) 14        (c) 28
    (d) 30        (e) 50        (f) 72
    (g) 226       (h) 300       (i) 496

16. Can a prime number ever be perfect? Deficient? Abundant?

17. When you test for divisibility on a calculator, what happens if the number is not divisible by the divisor?

18. The 135 county bus drivers are having a meeting.

    (a) Jim's job is to place chairs in the auditorium. He wants to have exactly one chair for each driver, yet he also wants each row to contain the same number of chairs. The space available could hold 10, 11, 12, 13, 14, or 15 rows. Testing for divisibility, how many rows should Jim have?

    (b) At the meeting the superintendent plans to divide the 135 drivers into study groups. If he wants equal numbers in each group with no more than 10 to a group, what are the possible group sizes?

19. Could Mr. Jones put the 2142 papers on his desk

    (a) into 6 equal piles?    (b) into 18 equal piles?
    (c) into 7 equal piles?    (d) into 8 equal piles?

20. John has 2144 pieces of candy. Can he divide them evenly among 8 friends? Why?

21. Without actually dividing, test each of these numbers for divisibility by 5, by 6, by 7, by 8, by 9, and by 11. Then check your answer with a calculator.

    (a) 6944        (b) 81,432      (c) 1,076,770
    (d) 50,177      (e) 1,161,914   (f) 162,122
    (g) 57,293      (h) 214,890     (i) 482,144
    (j) 150,024     (k) 23,606      (l) 63,732

22. Answer yes or no, and explain your answer.

    (a) Is $2^3 \cdot 3^2 \cdot 5^7$ a factor of $2^4 \cdot 3^9 \cdot 5^7$?
    (b) Is $3^7 \cdot 5 \cdot 17^4$ a factor of $3^7 \cdot 5^2 \cdot 17^3$?
    (c) Does $7^2 \cdot 3^5$ divide $7^5 \cdot 3^6$?
    (d) Is $7^4 \cdot 11^5 \cdot 5^4$ a multiple of $7^5 \cdot 11^3 \cdot 5$?
    (e) Formulate a rule for determining when a number written in terms of its prime factorization is divisible by another number written in the same way.

23. Leap years are divisible by 4. Which of the following are leap years? Years that end in two zeros must be divisible by 400 in order to be leap years.

    (a) 348     (b) 1492    (c) 1984    (d) 1776
    (e) 1940    (f) 1000    (g) 500     (h) 1024

24. For each of the following, give an example.

    (a) A composite number of the form $4n + 1$
    (b) A composite number of the form $3^p + 1$, where $p$ is a prime
    (c) A prime number of the form $n^2 + n$, where $n$ is a natural number
    (d) A prime number of the form $4n + 1$, where $n$ is a natural number
    (e) A prime number of the form $8n + 1$, where $n$ is a natural number

25. Classify the following as true or false. If false, give a counterexample. Let $a$, $b$, $c$, and $d$ be positive integers.

    (a) If $a \,|\, (b + c)$, then $a \,|\, b$ and $a \,|\, c$.
    (b) If $a \,|\, b$ and $a \,|\, c$, then $(a + b) \,|\, c$.
    (c) If $a \nmid b$ and $a \nmid c$, then $a \nmid bc$.
    (d) If $ab \,|\, c$, then $a \,|\, c$ and $b \,|\, c$.
    (e) If $a \,|\, bc$, then $a \,|\, b$ and $a \,|\, c$.
    (f) If $a \,|\, b$, $a \,|\, c$, $a \,|\, d$, and $a \,|\, e$, then $a \,|\, (b + c + d + e)$.
    (g) If $a \nmid b$ and $a \nmid c$, then $a \nmid (b + c)$.
    (h) $a \,|\, 0$, with $a \neq 0$

26. State the property that justifies each of the following.

    (a) $3 \,|\, 3$.
    (b) $3 \,|\, 6$, $6 \,|\, 18$; therefore, $3 \,|\, 18$.
    (c) $4 \,|\, 12$, $4 \nmid 3$; therefore, $4 \,|\, 36$.
    (d) $4 \,|\, 8$, $4 \nmid 9$; therefore, $4 \nmid (8 + 9)$.
    (e) $3 \,|\, 6$, $3 \,|\, 9$, $3 \,|\, 12$, and $3 \,|\, 15$; therefore, $3 \,|\, 42$.

27. Suppose that 3 new states were added to the United States, making a total of 53. Could the stars of our new flag be arranged in equal rows and columns? Design an arrangement for 53 stars.

28. (a) Without zeros, write a 5-digit number divisible by 9 and by 5.
    (b) Without zeros, write a 7-digit number divisible by 11 and by 6.
    (c) Without zeros, write a 6-digit number divisible by 7 and by 8.
    (d) Without zeros, write an 8-digit number divisible by 4 and by 3.

29. Write 726,664 in expanded notation, and verify that $8 \,|\, 726,664$ because $8 \,|\, 664$.

30. The symbol $n!$ is defined to be

$$n(n-1)(n-2) \cdots \cdot 3 \cdot 2 \cdot 1$$

Thus, for example, $4! = 4 \cdot 3 \cdot 2 \cdot 1 = 24$. Using the definition "$a \mid b$, if $b = ac$," find $c$ for the following.

(a) $7 \mid 7!$   (b) $5 \mid 6!$   (c) $6 \mid 5!$

31. Each basketball squad at Eastside Rec must consist of 11 players. 143 youngsters want to play. Will any be left out?

32. Classify as always, sometimes, or never true.

   (a) If a number is divisible by 2 and 3, it is divisible by 6.
   (b) If a number is divisible by 3, it is divisible by 9.
   (c) If a number is divisible by 8, it is divisible by 4.
   (d) $5 \nmid a$ and $7 \nmid a$. Yet $35 \mid a$.
   (e) If $d \mid ab$, then $d \mid a$ and $d \mid b$.
   (f) If $d \mid ab$, then $d \mid a$ or $d \mid b$.   (g) $a^3 b^2 c^4 \mid a^5 b^5 c^6$

33. Show that 1184 and 1210 are amicable numbers.

34. (a) Determine the first 20 numbers in the number system discussed in Example 13 of this section.
   (b) How many of these numbers are primes?
   (c) Is this system closed under multiplication?
   (d) Write 484 as a product of primes in 2 ways.

35. Will $n! + 2, n! + 3, \ldots, n! + n$ for $n \geq 2$ always be a sequence of $n - 1$ composite numbers? Why?

36. (a) Prove that every prime is deficient.
   (b) Prove that if $p$ is a prime, then $p^n$ is deficient for every natural number exponent $n$.
   This exercise shows that there are infinitely many deficient numbers. Why?

37. Consider the subset of numbers defined by

$$\{2x + 1 \mid x = 0, 1, 2, \ldots\}$$

   (a) Define a prime on this set of numbers; list the first ten primes.
   (b) Does the Fundamental Theorem of Arithmetic hold?

38. (a) Devise divisibility tests for 2 and 3 in base seven.
   (b) Devise divisibility tests for 2, 3, 4, and 5 in base six.

## REVIEW EXERCISES

39. Perform the indicated calculations without a calculator. Then check your answers on (a), (d), and (e) with a calculator.

   (a) $^-6(3 - {}^-5) - ({}^-7)$
   (b) $^-x + (2x + {}^-3x)$
   (c) $^-x({}^-z + y)$
   (d) $({}^-30 \div 6) \cdot {}^-8$
   (e) $^-4 + ({}^-16 \div {}^-8)$

40. Working from left to right, show that each of the following holds, and give reasons for each step.

   (a) $4({}^-3) + ({}^-4)(2) = 4({}^-3 + {}^-2)$
   (b) $(4 \cdot {}^-3) = 3 \cdot {}^-4$   (c) $^-4({}^-8 \cdot 2) = (8 \cdot {}^-2) \cdot {}^-4$

### ❖ PCR Excursion ❖

41. A palindrome is a number that reads the same forward and backward or that equals its reverse (digits reversed). For example, 1331, 2772, and 354453 are palindromes.

   (a) Start with any number, say 263. Reverse its digits and add. $(263 + 362 = 625)$ Repeat the process until a palindrome is obtained. Try some other examples. Will this process always result in a palindrome?
   (b) Which of the palindromes 232, 1331, 2772, 31513, and 354453 are divisible by 11?
   (c) Are all 4-digit palindromes divisible by 11? Why or why not?
   (d) Make up a 5-digit palindrome. Test for divisibility by 11. Are all 5-digit palindromes divisible by 11? Why or why not? Are any 5-digit palindromes divisible by 11?
   (e) What is characteristic of the 5-digit palindromes divisible by 11?
   (f) Are all palindromes with an even number of digits divisible by 11? Explain.
   (g) What is characteristic of all palindromes with an odd number of digits? Are such divisible by 11?

### ❖ PCR Excursion ❖

42. A string of 25 decorative lights are arranged on a mantel at Christmas. Timing devices are attached to switches on the lights so that the lighting pattern changes every minute. The sequence of changes repeats every 25 minutes. Switches are set so that at the end of the first minute, every light is turned on. At the end of the second minute, every second switch is reversed. At the end of the third minute, every third switch is reversed. At the end of the fourth minute, every fourth switch is reversed, and so on.

   (a) Label the light bulbs as 1, 2, 3, 4, 5, 6, ..., 25. Let an $x$ represent a bulb on and a $o$ the bulb off. Make several lines of a table showing with $x$'s and $o$'s whether the light bulbs are on or off. Let the first line represent 1 minute, the second line 2 minutes, the third line 3 minutes and so on.
   (b) Which lights are on at the end of the 25 minute cycle?

(c) Make a table showing the positive integer divisors (excluding 1) of the numbers 2 through 25.

(d) Compare the results in (b) and (c). Explain carefully the results from (b) in terms of the information from (c) about factors.

(e) Suppose there had been 64 lights on the mantel and a 64 minute cycle of turning lights off and on. Which lights would have been on at the end of 64 minutes?

# SECTION 4

# GREATEST COMMON DIVISOR AND LEAST COMMON MULTIPLE

**PROBLEM**

The cheerleaders for the Hardy College Bulldogs want to buy the same number of red pompons and black pompons. What is the least number of each they can buy if red ones are shipped in boxes of 8 and black ones are shipped in boxes of 6?

**OVERVIEW**

In this section, we discuss the greatest common divisor (or factor) and the least common multiple. Which of these two concepts can be used to solve the preceding problem? The greatest common divisor of two or more positive integers is the largest positive integer that will divide the numbers. The least common multiple of two positive integers is the least positive integer that both the numbers will divide. You may be surprised at the many applications of these two concepts.

**GOALS**

Illustrate the Number and Operations Standard, page 119.
Illustrate the Algebra Standard, page 4.
Illustrate the Connections Standard, page 147.
Illustrate the Reasoning and Proof Standard, page 23.
Illustrate the Communication Standard, page 78.

If a positive integer $d$ is a divisor of each of two positive integers $b$ and $c$, then $d$ is called a **common divisor** of $b$ and $c$.

**EXAMPLE 1**

The divisors of 18 are { 1, 2, 3, 6, 9, 18}.
The divisors of 30 are { 1, 2, 3, 5, 6, 10, 15, 30}.
The set of common divisors of 18 and 30 is {1, 2, 3, 6}.

In the set of common divisors of any two positive integers, there is a greatest number, called the **greatest common divisor** (or **highest common factor**). The greatest common divisor of 18 and 30 is 6.

**DEFINITION**

**Greatest Common Divisor**

The *greatest common divisor* (abbreviated g.c.d.) or *highest common factor* (h.c.f.) of two positive integers, $a$ and $b$, is the largest positive integer $d$ such that $d \mid a$ and $d \mid b$.

Thus, the greatest common divisor of two positive integers $a$ and $b$ is the largest positive integer $d$ that is a divisor of both numbers. It is denoted by

$$d = \text{g.c.d. } (a, b)$$

We saw previously that g.c.d. (18, 30) = 6.

Finding the greatest common divisor of two positive integers is easy when the numbers are small. In this case, we write the numbers as products of primes. We then note the powers of the primes that are common to both numbers. **The g.c.d. is the product of these common powers of primes.** This method is illustrated in Example 2.

**EXAMPLE 2**

Find g.c.d. (70, 90).

*SOLUTION*

$$70 = 2 \cdot 5 \cdot 7 \quad \text{and} \quad 90 = 2 \cdot 3 \cdot 3 \cdot 5 \quad \text{Fundamental Theorem of Arithmetic}$$

Notice that 2 and 5 are divisors of both 70 and 90 and that 2 and 5 are the only prime factors that 70 and 90 have in common. Thus,

$$\text{g.c.d. } (70, 90) = 2 \cdot 5 = 10$$

NOTE: The g.c.d. of two numbers is always less than or equal to each number.

**EXAMPLE 3**

Find g.c.d. (144, 180).

*SOLUTION*

$$144 = 2 \cdot 2 \cdot 2 \cdot 2 \cdot 3 \cdot 3 = 2^4 \cdot 3^2$$
$$180 = 2 \cdot 2 \cdot 3 \cdot 3 \cdot 5 = 2^2 \cdot 3^2 \cdot 5 \quad \text{Fundamental Theorem of Arithmetic}$$

Notice that two 2's and two 3's are divisors of both 144 and 180. Therefore,

$$\text{g.c.d. } (144, 180) = 2^2 \cdot 3^2 = 36$$

## Calculator Hint

In order to provide additional functions beyond those that can be provided on a small keyboard, many calculators use menus on which several functions are listed. Because those menus differ widely between calculators, we give a brief discussion in this text of the menus that appear on the TI 34 II, an inexpensive calculator with fraction arithmetic capabilities. For example, FracMode will be discussed in the next chapter, STATVAR will be discussed in Chapter 10, and a mathematics menu MATH will be discussed now. Our discussion of menus on the TI 34 II will help you navigate menus on other calculators. On some calculators the MATH menu contains entries such as *abs, round, ipart, fpart, min, max, lcm,* and *gcd.* After selecting the MATH menu, we use the cursor ▶ to underline the function we desire and then press =. Use the MATH menu to find the g.c.d. (144, 180).

MATH (underline *g.c.d.*) 144 , 180 ) =. Did you get 36?

The g.c.d. of three positive integers $a$, $b$, and $c$ can be found by pairing. First, find g.c.d. $(a, b) = e$. Then g.c.d. $(a, b, c) = $ g.c.d. $(e, c)$. Alternatively, the easiest way to proceed if the numbers are small is to write each number as a product of primes and then to find the primes common to all numbers.

**EXAMPLE 4**

Find g.c.d. (54, 72, 84).

*SOLUTION*

$$54 = 2 \cdot 3^3$$
$$72 = 2^3 \cdot 3^2 \qquad \text{Fundamental Theorem of Arithmetic}$$
$$84 = 2^2 \cdot 3 \cdot 7$$
$$\text{g.c.d. } (54, 72, 84) = 2 \cdot 3 = 6$$

# CALCULATOR HINT

Find g.c.d. (54, 72, 84) using a calculator.

[MATH] (underline *g.c.d*) [=] 54 [,] 72 [=] getting 18.

[MATH] (underline *g.c.d*) [=] 18 [,] 84 [=] getting 6.    g.c.d. (54, 72, 84) = 6.

**EXAMPLE 5**

A teacher decides to use discussion groups in 4 classes. There are 24 students in the first class, 32 in the second, 28 in the third, and 40 in the fourth. If the teacher has the same number of groups in every class and wants every group in a given class to have an equal number of students, what is the largest possible number of groups common to each class?

*SOLUTION*

The largest number of groups that can be used is g.c.d. (24, 32, 28, 40). We can factor these numbers as follows:

$$24 = 2^3 \cdot 3 \qquad 28 = 2^2 \cdot 7$$
$$32 = 2^5 \qquad 40 = 2^3 \cdot 5$$
$$\text{g.c.d. } (24, 32, 28, 40) = 2^2 = 4$$

Each class should have 4 groups.

Sometimes positive integers have no common factors except 1. For example, g.c.d. (9, 11) is 1. Consequently, 9 and 11 are said to be relatively prime.

---

**DEFINITION**

**Relatively Prime**    If the greatest common divisor of two positive integers $a$ and $b$ is 1, then $a$ and $b$ are *relatively prime*.

---

**EXAMPLE 6**

(a) 5 and 7 are relatively prime because g.c.d. (5, 7) = 1.
(b) 23 and 124 are relatively prime because g.c.d. (23, 124) = 1.
(c) 111 and 6 are not relatively prime because g.c.d. (111, 6) = 3.

All positive integers less than a prime number are relatively prime to that number. For instance, each of the numbers 1, 2, 3, 4, 5, and 6 is relatively prime to the prime number 7.

Writing a positive integer as a product of prime factors can be quite cumbersome when the numbers are large. Consequently, we need a more practical method for finding the g.c.d. for large numbers. The method we use at this time is based on the **division algorithm.** This algorithm states that, for positive integers $a$ and $b$, $a$ can always be written as

$$a = bq + r$$

where $q$ and $r$ are nonnegative integers with $0 \le r < b$. This representation is accomplished by division. If $a = 7$ and $b = 2$, then

$$\begin{array}{r} 3 \\ 2\overline{)7} \\ \underline{6} \\ 1 \end{array}$$

can be written as $7 = 2(3) + 1$, where $q = 3$ and $r = 1$ with (obviously) $0 \le 1 < 2$.

The following algorithm facilitates using the division algorithm to find the g.c.d. of two positive integers.

| **Euclidean Algorithm** | If $a$ and $b$ are two positive integers, with $a \ge b$, and if positive integer $r$ is the remainder when $a$ is divided by $b$, then $$\text{g.c.d. } (a, b) = \text{g.c.d. } (b, r)$$ |
| --- | --- |

The preceding algorithm enables you to find the g.c.d. of two numbers $a$ and $b$ by dividing several times, each time letting the divisor become the dividend and the remainder become the divisor. The last **nonzero remainder** is g.c.d. $(a, b)$.

**EXAMPLE 7**

Find g.c.d. (180, 144).

*SOLUTION*

**Computation**          **Verification**

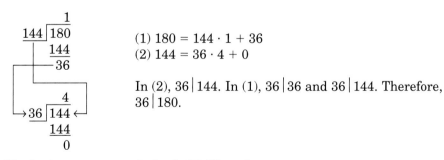

(1) $180 = 144 \cdot 1 + 36$
(2) $144 = 36 \cdot 4 + 0$

In (2), $36 \,|\, 144$. In (1), $36 \,|\, 36$ and $36 \,|\, 144$. Therefore, $36 \,|\, 180$.

The last nonzero remainder is 36. Therefore,

$$\text{g.c.d. } (180, 144) = 36$$

## CALCULATOR HINT

Using the Euclidean Algorithm find the g.c.d. (180, 144) on a calculator.

180 [INT÷] 144 [=] ; the remainder is 36.

144 [INT÷] 36 [=] ; the remainder is 0.        g.c.d (180,144) = 36

---

**EXAMPLE 8**

Find g.c.d. (3360, 4576).

**Computation**

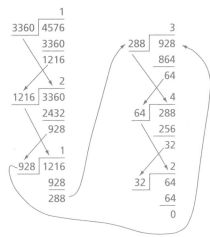

The last nonzero remainder is 32. Therefore,

$$\text{g.c.d. } (3360, 4576) = 32$$

## CALCULATOR HINT

Using the Euclidean Algorithm find the g.c.d. (3360, 4576) on a calculator.

4576 [INT÷] 3360 [=] ; the remainder is 1216

3360 [INT÷] 1216 [=] ; the remainder is 928.

1216 [INT÷] 928 [=] ; the remainder is 288.

928 [INT÷] 288 [=] ; the remainder is 64.

288 [INT÷] 64 [=] ; the remainder is 32.

64 [INT÷] 32 [=] ; the remainder is 0.        g.c.d. (4576, 3360) = 32.

# LEAST COMMON MULTIPLE

We are not only interested in the common divisors of two or more numbers; we are also interested in common multiples of numbers. If $b \mid c$, then $c$ is said to be a **multiple** of $b$. A number $m$ is a **common multiple** of two positive integers $b$ and $c$ if it is a positive multiple of both $b$ and $c$. How many common multiples can two integers have? Do you agree that the answer is infinitely many?

**EXAMPLE 9**

Multiples of 5 are

$$5, 10, 15, 20, 25, \ldots$$

Multiples of 2 are

$$2, 4, 6, 8, 10, 12, 14, 16, 18, 20, \ldots$$

Common multiples of 2 and 5 are

$$10, 20, 30, \ldots$$

The smallest of these numbers is 10; so the least common multiple of 5 and 2 is 10.

**EXAMPLE 10**

Examine the multiples of 6 and 9. The multiples of 6 are {6, 12, 18, 24, 30, 36, . . .}, and the multiples of 9 are {9, 18, 27, 36, . . .}. The common multiples of 6 and 9 are {18, 36, 54, 72, . . .}, and 18 is the least of these common multiples, so 18 is the least common multiple.

---

**DEFINITION**

**Least Common Multiple**

A positive integer $m$ is the *least common multiple* (abbreviated l.c.m.) of two positive integers $a$ and $b$ if $m$ is the least positive integer divisible by both $a$ and $b$. We write

$$m = \text{l.c.m. } (a, b)$$

---

**The least common multiple of 2 positive integers is therefore the smallest positive integer that is divisible by both numbers. It is never smaller than the larger number involved.** Thus, l.c.m. (3, 4) = 12; l.c.m. (4, 5) = 20; and l.c.m. (6, 15) = 30.

Because finding the l.c.m. by finding the set of common multiples and then selecting the least of these is a rather cumbersome process, additional methods must be considered. The first procedure uses prime factorization. **The l.c.m. of two numbers is the product of the highest powers of all the different prime factors that occur in factoring the numbers.**

**EXAMPLE 11**

(a) Find l.c.m. (8, 12).
(b) Find l.c.m. (120, 180).

*SOLUTION*

(a) The set of multiples of 8 is {8, 16, 24, 32, 40, . . .}. The set of multiples of 12 is {12, 24, 36, 48, . . .}. By inspection, we see that 24 is the least common multiple. Let's examine 8, 12, and 24 in factored form.

$$
\begin{array}{ccc}
2)\overline{24} & 2)\overline{8} & 2)\overline{12} \\
2)\overline{12} & 2)\overline{4} & 2)\overline{6} \\
2)\overline{6} & 2)\overline{2} & 3)\overline{3} \\
3)\overline{3} & 1 & 1 \\
1 & &
\end{array}
$$

$$24 = 2 \cdot 2 \cdot 2 \cdot 3 \qquad 8 = 2 \cdot 2 \cdot 2 \qquad 12 = 2 \cdot 2 \cdot 3$$

24 contains $2^3$, in order that the $2^3$ in 8 and the $2^2$ in 12 will divide it. Also, 24 contains a 3, so the 3 in 12 will divide it.

$$\text{l.c.m. } (8, 12) = 2^3 \cdot 3 = 24$$

(b) $120 = 2^3 \cdot 3 \cdot 5$ and $180 = 2^2 \cdot 3^2 \cdot 5$. Thus, l.c.m. $(120, \ 180) = 2^3 \cdot 3^2 \cdot 5 = 360$.

## CALCULATOR HINT

Some calculators have built in functions that enable us to easily compute the least common multiple. We will use the [MATH] menu that contains the l.c.m. to find the l.c.m. (120, 180).

[MATH] (underline *lcm*) 120 [ , ] 180 [ ) ] [ = ]. Did you get 360?

**EXAMPLE 12**

Find l.c.m. (24, 15, 20, 6).

*SOLUTION*

$$24 = 2^3 \cdot 3$$
$$15 = 3 \cdot 5$$
$$20 = 2^2 \cdot 5$$
$$6 = 2 \cdot 3$$
$$\text{l.c.m. } (24, 15, 20, 6) = 2^3 \cdot 3 \cdot 5 = 120$$

## COMMON ERROR

Sometimes students compute the least common multiple by using each prime factor the total number of times it occurs in all of the factorizations. Why is this incorrect?

# ⠿ DIVISION BY PRIMES

If the l.c.m. of several positive integers is to be found, a procedure involving division by primes is sometimes quicker than the methods we have used so far.

**EXAMPLE 13**

Find l.c.m. (12, 30).

First divide both 12 and 30 by 2, getting 6 and 15; then divide both 6 and 15 by 3, getting 2 and 5. Because 2 and 5 are primes, the process is complete. To obtain the l.c.m. by this procedure, we continue the division process until the row of answers (after division) consists of relatively prime numbers.

$$
\begin{array}{r|rr}
2 & 12 & 30 \\
3 & 6 & 15 \\
\hline
 & 2 & 5
\end{array}
\qquad \text{l.c.m. } (12, 30) = 2 \cdot 3 \cdot 2 \cdot 5 = 60
$$

Now suppose that we change the order of the division, dividing by 2, 2, and then 3:

$$
\begin{array}{r|rr}
2 & 12 & 30 \\
2 & 6 & 15 \\
3 & 3 & 15 \\
\hline
 & 1 & 5
\end{array}
\qquad \text{l.c.m. } (12, 30) = 2 \cdot 2 \cdot 3 \cdot 1 \cdot 5 = 60
$$

In the second line of this procedure, 2 does not divide 15; thus, we simply bring down the 15.

**EXAMPLE 14**

Find l.c.m. (24, 15, 20, 6).

*SOLUTION*

$$
\begin{array}{r|cccc}
2 & 24 & 15 & 20 & 6 \\
\hline
3 & 12 & 15 & 10 & 3 \\
\hline
5 & 4 & 5 & 10 & 1 \\
\hline
2 & 4 & 1 & 2 & 1 \\
\hline
 & 2 & 1 & 1 & 1
\end{array}
$$

l.c.m. $(24, 15, 20, 6) = 2 \cdot 3 \cdot 5 \cdot 2 \cdot 2 \cdot 1 \cdot 1 \cdot 1 = 120$

**PRACTICE PROBLEM**

Find the least common multiple of 16, 36, and 40.

*ANSWER*

720

## COMMON ERROR

Students often think that the greatest common divisor of two positive integers is *always* less than the least common multiple. Explain why this is not correct.

An interesting relationship exists between the l.c.m. and the g.c.d. of two numbers. From this relationship, if you know one of these quantities, you can find the other.

**Formula for Least Common Multiple**

The *least common multiple* of two positive integers $a$ and $b$ can be found from the greatest common divisor of $a$ and $b$ by computing the quotient

$$
\text{l.c.m. } (a, b) = \frac{ab}{\text{g.c.d. } (a, b)}
$$

**EXAMPLE 15**

Find l.c.m. (144, 180).

*SOLUTION*

$$
144 = 2^4 \cdot 3^2
$$

$$
180 = 2^2 \cdot 3^2 \cdot 5
$$

$$
\text{g.c.d. } (144, 180) = 2^2 \cdot 3^2
$$

$$
\text{l.c.m. } (144, 180) = \frac{(144)(180)}{\text{g.c.d. } (144, 180)}
$$

$$
= \frac{2^4 \cdot 3^2 \cdot 2^2 \cdot 3^2 \cdot 5}{2^2 \cdot 3^2}
$$

$$
= 2^4 \cdot 3^2 \cdot 5 = 720
$$

**CALCULATOR HINT**

[MATH] (underline *lcm*) 144 [ , ] 180 [ ) ] [ = ]. Did you get 720?

## LEAST COMMON DENOMINATOR

An important application of finding the least common multiple of two numbers is finding the least common denominator of two or more denominators of fractions. Let $\frac{a}{b}$ and $\frac{c}{d}$ be two fractions with denominators $b$ and $d$. The least common denominator of $b$ and $d$ (that is, the smallest positive integer that $b$ and $d$ will divide) is simply l.c.m. $(b, d)$.

**EXAMPLE 16**

Find the least common denominator for the 2 denominators 90 and 168 of $\frac{13}{90}$ and $\frac{11}{168}$.

*SOLUTION*

$$90 = 2 \cdot 3^2 \cdot 5$$
$$168 = 2^3 \cdot 3 \cdot 7$$

The least common denominator of 90 and 168 is

$$\text{l.c.m. } (90, 168) = 3^2 \cdot 2^3 \cdot 5 \cdot 7 = 2520$$

2520 is the smallest number that 90 and 168 will divide, and thus is the least common denominator.

The solution to the introductory problem is given in Example 17.

**EXAMPLE 17**

The cheerleaders of the Hardy College Bulldogs want to buy the same number of red pompons and black pompons. What is the least number of each they can buy if red ones are shipped in boxes of 8 and black ones are shipped in boxes of 6?

*SOLUTION*

**UNDERSTANDING THE PROBLEM** The number of red pompons purchased must be a multiple of 8, and the number of black pompons purchased must be a multiple of 6. Because the cheerleaders are purchasing the same number of each color, this number must be a multiple of both 6 and 8.

**DEVISING A PLAN** We want to obtain the least common multiple, which in this case is

$$\text{l.c.m. } (6, 8) = 24$$

**CARRYING OUT THE PLAN** Therefore, the cheerleaders will purchase 3 boxes of red pompons (or $3 \cdot 8 = 24$ red pompons) and 4 boxes of black pompons (or $4 \cdot 6 = 24$ black pompons).

# Just for Fun

Three neighborhood dogs barked consistently last night. Spot, Patches, and Lady began with a simultaneous bark at 11:00 P.M. Then Spot barked every 4 minutes, Patches every 2 minutes, and Lady every 5 minutes. When Mr. Jones awakened at 11:20 P.M., did he do so because it was unusually quiet at that moment or because all three dogs barked at once?

# EXERCISE SET 4

*R* 1. Find the greatest common divisor and the least common multiple for each pair of numbers without a calculator. Then check your answer with a calculator.

(a) 105 and 30  (b) 15 and 21  (c) 66 and 90
(d) 60 and 108  (e) 16 and 42  (f) 57 and 90
(g) 10 and 9   (h) 252 and 96

2. Find the greatest common divisor and the least common multiple for each of the following.

(a) 24, 30, 42       (b) 26, 36, 39
(c) 4600, 224, 228   (d) 45, 36, 24
(e) 15, 39, 30, 21, 70  (f) 42, 96, 104, 18

3. (a) Express the greatest common divisor g.c.d. $(a, b)$ as a product of primes, where

$$a = 3^3 \cdot 5^2 \cdot 2$$
$$b = 3 \cdot 5 \cdot 2^3$$

(b) Express g.c.d. $(a, b)$ as a product of primes, where $r$, $s$, and $t$ are distinct primes and where
$$a = r \cdot s^3 \cdot t^2$$
$$b = r^4 \cdot s^2 \cdot t^5$$

(c) Express the least common multiple of $a$ and $b$ in part (a) as a product of primes.

(d) Express l.c.m. $(a, b)$ in part (b) as a product of primes.

4. For the fractions $\frac{13}{162}$ and $\frac{11}{120}$ find the least common denominator of 162 and 120.

5. For the fractions $\frac{11}{200}$ and $\frac{19}{270}$ find the least common denominator of 200 and 270.

*T* 6. (a) When does l.c.m. $(a, b) = ab$?
(b) When does l.c.m. $(a, b) = a$?
(c) When does g.c.d. $(a, b) = ab$?
(d) When does g.c.d. $(a, b) = a$?

7. Find the least common multiple of the following by two different processes.

(a) 44 and 92   (b) 45 and 72
(c) 146 and 124  (d) 840 and 1800

8. Using the Euclidean algorithm, find the greatest common divisor. Then check with a calculator.

(a) 1122 and 105   (b) 4652 and 232
(c) 2244 and 418   (d) 735 and 850
(e) 220 and 315    (f) 486 and 522
(g) 912 and 19,656  (h) 7286 and 1684

9. (a) What are the g.c.d. and the l.c.m. of $a$ and $b$ if $a$ and $b$ are distinct primes?
(b) If $a \mid b$, compute g.c.d. $(a, b)$ and l.c.m. $(a, b)$.
(c) If $a$ is any natural number, compute g.c.d. $(a, a)$ and l.c.m. $(a, a)$.
(d) If $a$ and $b$ are composite, under what conditions will l.c.m. $(a, b) = ab$?

(e) What is the relationship between $a$ and $b$ if l.c.m. $(a, b) = a$?

(f) What are the l.c.m. and the g.c.d. of 1 and 4, of 1 and 101, and of 1 and $a$ $(a \neq 0)$?

(g) If g.c.d. $(a, b) = 1$, find l.c.m. $(a, b)$.

(h) Find g.c.d. $(a^2, a)$ and l.c.m. $(a^2, a)$, $(a \neq 0)$.

10. The l.c.m. of 2 numbers is $2^2 \cdot 3^4 \cdot 7 \cdot 11 \cdot 13$. The g.c.d. of the same 2 numbers is $2 \cdot 3 \cdot 7$. One of the numbers is $2^2 \cdot 3 \cdot 7 \cdot 11$. What is the other number?

11. Two people are running around a track in the same direction. One person does a mile in 4 min., but the second one takes 10 min. If they start at the same time and same place, how long will it take them to be at this place together if they continue to run?

12. First, 40 pupils are separated into teams, with each team having the same number of pupils. Later, 24 more pupils are distributed equally to the teams, with no one left over. What is the greatest number of teams possible?

13. An elementary art teacher has 3 art classes with 21, 35, and 28 students, respectively. The teacher wants to order some equipment that can be used by equal-sized groups in each class. What is the largest number of students in a group in each class so that each group has the same number of students?

14. Before checking with the caterer, a cook cuts a cake into 35 equal pieces and an identical cake into 42 equal pieces. The caterer, however, insists that the cakes be cut exactly alike. Into how many pieces must each cake now be cut?

15. Event A happens every 15 min. and event B happens every 18 min. If events A and B happen together at 6 P.M., when is the next time they will happen together?

16. Three cars are warming up at a circular race track, all going in the same direction. Car A goes around the track in 5 min., car B in 8 min., and car C in 12 min. If all 3 cars begin at the same moment, how many minutes will elapse before they are all again at the starting point at the same time?

17. Jane is planning a party at which she expects 16, 24, or 12 guests. Because she is unsure of the number of guests but wants to have available an equal number of hors d'oeuvres for each guest with none left over, how many should she order?

18. When Jill counts the pennies in her bank by 5's, there are 3 pennies left over. When she counts them by 3's, there are 2 left over. What is the least number of pennies in the bank?

C 19. Fifteen pears, 25 apples, and 35 oranges are to be packed in 2 or more baskets.

(a) What is the least number of baskets needed if each basket is to have the same numbers of pears, apples, and oranges?

(b) How many pears will be in each basket? How many apples? How many oranges?

20. Is l.c.m. $(a, b, c) = (a \cdot b \cdot c)/\text{g.c.d.} (a, b, c)$?

21. Consider finding the g.c.d. of two numbers as an operation. Let $aGb$ represent finding g.c.d. $(a, b)$. For example, $2G3 = 1$, $2G4 = 2$, $8G12 = 4$, and so on. Which of the following properties hold for this operation on the set of natural numbers?

(a) closure       (b) commutativity
(c) associativity    (d) an identity
(e) an inverse for each element

22. Let $aLb$ represent finding l.c.m. $(a, b)$. Answer the questions in Exercise 21.

23. Consider the operations defined in Exercises 21 and 22.

(a) Does L distribute over G?
(b) Does G distribute over L?

24. For any natural number $x$, prove that

(a) g.c.d. $(xa, xb) = x \cdot \text{g.c.d.} (a, b)$.
(b) l.c.m. $(xa, xb) = x \cdot \text{l.c.m.} (a, b)$.

## REVIEW EXERCISES

25. Verify for two cases that $2^{k+1} \cdot 3$, where $k$ is a positive integer, always produces an abundant number.

26. Euclid showed that if $2^p - 1$ is prime, then $2^{p-1}(2^p - 1)$ is perfect. Verify this theorem for four values of $p$.

27. Determine whether the following numbers are prime.

(a) 1     (b) 157     (c) 83     (d) 391

28. Factor the following numbers into primes.

(a) 126     (b) 525     (c) 2475     (d) 252

29. Is it possible to have exactly six consecutive composite numbers between two primes?

30. Answer the question in Exercise 29 for seven consecutive composite numbers.

31. Solve on the domain of integers.

(a) $^-2x = ^-4$      (b) $x + {}^-4 = 8$
(c) $^-2x - 5 = {}^-3$    (d) $^-3x + 4 = {}^-5$
(e) $x - 4 < {}^-2$

### ❖ PCR Excursion ❖

32. One warm-up procedure in basketball is to place players in a circle and pass the ball to different persons in the circle. For example, the coach might call, "Routine three," meaning pass the ball to every third person. Suppose the captain starts the warm-up. If the ball is passed to every person before it is passed back to the captain, we have what we call a complete cycle. For example, in the first figure we have eight players in the warm-up. If the ball is passed to every third person, observe that every person receives the ball before it is passed to the captain. Therefore we have a complete cycle. However, suppose as in the second figure there are nine players in the warm-up. Notice that the captain receives the ball before many players. This is not a complete cycle.

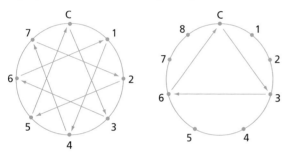

Consider the following warm-ups. In Routine 1, each player passes the ball to the adjacent player. In Routine 2, players pass the ball to the second person from them, and so on.

(a) What is the only routine that will ensure that a complete cycle occurs, no matter how many players are on the team?

(b) Which of Routines 2, 3, 4, 5, and 6 give complete cycles for a team of 7 players?

(c) For warm-ups consisting of 8, 9, 10, 11, and 12 players, determine which of Routines 2, 3, 4, . . . will give a complete cycle. Summarize your work from (b) and (c) in a table.

(d) Look at the table from (c). For which number of players was there the largest number of complete cycles?

(e) Describe verbally any relationship you have discovered between the number of players and the Routines (2, 3, 4, . . .) for which there are complete cycles.

(f) Find the number of routines that give complete cycles for 18 players. Use the relationship from (e) if possible.

(g) Which number of players has the largest number of complete cycles, 34 or 42?

(h) Verbally describe which routines give complete cycles for 13, 17, 19, and 23 players.

(i) Verbally describe the number of routines that give complete cycles if the number of players is a prime.

# MODULAR ARITHMETIC

| PROBLEM | If July 4 is on Friday this year and if next year is a leap year, then on what day will July 4 be next year? If Thanksgiving falls on November 28, then Christmas will be on what day of the week? If Valentine's Day is 47 days before Easter Sunday, then on what day of the week will it fall? |
| --- | --- |
| OVERVIEW | We now introduce mathematical systems with a finite number of elements. |
| GOALS | Illustrate the Number and Operations Standard, page 119.<br>Illustrate the Algebra Standard, page 4.<br>Illustrate the Connections Standard, page 147.<br>Illustrate the Problem Solving Standard, page 15.<br>Illustrate the Representation Standard, page 50. |

# TWELVE-HOUR CLOCK

**FIGURE 5-14**

To understand what characterizes a mathematical system, let us consider a normal 12-hour clock with only 12 numbers on it. The clock in Figure 5-14 is such a clock, modified by having its minute hand removed and by having the 12 replaced by 0.

When considering future time, we add a fixed number of hours to the present time. Suppose that the time is now 9:00 A.M. and that you have an appointment in 5 hours. Five hours after 9:00 A.M. is 2:00 P.M. (See Figure 5-15.) Thus, 9:00 A.M. $\oplus$ 5 hours = 2:00 P.M., where $\oplus$ is used as the symbol for clock addition.

**FIGURE 5-15**

In the same way, past events are denoted by subtraction. Suppose that we had breakfast 4 hours ago, and that the time is now 9:00 A.M. At what time did we have breakfast? (See Figure 5-16.) Letting the symbol $\ominus$ denote clock subtraction, we have 9 A.M. $\ominus$ 4 hours = 5 A.M. Thus we had breakfast at 5 A.M.

**FIGURE 5-16**

If we were unfamiliar with time as kept by a 12-hour clock, this new type of arithmetic might be difficult. However, we can easily verify that each of the following examples is correct.

**EXAMPLE 1**

Add 9 h. to 6 P.M.

*SOLUTION*

$6 \oplus 9 = 3$ A.M. Check against your watch.

**EXAMPLE 2**

Add 27 h. to 7 A.M.

*SOLUTION*

$7 \oplus 27 = 10$. The clock time is 10 A.M.

**EXAMPLE 3**

It takes 7 hours to drive to your house. You wish to arrive by 3 P.M. When should you start?

*SOLUTION*

$3 \ominus 7 = 8$, so you should start by 8 A.M.

Next, consider the numbers 0 through 11 as being the elements of a set, and consider addition as being based on counting in a clockwise direction on the face of a clock. A table of addition facts for a 12-hour clock (forgetting any A.M. or P.M. designations) is shown in Table 5-3. No matter where we start on the clock, we are always back at the starting position 12 hours later.

**TABLE 5-3**

| $\oplus$ | 0 | 1 | 2 | 3 | 4 | 5 | 6 | 7 | 8 | 9 | 10 | 11 |
|---|---|---|---|---|---|---|---|---|---|---|---|---|
| **0** | 0 | 1 | 2 | 3 | 4 | 5 | 6 | 7 | 8 | 9 | 10 | 11 |
| **1** | 1 | 2 | 3 | 4 | 5 | 6 | 7 | 8 | 9 | 10 | 11 | 0 |
| **2** | 2 | 3 | 4 | 5 | 6 | 7 | 8 | 9 | 10 | 11 | 0 | 1 |
| **3** | 3 | 4 | 5 | 6 | 7 | 8 | 9 | 10 | 11 | 0 | 1 | 2 |
| **4** | 4 | 5 | 6 | 7 | 8 | 9 | 10 | 11 | 0 | 1 | 2 | 3 |
| **5** | 5 | 6 | 7 | 8 | 9 | 10 | 11 | 0 | 1 | 2 | 3 | 4 |
| **6** | 6 | 7 | 8 | 9 | 10 | 11 | 0 | 1 | 2 | 3 | 4 | 5 |
| **7** | 7 | 8 | 9 | 10 | 11 | 0 | 1 | 2 | 3 | 4 | 5 | 6 |
| **8** | 8 | 9 | 10 | 11 | 0 | 1 | 2 | 3 | 4 | 5 | 6 | 7 |
| **9** | 9 | 10 | 11 | 0 | 1 | 2 | 3 | 4 | 5 | 6 | 7 | 8 |
| **10** | 10 | 11 | 0 | 1 | 2 | 3 | 4 | 5 | 6 | 7 | 8 | 9 |
| **11** | 11 | 0 | 1 | 2 | 3 | 4 | 5 | 6 | 7 | 8 | 9 | 10 |

**EXAMPLE 4**

Clock multiplication, denoted by $\odot$, is defined in terms of repeated addition. On a clock, find (a) $3 \odot 2$ and (b) $3 \odot 5$.

*SOLUTION*

(a) Start at 0 and add 2 hours three times:
$$3 \odot 2 = 2 \oplus 2 \oplus 2 = 6$$
(b) $3 \odot 5 = 3$    $5 \oplus 5 = 10$ and $10 \oplus 5 = 3$

**EXAMPLE 5**

Find $3 \odot 8$ on a 12-hour clock.

*SOLUTION*

$3 \odot 8 = 8 \oplus 8 \oplus 8 = 4 \oplus 8 = 0$    $8 \oplus 8 = 4$

**PRACTICE PROBLEM**

Find $5 \odot 4$ on a 12-hour clock.

*ANSWER*

$5 \odot 4 = 8$

# Four-minute Clock

A finite mathematical system can also be illustrated by a 4-minute clock. This type of clock might be used to time the rounds and intermissions in a boxing match. The sketch in Figure 5-17 represents the face of such a clock. Notice that 0 is used in place of 4; thus, at the end of 4 minutes, the clock starts over.

This mathematical system contains only 4 numbers—0, 1, 2, and 3. Addition in this system (denoted by $\oplus$) is defined and interpreted to mean that the hand of the clock moves the same number of positions as the number to be added. If the hand is at 1 and moves for 2 minutes, it will be at 3; so $1 \oplus 2 = 3$. Similarly, if the hand is at 2 and moves for 2 minutes, it will be at 0; so $2 \oplus 2 = 0$. Finally, if the hand is at 3 and moves for 2 minutes, it will be at 1; so $3 \oplus 2 = 1$.

The operation of addition for this system is traced in Table 5-4. Check the entries in this table by visualizing the movement of the hand of the 4-minute clock.

**TABLE 5-4**

| $\oplus$ | 0 | 1 | 2 | 3 |
|---|---|---|---|---|
| 0 | 0 | 1 | 2 | 3 |
| 1 | 1 | 2 | 3 | 0 |
| 2 | 2 | 3 | 0 | 1 |
| 3 | 3 | 0 | 1 | 2 |

**EXAMPLE 6**

(a) $2 \oplus 3 = 1$, from Table 5-4.
(b) $2 \ominus 3 = 3$, since Table 5-4 indicates that $3 \oplus 3 = 2$.

Our discussions of the arithmetic on a 12-hour clock and on a 4-minute clock illustrate finite mathematical systems. By a **finite mathematical system** we mean a system which is limited to a finite set of numbers such as the 12 numbers on a 12-hour clock. Although this type of number system may seem strange to you, it is the kind of mathematics used for many machines with dials or controls.

# Modular Arithmetic

Now let's do a similar discussion of mathematical systems where the set of numbers is not finite. That is, we will have an infinite set of numbers such as the set of whole numbers {0, 1, 2, 3, 4, 5, . . .}.

In modular arithmetic, emphasis is placed on the fact that two distinct numbers differ by a multiple of some natural number. For instance, 3 minutes and 7 minutes register the same way on a 4-minute clock because $7 - 3$ is a multiple of 4. Numbers that differ by multiples of a given natural number are said to be congruent modulo the given natural number.

**DEFINITION**

| Congruent Modulo $m$ | Two integers $a$ and $b$ are *congruent modulo m* (where $m$ is a natural number called the **modulus**) if and only if $m$ divides the difference of $a$ and $b$. |
|---|---|

The relationship is denoted by $a \equiv b \pmod{m}$ and is read "$a$ is congruent to $b \pmod{m}$." Thus two numbers are congruent modulo $m$ if their difference is divisible by $m$.

**EXAMPLE 7**

| | | |
|---|---|---|
| $17 \equiv 3 \pmod 7$ | because | $7 \mid (17 - 3)$ |
| $14 \equiv 8 \pmod 6$ | because | $6 \mid (14 - 8)$ |
| $21 \equiv 5 \pmod 4$ | because | $4 \mid (21 - 5)$ |
| $10^2 \equiv 2 \pmod 7$ | because | $7 \mid (10^2 - 2)$ |
| $6 \not\equiv 4 \pmod 3$ | because | $3 \nmid (6 - 4)$ |

Another criterion for determining whether natural numbers $a$ and $b$ are congruent is as follows.

| Requirement for Congruence | Positive integers $a$ and $b$ are **congruent modulo m** if their remainders are equal when divided by $m$. |
|---|---|

**EXAMPLE 8**

(a) $17 \equiv 35 \pmod 3$ because 17 divided by 3 leaves a remainder of 2, and 35 divided by 3 also produces a remainder of 2.

(b) $701 \equiv 7001 \pmod 7$ because, when the numbers are divided by 7, the remainder is 1 in each case.

## CALCULATOR HINT

Show that 462 and 230 are congruent mod 4.

462 [INT÷] 4 [=] ; the remainder is 2.

230 [INT÷] 4 [=] ; the remainder is 2.

Therefore, $462 \equiv 230 \pmod 4$.

In particular, $a \equiv 0 \pmod m$ means that $m \mid a$; conversely, if $m \mid a$, then $a \equiv 0 \pmod m$.

Addition in a system modulo $m$ is clearly the same as addition of whole numbers when the sum is less than $m$. When the sum is greater than or equal to $m$, we divide the sum by $m$ and use the remainder in place of the ordinary sum. For instance, Table 5-4 is a mod 4 addition table, and Table 5-5 is a mod 5 addition table. The following examples use the table for addition mod 5. Check the work by referring to Table 5-5.

| TABLE 5-5   Addition Mod 5 | | | | |
|---|---|---|---|---|
| **+** | **0** | **1** | **2** | **3** | **4** |
| **0** | 0 | 1 | 2 | 3 | 4 |
| **1** | 1 | 2 | 3 | 4 | 0 |
| **2** | 2 | 3 | 4 | 0 | 1 |
| **3** | 3 | 4 | 0 | 1 | 2 |
| **4** | 4 | 0 | 1 | 2 | 3 |

| TABLE 5-6   Multiplication Mod 5 | | | | |
|---|---|---|---|---|
| **·** | **0** | **1** | **2** | **3** | **4** |
| **0** | 0 | 0 | 0 | 0 | 0 |
| **1** | 0 | 1 | 2 | 3 | 4 |
| **2** | 0 | 2 | 4 | 1 | 3 |
| **3** | 0 | 3 | 1 | 4 | 2 |
| **4** | 0 | 4 | 3 | 2 | 1 |

**EXAMPLE 9**

$$2 + 1 \equiv 3 \ (\text{mod } 5) \qquad 2 + 3 \equiv 0 \ (\text{mod } 5)$$
$$2 + 2 \equiv 4 \ (\text{mod } 5) \qquad 2 + 4 \equiv 1 \ (\text{mod } 5)$$

**PRACTICE PROBLEM**

Find $5 + 4 \ (\text{mod } 7)$.

*ANSWER*

$5 + 4 \equiv 2 \ (\text{mod } 7)$

Subtraction in a system modulo $m$ is the same as subtraction for whole numbers, with one exception. If subtraction resulting in a nonnegative difference cannot be performed, change the numbers (modulo $m$) until such subtraction can be performed. Thus, $3 - 4 \ (\text{mod } 5)$ is equivalent to

$$(3 + 5) - 4 \equiv 8 - 4 = 4 \ (\text{mod } 5)$$

Multiplication in a system modulo $m$ is the same as multiplication of whole numbers, except when the product is greater than or equal to $m$; then it is reduced by the congruence relationship $a \equiv b \ (\text{mod } m)$ to a number less than $m$. Table 5-6 illustrates multiplication $(\text{mod } 5)$.

Division in a system modulo $m$ is the inverse of multiplication. If a multiplication table has been constructed, you can obtain from it an answer that, when multiplied by the divisor, gives the dividend. If a multiplication table has not been constructed, you can repeatedly increase the dividend by the modulus until the division is obvious (provided that an answer exists). For example, $3 \div 5 \equiv 2 \ (\text{mod } 7)$ because $(3 + 7) \div 5 \equiv 2 \ (\text{mod } 7)$.

**EXAMPLE 10**

(a) $3 \cdot 3 \equiv 4 \ (\text{mod } 5)$
(b) $2 \cdot 3 \equiv 1 \ (\text{mod } 5)$
(c) $2 \div 3 \equiv 4 \ (\text{mod } 5)$, since $3 \cdot 4 \equiv 2 \ (\text{mod } 5)$   $(2 + 10) \div 3 = 4$
(d) $4 \div 3 \equiv 3 \ (\text{mod } 5)$, since $3 \cdot 3 \equiv 4 \ (\text{mod } 5)$   $(4 + 5) \div 3 = 3$

According to Table 5-6, division by 3 $(\text{mod } 5)$ is always possible because every element occurs in the fourth row. Similarly, division by 1, 2, and 4 is always possible in mod 5 arithmetic. You will discover, however, that division is not always possible in other finite systems. For example, make a mod 4 multiplication table and look at division by 2.

Many properties of equality carry over to congruence.

| **Properties of Congruence** | If $a \equiv b \pmod{m}$ and $c \equiv d \pmod{m}$, then<br>1. $a + c \equiv b + d \pmod{m}$.<br>2. $a - c \equiv b - d \pmod{m}$, $a \geq c$ and $b \geq d$.<br>3. $ka \equiv kb \pmod{m}$, when $k$ is any natural number.<br>4. $ac \equiv bd \pmod{m}$ |
|---|---|

**EXAMPLE 11**

It is easy to verify using the definition of congruence that $16 \equiv 2 \pmod{7}$ and $38 \equiv 17 \pmod{7}$. Using these, without using the definition of $\pmod 7$, argue that

(a) $54 \equiv 19 \pmod{7}$    (b) $22 \equiv 15 \pmod{7}$    (c) $608 \equiv 34 \pmod{7}$

Of course, each of these is true by one of the properties just noted.

**EXAMPLE 12**

For each of the following, find a positive integer answer less than the modulus.

(a) $5 + 2 \pmod{3}$    (b) $18 + 5 \pmod{7}$    (c) $4 - 8 \pmod{5}$

*SOLUTION*

(a) $5 + 2 \equiv 1 \pmod{3}$ because $5 + 2 = 7 = 2(3) + 1$.
(b) $18 + 5 \equiv 2 \pmod{7}$ because $18 + 5 = 23 = 3(7) + 2$.
(c) $4 - 8 \equiv 1 \pmod{5}$ because $4 + 5 - 8 = 1$. (Notice that the modulus, 5, is added to make subtraction possible.)

| | | November | | | | |
|---|---|---|---|---|---|---|
| S | M | T | W | T | F | S |
| | | | | 1 | 2 | 3 |
| 4 | 5 | 6 | 7 | 8 | 9 | 10 |
| 11 | 12 | 13 | 14 | 15 | 16 | 17 |
| 18 | 19 | 20 | 21 | 22 | 23 | 24 |
| 25 | 26 | 27 | 28 | 29 | 30 | |

F I G U R E   5 - 1 8

We can apply the concepts of congruence mod 7 to solve calendar problems like the ones at the beginning of this section. A calendar for November is given in Figure 5-18.

Notice that the five Thursdays have dates 1, 8, 15, 22, and 29; they differ by multiples of 7. The same is true of any other day of the week. Two weeks after Friday the ninth will be Friday the

$$9 + 2 \cdot 7 = 23\text{rd} \qquad \text{The modulus is 7.}$$

**EXAMPLE 13**

What day of the week will be November 1 exactly 1 year after the calendar of Figure 5-18? (Exclude leap year.)

*SOLUTION*

A normal year has 365 days. Because $365 = (52)7 + 1$, it follows that $365 \equiv 1 \pmod{7}$. Dates that differ by a multiple of 7 fall on the same day of the week. A date exactly 52 weeks after a Thursday is a Thursday. One day later is a Friday. Therefore, November 1 a year later will be on a Friday.

**EXAMPLE 14**

Christmas is on Wednesday this year. On what day will it be next year if next year is a leap year?

*SOLUTION*

There are 366 days in a leap year. Therefore, we compute

$$366 \equiv 2 \pmod{7}$$

Consequently, Christmas will be 2 days after Wednesday, on Friday.

**EXAMPLE 15**    Mr. Jones buys 255 bottles of different sodas, and Mark is mixing the different brands in 6-pack containers with the understanding that he can have all the bottles that will not go in the containers. How many sodas are left for Mark?

SOLUTION    $255 \equiv 3$ (mod 6). Therefore, Mark will enjoy 3 sodas.

# Just for Fun

Can you write a schedule for 12 basketball teams so that each team will play every other team only once and no team will be idle? (Hint: Try mod 11 arithmetic to see if it is possible. How many games will each team play?)

# EXERCISE SET 5

*R* 1. Perform each of the following operations on a 12-hour clock.

    (a) $5 \oplus 8$      (b) $3 \ominus 7$      (c) $7 \ominus 11$
    (d) $9 \oplus 6$      (e) $4 \odot 3$      (f) $7 \odot 3$
    (g) $9 \odot 2$      (h) $8 \odot 5$

2. Perform each of the following operations on a 4-minute clock.

    (a) $2 \oplus 3$      (b) $2 \oplus 5$      (c) $2 \ominus 3$
    (d) $3 \odot 1$      (e) $2 \odot 3$      (f) $3 \odot 2$
    (g) $1 \odot 3$      (h) $3 \odot 3$

3. (a) If Mr. Smith gets to work at 8 A.M. and works 8 hours, at what time does his work day end?

(b) The Smiths are leaving on a car trip at 5 A.M. tomorrow. Assuming that the trip is made in exactly 17 hours, at what time will they arrive at their destination?

(c) Mr. Smith, exhausted from the long trip, fell into bed 1 hour after their arrival and slept for 14 hours. At what time did he awaken?

4. Find $t$ for the following, where the answer is some number on a 12-hour clock.

    (a) $t = 8 + 7$      (b) $t = 5 - 8$
    (c) $t = 9 - 11$      (d) $t = (4)(7)$
    (e) $t = (3)(9)$      (f) $t = 2 - 7$

5. Work Exercise 4 for a 24-hour clock.

*T* 6. The Fourth of July holiday occurs on Tuesday this year. On what day of the week will it occur next year if next year is not a leap year?

7. Suppose Christmas is on Friday this year. In how many years from now will it be on Sunday if no leap years are involved? If next year is a leap year?

8. Labor Day always falls on the first Monday in September. This year, it was on Monday, September 4. On what date will it be next year if next year is not a leap year? If next year is a leap year?

9. Find three positive integers congruent to the following integers.
   (a) $1 \equiv$ _____ (mod 4)
   (b) $3 \equiv$ _____ (mod 7)

10. If Thanksgiving is on November 28, Christmas will be on what day of the week?

11. If Valentine's Day is 47 days before Easter Sunday, on what day of the week will it fall?

12. (a) Is $4 + 7 \equiv 15 + 4$ (mod 8)?
    (b) Is $6 + 2 \equiv 31 + 19$ (mod 10)?
    (c) In what modulus is $3 + 8 \equiv 5$?
    (d) For what modulus is $5 + 9 \equiv 6$?

13. Label each of the following as either true or false.
    (a) $6 \cdot 5 \equiv 4$ (mod 6)    (b) $18 - 5 \equiv 1$ (mod 12)
    (c) $6 \equiv 5$ (mod 3)    (d) $79 \equiv 17$ (mod 4)
    (e) $7 + 4 \equiv 5$ (mod 6)    (f) $6 \equiv 3$ (mod 8)
    (g) $8 \equiv 7 + 1$ (mod 5)    (h) $34 \equiv 7$ (mod 13)

14. Without tables, perform the following operations. In each case reduce the answer to a whole number less than the modulus. Then check your answers to (k), (l), (m) and (n) with a calculator.
    (a) $2 + 3$ (mod 5)
    (b) $3 + 2$ (mod 5)
       (a) and (b) illustrate what property?
    (c) $1 - 5$ (mod 3)    (d) $15 - 2$ (mod 7)
    (e) $3 + (2 + 5)$ (mod 6)
    (f) $(3 + 2) + 5$ (mod 6)
       (e) and (f) illustrate what property?
    (g) $3(12)$ (mod 5)
    (h) $(12)(3)$ (mod 5)
       (g) and (h) illustrate what property?
    (i) $2 \cdot 4 + 2 \cdot 11$ (mod 7)
    (j) $2(4 + 11)$ (mod 7)
       (i) and (j) illustrate what property?
    (k) $5^3$ (mod 9)    (l) $4^3$ (mod 12)
    (m) $(22)(46)$ (mod 7)    (n) $(624)(589)$ (mod 9)

15. By inspection, find a number $x$ (positive integer less than the modulus) that will make each of these statements true.
    (a) $x \equiv 5$ (mod 3)    (b) $x + 10 \equiv 9$ (mod 4)
    (c) $x + 5 \equiv 7$ (mod 9)    (d) $x + 20 \equiv 13$ (mod 3)
    (e) $x + 4 \equiv 6$ (mod 7)    (f) $x^2 \equiv 1$ (mod 3)

16. If July 4 falls on a Friday this year and if next year is leap year, on what day of the week will July 4 fall next year?

*C* 17. Consider a light switch with four positions in the order off, dim, bright, brightest. If Sam starts with the position named first in the following, at what position will he be when he completes the number of turns given?
    (a) off, 4    (b) brightest, 5   (c) dim, 6
    (d) bright, 3 (e) off, 7    (f) dim, 23

18. Use your knowledge of the number of days in each month to answer the following questions.
    (a) If April 1 is on Saturday, on what day is May 30?
    (b) If Pat's birthday is exactly 6 weeks before Christmas, what is its date?
    (c) If Christmas is on Monday, on what day will Diane's birthday (January 29) be?
    (d) Do the 125th and 256th days of the year fall on the same day of the week? Why or why not?

19. Suppose that we agree to order the numbers in a 4-hour clock as
$$0 < 1 < 2 < 3$$
    (a) Does the transitive property for $<$ hold?
    (b) If $a < b$, is $a + c < b + c$?

## REVIEW EXERCISES

20. Find the named quantity without a calculator. Then check the answers with a calculator.
    (a) g.c.d. (126, 525)
    (b) l.c.m. (525, 2475)
    (c) g.c.d. (126, 252, 525)
    (d) l.c.m. (126, 252)

21. What is the greatest common divisor of 5734 and 12,862? What is the least common multiple? First work without a calculator. Then check your answers with a calculator.

22. Suppose that the least common multiple of two numbers is the same as their greatest common divisor. What can you say about the numbers?

23. (a) Find 5 consecutive composite numbers that are all less than 30.
    (b) Explain why $(25 \cdot 24 \cdots 3 \cdot 2) + 2$ and $(25 \cdot 24 \cdots 3 \cdot 2) + 3$ are composite.

24. For each natural number $n$,
    (a) show that $3 \,|\, (n^3 - n)$.
    (b) show that $5 \,|\, (n^5 - n)$.
    (c) does $9 \,|\, (n^9 - n)$?

25. Find the smallest natural number divisible by 2, 3, 4, 5, and 6.

## LABORATORY ACTIVITY
• • • • • • • • • • • • • • • • • • • • • • • • • • • • • • • •

26. Take any three-digit number, reverse its digits, and subtract. Repeat the process until something interesting happens. What do you get? Does this always happen?

### ❖ PCR Excursion ❖

27. We can use modulo arithmetic to create what we will call *modular designs*.

A. First we will draw what we call a 9 × 4 design. We will multiply 4 by all positive integers 1, 2, 3, 4, 5, 6, 7, 8, modulo 9; that is,

$$4 \cdot 1 \equiv 4 \ (\text{mod } 9), \qquad 4 \cdot 5 \equiv 2 \ (\text{mod } 9),$$
$$4 \cdot 2 \equiv 8 \ (\text{mod } 9), \qquad 4 \cdot 6 \equiv 6 \ (\text{mod } 9),$$
$$4 \cdot 3 \equiv 3 \ (\text{mod } 9), \qquad 4 \cdot 7 \equiv 1 \ (\text{mod } 9),$$
$$4 \cdot 4 \equiv 7 \ (\text{mod } 9), \qquad 4 \cdot 8 \equiv 5 \ (\text{mod } 9).$$

Next draw a circle and divide the circumference into 8 (1 less than the modulus) equal parts, labeling the points of division 1, 2, 3, 4, 5, 6, 7, and 8. Now connect with line segments the numbers multiplied by 4 and the answers: 1 and 4, 2 and 8, 3 and 3, 4 and 7, 5 and 2, 6

and 6, 7 and 1, and 8 and 5. Shade triangular regions to get the design below.

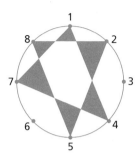

B. (a) Draw the simple design 5 × 3.
   (b) Now draw the 5 × 2 design. Discover why the designs for 5 × 3 and 5 × 2 are the same.

C. Draw the design
   (a) 11 × 5          (b) 13 × 3
   (c) 19 × 9          (d) 21 × 10

D. Can you discover for what value of *N* that
   (a) the design for 11 × *N* is the same as for 11 × 5?
   (b) The design for 13 × *N* is the same as for 13 × 3?
   (c) 19 × *N* is the same as 19 × 9?
   (d) 21 × *N* is the same as 21 × 10?

---

# SOLUTION TO INTRODUCTORY PROBLEM

**UNDERSTANDING THE PROBLEM**   When the trainer arranges the towels in 4 stacks, there is one left over; that is, if the number of towels is divided by 4, there is a remainder of 1. The same is true for division by 5 and by 6. When the trainer uses 7 stacks, there are no extra towels. Therefore, the number of towels is divisible by 7. We are to find the smallest possible number of towels.

**DEVISING A PLAN**   *Plan 1:* Write the multiples of 7, and divide each by 4, 5, and 6. Find the smallest multiple for which each of the three divisions produces a remainder of 1. This may be time consuming, but it will certainly produce the correct answer.

*Plan 2:* Let's use the fact that if $n \div 4$ leaves a remainder of 1, then $4 \,|\, (n-1)$. Likewise, $5 \,|\, (n-1)$ and $6 \,|\, (n-1)$. Since $7 \,|\, n$, then $n = 7k$.

$$4 \,|\, (7k - 1)$$
$$5 \,|\, (7k - 1)$$
$$6 \,|\, (7k - 1)$$

Now the smallest number that 4, 5, and 6 will divide is

$$\text{l.c.m. } (4, 5, 6) = 60$$

Since every common multiple is a multiple of the least common multiple,

$$60l = 7k - 1 \quad \text{or} \quad 7k = 60l + 1 \quad \text{or} \quad 7 \,|\, (60l + 1)$$

**CARRYING OUT THE PLAN**   Let *l* take on values 1, 2, 3, . . . until a $(60l + 1)$ is found that is divisible by 7.

$$7 \!\not|\, 61 \quad 7 \!\not|\, 121 \quad 7 \!\not|\, 181 \quad 7 \!\not|\, 241 \quad \text{but } 7 \,|\, 301$$

**LOOKING BACK**   Check to see if 301 towels is a valid answer.

# SUMMARY AND REVIEW

## THE INTEGERS

1. The set of integers $I$ consists of $\{\ldots, ^-3, ^-2, ^-1, 0, 1, 2, 3, \ldots\}$.

2. $I = \{$positive integers$\} \cup \{$negative integers$\} \cup \{0\}$.

3. For each integer $a$, there exists a unique *additive inverse* $^-a$ such that $a + ^-a = ^-a + a = 0$.

4. Addition is defined by the following cases, where $a$ and $b$ are whole numbers.

   (a) $a + b = n(A \cup B)$, where $a = n(A)$, $b = n(B)$, and $A \cap B = \varnothing$ for sets $A$ and $B$

   (b) $^-a + ^-b = ^-(a + b)$

   (c) $a + ^-b = ^-b + a = (a - b)$ if $a > b$

   (d) $a + ^-b = ^-b + a = 0$ if $a = b$

   (e) $a + ^-b = ^-b + a = ^-(b - a)$ if $a < b$

5. The subtraction of $y$ from $x$ is defined as $x - y = z$ if and only if $x = y + z$. In any case,

$$x - y = x + ^-y$$

6. Multiplication is defined by the following cases, where $a$ and $b$ are whole numbers.

   (a) $a \cdot b = n(A \times B)$, where $a = n(A)$ and $b = n(B)$ for sets $A$ and $B$

   (b) $^-a \cdot ^-b = ab$

   (c) $^-a \cdot b = b \cdot ^-a = ^-ab$

7. The division $a \div b = c$ only if $a = bc$.

   (a) If $a \neq 0$, then $0 \div a = 0$, because $a \cdot 0 = 0$.

   (b) If $a \neq 0$, then $a \div 0$ is undefined, because $a = 0 \cdot d$ is false for any $d$.

   (c) $0 \div 0$ is undefined, because there is no unique quotient.

## DIVISIBILITY

1. If $a$ and $b$ are integers, then $b$ divides $a$ (denoted $b \mid a$) if and only if there is a unique integer $c$ such that $a = bc$.

2. The following are basic divisibility properties for integers $x$, $y$, and $z$.

   (a) $1 \mid x$ and $x \mid x$.

   (b) If $x \mid y$ and $y \mid z$, then $x \mid z$.

   (c) If $x \mid y$ and $x \mid z$, then $x \mid (y + z)$ and $x \mid (y - z)$.

   (d) If $x \mid y$ and $x \nmid z$, then $x \nmid (y + z)$ and $x \nmid (y - z)$.

   (e) If $x \mid y$ and $x \mid (y + z)$ [or $x \mid (y - z)$], then $x \mid z$.

   (f) If $x \mid y$ or $x \mid z$, then $x \mid yz$.

3. Generalization of divisibility properties

   If $a$ divides the sum of $n$ positive integers and also divides each one of $n - 1$ of these positive integers, then $a$ divides the remaining positive integer.

## DIVISIBILITY TESTS (BASE TEN)

1. An integer is divisible by 2 or by 5 if and only if the units digit of the number is divisible by 2 or by 5.

2. An integer is divisible by 4 if and only if the last two digits of the number represent a number divisible by 4.

3. An integer is divisible by 8 if and only if the last three digits of the number represent a number divisible by 8.

4. An integer is divisible by 3 or by 9 if and only if the sum of its digits is divisible by 3 or by 9.

5. An integer is divisible by 6 if and only if it is divisible by both 2 and 3.

6. An integer is divisible by 7 if and only if it satisfies the following property. Subtract the double of the right-hand digit from the number represented by the remaining digits. If the difference is divisible by 7, then the original number is divisible by 7. Repeat the process until divisibility is obvious.

7. An integer is divisible by 10 if and only if it ends in 0.

8. An integer is divisible by 11 if and only if the difference between the sum of the digits in the odd-numbered positions and the sum of the digits in the even-numbered positions is divisible by 11.

## PRIMES AND COMPOSITE NUMBERS

1. A positive integer $p$, greater than 1, is *prime* if and only if the only divisors of $p$ are 1 and $p$.

2. A positive integer is *composite* if and only if it has a positive integer divisor other than itself and 1.

3. If $n$ is an integer greater than 1 and not divisible by any prime $p$ such that $p^2 \leq n$, then $n$ is prime.

4. When a composite number is written as a product of all of its prime divisors, the product is called a *prime factorization* of the number.

5. The Fundamental Theorem of Arithmetic: Every composite number has a unique prime factorization (except for the order of the factors).

## GREATEST COMMON DIVISOR (G.C.D.) AND LEAST COMMON MULTIPLE (L.C.M.)

1. The *greatest common divisor* of two positive integers $a$ and $b$ is the largest positive integer $d$ such that $d \mid a$ and $d \mid b$.

2. If the greatest common divisor of two positive integers $a$ and $b$ is 1, then $a$ and $b$ are *relatively prime*.

3. If $a$ and $b$ are two positive integers, with $a \geq b$, and if $r \neq 0$ is the remainder when $a$ is divided by $b$, then g.c.d. $(a, b) = $ g.c.d. $(b, r)$.

4. Euclidean algorithm: If the fact from 3 is applied repeatedly until a remainder of zero is obtained, the last nonzero remainder is g.c.d. $(a, b)$.

5. A positive integer $m$ is the *least common multiple* of two positive integers $a$ and $b$ if $m$ is the least positive integer divisible by both $a$ and $b$.

6. l.c.m. $(a, b) = \dfrac{ab}{\text{g.c.d. } (a, b)}$

### ADDITIONAL DEFINITIONS OF NUMBER THEORY

1. The proper divisors of a positive integer are all the positive integer divisors except the number itself.

2. If the sum of the proper divisors of a positive integer is equal to the positive integer, the number is called a *perfect* number; if this sum is less than the positive integer, it is called a *deficient* number; if it is greater than the positive integer, it is called an *abundant* number.

3. Two positive integers are said to be *amicable* if each is the sum of the proper divisors of the other.

### CLOCK AND MODULAR ARITHMETIC

1. Two integers $a$ and $b$ are congruent modulo $m$ if and only if $m \mid (a - b)$.

2. Positive integers $a$ and $b$ are congruent modulo $m$ if their remainders are equal when divided by $m$.

3. If $a \equiv b \pmod{m}$ and $c \equiv d \pmod{m}$, then

   (a) $a + c \equiv b + d \pmod{m}$
   (b) $a - c \equiv b - d \pmod{m}$
   (c) $ka \equiv kb \pmod{m}$, where $k$ is any natural number
   (d) $ac \equiv bd \pmod{m}$

# CHAPTER TEST

1. Make an addition table modulo 4.

2. Which of the following numbers are primes?
   (a) 167    (b) 177    (c) 187

3. Test 13,341 for divisibility by all counting numbers from 2 through 9.

4. Write 252 as a product of its prime factors.

5. Find the greatest common divisor for 114 and 90.

6. Find the least common multiple of 114 and 90.

7. Compute the following:
   (a) $^-4 + 7$    (b) $^-8 + \ ^-9$
   (c) $4 \cdot \ ^-5$    (d) $^-4 - \ ^-7$

8. Illustrate on a number line.
   (a) The additive inverse of $^-5$
   (b) $3 \cdot \ ^-6$    (c) $^-4 + \ ^-5$

9. Classify each of the following as true or false.
   (a) If a positive integer is divisible by 3, it must be divisible by 9.
   (b) $12 \mid 6$.
   (c) Every odd positive integer is a prime number.
   (d) If a positive integer is divisible by 8, it must be divisible by 4.
   (e) $7 \mid 7$.
   (f) If a positive integer is divisible by 2 and by 10, it must be divisible by 20.

10. Perform the indicated operations.
    (a) $^-1 - (^-3) - (^-4) - 17$    (b) $x(w + \ ^-y) + xw$

11. If November 30 falls on Wednesday, on what day will it fall next year if next year is not a leap year? If next year is a leap year?

12. If 1 is the greatest common divisor of two positive integers, what can you say about their least common multiple?

13. Use the Euclidean algorithm to find the greatest common divisor of 7286 and 1684.

14. Find the g.c.d. and the l.c.m. of 192, 96, and 288.

15. The Fourth of July occurs on Tuesday this year. Assuming that the next 2 years are not leap years, on what day of the week will the Fourth of July fall the year after next?

16. If May 19 is a Thursday, what day of the week is the Fourth of July of the same year?

17. Perform the following computation without using a table, and reduce the answer to a positive integer less than the modulus.

$$6 \cdot 3 \pmod 7$$

18. (a) When is g.c.d. $(a, b) = $ l.c.m. $(a, b)$?
    (b) What is the relationship between $a$ and $b$ if g.c.d. $(a, b) = a$?
    (c) What is the relationship between $a$ and $b$ if l.c.m. $(a, b) = a$?
    (d) If $a \mid b$, find g.c.d. $(a, b)$ and l.c.m. $(a, b)$.
    (e) Find g.c.d. $(b^2, b)$ and l.c.m. $(b^2, b)$.

# INTRODUCTION TO THE RATIONAL NUMBERS

# 6

A used car dealer was asked how many cars he had sold during the week. He replied, "Monday I sold one-half of my cars plus half a car. Tuesday I sold a third of what remained plus one-third of a car. Wednesday I sold a fourth of what remained plus one-fourth of a car. Thursday I sold a fifth of what remained plus one-fifth of a car. I have not sold any today, Friday. I now have 11 cars left on the lot." Of course he could not have divided a car, so how many cars did the salesman have on the lot at the beginning of the week?

It is impossible in the system of integers to divide 6 candy bars equally among 4 children or to express 20 laps as part of a 100-lap race. The system of integers lacks closure for division. In this chapter, we investigate some new numbers produced by extending the system of integers to a new number system, called rational numbers. Recall that we extended the system of whole numbers to the system of integers to provide closure for subtraction. We now extend the system of integers to the system of rational numbers to provide closure for division. The operations on rational numbers will be defined in a way that satisfies the same properties of operations that we studied for integers. In accordance with the Standards, we will use both models and estimations to improve understanding. The goal of this chapter is not only to give you a better understanding of fractions, but to improve your dexterity in using them. This chapter makes use of the following problem-solving strategies:  ▪ Discover a pattern. (280, 281, 292, 301, 311, 312)  ▪ Use reasoning. (289, 312)  ▪ Guess, test, and revise. (265, 278, 281)  ▪ Use an equation. (throughout the chapter)  ▪ Use properties of numbers. (265, 289)  ▪ Restate the problem. (298)  ▪ Design a model. (265, 280, 289, 291, 300, 311)

# SECTION 1  THE SET OF RATIONAL NUMBERS

| | |
|---|---|
| **PROBLEM** | Bob bragged that he and his girlfriend had eaten $\frac{18}{24}$ of a 24-inch pizza. Not to be outdone, Lee claimed to have eaten $\frac{3}{4}$ of a similar pizza by himself. Compare the amounts of pizza eaten by the couple and by Lee. |
| **OVERVIEW** | In this section, you will study the Fundamental Law of Fractions, which will enable you to solve the preceding problem. A fraction can be interpreted in several ways: as a division problem, as a part of a whole, as a relative amount, and so on. Although you have used fractions for many years, you may not have a full appreciation of this concept. Has it ever occurred to you that several fractions may represent the same number concept, but only one whole number represents a given number concept? This property of fractions should help you understand rational numbers. |
| **GOALS** | Illustrate the Number and Operations Standard, page 119.*<br>Illustrate the Representation Standard, page 50.*<br>Illustrate the Problem Solving Standard, page 15.*<br>Illustrate the Connections Standard, page 147.*<br><br>* The complete statement of the standard is given on this page of the book. |

## FRACTIONS

(a)

(b)

FIGURE 6-1

There are many experiences that cannot be fully quantified with whole numbers and integers. For example, try to describe with whole numbers the portion of the pizza that has been cut out in Figure 6-1(a) or the part of the glass of root beer that has been consumed in Figure 6-1(b). It should be obvious that we need to extend or enlarge our system of integers to handle these situations.

Probably your first encounter with fractions involved describing something. One-fourth of the pizza has been removed, and one-half of the root beer has been consumed. We denote these new numbers by $\frac{1}{4}$ and $\frac{1}{2}$, respectively, and we call them **fractions.** A fraction is a number of the form $\frac{a}{b}$, where $a$ and $b$ are any numbers ($b \neq 0$). In such a fraction, $a$ is called the **numerator,** and $b$ the **denominator.** Throughout this section, we will consider a subset of fractions in which $a$ and $b$ are required to be integers.

Historically, the first concept of a fraction involved ratios of parts-to-the-whole or parts-of-the-whole. If $a$ and $b$ are whole numbers ($b \neq 0$), then the fraction $\frac{a}{b}$ represents $a$ out of $b$ equivalent parts. Here *equivalent* means possessing a common attribute such as length, area, volume, or weight. Thus $\frac{5}{9}$ represents 5 parts out of 9 equivalent parts.

**EXAMPLE 1**

(a)

(b)

(c)

FIGURE 6-2

Consider the set of congruent regions (regions with same size and shape) as shown in each part of Figure 6-2. Certain regions in each part are shaded. After pairing the number of shaded regions with the total number of regions, write the result. Figure 6-2(a) shows that $\frac{5}{9}$ of the square is shaded; part (b) shows that $\frac{2}{4}$ of the circle is shaded; and part (c) shows that $\frac{1}{3}$ of the triangle is shaded. Intuitively, we recognize that these fractions represent parts of the whole.

John has a salary of $400 a week. The salary of his friend, Paul, is $500 a week. How does John's salary compare to Paul's? The fraction $\frac{400}{500}$ or $\frac{4}{5}$ gives this comparison. That is, John's salary is $\frac{4}{5}$ Paul's salary. Such a fraction is an example of a **ratio model.**

In addition to representing a part of the whole and a ratio, fractions also represent a number concept. Recall from Chapter 3 that the sets $\{a, b, c\}$, $\{x, y, z\}$, $\{1, 2, 3\}$, and $\{\triangle, \square, \bigcirc\}$ have a property of "threeness"; consequently, we used the cardinal number 3 to represent the number of elements in each of these sets. Similarly, an abstract number can be used to represent the common attribute of the figures in Figure 6-3.

The figures differ in size and shape, but 3 out of 8 equivalent parts are shaded. This attribute can be represented by the fraction $\frac{3}{8}$. Notice that in this interpretation we are restricting the $a$ and the $b$ in $\frac{a}{b}$ to being whole numbers ($b \neq 0$). Similarly $\frac{2}{5}$ describes a number concept of the relative amount associated with the set of blocks, the fraction bar, and the number line in Figure 6-4.

We need now to enlarge the set of fractions denoted by $\frac{a}{b}$, where $a$ and $b$ are whole numbers ($b \neq 0$), to a set of fractions that will provide solutions to equations such as

$$3x = -2$$

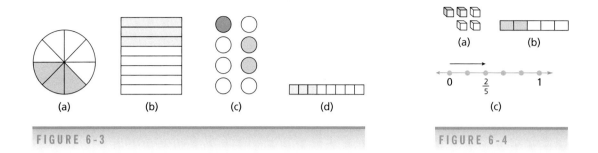

FIGURE 6-3

FIGURE 6-4

We represent this solution by $-2 \div 3$ or $\frac{-2}{3}$. Notice that the numerator is an integer. These new fractions, where the numerator and denominator can be integers (but the denominator must still be unequal to zero), are called **rational numbers.**

| DEFINITION | **Rational Number** | A *rational number* is a number that can be represented by an ordered pair of integers $a$ and $b$, where $b \neq 0$, and can be written as $$\frac{a}{b}, \qquad a/b, \qquad \text{or} \qquad a \div b$$ |
| --- | --- | --- |

A rational number may also be represented as a terminating or repeating decimal and in other ways, but it can always be represented as a fraction.

Now let's represent a point on a number line by a fraction. (See Figure 6-5). Consider the line segment from 0 to 3 on a number line. As best you can, divide this segment into 4 equal parts. (One procedure would be to cut a strip of paper the length of the segment and fold it in half twice.) Each part of the division would be of length $\frac{3}{4}$. Thus a point that is represented by $\frac{3}{4}$ can be located on a number line.

In using a number line to describe the concept of a fraction, we divide each unit interval into a given natural number of divisions of equal length. For example, if we divide each unit into 3 divisions, as in Figure 6-6, our fractions will have denominators of 3. The numerators will then count the number of subintervals to the left and right of 0.

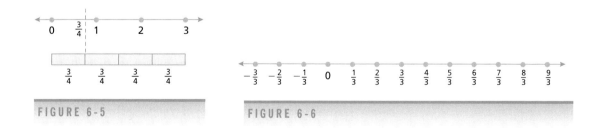

FIGURE 6-5

FIGURE 6-6

This description of fractions and the number line leads to the following property.

| Property of Rational Numbers | Every rational number corresponds to some unique point on a number line. |
|---|---|

Another interesting property of rational numbers is that many different fractions may be used to represent the same parts of the whole. Thus, the shaded region of Figure 6-7 is 1 part out of 3 (represented by $\frac{1}{3}$) or 2 parts out of 6 (represented by $\frac{2}{6}$) or 4 parts out of 12 (represented by $\frac{4}{12}$). Notice that $\frac{4}{12}$, $\frac{2}{6}$, and $\frac{1}{3}$ all represent the same shaded region of Figure 6-7.

Similarly, Figure 6-8 shows that a rational point on the number line is represented by many (in fact, infinitely many) different fractions. For example, $\frac{1}{3}$, $\frac{2}{6}$, $\frac{3}{9}$, and $\frac{4}{12}$ all represent the same point.

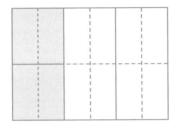

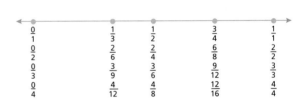

FIGURE 6-7                              FIGURE 6-8

### CALCULATOR HINT

Note that when quotients are entered using the backslash key $\boxed{/}$, the quotients are entered automatically as fractions. When quotients are entered using the $\boxed{\div}$ key, the fractions appear as decimal expressions. The $\boxed{\blacktriangleright F}$ key changes decimal expressions to fractions when possible. For example insert ‾17/31 into your calculator.

(a) $\boxed{(-)}$ 17 $\boxed{/}$ 31 $\boxed{=}$        (b) $\boxed{(-)}$ 17 $\boxed{\div}$ 31 $\boxed{\blacktriangleright \text{Fac}}$ $\boxed{=}$.

Did you get ‾17/31 in both cases?

## EQUIVALENT FRACTIONS

Fractions that represent the same rational number point on the number line are called **equivalent fractions.** Each of these fractions is said to be *equivalent* to the other fractions, and a set such as

$$\left\{ \cdots, \frac{-4}{-12}, \frac{-3}{-9}, \frac{-2}{-6}, \frac{-1}{-3}, \frac{1}{3}, \frac{2}{6}, \frac{3}{9}, \frac{4}{12}, \cdots \right\}$$

is called an **equivalence class** of fractions.

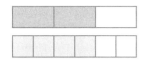

FIGURE 6-9

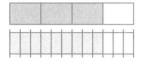

FIGURE 6-10

Fraction bars are very useful in demonstrating equivalent fractions. By placing two fraction bars side by side, we can see that 4 parts of 6 represents the same part of the whole as 2 parts of 3. (See Figure 6-9.) Figure 6-10 shows a fraction bar that has 3 parts shaded of 4. If we then subdivide each of the 4 parts of the fraction bar into 3 equal parts, the number of shaded parts is $3 \cdot 3$ and the total number of parts is $3 \cdot 4$. The shaded parts thus represent $\frac{3 \cdot 3}{3 \cdot 4}$ of the whole. But this amount was originally $\frac{3}{4}$. Therefore,

$$\frac{3}{4} = \frac{3 \cdot 3}{3 \cdot 4}$$

This intuitively evident property of fractions is expressed as the Fundamental Law of Fractions and will be used throughout this chapter to obtain equivalent fractions (or equal rational numbers).

| | |
|---|---|
| **Fundamental Law of Fractions** | Let $b \neq 0$. For any fraction $\dfrac{a}{b}$ and any natural number $c$, $$\frac{a}{b} = \frac{ac}{bc} \qquad \text{and} \qquad \frac{ca}{cb} = \frac{a}{b}$$ |

Because $c/c = 1$, and because we will learn later that $a/b \cdot c/c = ac/bc$, this property is simply an application of multiplying by the multiplicative identity.

$$\frac{a}{b} = \frac{a}{b} \cdot 1 = \frac{a}{b} \cdot \frac{c}{c} = \frac{ac}{bc}$$

**EXAMPLE 2**

The fraction $\frac{9}{15}$ can be written as $\frac{3 \cdot 3}{5 \cdot 3}$, because $9 = 3 \cdot 3$ and $15 = 5 \cdot 3$. Thus, by the Fundamental Law of Fractions,

$$\frac{9}{15} = \frac{3 \cdot 3}{5 \cdot 3} = \frac{3}{5} \qquad \text{Here, the } c \text{ in the Fundamental Law is 3.}$$

**EXAMPLE 3**

$$\frac{r^2}{r^5} = \frac{1 \cdot r^2}{r^3 \cdot r^2} = \frac{1}{r^3} \qquad r \neq 0 \qquad \begin{array}{l} r^2 = 1 \cdot r^2 \\ r^5 = r^3 \cdot r^2 \end{array}$$

## Simplifying Fractions

The answers in Example 2 and 3 are said to be in **lowest terms** or *simplest form.*

| | |
|---|---|
| **DEFINITION** **Simplest Form** | A fraction $\dfrac{a}{b}$ is in its *simplest form* when $b$ is positive and the largest positive integer that will divide both $a$ and $b$ is 1; that is, $a$ and $b$ are relatively prime. |

Using the Fundamental Law of Fractions, we can place a fraction in its simplest form by finding the greatest common divisor (g.c.d.) of the numerator and the denominator and writing each as the product involving the g.c.d. For instance,

$$\frac{120}{255} = \frac{15 \cdot 8}{15 \cdot 17} = \frac{8}{17}$$    The g.c.d. (120, 255) is 15.

**EXAMPLE 4**

Simplify $\frac{1020}{1380}$ to lowest terms.

*SOLUTION*

Because g.c.d. (1020, 1380) = 60, write

$$\frac{1020}{1380} = \frac{60 \cdot 17}{60 \cdot 23} = \frac{17}{23}$$    The common factor is 60.

**EXAMPLE 5**

Write $x^2y/xy^4$ in simplest form, where $x \neq 0$ and $y \neq 0$.

*SOLUTION*

$$\frac{x^2y}{xy^4} = \frac{\cancel{x} \cdot x \cdot \cancel{y}}{\cancel{x} \cdot y \cdot y \cdot y \cdot \cancel{y}} = \frac{x}{y^3}$$

## COMMON ERROR

How would you explain $\dfrac{1\cancel{6}}{\cancel{6}4} = \dfrac{1}{4}$   and   $\dfrac{1\cancel{9}}{\cancel{9}5} = \dfrac{1}{5}$?

Can you find other examples where this incorrect procedure gives a correct answer?

## CALCULATOR HINT

It is advisable to check the settings on the  FracMode  menu (*A b/c, d/e, Manual, Auto*) before doing fraction arithmetic. These settings sometimes affect whether you can use a procedure involving fractions and always affect the form of the answer. Underline the settings you want and press  =  for each setting.

When  FracMode  is set on *Manual* and *d/e*,  ▶ Simp  will simplify a fraction by dividing the numerator and denominator by the smallest prime that is a common factor. Keep pressing  ▶ Simp  until you get a reduced fraction. *Example:* Reduce 30/36.

30  /  36  ▶ Simp   =   ▶ Simp   = . Did you get 5/6 ?

If you select *Auto* in  FracMode , all answers will be reduced fractions.

30  /  36  = . Your answer is again 5/6.

**EXAMPLE 6**

Write each of the following in simplest form:

(a) $\dfrac{x^2 + x}{x + 1}$   (b) $\dfrac{2 + x^2}{2x}$

SOLUTION

(a) $\dfrac{x^2 + x}{x + 1} = \dfrac{x(x + 1)}{(x + 1)} = x$

(b) $\dfrac{2 + x^2}{2x}$ cannot be simplified because $2 + x^2$ and $2x$ have no common factors except 1.

In the next example, the Fundamental Law of Fractions is used to show that two fractions are equivalent.

**EXAMPLE 7**

Show that $\frac{35}{42}$ and $\frac{40}{48}$ are equivalent, by writing each fraction with a denominator that is the product of the two denominators.

SOLUTION

$$\frac{35}{42} = \frac{35 \cdot 48}{42 \cdot 48} = \frac{1680}{2016} \qquad \text{and} \qquad \frac{40}{48} = \frac{40 \cdot 42}{48 \cdot 42} = \frac{1680}{2016}$$

By carefully studying the preceding example, we can identify an interesting relationship. Because $35 \cdot 48 = 1680 = 40 \cdot 42$, the fractions $\frac{35}{42}$ and $\frac{40}{48}$ are equivalent. This fact can be generalized as follows.

**Equivalent Fractions**

The fractions $\dfrac{a}{b}$ and $\dfrac{c}{d}$ ($b, d \neq 0$) are *equivalent* (that is, the two rational numbers represented by $\dfrac{a}{b}$ and $\dfrac{c}{d}$ are equal) if and only if $ad = bc$.

**EXAMPLE 8**

Are the fractions $\frac{3}{8}$ and $\frac{6}{24}$ equivalent?

SOLUTION

$3 \cdot 24 = 72$ and $6 \cdot 8 = 48$. Because $72 \neq 48$, it follows that $\frac{3}{8} \neq \frac{6}{24}$.

**PRACTICE PROBLEM**

Are $\frac{24}{60}$ and $\frac{36}{90}$ equivalent?

ANSWER

Yes, because $24 \cdot 90 = 2160$ and $36 \cdot 60 = 2160$.

We can use three methods for determining whether fractions are equivalent (or, in other words, rational numbers are equal). Example 8 uses cross-multiplication. Example 9 demonstrates two other methods.

**EXAMPLE 9**

(a) Show that $\frac{24}{36}$ and $\frac{48}{72}$ are equivalent by writing each fraction in simplest form.
(b) Show that $\frac{10}{16}$ and $\frac{15}{24}$ are equivalent by changing each fraction to an equivalent fraction with a denominator that is the least common denominator of 24 and 16, l.c.m. (24, 16). (See p. 206.)

SOLUTION

(a) The g.c.d. (24, 36) = 12 and g.c.d. (48, 72) = 24.

$$\frac{24}{36} = \frac{12 \cdot 2}{12 \cdot 3} = \frac{2}{3} \qquad \text{and} \qquad \frac{48}{72} = \frac{24 \cdot 2}{24 \cdot 3} = \frac{2}{3}$$

So $\frac{24}{36}$ and $\frac{48}{72}$ are equivalent.

(b) The least common denominator of 16 and 24 equals l.c.m. (16, 24) = 48.

$$\frac{10}{16} = \frac{10 \cdot 3}{16 \cdot 3} = \frac{30}{48} \quad \text{and} \quad \frac{15}{24} = \frac{15 \cdot 2}{24 \cdot 2} = \frac{30}{48}$$

Because both fractions are equivalent to $\frac{30}{48}$, they are equivalent.

The concept of equivalent fractions naturally leads to issues of ordering and estimation.

## ORDERING OF RATIONAL NUMBERS

The concept of "less than" for rational numbers must be such that it holds for integers, because the integers are a subset of the rational numbers. This subset consists of

$$\left\{ \ldots, \frac{-4}{1}, \frac{-3}{1}, \frac{-2}{1}, \frac{-1}{1}, 0, \frac{1}{1}, \frac{2}{1}, \frac{3}{1}, \frac{4}{1}, \ldots \right\}$$

Recall that, for integers, the smaller of 2 numbers was to the left of the larger on the number line. As Figure 6-11 shows, $\frac{-5}{4}$ lies to the left of $\frac{-1}{4}$, and $\frac{1}{4}$ lies to the left of $\frac{5}{4}$; therefore,

$$\frac{-5}{4} < \frac{-1}{4} \quad \text{and} \quad \frac{1}{4} < \frac{5}{4}$$

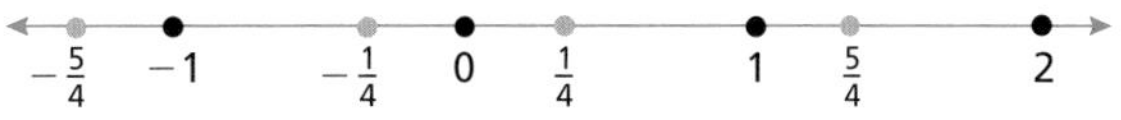

FIGURE 6-11

All of these fractions have the same denominator (4); but among the numerators, −5 < −1 and 1 < 5. This suggests the following definition.

| DEFINITION | **Less Than** | Let $\frac{a}{c}$ and $\frac{b}{c}$ be any fractions with $c > 0$. Then $$\frac{a}{c} < \frac{b}{c} \quad \text{if and only if} \quad a < b$$ |
|---|---|---|

Thus $\frac{3}{5} < \frac{4}{5}$ (because 3 < 4), $\frac{-2}{5} < \frac{2}{5}$ (because −2 < 2), and $\frac{-3}{5} < \frac{-2}{5}$ (because −3 < −2). To test whether $\frac{2}{3}$ is less than $\frac{3}{4}$, we use the Fundamental Law of Fractions to change $\frac{2}{3}$ and $\frac{3}{4}$ to equivalent fractions with the same (or a common) denominator:

$$\frac{2}{3} = \frac{2 \cdot 4}{3 \cdot 4} \quad \text{and} \quad \frac{3}{4} = \frac{3 \cdot 3}{3 \cdot 4}$$

Therefore,

$$\frac{2}{3} < \frac{3}{4} \qquad \text{Because } 2 \cdot 4 = 8 < 9 = 3 \cdot 3$$

This example suggests the following definition.

---

**Less Than for Rational Numbers**

If $\frac{p}{q}$ and $\frac{r}{s}$ are rational numbers expressed with positive denominators, then

$$\frac{p}{q} < \frac{r}{s} \qquad \text{if and only if} \qquad ps < qr$$

---

**EXAMPLE 10**

(a) $\frac{2}{3} < \frac{5}{4}$ because $2 \cdot 4 < 3 \cdot 5$     (b) $\frac{-4}{5} < \frac{1}{4}$ because $-4 \cdot 4 < 5 \cdot 1$

 ESTIMATION

The ordering of rational numbers and the Fundamental Law of Fractions both assist in estimating the size of rational numbers. For example, let's estimate and simplify $\frac{33}{64}$. Evidently, $\frac{33}{64}$ is slightly larger than $\frac{32}{64}$, which equals $\frac{1}{2}$. Thus, $\frac{33}{64}$ is slightly larger than $\frac{1}{2}$. By similar reasoning, $\frac{29}{90}$ is slightly smaller than $\frac{30}{90}$ which equals $\frac{1}{3}$. Therefore, $\frac{29}{90}$ is slightly smaller than $\frac{1}{3}$.

Now let's return to the introductory problem of this section.

**EXAMPLE 11**

Bob bragged that he and his girlfriend had eaten $\frac{18}{24}$ of a 24-in. pizza. Not to be outdone, Lee claimed to have eaten $\frac{3}{4}$ of a pizza by himself. Compare the amounts of pizza eaten by the couple and by Lee.

*SOLUTION*

They ate the same amount. By the Fundamental Law of Fractions,

$$\frac{18}{24} = \frac{3 \cdot 6}{4 \cdot 6} = \frac{3}{4}$$

We conclude this section with some important remarks. Early on, we made a distinction between a fraction and a rational number. A fraction is a numeral representing a number; in this chapter the number it represents is a rational number. Two fractions that represent the same rational number are said to be **equivalent,** whereas the corresponding rational numbers are said to be **equal.** However, once the distinction between rational numbers and fractions representing rational numbers is made, there is no need to belabor it; in general usage, the terms are interchangeable. However, we must note that some fractions do not represent rational numbers. For example, a fraction whose numerator or denominator is not an integer, like $\frac{\sqrt{2}}{3}$, may not be a rational number.

# Just for Fun

If you double $\frac{1}{6}$ of a fraction and then multiply by the fraction, the answer is $\frac{2}{17}$. What is the fraction?

# EXERCISE SET 1

*R* 1. What fraction does the shaded portion of each diagram illustrate?

(a)    (b)

(c)    (d)

(e)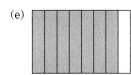

2. Write five fractions equivalent to each of the following fractions.

   (a) $\frac{3}{4}$   (b) $\frac{2}{3}$   (c) $\frac{0}{8}$

3. Which of the following are pairs of equivalent fractions?

   (a) $\frac{4}{1}$ and $\frac{8}{2}$   (b) $\frac{6}{1}$ and $\frac{18}{2}$

   (c) $\frac{0}{1}$ and $\frac{0}{2}$   (d) $\frac{7}{20}$ and $\frac{21}{60}$

   (e) $(3a + b)/3c$ and $(a + b)/c$, where $a$, $b$, and $c$ are any whole numbers with $b \neq 0$ and $c \neq 0$

   (f) $\frac{2}{0}$ and $\frac{1}{0}$

4. Write whole numbers that correspond to the following fractions.

   (a) $\frac{15}{5}$   (b) $\frac{16}{2}$   (c) $\frac{27}{9}$

5. Find the values of $x$ such that the following fractions are equivalent.

   (a) $\frac{1}{4} = \frac{2x}{24}$   (b) $\frac{7x}{10} = \frac{42}{20}$   (c) $\frac{0}{6} = \frac{x}{4}$

6. Represent with a fraction
   (a) the part of the collection of dots inside the rectangle.
   (b) the part of the collection of dots outside the rectangle.
   (c) the part of the collection of dots inside the triangle.
   (d) the part of the collection of dots inside the intersection of the interiors of the triangle and the rectangle.

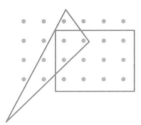

7. Without using a calculator write each of the following in simplest form. Assume that $r, s, t, x, y, z \neq 0$. In (c) and (i), use the g.c.d. of the numerator and the denominator. Then check (a), (b), (c), (d) and (i) with a calculator.

   (a) $\dfrac{162}{88}$   (b) $\dfrac{252}{210}$   (c) $\dfrac{308}{418}$

   (d) $\dfrac{14}{261}$   (e) $\dfrac{x^2 y}{y z^2}$   (f) $\dfrac{r^2 s^3 t^4}{r^2 s^4 t}$

   (g) $\dfrac{x^3 y z}{x^2 y^2 z^4}$   (h) $\dfrac{xy + xz}{x^2 y + x^2 z}$   (i) $\dfrac{-416}{512}$

8. Find the fraction $\frac{a}{b}$, with specified numerator or denominator, that equals the given fraction. Make sure that both $a$ and $b$ are whole numbers; observe that in some problems this is impossible. Check the answer you get with a calculator.

   (a) $\dfrac{7}{3}, b = 21$   (b) $\dfrac{2}{3}, b = 27$   (c) $\dfrac{4}{9}, b = 14$

   (d) $\dfrac{5}{3}, b = 10$   (e) $\dfrac{3}{4}, a = 30$   (f) $\dfrac{9}{5}, a = 27$

   (g) $\dfrac{5}{9}, a = 8$   (h) $\dfrac{7}{16}, a = 32$   (i) $\dfrac{5}{12}, a = 10$

   (j) $\dfrac{7}{3}, a = 14$   (k) $\dfrac{3}{5}, b = 10$   (l) $\dfrac{7}{3}, b = 12$

9. Write five elements of the set described, where $a$ and $b$ are whole numbers and $b \neq 0$.

   (a) $\left\{ x \mid x = \dfrac{a}{b} \text{ and } a + b = 5 \right\}$

   (b) $\left\{ x \mid x = \dfrac{a}{b} \text{ and } a + b < 7 \right\}$

   (c) $\left\{ x \mid x = \dfrac{a}{b} \text{ and } a - b < 2 \right\}$

   (d) $\left\{ x \mid x = \dfrac{a}{b} \text{ and } a - b < 4 \right\}$

   (e) $\left\{ x \mid x = \dfrac{a}{b} \text{ and } a = b \right\}$

10. A set of fractions equivalent to $a/b$ can be written as $\{ca/cb \mid a$ and $b$ are whole numbers with $b \neq 0$ and $c$ is any natural number$\}$. Write a set of fractions equivalent to each of the following fractions.

    (a) $\dfrac{3}{5}$   (b) $\dfrac{7}{9}$   (c) $\dfrac{x}{y}$

11. Without a calculator determine which of the following fractions are in simplest form? Then check your answer with a calculator.

    (a) $\dfrac{9}{17}$   (b) $\dfrac{2}{4}$   (c) $\dfrac{7}{14}$

    (d) $\dfrac{20}{49}$   (e) $\dfrac{3}{17}$   (f) $\dfrac{9}{14}$

12. Find a fraction equivalent to $\frac{1}{2}$ such that the product of the numerator and the denominator is 72.

13. Insert $>$, $=$, or $<$ to express the correct relationship between the following pairs of numbers.

    (a) $\dfrac{9}{10} \quad \dfrac{7}{11}$   (b) $\dfrac{33}{9} \quad \dfrac{21}{7}$   (c) $\dfrac{5}{7} \quad \dfrac{2}{5}$

    (d) $\dfrac{29}{3} \quad \dfrac{16}{2}$   (e) $\dfrac{17}{51} \quad \dfrac{6}{7}$   (f) $\dfrac{22}{73} \quad \dfrac{25}{71}$

14. Approximate the following by a fraction that can be simplified. Then give the simplified fraction.

    (a) $\dfrac{41}{80}$   (b) $\dfrac{17}{27}$   (c) $\dfrac{24}{71}$

15. Place each of the following between two simple fractions.

    *EXAMPLE:* For $\dfrac{45}{66}, \dfrac{2}{3} < \dfrac{45}{66} < \dfrac{3}{4}$

    (a) $\dfrac{12}{30}$   (b) $\dfrac{39}{48}$

16. Identify the following fractions as $-1$, $1$, $0$, or undefined.

    (a) $\dfrac{1}{0}$   (b) $\dfrac{0}{1}$   (c) $\dfrac{0}{0}$

    (d) $\dfrac{4}{0}$   (e) $\dfrac{0}{4}$   (f) $\dfrac{-1}{0}$

17. Compare the following numbers mentally, and then choose the larger number.

    (a) $\dfrac{13}{14}$ and $\dfrac{15}{16}$   (b) $\dfrac{-5}{8}$ and $\dfrac{1}{8}$

    (c) $\dfrac{-19}{20}$ and $\dfrac{-17}{18}$

18. Without using a calculator determine whether the following fractions are equal, by writing each in simplest form. Then check your answer with a calculator.

    (a) $\dfrac{750}{2000}, \dfrac{39}{104}$   (b) $\dfrac{42}{172}, \dfrac{51}{215}$

19. Find a replacement value for $x$.

    (a) $\dfrac{x}{9} = \dfrac{5}{3}$   (b) $\dfrac{4}{6} = \dfrac{x}{-12}$

    (c) $\dfrac{x}{72} = \dfrac{5}{8}$   (d) $\dfrac{3}{x} = \dfrac{4}{16}$

20. For the fractions in Exercise 18, determine which are equal by finding a least common denominator.

*T* 21. If $W$ is the set of whole numbers, $I$ is the set of integers, and $Q$ is the set of rational numbers, indicate which of the following are true and which are false.

    (a) $W \subset Q$   (b) $W \subset (Q \cup I)$
    (c) $W \subseteq I$   (d) $I \cap Q = W$
    (e) $W \cap Q = W \cap I$   (f) $W \cup Q = I \cup W$

22. (a) If $a = c$ and $d \neq 0$, is $a/d$ equal to $c/d$? Why?
    (b) If $b = d$, is $a/b$ equal to $c/d$ $(b, d \neq 0)$? Why?
    (c) If $(a/b) = (c/d)$ and $b = d$, what is true about $c$ and $a$?

23. Assume that each variable expression in a denominator is not equal to 0, and write each of the following in simplest form.

    (a) $\dfrac{2 \cdot 3^2 \cdot 5^3 \cdot 7}{2^2 \cdot 3 \cdot 5 \cdot 11}$   (b) $\dfrac{4a^6b^5}{8a^4b^2}$   (c) $\dfrac{4x^3y^7}{8xy^9}$

    (d) $\dfrac{3c + 6d}{9e}$   (e) $\dfrac{a(x + y)}{b^2(x + y)}$   (f) $\dfrac{a}{2a + ba}$

    (g) $\dfrac{a}{2a + b}$   (h) $\dfrac{ab}{a + b}$   (i) $\dfrac{ab(a + b)}{a - b}$

24. (a) Discuss $a$ and $b$ if $a/c = b/c$.
    (b) Discuss $b$ and $d$ if $a/b = a/d$.

25. Last season, Tom got 32 hits in 94 times at bat. Ramon had 27 hits in 87 times at bat. Who had the better batting average?

26. What is wrong with the following equations?

    (a) $\dfrac{6 + \cancel{4}}{7 + \cancel{4}} = \dfrac{6}{7}$   (b) $\dfrac{5 - \cancel{3}}{8 - \cancel{3}} = \dfrac{5}{8}$

    (c) $\dfrac{13}{3} = \dfrac{10 + \cancel{3}}{\cancel{3}} = 11$

    (d) If you use the word *canceling*, what must be true before you can cancel something from the numerator and the denominator of a fraction?

27. If $(a/b) < (c/d)$ is $(a/b) < (a + c)/(b + d) < (c/d)$ always true if $a, b, c, d > 0$? Explain.

28. $\frac{4}{5}$ of the people in the United States drink Lola Cola. This answer was obtained by polling 8000 people. How many of these indicated that they drink Lola Cola?

29. Determine whether or not the two rational numbers are equal by using the fact that $\frac{a}{b} = \frac{c}{d}$ if and only if $ad = bc$. Check your answer by simplifying the fractions with a calculator.

    (a) $\dfrac{9}{15}, \dfrac{1200}{2000}$   (b) $\dfrac{6}{24}, \dfrac{41}{164}$   (c) $\dfrac{33}{43}, \dfrac{62}{189}$

*C* 30. Two third-grade teachers at Ourtown Elementary gave their classes the same test. Of Mr. Brown's 27 students, 20 passed the test; and 22 of Ms Gray's 29 also passed. What part of each class passed the test?

31. (a) Show that the relationship "is equivalent to" is an equivalence relation on the set of fractions.
    (b) Find some equivalence classes for this equivalence relation.

32. Write in simplest form. Check the answer using a calculator by finding the value before and after simplifying.

    (a) $\dfrac{2^{10} - 2^9}{2^{11} - 2^{10}}$   (b) $\dfrac{3^{12} - 3^9}{3^{15} - 3^{12}}$

## LABORATORY ACTIVITIES

33. Represent $\frac{2}{5}$ with a fraction bar. Then draw lines to show that $\frac{2}{5}$ is equivalent to $\frac{6}{15}$.

34. For the fraction $\frac{4}{5}$, demonstrate a relative amount model using the following.

    (a) Sets of blocks   (b) Number line
    (c) Fraction bar

### ✧ PCR Excursion ✧

35. In Chapter 11 we will be studying plane figures such as

Octagon   Hexagon   Pentagon   Quadrilateral

Trapezoid   Rhombus   Triangle A   Triangle B

(a) The hexagonal region can be divided into non-overlapping copies of regions bounded by the trapezoid, the rhombus, and triangle A. What fractional part of the hexagonal region is each?

Trapezoids          Rhombi          Triangles

(b) In a similar manner divide the octagonal region into regions bounded by pentagons; then quadrilaterals; then Triangle B's. What fractional part is each of the whole?

(c) What fractional part is the triangular region A of the trapezoidal region? the region bounded by the rhombus?

(d) What fractional part is the triangular region B of the pentagonal region? the region bounded by the quadrilateral?

(e) What fractional part is the region bounded by the rhombus of the trapezoidal region?

(f) What fractional part is the quadrilateral region of the pentagonal region?

(g) What fractional part of 18 regions identical to the region bounded by triangle A is the hexagonal region? The trapezoidal region? The region bounded by the rhombus?

(h) What fractional part of 16 regions identical to the region bounded by Triangle B is the octagonal region? The pentagonal region? The quadrilateral region?

(i) What fractional part of 12,000 regions bounded by rhombi is the region bounded by the hexagon? The trapezoid? Triangle A?

(j) What fractional part of 12 quadrilateral regions is the region bounded by the octagon? The pentagon? Triangle B?

(k) What fractional part of the region bounded by 3 trapezoids is the region bounded by the hexagon? The rhombus? Triangle B?

(l) What fractional part of the region bounded by 6 pentagons is the region bounded by the octagon? The pentagon? Triangle B?

# SECTION 2 ADDITION AND SUBTRACTION OF RATIONAL NUMBERS

**PROBLEM**

Use four numbers—2, 4, 5, and 8—two as numerators and two as denominators of two fractions, such that each fraction is less than 1 and the difference in the two fractions is a maximum.

**OVERVIEW**

The preceding problem anticipates the introduction in this section of addition and subtraction of rational numbers. Two methods will be presented for adding rational numbers. The first method uses the product of the 2 denominators and the Fundamental Law of Fractions. The second (and more common) method uses a least common denominator (l.c.m.) of 2 denominators.

**GOALS**

Illustrate the Number and Operations Standard, page 119.*
Illustrate the Representation Standard, page 50.*
Illustrate the Problem Solving Standard, page 15.*
Illustrate the Algebra Standard, page 4.*

* The complete statement of the standard is given on this page of the book.

 ## ADDITION OF RATIONAL NUMBERS WITH LIKE DENOMINATORS

Because the integers are a subset of the rational numbers, the operations on rational numbers must be consistent with the operations on integers. Remember this as we define addition for rational numbers. To visualize the addition of two rational numbers, start with the number line as a model.

**EXAMPLE 1**

We already know that $3 + 4 = 7$ or, using the corresponding rational numbers,

$$\frac{3}{1} + \frac{4}{1} = \frac{7}{1}$$

as shown in Figure 6-12.

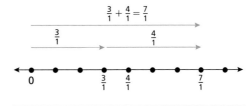

FIGURE 6-12

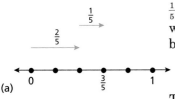

(a)

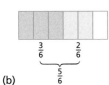

(b)

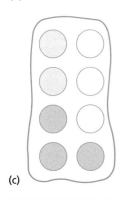

(c)

FIGURE 6-13

Suppose that it is $\frac{2}{5}$ mile from your dorm to the bookstore and another $\frac{1}{5}$ mile from the bookstore to your classroom. We know intuitively that the whole distance traveled from your dorm to your classroom by way of the bookstore is

$$\frac{2}{5} + \frac{1}{5} = \frac{3}{5} \text{ mile}$$

This is shown in Figure 6-13(a).

In another case, your chocolate bar is divided into 6 equal parts. You eat 3 parts and give your roommate 2 parts. Then

$$\frac{3}{6} + \frac{2}{6} = \frac{5}{6}$$

of the candy bar has been consumed. (See Figure 6-13(b).)

In a third case, 3 of 8 of your tennis balls are dead. After a hard game, 2 more balls are dead. What part of your 8 tennis balls are now dead? As illustrated in Figure 6-13(c), the answer is

$$\frac{3}{8} + \frac{2}{8} = \frac{5}{8}$$

Thus, the number-line model, the fraction-bar model, and the set model in Figure 6-13 all show that the sum of 2 fractions that have the same denominator can be found by adding the numerators. That is,

$$\left(\frac{a}{b}\right) + \left(\frac{c}{b}\right) = \frac{a + c}{b}$$

This can be interpreted as $a$ (of $b$ parts) plus $c$ (of $b$ parts), which is $a + c$ (of $b$ parts) or $\frac{a+c}{b}$, just as $a$ apples $+ c$ apples $= a + c$ apples.

## ADDITION OF RATIONAL NUMBERS WITH UNLIKE DENOMINATORS

Consider now the case where the denominators are unequal.

**EXAMPLE 2**

Add $\frac{1}{3}$ and $\frac{2}{5}$.

*SOLUTION* Using the fact that, for $c \neq 0$, $\frac{a}{b}$ and $\frac{ac}{bc}$ are equal, we see that

$$\frac{1}{3} = \frac{1 \cdot 5}{3 \cdot 5} \qquad \text{and} \qquad \frac{2}{5} = \frac{3 \cdot 2}{3 \cdot 5}$$

The denominators are now the same, and we add according to the preceding discussion. Thus

$$\frac{1}{3} + \frac{2}{5} = \frac{1 \cdot 5}{3 \cdot 5} + \frac{3 \cdot 2}{3 \cdot 5}$$

$$= \frac{(1 \cdot 5) + (3 \cdot 2)}{3 \cdot 5} \qquad \frac{a}{b} + \frac{c}{b} = \frac{a + c}{b}$$

$$= \frac{11}{15}$$

## CALCULATOR HINT

If you have a calculator that can do fraction arithmetic try this:

1 $\boxed{/}$ 3 $\boxed{+}$ 2 $\boxed{/}$ 5 $\boxed{=}$. Did you get 11/15?

The fraction-bar model shown in Figure 6-14 illustrates the preceding example.

 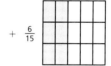

FIGURE 6-14

In general, let $\frac{a}{b}$ and $\frac{c}{d}$ represent rational numbers. The denominators can be made identical by applying the Fundamental Law of Fractions:

$$\frac{a}{b} = \frac{ad}{bd} \qquad \text{and} \qquad \frac{c}{d} = \frac{bc}{bd}$$

Therefore,

$$\frac{a}{b} + \frac{c}{d} = \frac{ad}{bd} + \frac{bc}{bd} = \frac{ad + bc}{bd}$$

Because of these considerations, we define addition of rational numbers in the following manner.

**DEFINITION**

**Addition of Rational Numbers**

(a) If $\frac{a}{b}$ and $\frac{c}{b}$ are *rational numbers* with equal denominators, then

$$\frac{a}{b} + \frac{c}{b} = \frac{a + c}{b}$$

(b) If $\frac{a}{b}$ and $\frac{c}{d}$ are *any two rational numbers,* then

$$\frac{a}{b} + \frac{c}{d} = \frac{ad + bc}{bd}$$

The equivalence of this two-part definition is demonstrated by adding $\frac{3}{7} + \frac{2}{7}$. By (a),

$$\frac{3}{7} + \frac{2}{7} = \frac{3+2}{7} = \frac{5}{7}$$

By (b),

$$\frac{3}{7} + \frac{2}{7} = \frac{(3 \cdot 7) + (7 \cdot 2)}{7 \cdot 7} = \frac{35}{49}$$

However,

$$\frac{5}{7} = \frac{35}{49}$$

## COMMON ERROR

$\frac{4}{5} + \frac{2}{3} \neq \frac{6}{8}$. You do not add numerators and denominators when adding fractions.

**EXAMPLE 3**

(a) Sue recorded that it rained $\frac{2}{3}$ of an inch yesterday and $\frac{4}{7}$ of an inch today. What is the total rainfall?

(b) Joe is a nurse. The temperature of one of his patients rose $\frac{5}{7}$ of a degree during the first hour after treatment and dropped $\frac{2}{4}$ of a degree the second hour. What was the change in temperature over the two hour period?

*SOLUTION*

(a) $\dfrac{2}{3} + \dfrac{4}{7} = \dfrac{2(7) + 3(4)}{3 \cdot 7} = \dfrac{14 + 12}{21} = \dfrac{26}{21}$

(b) $\dfrac{5}{7} + \dfrac{-2}{4} = \dfrac{5(4) + 7(-2)}{7 \cdot 4} = \dfrac{20 + -14}{28} = \dfrac{6}{28} = \dfrac{3}{14}$

## COMMON ERRORS

Explain what is wrong.

(a) $\dfrac{\cancel{3}}{4} + \dfrac{1}{\cancel{3}} = \dfrac{2}{5}$        (b) $\dfrac{3}{4+5} = \dfrac{3}{4} + \dfrac{3}{5}$

# LEAST COMMON DENOMINATOR

In our definition of addition of fractions, we found a common denominator for the pair of fractions, but we did not at all times find what is called the *least common denominator*. To find a least common denominator of two or more denominators we find the least common multiple (l.c.m.) of the denominators. (See Chapter 5.)

Finding the least common denominator allows us to work with smaller numbers. In this procedure, the Fundamental Law of Fractions is applied (as before) to each fraction to make the denominators identical.

**Least Common Denominator**

The *least common denominator* is the l.c.m. of the denominators of the fractions to be added.

**EXAMPLE 4**

Perform the following computations.

(a) $\dfrac{5}{6} + \dfrac{7}{15}$    (b) $\dfrac{5}{24} + \dfrac{11}{36}$    (c) $\dfrac{3}{16} + \dfrac{1}{4} + \dfrac{5}{8}$

*SOLUTION*

(a) The least common denominator of 6 and 15 is 30. Clearly, 6 must be multiplied by 5 to get 30, and 15 must be multiplied by 2 to get 30. Thus, by the Fundamental Law of Fractions,

$$\frac{5}{6} + \frac{7}{15} = \frac{5 \cdot 5}{6 \cdot 5} + \frac{7 \cdot 2}{15 \cdot 2} = \frac{25}{30} + \frac{14}{30} = \frac{39}{30} = \frac{13}{10}$$

(b) $\dfrac{5}{24} + \dfrac{11}{36} = \dfrac{5}{2^3 \cdot 3} + \dfrac{11}{2^2 \cdot 3^2}$     Least common denominator is $2^3 \cdot 3^2$.

$$= \frac{5 \cdot 3}{2^3 \cdot 3^2} + \frac{11 \cdot 2}{2^3 \cdot 3^2} = \frac{15 + 22}{2^3 \cdot 3^2} = \frac{37}{72}$$

(c) $\dfrac{3}{16} + \dfrac{1}{4} + \dfrac{5}{8} = \dfrac{3}{16} + \dfrac{1 \cdot 4}{4 \cdot 4} + \dfrac{5 \cdot 2}{8 \cdot 2}$     Least common denominator is 16.

$$= \frac{3}{16} + \frac{4}{16} + \frac{10}{16} = \frac{17}{16}$$

**EXAMPLE 5**

Find $\dfrac{5}{48} + \dfrac{7}{36} + \dfrac{1}{54}$.

*SOLUTION*

Factor each denominator into prime factors. Then the least common denominator can be obtained easily:

$$\frac{5}{2 \cdot 2 \cdot 2 \cdot 2 \cdot 3} + \frac{7}{2 \cdot 2 \cdot 3 \cdot 3} + \frac{1}{2 \cdot 3 \cdot 3 \cdot 3}$$

The least common denominator is $2 \cdot 2 \cdot 2 \cdot 2 \cdot 3 \cdot 3 \cdot 3 = 432$. Therefore,

$$\frac{5}{48} + \frac{7}{36} + \frac{1}{54} = \frac{5(3 \cdot 3)}{432} + \frac{7(2 \cdot 2 \cdot 3)}{432} + \frac{1(2 \cdot 2 \cdot 2)}{432}$$

$$= \frac{45 + 84 + 8}{432}$$

$$= \frac{137}{432}$$

## CALCULATOR HINT

5 $\boxed{/}$ 48 $\boxed{+}$ 7 $\boxed{/}$ 36 $\boxed{+}$ 1 $\boxed{/}$ 54 $\boxed{=}$.

Did you get $\dfrac{137}{432}$?

## COMMON ERROR

Explain what is wrong with the following:

(a) $2 = \dfrac{6}{3} = \dfrac{3+3}{3} = \dfrac{3}{3} + 3 = 4$

(b) $\dfrac{ab+c}{a} = b + c$

(c) $\dfrac{3 + \cancel{6}}{4 + \cancel{6}} = \dfrac{3}{4}$

(d) $\dfrac{1\cancel{3}}{\cancel{5}4} = \dfrac{1}{4}$

# MIXED NUMBERS

Sometimes we use what are called *mixed numbers,* that is, numbers made up of an integer and a fractional part of an integer. For example $4\frac{2}{3}$ denotes $4 + \frac{2}{3}$. The mixed number $7\frac{3}{8}$ means $7 + \frac{3}{8}$ or $\frac{7}{1} + \frac{3}{8} = \frac{59}{8}$. Fractions like $\frac{59}{8}$, in which the numerator is larger in absolute value than the denominator, are called *improper fractions.*

**Improper Fraction**    An *improper fraction* is one in which the absolute value of the numerator is larger than or equal to the absolute value of the denominator. A fraction in which this is not the case is called a **proper fraction.**

**EXAMPLE 6**

(a) Change $5\frac{3}{4}$ to an improper fraction and check with a calculator.
(b) Change $\frac{17}{3}$ to a mixed number and check with a calculator.
(c) Add $4\frac{1}{3}$ and $6\frac{1}{2}$.
(d) Change $-3\frac{5}{8}$ to an improper fraction.

*SOLUTION*

(a) $5\dfrac{3}{4} = 5 + \dfrac{3}{4} = \dfrac{5}{1} + \dfrac{3}{4} = \dfrac{20 + 3}{1 \cdot 4} = \dfrac{23}{4}$

(b) $\dfrac{17}{3} = \dfrac{15 + 2}{3} = \dfrac{15}{3} + \dfrac{2}{3} = \dfrac{5}{1} + \dfrac{2}{3} = 5 + \dfrac{2}{3} = 5\dfrac{2}{3}$    or    $3\overline{\smash{)}17}$ gives $5$ with $\dfrac{15}{2}$

(c) $4\dfrac{1}{3} + 6\dfrac{1}{2} = \left(4 + \dfrac{1}{3}\right) + \left(6 + \dfrac{1}{2}\right)$    or    $4\dfrac{1}{3} = 4\dfrac{2}{6}$

$\qquad = (4 + 6) + \left(\dfrac{1}{3} + \dfrac{1}{2}\right)$    $+ 6\dfrac{1}{2} = 6\dfrac{3}{6}$

$\qquad = 10 + \dfrac{5}{6} = 10\dfrac{5}{6}$    $10\dfrac{5}{6}$

(d) $-3\frac{5}{8} = -\left(3 + \frac{5}{8}\right) = -3 + \frac{-5}{8} = \frac{-24}{8} + \frac{-5}{8} = \frac{-29}{8}$

## CALCULATOR HINT

Mixed numbers are inserted in the calculator by using the $\boxed{\text{Unit}}$ key. You can specify in advance the form of the fraction in the answer by selecting either $A\ b/c$ or $d/e$ in the $\boxed{\text{FracMode}}$ menu. Or you can insert the mixed fraction, use $\boxed{=}$ to look at its form, and then use $\boxed{\text{A b/c} \blacktriangleleft \blacktriangleright \text{d/e}}$ (if needed), which changes mixed numbers to fractions and fractions to mixed numbers.

(a) 5 $\boxed{\text{Unit}}$ 3 $\boxed{/}$ 4 $\boxed{\text{A b/c} \blacktriangleleft \blacktriangleright \text{d/e}}$ $\boxed{=}$ 23/4   or   5 $\boxed{\text{Unit}}$ 3 $\boxed{/}$ 4 $\boxed{=}$ 23/4

(b) 17 $\boxed{/}$ 3 $\boxed{\text{A b/c} \blacktriangleleft \blacktriangleright \text{d/e}}$ $\boxed{=}$ 5 2/3   or   17 $\boxed{/}$ 3 $\boxed{=}$ 5 2/3

**PRACTICE PROBLEM**

*ANSWER*

Add $3\frac{2}{3}$ and $5\frac{1}{6}$.

$8\frac{5}{6}$

## CALCULATOR HINT

Some calculators can add the mixed fractions:   $4\frac{2}{3} + 6\frac{1}{4}$.

If $\boxed{\text{FracMode}}$ has been set on $A\ b/c$, 4 $\boxed{\text{Unit}}$ 2 $\boxed{/}$ 3 $\boxed{+}$ 6 $\boxed{\text{Unit}}$ 1 $\boxed{/}$ 4 $\boxed{=}$ to obtain $10\frac{11}{12}$.

# SUBTRACTION OF RATIONAL NUMBERS

Subtraction of rational numbers can be explained by using a takeaway approach, a missing-addend approach, or an additive inverse approach.

**EXAMPLE 7**

Find $\frac{4}{5} - \frac{3}{5}$, using the takeaway approach. (For example, suppose that $\frac{4}{5}$ of a pie is in the refrigerator. Coming home hungry from school, Carl eats $\frac{3}{5}$ of the pie. How much of the pie is left?)

*SOLUTION*

The fraction bar models in Figure 6-15 illustrate this approach.

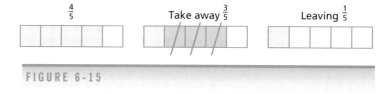

FIGURE 6-15

The preceding example suggests the following definition of the subtraction of rational numbers involving fractions with common denominators.

**DEFINITION**

| Subtraction of Rational Numbers with Common Denominators | Let $\dfrac{a}{b}$ and $\dfrac{b}{c}$ be any rational numbers. Then $$\frac{a}{c} - \frac{b}{c} = \frac{a-b}{c}$$ |

The missing-addend approach to subtraction is much the same for rational numbers as for integers and whole numbers.

$$\frac{a}{b} - \frac{c}{d} = \frac{e}{f} \quad \text{if} \quad \frac{a}{b} = \frac{c}{d} + \frac{e}{f}$$

**EXAMPLE 8**

$$\frac{7}{3} - \frac{5}{3} = \frac{2}{3} \quad \text{because} \quad \frac{7}{3} = \frac{5}{3} + \frac{2}{3}$$

To find the sum of two fractions, we find a common denominator, rewrite both fractions using the common denominator, and then add numerators. To subtract fractions, we again write them with a common denominator and then subtract numerators.

**DEFINITION**

| Subtraction of Rational Numbers | For rational numbers $\dfrac{a}{b}$ and $\dfrac{c}{d}$, $$\frac{a}{b} - \frac{c}{d} = \frac{ad - bc}{bd}$$ |

**EXAMPLE 9**

$$\frac{4}{5} - \frac{2}{3} = \frac{4(3) - 5(2)}{15} = \frac{2}{15}$$

In our discussion of subtraction of integers, we saw that subtracting an integer is equivalent to adding its additive inverse. The same is true for rational numbers. Thus, for rational numbers $\dfrac{a}{d}$ and $\dfrac{b}{e}$

$$\frac{a}{d} - \frac{b}{e} = \frac{a}{d} + \frac{-b}{e}$$

**EXAMPLE 10**

(a) $\dfrac{-4}{5} - \dfrac{2}{3} = \dfrac{-4}{5} + \dfrac{-2}{3} = \dfrac{-12 + -10}{15} = \dfrac{-22}{15}$

(b) $\dfrac{5}{4} - \dfrac{-1}{6} = \dfrac{5}{4} + \dfrac{1}{6} = \dfrac{15 + 2}{12} = \dfrac{17}{12}$   *l.c.d.* (4, 6) = 12

**EXAMPLE 11**

Complete the subtraction $17/12 - 8/9$.

*SOLUTION*

The least common denominator of 9 and 12 is 36.

$$\frac{17}{12} = \frac{51}{36} \quad \text{and} \quad \frac{8}{9} = \frac{32}{36}$$

$$\frac{17}{12} - \frac{8}{9} = \frac{51}{36} - \frac{32}{36} = \frac{19}{36}$$

**PRACTICE PROBLEM**

Complete the subtraction $\frac{7}{12} - \left(\frac{-3}{10}\right)$.

*ANSWER*

$\frac{53}{60}$

**EXAMPLE 12**

Alisha notes that his dogs can pull the empty sled to the store in $2\frac{5}{9}$ hours. When the sled is loaded to return home, it takes $4\frac{1}{3}$ hours. How much longer does it take to return home than to go to the store?

*SOLUTION*

$$4\frac{1}{3} - 2\frac{5}{9} = 4\frac{3}{9} - 2\frac{5}{9}$$

$$= 3 + \left(1 + \frac{3}{9}\right) - 2\frac{5}{9}$$

$$= 3 + \frac{12}{9} - 2\frac{5}{9}$$

$$= 1\frac{7}{9}$$

 ## CALCULATOR HINT

If [FracMode] has been set on $A\ b/c$ and *auto,* then

4 [Unit] 1 [/] 3 [−] 2 [Unit] 5 [/] 9 [=] $1\frac{7}{9}$.

 # ESTIMATION

We learned in the preceding section to approximate fractions by using fractions that can be simplified. For example, the fraction $\frac{61}{80}$ can be approximated by $\frac{60}{80}$ or $\frac{3}{4}$. The approximation $\frac{3}{4}$ is $\frac{1}{80}$ smaller than $\frac{61}{80}$. The following example illustrates one way to estimate the answer for the addition and subtraction of fractions using this technique.

**EXAMPLE 13**

Estimate answers to the following computations:

(a) $\frac{13}{24} + \frac{16}{45}$    (b) $\frac{13}{21} - \frac{10}{27}$

*SOLUTION*

(a) We see that $\frac{13}{24}$ is approximately $\frac{12}{24} = \frac{1}{2}$ and that $\frac{16}{45}$ is approximately $\frac{15}{45} = \frac{1}{3}$. Both fractions are slightly larger than their approximations. Therefore,

$$\frac{13}{24} + \frac{16}{45} \approx \frac{1}{2} + \frac{1}{3} = \frac{5}{6} \qquad \text{Smaller than actual answer}$$

(b) We see that $\frac{13}{21}$ is approximately $\frac{14}{21} = \frac{2}{3}$ and that $\frac{10}{27}$ is approximately $\frac{9}{27} = \frac{1}{3}$. Here $\frac{2}{3}$ is more than $\frac{13}{21}$, but $\frac{1}{3}$ is less than $\frac{10}{27}$. Thus,

$$\frac{13}{21} - \frac{10}{27} \approx \frac{2}{3} - \frac{1}{3} = \frac{1}{3} \qquad \text{Larger than actual answer}$$

**Range estimation** is a useful procedure when you are adding or subtracting mixed numbers. For example, suppose that you want to add $4\frac{5}{8} + 7\frac{1}{2}$. Now $4 < 4\frac{5}{8} < 5$ and $7 < 7\frac{1}{2} < 8$. Therefore, the sum $4\frac{5}{8} + 7\frac{1}{2}$ must be between $4 + 7 = 11$ and $5 + 8 = 13$.

**Estimation by rounding** is very similar to range estimation. To approximate the answer, you round each fraction to the nearest whole number. This time, suppose that you want to add $4\frac{5}{8} + 2\frac{1}{4}$. In this case, $4\frac{5}{8}$ rounds to 5 and $2\frac{1}{4}$ rounds to 2. Therefore,

$$4\frac{5}{8} + 2\frac{1}{4} \approx 5 + 2 = 7$$

For greater precision in rounding, you can round to the nearest $\frac{1}{2}$. Suppose this time that you want to add $4\frac{5}{8} + 3\frac{5}{6}$. Now $4\frac{5}{8}$ rounds to $4\frac{1}{2}$, and $3\frac{5}{6}$ rounds to 4. Thus,

$$4\frac{5}{8} + 3\frac{5}{6} \approx 4\frac{1}{2} + 4 = 8\frac{1}{2}$$

Probably the easiest way to estimate the addition and subtraction of mixed fractions is to find the exact answer for the whole part of the answer and to estimate the fraction part of the answer. Thus, for example,

$$6\frac{5}{8} - 2\frac{2}{3} = (6 - 2) + \left(\frac{5}{8} - \frac{2}{3}\right)$$

$$\approx 4 + \left(\frac{1}{2} - \frac{1}{2}\right) \qquad \text{Round to nearest } \frac{1}{2}$$

$$\approx 4$$

**EXAMPLE 14**

Estimate each of the following, and indicate whether the estimated answer is larger or smaller than the actual answer.

(a) $\dfrac{25}{48} + \dfrac{26}{75}$     (b) $\dfrac{35}{54} - \dfrac{10}{27}$

*SOLUTION*

(a) $\dfrac{1}{2} + \dfrac{1}{3} = \dfrac{5}{6}$    Smaller than actual answer

(b) $\dfrac{2}{3} - \dfrac{1}{3} = \dfrac{1}{3}$    Larger than actual answer

## ⋯ MENTAL ARITHMETIC

By breaking down a fraction into parts and regrouping, you can mentally perform additions and subtractions of fractions. For example, in the following addition, consider $\frac{4}{5}$ as $\frac{2}{5} + \frac{2}{5}$:

$$4\frac{3}{5} + 2\frac{4}{5} = (4 + 2) + \left(\frac{3}{5} + \frac{2}{5}\right) + \frac{2}{5}$$

$$= 6 + 1 + \frac{2}{5}$$

$$= 7\frac{2}{5}$$

In subtraction, the equal-addition procedure is sometimes helpful. For example, to subtract $4\frac{5}{7} - 2\frac{6}{7}$, add $\frac{1}{7}$ to both $4\frac{5}{7}$ and $2\frac{6}{7}$. Of course, the $\frac{1}{7}$ was selected to change $2\frac{6}{7}$ to 3. As a result, we have

$$4\frac{5}{7} - 2\frac{6}{7} = 4\frac{6}{7} - 3 = 1\frac{6}{7}$$

Let us return now to the problem at the beginning of this section.

**EXAMPLE 15**

Use four numbers—2, 4, 5, and 8—two as numerators and two as denominators of two fractions, such that each fraction is less than 1 and the difference in the two fractions is a maximum.

*SOLUTION*

To make a maximum difference in $a - b$ (where $a$ and $b$ arc fractions), we want $a$ to be as large as possible and $b$ to be as small as possible. For these 4 numbers, the biggest fraction is $\frac{4}{5}$ and the smallest fraction is $\frac{2}{8}$. We are in luck, because the biggest fraction and smallest fraction use different numbers. Therefore, the maximum is

$$\frac{4}{5} - \frac{2}{8} = \frac{22}{40} = \frac{11}{20}$$

# Just for Fun

Complete the magic square.

# EXERCISE SET 2

_R_ 1. Perform the following computations, using the definition of addition. Leave answers in simplest form.

(a) $\dfrac{3}{4} + 0$    (b) $\dfrac{7}{8} + \dfrac{2}{3}$    (c) $\dfrac{5}{9} + \dfrac{1}{4}$

(d) $\dfrac{2}{3} + \dfrac{3}{5}$    (e) $\dfrac{4}{x} + \dfrac{5}{2x}$    (f) $\dfrac{7}{xy} + \dfrac{4}{y}$

2. In Exercise 1, change each addition sign to a subtraction sign, and perform the computation.

3. Find the least common denominator of the following fractions, and then add the fractions.

(a) $\dfrac{5}{8}, \dfrac{3}{2}, \dfrac{1}{4}$    (b) $\dfrac{5}{32}, \dfrac{5}{4}, \dfrac{7}{8}$    (c) $\dfrac{7}{54}, \dfrac{4}{27}, \dfrac{5}{6}$

(d) $\dfrac{3}{4}, \dfrac{5}{3}, \dfrac{7}{5}$    (e) $\dfrac{5}{11}, \dfrac{11}{5}, \dfrac{3}{7}$    (f) $\dfrac{3}{3}, \dfrac{4}{3}, \dfrac{7}{9}$

4. Change the following improper fractions to mixed numbers, then check with a calculator.

(a) $\dfrac{58}{3}$    (b) $\dfrac{19}{4}$    (c) $\dfrac{49}{8}$    (d) $\dfrac{27}{4}$

5. Change the following mixed numbers to improper fractions, then check with a calculator.

(a) $3\frac{5}{6}$    (b) $4\frac{3}{4}$    (c) $16\frac{1}{3}$    (d) $50\frac{1}{8}$

6. The sum of two fractions is given in the following expressions. Determine the two fractions, writing each in lowest terms.

_Example:_ $\dfrac{15 + 24}{36} = \dfrac{15}{36} + \dfrac{24}{36} = \dfrac{5}{12} + \dfrac{2}{3}$

(a) $\dfrac{64 + 96}{512}$    (b) $\dfrac{4(5) + 12(9)}{9(5)}$

(c) $\dfrac{15 + 19}{4}$    (d) $\dfrac{4(4) + 3(5)}{5(4)}$

7. Find the sum and the difference (first minus second) for the following mixed numbers. Write your answer as a mixed number. Then check with a calculator

(a) $8\frac{4}{21}, 2\frac{3}{7}$    (b) $89\frac{1}{2}, 1\frac{12}{13}$

(c) $23\frac{3}{4}, 7\frac{7}{9}$    (d) $17\frac{1}{2}, 8\frac{9}{10}$

8. Compute (a) through (d).

(a) $\dfrac{7}{8} + \left(\dfrac{2}{3} + \dfrac{1}{2}\right)$    (b) $\left(\dfrac{7}{8} + \dfrac{2}{3}\right) + \dfrac{1}{2}$

(c) $\dfrac{7}{8} - \left(\dfrac{2}{3} - \dfrac{1}{2}\right)$    (d) $\left(\dfrac{7}{8} - \dfrac{2}{3}\right) - \dfrac{1}{2}$

(e) What do (a) and (b) demonstrate?

(f) What do (c) and (d) demonstrate?

9. Carry out the indicated operations and then check with a calculator.

(a) $5\frac{1}{2} + 4\frac{3}{4} + 1\frac{1}{3}$    (b) $4\frac{1}{2} - 2\frac{3}{4}$

(c) $3\frac{5}{9} + 2\frac{2}{3} + 5\frac{1}{2}$    (d) $2\frac{5}{9} + 4\frac{2}{3} + 3\frac{1}{4}$

(e) $\dfrac{5}{12} + 7\frac{2}{5}$    (f) $17\frac{7}{12} - 7\frac{2}{5}$

(g) $17\frac{4}{5} - 6\frac{3}{8}$    (h) $24\frac{5}{8} + 17\frac{4}{9}$

(i) $80\frac{17}{20} + 19\frac{1}{15}$    (j) $4\frac{1}{6} + 19\frac{1}{10}$

10. Using the fractions $\frac{3}{4}$ and $\frac{2}{5}$, demonstrate for rational numbers

(a) the commutative property of addition.

(b) with $\frac{2}{3}$ the associative property of addition.

(c) that 0/1 is the additive identity.

11. Use a number line to indicate the following additions.

(a) $\dfrac{2}{5} + \dfrac{1}{5}$    (b) $\dfrac{1}{3} + \dfrac{1}{4}$    (c) $\dfrac{1}{6} + \dfrac{1}{4}$

12. Calculate mentally, and explain what procedures you used.

(a) $\dfrac{4}{7} + \left(\dfrac{1}{2} + \dfrac{3}{7}\right)$    (b) $\dfrac{2}{5} + \left(\dfrac{5}{8} + \dfrac{3}{5}\right)$

(c) $4\dfrac{5}{8} + 2\dfrac{3}{8}$    (d) $3\dfrac{4}{5} + 2\dfrac{3}{5}$

(e) $4\dfrac{3}{8} - 2\dfrac{5}{8}$    (f) $6\dfrac{3}{5} - 3\dfrac{4}{5}$

13. Estimate the integral range of the answer for each of the following:

(a) $6\frac{7}{11} + 2\frac{5}{6} + 5\frac{1}{3}$    (b) $8\frac{1}{9} + 3\frac{1}{3} + 7\frac{2}{3}$

14. Estimate the answer in Exercise 13 by rounding to the nearest whole number.

15. Estimate the answer in Exercise 13 by rounding to the nearest one-half.

16. Does each of the following properties hold for subtraction of rational numbers? Demonstrate with examples or counterexamples.

(a) Closure

(b) Commutative

(c) Associative

(d) Inverse

(e) Identity

17. Solve the following equations:

(a) $x - \frac{2}{3} = \frac{1}{4}$    (b) $x + 2\frac{1}{2} = 3\frac{1}{3}$

(c) $3x - \frac{4}{3} = 2$    (d) $2x + \frac{1}{3} = 5$

18. Using the fact that $-\left(\frac{a}{b} + \frac{c}{d}\right) = \frac{-a}{b} + \frac{-c}{d}$, express each of the following in the form $\frac{a}{b}$, where $b$ is positive.

    (a) $-\left(\frac{3}{5} + \frac{7}{8}\right)$       (b) $-\left(\frac{-7}{5} - \frac{-3}{7}\right)$

    (c) $-\left(\frac{6}{5} + \frac{-7}{9}\right)$       (d) $-\left(\frac{7}{3} - \frac{6}{5}\right)$

*T* 19. Without actually finding the exact answer, state which of these answers is the best estimate of the given sum or difference.

    (a) $\frac{9}{16} + \frac{6}{10} - \frac{2}{9}$   $\left(\frac{1}{4} \text{ or } \frac{5}{6} \text{ or } \frac{13}{10}\right)$

    (b) $\frac{4}{15} + \frac{6}{61} - \frac{1}{100}$   $\left(\frac{19}{100} \text{ or } \frac{34}{100} \text{ or } \frac{94}{100}\right)$

20. Find the sum and the difference (first minus second) of each of the following, without using a calculator. Then check with a calculator.

    (a) $\frac{7}{39}, \frac{2}{91}$       (b) $\frac{13}{243}, \frac{5}{162}$

    (c) $\frac{7}{2^3 \cdot 3^2}, \frac{5}{2^3 \cdot 3^4}$       (d) $\frac{5}{2^2 \cdot 3^{10} \cdot 7^4}, \frac{8}{3^6 \cdot 5^2 \cdot 7^3}$

    (e) $41\frac{17}{23}, 16\frac{5}{11}$       (f) $74\frac{3}{40}, 8\frac{5}{36}$

21. Find the sum and difference (first minus second) for each pair.

    (a) $\frac{3yz}{xy^4z^2}, \frac{3x}{x^2y}$       (b) $\frac{6a^2b}{3xba}, \frac{5x^3b}{14ba^3}$

    (c) $\frac{11}{mn^2}, \frac{mnx}{12m^2}$       (d) $\frac{d}{a^3bc^2}, \frac{a}{ab^3c}$

22. For the following problems, formulate an open sentence (an equation) that serves as a model for a problem, and then find the solution.

    (a) Tricia purchased $\frac{1}{2}$ lb. of cheese, a $6\frac{1}{4}$-lb. ham, a $2\frac{3}{4}$-lb. roast, a $12\frac{1}{8}$-lb. turkey, and 1 lb. of butter.
        (i) What was the total weight of this purchase?
        (ii) If Tricia returned the roast, what was the weight of the remaining purchase?

    (b) About $\frac{3}{10}$ of the cars in the United States are less than 3 years old. About $\frac{1}{4}$ of the cars in the United States are at least 3 years old but less than 6 years old. What fraction of the cars in the United States are less than 6 years old?

    (c) Paul owns $\frac{1}{3}$ of the stock of the company of which he is president. His wife owns $\frac{1}{3}$ of the stock. What fractional part of the stock will his daughter, Brooke, need to purchase in order for the family to own $\frac{3}{4}$ of the stock?

    (d) An Alabama farmer uses $61\frac{1}{3}$ acres of his 100-acre farm for cattle grazing. He has $22\frac{4}{5}$ acres planted in pine trees. How many acres are left for growing hay?

    (e) On a recent cross-country trip in an airplane, Lane spent $4\frac{5}{8}$ hours on the first flight and $2\frac{1}{4}$ hours on the second flight. If there was a $1\frac{1}{3}$-hour wait between flights, how long did it take Lane to make the trip?

23. What is wrong with each argument that follows?

    (a) $\frac{(2+3)}{2} = \frac{2}{2} + 3 = 1 + 3 = 4$

    (b) $1 = \frac{8}{8} = \frac{8}{4+4} = \frac{8}{4} + \frac{8}{4} = 2 + 2 = 4$

    (c) $\frac{3}{4} + \frac{7}{9} = \frac{27 + 28}{36 + 36} = \frac{55}{72}$

24. Estimate the answers for the following addition problems, and state whether your answer is too large or too small.

    (a) $\frac{26}{36} + \frac{29}{45}$

    (b) $\frac{31}{48} + \frac{11}{60}$

25. Does $6\frac{2}{3} + 5\frac{3}{4} = 6\frac{3}{4} + 5\frac{2}{3}$? Explain.

26. Add the following fractions:

    (a) $\frac{1}{x^3y^4} + \frac{1}{xy^2}$       (b) $\frac{2}{x^5y^2} + \frac{2}{xy^4}$

    (c) $\frac{x}{x-y} + \frac{y}{x+y}$       (d) $\frac{x}{x^2-y^2} + \frac{-y}{x-y}$

27. By finding a counterexample, show for rational numbers that $a - (b - c - d) \neq (a - b - c) - d$.

28. (a) Find a pattern for the following sequence.

$$\frac{1}{2} + \frac{1}{3} = \frac{5}{6}$$
$$\frac{1}{3} + \frac{1}{4} = \frac{7}{12}$$
$$\frac{1}{4} + \frac{1}{5} = \frac{9}{20}$$
$$\vdots \quad \vdots \quad \vdots$$

    (b) Does the pattern work for $\frac{1}{12} + \frac{1}{13}$?

29. A unit fraction has a numerator of 1. Express the following as a sum of 2 or more unit fractions.

    (a) $\frac{1}{2}$   (b) $\frac{1}{11}$   (c) $\frac{1}{100}$

    (d) $\frac{2}{3}$   (e) $\frac{2}{5}$   (f) $\frac{2}{11}$

30. Check to see that

$$\left(1\tfrac{1}{2}\right)^2 = 1 \cdot 2 + \tfrac{1}{4}$$
$$\left(2\tfrac{1}{2}\right)^2 = 2 \cdot 3 + \tfrac{1}{4}$$
$$\left(3\tfrac{1}{2}\right)^2 = 3 \cdot 4 + \tfrac{1}{4}$$

(a) Without squaring, write down an expression for $\left(4\tfrac{1}{2}\right)^2$.
(b) Do the same for $\left(5\tfrac{1}{2}\right)^2$.
(c) Write a general expression for $\left(n + \tfrac{1}{2}\right)^2$.
(d) Can you justify your answer in (c)?

## REVIEW EXERCISES

31. Write 3 fractions equivalent to $\tfrac{3}{8}$.

32. Find the simplest form for each fraction. Then check (a) and (c) with a calculator.

(a) $\dfrac{24}{36}$

(b) $\dfrac{ax^2}{axy}$

(c) $\dfrac{45}{162}$

## LABORATORY ACTIVITIES

33. Illustrate the addition $\tfrac{2}{5} + \tfrac{1}{4}$, using the following models:

(a) Number line
(b) Fraction bars
(c) Set of blocks

34. Illustrate the subtraction $\tfrac{2}{3} - \tfrac{1}{2}$, using the following models.

(a) Takeaway approach
(b) Missing-addend approach
(c) Additive inverse approach on a number line

35. Complete the magic squares.

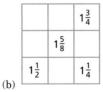

(a)                    (b)

36. On p. 267 we considered the plane figures

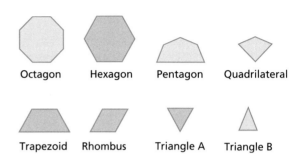

Octagon    Hexagon    Pentagon    Quadrilateral

Trapezoid    Rhombus    Triangle A    Triangle B

(a) Represent the number of hexagonal regions in 13 triangular regions (A) as a mixed number. The number of trapezoidal regions. The number of regions formed by rhombi.
(b) Represent the number of octagonal regions in 45 triangular regions (B) as a mixed number. The number of pentagonal regions. The number of quadrilateral regions.
(c) Represent the number of hexagonal regions in 17 regions formed by rhombi as a mixed number. The number of trapezoidal regions. The number of triangular regions (A).
(d) Represent the number of octagonal regions in 49 quadrilateral regions as a mixed number. The number of pentagonal regions. The number of triangular regions (B).
(e) Represent the number of hexagonal regions in 19 trapezoidal regions. The number of regions formed by rhombi. The number of triangular regions (A).
(f) Represent the number of octagonal regions in 51 pentagonal regions. The number of quadrilateral regions. The number of triangular regions (B).
(g) If 1 hexagon region has a value of a whole, represent 3 hexagons, 5 trapezoids, 7 rhombi, and 9 triangles (A) as a mixed number.
(h) If 1 octagon region has a value of a whole, represent 5 pentagons, 7 quadrilaterals, 9 octagons, and 17 triangles (B) as a mixed number.
(i) If 1 trapezoid region has a value of a whole, represent 5 hexagons, 7 rhombi, and 11 triangles (A).
(j) If 1 quadrilateral region has a value of a whole, represent 5 octagons, 7 pentagons, and 23 triangles (B).

# MULTIPLICATION AND DIVISION OF RATIONAL NUMBERS

**PROBLEM**

John gave away all of his apples to his two brothers. To Leo he gave half of his apples plus half an apple. To Ned he gave half of what he had left plus half an apple. John then had no apples left. How many apples did he have to begin with? (Notice that at no time were the apples cut, split, or divided.)

**OVERVIEW**

If we set up this problem as an open sentence, we must use multiplication and division to solve it. The operations in themselves won't be difficult because we've been multiplying and dividing rational numbers for many years. But the goal of this section is to help you understand these operations better. As a step in that direction, we begin our study by interpreting $\frac{1}{3} \cdot \frac{2}{5}$ as $\frac{1}{3}$ of $\frac{2}{5}$. We can then explain the answer geometrically.

**GOALS**

Illustrate the Number and Operations Standard, page 119.
Illustrate the Problem Solving Standard, page 15.
Illustrate the Representation Standard, page 50.
Illustrate the Connections Standard, page 147.
Illustrate the Algebra Standard, page 4.

## MULTIPLICATION OF RATIONAL NUMBERS

We want to extend the operation of multiplication from whole numbers and integers to rational numbers. We can easily accomplish this by considering multiplication as repeated addition. Thus, for example, $3 \cdot \frac{1}{4}$ is equal to $\frac{3}{4}$. This is shown on a number line in Figure 6-16(a) and with a fraction bar in Figure 6-16(b).

However, consider the following problem: Lane Park in Nashville, Tennessee is about $\frac{1}{3}$ of a block wide and about $\frac{2}{5}$ of a block long. The park is what part of a square block? Figure 6-16(c) shows the square block and Lane Park (shaded). You can see that the park is $\frac{2}{15}$ of the square block. Now the area of a rectangle is obtained by multiplying the width times the length or

$$\frac{1}{3} \cdot \frac{2}{5}$$

The answer $\frac{2}{15}$ can be obtained by multiplying the two numerators to obtain a numerator of 2 and multiplying the two denominators to obtain a denominator of 15. To further illustrate this concept consider the fraction bar in Figure 6-17(a), which represents $\frac{2}{5}$. In Figure 6-17(b), the region has been further divided into three equal parts, and one of these has been shaded to represent $\frac{1}{3}$. Thus the product $\frac{1}{3} \cdot \frac{2}{5}$ is indicated by the dark-shaded region, which covers 2 parts out of 15. Expressed mathematically,

$$\frac{1}{3} \cdot \frac{2}{5} = \frac{2}{15}$$

The multiplication $\frac{1}{3} \cdot \frac{2}{5}$ is often interpreted as $\frac{1}{3}$ of $\frac{2}{5}$. This interpretation can be illustrated by dividing a square into 15 like regions arranged in three columns and five rows. (See Figure 6-17(c).) Shade any 1 of the 3 columns, and then shade any 2 of the 5 rows. Because 2 parts have been shaded twice, we see that $\left(\frac{1}{3}\right) \cdot \left(\frac{2}{5}\right)$ is $\frac{2}{15}$. This result can also be obtained by multiplying the numerators and multiplying the denominators of the 2 fractions.

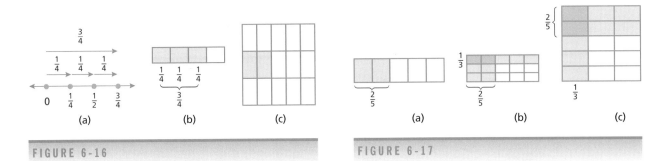

FIGURE 6-16          FIGURE 6-17

These examples illustrate a concept that is expressed in the following definition.

**DEFINITION**

**Multiplication of Rational Numbers**    If $\frac{a}{b}$ and $\frac{c}{d}$ are any rational numbers, then

$$\frac{a}{b} \cdot \frac{c}{d} = \frac{ac}{bd}$$

**EXAMPLE 1**    (a) $\frac{4}{5} \cdot \frac{1}{2} = \frac{4 \cdot 1}{5 \cdot 2} = \frac{4}{10}$    (b) $\frac{2}{3} \cdot \frac{7}{11} = \frac{2 \cdot 7}{3 \cdot 11} = \frac{14}{33}$

## COMMON ERROR

$\frac{1}{3} \cdot 2 = \frac{2}{6}$    is false.    Write 2 as $\frac{2}{1}$ and then multiply.

Leaving the numerator and denominator in factored form sometimes simplifies the process of simplifying the fraction.

**EXAMPLE 2**    (a) $\frac{7}{8} \cdot \frac{8}{15} = \frac{7 \cdot \overset{1}{\cancel{8}}}{\underset{1}{\cancel{8}} \cdot 15} = \frac{7}{15}$

The 8 is divided out, as prescribed by the Fundamental Law of Fractions.

(b) $\frac{2}{3} \cdot \frac{21}{4} = \frac{\overset{1}{\cancel{2}} \cdot \overset{7}{\cancel{21}}}{\underset{1}{\cancel{3}} \cdot \underset{2}{\cancel{4}}} = \frac{1 \cdot 7}{1 \cdot 2} = \frac{7}{2}$

In accordance with the Fundamental Law of Fractions, the 3 is divided out of 21 in the numerator and out of 3 in the denominator, and the 2 is divided out of 2 in the numerator and out of 4 in the denominator.

A word of caution is appropriate here. Instead of using the Fundamental Law of Fractions, students sometimes randomly cancel numbers and thus make errors. For example,

## COMMON ERRORS

$$\frac{1\!\!\!/7}{7\!\!\!/4} = \frac{1}{4}$$

Why is this incorrect?

$$\frac{5 + \overset{1}{2\!\!\!/}}{\underset{2}{4\!\!\!/}} = \frac{6}{2} = 3$$

Why is this incorrect?

$$\frac{34 + 3}{9 \cdot 2} = \frac{\overset{17}{3\!\!\!/4} + \overset{1}{3\!\!\!/}}{\underset{3}{9\!\!\!/} \cdot \underset{1}{2\!\!\!/}} = \frac{18}{3} = 6$$

Why is this incorrect?

## CALCULATOR HINT

A fraction calculator can be used to find products such as $\frac{3}{5} \cdot \frac{-4}{17}$.

$3 \boxed{/} 5 \boxed{\times} \boxed{(-)} \boxed{4} \boxed{/} 17 \boxed{=}$. The answer is, of course, $\frac{-12}{85}$.

The fraction $\frac{1}{1} = 1$ represents the multiplicative identity for rational numbers. It is easy to verify this fact because

$$\frac{c}{d} \cdot \frac{1}{1} = \frac{c \cdot 1}{d \cdot 1} = \frac{c}{d}$$

Similarly, $\frac{a}{a}$, where $a \neq 0$, represents the multiplicative identity because $\frac{1}{1} = \frac{a}{a}$.

The product of certain rational numbers always produces the multiplicative identity.

**EXAMPLE 3**

$$\left(\frac{5}{6}\right) \cdot \left(\frac{6}{5}\right) = 1$$

$$\left(\frac{-6}{3}\right) \cdot \left(\frac{3}{-6}\right) = 1$$

In general, for $x \neq 0$ and $y \neq 0$,

$$\frac{x}{y} \cdot \frac{y}{x} = \frac{xy}{yx} = 1$$

| DEFINITION | Multiplicative Inverses | If the product of 2 numbers equals the identity 1, then the 2 numbers are called *multiplicative inverses*. In the system of rational numbers, there exists a multiplicative inverse $y/x$ for every rational number $x/y$ if $x \neq 0$. However, the multiplicative inverse for $x/y$ where $x = 0$ does not exist. The multiplicative inverse of a rational number is also called its **reciprocal.** |

Multiplication involving mixed numbers may be performed in either of two ways. The mixed numbers may be changed to improper fractions, or the distributive property of multiplication over addition may be used. Both methods are illustrated in the following example.

**EXAMPLE 4**

Multiply $2\dfrac{3}{4} \cdot 7\dfrac{1}{8}$.

*SOLUTION*

Converting the mixed fractions to improper fractions, we get

$$2\frac{3}{4} \cdot 7\frac{1}{8} = \frac{11}{4} \cdot \frac{57}{8} = \frac{627}{32} = 19\frac{19}{32}$$

Using the distributive property, we get

$$2\frac{3}{4} \cdot 7\frac{1}{8} = \left(2 + \frac{3}{4}\right)\left(7 + \frac{1}{8}\right)$$

$$= 2\left(7 + \frac{1}{8}\right) + \frac{3}{4}\left(7 + \frac{1}{8}\right)$$

$$= 2(7) + 2\left(\frac{1}{8}\right) + \left(\frac{3}{4}\right)(7) + \left(\frac{3}{4}\right)\left(\frac{1}{8}\right)$$

$$= 14 + \frac{2}{8} + \left(5 + \frac{1}{4}\right) + \frac{3}{32}$$

$$= 19 + \left(\frac{2}{8} + \frac{1}{4} + \frac{3}{32}\right)$$

$$= 19\frac{19}{32}$$

**PRACTICE PROBLEM**

Compute $3\dfrac{5}{6} \cdot 2\dfrac{1}{4}$ and check with a calculator.

*ANSWER*

$8\dfrac{5}{8}$.   *Check*: Set $\boxed{\text{FracMode}}$ on $A\,b/c$.

$3\,\boxed{\text{Unit}}\,5\,\boxed{/}\,6\,\boxed{\times}\,2\,\boxed{\text{Unit}}\,1\,\boxed{/}\,4\,\boxed{=}$. Did you get $8\frac{5}{8}$?

## ⠿ DIVISION OF RATIONAL NUMBERS

The fraction bar models in Figure 6-18 illustrate the measurement model of division. Consider (a) $3 \div \frac{3}{4}$ and (b) $\frac{3}{4} \div \frac{1}{8}$.

(a) How many $\frac{3}{4}$'s are there in 3?    (b) How many $\frac{1}{8}$'s are there in $\frac{3}{4}$?

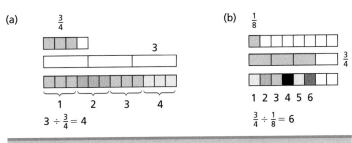

**FIGURE 6-18**

Larry has grown 3 inches in the last 2 years. His younger sister, Lucy, has grown only $\frac{3}{4}$ of an inch. Compare the two growths. In this problem we consider the comparison as a division or as a quotient,

$\frac{3}{\frac{3}{4}}$. Then $\dfrac{3 \cdot \frac{4}{3}}{\frac{3}{4} \cdot \frac{4}{3}} = \frac{4}{1}$  Multiply numerator and denominator by the reciprocal of $\frac{3}{4}$.

Larry's growth is four times that of Lucy.

**EXAMPLE 5**

How many pieces of ribbon $\frac{1}{8}$ inch wide can Marcia cut from a ribbon $\frac{3}{4}$ inch wide?

*SOLUTION*

$$\dfrac{\frac{3}{4}}{\frac{1}{8}} = \dfrac{\frac{3}{4} \cdot \frac{8}{1}}{\frac{1}{8} \cdot \frac{8}{1}} = \dfrac{\frac{24}{4}}{\frac{8}{8}} = \frac{6}{1} = 6$$

The preceding examples illustrate that we can replace a division problem with a multiplication problem by using the reciprocal of the divisor.

**DEFINITION**

**Division as a Multiplication**

If $\frac{c}{d} \neq 0$, then

$$\frac{a}{b} \div \frac{c}{d} = \frac{a}{b} \cdot \frac{d}{c}$$

where $\frac{d}{c}$ is the *multiplicative inverse* or *reciprocal* of $\frac{c}{d}$.

**EXAMPLE 6**

(a) $\dfrac{4}{7} \div \dfrac{3}{8} = \dfrac{4}{7} \cdot \dfrac{8}{3} = \dfrac{32}{21}$

(b) $\dfrac{-1}{4} \div \dfrac{-1}{3} = \dfrac{-1}{4} \cdot \dfrac{3}{-1} = \dfrac{-3}{-4} = \dfrac{3}{4}$

(c) $\dfrac{8}{15} \div \dfrac{4}{9} = \dfrac{\overset{2}{\cancel{8}}}{\underset{5}{\cancel{15}}} \cdot \dfrac{\overset{3}{\cancel{9}}}{\underset{1}{\cancel{4}}} = \dfrac{2 \cdot 3}{5 \cdot 1} = \dfrac{6}{5}$

**EXAMPLE 7**

Find $3\frac{3}{5} \div 2\frac{5}{8}$.

*SOLUTION*

$$3\frac{3}{5} \div 2\frac{5}{8} = \frac{18}{5} \div \frac{21}{8} = \frac{\overset{6}{\cancel{18}}}{5} \cdot \frac{8}{\underset{7}{\cancel{21}}} = \frac{48}{35} = 1\frac{13}{35}$$

**CALCULATOR HINT**

To get an answer as a mixed fraction, set ⌊FracMode⌋ on *A b/c* and *Auto*.

3 ⌊Unit⌋ 3 ⌊/⌋ 5 ⌊÷⌋ 2 ⌊Unit⌋ 5 ⌊/⌋ 8 ⌊=⌋ 1 13/35

**EXAMPLE 8**

Simplify $\dfrac{2+\dfrac{3}{4}}{1+\dfrac{1}{4}}$.

*SOLUTION*

$$\frac{2+\dfrac{3}{4}}{1+\dfrac{1}{4}} = \frac{\dfrac{2}{1}+\dfrac{3}{4}}{\dfrac{1}{1}+\dfrac{1}{4}} = \frac{\dfrac{8+3}{4}}{\dfrac{4+1}{4}} = \frac{\dfrac{11}{4}}{\dfrac{5}{4}} = \frac{11}{\cancel{4}_1} \cdot \frac{\cancel{4}^1}{5} = \frac{11}{5}$$

Because division is the inverse of multiplication, and because every rational number unequal to 0 has a multiplicative inverse, division of rational numbers is always possible except when the divisor is 0. Moreover, the answer is unique in that it is one of a number of equivalent fractions.

Because 0 plays such a special role in division, here is a word of caution: **Division of a fraction by 0** is not defined because

$$\frac{a}{b} \div 0 \qquad (a \neq 0)$$

cannot equal an answer $\dfrac{e}{f}$. This is because

$$\frac{e}{f} \cdot 0 = 0 \neq \frac{a}{b}$$

# MENTAL ARITHMETIC

Once again, you can perform mental arithmetic by appropriately grouping the factors. For example, to multiply $4 \cdot \frac{1}{3} \cdot \frac{1}{2}$, you can group the $\frac{1}{2}$ with the 4 to obtain

$$\left(4 \cdot \frac{1}{2}\right) \cdot \frac{1}{3} = 2 \cdot \frac{1}{3} = \frac{2}{3}$$

Now multiply $3\frac{1}{6} \cdot 18$, using the distributive property:

$$\left(3 + \frac{1}{6}\right) \cdot 18 = 3 \cdot 18 + \frac{1}{6} \cdot 18$$
$$= 54 + 3$$
$$= 57$$

For division, $18\frac{1}{4} \div 6$ is

$$\left(18 + \frac{1}{4}\right) \cdot \frac{1}{6} = \frac{18}{6} + \frac{1}{4} \cdot \frac{1}{6}$$
$$= 3 + \frac{1}{24}$$
$$= 3\frac{1}{24}$$

## CALCULATOR HINT

If [ FracMode ] is set on $A\ b/c$, 18 [ Unit ] 1 [ / ] 4 [ ÷ ] 6 [ = ] getting $3\dfrac{1}{24}$.

John Wallis was born a Protestant minister's son in the village of Ashford, England. Recognizing that John was more academically oriented than his 4 siblings, at age 14 his parents arranged for him to study with a special teacher at a school near Ashford. There, he studied Greek, Hebrew, and Latin–but not mathematics. He received the only mathematics instruction of his life from one of his brothers who had studied some arithmetic while learning a trade. As Wallis wrote in his autobiography, "Mathematicks were not, at that time, looked upon as Academical Learning."

The Civil War in England in 1649 led Wallis toward mathematics. Early on in the upheaval, he used his knowledge of mathematics to decipher secret coded messages for a political group called the Parliamentarians. In 1649, upon the ouster of a faculty member who was a Royalist, another political group, Wallis was appointed to a chair of geometry at Oxford University. From that time on, he was active in mathematical research. He was a charter member of the Royal Society, an organization he helped to form that endures as one of the world's oldest scientific organizations. Most of Wallis's work was in analytic geometry and in "infinite analysis," the precursor of calculus. Though analytic geometry was begun by Descartes, Wallis also had a great influence on its development. He arithemetized most of what was then known about geometry, using algebraic equations for lines, circles, ellipses, and the other common geometric curves. Wallis succeeded in marrying the topics of algebra and geometry.

Wallis was one of many mathematicians who extended the idea of adding or multiplying many fractions to include infinite collections of fractions. Some examples follow:

$$\frac{1}{2} + \frac{1}{4} + \frac{1}{8} + \frac{1}{16} + \frac{1}{32} + \frac{1}{64} + \cdots = 1$$

$$\frac{1}{1} + \frac{1}{3} + \frac{1}{6} + \frac{1}{10} + \frac{1}{15} + \frac{1}{21} + \frac{1}{28} + \cdots = 2$$

Wallis also explored ideas in mathematics that led to a much deeper understanding of how fractions are related to decimals.

We owe much of our algebraic notation to Wallis, including the use of negative exponents, the symbol $\infty$ for infinity, and the symbol $\div$ for division.

## ESTIMATION

Range estimation can be used in multiplication and division, as well as in addition and subtraction. Suppose that you want to multiply $4\frac{1}{3} \cdot 6\frac{7}{8}$. Now, because $4 < 4\frac{1}{3} < 5$ and $6 < 6\frac{7}{8} < 7$, we know that $4\frac{1}{3} \cdot 6\frac{7}{8}$ is between $4 \cdot 6 = 24$ and $5 \cdot 7 = 35$.

To illustrate estimation of multiplication using rounding, consider $7\frac{1}{8} \cdot 3\frac{5}{8}$. Rounded to whole numbers, $7\frac{1}{8}$ becomes 7 and $3\frac{5}{8}$ becomes 4. Thus,

$$7\frac{1}{8} \cdot 3\frac{5}{8} \approx 7 \cdot 4 = 28$$

This answer can be made somewhat more accurate by rounding to the nearest $\frac{1}{2}$:

$$7\frac{1}{8} \cdot 3\frac{5}{8} \approx 7 \cdot 3\frac{1}{2} = 24\frac{1}{2}$$

We now use the material of this section to solve the introductory problem.

**EXAMPLE 9**

John gave away all of his apples to his brothers. To Leo he gave half of his apples plus half an apple. To Ned he gave half of what he had left plus half an apple. John then had no apples left. How many apples did he have to begin with?

SOLUTION

Let $x$ be the number of apples at the beginning. Leo received
$$\left(\frac{1}{2} \text{ of } x\right) + \frac{1}{2} \text{ apple}$$
But by simple calculation, we see that
$$x - \left[\left(\frac{1}{2} \text{ of } x\right) + \frac{1}{2}\right] = \frac{x}{2} - \frac{1}{2}$$
This is what remains after the gift to Leo. Therefore, Ned received
$$\left(\frac{1}{2}\right)\left(\frac{x}{2} - \frac{1}{2}\right) + \frac{1}{2} = \frac{x}{4} + \frac{1}{4} \text{ apples}$$
The sum of the apples Leo received and the apples Ned received equals the total number of apples ($x$).
$$\left(\frac{x}{2} + \frac{1}{2}\right) + \left(\frac{x}{4} + \frac{1}{4}\right) = x$$
$$\frac{3x}{4} + \frac{3}{4} = x$$

$x = 3$ satisfies this equation.

Therefore, John had 3 apples originally and gave 2 to Leo and 1 to Ned. Look back at the problem and see if this solution makes sense.

# Just for Fun

In Ahab's will, $\frac{1}{2}$ of his camels were to go to his oldest child, $\frac{1}{3}$ to his second child, and $\frac{1}{9}$ to his youngest child. At his death, there were 17 camels. A clever lawyer borrowed a camel to make 18.

$$\frac{1}{2} \text{ of } 18 = 9$$

$$\frac{1}{3} \text{ of } 18 = 6$$

$$\frac{1}{9} \text{ of } 18 = 2$$

Total = 17

Then the lawyer returned the borrowed camel, and everybody was happy thereafter. Are you? Why not?

# EXERCISE SET 3

*R* 1. Use rectangular regions such as those in Figure 6-17(c) to illustrate the following computations.

(a) $\dfrac{1}{3} \cdot \dfrac{1}{3} = \dfrac{1}{9}$   (b) $\dfrac{1}{5} \cdot \dfrac{2}{3} = \dfrac{2}{15}$

2. Write the multiplicative inverse of each of the following, where possible, and perform the operations necessary to verify your answers.

(a) $\dfrac{7}{1}$   (b) $\dfrac{3}{5}$   (c) $\dfrac{0}{8}$

(d) $\dfrac{6}{11}$   (e) $\dfrac{w}{z}$   (f) $\dfrac{z}{w+y}$

3. Perform the following multiplications and divisions, and write your answer as a fraction. Then check your answers with a calculator.

(a) $\dfrac{3}{4}\left(\dfrac{1}{2} \cdot \dfrac{2}{3}\right)$   (b) $\dfrac{4}{3}\left(\dfrac{1}{3} \cdot \dfrac{1}{2}\right)$

(c) $\dfrac{4}{7}\left(\dfrac{5}{8} \cdot \dfrac{2}{3}\right)$   (d) $\left(\dfrac{2}{3} \cdot \dfrac{0}{1}\right)\left(\dfrac{13}{2}\right)$

(e) $4 \div \dfrac{1}{9}$   (f) $8 \div \dfrac{1}{4}$

(g) $\left(\dfrac{2}{3} + \dfrac{1}{5}\right) \div \dfrac{7}{10}$   (h) $\left(8 \div \dfrac{5}{11}\right)\left(\dfrac{2}{3} \div 9\right)$

(i) $\left(\dfrac{4}{3} \div \dfrac{7}{8}\right) \cdot \left(\dfrac{1}{2} \div \dfrac{3}{4}\right)$

(j) $\left(\dfrac{3}{2} + \dfrac{7}{12}\right) \div \left(\dfrac{2}{5} + \dfrac{3}{8}\right)$

(k) $\dfrac{16}{45} \cdot \dfrac{15}{72}$   (l) $\dfrac{7}{9} \div \dfrac{42}{5}$

(m) $\dfrac{9}{14} \cdot \dfrac{8}{30} \cdot \dfrac{7}{27}$   (n) $\dfrac{16}{27} \cdot \dfrac{3}{4} \cdot \dfrac{9}{32}$

4. Simplify each of the following expressions, using the same concepts as when simplifying numbers represented by numerals.

(a) $\dfrac{a}{b} \cdot \dfrac{abc}{ac}$   (b) $\dfrac{x+y}{x} \cdot \dfrac{x}{y}$

(c) $\dfrac{ab}{c}\left(\dfrac{a}{e} + \dfrac{b}{d}\right)$   (d) $2x \cdot \dfrac{3}{xy} \cdot \dfrac{y}{z}$

(e) $\dfrac{y^2}{x} \cdot \dfrac{x^3}{y^4}$   (f) $\dfrac{r^2 s^3}{t} \cdot \dfrac{t^2}{rs^2}$

5. Perform the following operations involving mixed numbers without using a calculator. Then check your answers with a calculator

(a) $7\dfrac{1}{8} \cdot 6\dfrac{1}{4}$   (b) $8\dfrac{1}{7} \cdot 3\dfrac{1}{2}$   (c) $2\dfrac{1}{10} \cdot 8\dfrac{1}{3}$

(d) $16\dfrac{2}{3} \cdot 4\dfrac{1}{4}$   (e) $7\dfrac{1}{2} \div 4\dfrac{1}{3}$   (f) $\dfrac{16}{3} \div 2\dfrac{1}{6}$

(g) $4\dfrac{7}{8} \div 1\dfrac{3}{4}$   (h) $6\dfrac{1}{4} \div \dfrac{4}{11}$

6. Without using a calculator express each as a fraction in simplest form. Then check (a) – (g) with a calculator

(a) $\dfrac{\frac{3}{8}}{\frac{4}{16}}$   (b) $\dfrac{\frac{1}{7}}{\frac{8}{21}}$   (c) $\dfrac{\frac{1}{3} + \frac{2}{5}}{\frac{5}{6} - \frac{3}{5}}$

(d) $\dfrac{\frac{1}{3} + \frac{1}{4}}{\frac{7}{12}}$   (e) $\dfrac{\frac{2}{3} + \frac{6}{7}}{\frac{13}{27}}$   (f) $\dfrac{1}{\frac{1}{2} + \frac{3}{4}}$

(g) $\dfrac{\frac{3}{4} + \frac{2}{7}}{\frac{3}{4} - \frac{1}{2}}$   (h) $\dfrac{x-y}{\frac{x+y}{x}}$   (i) $\dfrac{\frac{z+y}{z-y}}{\frac{z^3}{z-y}}$

*T* 7. Solve these equations on the domain of rational numbers.

(a) $\dfrac{1}{4}x = \dfrac{5}{8}$   (b) $\dfrac{3}{4}x = \dfrac{6}{5}$

(c) $\dfrac{1}{4} = \dfrac{7}{5}x$   (d) $x \div \dfrac{5}{4} = \dfrac{3}{5}$

(e) $\dfrac{3}{4} \div x = \dfrac{1}{4}$   (f) $\dfrac{5}{8} + \dfrac{1}{4}x = \dfrac{3}{4}$

(g) $\dfrac{2x}{3} - \dfrac{1}{6} = 1\dfrac{1}{6}$   (h) $x - \dfrac{1}{3} = \dfrac{13}{15} - \dfrac{x}{5}$

8. (a) Make up two examples to show that division for rational numbers is neither commutative nor associative.

(b) Do rational numbers have a division identity? Give reasons to support your answer.

(c) Does the distributive property
$$(a + b) \div c = (a \div c) + (b \div c)$$
hold true in the system of rational numbers? Demonstrate your answer with an example.

(d) Does the distributive property of multiplication over subtraction hold true for rational numbers? Demonstrate your answer with an example.

9. Find a range to estimate each of the following.

(a) $6\dfrac{5}{8} \cdot 4\dfrac{1}{3}$   (b) $8\dfrac{3}{5} \cdot 6\dfrac{7}{8}$

(c) $8\dfrac{1}{4} \div 3\dfrac{7}{8}$   (d) $9\dfrac{5}{8} \div 3\dfrac{1}{3}$

10. Find an approximate answer for each part of Exercise 9 by rounding to whole numbers.

11. Find an approximate answer for each part of Exercise 9 by rounding one factor to the nearest $\frac{1}{2}$ (the fraction closest to $\frac{1}{2}$) and the other to the nearest whole number.

12. Without actually multiplying or dividing, decide whether each of the following is greater than or less than 2.

    (a) $\frac{26}{14} \cdot \frac{15}{17}$   (b) $\frac{19}{9} \cdot \frac{3}{4}$   (c) $4 \div 2\frac{1}{16}$

    (d) $3 \div 2\frac{1}{8}$   (e) $8\frac{1}{3} \div 4\frac{3}{50}$   (f) $4\frac{7}{8} \div 2\frac{3}{13}$

13. Regroup the following numbers in order to perform the operations mentally.

    (a) $4 \cdot \left(\frac{1}{5} \cdot \frac{1}{2}\right)$   (b) $18 \cdot \left(\frac{3}{4} \cdot \frac{1}{6}\right)$   (c) $4\frac{7}{8} \cdot 16$

    (d) $8\frac{5}{9} \cdot 18$   (e) $24\frac{1}{2} \div 3$   (f) $25\frac{1}{3} \div 5$

*For Exercises 14 through 29, set up an equation as a model that represents each written statement, solve the equation, and then follow Polya's suggestion to look back to see if your answer makes sense.*

14. Sandy's mathematics class meets 2 days a week for $\frac{5}{4}$ hours each day. How many hours is that a week?

15. Jeff is $6\frac{1}{2}$ ft. tall. Ron is $5\frac{3}{4}$ ft. tall. How many inches taller than Ron is Jeff?

16. A parking lot holds 64 cars. The parking lot is $\frac{7}{8}$ filled. How many spaces remain in the lot?

17. A tailor uses $3\frac{1}{2}$ m² of material to make a suit. A bolt contains 60 m² of material. How many suits can be made from a bolt of material?

18. A machine can desalinate $12\frac{1}{3}$ L of water per hour. How long will it take to desalinate 100 L?

19. John can mow the yard in $\frac{6}{7}$ hours. Joel can do the same job in $\frac{5}{6}$ hours. How long will it take them together to mow the yard?

20. Two-thirds of the students like math. One-half of the students who like math like science. One-fourth of the students who like math and science like spelling. What part of the students like all three subjects?

21. Hardy College reduced its faculty by $\frac{1}{8}$ one year and by $\frac{1}{5}$ the next year. If 168 faculty members were left, how many were there originally?

22. At Joe's Pizza Parlor, a slice of pizza is $\frac{1}{12}$ of a whole pizza. Joe noted that there was $\frac{2}{3}$ of a pizza left. A customer asked for 9 slices. Was Joe able to serve the customer without baking a new pizza?

23. Aaron owns $\frac{3}{7}$ of the stock in a company. His sister Jodi owns $\frac{2}{3}$ as much as he owns. What part of the stock do the two own?

24. The Boy Scouts collected $2\frac{1}{2}$ times as many pounds of paper as the Girl Scouts and the soccer league did together. The soccer league collected 50 lb. If all three collected 490 lb. of paper, how much did the Girl Scouts and the Boy Scouts collect?

25. Together, Shirley and Linda weigh 190 lb. Shirley weighs $\frac{2}{3}$ as much as Linda. How much does each weigh?

26. Mary bought $2\frac{1}{3}$ yards of ribbon. How many bows can Mary make if each bow uses $\frac{3}{8}$ ft. of ribbon? How many whole bows does she have? What part of a bow is left over?

27. Use the formula $d = rt$ (distance equals rate times time) to solve these problems.

    (a) Dennis has been bicycling at an average speed of $2\frac{1}{2}$ km per h. for $3\frac{1}{3}$ h. How far has he traveled?

    (b) Jerry has driven $20\frac{3}{4}$ h. to cover $1724\frac{1}{2}$ km. What has been his average speed per hour?

28. A merchant sold a suit for $176, thereby gaining $\frac{1}{3}$ of the suit's cost. What was the cost of the suit?

29. Joan Makenny received a raise of $\frac{1}{10}$ of her salary. What was her salary if her new salary is $39,600?

30. (a) John had $\frac{3}{4}$ lb. of candy. He gave his friend Joel $\frac{1}{3}$ of the candy by weight. What fraction of a pound of the candy did he give Joel?

    (b) Katherine has eaten $\frac{5}{6}$ of a box of candy. If she gives Mary $\frac{1}{2}$ of the remainder, how much is left?

31. Here is a recipe for pancakes. (mL represents milliliter.)

    | | |
    |---|---|
    | 325 mL flour | 3 eggs |
    | 6 mL salt | 60 mL butter |
    | 45 mL sugar | 375 mL milk |
    | 8 mL baking powder | |

    (a) Rewrite this recipe to make 4 times as many pancakes.

    (b) Rewrite to make $\frac{1}{3}$ as many pancakes.

32. Simplify each computation.

    (a) $\frac{a^3}{b^5} \cdot \frac{b^2 a}{a^4}$   (b) $\frac{a^2 + b}{a^3 c} \cdot \frac{2a^4}{ac}$

    (c) $\frac{ca^4}{b^3} \div \frac{3c^2 a^3}{b^2}$   (d) $\frac{6a^3 b}{a^3 c} \div \frac{4ac}{a^4}$

33. Suppose that prices due to inflation increase at a constant rate of $\frac{1}{20}$ each year. In 5 years, what will be the price of a

    (a) $1.60 hamburger?

    (b) $20,000 automobile?

34. If Arn's salary increases by $\frac{1}{10}$ each year for the next 3 years, what will Arn's salary be at the end of that time if his salary today is

    (a) $40,000?   (b) $80,000?

35. Find a fraction that doubles when you add the denominator to the numerator.
36. Find a fraction that triples when you add the denominator to the numerator.
37. Express each expression as a simple fraction.

(a) $\dfrac{a}{b} + \dfrac{c}{d} + \dfrac{e}{f}$   (b) $\dfrac{a}{b} \cdot \dfrac{c}{d} + \dfrac{e}{f} \cdot \dfrac{g}{h}$

(c) $\dfrac{\dfrac{a}{b} + \dfrac{c}{d}}{\dfrac{e}{f} + \dfrac{g}{h}}$   (d) $\dfrac{\dfrac{a}{b} \div \dfrac{c}{d}}{\dfrac{e}{f} \div \dfrac{g}{h}}$

38. Sales at Acme Agency increased by $\frac{3}{10}$ in 1999 over sales in 1998, decreased by $\frac{2}{10}$ in 2000 over 1999, and increased by $\frac{1}{10}$ in 2001 over 2000. The sales in 2001 were what part of the sales in 1998?
39. Multiply the following.

(a) $\left( \dfrac{1}{4} \cdot \dfrac{3}{5} \cdot \dfrac{5}{6} \right)_{\text{seven}}$

(b) $\left( 4\dfrac{9}{\text{T}} \cdot 2\text{E}\dfrac{7}{8} \right)_{\text{twelve}}$

## REVIEW EXERCISES

40. Compute $x + y$ and $x - y$ for the following pairs of fractions.

(a) $x = \dfrac{4}{7}$ and $y = \dfrac{2}{5}$

(b) $x = \dfrac{11}{9}$ and $y = \dfrac{4}{5}$

(c) $x = \dfrac{9}{10}$ and $y = \dfrac{3}{4}$

(d) $x = \dfrac{4}{3}$ and $y = \dfrac{5}{6}$

41. If $x = 2$, determine the reduced form for each of these rational numbers.

(a) $\dfrac{6}{x^2 + 4}$   (b) $\dfrac{a - 1}{2a - x}$

(c) $\dfrac{2x^2 - 3x - 2}{3 + x^2}$   (d) $\dfrac{6x}{3 + (x - 2)}$

42. Replace $x$ with a whole number so that each statement is true.

(a) $\dfrac{0}{3} + \dfrac{2}{5} = \dfrac{2}{x}$   (b) $\dfrac{1}{8} + \dfrac{3}{5} = \dfrac{29}{x}$

(c) $\dfrac{3}{5} + \dfrac{x}{4} = \dfrac{17}{20}$   (d) $\dfrac{x}{17} - \dfrac{2}{3} = \dfrac{242}{51}$

(e) $\dfrac{4}{3} + \dfrac{x}{4} = \dfrac{25}{12}$   (f) $\dfrac{5}{8} + \dfrac{7}{3} = \dfrac{71}{x}$

## LABORATORY ACTIVITIES

43. Draw visual representations (rectangular models) of the following.

(a) $\dfrac{1}{2} \cdot \dfrac{2}{3}$

(b) $\dfrac{3}{4} \cdot \dfrac{3}{4}$

44. Use fraction bars to represent the following.

(a) $\dfrac{2}{3} \div \dfrac{1}{6}$

(b) $\dfrac{3}{4} \div \dfrac{3}{8}$

## ❖ PCR Excursion ❖

45. In each of the following sequences, find a pattern. Write the next 2 terms and an expression for the $n$th term.

(a) $1, \dfrac{1}{3}, \dfrac{1}{9}, \dfrac{1}{27}, \cdots$

(b) $1, \dfrac{1}{2}, \dfrac{1}{2^2}, \dfrac{1}{2^3}, \cdots$

(c) $\dfrac{1}{4}, \dfrac{-1}{4^2}, \dfrac{1}{4^3}, \dfrac{-1}{4^4}, \cdots$

(d) $\dfrac{2}{3}, \dfrac{2}{9}, \dfrac{2}{27}, \dfrac{2}{81}, \cdots$

(e) In (a), (b), (c), and (d), each term is multiplied by a common ratio to get the term that follows it. Find the common ratio for each. (Such sequences are called *geometric sequences*.)
(f) If the common ratio of a sequence is $\frac{-1}{2}$ and the first term is 3, find the next 4 terms. Write an expression for the $n$th term.
(g) If the common ratio of a geometric sequence is $\frac{-3}{2}$ and the eighth term is $-16$, find the fourth and fifth terms.
(h) In the PCR on p. 14, we found the sum $S$ of such sequences by multiplying $S$ by the common ratio $r$ and subtracting the equation for $S$ from the equation for $rS$. Find the sum of 10 terms of (a).
(i) Using the pattern of (h), find the sum of 100 terms of (b).

# RATIO AND PROPORTION

| | |
|---|---|
| **PROBLEM** | If 10 workers assemble 30 television sets in 8 h., how many television sets will 40 workers assemble in 4 h., assuming that they work at the same rate? |
| **OVERVIEW** | In this section, we give renewed emphasis to critical thinking and problem solving. Now, however, we shift our attention to the kind of thinking involved in everyday experience in the marketplace. |
| **GOALS** | Illustrate the Number and Operations Standard, page 119.<br>Illustrate the Algebra Standard, page 4.<br>Illustrate the Connections Standard, page 147.<br>Illustrate the Representation Standard, page 50. |

 RATIO

The concepts of ratio and proportion are closely related to the rational numbers. Let's begin with *ratio*. **A ratio expresses a relation in size between two entities.**

**EXAMPLE 1**   Suppose that there are 10 boys and 15 girls in our class. The ratio of the number of boys to the number of girls is 10 to 15, written 10:15 or 10/15.

**EXAMPLE 2**   On a cold day, the teacher notices that the ratio of the number of gloves to the number of scarves is 5 to 2, written 5:2 or 5/2.

---

**DEFINITION**

**Ratio**   A *ratio* is an ordered pair of numbers, designated by either $a/b$ or $a:b$, with $b \neq 0$.

---

**EXAMPLE 3**   Express each ratio as a fraction in simplest form.

(a) 31/10:3271/100        (b) 2 1/3:3 1/4

SOLUTION   (a) $\dfrac{31/10}{3271/100} = \dfrac{31(10)}{3271} = \dfrac{310}{3271}$        (b) $\dfrac{2\frac{1}{3}}{3\frac{1}{4}} = \dfrac{\frac{7}{3}}{\frac{13}{4}} = \dfrac{7}{3} \cdot \dfrac{4}{13} = \dfrac{28}{39}$

Ratios play an important role in everyday life, as the following example indicates.

**EXAMPLE 4**

A 120-g can of Lovely Mushrooms costs 84¢. A 160-g can of Earthy Mushrooms costs $1.28. Which is the better buy?

*SOLUTION*

We compute the ratio of cost to quantity for each brand. This is called **unit pricing.** The results are given in Table 6-1. Because Lovely Mushrooms have the lower cost-to-quantity ratio, this brand is the better buy.

| TABLE 6-1 | | |
|---|---|---|
| | **Lovely** | **Earthy** |
| Cost | 84¢ | 128¢ |
| Number of grams | 120 g | 160 g |
| Ratio (cost per gram) | $\frac{84}{120} = \frac{7}{10}$ | $\frac{128}{160} = \frac{8}{10}$ |

**PRACTICE PROBLEM**

A 15-oz. box of Oaty Flakes costs $1.89, whereas a 20-oz. box costs $2.79. Which is the better buy?

*ANSWER*

The 15-oz. box

Recall that two rational numbers $\frac{a}{b}$ and $\frac{c}{d}$ are equal if $ad = bc$. The equality for two ratios is determined in this same manner.

**DEFINITION**

**Equal Ratios (Cross-Multiplication)**

If $a$, $b$, $c$, and $d$ are integers with $b \neq 0$ and $d \neq 0$, two ratios $\frac{a}{b}$ and $\frac{c}{d}$ are *equal* if and only if $ad = bc$.

**EXAMPLE 5**

In mixing fuel for a 2-cycle motor, one must mix 2 parts oil with each 8 parts gasoline. How much oil is needed to mix with 24 liters (L) of gasoline?

*SOLUTION*

We need to identify the unknown carefully. Let $x$ = the number of liters of oil necessary to mix with 24 L of gasoline. Our information is summarized in Table 6-2.

Because the ratio 2/8 must equal the ratio x/24,

$$\frac{2 \text{ parts oil}}{8 \text{ parts gas}} = \frac{x \; L \text{ oil}}{24 \; L \text{ gas}}$$

$$2 \cdot 24 = 8 \cdot x$$

$$48 = 8x$$

$$6 = x$$

Thus, 6 L of oil are needed.

| TABLE 6-2 | | |
|---|---|---|
| | *Standard Mixture* | *New Mixture* |
| Oil | 2 parts | $x$ L |
| Gasoline | 8 parts | 24 L |

| TABLE 6-3 | | |
|---|---|---|
| Number of eggs | 3 | 2 |
| Number of cups of sugar | $1\frac{1}{2}$ | $x$ |
| Number of cups of flour | 2 | $y$ |

 **PROPORTION**

This discussion leads us to the idea of proportions. Ratios that are equal form proportions.

| DEFINITION | **Proportion** | For integers $a$, $b$, $c$, and $d$ ($b \neq 0$ and $d \neq 0$), the statement that the two ratios $\frac{a}{b}$ and $\frac{c}{d}$ are equal is a *proportion*. Thus, $\frac{a}{b} = \frac{c}{d}$ is a proportion. |
|---|---|---|

**EXAMPLE 6**

Part of a recipe calls for 3 eggs, $1\frac{1}{2}$ cups of sugar, and 2 cups of flour. If we have only 2 eggs, how should the amounts of sugar and flour be changed to keep the proportions the same?

SOLUTION

Suppose that we let $x$ = the number of cups of sugar and $y$ = the number of cups of flour now needed. Again, we arrange the given information in a table. (See Table 6-3.)

The ratio of the number of eggs to the number of cups of sugar should be the same in both columns. Hence, we have

$$\frac{3 \text{ eggs}}{\frac{3}{2} \text{ cup of sugar}} = \frac{2 \text{ eggs}}{x \text{ cups of sugar}}$$

$$3x = \left(\frac{3}{2}\right)(2) = 3$$

$$x = 1$$

The ratio of the number of eggs to the number of cups of flour should also be the same. Therefore,

$$\frac{3 \text{ eggs}}{2 \text{ cups of flour}} = \frac{2 \text{ eggs}}{y \text{ cups of flour}}$$

$$3y = 4$$

$$y = \frac{4}{3}$$

The new recipe should include 1 cup of sugar and $\frac{4}{3}$ cups of flour.

**EXAMPLE 7**

A mixture of dog food is composed of 5 parts meat and 3 parts cereal. How much cereal is included in 20 kilograms (kg) of dog food?

*SOLUTION*

Each 8 units of dog food contains 3 units of cereal. Therefore, the ratio of cereal to dog food is 3 to 8 or 3/8. Let $x$ represent the amount of cereal in 20 kg of dog food. (See Table 6-4.)

Then we compute the proportion as follows:

$$\frac{3 \text{ kg cereal}}{8 \text{ kg dog food}} = \frac{x \text{ kg cereal}}{20 \text{ kg dog food}}$$

$$8x = 60$$

$$x = 7\frac{1}{2} \text{ kg of cereal}$$

**TABLE 6-4**

| | | |
|---|---|---|
| Number of kg of cereal | 3 | x |
| Number of kg of dog food | 8 | 20 |

**TABLE 6-5**

| | Start | Final |
|---|---|---|
| Number of red balls | 6 | 10 |
| Number of green balls | 3 | 3 + x |

**PRACTICE PROBLEM**

A veterinarian's office is full of cats and dogs. If there are 50 animals in the room, and if the ratio of cats to dogs is 3 to 7, find the number of cats in the office.

*ANSWER*

15

**EXAMPLE 8**

A box contains 6 red balls and 3 green balls. If 4 red balls are added, how many green balls must be added in order to keep the ratio of red balls to green balls the same?

*SOLUTION*

$x$ is the number of green balls to be added. The relevant data are given in Table 6-5.

$$\frac{6}{3} = \frac{10}{3 + x}$$

$$6(3 + x) = 3(10)$$

$$18 + 6x = 30$$

$$6x = 12$$

$$x = 2$$

Two additional green balls are needed.

**EXAMPLE 9**

In the city of Ft. Thomas, property is taxed at a rate of $20 per $1000 of assessed value. This fact can be represented by the following proportion:

$$\frac{20}{1000} = \frac{\text{tax to be paid}}{\text{assessed value of property}}$$

Hence, if a home is assessed at \$256,000, the tax is computed by solving the following proportion for $x$:

$$\frac{20}{1000} = \frac{x}{256,000}$$

$$(20)(256,000) = 1000x$$

$$x = \$5120$$

Sometimes we use the expression **directly proportional.** For instance, the distance $d$ traveled in a given time $t$ is directly proportional to the speed or rate of travel $r$:

$$d = rt$$

where $t$ is the constant of proportionality. If the rate (or speed) $r$ is fixed, the distance $d$ is directly proportional to the time $t$. Thus, if $r$ is fixed at 65 miles per h., distance varies with time as shown in Table 6-6.

**TABLE 6-6**

| $d$ | 65 | 130 | 195 | 260 | 325 |
|-----|----|-----|-----|-----|-----|
| $t$ | 1 | 2 | 3 | 4 | 5 |

Notice that for each pair the value of the ratio is 65:

$$\frac{d}{t} = \frac{260}{4} = \frac{195}{3} = \frac{130}{2} = 65$$

In many cases, ratios do not help us determine the sizes of the parts involved without additional information.

**EXAMPLE 10**

Suppose that $\frac{1}{2}$ of the girls and $\frac{1}{4}$ of the boys made A's or B's on the last test. What part of the class made A's or B's?

*SOLUTION*

We are unable to work this problem because we do not know the number of girls and boys. But suppose that there are 24 girls and 16 boys. Then the part of the class that made A's or B's is

$$\frac{(\frac{1}{2})24 + (\frac{1}{4})16}{24 + 16} = \frac{16}{40} = \frac{2}{5}$$

So 2/5 of the class made A's or B's.

The proportion $\frac{a}{b} = \frac{c}{d}$ can be written in several different ways. In fact, if $\frac{a}{b} = \frac{c}{d}$, then four other statements are also true:

$$\frac{b}{a} = \frac{d}{c} \qquad \frac{d}{b} = \frac{c}{a} \qquad \frac{b}{d} = \frac{a}{c} \qquad ad = bc$$

(flip)          (switch)          (dial)          (cross-multiply)

Thus, for example, if $\frac{3}{4} = \frac{6}{8}$, then

$$\frac{4}{3} = \frac{8}{6} \qquad \frac{8}{4} = \frac{6}{3} \qquad \frac{4}{8} = \frac{3}{6} \qquad 4 \cdot 6 = 3 \cdot 8$$

(flip)          (switch)          (dial)          (cross-multiply)

# SCALE DRAWINGS

Ratio and proportion are very useful in making scale drawings. For example, 1 centimeter (cm) on a scale drawing may represent 10 m in actual length. Then the ratio is 1/10, which in this case relates centimeters to meters.

**EXAMPLE 11**

In a particular scale drawing, $\frac{1}{2}$ cm represents 1 m.
(a) What length is represented by 2 cm?
(b) 40 m is represented on the drawing by how many centimeters?

*SOLUTION*

(a) The ratio of the scale drawing to the actual length is $\frac{\frac{1}{2}}{1} = \frac{1}{2}$. This ratio is centimeters to meters. Let $x$ m correspond to 2 cm. Then

$$\frac{2 \text{ cm}}{x \text{ m}} = \frac{1 \text{ cm}}{2 \text{ m}} \qquad \text{or} \qquad x = 4 \text{ m}$$

(b) Let $y$ cm correspond to 40 m. Then

$$\frac{y \text{ cm}}{40 \text{ m}} = \frac{1 \text{ cm}}{2 \text{ m}}$$

$$2y = 40$$

$$y = 20 \text{ cm}$$

Let's consider now the introductory problem for this section.

**EXAMPLE 12**

If 10 workers assemble 30 television sets in 8 hours, how many television sets will 40 workers assemble in 4 hours, assuming that they work at the same rate?

*SOLUTION*

Use the strategy of restating the problem to find out how many television sets the workers can assemble in 1 hour. We know that 10 workers can assemble 30/8 sets in 1 hour. Let $x$ be the number that 40 workers would assemble in 1 hour. Then

$$\frac{10}{40} = \frac{\frac{30}{8}}{x}$$

$$10x = \frac{30}{8}(40) = 150$$

$$x = 15$$

Therefore, 40 workers would assemble 15 sets in 1 hour and $4 \cdot 15 = 60$ sets in 4 hours.

# Just for Fun

Look at the ratios of numbers in the circles
on the opposite ends of the line segments.
All ratios are proportional, with one
exception. Can you find the exception?

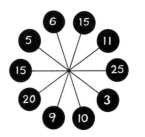

# EXERCISE SET 4

*R* 1. Express each ratio as a fraction in simplest terms.

   (a) 5 to 3    (b) $7:11$    (c) $2:7$

   (d) $21:72$    (e) $\dfrac{1}{100}:\dfrac{23}{10}$

2. Express in simplest form.

   (a) $3:27$    (b) $4\frac{1}{3}:5\frac{2}{3}$    (c) $8\frac{1}{2}:11\frac{1}{4}$

3. Solve for $x$ in each proportion.

   (a) $\dfrac{6}{x} = \dfrac{9}{12}$    (b) $\dfrac{x}{21} = \dfrac{3}{7}$    (c) $\dfrac{23}{27} = \dfrac{13}{x}$

   (d) $\dfrac{16}{x} = \dfrac{31}{4}$    (e) $\dfrac{25}{5} = \dfrac{x}{8}$    (f) $\dfrac{17}{21} = \dfrac{x}{13}$

4. A mathematics class consists of 16 women and 24 men.

   (a) What is the ratio of women to men?
   (b) What is the ratio of men to women?
   (c) What is the ratio of women to all students in the class?

5. Write fractions in simplest form representing the following ratios.

   (a) The mile-per-gallon ratio for a trip of 387 miles on 18 gal.
   (b) The student-to-faculty ratio at a college with 7527 students and a faculty of 215.
   (c) The administrator-to-employee ratio if there are 15 administrators and 85 employees.
   (d) The profit-per-lot ratio if 18 lots are sold for a total profit of $32,864.

6. Find the value of $x$ in each of the following proportions.

   (a) $x:4 = 6:20$    (b) $4:1 = 8:x$
   (c) $9:14 = x:8$    (d) $9:x = 56:30$

7. If the ratio of boys to girls in a class is $3:7$, find the following.

   (a) The ratio of girls to boys.
   (b) The ratio of boys to the total class.
   (c) The number of boys in the class if the total number in the class is 40.

8. Determine which item in each of the following pairs is the better buy. Make a table showing unit pricing.

   (a) A 900-g can of peaches for 49¢ or an 850-g can for 46¢.
   (b) A giant-size box (1500 g) of detergent for $1.51 (151 cents) or a king-size box (2400 g) for $2.47 (247 cents).

9. Determine which item in each of the following pairs is the better buy.

   (a) A 29-oz. can of beans for 98¢ or a 27-oz. can for 92¢.
   (b) A giant-size (49-oz.) box of corn flakes for $1.51 (151 cents), or a king-size box (84-oz.) for $2.47 (247 cents).
   (c) A 3-lb. canned ham for $5.49 (549 cents) or a 5-lb. canned ham for $9.49 (949 cents).
   (d) A 16-oz. can of corn for 58¢ or a 14-oz. can for 54¢.

(e) A 6-oz. box of spaghetti for 54¢ or an 8-oz. box for 60¢.

*T* 10. A recipe calls, in part, for 2 eggs, 1 teaspoon of vanilla, 3 cups of flour, and 1 3/4 cups of sugar. If one ingredient is changed to the amount given below, how must the other ingredients be changed to keep the proportions the same?

   (a) 1 egg
   (b) 2 cups of flour
   (c) $3\frac{1}{2}$ cups of sugar
   (d) 3 eggs

11. A distance of 1 cm on a map represents 50 km. What is the actual distance between two towns that are $2\frac{1}{2}$ cm apart on the map?

12. What would be the scale distance on the map in Exercise 11 for two towns 10 km apart?

13. The quantity of poison required to kill a mammal is approximately proportional to body weight. If 1/2 gram (g) will kill a rat weighing 400 g, how much poison is needed to kill a groundhog weighing 1500 g?

14. The ratio of boys to girls in a class is 8 : 7. If the class has 60 students, how many are girls?

15. A student can type 30 words per minute. How long will it take the student to type a 540-word report?

16. A blueprint has a scale of 1 in. representing 45 ft. How long is a side of a building represented by $2\frac{1}{2}$ in.?

*C* 17. A mixture of paint requires one part green, one part yellow, and four parts white. How much of each is needed to paint a house requiring 60 liters (L) of paint?

18. If it takes 10 minutes to saw a log into 3 pieces, how long does it take to saw it into 4 pieces? (Be careful!)

19. The ratio of boys to girls in two classes is 3 : 4 and 4 : 5. If the two classes have the same number of students, compare the number of boys in the two classes.

20. The electrical resistance of a wire is proportional to its length. The resistance of 4 ft. of a certain wire is $3\frac{1}{2}$ ohms. What resistance should be found in a 14-ft. piece of the same wire?

21. The capacities of two buckets having the same shapes are proportional to the cubes of the diameters of their tops. A bucket that is 10 in. across the top holds 2 gal. A second bucket shaped similarly has a diameter of 30 in. How many gallons will it hold?

22. The strength of a cable is proportional to the square of its diameter. If a $\frac{1}{2}$-in. cable will support 1200 lb., how much will a $\frac{5}{8}$-in. cable support?

23. Oxygen and hydrogen combine in a fixed ratio to form water. The ratio is about 8 parts of oxygen to 1 part of hydrogen, by weight. How many grams of oxygen are needed to form 32 g of water?

24. Two triangles are similar if and only if their corresponding sides are proportional. Show that the following triangles are similar.

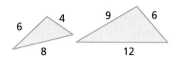

25. Find $x$ in the following figures, assuming in each part that the triangles are similar. (See Exercise 24.)

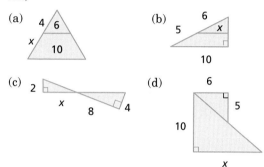

26. If Mark can paint a room in 6 hours, and Kate can paint the room in 8 hours, how long will it take the two of them to paint the room?

27. Suppose that the ratio of female students to male students in your class is 3 to 2. Find an expression for the

   (a) ratio of male students to female students.
   (b) fractional part of the class consisting of female students.
   (c) fractional part of the class consisting of male students.

28. If $a$, $b$, $c$, and $d$ are unequal to zero, prove algebraically that $a/b = c/d$ if and only if $ad = bc$.

29. In Exercise 28, prove that $(b/a) = (d/c)$.

30. In Exercise 28, prove that $(d/b) = (c/a)$.

31. In Exercise 28, prove that $(b/d) = (a/c)$.

32. If $\frac{a}{b} = \frac{c}{d}$, show that the following are true.

   (a) $\dfrac{a+b}{b} = \dfrac{c+d}{d}$

   (b) $\dfrac{d-b}{d} = \dfrac{c-a}{c}$

   (c) $\dfrac{a}{a+b} = \dfrac{c}{c+d}$

33. $(a/b) = c/d$. If $a$ and $b$ are each reduced by 3, will the resulting ratio equal $c/d$?

## REVIEW EXERCISES

• • • • • • • • • • • • • • • • • • • • • • • • • • • • • • • • • •

34. Compute $x \cdot y$ and $x \div y$ for the following pairs of fractions.

    (a) $x = \dfrac{4}{7}$ and $y = \dfrac{2}{5}$

    (b) $x = \dfrac{11}{9}$ and $y = \dfrac{4}{5}$

    (c) $x = \dfrac{9}{10}$ and $y = \dfrac{3}{4}$

    (d) $x = \dfrac{4}{3}$ and $y = \dfrac{5}{6}$

35. Solve the following equations on the domain of rational numbers.

    (a) $4x + 2 = -10$
    (b) $5y + -3 = 12$

36. A will provides that an estate is to be divided among a husband and 3 children. The husband is to receive $40,000, and the remainder is to be divided 4 ways, with the husband receiving twice each child's share. If the estate is valued at $240,000, how much does each receive?

### ∴ PCR Excursion ∴

37. Everybody thought that Dah was a genius because she found mentally the fraction answer for

$$\frac{1}{1 \cdot 2} + \frac{1}{2 \cdot 3} + \frac{1}{3 \cdot 4} + \cdots + \frac{1}{50 \cdot 51} =$$

    (a) What problem-solving strategy do you suspect that Dah used?
    (b) Find the sum of the first few partial sums such as

$$\frac{1}{1 \cdot 2}, \ \frac{1}{1 \cdot 2} + \frac{1}{2 \cdot 3}, \ \frac{1}{1 \cdot 2} + \frac{1}{2 \cdot 3} + \frac{1}{3 \cdot 4}$$

    and study your answers until you discover a pattern.
    (c) Find Dah's answer.
    (d) Write in words the pattern you discovered.
    (e) Find the sum of $n$ such fractions.
    (f) Use $\frac{3}{4}$, the sum of 3 fractions, and add to this the fourth fraction, $\frac{1}{4 \cdot 5}$. Does the sum equal what you found by the pattern?
    (g) Using your pattern, find the sum of

$$\frac{1}{1 \cdot 2} + \frac{1}{2 \cdot 3} + \frac{1}{3 \cdot 4} + \cdots + \frac{1}{(k-1)(k)}.$$

Now add $\dfrac{1}{(k)(k+1)}$. Does this verify the result you obtained in (e)?

---

# THE RATIONAL NUMBER SYSTEM

**PROBLEM**

Joyce joined the local weight-reducing club. She weighed $128\frac{1}{2}$ lb. when she joined. During the first 2 weeks, she lost $7\frac{1}{4}$ lb. During the next 4 weeks, she lost weight at a rate of $1\frac{1}{3}$ lb. per week. If she continues to lose weight at this rate, how long will it take for her to attain her goal of weighing 110 lb?

**OVERVIEW**

In this section, we investigate some of the most important properties of rational numbers. A select number of these properties are called *field properties*. Then we consider order for rational numbers, leading to the density property of rational numbers. Finally, we consider properties of equalities and properties of inequalities.

| GOALS | Illustrate the Number and Operations Standard, page 119.<br>Illustrate the Representation Standard, page 50.<br>Illustrate the Algebra Standard, page 4.<br>Illustrate the Reasoning and Proof Standard, page 23.<br>Illustrate the Communication Standard, page 78. |
|---|---|

# CHARACTERISTICS OF RATIONAL NUMBERS

A comparison of the three number systems—whole numbers, integers, and rational numbers—should start with the fact that integers have all the properties of whole numbers plus one additional property: For every integer, there exists a unique additive inverse. Similarly, the system of rational numbers has all the properties of integers plus the additional property that every rational number except $\frac{0}{1}$ (or 0) has a multiplicative inverse. Let's enumerate some of these properties.

The additive identity for rational numbers may be written as $\frac{0}{b}$ $(b \neq 0)$ because

$$\frac{a}{b} + \frac{0}{b} = \frac{0}{b} + \frac{a}{b} = \frac{a}{b}$$

Because $\frac{0}{b} = \frac{0}{1}, \frac{0}{1}$ or the corresponding integer 0 is usually used to represent the additive identity.

For each rational number $\frac{a}{b}$, we define $-\frac{a}{b}$ to be the additive inverse such that $\frac{a}{b} + -\frac{a}{b} = \frac{0}{b} = \frac{0}{1} = 0$. Observe that

$$\frac{a}{b} + \frac{-a}{b} = \frac{a + (-a)}{b} = \frac{0}{b} = 0$$

Thus, $\frac{-a}{b}$ is also an additive inverse of $\frac{a}{b}$. In like manner,

$$\frac{a}{b} + \frac{a}{-b} = \frac{a + (-a)}{b} = \frac{0}{b} = 0$$

so $\frac{a}{-b}$ is an additive inverse of $\frac{a}{b}$.

Because $-\frac{a}{b}, \frac{-a}{b}$, and $\frac{a}{-b}$ are all additive inverses for $\frac{a}{b}$, and because the additive inverse of $\frac{a}{b}$ is unique, these fractions all represent the same rational number.

| EXAMPLE 1 | The additive inverse of $\frac{2}{3}$ is $\frac{-2}{3}$ because $\frac{2}{3} + \frac{-2}{3} = \frac{2 + -2}{3} = \frac{0}{3} = 0.$ |
|---|---|

# MULTIPLICATIVE INVERSE

The existence of a multiplicative inverse $\frac{y}{x}$ for each $\frac{x}{y} \neq 0$ *almost* gives closure for the division of rationals. The one exception is that $\frac{0}{1}$ (or 0) has no multiplicative inverse.

Thus, the multiplicative inverse of $\frac{2}{3}$ is $\frac{3}{2}$, and the multiplicative inverse of $\frac{-4}{5}$ is $\frac{5}{-4}$ or $\frac{-5}{4}$ or $-\frac{5}{4}$. These numbers are sometimes called **reciprocals.** The reciprocal of $\frac{2}{3}$ is

$$\frac{1}{\frac{2}{3}} = \frac{1}{1} \cdot \frac{3}{2} = \frac{3}{2} \qquad \text{or} \qquad \frac{2}{3} \cdot \frac{3}{2} = 1$$

## CALCULATOR HINT

To find the multiplicative inverse on a calculator use the key $\boxed{x^{-1}}$.

Set $\boxed{\text{FracMode}}$ on $d/e.$ 2 $\boxed{/}$ 3 $\boxed{x^{-1}}$ $\boxed{=}$. Did you get 3/2?

Before summarizing the properties of rational numbers, let's emphasize that the integers are a subset of the rational numbers through the following correspondence:

... −4   −3   −2   −1   0   1   2   3   4 ...
   ↕    ↕    ↕    ↕    ↕    ↕    ↕    ↕    ↕
... $\frac{-4}{1}$  $\frac{-3}{1}$  $\frac{-2}{1}$  $\frac{-1}{1}$  $\frac{0}{1}$  $\frac{1}{1}$  $\frac{2}{1}$  $\frac{3}{1}$  $\frac{4}{1}$ ...

# FIELD PROPERTIES

The properties that we enumerate for the rational number system (see page 304) are called **field properties.** The rational number system is the first field (that is, the first system for which all the field properties are satisfied) that we have developed in this text. The real number system, to be discussed later, is also a field.

**EXAMPLE 2**

Review the field properties on page 304 to determine whether the number system defined by Table 6-7 (a) and (b) is a field.

(a)                                             (b)

**TABLE 6-7**

| $\oplus$ | 0 | 1 | 2 | 3 | 4 |
|---|---|---|---|---|---|
| **0** | 0 | 1 | 2 | 3 | 4 |
| **1** | 1 | 2 | 3 | 4 | 0 |
| **2** | 2 | 3 | 4 | 0 | 1 |
| **3** | 3 | 4 | 0 | 1 | 2 |
| **4** | 4 | 0 | 1 | 2 | 3 |

| $\odot$ | 0 | 1 | 2 | 3 | 4 |
|---|---|---|---|---|---|
| **0** | 0 | 0 | 0 | 0 | 0 |
| **1** | 0 | 1 | 2 | 3 | 4 |
| **2** | 0 | 2 | 4 | 1 | 3 |
| **3** | 0 | 3 | 1 | 4 | 2 |
| **4** | 0 | 4 | 3 | 2 | 1 |

*SOLUTION*

First, notice that there is closure for both addition and multiplication because the sum of any two numbers and the product of any two numbers are numbers in the set {0, 1, 2, 3, 4}. It is easy to check that commutativity holds. For example, $2 \oplus 4 = 1$ and $4 \oplus 2 = 1$. Likewise, $3 \odot 4 = 2$ and $4 \odot 3 = 2$. There are actually $5 \cdot 5 = 25$ additions and multiplications to check; however, at a glance you can see that numbers equidistant from the diagonal (upper left to lower right) are equal. The additive identity is 0, and

**System of Rational Numbers**   The system of rational numbers consists of the set $Q$ of all rational numbers and two binary operations, addition ($+$) and multiplication ($\cdot$), with the following properties for any rational numbers $\frac{a}{b}$, $\frac{c}{d}$, and $\frac{e}{f}$.

*Closure Properties*

1. $\frac{a}{b} + \frac{c}{d} = \frac{ad + bc}{bd}$ is a unique rational number in $Q$.

2. $\frac{a}{b} \cdot \frac{c}{d} = \frac{ac}{bd}$ is a unique rational number in $Q$.

*Commutative Properties*

3. $\frac{a}{b} + \frac{c}{d} = \frac{c}{d} + \frac{a}{b}$

4. $\frac{a}{b} \cdot \frac{c}{d} = \frac{c}{d} \cdot \frac{a}{b}$

*Associative Properties*

5. $\left( \frac{a}{b} + \frac{c}{d} \right) + \frac{e}{f} = \frac{a}{b} + \left( \frac{c}{d} + \frac{e}{f} \right)$

6. $\left( \frac{a}{b} \cdot \frac{c}{d} \right) \cdot \frac{e}{f} = \frac{a}{b} \cdot \left( \frac{c}{d} \cdot \frac{e}{f} \right)$

*Distributive Property of Multiplication Over Addition*

7. $\frac{a}{b} \cdot \left( \frac{c}{d} + \frac{e}{f} \right) = \frac{a}{b} \cdot \frac{c}{d} + \frac{a}{b} \cdot \frac{e}{f}$

8. $\left( \frac{c}{d} + \frac{e}{f} \right) \cdot \frac{a}{b} = \frac{c}{d} \cdot \frac{a}{b} + \frac{e}{f} \cdot \frac{a}{b}$

*Identity Elements*

9. There exists a unique rational number $\frac{0}{1}$ such that
$$\frac{a}{b} + \frac{0}{1} = \frac{0}{1} + \frac{a}{b} = \frac{a}{b}$$

10. There exists a unique rational number $\frac{1}{1}$ such that
$$\frac{a}{b} \cdot \frac{1}{1} = \frac{1}{1} \cdot \frac{a}{b} = \frac{a}{b}$$

*Inverse Elements*

11. For each $\frac{a}{b}$ there exists a unique additive inverse $\frac{-a}{b}$ such that
$$\frac{a}{b} + \frac{-a}{b} = \frac{-a}{b} + \frac{a}{b} = \frac{0}{1}$$

12. For each $\frac{a}{b} \neq 0$, there exists a unique multiplicative inverse $\frac{b}{a}$ such that
$$\frac{a}{b} \cdot \frac{b}{a} = \frac{b}{a} \cdot \frac{a}{b} = \frac{1}{1}$$

the multiplicative identity is 1. There is also an additive inverse for each element. For example, 3 is the additive inverse of 2 because $2 \oplus 3 = 0$. (Can you find the additive inverses of the other four numbers?) Likewise, each element (except 0) has a multiplicative inverse. For example, 2 is the multiplicative inverse of 3 because $3 \odot 2 = 1$. Find the multiplicative inverses of the other three numbers. Then pick some numbers and show that the associative and distributive properties hold for them. Eventually, you would check and confirm all of the $5 \cdot 5 \cdot 5 = 125$ possibilities. Therefore, this number system satisfies the properties of a field. Did you notice that this is the addition and multiplication as defined on a five-hour clock in Chapter 5?

## MENTAL ARITHMETIC

In earlier sections we used some of the field properties of rational numbers in mentally computing the answers to certain problems. As we return to mental arithmetic, we investigate why we can arrange a problem in convenient form for computation. Notice the use of field properties here as we group terms for mental arithmetic:

1. $(16 \cdot 13) \cdot \dfrac{1}{4} = (13 \cdot 16) \cdot \dfrac{1}{4}$      Commutative property of multiplication

$$= 13\left(16 \cdot \dfrac{1}{4}\right)$$      Associative property of multiplication

$$= 13 \cdot 4$$

$$= 52$$

2. $\left(\dfrac{1}{4} + \dfrac{2}{3}\right) + \dfrac{3}{4} = \left(\dfrac{2}{3} + \dfrac{1}{4}\right) + \dfrac{3}{4}$      Commutative property of addition

$$= \dfrac{2}{3} + \left(\dfrac{1}{4} + \dfrac{3}{4}\right)$$      Associative property of addition

$$= \dfrac{2}{3} + 1 = 1\dfrac{2}{3}$$

3. $3\dfrac{1}{4} \cdot 16 = \left(3 + \dfrac{1}{4}\right)16$

$$= 3 \cdot 16 + \dfrac{1}{4} \cdot 16$$      Distributive property of multiplication over addition

$$= 48 + 4 = 52$$

## DENSITY PROPERTY

We turn once again to inequalities to introduce an important property of rational numbers. First, given any two rational numbers, either the numbers are equal or one is less than the other.

| **Trichotomy Property for Rational Numbers** | If $\dfrac{a}{b}$ and $\dfrac{c}{d}$ are any rational numbers, then exactly one of the following is true: $$\dfrac{a}{b} < \dfrac{c}{d} \qquad \dfrac{a}{b} = \dfrac{c}{d} \qquad \dfrac{a}{b} > \dfrac{c}{d}$$ |
| --- | --- |

The average of two rational numbers $\frac{a}{b}$ and $\frac{c}{d}$ $\left(\text{where } \frac{a}{b} < \frac{c}{d}\right)$ is

$$\frac{\frac{a}{b} + \frac{c}{d}}{2} = \frac{ad + bc}{2bd}$$

which is a number between $\frac{a}{b}$ and $\frac{c}{d}$. For example, if $\frac{a}{b}$ is $\frac{2}{3}$ and $\frac{c}{d}$ is $\frac{3}{4}$, then $\left(\frac{1}{2}\right)\left(\frac{2}{3} + \frac{3}{4}\right) = \frac{17}{24}$. It is easy to verify that

$$\frac{2}{3} < \frac{17}{24} < \frac{3}{4}$$

Thus, another rational number exists between any two rational numbers. We express this generalization as the density property of rational numbers.

| | |
|---|---|
| **Density Property of Rational Numbers** | For any two rational numbers, it is always possible to find another rational number between them. |

In the preceding example, for instance, the average of $\frac{2}{3}$ and $\frac{17}{24}$ is

$$\frac{\frac{2}{3} + \frac{17}{24}}{2} = \frac{33}{48}$$

which is a rational number between $\frac{2}{3}$ and $\frac{17}{24}$. Thus we have found 2 rational numbers between $\frac{2}{3}$ and $\frac{3}{4}$. By the same method, we can find a rational number between $\frac{2}{3}$ and $\frac{33}{48}$. By continuing this procedure indefinitely, we can demonstrate that infinitely many rational numbers lie between any 2 rational numbers on the number line, so there exists no "next" rational number in the sense that 3 is the next positive integer after 2.

# SOLVING EQUATIONS AND INEQUALITIES

We now use the properties of rational numbers to solve equations.

| | |
|---|---|
| **Properties of Equalities** | Let $\frac{a}{b}$, $\frac{c}{d}$, and $\frac{e}{f}$ be rational numbers with $\frac{a}{b} = \frac{c}{d}$. Then<br><br>1. $\dfrac{a}{b} + \dfrac{e}{f} = \dfrac{c}{d} + \dfrac{e}{f}$<br><br>2. $\dfrac{a}{b} \cdot \dfrac{e}{f} = \dfrac{c}{d} \cdot \dfrac{e}{f}$ |

**EXAMPLE 3**  Solve $x + 4 = 17$ when the domain is the set of rational numbers.

*SOLUTION*    To solve this equation, we wish to isolate $x$ on one side of the equation. To do this, we must remove the 4 from the left-hand side. Hence, we add the additive inverse of 4, namely −4, to both sides.

| | |
|---|---|
| $x + 4 = 17$ | **Given** |
| $(x + 4) + -4 = 17 + -4$ | **If $a/b = c/d$, then $a/b + e/f = c/d + e/f$.** |
| $(x + 4) + -4 = 13$ | **Addition** |
| $x + (4 + -4) = 13$ | **Associative property of addition** |
| $x + 0 = 13$ | **Additive inverse** |
| $x = 13$ | **Additive identity** |

Thus, 13 is the solution of the equation. We say that the solution set is {13}. A solution can be checked by substituting the solution into the original equation:

$$13 + 4 = 17$$

**EXAMPLE 4**    Solve $3x + 4 = -5$ on the domain of rational numbers.

*SOLUTION*    Again, we wish to isolate $x$ on one side. To accomplish this, we add −4 to both sides. Since $3x$ remains, we multiply both sides of the equation by $\frac{1}{3}$ to undo multiplication by 3. The solution set is {−3}.

| | |
|---|---|
| $3x + 4 = -5$ | **Given** |
| $(3x + 4) + -4 = -5 + -4$ | **If $a/b = c/d$, then $a/b + e/f = c/d + e/f$.** |
| $(3x + 4) + -4 = -9$ | **Addition** |
| $3x + (4 + -4) = -9$ | **Associative property of addition** |
| $3x + 0 = -9$ | **Additive inverse** |
| $3x = -9$ | **Additive identity** |
| $\left(\dfrac{1}{3}\right)(3x) = \left(\dfrac{1}{3}\right)(-9)$ | **If $a/b = c/d$, then $(a/b)(e/f) = (c/d)(e/f)$.** |
| $\left(\dfrac{1}{3} \cdot 3\right)x = \left(\dfrac{1}{3} \cdot -9\right)$ | **Associative property of multiplication** |
| $1 \cdot x = -3$ | **Multiplication** |
| $x = -3$ | **Multiplicative identity** |

Similar properties are used to solve inequalities.

**Properties of Inequalities**

Let $\dfrac{a}{b}, \dfrac{c}{d}$, and $\dfrac{e}{f}$ be rational numbers with $\dfrac{a}{b} < \dfrac{c}{d}$. Then

1. $\dfrac{a}{b} + \dfrac{e}{f} < \dfrac{c}{d} + \dfrac{e}{f}$

2. $\dfrac{a}{b} \cdot \dfrac{e}{f} < \dfrac{c}{d} \cdot \dfrac{e}{f}$    if    $\dfrac{e}{f} > 0$

3. $\dfrac{a}{b} \cdot \dfrac{e}{f} > \dfrac{c}{d} \cdot \dfrac{e}{f}$    if    $\dfrac{e}{f} < 0$

**EXAMPLE 5**

Solve $x + 3 < -5$ when the domain is rational numbers.

SOLUTION

To isolate $x$ on one side, add $-3$ to both sides of the inequality.

$$x + 3 < -5 \qquad \text{Given}$$

$$(x + 3) + -3 < -5 + -3 \qquad \text{If } a/b < c/d, \text{ then } a/b + e/f < c/d + e/f.$$

$$(x + 3) + -3 < -8 \qquad \text{Addition}$$

$$x + (3 + -3) < -8 \qquad \text{Associative property of addition}$$

$$x + 0 < -8 \qquad \text{Additive inverse}$$

$$x < -8 \qquad \text{Additive identity}$$

The solution set is $\{x \,|\, x \text{ is a rational number less than } -8\}$.

**EXAMPLE 6**

Solve $\frac{-2x}{3} + -3 \leq 5$ on the rational numbers.

SOLUTION

To isolate $x$, we add 3 to both sides of the inequality. Because $\frac{-2x}{3}$ remains, we multiply by $-\frac{3}{2}$ to undo multiplication by $\frac{-2}{3}$. Remember that multiplying by a negative number reverses the inequality.

$$\frac{-2x}{3} + -3 \leq 5 \qquad \text{Given}$$

$$\left(\frac{-2x}{3} + -3\right) + 3 \leq 5 + 3 \qquad \text{If } a/b < c/d, \text{ then } a/b + e/f < c/d + e/f.$$

$$\frac{-2x}{3} + (-3 + 3) \leq 5 + 3 \qquad \text{Associative property of addition}$$

$$\frac{-2x}{3} \leq 8 \qquad \text{Addition}$$

$$\left(-\frac{3}{2}\right)\left(\frac{-2x}{3}\right) \geq \left(-\frac{3}{2}\right)(8) \qquad \text{If } a/b < c/d, (a/b)(e/f) > (c/d)(e/f) \text{ if } \frac{e}{f} < 0.$$

$$x \geq -12 \qquad \text{Associative property of multiplication and multiplication}$$

$\{x \,|\, x \text{ is a rational number} \geq -12\}$

Now let's solve the introductory problem.

**EXAMPLE 7**

Joyce joined the local weight-watching club. She weighed $128\frac{1}{2}$ lb when she joined. During the first 2 weeks she lost $7\frac{1}{4}$ lb. For the next 4 weeks, she lost weight at a rate of $1\frac{1}{3}$ lb per week. If she continues to lose weight at this rate, how long will it take for her to attain her goal of weighing 110 lb?

SOLUTION

At the start, Joyce weighed $128\frac{1}{2}$ lb. She has since lost $7\frac{1}{4} + 4(1\frac{1}{3}) = 12\frac{7}{12}$ lb. Therefore, her weight at the end of 6 weeks is

$$128\frac{1}{2} - 12\frac{7}{12} = 115\frac{11}{12} \text{ lb}$$

If she continues to lose weight at a rate of $1\frac{1}{3}$ lb per week for $x$ additional weeks, her new weight will be

$$115\frac{11}{12} - x\left(1\frac{1}{3}\right)$$

Thus, if her goal is a weight of 110 lb, we have the equation

$$115\frac{11}{12} - x\left(\frac{4}{3}\right) = 110$$

$$x\left(\frac{4}{3}\right) = \frac{71}{12}$$

$$x = \frac{71}{12} \cdot \frac{3}{4}$$

$$= \frac{71}{16} \quad \text{or} \quad 4\frac{7}{16} \text{ weeks}$$

## Just for Fun

Professor Abstract gave his report on the grades of the last test in the following manner: One-fifth made A's, one-fourth made B's, one-third made C's, one-sixth made D's, and 12 made F's. How many are in this class?

A = ⅕   D = ⅙
B = ¼   F = 12
C = ⅓

# EXERCISE SET 5

**R** 1. Insert > , = , or < to express the correct relationship between the following pairs of numbers.

(a) $\dfrac{9}{10}, \dfrac{7}{11}$   (b) $\dfrac{33}{9}, \dfrac{21}{7}$   (c) $\dfrac{5}{7}, \dfrac{2}{5}$

(d) $\dfrac{29}{3}, \dfrac{16}{2}$   (e) $\dfrac{17}{51}, \dfrac{6}{7}$   (f) $\dfrac{22}{73}, \dfrac{25}{71}$

2. Arrange the following sets of rational numbers in increasing order, such as $\frac{1}{2} < 2 < \frac{7}{2}$.

(a) $\dfrac{71}{100}, \dfrac{3}{2}, \dfrac{23}{30}$   (b) $\dfrac{9}{2}, 4, \dfrac{165}{41}$

(c) $\dfrac{2}{3}, \dfrac{11}{18}, \dfrac{16}{27}, \dfrac{67}{100}$   (d) $\dfrac{25}{28}, \dfrac{27}{20}, \dfrac{14}{16}, \dfrac{79}{90}$

(e) $\dfrac{22}{7}, \dfrac{7}{2}, \dfrac{10}{3}, \dfrac{156}{50}$   (f) $\dfrac{51}{95}, \dfrac{19}{36}, \dfrac{17}{30}, \dfrac{14}{29}$

3. (a) For what integral values of $x$ will $\frac{x}{8}$ be less than $\frac{1}{4}$?

   (b) For what integral values of $x$ will $\frac{x}{9}$ be greater than $\frac{1}{3}$?

4. Classify as true or as false.

   (a) Every rational number is an integer.
   (b) Every whole number is a rational number.
   (c) The rationals are a subset of the integers.
   (d) Some whole numbers are not rational numbers.

5. Let $W$ be the whole numbers, $F$ the nonnegative rational numbers, $Q$ all the rationals, $I$ the integers, and $N$ the negative integers. Which of these sets have the following properties?

   (a) $-2$ is a member of the set.
   (b) $3/-7$ is a member of the set.
   (c) The set has an additive inverse for each element.
   (d) The set has a multiplicative inverse for each element.
   (e) The set is closed under addition.
   (f) The set is closed under multiplication.
   (g) The set is closed under subtraction.
   (h) The set is closed under division.

6. State the property that is used in each of (a)–(g).

   (a) $\frac{3}{7} + \frac{9}{14} < \frac{4}{7} + \frac{9}{14}$

   (b) $\frac{3}{7} + \left(\frac{3}{8} + \frac{1}{4}\right) = \frac{3}{7} + \left(\frac{1}{4} + \frac{3}{8}\right)$

   (c) $\frac{5}{13} \cdot \frac{4}{9} < \frac{6}{13} \cdot \frac{4}{9}$

   (d) $\frac{1}{18} + \left(\frac{3}{11} + \frac{5}{9}\right) = \left(\frac{1}{18} + \frac{3}{11}\right) + \frac{5}{9}$

   (e) $\frac{5}{9} \cdot \frac{-3}{4} > \frac{6}{9} \cdot \frac{-3}{4}$

   (f) $\frac{2}{3} \cdot \frac{1}{4} + \frac{3}{11} \cdot \frac{1}{4} = \left(\frac{2}{3} + \frac{3}{11}\right) \cdot \frac{1}{4}$

   (g) $\frac{5}{9} \cdot \left(\frac{1}{9} \cdot \frac{1}{7}\right) = \frac{5}{9}\left(\frac{1}{7} \cdot \frac{1}{9}\right)$

7. Given the first relationship in each item that follows, is the conclusion necessarily true?

   (a) If $\frac{6}{7} + \frac{1}{3} < \frac{29}{23}$, then $\frac{6}{7} < \frac{29}{23} - \frac{1}{3}$.

   (b) If $\frac{11}{12} > \frac{2}{5}$, then $\frac{-11}{12} > \frac{-2}{5}$.

   (c) If $\frac{-63}{87} < \frac{1}{3}$, and $\frac{1}{3} \le \frac{23}{69}$, then $\frac{-63}{87} < \frac{23}{69}$.

   (d) If $\frac{a}{b} < 0$ and $\frac{c}{d} > 0$, then $\frac{a}{b} \cdot \frac{c}{d} > 0$.

   (e) If $\frac{a}{b} < \frac{c}{d}$ and $x < 0$, then $\left(\frac{a}{b}\right)x < \left(\frac{c}{d}\right)x$.

8. Using the average, or midpoint, find 3 rational numbers between $\frac{1}{6}$ and $\frac{3}{4}$.

9. If $a/b$ and $c/d$ are unequal rational numbers, is the rational number $(a + c)/(b + d)$ between them?

10. Find the solution set in each case. The domain for the variable is the set of rational numbers.

    (a) $2y - 3 = 11$
    (b) $5y + 4 < -6$
    (c) $3z + (-5) \ge 7$
    (d) $2x + -2 = 20$
    (e) $6x = 4x + 4$
    (f) $2m + 3 = m + 7$

    (g) $\frac{x}{3} + (-2) < -5$

    (h) $4x + (-3) < -7$

    (i) $\frac{2x}{3} - \frac{1}{4} \le \frac{1}{6}$

    (j) $\frac{-3x}{4} + \frac{x}{7} \le 2$

11. (a) Find the smallest rational number greater than 0.
    (b) Find the largest rational number less than 2.

*T* 12. Solve each of the following equations on the domain of rational numbers and give reasons for performing each step.

    (a) $x + 8 = 15$    (b) $t - 15 = 6$
    (c) $2x = 12$        (d) $2x + 4 = 6$
    (e) $y + 7 = 8$      (f) $3x + 2 = 7$
    (g) $5x + 7 = 12$    (h) $2r + 3 = 5$

13. Rearrange the following problems for mental arithmetic, and state the field property of rational numbers used in each step.

    (a) $(18 \cdot 16) \cdot \left(\frac{1}{3}\right)$    (b) $\left(\frac{1}{3} \cdot 5\right) \cdot 9$
    (c) $\left(\frac{1}{3} + \frac{3}{4}\right) + \frac{2}{3}$    (d) $\left(\frac{5}{6} + \frac{3}{8}\right) + \frac{1}{6}$
    (e) $2\frac{5}{8} \cdot 16$    (f) $4\frac{1}{3} \cdot 9$

14. Solve each of the sets of inequalities on the domain of rational numbers and give reasons for performing each step.

    (a) $x + 2 < 6$
    (b) $-x + 1 \ge 3$
    (c) $-2x - 4 < -6$
    (d) $3x + 5 \le -1$

15. Which field properties of rational numbers fail if you remove the following from the rational numbers?

    (a) 0    (b) 1    (c) $-5$    (d) $\frac{1}{4}$

*For Exercises 16 through 20, set up an equation as a model for the problem. Then solve the equation.*

16. A man leaves $\frac{1}{2}$ of his estate to his wife, $\frac{1}{8}$ plus $1000 to each of 3 children, and the remainder to his pet dog. If he has a $1,000,000 estate, how much does the dog get?

17. Mrs. Smith's electric bill was 6 times Mrs. Jones's bill. The two bills totaled $168. What was the cost of each woman's bill?

18. The difference between two integers is 21. The larger integer is equal to twice the smaller integer plus 20. What are the 2 integers?

19. Three times Ralph's weight added to 54 kg is equal to 300 kg. How much does Ralph weigh?

20. Jerry's age is 4 times the age Diane will be in 2 years. The sum of their ages is 78. How old are Jerry and Diane?

*C* 21. Prove that the positive rational numbers with operations + and · do not form a field.

22. Prove that if $p/q < r/s$, then

(a) $\dfrac{p}{q} + \dfrac{t}{v} < \dfrac{r}{s} + \dfrac{t}{v}$.

(b) $\dfrac{p}{q} \cdot \dfrac{t}{v} < \dfrac{r}{s} \cdot \dfrac{t}{v}$   if   $\dfrac{t}{v} > 0$.

(c) $\dfrac{p}{q} \cdot \dfrac{t}{v} > \dfrac{r}{s} \cdot \dfrac{t}{v}$   if   $\dfrac{t}{v} < 0$.

23. Which of these systems are fields?

(a)

| △ | 0 | 1 |
|---|---|---|
| 0 | 0 | 1 |
| 1 | 1 | 0 |

| △ | 0 | 1 |
|---|---|---|
| 0 | 0 | 0 |
| 1 | 0 | 1 |

(b)

| ⊞ | 0 | 1 | 2 |
|---|---|---|---|
| 0 | 0 | 1 | 2 |
| 1 | 1 | 2 | 0 |
| 2 | 2 | 0 | 1 |

| ⊡ | 0 | 1 | 2 |
|---|---|---|---|
| 0 | 0 | 0 | 0 |
| 1 | 0 | 1 | 2 |
| 2 | 0 | 2 | 1 |

24. (a) Prove that if $p/q + t/v < r/s + t/v$, then $p/q < r/s$.

(b) Prove that if $p/q \cdot t/v < r/s \cdot t/v$ where $t/v > 0$, then $p/q < r/s$.

## REVIEW EXERCISES
• • • • • • • • • • • • • • • • • • • • • • • • • • • • • • •

25. Two backpackers start toward each other from 2 points on the Appalachian Trail. Initially they are 12 km apart. If one walks 4 km per h. and the other walks 2 km per h., how long will it be until they meet?

26. Perform each of the following computations.

(a) $\dfrac{3}{16} \div \dfrac{4}{6}$    (b) $4\frac{1}{3} \cdot 5\frac{2}{5}$

(c) $(6\frac{1}{4} - 2\frac{1}{3}) \cdot 7\frac{1}{4}$    (d) $(\frac{7}{8} \div \frac{-1}{4}) \div \frac{2}{3}$

27. Simplify $\dfrac{2/3 - 1/4}{5/6 - 7/8}$

28. Find the rational-number solution for each of the following:

(a) $\dfrac{x}{3} = \dfrac{-18}{54}$

(b) $2x + {-4} = 16$

(c) $-3x - 7 = 41$

(d) $4x - 3 = 5$

(e) $\dfrac{x}{100} = \dfrac{480}{54}$

(f) $-3x + 7 = -5$

### ∴ PCR Excursion ∴

29. Consider $n \times m$ rectangles consisting of $nm$ unit squares where $n$ and $m$ are relatively prime. Draw 1 diagonal, and count the number of unit squares that the diagonal passes through.

3 × 4 Rectangle

Diagonal passes through 6 squares.

(a) First draw a number of $2 \times m$ rectangles where g.c.d. $(2, m) = 1$ and count the squares cut by a diagonal. Generalize for a $2 \times m$ rectangle.

(b) Do the same for $3 \times m$ rectangles.

(c) Do the same for $4 \times m$ rectangles.

(d) Do the same for $5 \times m$ rectangles.

(e) Formulate a rule for determining the number of unit squares the diagonal passes through for a $n \times m$ rectangle where $n$ and $m$ are relatively prime.

(f) How many unit squares are cut by a diagonal of a $31 \times 43$ rectangle?

# SOLUTION TO INTRODUCTORY PROBLEM

UNDERSTANDING THE PROBLEM. The number $N$ of cars that the salesman had at the beginning of the week must be such that the cars sold each day (or the cars remaining on the lot at the end of each day) is a whole number. Second, $N$ must be such that the number of cars in the lot on Friday is 11.

DEVISING A PLAN. The plan for solving this problem is to express the number of cars on the lot at the end of each day in terms of $N$. In the process, we shall look for a pattern to save time in computation. Then we shall set the expression for the number of cars on the lot on Friday evening equal to 11 and solve the equation.

CARRYING OUT THE PLAN. Number sold on Monday:

$$\frac{N}{2} + \frac{1}{2}$$

Number in the lot at the end of the day on Monday:

$$N - \left(\frac{N}{2} + \frac{1}{2}\right) = \frac{N}{2} - \frac{1}{2}$$

Number sold on Tuesday:

$$\frac{1}{3} \cdot \left(\frac{N}{2} - \frac{1}{2}\right) + \frac{1}{3}$$

Number in the lot at the end of the day on Tuesday:

$$\frac{N}{2} - \frac{1}{2} - \frac{1}{3}\left(\frac{N}{2} - \frac{1}{2}\right) - \frac{1}{3} = \frac{N}{3} - \frac{2}{3}$$

Do you suppose we have discovered a pattern? Check to see if the number in the lot at end of the day on Wednesday is

$$\frac{N}{4} - \frac{3}{4}$$

If this is true, then on Friday

$$\frac{N}{5} - \frac{4}{5} = 11 \quad \text{or} \quad \frac{N}{5} - \frac{4}{5} = \frac{55}{5} \quad \text{or} \quad N = 59$$

LOOKING BACK. Is $\dfrac{N}{2} - \dfrac{1}{2}$

a whole number of cars? Yes,

$$\frac{59}{2} - \frac{1}{2} = 29$$

Is $\dfrac{N}{3} - \dfrac{2}{3}$

a whole number of cars? Yes,

$$\frac{59}{3} - \frac{2}{3} = 19$$

Is $\dfrac{N}{4} - \dfrac{3}{4}$

a whole number of cars? Yes,

$$\frac{59}{4} - \frac{3}{4} = 14$$

Finally,

$$14 - \left(\frac{1}{5} \text{ of } 14 + \frac{1}{5}\right) = 11$$

# SUMMARY AND REVIEW

## THE SET OF RATIONAL NUMBERS

1. Fractions such as $a/b$ are used to represent the following:
   (a) a part of a whole    (b) a relative amount
   (c) a division problem or a solution to a multiplication problem
   (d) a comparison between two amounts

2. Fractions represent numbers called *rational numbers*.

3. Two fractions are equivalent if they represent the same (or equal) rational number(s).

4. Fundamental Law of Fractions: $a/b$ is equivalent to $ac/bc$ (or rational numbers $a/b$ and $ac/bc$ are equal).

5. The fraction $a/b$ is in simplest form when $b > 0$ and the largest natural number that will divide both $a$ and $b$ is 1.

6. The fractions $a/b$ and $c/d$ ($b, d \neq 0$) are equivalent (or the rational numbers represented by $a/b$ and $c/d$ are equal) if and only if $ad = bc$.

## ADDITION AND SUBTRACTION OF RATIONAL NUMBERS

1. If $a/b$ and $c/b$ are rational numbers (note equal denominators), then

$$\frac{a}{b} + \frac{c}{b} = \frac{a+c}{b}$$

2. If $a/b$ and $c/d$ are any two rational numbers, then

$$\frac{a}{b} + \frac{c}{d} = \frac{ad+bc}{bd}$$

3. The least common denominator of two or more denominators of fractions is the l.c.m. of the denominators.

4. An *improper fraction* is a fraction with positive denominator in which the absolute value of the numerator is larger than or equal to the denominator. A fraction in which the absolute value of the numerator is smaller than the denominator is called a *proper fraction*. Improper fractions can be converted to integers or to mixed numbers such as $3\frac{1}{4}$.

5. Subtraction of rational numbers: $\dfrac{a}{b} - \dfrac{c}{d} = \dfrac{ad-bc}{bd}$

## MULTIPLICATION AND DIVISION OF RATIONAL NUMBERS

1. $\dfrac{a}{b} \cdot \dfrac{c}{d} = \dfrac{ac}{bd}$

2. $\left(\dfrac{p}{q}\right) \div \left(\dfrac{r}{s}\right) = \left(\dfrac{p}{q}\right) \cdot \left(\dfrac{s}{r}\right)$, where $\dfrac{s}{r}$ is the multiplicative inverse of $\dfrac{r}{s}$.

## PROPERTIES OF RATIONAL NUMBERS WITH OPERATIONS + AND · (CALLED FIELD PROPERTIES)

*Closure properties*

1. $\dfrac{a}{b} + \dfrac{c}{d} = \dfrac{ad+bc}{bd}$ is a unique rational number.

2. $\dfrac{a}{b} \cdot \dfrac{c}{d} = \dfrac{ac}{bd}$ is a unique rational number.

*Commutative properties*

3. $\dfrac{a}{b} + \dfrac{c}{d} = \dfrac{c}{d} + \dfrac{a}{b}$    4. $\dfrac{a}{b} \cdot \dfrac{c}{d} = \dfrac{c}{d} \cdot \dfrac{a}{b}$

*Associative properties*

5. $\left(\dfrac{a}{b} + \dfrac{c}{d}\right) + \dfrac{e}{f} = \dfrac{a}{b} + \left(\dfrac{c}{d} + \dfrac{e}{f}\right)$

6. $\left(\dfrac{a}{b} \cdot \dfrac{c}{d}\right) \cdot \dfrac{e}{f} = \dfrac{a}{b} \cdot \left(\dfrac{c}{d} \cdot \dfrac{e}{f}\right)$

*Distributive property of multiplication over addition*

7. $\dfrac{a}{b} \cdot \left(\dfrac{c}{d} + \dfrac{e}{f}\right) = \dfrac{a}{b} \cdot \dfrac{c}{d} + \dfrac{a}{b} \cdot \dfrac{e}{f}$

8. $\left(\dfrac{c}{d} + \dfrac{e}{f}\right) \cdot \dfrac{a}{b} = \dfrac{c}{d} \cdot \dfrac{a}{b} + \dfrac{e}{f} \cdot \dfrac{a}{b}$

*Identity elements*

9. There exists a unique rational number $\frac{0}{1}$ (or 0) such that

$$\frac{a}{b} + \frac{0}{1} = \frac{0}{1} + \frac{a}{b} = \frac{a}{b}$$

10. There exists a unique rational number $\frac{1}{1}$ (or 1) such that

$$\frac{a}{b} \cdot \frac{1}{1} = \frac{1}{1} \cdot \frac{a}{b} = \frac{a}{b}$$

*Inverse elements*

11. For each $a/b$, there exists a unique additive inverse $-a/b$ such that

$$\frac{a}{b} + \frac{-a}{b} = \frac{-a}{b} + \frac{a}{b} = \frac{0}{1}$$

and a unique multiplicative inverse $b/a$ $(a/b \neq 0)$ such that

$$\frac{a}{b} \cdot \frac{b}{a} = 1$$

## RATIO AND PROPORTION

1. *Ratio* is a name used for the quotient of two numbers when this quotient is used to compare the sizes of two quantities.

2. The ratio of cost to quantity is called the *unit price ratio* and is used in comparison shopping.

3. Two ratios $\frac{a}{b}$ and $\frac{c}{d}$ are equal if and only if $ad = bc$.

4. A statement that two ratios are equal is called a *proportion*.

# CHAPTER TEST

1. (a) Draw a diagram to represent $\frac{3}{8}$.
   (b) Write three fractions equivalent to $\frac{2}{3}$.

2. (a) Change $18\frac{2}{3}$ to an improper fraction.
   (b) Change $\frac{326}{6}$ to a mixed number in lowest terms.

3. Subtract and simplify your answer: $4\frac{5}{6} - 2\frac{7}{8}$

4. Using a least common denominator, find

$$\frac{3}{8} - \frac{1}{6} + \frac{11}{18} - \frac{7}{24}$$

5. Find the additive inverse of $\frac{3}{5}$.

6. Find the multiplicative inverse of each of the following.

   (a) $\frac{5}{7}$   (b) 15   (c) 0   (d) $\frac{1}{3} - \frac{1}{4}$

7. Divide and leave answer in simplest form: $3\frac{1}{2} \div 5\frac{1}{4}$

8. Multiply the following, and leave the answer as a mixed number in simplest form: $5\frac{1}{2} \cdot 6\frac{7}{22}$

9. Solve for $x$.

   (a) $2 : 16 = x : 72$   (b) $\frac{16}{x} = \frac{10}{4}$

10. Evaluate:

$$\left(-4\frac{1}{6} + 6\frac{1}{3}\right) \div \frac{-9}{4}$$

11. Pam is holding 55¢ in her hand. If she has twice as many nickels as quarters and one more dime than quarters, how many coins does she hold? How many coins of each type does she have?

12. Reduce the following to simplest form.

   (a) $\frac{2x^3y}{x^4y^2}$   (b) $\frac{a^2 + ax}{a + x}$

13. Express as a fraction in simplest form:

$$\frac{\frac{3}{4} + \frac{2}{7}}{\frac{3}{4} - \frac{1}{2}}$$

14. The ratio of girls to boys in Joy's class is 3 to 2. Of the 30 students in the class, how many are girls?

15. Canoeing enthusiasts find that they can paddle upstream at 4 km per h. and downstream at 12 km per h. How long can they paddle upstream if they must return to the same spot 4 h. after departing?

16. Solve on the domain of rational numbers.

   (a) $\frac{-3x}{2} + 6 = \frac{1}{2}x - \frac{1}{3}$

   (b) $\frac{-8}{5} - \frac{3x}{10} = \frac{-3}{5}x + \frac{7}{10}$

17. List one field property that is
   (a) satisfied on the rationals but not on the integers.
   (b) satisfied on the integers but not on the whole numbers.
   (c) satisfied on the rationals but not on the whole numbers.
   (d) satisfied on the counting numbers, whole numbers, integers, and rational numbers.

18. Suppose that addition for rational numbers is defined as $\frac{x}{y} \oplus \frac{w}{z} = \frac{x + w}{y + z}$.
   (a) Is there closure for $\oplus$?
   (b) Is $\oplus$ commutative?
   (c) If there is an additive identity, find it.
   (d) If there is an additive inverse for $\frac{a}{b}$, find it.

# FROM RATIONAL NUMBERS TO REAL NUMBERS

# 7

Find the numbers to complete the magic square if the sum of the rows, columns, and diagonals is the same.

| | | |
|---|---|---|
| | | $1.\overline{6}$ |
| | $2.08\overline{3}$ | 1.25 |
| 2.5 | | |

The numbers we use most of the time are called real numbers. In this chapter, we extend the rational numbers to real numbers. To do this, we first write rational numbers as decimals. Next rational numbers are extended to real numbers by writing rational numbers as infinite decimals and then defining an infinite decimal that does not represent a rational number as an irrational number. The real numbers consist of the rational numbers and the irrational numbers. With this extension we will have completed our extension of number systems from whole numbers to real numbers. Problem solving strategies include:

- Use a table. (throughout Section 3)  - Use a variable. (338, throughout Section 3)  - Guess, test, and revise. (315, 327)  - Look for a pattern. (318, 319, 330, 338, 360, 363)  - Use an equation. (throughout Section 3)  - Draw a diagram. (throughout Section 3)  - Use indirect reasoning. (352, 354, 355)  - Validate all possibilities. (315)

# INTRODUCTION TO DECIMALS

| PROBLEM | Two trains start out 53.6 km apart. One train travels 47.3 km away from the other train. The second train travels 61.4 km toward the first train. How far apart are the 2 trains? |
| --- | --- |
| OVERVIEW | The preceding problem involves decimals. Why study decimals? Although considerably more useful than whole numbers alone, fractions are not always adequate, and they often lead to awkward manipulative procedures. You will learn in Section 2 that all fractions can be expressed as decimals, but in this section you will express a subset of fractions as terminating decimals. |
| GOALS | Illustrate the Number and Operations Standard, page 119.*<br>Illustrate the Algebra Standard, page 4.*<br>Illustrate the Representation Standard, page 50.*<br><br>* The complete statement of the standard is given on this page of the book. |

## DECIMALS

Decimal representation of rational numbers is the natural extension of the base ten place-value representation of whole numbers. Decimal fractions are constructed by placing a dot, called a **decimal point,** after the units digit and letting the digits to the right of the dot denote the number of tenths, hundredths, thousandths, and so on. If there is no whole number part in a given numeral, a 0 is usually placed before the decimal point (for example, 0.24).

The digits of a decimal fraction are named as indicated in Figure 7-1. There, the number being described is 2462.3185.

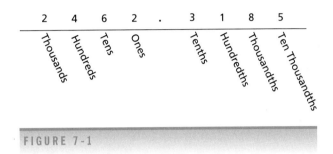

FIGURE 7-1

Decimals can be illustrated graphically by the color shading shown in Figure 7-2.

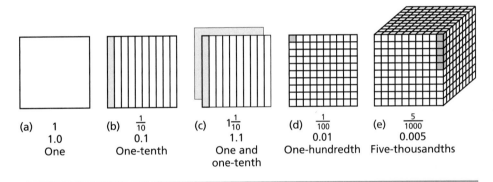

FIGURE 7-2

Notice that each large square in Figure 7-2 (a-d) represents 1, as does the cube in 7-2 (e). Because there are 10 strips in each square of Figure 7.2 (b-d), each strip represents 0.1. In Figure 7-2 (c), the diagram represents 1.1 (one square and one strip out of ten). In 7-2 (d), each small square is $\frac{1}{100}$ of the total number of squares; and in 7-2 (e), each block is $\frac{1}{1000}$ of the total cube.

**EXAMPLE 1**    Mark purchased 2.61 pounds of hamburger meat. Illustrate the number of pounds with decimal squares.

*SOLUTION*    The answer is given in Figure 7-3.

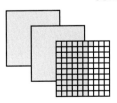

If the denominator of a fraction is 10, there is one digit to the right of the decimal point in the decimal fraction. If the denominator is 100, there are two digits to the right of the decimal point. If the denominator is 1000, there are three; and so on.

FIGURE 7-3

EXAMPLE 2

$$\frac{31}{100} = 0.31 \qquad \frac{191}{100} = 1.91 \qquad \frac{1654}{100} = 16.54$$

$$\frac{426}{1000} = 0.426 \qquad \frac{1833}{1000} = 1.833 \qquad \frac{17{,}927}{1000} = 17.927$$

# EXPONENTIAL NOTATION

Since decimals involve a power of 10 in the denominator of a fraction, we need to extend to fractions the discussion of exponents (found in Chapter 4).

EXAMPLE 3

$$\frac{6^5}{6^3} = \frac{6 \cdot 6 \cdot 6 \cdot 6 \cdot 6}{6 \cdot 6 \cdot 6} = 6^2 = 6^{5-3}$$

$$\frac{a^2}{a^7} = \frac{\cancel{a} \cdot \cancel{a}}{\cancel{a} \cdot \cancel{a} \cdot a \cdot a \cdot a \cdot a \cdot a} = \frac{1}{a^5} = \frac{1}{a^{7-2}}$$

In general,

$$\frac{a^m}{a^n} = a^{m-n} \qquad \text{or} \qquad \frac{a^m}{a^n} = \frac{1}{a^{n-m}} \qquad (a \neq 0)$$

The expression on the left is usually used when $m > n$; the one on the right is used when $m < n$.

Remember that

$$(a^m)^n = a^{mn} \qquad \text{and} \qquad \left(\frac{a}{b}\right)^m = \frac{a^m}{b^m} \qquad (b \neq 0)$$

Again, because any nonzero number divided by itself is 1,

$$\frac{a^3}{a^3} = 1 \qquad (a \neq 0)$$

But by the pattern we observed earlier,

$$\frac{a^3}{a^3} = a^{3-3} = a^0$$

Therefore, $a^0$ is defined to be 1.

EXAMPLE 4

$10^0 = 1; \qquad 4^0 = 1; \qquad n^0 = 1 \qquad (n \neq 0)$

The same pattern can be followed to infer that

$$\frac{1}{3^4} = \frac{3^0}{3^4} = 3^{0-4} = 3^{-4}$$

Thus,

$$3^{-4} = \frac{1}{3^4}$$

In general, $a^{-m}$ is defined to be $1/a^m$, when $a \neq 0$.

| | |
|---|---|
| **Properties of Exponents** | If $a$ and $b$ are any numbers unequal to 0, and $m$ and $n$ are any exponents, then the following properties hold. |

$$a^m \cdot a^n = a^{m+n} \qquad\qquad a^{-m} = \frac{1}{a^m}$$

$$\frac{a^m}{a^n} = a^{m-n} = \frac{1}{a^{n-m}} \qquad\qquad (a^m)^n = a^{m \cdot n}$$

$$a^0 = 1 \qquad\qquad (ab)^m = a^m b^m$$

**EXAMPLE 5**

$$10^8 \cdot 10^{-5} = 10^8 \cdot \frac{1}{10^5} = 10^{8-5} = 10^3$$

$$10^{-6} \cdot 10^{-3} = \frac{1}{10^6} \cdot \frac{1}{10^3} = \frac{1}{10^6 \cdot 10^3} = \frac{1}{10^9} = 10^{-9}$$

**EXAMPLE 6**

$$\left(\frac{2}{3}\right)^2 = \frac{2^2}{3^2} = \frac{4}{9}$$

$$\left(\frac{1}{10}\right)^2 = \frac{1^2}{10^2} = \frac{1}{100}$$

$$\left(\frac{1}{10}\right)^3 = \frac{1^3}{10^3} = \frac{1}{1000}$$

# FRACTIONS AND DECIMALS

The following example illustrates important relationships between fractions and decimals.

**EXAMPLE 7**

$$0.46 = \frac{46}{100} = \frac{4(10) + 6(1)}{100}$$

$$= 4\left(\frac{10}{100}\right) + 6\left(\frac{1}{100}\right) = 4\left(\frac{1}{10}\right) + 6\left(\frac{1}{10}\right)^2 = 4(10)^{-1} + 6(10)^{-2}$$

$$6.895 = \frac{6895}{1000} = \frac{6(1000)}{1000} + \frac{8(100)}{1000} + \frac{9(10)}{1000} + \frac{5(1)}{1000}$$

$$= 6 + 8\left(\frac{1}{10}\right) + 9\left(\frac{1}{100}\right) + 5\left(\frac{1}{1000}\right)$$

$$= 6 + 8\left(\frac{1}{10}\right) + 9\left(\frac{1}{10}\right)^2 + 5\left(\frac{1}{10}\right)^3$$

$$= 6 + 8(10)^{-1} + 9(10)^{-2} + 5(10)^{-3}$$

## TERMINATING DECIMALS

The decimals we discussed at the beginning of this section are called **terminating decimals.** In every example of terminating decimals, the decimal can be written as a fraction with a denominator as a power of 10. Note that $10 = 2 \cdot 5$; $100 = 10^2 = 2^2 \cdot 5^2$; $1000 = 10^3 = 2^3 \cdot 5^3$; and, in general, $10^n = 2^n \cdot 5^n$. This suggests the following property.

| | |
|---|---|
| **Terminating Decimals** | The only fractions in simplest form that can be written as terminating decimals are those in which the denominators can be factored into powers of 2 and 5. |

**EXAMPLE 8**

Change to terminating decimals (a) $\frac{3}{20}$ and (b) $\frac{5}{6}$.

*SOLUTION*

(a) To change this fraction to a decimal, we need a power of 10 in the denominator. Therefore, multiply the numerator and the denominator by 5, to get

$$\frac{3 \cdot 5}{20 \cdot 5} = \frac{15}{100} = 0.15$$

(b) $\frac{5}{6} = \frac{5}{3 \cdot 2}$. Since the denominator contains a 3, this fraction cannot be changed to a terminating decimal.

Because all terminating decimals can be expressed as fractions with denominators of powers of 10, interesting properties can be developed for multiplying and dividing a decimal by powers of 10.

**EXAMPLE 9**

$$14.37(10) = \left[ 1(10) + 4 + 3\left(\frac{1}{10}\right) + 7\left(\frac{1}{10}\right)^2 \right] \cdot 10$$

$$= 1(10)^2 + 4(10) + 3 + 7\left(\frac{1}{10}\right) = 143.7$$

$$2.615(100) = \left[ 2 + 6\left(\frac{1}{10}\right) + 1\left(\frac{1}{10}\right)^2 + 5\left(\frac{1}{10}\right)^3 \right] \cdot 10^2$$

$$= 2(10)^2 + 6(10) + 1 + 5\left(\frac{1}{10}\right) = 261.5$$

Do you see a pattern? Now let's divide by powers of 10.

**EXAMPLE 10**

Find the following quotients.
(a) $4.2 \div 10$     (b) $4.2 \div 100$     (c) $4.2 \div 1000$

*SOLUTION*

(a) $\dfrac{4.2}{10} = \dfrac{(4.2)(10)}{10(10)} = \dfrac{42}{100} = 0.42$

(b) $\dfrac{4.2}{100} = \dfrac{(4.2)(10)}{(100)(10)} = \dfrac{42}{1000} = 0.042$

(c) $\dfrac{4.2}{1000} = \dfrac{(4.2)(10)}{(1000)(10)} = \dfrac{42}{10,000} = 0.0042$

The preceding discussion suggests the following statement.

**Powers of 10**

To **divide** a number $N$ by $10^k$, where $k$ is a whole number, move the decimal point (in $N$) $k$ places to the left. To **multiply** a number $N$ by $10^k$, move the decimal point (in $N$) $k$ places to the right.

**EXAMPLE 11**

$176.2 \div 10^3 = 0.1762$

$0.163 \cdot 10^4 = 1630.$

# ROUNDING DECIMALS

On page 171 we discussed rounding techniques. These apply to decimals as well as to whole numbers.

**EXAMPLE 12**

Round off the following numbers to two decimal places.

$3.1671 \approx 3.17 \qquad 5.315 \approx 5.32$

$8.134 \approx 8.13 \qquad 7.245 \approx 7.25$

$7.1451 \approx 7.15 \qquad 3.1416 \approx 3.14$

## CALCULATOR HINT

Calculators often have built-in capability to round. We select the *round* command under the MATH menu. Suppose we wish to round 3.1671 to two decimal places.

MATH (underline *round*) 3.1671 ⌐,⌐ 2 ⌐=⌐. Did you get 3.17? Did you notice the 2 that followed the 3.1671? What happens if you replace the 2 with a 3 in this sequence of commands?

If the approximation 5.84 is a rounded-off representation of the exact number $N$, then

$$5.835 \leq N < 5.845$$

One-half of the difference between 5.835 and 5.845 is called the approximation's **greatest possible error.** Thus the greatest possible error of 5.84 is 0.005.

**EXAMPLE 13**

Find the greatest possible error of each of the following.
(a) 7.03    (b) 2.1    (c) 2.111

*SOLUTION*

(a) 0.005    (b) 0.05    (c) 0.0005

Scientists use exponential notation to express extremely large numbers and extremely small numbers, such as those involved in the velocity of light, the velocity of sound, the distance of a light year, or the mass of an electron. Expressing exponents in *scientific notation* allows us to present large and small numbers in shortcut forms that are easy to write, read, and understand.

**DEFINITION**

**Scientific Notation**   In *scientific notation*, a number $N$ is written in the form

$$N = A \cdot 10^n$$

where $1 \leq A < 10$ (such as 2.61) and $n$ is an integral exponent.

To write a number in scientific notation, we first move the decimal so as to place the number between 1 and 10. Then, by inspection, we identify the power of 10 that makes the two expressions equal.

**EXAMPLE 14**

Write 432.7 in scientific notation.

*SOLUTION*

$$1 \leq 4.327 < 10$$

$$432.7 = 4.327 \cdot 10^x$$

We must multiply 4.327 by 100 to get 432.7. Thus,

$$432.7 = 4.327 \cdot 10^2$$

## CALCULATOR HINT

Calculators often provide a special key such as $\boxed{EE}$ that allows us to enter a number in scientific notation. The number following the EE is the power of 10 in scientific notation. Insert $4.32 \cdot 10^{12}$ into your calculator.

$4.32 \boxed{EE} 12 \boxed{=}$. Did you get $4.32\ 10^{12}$?

To change a number from scientific notation to standard form, multiply by the power of 10.

**EXAMPLE 15**

Change $2.174 \cdot 10^{-3}$ to standard form.

*SOLUTION*

$$2.174 \cdot 10^{-3} = 2.174 \cdot \frac{1}{10^3} = \frac{2.174}{1000} = 0.002174$$

**EXAMPLE 16**

Change the following numbers from scientific notation to standard notation. To do so use your knowledge of the pattern of multiplying by powers of 10, rather than direct multiplication.

(a) $2.67 \cdot 10^4$     (b) $2.6705 \cdot 10^3$     (c) $9.2347 \cdot 10^1$
(d) $3.457 \cdot 10^{-7}$     (e) $5.55555 \cdot 10^5$     (f) $6.233 \cdot 10^{-2}$

*SOLUTION*

(a) 26,700     (b) 2670.5     (c) 92.347
(d) 0.0000003457     (e) 555,555     (f) 0.06233

**EXAMPLE 17**

Express the following numbers in scientific notation.

(a) 0.00012     (b) 467,000     (c) $453 \cdot 10^{-2}$

*SOLUTION*

(a) $1.2 \cdot 10^{-4}$     (b) $4.67 \cdot 10^5$     (c) $4.53 \cdot 10^0$

# ADDITION AND SUBTRACTION OF DECIMALS

The models in Figure 7-4 introduce the addition of numbers involving decimals (in this case, $2.41 + 1.23$).

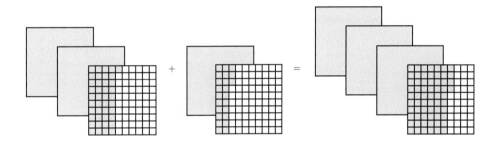

Place-value chart

| Ones | Tenths | Hundredths | | |
|------|--------|------------|---|---|
| 2 | 4 | 1 | | 2.41 |
| 1 | 2 | 3 | or | + 1.23 |
| 3 | 6 | 4 | | 3.64 |

FIGURE 7-4

In Example 18, addition is performed in four different ways: as fractions; in expanded notation using $\frac{1}{10}$, $\frac{1}{100}$, and so on; in powers of 10; and finally with an algorithm similar to that used for whole numbers.

**EXAMPLE 18**

Perform the addition $0.253 + 0.14$ in 4 different ways.

*SOLUTION*

(a) First, we interpret the decimals as fractions:

$$0.253 = \frac{253}{1000} \quad \text{and} \quad 0.14 = \frac{14}{100}$$

Hence, we can compute

$$\frac{253}{1000} + \frac{14}{100} = \frac{253}{1000} + \frac{140}{1000} = \frac{393}{1000} = 0.393$$

(b) Next, we state the decimals as fractions in expanded notation:

$$0.253 = \quad 2\left(\frac{1}{10}\right) + 5\left(\frac{1}{100}\right) \quad + 3\left(\frac{1}{1000}\right)$$

$$0.14 = \quad 1\left(\frac{1}{10}\right) + 4\left(\frac{1}{100}\right) \quad + 0\left(\frac{1}{1000}\right)$$

$$\overline{3\left(\frac{1}{10}\right) + 9\left(\frac{1}{100}\right) + 3\left(\frac{1}{1000}\right)} = 0.393$$

(c) Next, we convert the decimals to powers of 10:

$$0.253 = 2(10)^{-1} + 5(10)^{-2} + 3(10)^{-3}$$
$$+\ \underline{0.14\ \ = 1(10)^{-1} + 4(10)^{-2}}$$
$$= 3(10)^{-1} + 9(10)^{-2} + 3(10)^{-3}$$
$$= 0.393$$

(d) Finally, we use a standard algorithm:

$$\begin{array}{r} 0.253 \\ +\ 0.14 \\ \hline 0.393 \end{array}$$

These examples suggest the following algorithm.

| **Addition and Subtraction of Decimals** | The addition and subtraction of decimals are accomplished by lining up the decimal points in a vertical column and using an algorithm similar to that used for whole numbers. |
|---|---|

## CALCULATOR HINT

On many calculators are found keys such as $\boxed{\blacktriangleright F}$ that convert decimals to fractions if such is possible. Find $31.93 + 8.275$ and convert to a fraction if possible. If $AU\ b/c$ is underlined in $\boxed{\text{FracMode}}$,

$$31.93\ \boxed{+}\ 8.275\ \boxed{\blacktriangleright F}\ \boxed{=}.\ \text{Did you get } 40\ 41/200?$$

**EXAMPLE 19**

Find $32.4 - 6.73$.

*SOLUTION*

$$32.4\ = \frac{324}{10} = \frac{3240}{100}$$

$$-\ 6.73 = \frac{673}{100} = \frac{673}{100}$$

$$\overline{\phantom{xxx}}$$

$$25.67 \qquad\qquad \frac{2567}{100} = 25.67$$

### CALCULATOR HINT

Find $23.54 - 9.027$ and convert to a fraction if possible. If $AU\ b/c$ is underlined in $\boxed{\text{FracMode}}$,

$$23.54 \; \boxed{-} \; 9.027 \; \boxed{\blacktriangleright F} \; \boxed{=}. \text{ Did you get } 14\ 513/1000?$$

## MENTAL MATHEMATICS

The regrouping process works with decimals in the same way as it does with whole numbers:

$$2.8 + (5.1 + 1.2) = (2.8 + 1.2) + 5.1$$
$$= 4 + 5.1$$
$$= 9.1$$

Some people do mental arithmetic by splitting or separating the numbers into whole numbers plus decimals:

$$4.2 + 2.5 = (4 + 2) + (.2 + .5)$$
$$= 6 + .7$$
$$= 6.7$$

Adding or subtracting numbers to facilitate mental arithmetic works with decimals as well as with whole numbers. For example,

$$4.74 + 2.98 = (4.74 - .02) + (2.98 + .02) \qquad \text{Add and subtract.}$$
$$= 4.72 + 3$$
$$= 7.72$$

## APPROXIMATE NUMBERS

Addition and subtraction with numbers that are rounded numbers or approximations can be no more accurate than the least accurate number in the computation. This result may be stated as follows.

| Addition of Approximate Numbers | When *approximate numbers are added* (or subtracted), their sum (or difference) should be rounded to the same decimal place as the term with the greatest possible error. |
|---|---|

**EXAMPLE 20**

When we add $3.15 + 0.732$ on a calculator, we obtain $3.882$. But because $3.15$ has the greatest possible error, we round $3.882$ to $3.88$. Now consider

$$23.4341 - 6.112 = 17.322$$

Why?

Inevitably, you will, on occasion, enter the wrong information into a calculator. Such a mistake will often yield a large error in the display. For this reason, an important skill in work with a calculator is the ability to estimate answers mentally in order to recognize large mistakes.

This idea is sometimes formalized in a practice of rounding off the larger numbers to one or two digits and ignoring numbers that are small in comparison to the others. The computation can then be performed mentally (or quickly on paper) to check whether the order of magnitude of the answer displayed on the machine is reasonable.

**EXAMPLE 21**

|  | **Subtract** | **Check by rounding off to the nearest 100** |
|---|---|---|

$$
\begin{array}{r}
4137.6 \\
-\ \ 861.74 \\
\hline
3275.86
\end{array}
\qquad
\begin{array}{r}
4100 \\
-\ \ 900 \\
\hline
3200
\end{array}
$$

The answer seems reasonable.

**EXAMPLE 22**

|  | **Add** | **Check by rounding off to the nearest 1000** |
|---|---|---|

$$
\begin{array}{r}
4373 \\
79 \\
5 \\
7580 \\
\hline
7037
\end{array}
\qquad
\begin{array}{r}
4000 \\
00 \\
0 \\
8000 \\
\hline
12{,}000
\end{array}
$$

Undoubtedly, we have made a mistake. Can you find it?

## CALCULATOR HINT

When we use the $\boxed{\div}$ key in entering fractions, the results are given in terms of decimal fractions. However with a change to fraction key $\boxed{\blacktriangleright F}$ the result is a fraction. For example, $2 \boxed{\div} 3 \boxed{+} 7 \boxed{\div} 8 \boxed{\blacktriangleright F} \boxed{=} 37/24$ or $1\ 13/24$.

The algorithm we have used for addition and subtraction in base ten can also be used with other bases.

**EXAMPLE 23**

| **Base seven** | **Base two** |
|---|---|

$$
\begin{array}{r}
46.143 \\
+\ 24.35 \\
\hline
103.523
\end{array}
\qquad
\begin{array}{r}
111.0101 \\
+\ \ \ 11.1011 \\
\hline
1011.0000
\end{array}
$$

However, it is a bit trickier to change such expressions to base ten.

**EXAMPLE 24**

Change $42.14_{\text{five}}$ to base ten.

$$
42.14_{\text{five}} = \left[ 4(5) + 2 + 1\left(\frac{1}{5}\right) + 4\left(\frac{1}{5}\right)^2 \right]_{\text{ten}}
$$

$$
= 20 + 2 + 0.2 + 0.16 = 22.36
$$

As you may have guessed from the preceding procedure, only decimals in bases two and five can always be changed to a terminating decimal in base ten.

# Just for Fun

Complete the subtraction magic square. In a subtraction magic square, you sum the numbers on the ends and subtract the middle. Thus, for example,

$$0.06 + 0.18 - 0.09 = 0.15$$

# EXERCISE SET 1

*R* 1. Write each of the following in expanded form.
   (a) 14.72   (b) 4.0016   (c) 0.146

2. Rewrite the following as decimals.
   (a) $4(10) + 3(10)^0 + 2\left(\frac{1}{10}\right) + 6\left(\frac{1}{10}\right)^2$
   (b) $4\left(\frac{1}{10}\right)^2 + 5\left(\frac{1}{10}\right)^3$
   (c) $7(10)^0 + 4\left(\frac{1}{10}\right)^3 + 2\left(\frac{1}{10}\right)^4$

3. Express each of the following decimal fractions as a fraction.
   (a) 0.085   (b) 3.25   (c) 12.74
   (d) 0.84    (e) 2.75   (f) 0.00025

4. Express the following fractions as decimal fractions. Then check your answers with a calculator.
   (a) $\frac{316}{10}$   (b) $\frac{602}{10,000,000}$   (c) $\frac{18}{15}$
   (d) $\frac{63}{70}$   (e) $\frac{651}{120}$   (f) $\frac{91}{140}$
   (g) $\frac{3}{8}$   (h) $\frac{5}{16}$   (i) $\frac{126}{105}$

*Estimate the sums or differences in Exercises 5 through 10. Then perform the operation, and see if your calculator answer is reasonably close to your estimate.*

5. $3909 + 161.79$   6. $5986 - 2653$
7. $0.08469 - 0.00104$   8. $631,428 - 241,586$
9. $425,631 + 27,408$   10. $726,410 - 14,190$

11. Without using a calculator, determine which of these fractions can be expressed as terminating decimal fractions?
    (a) $\frac{36}{15}$   (b) $\frac{21}{70}$   (c) $\frac{48}{36}$
    (d) $\frac{27}{60}$   (e) $\frac{10}{30}$   (f) $\frac{21}{36}$

12. Where possible, write the fractions in Exercise 11 as terminating decimals.

13. Multiply each number in Exercise 3 by $10^4$.

14. Multiply each number in Exercise 3 by $10^{-3}$.

15. Write the following numbers in words.
    (a) 2.036   (b) 3.0071
    (c) 0.0016   (d) 0.0007

16. Calculate the following mentally. Describe the process by which you get the answer.
    (a) $2.01 + 3.56$   (b) $3.98 + 4.61$
    (c) $14.35 - 8.98$   (d) $31.62 - 5.96$

17. Round off the numerals shown to the nearest hundredth. Then round them off to the nearest tenth, and finally to three digits. When rounding to the nearest hundreth, check your answer with a calculator.

    (a) 91.4833    (b) 1.6651    (c) 0.8535
    (d) 1400.176   (e) 273.871   (f) 436.436

18. For each of the following rounded numbers, what is the greatest possible original number? The least possible original number?

    (a) 4000 (rounded to thousands)
    (b) 4700 (rounded to hundreds)
    (c) 2.1
    (d) 3.25

19. Write the following numbers in scientific notation. Check your answers with a calculator.

    (a) 11 million      (b) 33 billion
    (c) 0.0000033       (d) 67,300,000,000,000
    (e) 0.036754        (f) 3891

20. Change these numbers from scientific notation to standard notation that does not use powers of 10 without using a calculator. Then check your answers with a calculator.

    (a) $4.6 \cdot 10^{-1}$      (b) $3.321 \cdot 10^{6}$
    (c) $6 \cdot 10^{-6}$        (d) $7.12 \cdot 10^{-2}$
    (e) $2.89 \cdot 10^{10}$
    (f) $2.99776 \cdot 10^{10}$ cm per sec (velocity of light)

21. Perform the following operations without using a calculator. Then check your answers with a calculator.

    (a) $28.32 + 7.521$      (b) $0.56 + 0.006$
    (c) $0.3 + 5.00311$      (d) $84.5 - 9.72$
    (e) $354.51 - 38.64$     (f) $389.27 - 63.99$

22. Change each decimal in Exercise 21 to a fraction, perform the operation, and then change each answer back to a decimal. Do you get the same answer?

23. Write each decimal in Exercise 21 in expanded form, using negative exponents, and perform the operation. Then change each answer back to a decimal and check it against the answer in Exercise 21.

24. In the same manner as you worked with fractions on p. 277, find the whole number range for the following computations.

    (a) $4.11 + 7.68$      (b) $3.96 + 8.54$
    (c) $8.243 + 3.665$    (d) $11.611 + 8.213$

25. Find the approximate values of the sums in Exercise 24 by rounding to the nearest whole number.

26. Estimate by rounding to hundreds.

    (a) $162.4 + 1152.3$      (b) $5462.1 - 843.7$

27. Find the whole-number range of each of the following. (See Exercise 24.)

    (a) $8.243 - 3.665$
    (b) $11.611 - 8.213$

28. Find the approximate values in Exercise 27 by rounding to the nearest whole number.

29. Perform the following additions and subtractions of approximate numbers, rounding answers to the appropriate number of decimal places.

    (a) $2.036 + 2.21 - 0.0072$
    (b) $40.672 - 0.05 + 31.6$

30. Calculate mentally.

    (a) $172.41 - 19.98$
    (b) $3.4 + 6.92$
    (c) $8.27 + 5.89$
    (d) $141.6 - 9.95$

*For Exercises 31 through 35, set up an equation to represent the problem, solve the equation, and then check the answer with data given in the problem.*

31. Three students in the mathematics club participated in a bike-a-thon to raise money. John rode 16.2 miles, Jim rode 13.5 miles, and Sue rode enough so that they collected on 40 miles. How many miles did Sue ride?

32. Chang had $142.36 in his checking account. He writes two checks for $25 each and one check for $13.42. How much money is left in his account?

33. Sue's gasoline tank on her new car holds 20 gallons. She fills the car by adding 14.6 gallons. How much gasoline was in the tank already?

34. Robert has agreed with his parents to pay for his long-distance calls out of his allowance. Last month, Robert made all the long-distance calls; the total phone bill for the family was $50, excluding tax. How much will Robert pay if the monthly cost of the telephone is $18.75?

35. *(Introductory problem)* Two trains start out 53.6 km apart. One train travels 47.3 km away from the other train. The second train travels 61.4 km toward the first train. How far apart are the two trains?

36. Perform the following operations on a calculator. Check by estimating (rounding each part to two digits).

    (a) $16.1756 + 2.71419$    (b) $452.7613 - 123.75$
    (c) $42.7131 + 14.7615$    (d) $642.95 - 43.71$

37. Find the value of the following.

    (a) $\dfrac{10^{-1}4^{0}}{10^{-2}}$      (b) $\dfrac{10^{0}4^{-1}}{4^{-3}}$

    (c) $(2^{-1})^{0}$      (d) $(3^{-1})^{-2}$

C 38. Perform the following operations.
 (a) $11.001_{two} + 1.110_{two}$  (b) $12.21_{three} - 2.222_{three}$
 (c) $0.043_{seven} + 6.45_{seven}$  (d) $51.35_{six} + 23.42_{six}$

39. Convert the following representations to base ten decimal fractions.
 (a) $0.22_{five}$  (b) $1.24_{five}$
 (c) $1.101_{two}$  (d) $0.011_{two}$

## LABORATORY ACTIVITIES

40. Represent the following with decimal squares.
 (a) 4.72
 (b) 2.016

41. Illustrate $2.31 + 4.83$ using decimal squares.

# SECTION 2 THE ARITHMETIC OF DECIMALS

| | |
|---|---|
| **PROBLEM** | Bill has the highest batting average of the Samford Bulldogs, .400 after 30 times at bat. In a slump, Bill gets only 2 hits in his next 8 times at bat. What is his new batting average? |
| **OVERVIEW** | Solving this problem requires an understanding of both multiplication and division of decimals. This section reviews these basic operations and presents algorithms that illustrate the general rules for placing decimal points.<br><br>Because the calculator and the computer play such important roles in our lives, we need to be able to estimate whether the answers that appear on the screen are in any way reasonable. This section explains how to estimate answers and how to appraise the accuracy of answers when working with approximate numbers. |
| **GOALS** | Illustrate the Number and Operations Standard, page 119.<br>Illustrate the Algebra Standard, page 4.<br>Illustrate the Representation Standard, page 50. |

## MULTIPLYING DECIMALS

We shall use two procedures to demonstrate multiplication of decimals: the fraction square model, and converting to common fractions before multiplying.

Consider the multiplication $0.5 \cdot 0.3$ as depicted by the fraction square model in Figure 7-5(a). Vertically, 0.3 is represented by 3 strips of 10, and these are shaded.

Horizontally, 0.5 is represented by 5 strips of 10, and these are shaded. The double-shaded region (15 blocks of 100) in Figure 7-5(a) represents 0.15. Thus,

$$0.5 \cdot 0.3 = 0.15$$

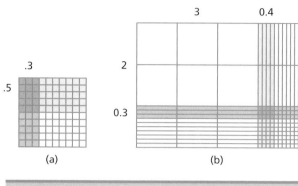

FIGURE 7-5

This idea can be extended to the multiplication $2.3 \cdot 3.4$. (See Figure 7-5(b).) Vertically, the shaded region represents .4 (4 strips of 10); horizontally, the shaded region represents .3 (3 strips of 10). As in whole-number multiplication, $3 \cdot 2 = 6$ unshaded blocks; however, 2(.4) is represented by 8 shaded rectangular strips, or .8; $.3 \cdot 3$ is represented by 9 shaded rectangular strips, or .9; and .3(.4) is represented by 12 double-shaded blocks of 100, or .12. Thus,

$$(2.3)(3.4) = 6 + .8 + .9 + .12 = 7.82$$

In both illustrations one digit lies to the right of the decimal in each factor, and this results in two digits appearing to the right of the decimal in the answer.

To help identify the multiplication algorithm for decimals, let's change the decimals to fractions before multiplying. For example, consider multiplying 32.4 by 0.12:

$$(32.4) \cdot (0.12) = \frac{324}{10} \cdot \frac{12}{100} = \frac{3888}{1000} = 3.888$$

Notice that the answer is obtained by multiplying the numbers 324 and 12 without decimals and then dividing the result by 1000 or $10^3$.

One digit lies to the right of the decimal point in 32.4, leading to a 10 in the denominator of the equivalent fraction; meanwhile, two digits lie to the right of the decimal point in 0.12, leading to a $10^2$ or 100 in the denominator of the equivalent fraction. Because $10 \cdot 10^2 = 10^3$ in the denominator of the answer, the answer as a decimal has three digits to the right of the decimal point. Do you see the pattern? This discussion leads to the following algorithm for multiplying decimals.

---

**Decimal Places in Multiplication**

If $r$ digits lie to the right of the decimal point in one factor and $s$ digits lie to the right of the decimal point in a second factor, multiply the two factors, ignoring decimal points, and place the decimal point so that $r + s$ digits lie to the right of the decimal point in the answer.

---

**EXAMPLE 1**

$$
\begin{array}{r}
32.4 \\
\cdot\, 0.12 \\
\hline
648 \\
324\phantom{0} \\
\hline
3.888
\end{array}
$$

32.4    1 digit to the right of the decimal point

·0.12    2 digits to the right of the decimal point

3.888    3 digits to the right of the decimal point

This algorithm can be further illustrated by expressing the multiplication in expanded form.

$$3(10) + 2(10)^0 + 4(10)^{-1}$$
$$\cdot\,[1(10)^{-1} + 2(10)^{-2}]$$
$$\overline{6(10)^{-1} + 4(10)^{-2} + 8(10)^{-3}} \quad \text{Multiply by } 2(10)^{-2}$$
$$\underline{3(10)^0 + 2(10)^{-1} + 4(10)^{-2}} \quad\quad \text{Multiply by } 1(10)^{-1}$$
$$3(10)^0 + 8(10)^{-1} + 8(10)^{-2} + 8(10)^{-3} = 3.888$$

Multiply $3.41_{\text{five}} \cdot 2.3_{\text{five}}$

*SOLUTION*

$$
\begin{array}{rl}
3.41 & \text{2 places} \\
\underline{\cdot\ \ 2.3} & \text{1 place} \\
2123 & \\
\underline{1232\ \ } & \\
14.443_{\text{five}} & \text{3 places}
\end{array}
$$

##  ORDER

Some people have difficulty ordering numbers that involve decimals. With whole numbers, a two-digit number is always greater than a one-digit number. The same is not true for decimals.

NOTE: **One suggestion for comparing the size of two decimal numbers is to place one just above the other so that the decimal points line up vertically.** If the whole parts are equal, scan both decimals from left to right until you note a place where corresponding digits differ.

For instance, $2.325176 < 2.3255$, because for

2.325176
2.3255

1 is less than 5 in the fourth decimal place.

### 📱 CALCULATOR HINT

When an answer to an operation is too large for the display, many calculators display the answer in scientific notation. Multiply $4{,}324{,}756 \cdot 1{,}462{,}375$.

$$4{,}324{,}756 \ \boxed{\times} \ 1{,}462{,}375 \ \boxed{=} \quad 6.324415056 \quad 12$$

The 12 on the right-hand margin of the display represents $10^{12}$.

## DIVISION OF DECIMALS

We investigate the division of decimal numbers by using three procedures: estimating the whole-number part of the answer, and thereby knowing where to place the decimal; placing the decimal point when dividing by a whole number; and converting decimal division problems into problems where the divisor is a whole number.

**EXAMPLE 3**

Place the decimal point in $302.46 \div 42.6$ by approximating the decimal numbers with whole numbers.

*SOLUTION*

$302.46 \approx 300$ and $42.6 \approx 40$. Now, $300 \div 40 \approx 7$. Therefore, we can perform the division on the numbers without decimal points and then place the decimal point after the first number (which should be approximately a 7):

$$
\begin{array}{r}
71 \\
426\overline{)30246} \\
\underline{2982} \\
426 \\
\underline{426}
\end{array}
$$

Because we know that the answer is approximately 7, we can now place the decimal point so that the exact answer is 7.1.

**EXAMPLE 4**

Compute $4.26 \div 2$ by converting 4.26 into a common fraction.

*SOLUTION*

Write 4.26 as $\frac{426}{100}$. Then

$$
\frac{426}{100} \div 2 = \frac{\overset{213}{\cancel{426}}}{100} \cdot \frac{1}{\underset{1}{\cancel{2}}} = \frac{213}{100} = 2.13
$$

Think of this division as

$$
\underset{\text{divisor}}{\overline{)\,\text{dividend}}}\overset{\text{quotient}}{} \qquad \text{or} \qquad 2\overline{)4.26}\overset{2.13}{}
$$

The preceding discussion suggests the following property of division of decimals.

---

**Division of Decimals**

When a decimal is divided by a whole number, the decimal point in the quotient is placed directly above the decimal point in the dividend.

---

Any division of decimals may be changed into a division in which the divisor is a whole number.

**EXAMPLE 5**

Divide 106.08 by 1.7.

*SOLUTION*

$106.08 \div 1.7$ can be expressed as

$$
\frac{106.08}{1.7}
$$

Then, by the Fundamental Law of Fractions,

$$
\frac{106.08}{1.7} = \frac{(106.08)(10)}{(1.7)(10)} = \frac{1060.8}{17} \qquad \text{or} \qquad
\begin{array}{r}
62.4 \\
17\overline{)1060.8} \\
\underline{102} \\
40 \\
\underline{34} \\
68 \\
\underline{68}
\end{array}
$$

In the preceding example, to convert $106.08 \div 1.7$ into division by a whole number, we multiplied both 106.08 and 1.7 by 10. This is equivalent to moving the decimal point one place to the right in both the dividend and the divisor. **In general, if there are $r$ decimal places in the divisor, we move the decimal point $r$ places to the right in the dividend and then divide by the divisor as a whole number.**

| | |
|---|---|
| **PRACTICE PROBLEM** | Divide 111.32 by 2.3. |
| *ANSWER* | 48.4 |

## ⠿ MENTAL ARITHMETIC

Knowing the fractional equivalents of decimals facilitates mental arithmetic:

$$\frac{1}{10} = 0.1 \qquad \frac{1}{4} = 0.25 \qquad \frac{1}{8} = 0.125 \qquad \frac{7}{8} = 0.875$$

$$\frac{1}{20} = 0.05 \qquad \frac{1}{2} = 0.5 \qquad \frac{3}{8} = 0.375$$

$$\frac{2}{10} = 0.2 \qquad \frac{3}{4} = 0.75 \qquad \frac{5}{8} = 0.625$$

We use these equivalents in the following example.

| | |
|---|---|
| **EXAMPLE 6** | Using mental arithmetic, find the following. |
| | (a) 0.75(84)  (b) 0.625(16)  (c) 0.125(24) |
| *SOLUTION* | (a) $0.75(84) = \dfrac{3}{4}(84) = 63$ |
| | (b) $0.625(16) = \dfrac{5}{8}(16) = 10$ |
| | (c) $0.125(24) = \dfrac{1}{8}(24) = 3$ |

## ⠿ APPROXIMATE NUMBERS

Let us now look at the accuracy of answers obtained by multiplying (or dividing) approximate numbers. In the preceding section, we learned to express numbers in scientific notation; that is, we wrote numbers in the form

$$N = A \cdot 10^n$$

where $1 \le A < 10$.

The digits in $A$ in the definition of scientific notation are called **significant digits** of the number.

**EXAMPLE 7**

$1.40 \cdot (10)^5$ has three significant digits.
$1.4 \cdot (10)^2$ has two significant digits.
$1.041 \cdot (10)^7$ has four significant digits.
$0.05$ has only one significant digit.

The following procedure uses the number of significant digits in the parts of a product (or quotient) as a basis for rounding the answer.

| Product of Approximate Numbers | When *approximate numbers are multiplied* (or divided), their product (or quotient) should be rounded off to have the same number of significant digits as the factor with the smallest number of significant digits. |
|---|---|

**EXAMPLE 8**

(a) $(31.45) \cdot (56.2) = 1767.49$ rounds to 1770 with three significant digits because 56.2 has only three significant digits.
(b) $(0.023) \cdot (0.45) = 0.01035$ rounds to 0.010 with two significant digits because 0.45 has only two significant digits.
(c) $635 \div 0.27 = 2351.85$ rounds to 2400 with two significant digits because 0.27 has only two significant digits.

#  ESTIMATION

Estimation is important at all times. In work with a calculator, however, it is essential. A good feel for the answer will sharpen your sensitivity to mistakes made when you hit a wrong key.

**EXAMPLE 9**

Find the range for the multiplication $3.81 \cdot 4.12$.

*SOLUTION*

Because $3 < 3.81 < 4$ and $4 < 4.12 < 5$, we know that

$$3 \cdot 4 < 3.81 \cdot 4.12 < 4 \cdot 5$$
$$12 < 3.81 \cdot 4.12 < 20$$

**EXAMPLE 10**

Find the range for the division $8.8 \div 3.6$.

*SOLUTION*

Clearly, $8 < 8.8 < 9$ and $3 < 3.6 < 4$. Therefore,

$$\frac{8}{4} < (8.8) \div (3.6) < \frac{9}{3} \qquad \text{or}$$
$$2 < (8.8) \div (3.6) < 3$$

Notice that, on the left of the inequality, we used the smallest possible value in the numerator and the largest possible value in the denominator, while on the right we used the largest possible value in the numerator and the smallest possible value in the denominator.

**EXAMPLE 11** | Multiply 12.47 and 0.623. Check by rounding each number to one significant digit.

## CALCULATOR HINT

On a calculator, 12.47 $\boxed{\times}$ 0.623 = 7.76881. For the check, $10 \cdot 0.6 = 6$. The answer seems reasonable.

**EXAMPLE 12**

| Divide | Check |
|--------|-------|
| 4324 $\boxed{\div}$ 2.12 = 196.742 | 4000 ÷ 2 = 2000 |

The calculated answer is evidently incorrect.

Finally, we return to the introductory problem for this section.

**EXAMPLE 13** | Bill has the highest batting average of the Samford Bulldogs, .400 after 30 times at bat. In a slump, Bill gets only 2 hits in his next 8 times at bat. What is his new batting average?

SOLUTION | Of 30 times a bat, Bill has

$$.400(30) = 12 \text{ hits}$$

He now has 12 + 2 = 14 hits of

$$30 + 8 = 38 \text{ times at bat}$$

Therefore,

$$14 \div 38 = .368 \text{ (Bill's new average)}$$

# REPEATING DECIMALS

Now let us turn to the problem of finding the decimal representation of a fraction.

**EXAMPLE 14** | Change the following to decimals.

(a) $\dfrac{3}{8}$  (b) $\dfrac{1}{3}$  (c) $\dfrac{3}{11}$

SOLUTION

(a)
```
        0.375
   8 ) 3.000
       2 4
       ─────
         60
         56
       ─────
         40
         40
       ─────
          0
```

(b)
```
        0.333
   3 ) 1.000
        9
       ────
        10
         9
       ────
        10
         9
       ────
         1
```

(c)
```
         0.2727
   11 ) 3.0000
        2 2
       ──────
         80
         77
       ──────
          30
          22
       ──────
          80
          77
       ──────
           3
```

Notice that in part (a) of Example 14 a remainder of 0 is obtained. For this reason, the decimal is said to **terminate.** In part (b), because a remainder of 1 repeats, and in part (c), because alternate remainders of 3 and 8 repeat, a remainder of 0 will never be attained. These decimal fractions, which are called **nonterminating decimals,** have interesting properties. Notice that in the quotient of part (b) the 3 repeats, whereas in the quotient of part (c) the pair of digits 27 repeats. Such decimals are called *repeating decimals.*

---

**DEFINITION**

**Repeating Decimal**    A *repeating decimal* is a nonterminating decimal in which (after a certain point) the same pattern of numbers repeats endlessly.

---

**EXAMPLE 15**

0.36 is a terminating decimal.
0.846333 is a terminating decimal.
0.242424 . . . is a repeating decimal.
0.4271271 . . . is a repeating decimal.
0.398210321032103 . . . is a repeating decimal.

Notice that each of the last three decimal representations in the preceding example contains repeating blocks of digits.

NOTE: **The fact that a block of digits repeats indefinitely is denoted by a bar above the repeating digits.**

**EXAMPLE 16**

(a) $\dfrac{2}{3} = 0.6666\ldots = 0.\overline{6}$

(b) $\dfrac{1}{7} = 0.142857142857\ldots = 0.\overline{142857}$

(c) $\dfrac{19}{444} = 0.04279279\ldots = 0.04\overline{279}$

If the division process does not convert a fraction into a terminating decimal, it will convert it into a repeating decimal. For example, perform the long division of 1.0000000 by 7. The first remainder after subtraction is 3, then 2, then 6, then 4, then 5, then 1, then 3. In fact, here only 6 digits can occur before a repetition. And for $a/b$, in simplest or reduced form, the largest number of possible remainders without a repetition is $b - 1$. Beyond this, the answer is a repetition of previous remainders. Therefore, if the division process does not yield a terminating decimal, it will yield a repeating decimal.

---

**Decimal Representation**    Every fraction can be represented by an integer or by a decimal that either terminates or eventually repeats; conversely, every decimal that either terminates or eventually repeats represents a fraction.

## CALCULATOR HINT

Write $\frac{2}{11}$ as a decimal.

On most calculators $2 \div 11$ yields .181818182. This looks like the repeating decimal .18 after it has been rounded. On many calculators we can check this assumption by converting the repeating decimal back to a fraction For example

$$.181818181818 \; \boxed{\blacktriangleright F} \; \boxed{=} \; 2/11. \text{ So } 2/11 = .\overline{18}$$

**EXAMPLE 17**

Find a reduced fraction that represents each of the following repeating decimals: (a) $.\overline{27}$ (b) $.01\overline{2}$

*SOLUTION*

(a) Let $N = 0.272727 \ldots$. We shall form a second equation (obtained from the first) so that, when we subtract the first equation from the second, the repeating decimal part will become zeros. The block of repeating digits consists of "27." Because two digits are in the block, we multiply by 100 to accomplish our goal in subtraction. (If there were 3 repeating digits, we would multiply by 1000; and so on.) Multiplying by 100 shifts the decimal point two places to the right, and we obtain

$$\begin{array}{r} 100N = \phantom{0}27.272727 \ldots \\ -N = -0.272727 \ldots \\ \hline 99N = 27 \end{array} \quad \text{or} \quad N = \frac{27}{99} = \frac{3}{11}$$

(b) Let $N = 0.012222 \ldots$.

$$\begin{array}{r} 10N = 0.12222 \ldots \\ -N = -0.01222 \ldots \\ \hline 9N = .11 \end{array} \quad \text{or} \quad N = \frac{.11}{9}$$

Multiplying numerator and denominator by 100 gives a whole number numerator, $N = \frac{11}{900}$.

## CALCULATOR HINT

Find a fraction representation for $0.7\overline{4}$.

$$.744444444 \; \boxed{\blacktriangleright F} \; \boxed{=} \qquad \text{The answer is 67/90.}$$

Interestingly, every terminating decimal can be written as a repeating decimal. To do this, simply subtract 1 from the last digit of the number and annex an infinite number of 9's.

**EXAMPLE 18**

$$26.4 = 26.3\overline{9}$$

$$0.005 = 0.004\overline{9}$$

To show that the preceding manipulation is correct, verify that $1 = 0.\overline{9}$. Let

$$\begin{array}{r} N = \phantom{0}0.9999 \ldots \\ 10N = \phantom{0}9.9999 \ldots \\ -N = -0.9999 \ldots \\ \hline 9N = 9 \\ N = 1 \end{array}$$

Therefore,

$$1 = 0.\overline{9}$$

# Just for Fun

Select an unusual number (say, a decimal such as 26.2347). Enter this number in your calculator. Multiply the number by 0.625; add 1.375; multiply by 3.68; divide by 2.3; add 2.8; subtract the number you started with. Your final result will be 5. Can you explain this trick?

# EXERCISE SET 2

R 1. Without using a calculator perform the following operations, using the algorithms of this section. (Do not round.) Then check your answers using a calculator.

 (a) $7.62 \cdot 0.021$ (b) $4002 \cdot 7.61$
 (c) $32.736 \div 12.4$ (d) $1.504 \div 0.32$

2. In Exercise 1, change each decimal to a fraction, perform the computation, and change the answer back to a decimal. Then check your answers against those for Exercise 1.

3. Perform the following divisions by first writing each division as a fraction that has a whole-number denominator. Then check your answers with a calculator.

 (a) $4.1\overline{)66.83}$ (b) $0.42\overline{)4.116}$
 (c) $0.31\overline{)5.301}$ (d) $2.6\overline{)445.64}$

4. Determine which of the following division problems have the same quotients, without actually dividing. How do you know which quotients will be identical?

 (a) $3.2\overline{)36}$ (b) $32\overline{)360}$
 (c) $0.320\overline{)3.06}$ (d) $0.032\overline{)3.06}$
 (e) $0.32\overline{)3.6}$ (f) $0.032\overline{)0.36}$

5. Without using a calculator perform the following operations. (Do not round.) Now check your answers with a calculator.

 (a) $(7.05)(0.006)$ (b) $(0.04)(6.011)$

 (c) $(2.04)(3.25)$ (d) $(21.7)(74.65)$

6. Use mental arithmetic to find the following.

 (a) $(0.375)(32)$ (b) $(0.75)(17)$
 (c) $(0.05)(18 + 2)$ (d) $(0.875)(56)$

7. Calculate mentally.

 (a) $3.16 \cdot 10^3$ (b) $3.16 \cdot 10^{-3}$
 (c) $4(5.5)$ (d) $8.5 \div 4$
 (e) $7.5 \cdot 12$ (f) $(0.5)(39)(4)$

8. Find the integral range for the following computations.

 (a) $(8.576) \cdot (6.459)$ (b) $(9.061) \cdot (7.821)$
 (c) $12.76 \div 3.41$ (d) $9.74 \div 4.12$

9. In Exercise 8, find the approximate value of each by rounding each factor to 1 significant digit.

10. Place one of the symbols $<$ , $>$ , or $=$ in each □ to make a true statement.

 (a) $31.216$ □ $31.20$ (b) $0.6$ □ $0.60$
 (c) $9.301$ □ $9.30$ (d) $2\frac{1}{4}$ □ $2.251$
 (e) $641.7125$ □ $641.7131$

*Estimate each quotient in Exercises 11 through 16, by rounding each to one significant digit. Then perform the operation, and see if your answer is reasonably close to your estimate.*

11. $84,967 \div 25$ 12. $0.0012 \div 468$

13. $0.016 \div 74$ 14. $95,640 \div 0.018$

15. $14,625 \div 0.0031$ 16. $0.0321 \div 0.0016$

17. Use the techniques discussed in this section to classify the following answers as probably correct or incorrect, without actually performing the computations.
    (a) $(7764) \cdot (1172) = 9,099,408$
    (b) $76 \div 0.011 = 2.2727273$
    (c) $(41) \cdot (4444) = 18,204$
    (d) $777 \div 12 = 8.333333$
    (e) $(42.167) \cdot (21.233) = 8.4231681$

18. Determine the number of significant digits in each of the following.
    (a) 23.41      (b) $2.7 \cdot 10^{-3}$
    (c) $0.301(10)^4$      (d) 300.4
    (e) $168(10)^3$      (f) 0.0012

19. Estimate the approximate size of each of the following computations by rounding each to one significant digit. Then find the answer by using a calculator.
    (a) $7.223 \cdot 4.2$      (b) $215.26 \div 24.6$
    (c) $0.66 \div 0.0022$      (d) $8.1 \cdot 0.0012$

20. Place the decimal point in each by estimating.
    (a) $(15.2)(3.8) = 5776$
    (b) $(7.32)(4.94) = 361608$
    (c) $(.085)(.07) = 595$      (d) $(13.4)(0.003) = 402$
    (e) $339.48 \div 0.36 = 943$      (f) $1.28 \div 3.2 = 4$

*T* 21. Change each factor to scientific notation, and then multiply or divide the approximate numbers.
    (a) $0.00023 \cdot 526,000$      (b) $682,000 \div 0.00024$
    (c) $2.71 \cdot (10)^{-3} \div 0.0002$      (d) $271 \cdot (10)^{-3} \cdot 1.6$

22. Write the following as repeating decimals without a calculator. Then check your answers with a calculator.
    (a) $\dfrac{1}{7}$    (b) $\dfrac{2}{11}$    (c) $\dfrac{3}{13}$

23. Write each decimal expression as a fraction in simplest form without using a calculator; then check with a calculator.
    (a) $0.346$      (b) $1.\overline{9}$      (c) $0.\overline{18}$
    (d) $3.8\overline{45}$      (e) $16.45\overline{9}$      (f) $146.\overline{1}$
    (g) $1.2\overline{54}$      (h) $0.01\overline{79}$      (i) $0.00\overline{9}$
    (j) $3.2\overline{5}$      (k) $0.01\overline{32}$      (l) $4.\overline{247}$

24. One procedure for ordering fractions is to compare decimal equivalents. Order the following fractions from smallest to largest.
    (a) $\dfrac{6}{11}, \dfrac{19}{34}, \dfrac{5}{9}, \dfrac{26}{41}$    (b) $\dfrac{15}{23}, \dfrac{32}{41}, \dfrac{38}{52}, \dfrac{18}{25}$

25. Is the decimal expansion of 131/5,297,323 terminating or nonterminating?

*In Exercises 26 through 29, first find an equation, then solve the equation, and then check your answer by looking back at the problem.*

26. 7.05 gal. of a certain liquid weigh 40 lb. How much does 1 gal. weigh?

27. Kelley's book satchel with books weighs 5.46 kg. It contains 2 books weighing 1.03 kg each and 3 notebooks weighing 0.72 kg each. How much does the satchel weigh?

28. The peel of an orange usually weighs $\frac{1}{20}$ of the total weight of the orange. You buy 50 pounds of oranges for $19.60. How much do you pay for the peel?

29. You travel 120 miles in 2 hours and 20 minutes, stop 30 minutes for lunch, and then travel 180 miles in 3 hours and 10 minutes. What is your average speed for the trip?

30. A seamstress purchased 2.40 m of $2.90-per-meter material; 5.40 m at $3.90 per meter; 7.5 meters at $2 per meter; 25.20 m of cording at 20¢ per meter; two spools of thread at $1.00 per spool; and 2.2 m of pleater tape at 80¢ per meter.
    (a) What was the total purchase price for these goods?
    (b) With a sales tax of 5¢ per dollar, what was her total bill?

31. At the candy counter, Gertrude weighed all the candy and nuts she had in stock. She found that she had 6.25 kg of divinity at 50¢ per $\frac{1}{4}$ kg, 4.3 kg of fudge at 80¢ per $\frac{1}{4}$ kg, 1.8 kg of chocolate-covered peanuts at 90¢ per $\frac{1}{2}$ kg, and 5.27 kg of mixed nuts at 130¢ per $\frac{1}{2}$ kg.
    (a) What was the total weight of the candy and nuts?
    (b) What was the monetary value of the supply of each type of candy and nuts?
    (c) What was the monetary value of all the candy and nuts?

32. Diane needs $2\frac{3}{4}$ m of material at $2.98 per meter, $4\frac{3}{4}$ m of material at $2.59 per meter, 5 spools of thread at $1.00 each, and 2 zippers at $1.50 each. Will a $20 bill cover the purchase?

33. Arrange in order, from smallest to largest.
    (a) 16.3, 16.259, 16, 16.2591
    (b) 0.01365, 0.013, 0.0136, 0.013654
    (c) 0.603, 0.063, 0.306, 0.036
    (d) 2.402, 2.204, 2.420, 2.024

34. Solve.
    (a) $2.16x + 7.3 = 9.05$
    (b) $0.02x - 17.6 = 41.5$

35. Perform the following operations (without using a calculator), leaving your answers in terms of decimals. (Do not round.) Then check your answers with a calculator.

    (a) $0.175 + 2\frac{3}{4}$     (b) $1\frac{1}{4} - 0.175$

    (c) $6\frac{3}{8} - 2.61$     (d) $-5.11 + 3\frac{7}{8}$

    (e) $2\frac{1}{2} \cdot 0.55$     (f) $-2.1 \cdot 3\frac{1}{4}$

    (g) $\dfrac{0.21 \cdot 2\frac{1}{4}}{1\frac{1}{2}}$     (h) $\dfrac{14.1 - 2\frac{1}{2}}{0.5}$

36. Arrange the following numbers from smallest to largest.

    (a) 6.71, 6.705, 6.715, 6.7015
    (b) 0.0036, 0.0035, 0.003, 0.00359
    (c) −0.46, −0.4, −0.45, −0.465

37. Multiply the following approximate numbers in scientific notation without using a calculator. Leave your answers in scientific notation. Then check your answers using a calculator.

    (a) $(3.4 \cdot 10^3) \cdot (2 \cdot 10^5)$
    (b) $(-1.6 \cdot 10^{-2}) \cdot (3 \cdot 10^{-3})$
    (c) $(1.61 \cdot 10^{-4}) \cdot (-2.1 \cdot 10^2)$
    (d) $(-1.811 \cdot 10^{-3}) \cdot (2.11 \cdot 10^{-2})$

38. Perform the divisions of the following approximate numbers in scientific notation without using calculator. Leave your answers in scientific notation. Now check your answers using a calculator.

    (a) $\dfrac{3.4 \cdot 10^{-3}}{2 \cdot 10^{-5}}$     (b) $\dfrac{6.21 \cdot 10^2}{3 \cdot 10^{-3}}$

    (c) $\dfrac{-6.541 \cdot 10^{-6}}{2.11 \cdot 10^{-4}}$     (d) $\dfrac{-6.522 \cdot 10^{-4}}{-2.1 \cdot 10^{-2}}$

39. Determine a fraction $\frac{a}{b}$ that corresponds to each of the following repeating decimals without a calculator. Check your answers with a calculator.

    (a) $14.\overline{23}$     (b) $0.072\overline{01}$
    (c) $30.0\overline{12}$     (d) $0.002\overline{46}$
    (e) $0.017\overline{9}$     (f) $6.5\overline{0}$

40. (a) Write $\frac{2}{3}$ as a repeating decimal.
    (b) Write $\frac{4}{6}$ as a repeating decimal.
    (c) Make a conjecture about repeating-decimal representations of equivalent fractions.

41. Add and subtract the following approximate numbers written in scientific notation without a calculator. Check your answers with a calculator.

    (a) $3.6(10)^3 - 2.1(10)^3 + 2.61(10)^3$
    (b) $7.8(10)^2 - 2.1(10)^2 - 2.61(10)^2$
    (c) $4.6(10)^3 + 1.02(10)^2 + 1.2(10)^4$
    (d) $3.4(10)^2 - 1.02(10)^3 - 1.4(10)$

42. "The product of two positive decimals (less than 1) is always greater than the smaller." Prove or disprove.

43. "The quotient of two positive decimals (less than 1) is always smaller than the numerator (when the quotient is expressed as a fraction)." Prove or disprove.

44. A share of OBM stock increased in value from $\$48\frac{1}{4}$ to $\$49\frac{3}{8}$. Joan owns 300 shares. How much did the value of her stock increase?

45. Perform the following operations.

    (a) $4.2_{\text{five}} \cdot 3.1_{\text{five}}$     (b) $7.\text{ET}_{\text{twelve}} \cdot 3.\text{T}_{\text{twelve}}$
    (c) $514.6_{\text{seven}} \cdot 2.43_{\text{seven}}$     (d) $10.101_{\text{two}} \cdot 0.011_{\text{two}}$
    (e) $1.01_{\text{two}} \cdot 1.11_{\text{two}}$     (f) $102.2_{\text{three}} \cdot 0.21_{\text{three}}$

## REVIEW EXERCISES

46. Perform the following computations on the approximate numbers, and round off answers to the appropriate number of decimal places.

    (a) $3.015 + 2.2 - 0.0074$
    (b) $14.671 - 3.1 + 0.0007$

47. Without a calculator compute $7 - 3.285 - 0.06$. (Do not round.) Check your answer with a calculator.

## LABORATORY ACTIVITIES

48. Illustrate the following multiplications, using a decimal square.

    (a) $(0.2) \cdot (0.6)$     (b) $(0.7) \cdot (0.8)$

49. Illustrate the following multiplications, using a rectangle.

    (a) $(1.3) \cdot (3.2)$     (b) $(0.05) \cdot (2.3)$

### ❖ PCR Excursion ❖

50. (a) Convert $0.\overline{1}$ to a fraction. Note the pattern for obtaining the numerator and the denominator, and then mentally write the fraction equivalent of $0.\overline{3}$, $2.\overline{6}$, and $3.\overline{7}$.

    (b) Convert $0.\overline{01}$ to a fraction and note the pattern. Then mentally convert to fractions $0.\overline{07}$, $0.\overline{23}$, and $4.\overline{31}$.

    (c) Convert $0.\overline{001}$ to a fraction and note the pattern. Then mentally convert to fractions $0.\overline{007}$, $0.\overline{023}$, $0.\overline{174}$, and $3.\overline{521}$.

    (d) Find a fraction to represent each of the following repeating decimals: $0.\overline{x}$, $0.\overline{xy}$, and $0.\overline{xyz}$.

    (e) Explain in words how you could obtain the fraction representation of numbers of the form $0.\overline{xyzw}$ (assuming no repetition of less than 4 digits).

(f) Explain in words how you would obtain the fraction representation of $0.\overline{xyzwv}$ (assuming no repetition of less than 5 digits).

(g) Find the 7th digit to the right of the decimal in the decimal expansion of $\frac{1}{11}$, the 99th digit, and the 1,147,922nd digit.

(h) Find the 15th digit to the right of the decimal in the decimal expansion of $\frac{1}{7}$, the 101st digit, and the 11,247,511th digit.

## ∵ PCR Excursion ∵

51. To convert a base ten fraction to its decimal representation, we divide the numerator by the denominator. In this way we can determine if the decimal representation is a terminating or repeating decimal. We now investigate how to answer similar questions in bases other than base ten.

A. (a) One method for changing a fraction to a representation involving a decimal point in another base requires that we rewrite the fraction in that base and then divide. For instance, to change $\frac{1}{6}$ and $\frac{3}{10}$ to representations involving a decimal point in base six we note that

$$\left(\frac{1}{6}\right)_{\text{ten}} = \left(\frac{1}{10}\right)_{\text{six}} \quad \text{and} \quad \left(\frac{3}{10}\right)_{\text{ten}} = \left(\frac{3}{14}\right)_{\text{six}}.$$

Dividing

Thus $\left(\dfrac{1}{6}\right)_{\text{ten}} = .1_{\text{six}}$ and $\left(\dfrac{3}{10}\right)_{\text{ten}} = 0.1\overline{4}_{\text{six}}$.

Do fractions that have terminating decimal representations in base ten necessarily have terminating decimal representations in base six?

(b) Find a representation using a decimal point in base six for the following base ten fractions.

(i) $\dfrac{1}{2}$    (ii) $\dfrac{2}{3}$    (iii) $\dfrac{3}{5}$    (iv) $\dfrac{1}{7}$

(c) Use the work from (a) and (b) to conjecture a rule for when a fraction has a terminating representation involving a decimal point in base six. (HINT: Study the base ten rule on p. 320.)

B. (a) Find a base seven representation involving a decimal point for each of the following fractions.

(i) $\dfrac{3}{7}$    (ii) $\dfrac{3}{10}$    (iii) $\dfrac{1}{14}$

(b) Conjecture a rule for when a fraction has a terminating representation involving a decimal point in base seven.

C. (a) When does a fraction have a terminating representation involving a decimal point in base eight? Demonstrate your answer by finding a fraction with a terminating representation and a fraction with a repeating representation.

(b) When does a fraction have a terminating representation involving a decimal point in base five? Demonstrate your answer by finding a fraction with a terminating representation and a fraction with a repeating representation.

# SECTION 3   THE LANGUAGE OF PERCENT

| PROBLEM | M-Mart plans a Fourth of July sale in which it will mark down all garden tools by 18%. What will be the sale price of a $50 wheelbarrow? |
|---|---|
| OVERVIEW | Percent is a familiar but elusive concept, one that many people have failed to master fully. Given that the term *percent* is used frequently in critical thinking and decision making—in addition to being a fundamental word in the language of the marketplace—it merits further study. Percents are used to compute taxes, to report unemployment and inflation rates, to determine sale prices and amounts of interest due on loans, and to perform a multitude of other critical tasks. |

**GOALS**

Illustrate the Number and Operations Standard, page 119.
Illustrate the Problem Solving Standard, page 15.
Illustrate the Representation Standard, page 50.
Illustrate the Connections Standard, page 147.

The word *percent* is derived from the Latin phrase *per centum* and means "by the hundred." Thus 46% could be correctly stated as "46 hundredths."

**DEFINITION**

**Percent**

A *percent* is a ratio giving the number of parts per hundred; thus, $n\% = n/100$.

As this definition suggests, percent is a fraction with a denominator of 100. Thus, to express $\frac{2}{5}$ as a percent, we must find the numerator of a fraction that is equivalent to $\frac{2}{5}$ and has a denominator of 100:

$$\frac{2}{5} = \frac{x}{100} \qquad \text{or} \qquad \frac{2}{5} = \frac{40}{100}$$

Consequently, $\frac{2}{5}$ expressed as a percent is 40%.

To convert a decimal fraction such as 0.016 into a percent, we must rewrite

$$0.016 = \frac{16}{1000}$$

as a fraction with a denominator of 100. We can do so by dividing the numerator and the denominator by 10:

$$0.016 = \frac{16}{1000} = \frac{1.6}{100}$$

Thus 0.016 can be written as 1.6%.

The preceding discussion illustrates that writing a decimal fraction as a percent merely involves moving the decimal point two places to the right and writing the symbol %.

**EXAMPLE 1**

Convert to percents: 0.047, 0.769, 3.56, and 0.00071.

*SOLUTION*

$$0.047 = 4.7\% \qquad\qquad 0.769 = 76.9\%$$
$$3.56 = 356\% \qquad\qquad 0.00071 = 0.071\%$$

**EXAMPLE 2**

Express as decimals.

(a) 1.7%   (b) 215%

*SOLUTION*

(a) $\dfrac{1.7}{100} = \dfrac{1.7(10)}{100(10)} = \dfrac{17}{1000} = 0.017$   (b) $\dfrac{215}{100} = 2.15$

| **EXAMPLE 3** | Convert the following to decimals: 2.12%, 0.012%, 324%. |
|---|---|
| *SOLUTION* | 2.12% = 0.0212; 0.012% = 0.00012; 324% = 3.24 |

## Calculator Hint

Many calculators have a key such as ▶% that allows us to express decimal answers as percents and a key % which converts percents to decimals.

(a) Convert $\dfrac{6}{1200}$ to a percent. (b) Convert $\dfrac{2\ 3/4}{6\ 7/8}$ to a percent.

(a) 6 [ / ] 1200 [ ▶% ] [ = ].   Did you get .5%?

(b) [ ( ] 2 [ Unit ] 3 [ / ] 4 [ ) ] [ ÷ ] [ ( ] 6 [ Unit ] 7 [ / ] 8 [ ) ] [ ▶% ] [ = ].
Did you get 40%?

Percent problems are worked easily by utilizing ratios. The procedure can be summarized as

$$\frac{\text{something is}}{\text{of a quantity}} = \frac{\text{percent}}{100}.$$

We also use the percent scale to assist in forming the ratios.

| **EXAMPLE 4** | Find 20% of 60. |
|---|---|

*SOLUTION*

Since 20% means 20/100, we want to find a number that compares to 60 as 20 compares to 100. Call this number $x$. (See Figure 7-6.) Then our proportion is

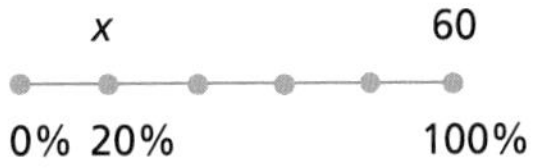

$$\frac{x}{60} = \frac{20}{100} \qquad \text{Ratio of numbers = ratio of percents}$$

$$100x = 1200 \qquad \text{If } \frac{a}{b} = \frac{c}{d}, \text{ then } ad = bc.$$

$$x = 12$$

x        60

0%  20%        100%

**FIGURE 7-6**

| **EXAMPLE 5** | What percent of 60 is 15? |
|---|---|

*SOLUTION*

Look at the percent scale model in Figure 7-7. Note that "15 is" and "of 60." Thus our proportion is

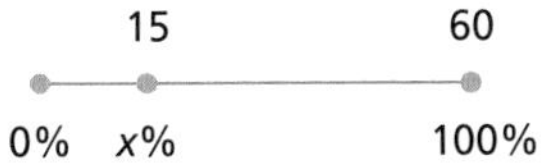

$$\frac{15}{60} = \frac{x}{100} \qquad \text{Ratio of numbers = ratio of percents}$$

$$60x = 1500 \qquad \text{If } \frac{a}{b} = \frac{c}{d}, \text{ then } ad = bc.$$

$$x = 25 \qquad \text{or} \qquad 25\%$$

15        60

0%   x%        100%

FIGURE 7-7

**EXAMPLE 6**

18 is 30% of what number?

*SOLUTION*

Look at the percent scale model in Figure 7-8. Here our proportion is

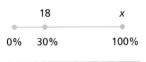

0%   30%   100%

**FIGURE 7-8**

$$\frac{18}{x} = \frac{30}{100} \qquad \text{Ratio of numbers = ratio of percents}$$

$$30x = 1800$$

$$x = 60$$

Another interpretation of a percent of a number is a decimal times the number. For instance, 8% of 40 can be interpreted as 0.08(40). Under this approach, we would write the same 3 equations that we used in the preceding exercises, where $x$ is the unknown. Notice that this approach answers directly the following 3 questions.

1. What is $p$ percent of $B$?
2. What percent is $A$ of $B$?
3. If $A$ is $p$ percent of $B$, then what is $B$?

**EXAMPLE 7**

(a) What is 20% of 60?   (b) What percent of 60 is 15?
(c) 18 is 30% of what number?

*SOLUTION*

(a) $x = 0.20(60) = 12$        (b) $15 = x \cdot 60$;    $x = 0.25 = 25\%$
(c) $18 = 0.30(x)$;    $x = 60$

**EXAMPLE 8**

Raymond, who is currently making $25,000 a year, receives a 5.5% raise. What is his new salary?

*SOLUTION*

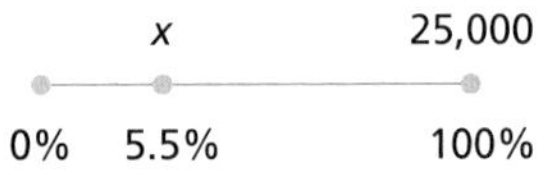

$x$   25,000

0%   5.5%   100%

**FIGURE 7-9**

First we compute his salary increase. Look at the percent scale in Figure 7-9.

$$\text{Increase in salary} = x = 0.055(25{,}000)$$

$$= \$1375$$

Using ratios, we tabulate the information as shown in Table 7-1.

Therefore, our proportion is

$$\frac{x}{\$25{,}000} = \frac{5.5}{100}$$

$$x = \frac{5.5(\$25{,}000)}{100}$$

$$= 0.055(\$25{,}000) = \$1375$$

**TABLE 7-1**

| Dollars | $x | $25,000 |
|---|---|---|
| Percent | 5.5% | 100% |

Thus, Raymond's new salary is

$$\$25{,}000 + \$1375 = \$26{,}375.00$$

Other problems that use percent involve solving simple equations.

**EXAMPLE 9**

Tom weighed 65 kg in January and lost 13 kg by April. What percent of his body weight had he lost?

*SOLUTION*

First look at the percent scale in Figure 7-10. Notice that $x$ is the unknown rate:

$$13 = x \cdot 65$$

$$x = \frac{13}{65} = 0.20 = 20\%$$

**FIGURE 7-10**

To use ratios, we would set up the data as shown in Table 7-2.

Thus, our proportion would be

| TABLE 7-2 | | |
|---|---|---|
| | *Loss* | *Total* |
| *Weight* | 13 | 65 |
| *Percent* | *x%* | 100% |

$$\frac{13}{65} = \frac{x}{100}$$

$$x = \frac{1300}{65} = 20\%$$

Tom lost 20% of his body weight between January and April.

**EXAMPLE 10**

Of the freshmen at Star College, 18% made A's in English. If 396 freshmen made A's in English, how many freshmen are enrolled in Star College?

*SOLUTION*

From the percent scale in Figure 7-11, we see that

$$0.18x = 396$$

$$x = \frac{396}{0.18} = 2200 \text{ freshmen}$$

**FIGURE 7-11**

Making a table and using ratios, we begin with the information in Table 7-3. Then we have as our proportion

$$\frac{18}{100} = \frac{396}{x}$$

$$x = \frac{396(100)}{18} = 2200$$

It follows that 2200 freshmen are enrolled in Star College.

| TABLE 7-3 | | |
|---|---|---|
| | *Made A's* | *Total* |
| *Number of Students* | 396 | *x* |
| *Percent* | 18% | 100% |

**PRACTICE PROBLEM**

A football game played in a stadium with a seating capacity of 65,000 is attended by 52,000 people. What percent of the seats are empty?

*ANSWER*

20%

**EXAMPLE 11**

92 is what percent of 400?

*SOLUTION*

From the percent scale in Figure 7-12, we see that

$$92 = x(400)$$

$$x = \frac{92}{400} = 0.23 = 23\%$$

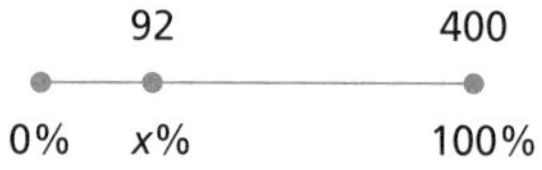

92        400

0%    x%        100%

**FIGURE 7-12**

By the alternative approach of using ratios, we begin with Table 7-4. The proportion we should use is therefore

$$\frac{92}{400} = \frac{x}{100}$$

$$x = \frac{9200}{400} = 23\%$$

**TABLE 7-4**

| Number | 92 | 400 |
|---|---|---|
| Percent | x% | 100% |

Consequently, 92 is 23% of 400.

In everyday experience, reference is frequently made to a change in percent.

**EXAMPLE 12**

The number 36 is what percent less than 48?

*SOLUTION*

First, find the change.

$$\text{Change} = 48 - 36 = 12$$

12        48

0%    x%        100%

**FIGURE 7-13**

Then from Figure 7-13, we have

$$12 = x \cdot 48$$

$$x = \frac{12}{48} = 0.25 = 25\%$$

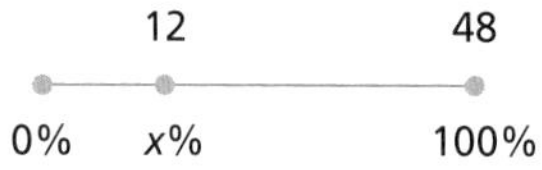

Using ratios, we start with Table 7-5. Our proportion in this case is

$$\frac{12}{48} = \frac{x}{100}$$

$$x = \frac{1200}{48} = 25\%$$

**TABLE 7-5**

| Number | 12 | 48 |
|---|---|---|
| Percent | x% | 100% |

Thus, 36 is 25% less than 48.

# BUSINESS TERMINOLOGY

A **commission** is defined as the commission rate times the sales price. A **markup** is usually the markup rate times cost. A **selling price** equals cost plus markup. A **markdown** is subtracted from a selling price to get a new selling price. The **markdown rate** is based on the original selling price. The **markup rate** can be based on the selling price, but this needs to be stated in a problem. We use these terms in the following examples.

**EXAMPLE 13**

A real estate agent receives a commission of 6% of the sales price of the property she sells. If a lot is sold for $38,000, what is the agent's commission?

*SOLUTION*

$$\text{Commission} = (\text{rate})(\text{sales price})$$
$$= (0.06)(38,000) = 2280$$

The commission is $2280. The seller of the lot receives $38,000 − $2280 = $35,720.

**EXAMPLE 14**

The Korner Drugstore marks up its prescription drugs by 42% of cost. If a certain drug costs $3 from the wholesale distributor, what will be its selling price at the store?

*SOLUTION*

$$\text{Markup} = (\text{rate})(\text{cost})$$
$$= (0.42)(3)$$
$$= \$1.26$$
$$\text{Retail price} = \$3.00 + \$1.26 = \$4.26$$

We now take up the section introductory problem.

**EXAMPLE 15**

M-Mart plans a Fourth of July sale in which it will mark down all garden tools by 18%. What will be the sale price of a $50 wheelbarrow?

*SOLUTION*

The wheelbarrow ordinarily sells for $50.00. Therefore,

$$\text{Markdown} = (0.18)(50) = \$9.00$$
$$\text{New sale price} = \$50.00 - \$9.00 = \$41.00$$

**PRACTICE PROBLEM**

Betty's Boutique marks up its merchandise by 65%. If a blouse costs $44 from the factory, what will be the selling price at the shop?

*ANSWER*

$72.60

**EXAMPLE 16**

The selling price of a suit of clothes is $200. If the cost of the article is $150, what is the percent of markup based on selling price?

*SOLUTION*

$$\text{Markup} = \text{selling price} - \text{cost}$$
$$= \$200 - \$150$$
$$= \$50$$
$$\text{Rate of markup} = \frac{50}{200} = 0.25$$

The percent of markup based on selling price is 25%.

# Just for Fun

The oil boom is on in the isolated frontier territory. Last year, 25% of the women in the territory got married, whereas 2.4% of the 1250 men married these local ladies. Assuming that no bigamy occurred, how many women are in the territory?

# EXERCISE SET 3

R 1. Convert to decimals.

(a) 5.5%    (b) 0.012%    (c) 0.31%
(d) 426%    (e) 43.6%    (f) 18.4%

2. Convert to percents without a calculator. Then check your answers with a calculator.

(a) 0.147    (b) 0.4472    (c) 0.005
(d) 0.086    (e) 8.61    (f) 23.14

3. Convert to percents without a calculator; then check with a calculator.

(a) $\dfrac{18}{15}$    (b) $\dfrac{33}{150}$    (c) $\dfrac{6}{15}$

(d) $\dfrac{18}{4}$    (e) $\dfrac{11}{40}$    (f) $\dfrac{143}{80}$

4. Using the fact that percent is a ratio, find the following.

(a) 17% of 200    (b) 23% of $1200
(c) 15% of 823    (d) 125% of 820
(e) 0.06% of 1100    (f) 6% of 88

5. Solve each part of Exercise 4 without using ratios.

6. A saleswoman makes a commission of 9% on all sales. If her monthly sales amount to $20,000, what is her commission that month?

7. Find the markup and retail price on the following goods if the percent of the markup is to be 30%.

| Item | Wholesale Cost |
| --- | --- |
| Paring knife | $1.20 |
| Electric clock | $8.40 |
| Can opener | $1.70 |
| Waste basket | $3.60 |

8. (a) What percent of 48 is 12?
(b) 24 is what percent of 96?
(c) Find 16% of 200.
(d) Find 0.01% of 1.632.
(e) What is 78% of 16?
(f) What percent of 150 is 6?
(g) What percent of 18 is 54?
(h) What number is 130% of 96?

9. Fill in the missing entries.

|   | Cost | % Markup Based on Cost | Selling Price |
|---|---|---|---|
| (a) | $1400 | 24 | — |
| (b) | $50 | — | $62.50 |
| (c) | — | 20 | $84 |

*T* 10. In some areas of business, the percent of markup is applied to the selling price rather than to the cost. In Exercise 9, compute the percent markup based on selling price for each article.

11. There are 23 women and 19 men in a freshman history class at Simmons College. Seven people failed the first test. What percent of the class failed the first test?

12. If your automobile payments are $400 per month, what percent of your $1900-per-month salary must be set aside to pay for your automobile?

13. In Exercise 12, suppose that you receive a 4% raise. What is your new monthly salary? What percent of your salary now must be set aside to pay for your automobile?

14. Of the 80,000 football seats in the stadium, 48,640 were filled. What percent of the stadium was filled?

15. If you spend 65% of your leisure time reading, how much leisure time do you have if you spend 13 hours per week reading?

16. A merchant sold a suit for $287, thereby gaining 30% above the suit's cost. What was the cost of the suit?

17. Mrs. Taylor considers buying a dress that costs $176, but she decides to have the dress made if she can save at least 25% of the cost. The materials for the dress cost $112. She finds out the dressmaker's fee and decides to buy the dress instead. What is the maximum amount of the fee?

18. Joan Makenny received a 12% raise on her salary of $21,500. What is her new salary?

19. U.S. oil imports from one nation increased from 22 million barrels to 29 million barrels in a 2-year period. What percent does this increase represent?

20. A newspaper reports that the Amalgamated Sweepers Union negotiated a raise of $0.40 per hour. If they were previously making $11.00 an hour, what is the percent of increase?

21. One serving of Wheats cereal contains 4% of the recommended daily allowance of protein. If one serving of Wheats contains 8 grams of protein, what is the recommended daily allowance for protein?

22. In a presidential election, only 54% of the registered voters voted. If 42 million went to the polls, how many registered voters were there?

23. The average white-collar government employee earned $17,350 in 1987 and $30,450 in 2000. What was the percent of increase?

24. (a) An airline ticket costs $284. A federal tax of 12% of this amount is added. What is the total amount of the ticket?
    (b) If the total cost of an airline ticket is $344, what is the amount of federal tax if the tax rate is 12%?

25. Sara Friedman makes $1200 a month plus 3% of all contracts she negotiates. If in a given month she generates $12,600 of business, what is her pay that month?

26. Fred Moss wants to sell T-shirts for $7.25. If he marks up his items by 35% over cost, what is the greatest amount he can pay for the shirts?

27. Thomas Sporting Goods buys boots for $69.95 and sells them for $110.50. What is the percent of markup based on cost?

28. What percent of markdown was used in each of the following sale prices?

**Handbag**
## Cut $2 to $11⁹⁹
Was **$13.99**

| Encyclopedia Bookcase | Highboy Bookcase |
|---|---|
| **Cut $15** | **Cut $20** |
| Now **$59⁸⁸** | Now **$99⁸⁸** |

*C* 29. What was the presale price on the lighting fixture advertised as follows?

**Cut 25%**
Now **$14⁹⁹**

30. Thomas Sporting Goods buys a racket at a wholesale price of $34.80. The racket is marked up by 45%, and then during the March sale it is marked down by 30%.
    (a) What is the sale price?
    (b) What percent of the markup on the wholesale cost does the sale price represent?

31. George Moore was hired at a yearly salary of $22,500. He received a raise of 11% and then received another of 6%. What is his current annual salary?

32. Sales at the Acme Agency increased by 30% in 1996, decreased by 20% in 1997, and increased by 10% in 1998. The sales in 1998 were what percent of sales in 1996?

33. A cotton-picking machine picks 70% of the cotton each time it goes over a field. If a farmer runs his machine over a field twice, what percentage of the cotton will be harvested?

34. Rahab receives a 12% salary increase. If his new salary is $3000 per month, what was his old salary?

35. Lai has decreased her weight by 8% in two weeks. If her weight now is 138 pounds, how much did she weigh two weeks ago?

36. June's salary increased from $22,000 to $33,080. What percent raise did she get?

## REVIEW EXERCISES

 37. Perform the following operations without a calculator. (Do not round.) Then check your answers with a calculator.
   (a) $2.001 \cdot 3.6$
   (b) $0.1266 \div 2.11$

 38. Without a calculator express $3.\overline{413}$ as a fraction. Then check your answer with a calculator.

---

 **SECTION 4**

# THE REAL NUMBER SYSTEM

| | |
|---|---|
| **PROBLEM** | A certain number, approximated to four decimal places, is 2.3135. Can you tell whether the original number is a rational number? |
| **OVERVIEW** | Do you know of decimals that neither terminate nor repeat? If so, then you are aware of numbers that are not rational. In fact you have discovered what we call *irrational numbers.* The rational numbers and the irrational numbers make up what we call the *real number system.* In the first part of this section, we encounter the concept of irrational numbers. Then we explore some interesting properties of real numbers. |
| **GOALS** | Illustrate the Number and Operations Standard, page 119.<br>Illustrate the Algebra Standard, page 4.<br>Illustrate the Representation Standard, page 50. |

## IRRATIONAL NUMBERS

In sections 1 and 2 of this chapter we learned that every rational number (a number that can be expressed as a fraction) can be represented by either a repeating decimal or by a terminating decimal. This implies that a number in decimal form that neither terminates nor repeats must represent a number other than a rational number. A decimal of this type represents what is called an **irrational number.**

If we construct an infinite decimal with no repeating cycle of digits, we have an irrational number. There are many such numbers. Consider

$$0.7377377737773777773 \ldots$$

where the number of 7's between successive 3's increases by 1 each time. This number is irrational because it is a nonterminating, nonrepeating decimal. It is nonrepeating because each pair of consecutive 3's has one more 7 between them than the preceding pair.

Probably the most famous irrational number is pi ($\pi$). Pi is the ratio of the circumference of a circle to its diameter. No matter what size circle we choose, if we measure the length around it and the length of the diameter, the ratio of these measures will be the same number, $\pi$. Hence, for any circle,

$$\pi = \frac{\text{circumference}}{\text{diameter}}$$

The number $\pi$ is not $3\frac{1}{7}$ or 3.1416; $\pi$ is an unending, nonrepeating decimal whose first few digits are given by 3.14159265358979323846.

The ancient Greeks spent a long time debating whether numbers other than rational numbers existed. In fact, the Pythagoreans, a group of religious mystics and mathematicians, long believed that all numbers were rational. But at some point they began to ponder the question: How long is the side of a square whose area is 2? (See Figure 7-14.) To investigate this question we, like the ancient Greeks, first need to define square root.

Define $\sqrt{3}$ to be the positive number that, when multiplied by itself, is equal to 3. In symbolic form,

$$\sqrt{3} \cdot \sqrt{3} = 3$$

Remember that the properties we developed in Chapter 4 for integral exponents hold when the exponents are any real numbers. For example,

$$3^{\frac{1}{2}} \cdot 3^{\frac{1}{2}} = 3^{\frac{1}{2} + \frac{1}{2}} \qquad \text{Add exponents.}$$
$$= 3^1 = 3$$

Thus $\sqrt{3}$ and $3^{\frac{1}{2}}$ represent the same number. In general,

$$\sqrt{a} = a^{\frac{1}{2}} \qquad (a \geq 0)$$

Note that $\sqrt{-1}$ does not exist because there is no real number $b$ such that $b = \sqrt{-1}$. If there were, then

$$b^2 = -1$$

This is impossible because $b^2 = b \cdot b$, and the product of any 2 positive numbers or any 2 negative numbers is always positive. Likewise, $\sqrt{-9}$, $\sqrt{-16}$, and $\sqrt{-25}$ do not exist in the real numbers. This discussion is summarized in the following definition.

Area = 2

$x$

FIGURE 7-14

---

**DEFINITION**

**Square Root**

(a) $b$ is the square root of $a$ if
$$b^2 = a \qquad (b \geq 0)$$

(b) $\sqrt{a}$ is the square root of $a$ with the property that
$$\sqrt{a} \cdot \sqrt{a} = a \qquad (a \geq 0)$$

(c) $\sqrt{a}$ can be written as $a^{\frac{1}{2}}$.

(d) $\sqrt{-a}$ where $a$ is a positive number is not a real number.

(e) $\sqrt{0} = 0$

For instance, $\sqrt{9} = 3$ because $3 \geq 0$ and $3^2 = 9$. Of course, $(-3)^2$ is also equal to 9, but $\sqrt{9} \neq -3$. Thus, $\sqrt{a}$ $(a \geq 0)$ always represents a positive number. When we write $\sqrt{25} = 5$, the 5 is sometimes called the **principal square root.** The symbol $\sqrt{\phantom{a}}$ is called a **radical sign,** and the $a$ in $\sqrt{a}$ is called a **radicand.**

**EXAMPLE 1**

Evaluate if possible.
(a) $\sqrt{16}$      (b) $-\sqrt{25}$      (c) $\sqrt{-9}$      (d) $\sqrt{0}$

SOLUTION

(a) 4      (b) –5      (c) Not a real number      (d) 0

**PRACTICE PROBLEM**

Evaluate.
(a) $\sqrt{36}$      (b) $\sqrt{3^4}$

ANSWER

(a) 6      (b) $3^2 = 9$

The area of a square is found by multiplying the length of a side times itself; therefore, turning back to Figure 7-14, $x \cdot x = 2$. The question is: What number $x$ multiplied by itself is equal to 2 (or what is the square root of 2, denoted by $\sqrt{2}$)? The Pythagoreans proved that $\sqrt{2}$ is not a rational number (and the proof was recorded in Euclid's *Elements*).

**EXAMPLE 2**

Prove that $\sqrt{2}$ is an irrational number.

SOLUTION

Let us assume that $\sqrt{2}$ is a rational number and then show that this leads to a contradiction, thus proving that $\sqrt{2}$ must be irrational. Assume that $\sqrt{2}$ is a rational number. Then it can be written as a quotient of two integers $a \div b$. If

$$\frac{a}{b} = \sqrt{2}$$

then

$$\left(\frac{a}{b}\right)^2 = 2 \quad \text{and} \quad a^2 = 2b^2$$

Factor $a$ into prime factors. One of these prime factors will be 2 because $a^2 = 2b^2$. Each factor of $a$ occurs twice in $a^2$, so $a^2$ contains an even number of 2's as factors. At the same time, $2b^2$ must contain an odd number of 2's as factors. But the fundamental theorem of arithmetic states that the factorization must be unique. Because we have reached an obvious contradiction, our hypothesis that $\sqrt{2}$ is a rational number must be false. Thus, $\sqrt{2}$ is irrational.

Using a similar argument, it can be proven that the square roots of all positive integers except the square numbers $\{1, 4, 9, 16, \ldots\}$ are irrational.

## Calculator Hint

Approximate values of square roots may be obtained on a calculator by using either the $\boxed{\sqrt{\phantom{a}}}$ key or the $\boxed{\wedge}$ key. For instance, to find $\sqrt{18}$, press

## Pythagoras
### 580?–497 B.C.

**T
H
E
N**

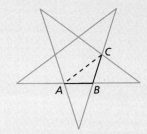

**N
O
W**

The Pythagorean Theorem that you will use in this section is one of the preeminent theorems in mathematics. Yet we only know about Pythagoras through his-torical accounts, for nothing that he wrote exists. Born on the Greek island of Samos, he traveled widely and studied mathematics and religion. He founded a secretive society based upon mathematical ideas, whose credo was "all is number." It is believed that the members of the cult, the Pythagoreans, coined the words *philosophy* (meaning love of wisdom) and *mathematics* (meaning that which is learned).

The Pythagoreans held their property in common and ate a strict vegetarian diet (but allowed no beans). Their symbol was the pentagram, which is known for its many geometric properties, such as the Golden Ratio, and its relevance to the biological and physical worlds.

The "all is number" philosophy of the Pythagoreans was just as sweeping as it sounds. They literally believed that *everything* could be explained by means of whole numbers and fractions with whole-number numerators and denominators.

It followed from their philosophy, then, that in a pentagon contained in a pentagram, the ratio of the diagonal AC to the side AB must be such a fraction. But it is not; it is, in fact $(\sqrt{5} + 1)/2$ in modern notation. (This number, approximately 1.618, is known as the Golden Ratio.) In your study of this section, you will recognize this fraction to be an irrational number. It was eventually proved after Pythagoras had died that this fraction could not be written as a fraction with a whole-number numerator and denominator. Legend holds that the man responsible for this proof was expelled from the Pythagorean society and put to death for his attack upon their symbol.

The Pythagorean Theorem you will use in this section is among the most widely used theorems in mathematics. It will be proved in Chapter 14 and used throughout the book.

$$\boxed{\sqrt{\phantom{x}}}\ 18\ \boxed{=},$$

which gives 4.24264. Alternatively, $\sqrt{18}$ can be written as $18^{\frac{1}{2}}$.

$$18\ \boxed{\wedge}\ .5\ \boxed{=},\ \text{which is } 4.24264$$

## THE REAL NUMBER SYSTEM

We have stated that every rational number can be represented by either a terminating decimal or a repeating decimal and that an irrational number can be represented by a nonrepeating decimal. The union of these two sets forms the real number system.

DEFINITION

**Real Number**

A *real number* is any number that is either a rational number or an irrational number. Thus,

Real numbers = {rational numbers} ∪ {irrational numbers}

We have previously shown that the rational numbers are dense and that infinitely many rational numbers lie between any two given rational numbers. Thus, it seems logical that the rational numbers fill up the number line. But this is not correct. In fact, we can demonstrate that the irrational numbers are also represented by points on the number line. To do this we introduce the Pythagorean Theorem for a right triangle (a triangle with a right or 90° angle). The Pythagorean Theorem will be discussed in greater detail in Chapter 12.

**Pythagorean Theorem**

In a right triangle, the square of the **hypotenuse** (the side opposite the right angle) is equal to the sum of the squares of the other two sides of the triangle. Expressed symbolically,

$$c^2 = a^2 + b^2$$

where $c$ is the hypotenuse and $a$ and $b$ are the other two sides.

In Figure 7-15 note that the hypotenuse $c$ is 5 and the other two sides of the right triangle $a$ and $b$ are 3 and 4. The Pythagorean Theorem states that

$$c^2 = a^2 + b^2 \quad \text{or}$$
$$5^2 = 3^2 + 4^2 \quad \text{or}$$
$$25 = 9 + 16 = 25$$

**FIGURE 7-15**

Now place one side of a square that has sides of length 1 (called a *unit square*) on a number line (as indicated in Figure 7-16(a)), draw a diagonal $c$, and consider the upper right triangle. By the Pythagorean Theorem,

$$c^2 = 1^2 + 1^2 = 2 \quad \text{or} \quad c = \sqrt{2}$$

Now rotate this diagonal clockwise about the point 0 until it lies on the number line. The end will mark a point on the number line that lies at a distance of $\sqrt{2}$ from the origin, and this point represents the irrational number $\sqrt{2}$. Similarly, if we rotate the diagonal counterclockwise about the origin, we can mark the point $-\sqrt{2}$. Thus, there exist points on the number line ($\sqrt{2}$ and $-\sqrt{2}$, for example) that cannot be represented by rational numbers.

(a)

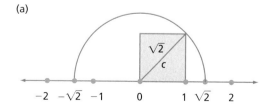

(b)

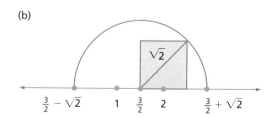

**FIGURE 7-16**

Next, move the unit square so that its lower left-hand corner is at $\frac{3}{2}$. (See Figure 7-16(b).) Then rotate the diagonal in both directions until it coincides with the number line. The points located at the ends of the diagonal correspond to $\frac{3}{2} + \sqrt{2}$ and $\frac{3}{2} - \sqrt{2}$. Is $\frac{3}{2} + \sqrt{2}$ a rational number or an irrational number? If we assume it to be a rational number $\frac{p}{q}$, then $\frac{3}{2} + \sqrt{2} = \frac{p}{q}$ or $\sqrt{2} = \frac{p}{q} - \frac{3}{2}$. But because the rational numbers are closed under the operations of addition and subtraction, $\frac{p}{q} - \frac{3}{2}$ must be a rational number, which implies that $\sqrt{2}$ is a rational number. This, of course, is a contradiction; consequently, $\frac{3}{2} + \sqrt{2}$ is an irrational number. This example illustrates that the sum of an irrational number and a rational number is an irrational number.

A similar argument can be used to show that the product of a nonzero rational number and an irrational number is irrational. Consider the product of 5 and $\sqrt{2}$. If we assume that this product is rational, it can be written as

$$5\sqrt{2} = \frac{a}{b}$$

$$\frac{1}{5}(5\sqrt{2}) = \frac{1}{5}\left(\frac{a}{b}\right) \qquad \text{where } a \text{ and } b \text{ are integers}$$

$$\sqrt{2} = \frac{a}{5b}$$

Now $\frac{a}{5b}$ is a rational number, which means that $\sqrt{2}$ is rational. This is a contradiction. Thus, the assumption is incorrect. Therefore, the product of a nonzero rational number and an irrational number is irrational.

**EXAMPLE 3**

Show that $\dfrac{3}{\sqrt{2}}$ is irrational.

*SOLUTION*

When we multiply the numerator and the denominator by $\sqrt{2}$, we get

$$\frac{3}{\sqrt{2}} = \frac{3\sqrt{2}}{\sqrt{2} \cdot \sqrt{2}} = \frac{3\sqrt{2}}{2} = \left(\frac{3}{2}\right)\sqrt{2}$$

which is irrational because the product of a nonzero rational and an irrational number is irrational.

The geometric process of placing the lower left-hand corner of the unit square at a rational number $\frac{a}{b}$ and then finding $\frac{a}{b} + \sqrt{2}$ and $\frac{a}{b} - \sqrt{2}$ always yields an irrational number. Therefore, there are as many irrational numbers of the form $\frac{a}{b} + \sqrt{2}$ or $\frac{a}{b} - \sqrt{2}$ as there are rational numbers $\frac{a}{b}$. The same argument could be used to show that $\frac{a}{b} + \pi, \frac{a}{b} + \sqrt{3}, \frac{a}{b} + \sqrt{5}$, and so on are irrational numbers. What does this say about the number of irrational numbers? It suggests that the irrational numbers are at least as numerous as the rationals.

Irrational numbers do not possess some important properties of rational numbers. For example, they are not closed under multiplication because

$$\sqrt{3} \cdot \sqrt{3} = 3$$

which is a rational number. Likewise, they do not have either an additive identity or a multiplicative identity because 0 and 1 are rational numbers.

In the system of real numbers, all numbers can be placed in one-to-one correspondence with all points on the number line; that is, every point on the number line represents a real number, and every real number can be represented by a point on the number line. The system of rational numbers does not have this characteristic because some points on the number line cannot be represented by rational numbers. Because every point on the number line represents a real number, we say that the real numbers are **complete,** in the sense that they account for every point on the entire number line.

Additional irrational numbers can be obtained by considering cube roots. For instance, $\sqrt[3]{27} = 3$ (rational) because $3 \cdot 3 \cdot 3 = 27$, but $\sqrt[3]{4}$ is irrational. Now $\sqrt[3]{27} = 27^{1/3}$ because

$$27^{1/3} \cdot 27^{1/3} \cdot 27^{1/3} = 27^{1/3+1/3+1/3} = 27$$

The preceding discussion leads to the following definition.

| DEFINITION | **Fractional Exponents** | For any real number $x$ and any natural number $n$, $$x^{1/n} = \sqrt[n]{x}$$ whenever $\sqrt[n]{x}$ is meaningful on the set of reals. |

**EXAMPLE 4**

Find the value.
(a) $8^{1/3}$    (b) $25^{-1/2}$

SOLUTION

(a) $8^{1/3} = \sqrt[3]{8} = 2$    (b) $25^{-1/2} = \dfrac{1}{25^{1/2}} = \dfrac{1}{\sqrt{25}} = \dfrac{1}{5}$

An irrational number that involves a radical is said to be in **simplest form** if all possible whole number factors have been removed from the radicand. Because we can easily compute the square roots of square numbers, in the next example look for numbers such as 4, 9, 16, 25, 36, 49, 64, 81, and 100. In each case we write the number under the radical as a product of a square number times another number.

**EXAMPLE 5**

Write in simplest form.
(a) $\sqrt{300}$    (b) $\sqrt{128}$

SOLUTION

(a) $\sqrt{300} = \sqrt{100 \cdot 3} = \sqrt{100} \cdot \sqrt{3} = 10\sqrt{3}$    $\sqrt{ab} = \sqrt{a} \cdot \sqrt{b}$
(b) $\sqrt{128} = \sqrt{64 \cdot 2} = \sqrt{64} \cdot \sqrt{2} = 8\sqrt{2}$

**EXAMPLE 6**

Simplify $\sqrt[3]{54}$.

SOLUTION

$$\sqrt[3]{54} = \sqrt[3]{27 \cdot 2} \qquad \text{Why did we select 27?}$$

$$= \sqrt[3]{27} \cdot \sqrt[3]{2} \qquad \sqrt[3]{ab} = \sqrt[3]{a} \cdot \sqrt[3]{b}$$

$$= 3\sqrt[3]{2}$$

A fraction involving radicals is said to be in simplest form if the denominator contains no radical expression. When we remove radicals from denominators, we say that the denominator is **rationalized.**

**EXAMPLE 7**

Rationalize the denominator of $\dfrac{5}{\sqrt{2}}$.

*SOLUTION*

We need to multiply the numerator and the denominator by a number that will make the quantity under the radical sign in the denominator a square number. Thus we select $\sqrt{2}$ for this purpose:

$$\frac{5}{\sqrt{2}} = \frac{5\sqrt{2}}{\sqrt{2}\sqrt{2}} = \frac{5\sqrt{2}}{\sqrt{4}} = \frac{5\sqrt{2}}{2} \qquad \sqrt{a}\cdot\sqrt{b} = \sqrt{ab}$$

**EXAMPLE 8**

Rationalize the denominator of $\dfrac{7}{\sqrt{8}}$.

*SOLUTION*

$$\frac{7}{\sqrt{8}} = \frac{7\sqrt{2}}{\sqrt{8}\sqrt{2}} = \frac{7\sqrt{2}}{\sqrt{16}} = \frac{7\sqrt{2}}{4} \qquad \sqrt{a}\cdot\sqrt{b} = \sqrt{ab}$$

The arithmetic operations of addition, subtraction, multiplication, and division can be performed with irrational numbers by using the properties listed in this section. Using the distributive properties of multiplication over addition, we can add (or subtract) numbers with the same radical and radicand by adding their coefficients.

**EXAMPLE 9**

Add $3\sqrt{5} + 2\sqrt{5}$.

*SOLUTION*

$$3\sqrt{5} + 2\sqrt{5} = (3 + 2)\sqrt{5} = 5\sqrt{5}$$

## Calculator Hint

Check your answer in Example 9 with a calculator.

$3 \;\boxed{\times}\; \boxed{\sqrt{}}\; 5 \;\boxed{)}\; \boxed{+}\; 2 \;\boxed{\times}\; \boxed{\sqrt{}}\; 5 \;\boxed{)}\; \boxed{=}$; and

$5 \sqrt{5}$ is $5 \;\boxed{\times}\; \boxed{\sqrt{}}\; 5 \;\boxed{=}$. Both of these equal 11.18033989.

**EXAMPLE 10**

Write an irrational number that is greater than $\frac{2}{3}$ and less than $\frac{3}{4}$.

*SOLUTION*

First, we write the two rational numbers as decimals.

$$\frac{2}{3} = 0.66666\overline{6}$$

$$\frac{3}{4} = 0.75$$

The irrational number 0.6676776777 . . . satisfies the requirement. Now, $\frac{3}{4} - \frac{2}{3} = \frac{1}{12}$, and $\sqrt{2}/1000$ is obviously less than $\frac{1}{12}$. Therefore, $\frac{2}{3} + \sqrt{2}/1000$ also satisfies our requirement.

In the preceding example, if we wished to place 10 irrational numbers between $\frac{2}{3}$ and $\frac{3}{4}$, we could write them as

$$\frac{2}{3} + \frac{\sqrt{2}}{1000}, \frac{2}{3} + \frac{2\sqrt{2}}{1000}, \frac{2}{3} + \frac{3\sqrt{2}}{1000}, \cdots \frac{2}{3} + \frac{10\sqrt{2}}{1000}$$

## CALCULATOR HINT

One other calculator operation that could help you to understand irrational numbers is $\boxed{\sqrt[x]{\phantom{x}}}$, which gives you the $x$th root of a number.

(a) Find the $\sqrt[3]{5}$ and $5^{1/3}$ and compare answers.

(b) Find $\sqrt[7]{2}$ and $2^{1/7}$ and compare answers.

(a) $3 \boxed{\sqrt[x]{\phantom{}}} 5 \boxed{=} 1.70998$ and $5 \boxed{\wedge} \boxed{(} 1 \boxed{/} 3 \boxed{)} \boxed{=} 1.70998$. You have demonstrated that $\sqrt[3]{5} = 5^{1/3}$.

(b) $7 \boxed{\sqrt[x]{\phantom{}}} 2 \boxed{=} 1.104090$ and $2 \boxed{\wedge} \boxed{(} 1 \boxed{\div} 7 \boxed{)} \boxed{=} 1.04090$. Likewise, you have demonstrated that $\sqrt[7]{2} = 2^{1/7}$.

By similar reasoning, it follows that the number of irrational numbers between each pair of rational numbers is infinite.

# PROPERTIES OF THE REAL NUMBER SYSTEM

In our extension of the system of rational numbers to the system of real numbers, we demonstrated that some numbers, called *irrational numbers,* cannot be represented as fractions. Nonetheless, these numbers obey the same rules for the operations of addition, subtraction, multiplication, and division as rational numbers because both were combined into the system of real numbers through the use of unending decimals. This presentation, along with geometric plausibility, suggests that the real number system satisfies the following properties.

**Properties of the Real Number System**

1. It is closed under addition.
2. It is closed under subtraction.
3. It is closed under multiplication.
4. It is closed under division except for division by 0.
5. The elements of the set satisfy the commutative and associative properties of addition and multiplication.
6. The elements of the set satisfy the distributive property of multiplication over addition.
7. There exist an additive identity, additive inverses, a multiplicative identity, and multiplicative inverses (except for 0).
8. It is an ordered set.

The properties of equality and inequality that hold for rational numbers also hold for real numbers. Because each point on the number line corresponds to a real number, the number line provides a good way to illustrate the real numbers that satisfy inequalities.

**EXAMPLE 11**

Show the values of $x$ that satisfy $x < 5$ on a number line.

SOLUTION

In Figure 7-17, the circle around 5 indicates that $x = 5$ is not part of the solution set.

**EXAMPLE 12**

Graph the solution set of $x + 5 \geq 2$.

SOLUTION

We are given that $x + 5 \geq 2$, so

$$(x + 5) - 5 \geq 2 - 5$$

$$x \geq -3$$

The graph in Figure 7-18 shows −3 as a solid dot, since it is part of the solution set.

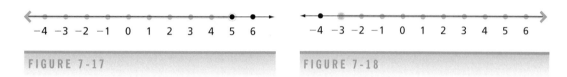

$$\begin{array}{ccccccccccc} -4 & -3 & -2 & -1 & 0 & 1 & 2 & 3 & 4 & 5 & 6 \end{array}$$

FIGURE 7-17

$$\begin{array}{ccccccccccc} -4 & -3 & -2 & -1 & 0 & 1 & 2 & 3 & 4 & 5 & 6 \end{array}$$

FIGURE 7-18

Figure 7-19 indicates that every negative integer is also an integer, a rational number, and a real number. Rational numbers that are not integers make up a subset of the rational numbers, which is a subset of the real numbers. Unless stated otherwise, all computations in the remainder of this book will be on the system of real numbers.

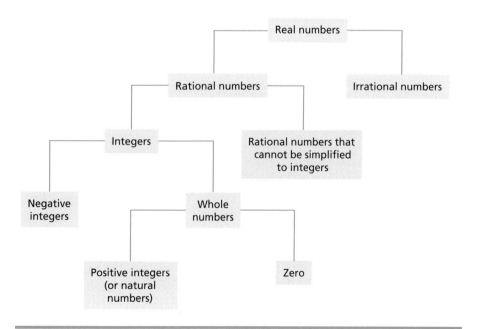

FIGURE 7-19

# Just for Fun

A rubber ball is known to rebound half the height it drops. If the ball is dropped from a height of 100 feet, how far will it have traveled by the time it hits the ground for
1. The first time?
2. The second time?
3. The third time?
4. The fourth time?
5. The fifth time?
6. The $n$th time? Look for a pattern.

# EXERCISE SET 4

1. Classify the following numbers as rational or irrational.
   (a) $\sqrt{3} - 1$    (b) $6 \cdot \sqrt{3}$    (c) $2 \div \sqrt{2}$
   (d) $\dfrac{\sqrt{2}}{4\sqrt{2}}$    (e) $\dfrac{\sqrt{2}}{\sqrt{3}}$    (f) $\dfrac{57}{1001}$
   (g) 0.253125312531 . . .
   (h) 26.1311311131111 . . .
   (i) $5\pi$    (j) 0    (k) 3.1416

2. If $R$ = {real numbers}, $I$ = {integers}, $Q$ = {rationals}, $W$ = {whole numbers}, and $Z$ = {irrationals}, classify the following as true or false.
   (a) $R \cap I = R$    (b) $R \cup Q = Q$
   (c) $I \cap Q = Q$    (d) $Q \cap W = W$
   (e) $Q \cap Z = \varnothing$    (f) $Q \subset R$
   (g) $Q \subset I$    (h) $I \subset Q$
   (i) $R \subset Z$    (j) $I \subset (Q \cap R)$
   (k) $I \cap (Q \cup Z) = I$
   (l) A repeating, nonterminating decimal is a rational number.
   (m) A nonrepeating, nonterminating decimal is an irrational number.
   (n) $Q \cup Z = R$    (o) $Z$ is closed under addition.
   (p) $Z$ is closed under multiplication.
   (q) $R$ is closed under both addition and multiplication.

3. Complete the following chart where A is a whole number, B an integer, C a rational number, D an irrational number, and E a real number.

| Number | A | B | C | D | E |
|---|---|---|---|---|---|
| $-5$ | No | Yes | Yes | No | Yes |
| $2.\overline{71}$ | No | No | Yes | No | Yes |
| 0.717117111 . . . | | | | | |
| 0 | | | | | |
| $5\sqrt{2}$ | | | | | |
| $3\pi$ | | | | | |
| $0.\overline{45}$ | | | | | |
| 4 | | | | | |

$T$  4. (a) Is there a greatest whole number less than 8?
   (b) Is there a greatest integer less than 8?
   (c) Is there a greatest rational number less than 8?
   (d) Is there a greatest irrational number less than 8?
   (e) Is there a greatest real number less than 8?
   (f) What property is involved in these questions?

5. A certain number, approximated to 4 decimal places, is 2.3135. Can you tell whether the original number is rational?

6. Show with examples each of the following:

   (a) The product of irrational numbers may be rational or irrational.

   (b) The sum of irrational numbers may be rational or irrational.

   (c) The difference of two irrational numbers may be rational or irrational.

   (d) The quotient of two irrational numbers may be rational or irrational.

7. Demonstrate with an example that the sum of an irrational number and a rational number is irrational.

8. Demonstrate with an example that the product of an irrational number and a nonzero rational number is irrational.

9. (a) How many integers are between $-2$ and 3?

   (b) How many rational numbers are between $-2$ and 3? Name three of them.

   (c) How many irrational numbers are between $-2$ and 3? Name three of them.

10. (a) How many counting numbers are between 0 and 10?

    (b) How many integers are between 0 and 10?

    (c) How many rational numbers are between 0 and 10?

    (d) How many irrational numbers are between 0 and 10?

    (e) How many real numbers are between 0 and 10?

11. Arrange the following real numbers from smallest to largest.

    $$0.75\overline{1}, 0.\overline{75}, 0.7510110111\ldots, 0.755, \frac{3}{4}$$

12. Arrange the following real numbers from smallest to largest.

    $$0.45, 0.\overline{4}, 0.44, 0.454454445\ldots, 0.\overline{45}, 0.455$$

13. Write without radicals.

    (a) $\sqrt[3]{64}$     (b) $\sqrt[4]{81}$

    (c) $\sqrt[5]{32}$     (d) $\sqrt{225}$

14. Classify as true or false.

    (a) $\sqrt{x} = -8$ for some real $x$

    (b) $\sqrt{17} = 4.12\overline{3}$

    (c) $\pi = 22/7$

    (d) $\sqrt{-x} = 4$ for some real $x$

15. Find 2 irrational numbers between $0.\overline{46}$ and $0.\overline{47}$.

16. Find 2 irrational numbers between 0.24 and 0.25.

17. Find 3 irrational numbers between 4.5 and 4.6.

18. Let $a = 0.565\overline{2}$ and $b = 0.5652020020002\ldots$.

    (a) Find a rational number between $a$ and $b$.

    (b) Find an irrational number between $a$ and $b$.

19. Evaluate the following:

    (a) $\sqrt{49}$     (b) $\sqrt{121}$

    (c) $\sqrt[3]{27}$     (d) $\sqrt[4]{16}$

    (e) $-\sqrt{81}$     (f) $-\sqrt{196}$

20. Simplify the following radicals.

    (a) $\sqrt{32}$     (b) $\sqrt{75}$

    (c) $\sqrt{125}$     (d) $\sqrt{162}$

21. Perform the indicated operation without a calculator. Then check with a calculator.

    (a) $\sqrt{5} + 2\sqrt{5}$     (b) $\sqrt{7} - 4\sqrt{7}$

    (c) $3\sqrt{18} - 5\sqrt{2}$     (d) $4\sqrt{6} - \sqrt{24}$

    (e) $4\sqrt{12} + 3\sqrt{48}$     (f) $\sqrt{20} + \sqrt{125}$

22. Evaluate without a calculator. Then check with a calculator.

    (a) $36^{1/2}$     (b) $9^{-1/2}$

    (c) $25^{-1/2}$     (d) $(-27)^{-1/3}$

    (e) $(-8)^{1/3}$     (f) $(-27)^{1/3}$

23. Perform the indicated operations without a calculator. Then check with a calculator.

    (a) $\sqrt{50}$     (b) $\sqrt{384}$

    (c) $\sqrt{\dfrac{147}{36}}$     (d) $\sqrt{\dfrac{450}{169}}$

24. Solve each of the following, and locate approximately the solution on a number line.

    (a) $4x + \sqrt{2} = 8$     (b) $\sqrt{2}x - 3 = 5$

    (c) $2y + 3 = 4\sqrt{3}$     (d) $\sqrt{3}y + 2 = 7$

    (e) $\pi x - \sqrt{5} = 2$     (f) $3\pi x - \sqrt{5} = 2\pi x - 1$

25. Rationalize the denominator of each.

    (a) $\dfrac{5}{\sqrt{2}}$     (b) $\dfrac{7}{\sqrt{5}}$     (c) $\dfrac{\sqrt{7}}{\sqrt{2}}$

    (d) $\dfrac{\sqrt{14}}{\sqrt{6}}$     (e) $\dfrac{\sqrt{15}}{\sqrt{7}}$     (f) $\dfrac{\sqrt{10}}{\sqrt{6}}$

    (g) $\dfrac{\sqrt{3}}{\sqrt{8}}$     (h) $\dfrac{\sqrt{5}}{\sqrt{18}}$

26. Without a calculator, simplify as much as possible. Then check with a calculator.

    (a) $\dfrac{4^{-2} \cdot 5^{-1}}{2}$     (b) $(10^{1/3})^{-6}$

    (c) $\dfrac{32^{-3/5}}{16^{-1/4}}$     (d) $\dfrac{16^{-1/2}}{4^{-3}}$

27. To 8 decimal places, $\pi = 3.14159265\ldots$. In what decimal place does the decimal representation for each fraction first differ from $\pi$?

    (a) $\dfrac{22}{7}$     (b) $\dfrac{333}{106}$     (c) $\dfrac{355}{113}$

    (d) $\dfrac{31,412}{9999}$     (e) $\dfrac{1046}{333}$

28. If $x$ and $y$ are positive integers such that $x > y$, which one of the following is true?

$$\frac{1}{x} > \frac{1}{y}, \frac{1}{x} = \frac{1}{y}, \frac{1}{x} < \frac{1}{y}$$

Why?

29. Simplify the following.
    (a) $\sqrt[3]{-81}$     (b) $\sqrt[4]{64}$     (c) $\sqrt[3]{-24}$
    (d) $\sqrt[5]{-32}$     (e) $-\sqrt[3]{-8}$     (f) $\sqrt{\dfrac{-16}{-2}}$

C 30. Place four irrational numbers between 0.1 and 0.2.

31. Place ten irrational numbers between $\frac{1}{4}$ and $\frac{1}{3}$.

32. Show the solution set of $|x| \geq 2$ on a number line.

33. Show the solution set of $|x| \leq 2$ on a number line.

34. Show the solution set of $|x - 2| \leq 4$ on a number line.

35. Show the solution set of $|x - 3| \geq 9$ on a number line.

36. Prove or disprove that $\sqrt{p}$ is always irrational, where $p$ is any prime.

37. Use your calculator to find approximations of the following.
    (a) $\sqrt{268.21}$
    (b) $\sqrt{0.023}$
    (c) $\sqrt{0.002175}$
    (d) $\sqrt{4682.1}$

38. Find the square roots expressed in scientific notation of the following numbers.
    (a) $\sqrt{2.16(10)^4}$
    (b) $\sqrt{3.04 (10)^{-4}}$
    (c) $\sqrt{1.67(10)^{-3}}$
    (d) $\sqrt{8.51(10)^3}$

39. Use your calculator to discover a pattern for the following.

$$\sqrt{16.4125}, \sqrt{1641.25}, \sqrt{164{,}125}$$

Then approximate $\sqrt{.164125}$.

40. Prove that $\sqrt[3]{2}$ is irrational.

41. Show that

$$b + \frac{1}{b} \geq 2$$

   (a) for $b = 2$
   (b) for $b = 0.5$
   (c) for $b = 4$
   (d) for $b = n$, a positive integer

42. Prove that $\sqrt{3}$ is irrational.

43. Prove or disprove that

$$\sqrt{a + b} = \sqrt{a} + \sqrt{b}$$

44. Prove that each of the following is an irrational number.
    (a) $3\sqrt{2}$
    (b) $3 + \sqrt{2}$
    (c) $\dfrac{3}{\sqrt{2}}$
    (d) $\sqrt{2} - 3$
    (e) $3 - 2\sqrt{2}$
    (f) $\dfrac{3}{2} + \sqrt{2}$
    (g) $2 + \sqrt{3}$

45. Add $2.\overline{4} + 1.\overline{31}$, and give a fraction answer.

46. Use base three notation to prove that $\sqrt{2}$ is not rational. (HINT: The definition of a rational number still holds in base three. Now look at the units digit of the square of any integer in base three.)

47. Indicate whether each statement is true or false. If false, explain why.
    (a) 4 is an irrational number.
    (b) Every rational number is a real number.
    (c) Every point on a number line represents a real number.
    (d) Every real number is a rational number.
    (e) The irrational numbers are closed under multiplication.
    (f) Every repeating decimal represents a real number.
    (g) Every irrational number can be represented by a repeating decimal.
    (h) $\frac{a}{0}$ is a rational number.
    (i) If $a$ and $b$ are rational numbers with $ab = 0$ and $b = 0$, then $a = 0$.
    (j) Division is always possible in the rational number system.
    (k) $\frac{2}{3}$ is the largest rational number less than 3.
    (l) 1 is the smallest positive integer as well as the smallest positive rational number.
    (m) 8 is the largest integer less than 9.
    (n) 0 is a positive rational number.
    (o) The set of rational numbers between 2 and 3 is dense.
    (p) If $a$ is any rational number, $a^2 > 0$.
    (q) $-1$ is the largest negative rational number.

## REVIEW EXERCISES

48. Compute the following.
    (a) 0.5% of $660     (b) 105% of 77
    (c) $\frac{9}{8}$ as a percent     (d) 1.007 as a percent

49. (a) An $89 coat is sold for $65. What is the percent of markdown?
    (b) John earns a commission of 4% on all sales in excess of $10,000. What is his commission in a month in which he sells $28,000 of merchandise?
    (c) A hat has a retail price of $57.75. What was the cost if the hat was marked up 65% based on cost?

50. Find the value in base ten of $3.02_{four}$.

## LABORATORY ACTIVITIES

51. Geometrically construct a representation of $\sqrt{3}$ on the number line.

52. Geometrically construct a representation of $\sqrt{4}$ on the number line, and note that it is 2.

# SOLUTION TO INTRODUCTORY PROBLEM

**UNDERSTANDING THE PROBLEM.** The numbers in the square are repeating decimals and terminating decimals.

**DEVISING A PLAN.** The addition involves terminating decimals and repeating decimals. Find the sum of the three numbers on the diagonal. If you have difficulty, you can express each as a fraction. Once this sum is attained, one can complete the square.

**CARRY OUT THE PLAN.**

| $.8\overline{3}$ | 3.75 | $1.\overline{6}$ |
|---|---|---|
| $2.91\overline{6}$ | $2.08\overline{3}$ | 1.25 |
| 2.5 | $0.41\overline{6}$ | $3.\overline{3}$ |

**LOOKING BACK.** Each sum should be 6.25.

# SUMMARY AND REVIEW

## PROPERTIES OF DECIMALS

1. To divide a number $N$ by $10^k$, where $k$ is a natural number, move the decimal point (in $N$) $k$ places to the left. In like manner, to multiply by $10^k$, move the decimal point $k$ places to the right.

2. An algorithm for adding (or subtracting) decimals consists of lining up the decimal points of the numbers in a vertical column, adding without reference to the decimal point, and then placing the decimal point in the answer directly underneath the given decimal points.

3. Multiplication of decimals: If there are $r$ decimal places in one factor and $s$ decimal places in a second factor, then there will be $r + s$ decimal places in the product.

4. When a decimal is divided by a whole number, the decimal point in the quotient is placed directly above the decimal point in the dividend.

5. A rational number $\frac{a}{b}$ (in reduced form) can be expressed as a terminating decimal fraction if and only if the prime factorization of $b$ is of the form $b = 2^x \cdot 5^y$, where $x$ and $y$ are whole numbers.

6. A *terminating decimal* is one that repeats zeros after a specific number of decimal places.

7. A *repeating decimal* is a nonterminating decimal in which (after a certain point) the same pattern of digits repeats endlessly.

8. Every rational number can be represented by an integer or by a decimal fraction that either terminates or eventually repeats; conversely, every decimal that either terminates or eventually repeats represents a rational number.

## ROUNDING NUMBERS

1. If the digit to the right of the last digit to be retained is less than 5, leave the last digit to be retained unchanged.

2. If the digit to the right of the last digit to be retained is 5 or greater, increase the last digit to be retained by 1.

3. In scientific notation, a number $N$ is written $N = A \cdot 10^n$, where $1 \leq A < 10$ and $n$ is an integral exponent. The necessary digits of $A$ are called *significant digits*.

4. For an approximate number with 1 decimal place, the greatest possible error is 0.05; for 2 decimal places, 0.005; for 3 decimal places, 0.0005.

## OPERATIONS WITH APPROXIMATE NUMBERS

1. When adding (or subtracting) approximate numbers, round the answer to the same decimal place as the term with the greatest possible error.
2. When multiplying (or dividing) approximate numbers, give the answer the same number of significant digits as the factor with the smallest number of significant digits.

## PERCENT

Percent is a ratio giving the number of parts per hundred.

## THE REAL NUMBERS

1. A number in decimal form which neither terminates nor repeats is an irrational number.
2. The real number system consists of rational numbers and irrational numbers.
3. Real numbers in exponential or radical form satisfy the following properties:
   (a) $\sqrt{x} \cdot \sqrt{x} = x$ and $\sqrt{x} = x^{1/2}$
   (b) $\sqrt[3]{x} \cdot \sqrt[3]{x} \cdot \sqrt[3]{x} = x$ and $\sqrt[3]{x} = x^{1/3}$
   (c) $\sqrt[n]{x} = x^{1/n}$

# CHAPTER TEST

1. Multiply $1.5(0.003) \cdot 5.2$. (Do not round.)
2. Simplify.
   (a) $\sqrt{192}$
   (b) $\dfrac{27^{-1/3}}{3^{-2}}$
3. Determine the fraction that corresponds to each of the following.
   (a) $3.66\overline{0}$
   (b) $0.24\overline{9}$
4. Find two irrational numbers between $2.\overline{43}$ and $2.\overline{44}$.
5. Classify as rational or irrational.
   (a) $\dfrac{3\sqrt{2}}{2\sqrt{2}}$
   (b) $16.0314314431444\ldots$
   (c) $5\sqrt{2} - \sqrt{2}$
6. Round the following numbers to 2 decimal places.
   (a) $3.1656$
   (b) $7.2450$
   (c) $13.235$
7. Write the following numbers in scientific notation.
   (a) $0.00021$
   (b) $467.300$

8. Multiply the following approximate numbers.

$$31.46 \cdot 0.21$$

9. Subtract the following approximate numbers.

$$431.6 - 3.271$$

10. 120% of what number is 60?
11. Divide $1.216 \div 0.04$. (Do not round.)
12. (a) Multiply $31.06 \cdot 4.612$. (Do not round.)
    (b) Change each decimal fraction in (a) to a common fraction, and multiply.
    (c) Show that the answers in (a) and (b) are the same.
13. Write $1.2\overline{54}$ as a reduced fraction.
14. Solve $\frac{x}{2} - 4 < 8$ on the real numbers, and draw a graph on a number line showing your solution.
15. Todd's salary was increased by 15%. If his salary now is \$2400 a month, what was his original salary?
16. Solve for $x$:

$$\pi x - 3 = \sqrt{5}x - \sqrt{3}$$

# CONSUMER MATHEMATICS

8

**T**hree men attending a convention rented a room at a nearby hotel for $30, so they paid $10 each. However, the desk clerk realized that the room cost only $25, so he sent the bellboy upstairs with $5 to return to the 3 men. The bellboy, rewarding himself with a $2 tip, returned only $3 to the men. Yet if each of the men received $1, then each man paid $9 for the room. But 3($9) + $2 = $27 + $2 = $29. What happened to the extra dollar?

Mathematics has been part of the language of the marketplace ever since the first nomad tried to decide whether to trade two goats for three woolen robes or three goats for four woolen robes. The abacus and other simple computing devices were inspired by the needs of problem solving in the marketplace. As a citizen of the modern world, you have replaced the abacus with a hand calculator or a computer, but you must still master the language of the marketplace—discount, percent, markup, finance charge, simple interest, compound interest, annuities, and so on. Some of these concepts were introduced in Chapter 7. We use a calculator approach to explore the remaining topics in this chapter. The calculator approach is significant for two reasons: you can find answers for percentages not usually given in tables, and you can find answers when tables are not available. You are encouraged to use the calculator on most problems. For this reason we have not designated certain problems as calculator problems.

In this chapter, we use the following problem-solving strategies:  ▪ Guess, test, and revise. (375) ▪ Work backward. (375)  ▪ Draw a diagram. (sixteen line diagrams)  ▪ Make a chart or a table. (382, 388)  ▪ Use reasoning. (381, 387, 388)  ▪ Find a pattern. (372, 378, 383, 387, 388)

# SECTION 1 SOME COMPARISONS OF INTEREST RATES

**PROBLEM**

Kate deposited $100 in a savings account that paid 6% per year compounded quarterly. At the end of the year, her statement indicated that she had $106 in her account. She believes there is a mistake. Is she right or wrong?

**OVERVIEW**

The phrase *compounded quarterly* indicates that this problem involves compound interest. What is the difference between compound and simple interest? You will learn the difference in this section and use both concepts to solve a variety of problems. To assist you in solving problems, we will use models called *time diagrams*.

**GOALS**

Illustrate the Algebra Standard, page 4.*
Illustrate the Connections Standard, page 147.*
Illustrate the Representation Standard, page 50.*
Illustrate the Problem Solving Standard, page 15.*

* The complete statement of the standard is given on this page of the book.

## SIMPLE INTEREST

The cost of borrowing money is called **interest.** Interest is the "rent" we pay the bank for using its money to purchase things such as automobiles. The bank pays us interest on money in our savings accounts. Simple interest is computed as a constant percentage of the money borrowed for a specific time (usually a single year or less) and is paid at the end of the specified time.

The sum borrowed is called the **principal** $P$ or sometimes the **present value;** $r$ denotes the rate of interest, which is usually expressed in percent per year; and $t$ is time, expressed in years or fractions of years. By definition, simple interest $I$ equals principal multiplied by the interest rate multiplied by the time in years.

$$I = Prt$$

**EXAMPLE 1**

A sum of \$400 is borrowed at 4% simple interest for 3 years. What is the interest?

$$I = (\$400)(0.04)(3) = \$48 \qquad \text{4\% as a decimal is 0.04.}$$

At the end of the term of the loan, the borrower must pay not only the interest but also the sum that was originally borrowed, the principal. Hence, the amount due at the end of the term is given by

$$Amount = principal + interest$$
$$A = P + I = P + Prt$$

Using the distributive property, we can write this formula as

$$A = P(1 + rt)$$

---

**Simple Interest and Amount Due**

The formulas for *simple interest I* and *amount due A* are

$$I = Prt$$
$$A = P + Prt$$
$$= P(1 + rt)$$

where $P$ = principal or present value, $r$ = annual simple-interest rate, $t$ = time (in number of years), and $A$ = amount or future value.

---

**EXAMPLE 2**

A loan of \$1000 is made for 6 months at a simple interest rate of 8%. How much does the borrower owe at the end of 6 months?

*SOLUTION*

In the formula replace $P$, $r$, and $t$ with their values:

$$P = \$1000$$
$$r = 0.08$$
$$t = \frac{1}{2} \qquad \text{Convert 6 months to years; } \frac{6}{12} = \frac{1}{2} \text{ year.}$$
$$A = P(1 + rt) = \$1000\left[1 + (0.08)\left(\frac{1}{2}\right)\right]$$

In this example and many that follow we will complete the computations in the Calculator Hint.

## CALCULATOR HINT

$$A \;=\; 1000 \;\boxed{\times}\;\boxed{(}\; 1 \;\boxed{+}\; .08 \;\boxed{\times}\; .5 \;\boxed{)}\;\boxed{=}$$

The borrower owes \$1040.

The relationship between principal (or present value) and amount (or future value) for this example is shown in the time diagram of Figure 8-1.

$r = 0.08$

$1000
Principal, or
present value

$1040
Amount, or
future value

0                    $\frac{1}{2}$ Year

FIGURE 8-1

**PRACTICE PROBLEM**

Steve Jones borrows $5000 to complete his senior year at Roebuck College. If the bank charges 8% simple interest, how much will Steve owe in 3 years?

*ANSWER*

$$A = P(1 + rt)$$

$$= 5000(1 + 0.08 \cdot 3) = \$6200$$

To find the present value of an amount at simple interest rate $r$, replace $A$, $r$, and $t$ with the given values and then solve for $P$.

**EXAMPLE 3**

Compute the present value of $1000 due in 3 months at a simple interest rate of 12% annually.

*SOLUTION*

$$\$1000 = P\left[1 + (0.12)\left(\frac{3}{12}\right)\right] \qquad A = \$1000$$
$$r = 0.12 \quad t = 3/12$$

$$P = 1000 \div ( 1 + .12 \times 3 \div 12 ) =$$

In this chapter, we will round all final answers (but not intermediate steps) to 2 decimal places (nearest cent). Hence, the present value is $970.87. The time diagram for this example is shown in Figure 8-2.

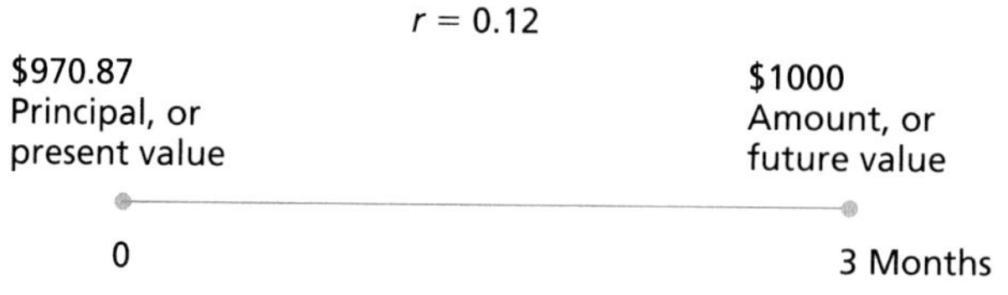

$r = 0.12$

$970.87
Principal, or
present value

$1000
Amount, or
future value

0                    3 Months

FIGURE 8-2

**EXAMPLE 4**

A sum of $1000 is borrowed for 2 years. At the end of that time, $1120 is repaid. What percent of simple interest was charged?

*SOLUTION*

$$A = P(1 + rt) \qquad \begin{array}{l} A = \$1120 \\ P = \$1000 \\ t = 2 \text{ years} \end{array}$$

$$\$1120 = \$1000(1 + 2r)$$

$$= \$1000 + \$2000r$$

## CALCULATOR HINT

Two procedures are given for solving this problem on a calculator. The first procedure uses the last line in Example 4, and the second procedure uses the next to the last line.

$$r = \boxed{(} \; 1120 \; \boxed{-} \; 1000 \; \boxed{)} \; \boxed{\div} \; 2000 \; \boxed{\blacktriangleright\%} \; \boxed{=} \quad \text{or}$$

$$r = \boxed{(} \; 1120 \; \boxed{\div} \; 1000 \; \boxed{-} \; 1 \; \boxed{)} \; \boxed{\div} \; 2 \; \boxed{\blacktriangleright\%} \; \boxed{=} .$$

In both cases $r = 6\%$ simple interest.

Now let's work Example 4 in a different way, after examining the time diagram. (See Figure 8-3.) Notice that the difference between \$1120 and \$1000 is equal to the interest.

$$I = \$1120 - \$1000 = \$120 \quad \text{and} \quad I = Prt$$

Therefore,

$$\$120 = \$1000(r)(2)$$

$$r = 0.06 = 6\% \text{ simple interest rate}$$

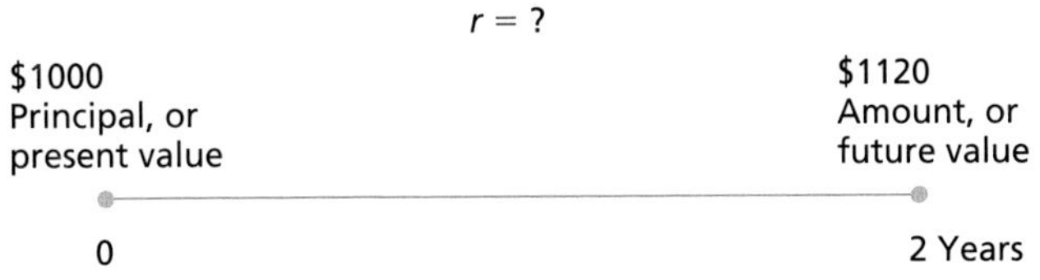

$$r = ?$$

| \$1000 | \$1120 |
|---|---|
| Principal, or present value | Amount, or future value |

| 0 | 2 Years |

FIGURE 8-3

Simple interest is not always as straightforward to calculate as it appears. Consider the story of Jill, who anticipated an IRS tax refund of \$300 in early May. Being short on cash, she decided on January 19 to borrow \$300 until May 10. Her banker agreed to give her the loan at 7.5% simple interest. First the banker calculated the number of days she would have the money.

| | | |
|---|---|---|
| Days remaining in January | $31 - 19 =$ | 12 |
| Days in February | | 28 |
| Days in March | | 31 |
| Days in April | | 30 |
| Days in May | | 10 |
| | | 111 |

She was comfortable with this computation, but she was quite surprised when the banker computed her interest in the following way:

$$I = Prt$$
$$= (300)(0.075)\left(\frac{111}{360}\right)$$
$$= \$6.94$$

The banker pointed out that the fraction $\frac{111}{360}$ represented the portion of the year that she would have the money. Bankers often compute interest on the basis of a 360-day year (called a *banker's year*). In the days prior to widespread use of calculators, this convention made computations easier; Jill also noted that a banker's year made more money for the bank than the 365-day year.

## COMMON ERROR

Suppose you must pay $6000 at the end of 1 year. If the bank charges 6% sim-ple interest, how much did you borrow? ANSWER: Computing $6000(1 + 0.06 \cdot 1)$ is incorrect because this gives the accumulated amount for a principle of $6000 invested for a year rather than a present value. The correct answer is $P = 6000 \div (1 + 0.06 \cdot 1)$.

| **PRACTICE PROBLEM** | In 5 years, you wish to have $10,000 to use as a down payment on a condominium. How much must you deposit today if the Secor Savings and Loan will pay you 8% simple interest? |
| --- | --- |
| *ANSWER* | $P = \$7142.86$ |

| **PRACTICE PROBLEM** | You borrow $1000 and the Easy Loan Company requires that you pay $1240 in 2 years. What simple-interest rate is the loan company charging? |
| --- | --- |
| *ANSWER* | $r = 0.12 = 12\%$ |

# COMPOUND INTEREST

We have learned that simple interest $I$ is found by using the formula $I = Prt$, where $P$ represents the principal, $r$ the rate, and $t$ the time. When interest is computed by this formula, the principal always remains the same. If instead the interest is added to the principal at the end of each interest period, so that the principal is increased, the interest is said to be **compounded.** The sum of the original principal and all the interest is called the **compound amount,** and the difference between the compound amount and the original principal is called the **compound interest.** We compare simple and compound interest in the following example.

| **EXAMPLE 5** | Find the simple interest on $1000 for 3 years at 6%. Then find the compound interest on $1000 for 3 years at 6% compounded annually. |
| --- | --- |
| *SOLUTION* | To emphasize the difference between simple and compound interest, we will compute them year by year. At the end of the first year, the interest for both will be |

$$I = \$1000(0.06) = \$60$$

For compound interest, this amount is added to the principal, and the new principal becomes $1000 + $60 = $1060. As a result, interest for the second year is

$$I = \$1060(0.06) = \$63.60$$

Now, how do you get the principal for the third year? Table 8-1 shows the computation.

## Isaac Newton
## 1642–1727

**T**
**H**
**E**
**N**

*and*

**N**
**O**
**W**

If you are working to help pay for your college education, you are in prestigious company: So did Isaac Newton. Born on Christmas Day of 1642 in England, Isaac had an unfortunate childhood. His father, though not a poor man, was illiterate and died before Isaac was born. His mother remarried when Isaac was 3, and at the urging of her new husband, abandoned her son to the care of her mother and brother. Newton was a "late bloomer" as far as his studies were concerned, but once he put his mind to it, he became the best student in his school. He went on to Trinity College in Cambridge at age 18, where he became a *sizar*, a student who earns his way by being a servant to richer students.

Newton's favorite subject at Trinity was chemistry, a subject that interested him his entire life. As far as we know, his first deep attraction to mathematics came from studying Descartes' geometry. (See Chapter 13.) After delving deeply into mechanics, optics, astronomy, and light, and studying the works of others such as Fermat, Galileo, and Huygens, he began his own mathematical research in 1664. (He later penned the famous quote, "If I have seen farther than Descartes, it is because I have stood on the shoulders of giants.")

By 1666 Newton had developed his theory of universal gravitation, wherein he described the motions of the planets. Newton also is credited with developing calculus, a cornerstone of the college mathematics curriculum. For these and many other works, Newton is known as one of the greatest of all scientific geniuses.

Newton's major publication appeared in 1687 entitled *Philosophiae Naturalis Principia Mathematica,* in which he described the system of the universe through his laws of motion and gravitation. "Newtonian physics" reigned supreme in science until the time of Einstein in the early 20th century.

Newton was elected to the Royal Society in 1672 and to the British Parliament in 1692. In 1696, he accepted the position of Warden of the Mint, becoming Master of the Mint 3 years later. British coins were at that time minted in the Tower of London, where Newton worked and lived. It was an arduous job, for during this time the country's currency was completely revamped. You can imagine how new currency at that time would affect all the commerce in England as well as abroad. He was well received by the financiers and the brokers in London and also amassed a personal fortune from astute financial dealings.

In 1703, Newton was elected president of the Royal Society, a position he held for life. He was knighted by Queen Anne in 1705, and he died in 1727. Sir Isaac Newton is buried in Westminster Abbey.

Today historians of mathematics identify Newton as a master problem solver. The economist John Maynard Keynes said of Newton that he "could hold a problem in his mind for hours and days and weeks until it surrendered to him its secret."

| TABLE 8-1 | | |
|---|---|---|
| | **Simple Interest** | **Compound Interest** |
| For the first year, | $I = \$1000(0.06) = \ \$60$ | $I = \$1000.00(0.06) = \ \$60$ |
| For the second year, | $I = \$1000(0.06) = \ \$60$ | $I = \$1060.00(0.06) = \ \$63.60$ |
| For the third year, | $I = \$1000(0.06) = \ \underline{\$60}$ | $I = \$1123.60(0.06) = \ \underline{\$67.42}$ |
| | $\$180$ | $\$191.02$ |

Thus, the simple interest for 3 years totals $180, whereas the compound interest totals $191.02. The compound amount is $1191.02. Notice that the principal changes each year when interest is compounded, but it remains the same when simple interest is used.

Now let's compute the compound amount at the end of each year in another way. At the end of the first year, the compound amount is

$$\$1000 + \$1000(0.06) = \$1000(1 + 0.06) = \$1000(1.06)$$

During the second year, the principal is $1000(1.06). Thus, at the end of the second year, the compound amount is

$$\$1000(1.06) + \$1000(1.06)(0.06) = \$1000(1.06)(1 + 0.06)$$
$$= \$1000(1.06)(1.06)$$
$$= \$1000(1.06)^2$$

The pattern described in the examples extends easily to the most general case. Suppose that $P$ dollars are deposited at an interest rate of $i$ per period for $n$ periods. Then Table 8-2 computes the amount that has accrued at the end of that time.

**TABLE 8-2**

| Period | Principal | + Interest | = Amount (at end of period) |
|--------|-----------|------------|------------------------------|
| 1 | $P$ | $+ Pi$ | $= P(1 + i)$ |
| 2 | $P(1 + i)$ | $+ P(1 + i)\,i$ | $= P(1 + i)^2$ |
| 3 | $P(1 + i)^2$ | $+ P(1 + i)^2 i$ | $= P(1 + i)^3$ |
| ⋮ | ⋮ | ⋮ | ⋮ |
| $n$ | $P(1 + i)^{n-1}$ | $+ P(1 + i)^{n-1}i$ | $= P(1 + i)^n$ |

Table 8-2 suggests that the compound amount can be found by multiplying the principal by $(1 + i)^n$, where $i$ is the interest rate per period (or $i = r/k$, where $r$ is the annual rate and $k$ is the number of interest periods per year) and $n$ is the number of interest periods. For example, for a rate quoted as 12% compounded monthly, $r = 0.12$, $k = 12$, and $i = r/k = 0.12/12 = 0.01$.

**Compound Interest**

The formula for *compound interest* is

$$A = P\left(1 + \frac{r}{k}\right)^{kt} = P(1 + i)^n$$

where $A$ = compound amount after $n$ periods, $P$ = principal invested (present value), $r$ = annual interest rate, $k$ = number of compound interest periods per year, $t$ = time in years, $i = r/k$, and $n = kt$.

**EXAMPLE 6**

Find the compound amount that results from investing $500 at 8% compounded annually for 4 years.

*SOLUTION*

We first examine the time diagram in Figure 8-4.

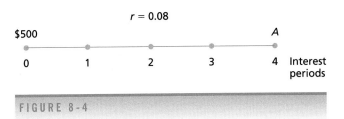

FIGURE 8-4

Substituting $P = \$500$, $r = 0.08$, $t = 4$, and $k = 1$ in the formula $A = P[1 + (r/k)]^{kt}$ yields

$$A = \$500(1 + 0.08)^4 = 500(1.08)^4$$

## CALCULATOR HINT

Use the $y^x$ key or the $\wedge$ key on your calculator.

$$500 \times 1.08 \wedge 4 =.$$

The result is $A = \$680.24$.

**EXAMPLE 7**

*SOLUTION*

Find the compound amount obtained from an investment of $2000 at 6% compounded quarterly for 5 years.

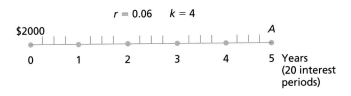

FIGURE 8-5

Figure 8-5 indicates that $r = 0.06$, $k = 4$, $t = 5$, and $n = kt = 4 \cdot 5 = 20$.
   Thus,

$$A = P\left(1 + \frac{r}{k}\right)^{kt}$$

$$= \$2000(1.015)^{20} = \$2693.71$$

The compound interest is

$$\$2693.71 - 2000 = \$693.71$$

Once we have established the relationship $A = P[1 + (r/k)]^{kt}$, we can use this formula to change the direction of our thinking. Consider the question, "How much principal must we invest now at 8% per year compounded quarterly in order to have the $6000 we need to buy a used car in 4 years?" When the question is asked in this way, the principal for which we are searching is called the *present value*. Thus, we are looking for the present value of the compound amount $6000 due in 4 years at 8% per year compounded quarterly. The time diagram for these conditions is shown in Figure 8-6.

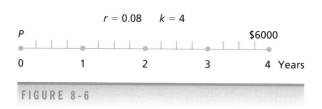

FIGURE 8-6

Inserting the relevant information in the formula for compound interest, we get

$$A = P\left(1 + \frac{r}{k}\right)^{kt}$$

$$\$6000 = P(1 + 0.02)^{16}$$

$$P = \frac{\$6000}{(1.02)^{16}}$$

$A = \$6000$
$r = 0.08$
$k = 4$
$\frac{r}{k} = 0.02$
$n = kt = (4)(4) = 16$

## CALCULATOR HINT

Do you have a reciprocal key $\boxed{1/x}$ or $\boxed{x^{-1}}$ on your calculator? This key enables you to divide by $x$ by multiplying by $1/x$.

$$1.02 \boxed{\wedge} 16 \boxed{=} \boxed{x^{-1}} \boxed{\times} 6000 \boxed{=}$$

to get $4370.67. Thus, the present value of $6000 due in 4 years at 8% compounded quarterly is $4370.67.

**EXAMPLE 8**

How much money should a person deposit in a savings and loan association account that pays 6% interest compounded quarterly in order to have $3000 in 5 years?

*SOLUTION*

The time diagram for these conditions appears in Figure 8-7.

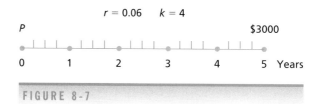

FIGURE 8-7

Inserting these values in the compound interest formula, we get

$$A = P\left(1 + \frac{r}{k}\right)^{kt}$$

$$\$3000 = P(1 + 0.015)^{20}$$

$$P = \frac{\$3000}{(1.015)^{20}}$$

$i = .06/4 = 0.015$
$n = kt = (4)(5) = 20$

Using a calculator, we find $P = \$2227.41$. Thus, a person must deposit $2227.41 now in a savings and loan association account in order to have $3000 in 5 years.

**PRACTICE PROBLEM**

You want to borrow $1000 for 6 years. Which of the following is a better interest rate for the loan?

(a) $8\frac{1}{4}\%$ simple interest
(b) 8% compounded quarterly

*ANSWER*

(a) $A = 1000(1 + 0.0825 \cdot 6) = \$1495$
(b) $A = 1000(1 + 0.02)^{6 \cdot 4} = \$1608.44$. The loan at $8\frac{1}{4}\%$ simple interest is the better rate.

**EXAMPLE 9**

How long will it take a dollar to double at 8% compounded semiannually?

*SOLUTION*

Suppose that we start with $P = \$1$. This amount is to double giving us $A = \$2$, with $r = 0.08$ and $k = 2$ (semiannually). Substituting these values in the compound interest formula, we get

$$A = P\left(1 + \frac{r}{k}\right)^{kt}$$
$$2 = (1.04)^n$$

## CALCULATOR HINT

Use the $\boxed{\wedge}$ key of your calculator and the very useful problem-solving strategy of "guessing and then testing" to get an approximate answer. Try any number for $n$. Suppose you select a 5.

$1.04 \boxed{\wedge} 5 \boxed{=} 1.217$     Much too small
$1.04 \boxed{\wedge} 10 \boxed{=} 1.480$     Still too small
$1.04 \boxed{\wedge} 20 \boxed{=} 2.191$     A bit too large
$1.04 \boxed{\wedge} 17 \boxed{=} 1.948$
$1.04 \boxed{\wedge} 18 \boxed{=} 2.026$

Thus, in 18 interest periods (or 9 years), the principal of $1 will more than double.

# Just for Fun

A man goes into a store and says to a salesperson, "Give me as much money as I have with me and I will spend $10 with you." It is done. The operation is repeated in a second and a third store, after which he has no money left. How much did he have originally?

# EXERCISE SET 1

*R* 1. Compute the simple interest earned when the principal, rate, and time of the loan are as given.

   (a) $P = \$500$, $r = 0.08$, $t = 2$ years
   (b) $P = \$300$, $r = 0.03$, $t = 4$ years
   (c) $P = \$500$, $r = 0.04$, $t = 5$ years

2. What is the amount to be repaid in (a), (b), and (c) of Exercise 1?

3. Find the interest and the amount of a loan for $3000 borrowed for 2 years at 6% simple interest.

*Find the compound interest and compound amount for the investments in Exercises 4 through 6.*

4. (a) $5000 at 4% compounded annually for 3 years
   (b) $5000 at 6% compounded semiannually for 3 years
   (c) $5000 at 6% compounded quarterly for 3 years

5. (a) $2000 at 4% compounded annually for 4 years
   (b) $2000 at 4% compounded semiannually for 4 years
   (c) $2000 at 4% compounded quarterly for 4 years

6. (a) $3000 at 5% compounded annually for 6 years
   (b) $3000 at 5% compounded semiannually for 6 years
   (c) $3000 at 5% compounded quarterly for 6 years

7. Compute the amount to be repaid when the principal, simple-interest rate, and time of the loan are as given.

   (a) $P = \$4000$, $r = 0.05$, $t = 2$ years
   (b) $P = \$3000$, $r = 0.06$, $t = 6$ months
   (c) $P = \$100$, $r = 0.08$, $t = 3$ years

8. Find the simple-interest rate on the loan when the principal, the amount repaid, and the term of the loan are as given. (HINT: Use a 360-day year.)

   (a) Principal, $3500; amount repaid in 2 years, $4130
   (b) Principal, $1000; amount repaid in 120 days, $1046.67
   (c) Principal, $500; amount repaid in 45 days, $510

*Find the present value of the money in Exercises 9 through 12.*

*T* 9. $5000 due in 5 years, if the money is worth 6% compounded annually

10. $6000 due in 5 years 6 months, if the money is worth 8% compounded semiannually

11. $7000 due in 4 years, if the money is worth 6% compounded semiannually

12. $8000 due in 7 years, if the money is worth 7% compounded annually

*Solve each formula for the indicated variable.*

13. $I = Prt$    $(r)$

14. $A = P + Prt$    $(P)$

*C* 15. How long will it take for $125 to amount to $375 at 6% interest compounded quarterly?

16. How many years will it take to double $1000 at 4% interest compounded semiannually?

17. How much interest will you owe on a $1000 loan from March 3 to August 7 at 5% simple interest? (Use a 360-day year.)

18. What is the interest on a $1500 loan from June 15 to September 11 at a simple-interest rate of 4%? (Use a 360-day year.)

19. Some lending institutions have a minimum amount of interest they must collect on any loan. An institution with a minimum charge of $5 will expect to receive $5 interest on a loan, even though actual interest charges are only $3. How long must $1000 be borrowed at 8% interest to reach the minimum service charge of $5? (Use a banker's year of 360 days in the computation.)

20. A sum of $1000 was deposited in a bank at an interest rate of 6% compounded semiannually. Five years later the rate increased to 8% compounded semiannually. If the money was not withdrawn, how much was in the account at the end of 6 years?

*The effects of inflation can be obtained by using the formula for compound interest, where* i *becomes the inflation rate per period.*

21. Find the cost of the following items in 10 years at an annual inflation rate of 3%.

   (a) A $100,000 house    (b) A $1.50 hamburger
   (c) A $5 movie    (d) A $14.65 hourly labor rate

22. Rework Exercise 21 for a 6% annual inflation rate.

23. Rework Exercise 21 for an 8% annual inflation rate.

24. How long will it take a price to double at the following average annual inflation rates?

   (a) 3%    (b) 4%    (c) 5%    (d) 6%

25. The population of a city of 60,000 is expected to increase at a rate of 4% per year for the next 10 years. What will the population be at the end of 10 years?

26. The number of fish in a lake is expected to increase at a rate of 12% per year for 5 years. How many fish will be in the lake in 5 years if 10,000 are placed in the lake today?

# 2 CALCULATOR APPROACH TO ANNUITIES

**PROBLEM**

Suppose you deposit $500 at the end of each quarter in a savings and loan association account that pays interest at a rate of 4% compounded quarterly. Will you have enough money in your savings account in 5 years to purchase a Honda Civic or a Mercedes?

**OVERVIEW**

In this section, we compute the value of sequences of interest-drawing payments or deposits in order to find the value immediately after the last deposit or the present value one period before the first deposit. The value of a savings account into which periodic deposits are made is a typical illustration of finding the value at the end of a sequence of payments.

**GOALS**

Illustrate the Algebra Standard, page 4.
Illustrate the Representation Standard, page 50.
Illustrate the Problem Solving Standard, page 15.

In this section it will be very important to remember some material from Chapter 1 of this text. Remember that sequences of the form $a$, $ar$, $ar^2$, $ar^3$, ... in which each term is obtained from its predecessor by multiplying by a fixed number, $r$, are called *geometric sequences*. Remember further that in a PCR Excursion in Chapter 1 we computed the sum of the first $n$ terms of a geometric sequence.

The sum of the first $n$ terms of a geometric sequence is given by the formula

$$a + ar + ar^2 + ar^3 + \cdots + ar^{n-1} = \frac{a - ar^n}{1 - r} = \frac{a(1 - r^n)}{1 - r}$$

where $a$ is the first term of the sequence, $r$ is the common ratio, and $n$ is the number of terms of the sequence.

## AMOUNT OF AN ANNUITY

An **annuity** is a sequence of equal payments made at equal time intervals. For the formula we shall develop, these payments (denoted by $R$) are made at the *ends* of equal successive payment periods, and the interest periods are the same as the payment periods. The sum of all payments $R$ plus their interest is called the **amount of an annuity,** as illustrated in the following example.

**EXAMPLE 1**

Suppose that you receive $100 at the end of each year and immediately invest it at 6% compounded annually. How much will you have at the end of 5 years?

*SOLUTION*

Your first deposit, made at the end of the first year, accumulates compound interest for only 4 years. The value of this deposit at the 5-year mark is $100(1.06)^4$. (See Figure 8-8.) The second deposit will accumulate interest for $5 - 2 = 3$ years and has a value of $\$100(1.06)^3$. At the 5-year mark the third deposit has a value of $\$100(1.06)^2$, the fourth deposit $\$100(1.06)^1$, and finally the last deposit is made on the 5-year mark and thus has a value of $\$100$. Hence you have in the bank at the end of 5 years the sum of the 5 accumulations.

$$S = \$100(1.06)^4 + \$100(1.06)^3 + \$100(1.06)^2 + \$100(1.06) + 100$$

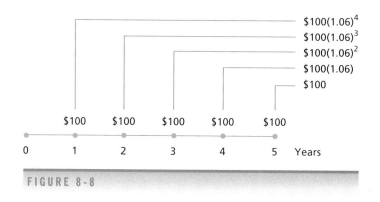

FIGURE 8-8

If these terms are rearranged from smallest to largest as

$$S = 100 + 100(1.06) + 100(1.06)^2 + 100(1.06)^3 + 100(1.06)^4$$

you can recognize this expression as the sum of a geometric sequence with ratio $r = 1.06$ and first term $a = 100$. Thus, the sum is

$$S = 100\left[\frac{1 - (1.06)^5}{1 - 1.06}\right] \quad \text{or} \quad 100\left[\frac{(1.06)^5 - 1}{0.06}\right]$$

This example suggests the following formula for the sum or amount of an annuity. Notice that, in Example 1, $R = 100$ and $i = 0.06$.

**Amount of an Annuity**

The formula for the *amount of an annuity* $S$ is

$$S = R\left[\frac{(1 + i)^n - 1}{i}\right]$$

where $S$ = amount of the annuity, $R$ = periodic payment of the annuity, $i$ = rate per period, and $n$ = number of payments (periods). (Payments are made at the end of each period.)

**EXAMPLE 2**

Kate deposits $300 at the end of each year in a savings account that pays 4% interest compounded annually. How much money does she have immediately after making the fifth deposit?

FIGURE 8-9

*SOLUTION*

Figure 8-9 shows the time diagram for these conditions. Remember that the formula above gives the amount just after a payment is made. Hence, this formula can be used to solve the problem.

$$S = R\left(\frac{(1 + i)^n - 1}{i}\right) \qquad \begin{array}{l} i = 0.04 \\[4pt] n = 5 \\[4pt] R = \$300 \end{array}$$

$$= \$300\left(\frac{(1 + 0.04)^5 - 1}{0.04}\right)$$

## CALCULATOR HINT

To determine the value of this expression on a calculator, try pressing

$$300 \;\boxed{\times}\; \boxed{(}\; 1.04 \;\boxed{\wedge}\; 5 \;\boxed{-}\; 1 \;\boxed{)}\; \boxed{\div}\; .04 \;\boxed{=}$$

Did you get $1624.90?

Now return to Example 1, and use your calculator to find the amount of money you would have at the end of 5 years. Did you get $563.71?

**EXAMPLE 3**

Mark deposits $1000 in the First Savings and Loan at the end of each quarter for 10 years. How much money does he have at the end of 10 years if the savings and loan pays 4% interest compounded quarterly?

*SOLUTION*

Figure 8-10 shows the relevant time diagram.
  Thus, we have

$$S = R\left(\frac{(1 + i)^n - 1}{i}\right)$$

$$= \$1000\left[\frac{(1 + .01)^{40} - 1}{0.01}\right] \qquad \begin{array}{l} i = \dfrac{r}{k} = \dfrac{0.04}{4} = 0.01 \\[6pt] n = kt = 4(10) = 40 \\[4pt] R = \$1000 \end{array}$$

$$= \$48,886.37$$

Mark has $48,886.37 in his savings account at the end of 10 years.

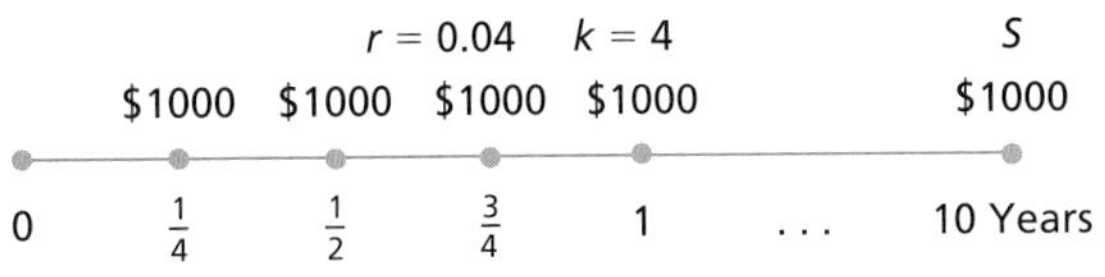

| | $r = 0.04$ | $k = 4$ | | $S$ |
|---|---|---|---|---|
| $1000 | $1000 | $1000 | $1000 | $1000 |

| 0 | $\frac{1}{4}$ | $\frac{1}{2}$ | $\frac{3}{4}$ | 1 | ... | 10 Years |

**FIGURE 8-10**

**PRACTICE PROBLEM**

If you deposit $100 a month in a savings and loan paying 6% compounded monthly, how much money do you have at the end of 5 years?

*ANSWER*

$$\$100\left(\frac{(1.005)^{60} - 1}{.005}\right) = \$6977$$

**EXAMPLE 4**

The Wilson family decides to make monthly deposits into a college-education fund for their daughter Brooke so she will have $20,000 at the end of 8 years. They locate a bond fund that pays 12% compounded monthly. How much must the Wilson family deposit each month to achieve the desired result?

*SOLUTION*

The time diagram is given in Figure 8-11.
 In this case, we have

$$S = R\left[\frac{(1 + i)^n - 1}{i}\right]$$

$$20,000 = R\left[\frac{(1 + 0.01)^{96} - 1}{0.01}\right]$$

$i = r/k = 0.12/12 = .01$
$n = kt = 12(8) = 96$
$S = \$20,000$

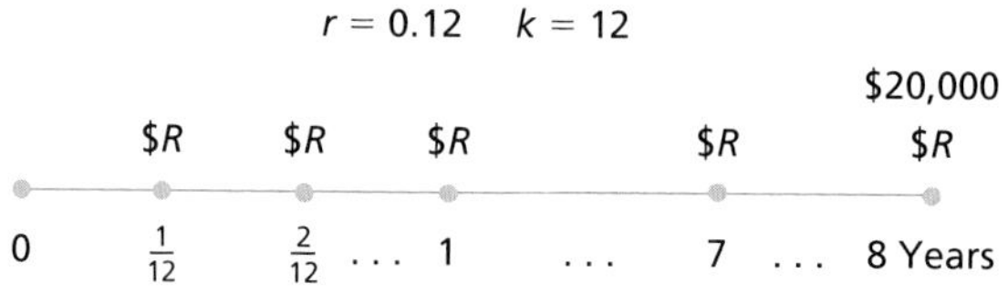

$r = 0.12$    $k = 12$

$20,000

| | \$R | \$R | \$R | | \$R | \$R |

0    $\frac{1}{12}$    $\frac{2}{12}$  ...  1    ...    7  ...  8 Years

FIGURE 8-11

## CALCULATOR HINT

If your calculator has a ⌨ $x^{-1}$ key, find $R$ as follows:

 ( 1.01 ^ 96 − 1 ) ÷ .01 = $x^{-1}$ × 20,000 = .

The result is $R = \$125.06$, which means that the Wilson family must deposit $125.06 each month in order for Brooke to have $20,000 at the end of 8 years.

**EXAMPLE 5**

Mr. Wilkins makes equal semiannual deposits in his savings account in order to have $1259.71 in 3 years. What is the amount of his semiannual deposit at 4% compounded semiannually?

*SOLUTION*

Semiannual deposits of $R each are shown in Figure 8-12.
 Substituting the relevant data in the annuity formula, we get

$$S = R\left[\frac{(1 + i)^n - 1}{i}\right]$$

$$1259.71 = R\left[\frac{(1 + 0.02)^6 - 1}{0.02}\right]$$

$i = r/k = 0.04/2 = 0.02$
$n = kt = 2(3) = 6$
$S = \$1259.71$

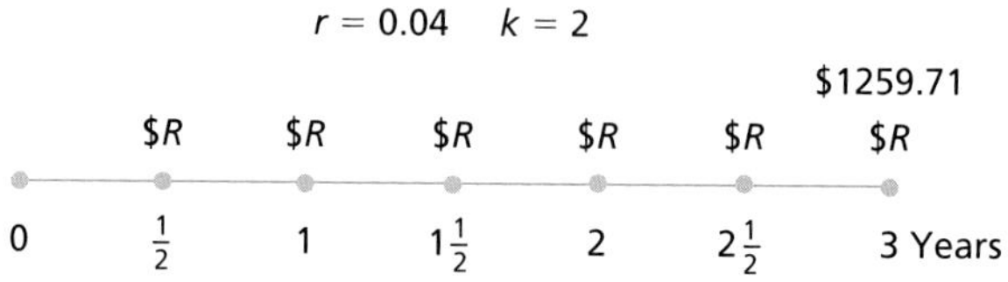

$r = 0.04$    $k = 2$

$1259.71

| \$R | \$R | \$R | \$R | \$R | \$R |

0    $\frac{1}{2}$    1    $1\frac{1}{2}$    2    $2\frac{1}{2}$    3 Years

FIGURE 8-12

**CALCULATOR HINT**

$R =$ ⎡ ( ⎤ 1.02 ⎡ ^ ⎤ 6 ⎡ − ⎤ 1 ⎡ ) ⎤ ⎡ ÷ ⎤ .02 ⎡ = ⎤ ⎡ x⁻¹ ⎤ ⎡ × ⎤ 1259.71 ⎡ = ⎤.

$R = \$199.70$

# Just for Fun

What is the temperature?

0

B.S.

M.S.

Ph.D.

(Look for a trick.)

## EXERCISE SET 2

*With a calculator, evaluate the expressions in Exercises 1 through 4. Interpret what you have found, relative to annuities.*

**R** 1. $100\left(\dfrac{(1.06)^{10} - 1}{0.06}\right)$     2. $50\left(\dfrac{(1.01)^{100} - 1}{0.01}\right)$

   3. $200\left(\dfrac{(1.08)^{14} - 1}{0.08}\right)$     4. $1\left(\dfrac{(1.12)^{17} - 1}{0.12}\right)$

*Find the amount of the annual annuities in Exercises 5 and 6. Interest is compounded annually.*

5. $R = \$100$, $i = 6\%$, $n = 10$

6. $R = \$1000$, $i = 8\%$, $n = 12$

*Find the amount of each of the annuities in Exercises 7 through 10. Payments or deposits are made at the end of each interest period.*

**T** 7. $R = \$100$, $r = 6\%$ compounded semiannually, $n = 10$ years

8. $R = \$1000$, $r = 5\%$ compounded quarterly, $n = 5$ years

9. $R = \$2000$, $r = 5\%$ compounded monthly, $n = 4$ years

10. $R = \$5000$, $r = 3.5\%$ compounded monthly, $n = 10$ years

11. Find the amount of the following annuities.

   (a) $1000 per year for 20 years at 8% interest compounded annually

   (b) $500 per quarter for 6 years at 3.5% compounded quarterly

   (c) $600 per half year for 5 years at 4% compounded semiannually

12. Suppose that you deposit $500 every 6 months in a credit union that pays 4% interest compounded semiannually. How much would you have after 5 years?

13. Juan is to receive $1000 at the end of each year for 5 years. If he invests each year's payment at 8% compounded annually, how much will he have at the end of the 5 years?

**C** 14. You place 1000 fish in your lake at the end of each year for 5 years. If the fish increase at a rate of 5%

per year, how many will you have at the end of 5 years?

15. John Sizemore makes monthly deposits in Burbank Savings and Loan, which pays 3% interest compounded monthly. What size deposit must John make in order to have the $10,000 he needs in 3 years to purchase an automobile?

16. What annual deposit is required to accumulate $50,000 after 20 years at 8% interest compounded annually?

17. What quarterly deposit is required to accumulate $7000 after 6 years at 3.5% compounded quarterly?

### REVIEW EXERCISES

18. Find the simple interest due on a loan of $5000 at 12% interest for 3 months.

19. Find the compound interest and compound amount for an investment of $4000 at 6% interest compounded semiannually for 10 years.

---

### ∵ PCR Excursion ∵

20. Mr. Wilkins wishes to make equal semiannual deposits in his savings account in order to have $2500 in 3 years.

(a) How large should his semiannual deposits be if each deposit grows at an interest rate of 4% compounded semiannually?

(b) Now investigate how Mr. Wilkins's savings fund grows each 6 months. To accomplish this goal, study the first line of the following table and complete the second line.

| End of Period | Interest | Semiannual Deposit | Amount in Fund |
|---|---|---|---|
| 1 | 0 | $396.31 | $396.31 |
| 2 | $7.93 | 396.31 | |

Did you get $800.55 in the fund at the end of the second period?

(c) Complete three more lines of the table.

(d) Explain in words how the fund accumulates.

(e) How much will be in the fund at the end of the sixth period?

---

SECTION **3** 

# PRESENT VALUE OF AN ANNUITY AND APR

| | |
|---|---|
| **PROBLEM** | Jodi agrees to pay $2500 down and $250 per month for 5 years for her new automobile. If the finance rate is 8% compounded monthly, what is the cash value of her car? |
| **OVERVIEW** | Problems like the one given are considered in this section. |
| **GOALS** | The goals are the same as those listed on page 377. |

## PRESENT VALUE OF AN ANNUITY

The **present value of an annuity** is the sum of the present values of each of a sequence of payments made at the end of equal periods. We will use the formulas for the sum of a geometric sequence to find the present value of the annuity in Example 1.

**EXAMPLE 1**

Compute the present value of an annuity of $100 per year for 5 years at 6% compounded annually. In other words, find the amount of money that you must invest now at 6% compounded annually so that payments of $100 per year can be made from this investment for 5 years.

*SOLUTION*

The $100 per year payments have been placed on the time diagram in Figure 8-13. Notice that the first payment is at the end of year 1. We learned in Section 1 that the present value of this payment (at time 0) is $100(1.06)^{-1}$. The second $100 payment is at the end of year 2. The present value of this payment (at time 0) is $100(1.06)^{-2}$. The present value of the third payment is $100(1.06)^{-3}$, the fourth $100(1.06)^{-4}$, and finally the present value of the fifth payment is $100(1.06)^{-5}$. The present value of the annuity is the sum of these terms.

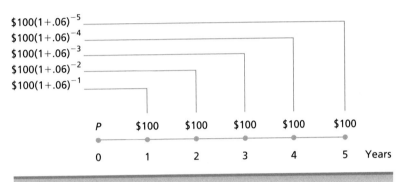

**FIGURE 8-13**

Thus,

$$P = 100(1.06)^{-1} + 100(1.06)^{-2} + 100(1.06)^{-3} + 100(1.06)^{-4} + 100(1.06)^{-5}$$

$$= \frac{100}{(1.06)^{1}} + \frac{100}{(1.06)^{2}} + \frac{100}{(1.06)^{3}} + \frac{100}{(1.06)^{4}} + \frac{100}{(1.06)^{5}}$$

You probably recognize this as the sum of a geometric sequence with ratio $1/1.06$ and first term $\frac{100}{1.06}$. Review page 377 to get the sum of this sequence to be

$$\frac{100}{1.06}\left[\frac{1 - \left(\frac{1}{1.06}\right)^{5}}{1 - \frac{1}{1.06}}\right] = 100\left[\frac{1 - (1.06)^{-5}}{.06}\right]$$

**Present Value of an Annuity**

The formula for the *present value of an annuity P* is

$$P = R\left[\frac{1 - (1 + i)^{-n}}{i}\right]$$

where $P$ = present value of an annuity, $R$ = periodic payment, $i$ = rate per period, and $n$ = number of payments (periods). Note that $P$ is the present value at the beginning of the first period; payments are made at the end of each period.

| EXAMPLE 2 | Compute the present value of an annuity that pays $100 each month for 4 years at 6% interest compounded monthly. |

*SOLUTION*

The time diagram for these conditions in shown in Figure 8-14.

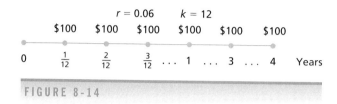

FIGURE 8-14

Substituting in the present value formula, we get

$$P = R\left[\frac{1 - (1 + i)^{-n}}{i}\right]$$

$$= 100\left[\frac{1 - (1.005)^{-48}}{0.005}\right]$$

$i = \dfrac{r}{k} = 0.06/12 = 0.005$

$n = kt = 12 \cdot 4 = 48$

$R = \$100$

## CALCULATOR HINT

Compute

$\boxed{(}\ 1\ \boxed{-}\ 1.005\ \boxed{\wedge}\ \boxed{(-)}\ 48\ \boxed{)}\ \boxed{\div}\ .005\ \boxed{\times}\ 100\ \boxed{=}$.

Thus, we find that $P = \$4258.03$

Now use your calculator to find the present value of the annuity in Example 1. Did you get $421.24?

| EXAMPLE 3 | Kamilla purchased a refrigerator for $150 down and $30 a month for 12 months. If the interest charge is 12% compounded monthly, find the cash price. |

*SOLUTION*

The payment schedule is depicted in Figure 8-15.

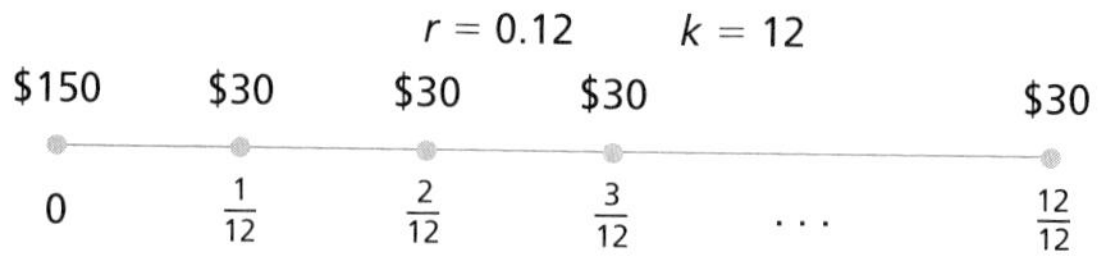

FIGURE 8-15

Therefore, the cash price can be expressed as

$$C = 150 + P$$

$$= 150 + 30\left[\frac{1 - (1.01)^{-12}}{0.01}\right]$$

$i = \dfrac{r}{k} = 0.12/12 = 0.01$

$n = kt = 12(1) = 12$

$R = \$30$

## CALCULATOR HINT

The cash price of Kamilla's refrigerator is $487.65.

**PRACTICE PROBLEM**

*ANSWER*

Find the cash value of Jodi's car in the introductory problem.

$$\text{Cash value} = \$2500 + \text{present value of a } \$250 \text{ annuity}$$

$$= 2500 + 250\left[\frac{1 - (1 + .08/12)^{-60}}{0.08/12}\right]$$

$$= \$14{,}829.61$$

**EXAMPLE 4**

Find the payment needed each month for 1 year to pay off a present debt of $1000 at 12% compounded monthly.

*SOLUTION*

Figure 8-16 shows the time diagram for this example. Let $R$ represent the payment. Thus, we have

$$P = R\left[\frac{1 - (1 + i)^{-n}}{i}\right]$$

$$1000 = R\left[\frac{1 - (1.01)^{-12}}{0.01}\right]$$

$i = 0.12/12 = 0.01$
$n = kt = 12(1) = 12$
$P = \$1000$
**Divide by coefficient of $R$.**

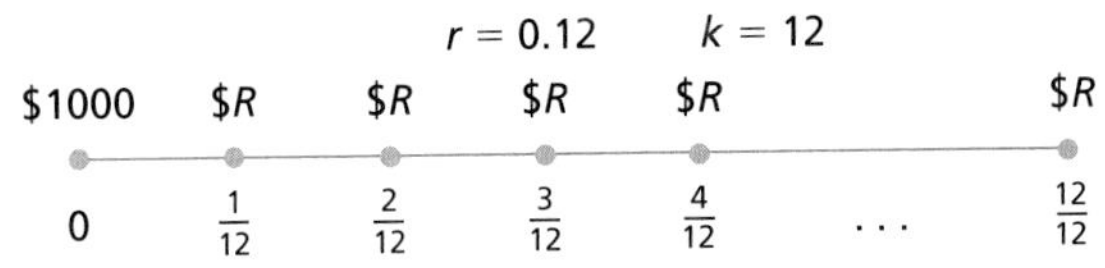

$r = 0.12$      $k = 12$

| $1000 | $R | $R | $R | $R | | $R |

| 0 | $\frac{1}{12}$ | $\frac{2}{12}$ | $\frac{3}{12}$ | $\frac{4}{12}$ | $\cdots$ | $\frac{12}{12}$ |

**FIGURE 8-16**

## CALCULATOR HINT

Compute

$$\boxed{(} \; 1 \; \boxed{-} \; 1.01 \; \boxed{\wedge} \; \boxed{(-)} \; 12 \; \boxed{)} \; \boxed{\div} \; .01 \; \boxed{=} \; \boxed{x^{-1}} \; \boxed{\times} \; 1000 \; \boxed{=}.$$

Therefore, $R$ is equal to $88.85.

## COMMON ERRORS

Students sometimes use the present value of an annuity when they should use amount, and vice versa. Read each problem carefully to determine which you should use.

**EXAMPLE 5**

You purchase a house for $100,000, pay 20% down, and to pay off your debt with monthly payments for 30 years. What is your monthly payment if the interest on your loan is 9% compounded monthly?

*SOLUTION*   $100,000 - 0.20($100,000) = $80,000$ (still owed on the house). We wish to find a sequence of equal payments whose present value is $80,000.

$$\$80{,}000 = R\left[\frac{1 - (1 + 0.0075)^{-360}}{0.0075}\right]$$

## CALCULATOR HINT

$$R = \$643.70$$

In circumstances like those examined in Examples 4 and 5, when a sequence of equal payments R is used to pay off a debt, we say that the payment R **amortizes** the debt. In Example 4, the payment of $88.85 amortizes the loan of $1000. For this reason, when the loan is used to buy a house, the loan is called a **mortgage.**

## ⠿ ANNUAL PERCENTAGE RATE

In the nineteen sixties, many different schemes for computing interest on installment loans were in vogue. Because most of these interest schemes were very misleading to consumers, in 1969 the federal Truth in Leading Act was passed. It requires that, for each loan, the lending institution quote the **Annual Percentage Rate (APR),** regardless of what scheme was used to compute the finances charges for the loan. **The APR is an uniform interest rate that gives you a basis of comparison when shopping for an installment loan.** It gives an accurate representation of how much rent you are paying for the money you are borrowing. More specifically, if you regard your sequence of installment payments as amortizing your loan, the APR is the interest rate at which the loan is amortized.

To compute the APR, most loan officers rely on tables compiled by the Federal Reserve Bank. Use of these tables requires that you know the finance charges for the loan. The **finance charges** are simply the total of all interest and fees charged by the lending institution to obtain the loan. Sample pages from these tables are given in the table in the appendix. The following steps describe in detail how the table is used.

**Steps in Using an APR Table**

1. Compute (finance charge · 100) ÷ (amount financed).
2. Look in the row labeled by the number of payments to be made, and find the entry closest to the number found in step 1.
3. Find the percentage rate at the top of the column in which this entry is found. This is the APR rounded to the nearest $\frac{1}{4}\%$.

**EXAMPLE 6**

Find the APR for a loan of $200 to be repaid with 15 monthly payments of $14.56.

*SOLUTION*

$$\text{Finance charges} = 15(14.56) - (200) = \$18.40$$

Therefore,

1. $\dfrac{(\$18.40)(100)}{\$200} = \$9.20$

2. Locate the row in the table in the appendix for 15 payments and find the entry in that row nearest 9.20. The nearest entry is 9.23.
3. The percentage rate at the top of the column is 13.50%. This is the APR (rounded to the nearest $\frac{1}{4}\%$).

# Just for Fun

Deposit $100 in the bank and then make withdrawals in the following manner.

|  | Withdraw | | Leaving a Balance of |
|---|---|---|---|
|  | $40 | | $60 |
|  | $30 | | $30 |
|  | $18 | | $12 |
| Adding, | $12 | | $ 0 |
| we have | $100 | and | $102 |

The total withdrawal is $100, whereas the total of the balances is $102. Can you now go to the bank to demand your extra $2?

# EXERCISE SET 3

*With a calculator, compute the values in Exercises 1 through 4. Interpret what you have found, relative to annuities.*

*R* 1. $50\left[\dfrac{1 - (1.06)^{-10}}{0.06}\right]$     2. $50\left[\dfrac{1 - (1.01)^{-100}}{0.01}\right]$

   3. $200\left[\dfrac{1 - (1.08)^{-14}}{0.08}\right]$     4. $1\left[\dfrac{1 - (1.12)^{-7}}{0.12}\right]$

*For Exercises 5 and 6, find the present value of each of the annuities. Interest is compounded annually.*

   5. $R = \$100$, $i = 6\%$, $n = 10$

   6. $R = \$1000$, $i = 8\%$, $n = 12$

*T* 7. Blueport National Bank offers to lend Bill $500 to be repaid in 18 monthly payments of $30.83.

   (a) What are the finance charges?
   (b) What is the APR (rounded to the nearest $\frac{1}{4}\%$)?

8. CIP Finance Company offers to lend Bill $508 for 18 months, to be repaid in 18 monthly payments of $31.

   (a) What are the finance charges?
   (b) What is the APR (rounded to the nearest $\frac{1}{4}\%$)?

9. Find the present value of the following annuities.

   (a) $1000 per year for 20 years at 8% interest compounded annually
   (b) $500 per quarter for $6\frac{1}{2}$ years at 8.5% interest compounded quarterly
   (c) $600 per half year for 5 years at 8% interest compounded semiannually

10. Find the monthly payments necessary to amortize a debt of $6000.

   (a) for 8 years at 6% interest compounded monthly
   (b) for 3 years at 6.4% interest compounded monthly

11. A businessperson wants to receive $6000 (including principal) from a fund at the end of each year for 10 years. How should she compute her required initial investment at the beginning of the first year if the fund earns 6% compounded annually?

12. A contract pays $200 at the end of each quarter for 4 years and an additional $2000 at the end of the last quarter. What is the present value of the contract at 8% interest compounded quarterly?

13. What should be the semiannual deposit in a fund established to pay off a loan of $300 at 6% interest compounded annually in 3 years, if the fund pays 6% interest compounded semi-annually?

*C* 14. Compute the payment necessary to finance a used car for $3500 at 6% interest compounded monthly for 3 years.

15. Find the payment necessary each quarter for 2 years to amortize a debt of $2000 at 6% interest compounded quarterly.

16. Henry accepted a $35,000 early retirement bonus from his company. He wishes to deposit some of it into an account that provides a payment of $300 at the end of each month for ten years. If the account pays 6% interest compounded monthly, how much of his $35,000 does he need to deposit?

17. Monica wishes to borrow $50,000 to purchase a house. If she obtains a 15-year loan, what will her monthly payments need to be if she pays 7.2% compounded monthly?

## REVIEW EXERCISES

18. Compute the compound interest and compound amount for an investment of $4000 at 8% compounded quarterly for 10 years.

19. Find the interest on $3000 at 12% simple interest for 6 months.

20. Compute the present value and the amount of an annuity of $400 per year for 10 years at 8% interest compounded annually.

### ❖ PCR Excursion ❖

21. In Example 4 we learned that a monthly payment of $88.85 would pay off a debt of $1000 in 1 year at an interest rate of 12% compounded monthly. Find the interest on $1000 for the first month. Subtract this from $88.85 to get the amount of principal repaid in the first month. Then complete the first line of the table.

| End of Month | Principal at Beginning of Month | Interest on Debt | Payment at End of Month | Principal Repaid |
|---|---|---|---|---|
| 1 | $1000 | $10 | $88.85 | |
| 2 | | | | |

   (a) What is the principal at the beginning of the second month? Did you get $921.15?
   (b) Complete four more lines of the table.
   (c) Explain in words how each payment reduces the debt.
   (d) What will the debt be at the end of 4 months?
   (e) What will the debt be at the end of 12 months?

# SOLUTION TO INTRODUCTORY PROBLEM

**UNDERSTANDING THE PROBLEM**   We first summarize the facts. The men thought that the rent for their hotel room was $30. Consequently, each of the three men paid $10. The rent for the room was only $25. The bellboy returned $3 to the three men and kept a $2 tip.

**DEVISING A PLAN**   As good consumers, we must find exactly how the books must balance. Specifically, the amount received by the motel ($25) and the bellboy ($2) must equal the amount paid by the guests.

**CARRYING OUT THE PLAN**

There is no missing dollar. There is no reason why 30 must equal 3(9) + 2 because 3(9) + 2 is the sum of the amount paid by the guests and the amount received by the bellboy. $30 is an irrelevant sum.

| Amount Received by Motel and Bellboy | | Amount Paid by Guests |
|---|---|---|
| $25 + $2 | = | 3($9) |

# SUMMARY AND REVIEW

### INTEREST

1. Simple interest is computed by the formula $I = Prt$, where $P$ is the principal, $r$ is the simple-interest rate, and time $t$ is expressed in years.

2. Amount $A = P + I$    or    $A = P(1 + rt)$

3. For compound interest, $A = P(1 + i)^n$, where $i$ is the interest rate per period and $n$ is the number of interest periods.

4. The APR rate is a uniform interest rate that can be used as a basis for comparison. We obtain the APR from a table in the appendix.

### ANNUITIES

1. Amount    $A = R\left[\dfrac{(1 + i)^n - 1}{i}\right]$

2. Present value    $P = R\left[\dfrac{1 - (1 + i)^{-n}}{i}\right]$

# CHAPTER TEST

1. Find the present value of $1000 due in 3 years at 8% compounded semiannually.

2. How much money will you have in the bank at the end of 3 years if you deposit $500 at the end of each month for the 3 years? Suppose that you draw 6% interest compounded monthly.

3. If $1000 accumulates to $1200 in 6 months, what simple-interest rate is being charged?

4. You borrow $2000 from the E-Z Loan Company. How much do you pay back in 2 years if they charge 12% compounded monthly?

5. For a used automobile, you agree to pay $100 every 6 months for 5 years. If you are charged interest at a rate of 8% compounded semiannually, what is the cash value of the automobile?

6. You agree to pay $2320 at the end of 2 years. If a bank charges 8% simple interest, how much do you obtain?

7. You place your savings in a bank that pays a simple-interest rate of 5%. If you have $2400 at the end of 4 years, how much did you deposit?

8. How much do you owe at the end of 90 days, if you borrow $2000 at a simple-interest rate of 18%? Use a 360-day year.

9. You purchase a house for $120,000, pay $20,000 down, and amortize the debt by monthly payments for 10 years. What are your monthly payments if money is worth 8% compounded monthly?

10. You borrow $10,000 for 10 years at an interest rate of 12% compounded monthly. Find the interest rate compounded yearly that would result in the same amount owed as the 12% compounded monthly. Would the amount borrowed make any difference in your answer?

11. How long will it take for a price to double at an annual inflation rate of 2%?

12. You buy a house for $200,000. Your monthly payments are $1200. How much do you owe at the end of four months, if you pay 6% compounded monthly?

# INTRODUCTION TO PROBABILITY THEORY

# 9

A couple wants to have 3 or 4 children, including exactly 2 girls. Is it more likely that they will get their wish with 3 children or with 4?

Archaeological artifacts indicate that many early peoples played some version of dice, either for recreation or to determine the will of the tribal deity. As more elaborate games were developed, the players began to observe certain patterns in the results, but they did not have the language of probability with which to describe and analyze them.

One of the first people to apply mathematical analysis to games of chance was the 16th-century Italian mathematician and gambler, Girolamo Cardano. Probability was firmly established as a legitimate field of inquiry by the work of Blaise Pascal and Pierre Fermat, who attempted to answer questions concerning a dice game. Pascal corresponded with Fermat on this problem, and they each arrived at a solution by different methods. Their interest in these questions generated widespread interest in probability. As time passed, it became clear that probability was much more than a plaything for gamblers. In the 19th and 20th centuries, probability has been used in many different applications. Physicists use probability in studying the various gas and heat laws as well as in atomic theory. Biologists apply the techniques of probability in genetics, in the theory of natural selection, and in learning theory. Managers in government and industry use probabilistic techniques in developing decision-making processes. Furthermore, probability is the theoretical basis of statistics, a discipline that permeates modern thinking in virtually all disciplines.

In this section, you will practice using the following problem-solving strategies:  ▪ Draw a diagram. (throughout chapter)  ▪ Construct a table or chart. (427, 432)  ▪ Guess, test, and revise. (400)  ▪ Look for a pattern. (427)  ▪ Use reasoning. (400,408)  ▪ Restate the problem. (400, 408)  ▪ Do a simulation. (411–413)

## SECTION 1  THE LANGUAGE OF PROBABILITY

| **PROBLEM** | A card is dealt from a shuffled deck of 52 cards. What is the probability that the card is an ace of hearts? |
| --- | --- |
| **OVERVIEW** | As this problem suggests, we shall be studying the language of uncertainty in this chapter. One significant characteristic of our increasingly complex society is that we must deal with questions for which there is no known answer but instead one or more probable (or improbable) answers. |
| **GOALS** | Use the Data Analysis and Probability Standard: ● formulate questions that can be addressed with data and collect, organize, and display relevant data to answer them; ● select and use appropriate statistical methods to analyze data; ● develop and evaluate inferences and predictions that are based on data; ● understand and apply basic concepts of probability.<br>Illustrate the Representation Standard, page 50.*<br>Illustrate the Problem Solving Standard, page 15.*<br><br>* The complete statement of the standard is given on this page of the book. |

## OUTCOMES

In this section, we discuss procedures for assigning probabilities, rules that govern these probabilities, and conclusions that can legitimately be deduced from a probability once it is assigned. Because probability is a language of uncertainty, any discussion of probability presupposes a process of observation or measurement in which the outcomes are not certain. Such a process is called an **experiment.** A result of an experiment is called an **outcome.** A list of all possible outcomes in an experiment is called a **sample space.**

**EXAMPLE 1**

*Experiment:* A coin is tossed.
*Possible outcomes:* Heads (H) or tails (T). (See Figure 9-1.)
*Sample space:* {H, T}.

**EXAMPLE 2**

*Experiment:* A die is tossed.
*Possible outcomes:* The top side of the die shows 1, 2, 3, 4, 5, or 6 dots. (See Figure 9-2.)
*Sample space:* {1, 2, 3, 4, 5, 6}.

Assuming that the coin is fair, the outcomes (getting heads or getting tails) in the first experiment are said to be **equally likely,** because one outcome has the same chance of occurring as the other. Likewise, the outcomes in the second experiment (getting a 1, 2, 3, 4, 5, or 6 in the roll of a die) are equally likely.

**EXAMPLE 3**

Suppose that we have a spinner whose needle is as likely to stop at one place as another. Suppose further that the spinner is divided into three sections labeled *R, B* and *W*. (See Figure 9-3.) Let an experiment consist of spinning the needle and observing the label of the region where it stops. A sample space for this experiment would be the list of labels:

$$\{R, W, B\}$$

Note that since the portion of the spinner labeled *W* is much larger than the portions of the circle labeled *R* or *B,* the outcomes of this experiment are not equally likely.

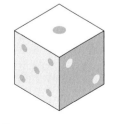

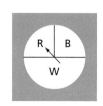

FIGURE 9-1          FIGURE 9-2          FIGURE 9-3

# TREE DIAGRAMS

Tree diagrams often serve to help list all possible outcomes of a multistep experiment. The procedure for drawing a tree diagram is illustrated in the following example.

**EXAMPLE 4**

Joe must guess at both questions on a quiz. The first question is a True/False question and the second is a multiple choice question with possible answers a through d. What are the outcomes of this experiment?

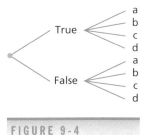

FIGURE 9-4

*SOLUTION*

From a single point, draw a line to each of the possible choices on the True/False question. From each of these answers draw four lines to each possible answer on the multiple choice question. Do you see that the tree diagram in Figure 9-4 shows eight possible outcomes?

# CHOOSING A SAMPLE SPACE

A first step in analyzing an experiment is the selection of a sample space. Different sample spaces can result from the same experiment, depending on how the observer chooses to record the outcomes.

**EXAMPLE 5**

A coin is flipped twice. Find three different sample spaces.

*SOLUTION*

The sample spaces for this experiment can best be understood with the aid of a diagram (Figure 9-5) in which we record possible results of each flip.

*Sample space A:* One complete listing of the outcomes is

$$S = \{(H, H), (H, T), (T, H), (T, T)\}$$

The letter listed first in each pair indicates the result of the first flip, and the letter listed second gives the result of the second flip. Each outcome of this sample space is equally likely.

*Sample space B:* An alternative way to list the outcomes is to ignore the order in which the heads and tails occur and to record only how many of each appear. Thus, another possible sample space is

$$S = \{(2H), (1H \text{ and } 1T), (2T)\}$$

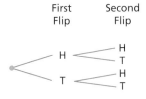

FIGURE 9-5

These outcomes are not equally likely. Because (1H and 1T) can occur in two ways (as HT and as TH), it is more likely to occur than the other two outcomes.

*Sample space C:* Another way to tabulate the same outcomes is to list the number of heads that occur: {0, 1, 2}. Again, the outcomes of this sample space are not equally likely. Zero heads can occur in only one way (as tails on one coin and tails on the other coin). Likewise, two heads can occur in only one way. But a result of heads on one coin and tails on the other can occur in two ways (as heads on the first coin and tails on the second, or as tails on the first flip and heads on the second).

Although the three sample spaces in the previous example are different, they share certain properties. Each set of outcomes classifies completely what can happen if the experiment is performed. Further, the members of each set of outcomes are distinct; that is, they do not overlap. In other words, each possible result of the experiment is represented by exactly one member of the set. This discussion suggests the following definition of a sample space.

| DEFINITION | **Sample Space** | A *sample space* (denoted by $S$) is a list of the possible outcomes of an experiment, constructed in such a way that it has the following characteristics:<br>1. The categories do not overlap.<br>2. No result is classified more than once.<br>3. The list is complete (exhausts all the possibilities). |
|---|---|---|

 **PROBABILITY**

To each outcome of a sample space we can assign a number, called its **probability,** that measures how likely that outcome is to occur. The probability of an outcome is a number between 0 and 1 inclusive. If the probability of an outcome is near 1, that outcome will likely occur when the experiment is performed. If the probability of an outcome is near 0, it is unlikely that that outcome will occur when the experiment is performed. A probability of zero indicates there is no outcome, and a probability of 1 indicates the outcome is the entire sample space.

| Properties of Probability | A **probability assignment on a sample space** must satisfy two properties:<br>1. If $A$ is a possible outcome, then its probability $P(A)$ is between 0 and 1 inclusive.<br>2. The sum of the probabilities of all outcomes in the sample space equals 1. |
|---|---|

For some sample spaces we are able to assign probabilities to outcomes based on our understanding of the properties of the sample space. Such assignments are called **theoretical probabilities.**

**EXAMPLE 6**

In the experiment of tossing a fair coin the outcomes of a head and a tail are equally likely to occur. Hence we will assign probabilities $P(H) = \frac{1}{2}$ and $P(T) = \frac{1}{2}$. Note that these assignments satisfy the properties of probability:

(a) $0 \le P(T) \le 1$    and    $0 \le P(H) \le 1$

(b) $P(S) = P(H) + P(T) = \frac{1}{2} + \frac{1}{2} = 1$

| **EXAMPLE 7** | Consider the experiment of spinning the spinner in Figure 9-3. A sample space for this experiment would be {R, B, W}. If we believe that the spinner is twice as likely to stop on region $W$ as on regions $R$ or $B$, then we would assign a probability of $\frac{1}{2}$ to outcome $W$ and probabilities of $\frac{1}{4}$ to outcomes $R$ and $B$. Again, |

(a) $0 \le P(R) \le 1$, $0 \le P(W) \le 1$, $0 \le P(B) \le 1$

(b) $P(R) + P(W) + P(B) = 1$

## UNIFORM SAMPLE SPACE

There is one whole class of sample spaces, called *uniform sample spaces,* whose probability assignments are particularly easy to determine.

| **DEFINITION** | **Uniform Sample Space** | If each possible outcome of the sample space is equally likely to occur, the sample space is called a *uniform sample space.* |

Suppose that a uniform sample space consists of $m$ possible outcomes {$A_1, A_2, \ldots, A_m$}. Because each of the outcomes is equally likely, it seems reasonable to assign to each outcome $A_i$ the same probability, denoted by $P$. Because the sum of the probabilities of the $m$ individual outcomes must be 1, it follows that

$$P(A_1) + P(A_2) + P(A_3) + \ldots + P(A_m) = 1$$
$$\underbrace{P + P + P + \ldots + P}_{m \text{ times}} = 1$$

$$mP = 1 \quad \text{or} \quad P = \frac{1}{m}$$

Thus, each of the $m$ outcomes has probability $1/m$.

| **Equal Probabilities** | In a uniform sample space with $m$ possible outcomes, each outcome has probability $1/m$. This is sometimes written as $$P(A) = \frac{1}{n(S)} = \frac{1}{m}$$ where $A$ represents one outcome and $n(S)$ represents the number of possible outcomes in the sample space. |

| **EXAMPLE 8** | Eight identical balls numbered 1 to 8 are placed in a box. (See Figure 9-6.) Find a sample space and probability assignments describing the experiment of randomly drawing one of them from the box. |

*SOLUTION*

A suitable sample space is {1, 2, 3, 4, 5, 6, 7, 8}, each number representing one of the 8 balls. Because each ball is equally likely to be drawn, we assign a probability of $\frac{1}{8}$ to each outcome:

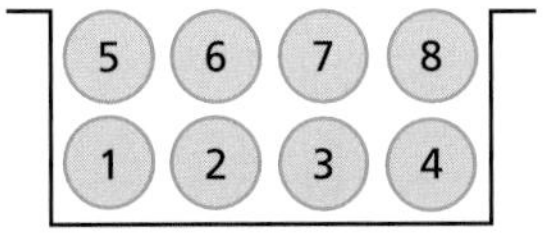

$$P(1) = \frac{1}{8}, \quad P(2) = \frac{1}{8}, \ldots, P(8) = \frac{1}{8}$$

FIGURE 9-6

---

**EXAMPLE 9**

A card is drawn from a shuffled deck (52 cards with four suits: clubs, diamonds, hearts, and spades; each suit with 13 cards: 2–10, jack, queen, king, and ace). What is the probability of drawing the ace of hearts? (See Figure 9-7.)

FIGURE 9-7

*SOLUTION*

A uniform sample space for this experiment would consist of a listing of all 52 cards. Hence each outcome would be assigned probability 1/52. In particular, the probability of drawing the ace of hearts is 1/52.

## EMPIRICAL PROBABILITY

An alternative way to assign probabilities to outcomes in an experiment involves performing the experiment many times and looking at the empirical data. Consider the following example.

**EXAMPLE 10**

A fair die is rolled 10,000 times. Table 9-1 itemizes the number of times a 1 has occurred at various stages of the process.

Notice that, as $N$ (number of rolls) becomes larger, the relative frequency ($m/N$) stabilizes in the neighborhood of $0.166 \approx 1/6$. Thus, we are willing to assign the probability

$$P(1) = \frac{1}{6}$$

| TABLE 9-1 | | |
|---|---|---|
| **Number of Rolls (N)** | **Number of 1's Occuring (m)** | **Relative Frequency (m/N)** |
| 10 | 4 | .4 |
| 100 | 20 | .2 |
| 1000 | 175 | .175 |
| 3000 | 499 | .166333 . . . |
| 5000 | 840 | .168 |
| 7000 | 1150 | .164285714 . . . |
| 10,000 | 1657 | .1657 |

Compare this thinking with our previous method of assigning probabilities. Assume that the die is constructed so that the outcomes are equally likely. Because the sample space $S = \{1, 2, 3, 4, 5, 6\}$ is a uniform sample space, the probability of obtaining a 1 in a sample space of 6 outcomes is 1/6, the same answer we obtained by looking at the empirical data.

In the previous example, we assigned probability to an outcome by assigning the fraction of times that the outcome occurred when the experiment was performed a large number of times. Similarly, suppose that a thumbtack lands with its point up 1000 times out of 10,000 trials. The relative frequency is $\frac{1000}{10,000} = \frac{1}{10}$. If we repeat the experiment 10,000 more times and find that the ratio is still approximately $\frac{1}{10}$, we will agree to assign this number as a measure of our degree of belief that it will land point up on the next toss. These examples suggest the following definition.

**DEFINITION**

| **Empirical Probability** | If an experiment is performed $N$ times, where $N$ is a very large number, the probability of an outcome $A$ should be approximately equal to the following ratio: $$P(A) = \frac{\text{number of times } A \text{ occurs}}{N}$$ |
|---|---|

**EXAMPLE 11**

A loaded die (one for which the outcomes are not equally likely) is thrown a number of times with results as shown in Table 9-2.
(a) How many possible outcomes are there?
(b) What are the possible outcomes?
(c) Using the frequency table (Table 9-2), assign a probability to each of the outcomes.

| TABLE 9-2 | | | | | | |
|---|---|---|---|---|---|---|
| **Outcome** | 1 | 2 | 3 | 4 | 5 | 6 |
| **Frequency** | 967 | 843 | 931 | 1504 | 1576 | 1179 |

## Blaise Pascal
### 1623–1662

Blaise Pascal was a frail youth whose mother died when he was 3 and whose upbringing fell to his father, Etienne, a French lawyer and mathematician. Blaise and his two sisters were educated at home by their father, and Blaise never attended a school or university.

His attraction to mathematics was evident by the time he was 12. At age 16, he had already published an original treatise on conic sections, a profound 1-page geometric result known today as Pascal's Theorem. Such a promising first paper showed indication of a brilliant career in geometry, but Pascal quickly turned to other mathematical interests. At age 18, he embarked on the unusual enterprise of designing, building, and selling mechanical calculating machines. Pascal's friend, the Chevalier de Mèrè, posed a gambling problem, whereupon Pascal—with Pierre de Fermat—originated the modern mathematical theory of probability. In this work, Pascal made use of the array that follows. Though this array was known by Jiǎ Xiàn in China in the 11th century, it is known as Pascal's Triangle because it was Pascal who developed its properties and showed its connections with other facets of mathematics.

Pascal's name first appears these days in our mathematics curriculum by way of Pascal's Triangle, usu-

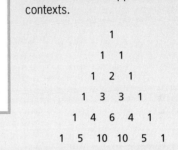

ally in grades 7 and 8. Students are asked to write down the numbers in the next row by looking for the pattern in the first few rows. Then they are asked to find other patterns within Pascal's Triangle, or to find the sum of each row, and so on. Then, throughout the first 2 or 3 courses of algebra, Pascal's work reappears in various contexts.

```
                1
              1   1
            1   2   1
          1   3   3   1
        1   4   6   4   1
      1   5  10  10   5   1
    1   6  15  20  15   6   1
```

We owe to Pascal appreciation for his innovation in the science of computation (IBM has one of his mechanical calculators in its museum). A popular and widely used computer language is named PASCAL. A great deal of attention is now being paid to the subject of artificial intelligence, and Pascal pondered the question of whether machines could think three centuries ago.

*SOLUTION*

There are 6 possible outcomes to this experiment: The die may show a 1, 2, 3, 4, 5, or 6. By adding the frequencies of the 6 outcomes, we see that the experiment was performed 7000 times. Thus the relative frequency of the outcome of 1 is 967/7000 or approximately 0.14. Similarly, the relative frequency of a 2 is 843/7000 or approximately 0.12. Continuing in this way, we make the following probability assignments:

$$P(1) = .14 \qquad P(2) = .12 \qquad P(3) = .13$$
$$P(4) = .21 \qquad P(5) = .23 \qquad P(6) = .17$$

# Just for Fun

Try to place the pennies in four equal columns subject to the following rules:

1. Each penny can be moved only once.
2. Each penny must be jumped over exactly two other pennies or stacks.
3. If an empty space occurs, ignore it.

Good luck.

# EXERCISE SET 1

*R* 1. A box contains a yellow, a green, a black, and a blue marble.

(a) How many outcomes are possible if you pick one marble from the box?
(b) List the possible outcomes.
(c) What is the probability of picking a blue marble?
(d) What is the probability of picking a green marble?
(e) Are the probabilities the same for each color of marble in the box?

2. Each letter of the word HOPEFUL is placed on a card, and the cards are placed in a basket. Sam draws a card without looking.

(a) List the possible outcomes.
(b) Are the outcomes equally likely?
(c) What is the probability of drawing a *P*?
(d) An *E*?

3. Yute has four belts of different colors (one of which is brown) in his drawer. He reaches into the drawer and picks a belt without looking. What is the probability that Yute has picked the brown belt?

4. The probability that Marsha will pick a blue button out of a jar is 1 of 2 or $\frac{1}{2}$. There are 16 buttons in the jar. How many of the buttons are blue?

5. Mark tossed a letter cube (with sides A, B, C, D, E, and F) and recorded the number of times (frequency) each letter occurred.

| Letter | Frequency |
|--------|-----------|
| A | 8 |
| B | 10 |
| C | 9 |
| D | 11 |
| E | 12 |
| F | 10 |

(a) How many possible outcomes are there?

(b) What are the possible outcomes?
(c) From the frequency table, assign a probability of tossing a D.
(d) Assign a probability to the outcome *D* based on the observation that each outcome is equally likely.
(e) Why is there a difference?
(f) Would the accuracy have improved if Mark had tossed the letter cube 100 times? 1000 times?

6. Suppose that you toss a coin 40 times and get 19 heads.

(a) On the basis of this experiment, assign a probability of getting heads in one toss of a coin.
(b) Using the fact that heads and tails are equally likely, compute the probability of getting heads in one toss.
(c) Why do the two answers differ?

7. Give the sample space for each of the four spinners.

(a)

(b)

(c)

(d)

8. Assign a probability to each outcome in the sample spaces of Exercise 7.

9. What is the probability of getting heads when a two-headed coin is tossed?

10. In Exercise 9, what is the probability of getting tails?

11. (a) A balanced coin is tossed 10,000 times. Approximately how many heads can one expect to occur?
    (b) A balanced coin is tossed 8 times and only 2 heads occur. Should we doubt that the coin is fair?

*In Exercises 12 through 15, tabulate a sample space and make probability assignments on the basis of the information given.*

12. Ten identically sized cards numbered 1 through 10 are placed in a box, and a single card is drawn from the box.

13. A die is found in a back alley and then rolled 1000 times with the following results.

| Observed | Frequency |
| --- | --- |
| 1 | 181 |
| 2 | 152 |
| 3 | 144 |
| 4 | 138 |
| 5 | 156 |
| 6 | 229 |

14. The U-POT-EM manufacturing company makes clay pots. Of the last 1200 crates of pots shipped, 440 contained a broken pot when received. Muriel's Flower Shop receives a crate and prepares to open it.

15. Of the last 850 fish caught at White Lake in the traps of the Game and Wildlife Department, 200 have been bass, 278 have been bream, 122 have been carp, and the rest have been catfish.

16. Which of the following could not be a probability? Why?
    (a) $-\frac{1}{2}$   (b) $\frac{17}{16}$   (c) .001   (d) 0
    (e) 1.03   (f) .01   (g) $\frac{5}{4}$   (h) 1

*T* 17. (a) Use the tree diagram below to tabulate a uniform sample space for the experiment of tossing a coin three times. Express each outcome as an ordered triple, such as HHH or HTT.

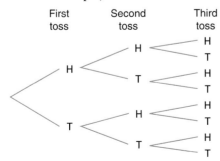

First toss   Second toss   Third toss

(b) What probability should be assigned to each outcome in this sample space?

*In Exercises 18 through 20 you may find a tree diagram to be helpful.*

18. A box contains 4 balls numbered 1 to 4. Record a sample space for the following experiments.
    (a) A ball is drawn and the number is recorded. Then the ball is returned, and a second ball is drawn and recorded.
    (b) A ball is drawn and recorded. Without replacing the first ball, the experimenter draws and records a second ball.

19. List the elements in a sample space for the simultaneous tossing of a coin and drawing of a card from a set of 6 cards numbered 1 through 6.

*C* 20. A box contains 3 red balls and 4 black balls. Let R represent a red ball and B a black ball. Tabulate a sample space if
    (a) one ball is drawn at a time.
    (b) two balls are drawn at a time.
    (c) three balls are drawn at a time.
    (d) Are these sample spaces uniform?

21. Three coins are tossed, and the number of heads is recorded. Which of the following sets is a sample space for this experiment? Why do the other sets fail to qualify as sample spaces?
    (a) {1, 2, 3}   (b) {0, 1, 2}
    (c) {0, 1, 2, 3, 4}   (d) {0, 1, 2, 3}

22. Three coins are tossed, and the number of heads is recorded. Which of the following sets are sample spaces for this experiment? If a set fails to qualify as a sample space, give the reason.
    (a) {0, 2, an odd number}
    (b) $\{x \mid x$ is a whole number and $x < 4\}$

23. If you flipped a fair coin 15 times and got 15 heads, what would be the probability of getting a head on the 16th toss?

## LABORATORY ACTIVITIES

• • • • • • • • • • • • • • • • • • • • • • • • • • • •

24. Carefully label two pennies as Penny 1 and Penny 2. Flip them and classify each outcome by recording the results for each coin. More precisely, if Penny 1 shows a head and Penny 2 shows a tail, classify the outcome as HT. Similarly, if Penny 1 shows a tail and Penny 2 shows a head, classify the outcome as TH. All outcomes can be classified as HH, HT, TH, or TT. Perform this experiment 200 times and assign probabilities by using the empirical data. Do you have evidence that this is a uniform sample space?

25. Flip 2 pennies 200 times and classify each toss as HH, H&T, or TT depending on what shows. When both a head and tail occur on the same toss, make no effort to determine which coin shows a head and which coin shows a tail. Assign probabilities to the outcomes HH, H&T, and TT on the basis of the empirical data. Does this appear to be a uniform sample space? Compare to the results of Exercise 24.

26. Toss a paper cup into the air 100 times. After each toss record whether the cup lands on its bottom, upside down on its top, or on its side. Assign probabilities to these outcomes by using empirical probabilities.

27. Roll a pair of dice 300 times. How many times do you get a pair of 1's? What probability would you assign to this outcome?

# SECTION 2 PROBABILITY OF EVENTS AND PROPERTIES OF PROBABILITY

| PROBLEM | (a) Two coins are tossed. What is the probability of getting at least one head?<br>(b) Of the freshmen who entered Loren College last year, 12% made A's in freshman English, 8% made A's in history, and 4% made A's in both English and history. An admissions counselor would like to know what percent made A's in English or history. |
| --- | --- |
| OVERVIEW | In the previous section we computed probabilities of a single outcome from an experiment. In this section we will examine the problem of computing the probability of an event where an event is a subset of outcomes from the sample space. Because an event is a subset of the sample space, we will find that many of the concepts and much of the language of Chapter 3 on sets will be useful. |
| GOALS | Illustrate the Data Analysis and Probability Standard, page 392.<br>Illustrate the Representation Standard, page 50.<br>Illustrate the Problem Solving Standard, page 15. |

We now extend our definition of probability to events.

| DEFINITION | **Event** | An *event* is a subset of a sample space. |
| --- | --- | --- |

**EXAMPLE 1**

In the previous section we found a uniform sample space for the experiment of tossing a pair of coins. Tabulate the event "at least one coin shows a head."

*SOLUTION*

| Sample Space | Event |
| --- | --- |
| {HH, HT, TH, TT} | {HH, HT, TH} |

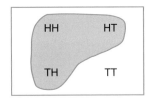

| HH | HT |
| TH | TT |

**FIGURE 9-8**

The event consists of the possible outcomes in the sample space that include at least one occurrence of heads, circled in Figure 9-8.

In the case of an event from a uniform sample space, the probability of an event is particularly easy to compute.

---

**DEFINITION**

**Probability of an Event**

Let $S$ represent a uniform sample space with $n(S)$ equally likely outcomes, and let $A$ be an event in $S$. If $A$ is an event consisting of $n(A)$ outcomes, then $P(A)$ is given by

$$P(A) = \frac{n(A)}{n(S)}$$

---

This rule is the classical definition of probability. Suppose that there are $N$ equally likely possible outcomes of an experiment. If $r$ of these outcomes have a particular characteristic so that they can be classified as a success, then the probability of a success is defined to be $r/N$.

**EXAMPLE 2**

Suppose that a card is drawn from a set of 6 (numbered 1 through 6). There are 6 equally likely possible outcomes of the experiment. Let 2 of these, a 3 and a 6, represent a success $E$. Then

$$P(E) = \frac{n(E)}{n(S)} = \frac{2}{6} = \frac{1}{3}$$

**PRACTICE PROBLEM**

What is the probability of drawing an ace from a standard deck?

*ANSWER*

4/52 or 1/13

**EXAMPLE 3**

Return to the experiment of tossing a pair of coins. What is the probability of tossing
(a) at least one head?
(b) exactly one head?

*SOLUTION*

Using the uniform sample space

$$S = \{HH, HT, TH, TT\}$$

we can tabulate the events in question.
(a) $A = \{HH, HT, TH\}$          (b) $B = \{HT, TH\}$

Thus,

(a) $P(A) = \dfrac{n(A)}{n(S)} = \dfrac{3}{4}$          (b) $P(B) = \dfrac{n(B)}{n(S)} = \dfrac{1}{2}$

**EXAMPLE 4**

A poll is taken of 500 workers to determine whether they want to go on strike. Table 9-3 indicates the results of this poll.
(a) What is the probability that a worker selected at random is in favor of a strike?
(b) What is the probability that such a worker has no opinion?

*SOLUTION*

One sample space for this experiment would consist of a list of the 500 workers. To choose a worker *at random* means that each employee has the same chance of being selected; hence the sample space is uniform.
(a) Let $E$ be the event "worker is in favor of the strike." Then

$$P(E) = \frac{n(E)}{n(S)} = \frac{280}{500} = \frac{14}{25}$$

(b) Similarly, $P(\text{no opinion}) = \frac{20}{500} = \frac{1}{25}$

**TABLE 9-3**

| In Favor of a Strike | Against a Strike | No Opinion |
|---|---|---|
| 280 | 200 | 20 |

Because events are subsets, we can, of course, form unions, intersections, and complements of events. The properties that govern unions, intersections, and complements of events are very important.

**DEFINITION**

**And, Or, and Complement**

1. The event $A \cup B$ ($A$ or $B$) is the collection of all outcomes that are in $A$ or in $B$ or in both $A$ and $B$.
2. The event $A \cap B$ ($A$ and $B$) is the collection of all outcomes that are in both $A$ and $B$.
3. The **complement** of an event $A$, denoted by $\overline{A}$, is the collection of all outcomes that are in the sample space and are not in $A$.

To illustrate these concepts, we consider examples involving the roll of a die.

**EXAMPLE 5**

In the rolling of a fair die, what is the probability that the result will be an odd number or a 4?

*SOLUTION*

We let $O$ represent an odd number and $F$ represent a 4, and then we seek $P(O \cup F)$. In Figure 9-9(b), using Venn diagrams, we see that

$$P(O \cup F) = \frac{n(O \cup F)}{n(S)} = \frac{4}{6} = \frac{2}{3}$$

FIGURE 9-9

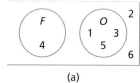

(a)

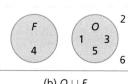

(b) $O \cup F$

Notice in Figure 9-9(a) that

$$P(O) = \frac{3}{6} \quad \text{and} \quad P(F) = \frac{1}{6}$$

Furthermore, $\qquad P(O \cup F) = P(O) + P(F)$

because $\qquad \frac{4}{6} = \frac{3}{6} + \frac{1}{6}$

**EXAMPLE 6**

*SOLUTION*

In the rolling of the same fair die, what is the probability of getting either an even number or a multiple of 3?

Let $E$ represent an even number, and let $M$ represent a multiple of 3. We seek $P(E \cup M)$. In Figure 9-10(b) we see that

$$P(E \cup M) = \frac{n(E \cup M)}{n(S)} = \frac{4}{6} = \frac{2}{3}$$

Notice in Figure 9-10(a) that

$$P(E) = \frac{3}{6} \quad \text{and} \quad P(M) = \frac{2}{6}$$

Here, however,

$$P(E \cup M) \neq P(E) + P(M)$$

because $\qquad \frac{4}{6} \neq \frac{3}{6} + \frac{2}{6}$

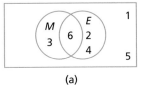

(a)

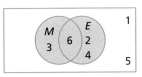

(b) $E \cup M$

FIGURE 9-10

What is the difference between the problems in the two previous examples? For $P(O \cup F)$, $F$ and $O$ had no points in common. For $P(E \cup M)$, $E$ and $M$ overlapped. This discussion suggests the following definition and property of probability.

**Mutually Exclusive Events**

1. Events $A$ and $B$ are *mutually exclusive* if they have no outcomes in common.
2. If events $A$ and $B$ are mutually exclusive,

$$P(A \cup B) = P(A) + P(B)$$

**EXAMPLE 7**

*SOLUTION*

From a standard deck of cards, you draw one card. What is the probability of your getting a spade or a red card?

Verify that $\qquad P(S) = \frac{13}{52} \quad \text{and} \quad P(R) = \frac{26}{52}$

Because the outcome of getting a spade and the outcome of getting a red card are mutually exclusive,

$$P(S \cup R) = P(S) + P(R)$$
$$= \frac{13}{52} + \frac{26}{52} = \frac{39}{52} = \frac{3}{4}$$

Now let's return to Example 6, where we noted that $P(E \cup M) = \frac{4}{6}$, $P(E) = \frac{3}{6}$, and $P(M) = \frac{2}{6}$. The reason $P(E \cup M) \neq P(E) + P(M)$ is that the outcome 6 is in both $E$ and $M$ and thus is counted twice in $P(E) + P(M)$. The probability of getting an outcome that is in both $E$ and $M$ is

$$P(E \cap M)$$

Because $E \cap M$ is included twice in $P(E) + P(M)$, we subtract one of these and note that

$$P(E \cup M) = P(E) + P(M) - P(E \cap M)$$

or $$\frac{4}{6} = \frac{3}{6} + \frac{2}{6} - \frac{1}{6}$$

We can generalize this concept by realizing that in set theory the number of outcomes in event $A$ *or* in event $B$ is the number in $A$ plus the number in $B$ less the number in $A \cap B$, which has been counted in both $A$ and $B$. (See Figure 9-11.) Thus,

$$n(A \cup B) = n(A) + n(B) - n(A \cap B)$$

In a uniform sample space we can divide both sides of the equation by $N$, the number of elements in a sample space, to obtain probabilities

$$\frac{n(A \cup B)}{N} = \frac{n(A)}{N} + \frac{n(B)}{N} - \frac{n(A \cap B)}{N}$$

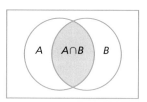

FIGURE 9-11

Thus, $$P(A \cup B) = P(A) + P(B) - P(A \cap B)$$

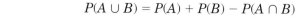

**Probability of A or B**

For any two events $A$ and $B$, the *probability of A or B* is given by

$$P(A \cup B) = P(A) + P(B) - P(A \cap B)$$

We return now to the second of our introductory problems.

**EXAMPLE 8**

Of the freshmen who entered Loren College last year, 12% made A's in English, 8% made A's in history, and 4% made A's in both English and history. What percent made A's in English or history?

*SOLUTION*

$$\begin{aligned} P(E) &= .12 \\ P(H) &= .08 \\ P(E \cap H) &= .04 \\ P(E \cup H) &= P(E) + P(H) - P(E \cap H) \\ &= .12 + .08 - .04 \\ &= .16 \end{aligned}$$

16% made A's in English or history.

**EXAMPLE 9**

In drawing a card from 8 cards numbered 1 through 8, what is the probability of getting an even number or a number less than 5?

SOLUTION

Let $A$ represent the event of getting a number less than 5, and let $B$ represent the event of getting an even number. Then

$$P(A) = \frac{4}{8} = \frac{1}{2}$$

$$P(B) = \frac{4}{8} = \frac{1}{2}$$

But 2 and 4 are both even and less than 5; consequently,

$$P(A \cap B) = \frac{2}{8} = \frac{1}{4}$$

Therefore,        $P(A \cup B) = P(A) + P(B) - P(A \cap B)$

$$= \frac{1}{2} + \frac{1}{2} - \frac{1}{4} = \frac{3}{4}$$

**PRACTICE PROBLEM**

A card is drawn from a standard deck of cards. What is the probability that it is either an ace or a spade?

ANSWER

$\dfrac{4}{13}$

The complement of event $A$ is everything in the sample space except $A$, denoted by $\overline{A}$. Because $A \cup \overline{A} = S$, and $A \cap \overline{A} = \varnothing$, $P(A \cup \overline{A}) = 1$ and $P(A \cap \overline{A}) = 0$, then $P(A \cup \overline{A}) = P(A) + P(\overline{A}) = 1$. Therefore, $P(\overline{A}) = 1 - P(A)$.

---

**Probability of the Complement of A**

If $\overline{A}$ is the complement of $A$, then

$$P(\overline{A}) = 1 - P(A).$$

---

**EXAMPLE 10**

What is the probability of not getting an ace in the drawing of a card from a standard deck of cards?

SOLUTION

$$P(\text{no ace}) = 1 - P(\text{ace}) = 1 - \frac{4}{52} = \frac{12}{13}$$

Sometimes probability statements are given in terms of odds, which is actually a comparison of the probability that event $E$ will occur and the probability that event $E$ will not occur (denoted by $\overline{E}$).

---

**DEFINITION**

**Odds**

The odds in favor of event $E$ equal the ratio

$$P(E):P(\overline{E}) \qquad \text{or} \qquad P(E):1 - P(E)$$

and the odds against event $E$ equal

$$P(\overline{E}):P(E) \qquad \text{or} \qquad 1 - P(E):P(E)$$

**EXAMPLE 11**

Find the odds in favor of rolling a 6 with a single die.

*SOLUTION*

$$P(6) = \frac{1}{6} \qquad 1 - P(6) = \frac{5}{6}$$

$$\text{Odds} = P(6):1 - P(6) \quad \text{or} \quad \frac{1}{6} \text{ to } \frac{5}{6} \quad \text{or} \quad 1 \text{ to } 5$$

Thus, the odds in favor of rolling a 6 are 1 to 5.

At times we are given the odds in favor of an event, and from the odds we obtain the probability that the event will occur.

---

**Probability from Odds**

If the odds favoring an event $E$ are $m$ to $n$, then

$$P(E) = \frac{m}{m + n} \quad \text{and} \quad P(\overline{E}) = \frac{n}{m + n}$$

---

**EXAMPLE 12**

The odds that it will rain today are 1 to 3. What is the probability that it will rain?

*SOLUTION*

For the given odds, $m$ can be taken as 1 and $n$ as 3. Therefore,

$$P(R) = \frac{1}{1 + 3} = \frac{1}{4}$$

**PRACTICE PROBLEM**

What are the odds against 2 heads on 2 tosses of a coin?

*ANSWER*

The odds against 2 heads are 3 to 1.

# Just for Fun

In a casino, a customer places quarters on numbers, and the house places silver dollars on numbers. The first to put money on any 3 squares that total 15 wins what the other has on the board.

EXAMPLE: Customer, 25¢ on 1, the house, $1 on 5; customer, 25¢ on 9, and the house $1 on 7; customer blocks with 25¢ on 3, and the house places $1 on 6; the customer blocks with 25¢ on 4 and the house places $1 on 2 and wins 4 · 25¢ = $1 because 7 + 6 + 2 = 15.

Show that this game is similar to tic-tac-toe and a 3-by-3 magic square with totals of 15. Can the house always win?

# EXERCISE SET 2

*R* 1. Use the accompanying spinner to find the probability that the needle will stop on

(a) a 6 or 3 wedge.
(b) a 5 or 4 wedge.

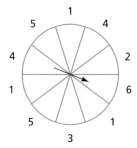

2. The letters of the word *MATHEMATICS* are placed on cards. Aaron draws a card without looking.

(a) What is the probability that it is an M, H, or T?
(b) What is the probability it is not an A?

3. Suppose that there is an equally likely probability that the spinner will stop at any one of the six sections for the given spinner.

(a) What is the probability of stopping on an even number?
(b) What is the probability of stopping on a multiple of 3?
(c) What is the probability of stopping on an even number or a multiple of 3?
(d) What is the probability of stopping on an even number or an odd number?
(e) What is the probability of stopping on a number less than 3?

4. A card is drawn from an ordinary deck. What is the probability of getting

(a) a heart?
(b) an ace?
(c) the jack of spades?
(d) a red card?
(e) a red ace?

5. A multiple-choice question has five possible answers. You haven't studied and hence have no idea which answer is correct, so you randomly choose one of the five answers. What is the probability that you select the correct answer? An incorrect answer?

6. A survey course in history contains 20 freshmen, 8 sophomores, 6 juniors, and 1 senior. A student is chosen at random from the class roll.

(a) Describe in words a uniform sample space for this experiment.
(b) What is the probability that the student is a freshman?

(c) What is the probability that the student is a senior?
(d) What is the probability that the student is a junior or a senior?
(e) What is the probability that the student is a freshman or a sophomore or a junior or a senior?

7. A box contains four black balls, seven white balls, and three red balls. If a ball is drawn, what is the probability of getting the following colors?

(a) Black          (b) Red
(c) White          (d) Red or white
(e) Black or white (f) Red or white or black

8. A number $x$ is selected at random from the set of numbers $\{1, 2, 3, \ldots, 8\}$. What is the probability that

(a) $x$ is less than 5?
(b) $x$ is even?
(c) $x$ is less than 5 and even?
(d) $x$ is less than 5 or a 7?

9. An experiment consists of tossing a coin seven times. Describe in words (without using the word *not*) the complement of each of the following.

(a) Getting at least two heads
(b) Getting three, four, or five tails
(c) Getting one tail          (d) Getting no heads

10. If $A$ and $B$ are events with $P(A) = .6$, $P(B) = .3$, and $P(A \cap B) = .2$, find $P(A \cup B)$.

11. At Brooks College, 30% of the freshmen make A's in mathematics, 20% make A's in English, and 15% make A's in both mathematics and English. What is the probability that a freshman makes an A in mathematics or English?

12. A single card is drawn from a 52-card deck. What is the probability that it is

(a) either a heart or a club?
(b) either a heart or a king?
(c) not a jack?
(d) either red or black?

13. In the experiment of tossing a pair of coins compute the probabilities of these events.

(a) all heads          (b) no heads

14. In Exercise 13, compute the probability of:

(a) at least one tail          (b) exactly one tail

15. The employment status of the residents in a certain town is given in the following table.

| Gender | Employed | Unemployed |
|--------|----------|------------|
| Male   | 1000     | 40         |
| Female | 800      | 160        |

Assign a probability that each of the following is true of a randomly selected person.

(a) Person is female.    (b) Person is male.
(c) Person is unemployed.
(d) Person is employed.

*T* 16. A sociology class made a study of the relationship between an employee's age and the number of on-the-job accidents. The following table summarizes the findings.

|  | **Number of Accidents** | | | |
| --- | --- | --- | --- | --- |
| **Age Group** | **0** | **1** | **2** | **3 or More** |
| Under 20 | 18 | 22 | 8 | 12 |
| 20–39 | 26 | 18 | 8 | 10 |
| 40–59 | 34 | 14 | 8 | 6 |
| 60 and over | 42 | 10 | 12 | 2 |

An employee is selected at random.

(a) What is the probability that the employee is in the 20–39 age group?
(b) What is the probability that the employee has had 2 accidents?
(c) What is the probability that the employee has had more than 2 accidents?
(d) What is the probability that the employee has had at least 1 accident?

17. Find the odds in favor of drawing a heart from an ordinary deck of 52 cards.

18. A die is rolled. What are the odds that a 2 will turn up?

19. What are the odds against selecting an ace in drawing one card from a deck of cards?

20. Leroy has a chance for 5 different summer jobs, 3 of which are at resort areas. If he selects a job at random, find the odds against its being at a resort area.

21. If $A$ and $B$ are events in a sample space such that $P(A) = .6$, $P(B) = .2$, and $P(A \cap B) = .1$, compute each of the following.

(a) $P(\overline{A})$
(b) $P(\overline{B})$
(c) $P(A \cup B)$
(d) $P(\overline{A \cap B})$

*C* 22. If $A$ and $B$ are events with $P(A \cup B) = \frac{5}{8}$, $P(A \cap B) = \frac{1}{3}$, and $P(\overline{A}) = \frac{1}{2}$, compute the following.

(a) $P(A)$
(b) $P(B)$
(c) $P(\overline{B})$
(d) $P(\overline{A \cup B})$

23. In a survey, families with children were classified as $C$, and those without children as $\overline{C}$. At the same time, families were classified according to $D$, husband and wife divorced, and $\overline{D}$, not divorced. For 200 families surveyed, the following results were obtained.

|  | **C** | **C̄** | **Total** |
| --- | --- | --- | --- |
| **D** | 60 | 20 | 80 |
| **D̄** | 90 | 30 | 120 |
| **Total** | 150 | 50 | 200 |

What is the probability that a family selected at random

(a) has children?
(b) has husband and wife who are not divorced?
(c) has children or husband and wife who are divorced?
(d) has no divorce or no children?

24. A recent survey found that 60% of the people in a given community drink Lola Cola and 40% drink other soft drinks; 15% of the people interviewed indicated that they drink both Lola Cola and other soft drinks. What percent of the people drink either Lola Cola or other soft drinks?

25. A coin is tossed 4 times. Let event $A$ be that exactly one head appears. Compute $P(A)$ and $P(\overline{A})$, and verify that $P(A) = 1 - P(\overline{A})$.

26. After 1000 rolls of a loaded die, the following probability assignments are made. Compute the probability of the following events.

| Outcome | 1 | 2 | 3 | 4 | 5 | 6 |
| --- | --- | --- | --- | --- | --- | --- |
| Probability | .15 | .3 | .3 | .05 | .05 | .15 |

(a) The roll is even.
(b) The roll is $> 4$.
(c) The roll is even or divisible by 3.
(d) The roll is even and divisible by 3.

27. Four coins are tossed. What is the probability of getting

(a) 4 tails?
(b) exactly 2 tails?
(c) at least 3 heads?
(d) exactly 1 head?

## REVIEW EXERCISES

28. A family plans to have three children.

(a) Use a tree diagram to help list all possible outcomes of this activity by classifying each child as either a boy or a girl.
(b) If each of these outcomes is equally likely, what probability should be assigned to each outcome?

SECTION

# 3  MULTISTEP EXPERIMENT, EXPECTED VALUES, AND SIMULATION

**PROBLEM**

A nationwide promotion promises a first prize of $25,000, 2 second prizes of $5000, and 4 third prizes of $1000. A total of 950,000 persons enter the lottery. Is entering worth the price of a stamp required to mail the lottery form?

**OVERVIEW**

In this section we examine three important problem-solving tools that are used for resolving uncertain situations: We find a use for tree diagrams in multistep experiments; we discuss the importance of expected value in determining how much a decision maker stands to gain or lose in an uncertain situation; and we look at the process of simulation as a way of exploring probability.

**GOALS**

Illustrate the Data Analysis and Probability Standard, page 392.
Illustrate the Representation Standard, page 50.
Illustrate the Connections Standard, page 147.
Illustrate the Algebra Standard, page 4.

Some experiments are performed as sequences of consecutive steps and are called *multistep experiments*. The probabilities of outcomes of multistep experiments are easily computed with tree diagrams. Consider the following example.

**EXAMPLE 1**

A basket contains two red balls and three black balls. A ball is drawn, set aside, and its color is noted. Then a second ball is drawn. Find a sample space for this experiment and assign probabilities.

*SOLUTION*

This is an example of a multistep experiment. The simple tree diagram in Figure 9-12 makes it easy to determine the sample space, but it is clear that this is not a uniform sample space. The outcome RR (red, then red) is certainly less likely than BB (black, then black).

The probability of obtaining a red ball on the first draw is 2/5, but, because the first ball is laid aside, the probability of the second draw producing a red ball is 1/4. Thus, only 1/4 of the times that the first ball is red will the second ball be red. Hence, it is reasonable to assign a probability of $\frac{2}{5} \cdot \frac{1}{4} = \frac{1}{10}$ to the outcome RR. Label each branch of the tree with the probability that the outcome on that branch will occur on the next step. (See Figure 9-12.) Notice that the probability of RR we computed is just the product of the probabilities along the path RR in the tree diagram. Similarly, the probability of RB is $\frac{2}{5} \cdot \frac{3}{4} = \frac{3}{10}$, the probability of BR is $\frac{3}{5} \cdot \frac{2}{4} = \frac{3}{10}$, and BB is $\frac{3}{5} \cdot \frac{2}{4} = \frac{3}{10}$.

The scheme used in Example 1 to compute probabilities in a multistep experiment can be used in general. Suppose a tree diagram is used to find a sample space for a multistep experiment.

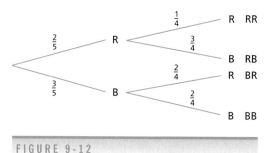

FIGURE 9-12

| **Probability Using a Tree Diagram** | The probability of an outcome described by a path through a tree diagram is equal to the product of the probabilities along that path. |

---

**EXAMPLE 2**

Let us use the basket from Example 1 that contains 2 red balls and 3 black balls. This time, after we draw the first ball and record the results, we will place it back in the basket. (See Figure 9-13.) In this circumstance, what are the probabilities of the outcomes RR and RB in the sample space?

*SOLUTION*

Do you see that the probability of RR is $\frac{2}{5} \cdot \frac{2}{5} = \frac{4}{25}$ while the probability of RB is $\frac{2}{5} \cdot \frac{3}{5} = \frac{6}{25}$?

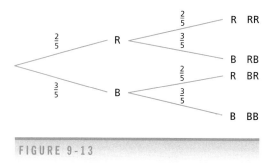

FIGURE 9-13

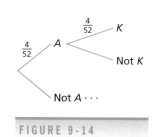

FIGURE 9-14

**PRACTICE PROBLEM**

Find the probability of BR and BB in Example 2.

*ANSWER*

$\frac{6}{25}, \frac{9}{25}$

**EXAMPLE 3**

A card is drawn from a deck of cards. Then the card is replaced, the deck is reshuffled, and a second card is drawn. What is the probability of getting an ace on the first draw and a king on the second?

*SOLUTION*

We do not need to draw the whole tree for the experiment of drawing two cards from a deck to compute this probability. Consider only the path of interest in the problem shown in Figure 9-14. The probability of an ace followed by a king is $\frac{4}{52} \cdot \frac{4}{52} \approx .0059$

**PRACTICE PROBLEM**

In Example 3 compute the probability if the first card drawn is not replaced.

*ANSWER*

$$\frac{4}{52} \cdot \frac{4}{51} = .006$$

# EXPECTED VALUE

An important property associated with probability is that of expectation or expected value. If we toss a fair coin 100 times, we expect to get heads approximately

$$100\left(\frac{1}{2}\right) = 50 \text{ times}$$

If we spin a fair spinner with 10 equal sectors 1000 times, we expect the spinner to stop on any given sector approximately

$$1000\left(\frac{1}{10}\right) = 100 \text{ times}$$

The concept is perhaps most easily explored in the analysis of a simple game of chance. Suppose that some poor benighted soul were persuaded to play the following game with us: A fair coin is flipped. If a head appears, we receive \$5; if a tail appears, our demented opponent receives \$2 (that is, we win –\$2).

What happens if we play the game 100 times? In 100 flips of the coin we can expect approximately 50 heads and 50 tails. Hence we can expect a pay-off of approximately 50(\$5) from the heads and a payoff of (50)(–\$2) from the tails. Our net profit will thus be (50)(\$5) + 50(–\$2) = \$150. Because the game is played 100 times, our average profit per game will be 150/100 = \$1.50.

Now observe this alternative way of computing average gain per game:

$$P(\text{winning \$5}) = P(H) = \frac{1}{2} \qquad P(\text{payoff of } - \$2) = P(T) = \frac{1}{2}$$

$$(\$5) \cdot P(\text{winning \$5}) + (-\$2) \cdot P(\text{payoff of -\$2}) = \frac{1}{2} \cdot \$5 + \frac{1}{2} \cdot -\$2$$

$$= \$2.50 - \$1 = \$1.50$$

This second set of computations suggests the following definition of expected value.

| DEFINITION | **Expected Value** | Suppose that there are $n$ payoff values in a given experiment: $A_1, A_2, A_3, \ldots, A_n$. The *expected value* for this experiment is<br><br>$$A_1 \cdot P(A_1) + A_2 \cdot P(A_2) + \cdots + A_n \cdot P(A_n)$$<br><br>where $P(A_1) + P(A_2) + \cdots + P(A_n) = 1$. |
| --- | --- | --- |

**EXAMPLE 4**

Let $x$ be a variable representing the number of tails that can appear in the toss of three coins. Of course, $x$ can assume the value 0, 1, 2, or 3. Tabulating the results as in Table 9-4 assists us in computing the expected value of $x$. The expected number of tails is $\frac{3}{2}$.

One interpretation of expected value is that it is the average payoff per experiment when the experiment is performed a large number of times.

**EXAMPLE 5**

A nationwide promotion promises a first prize of $25,000, two second prizes of $5000, and four third prizes of $1000. A total of 950,000 persons enter the lottery.
(a) What is the expected value if the lottery costs nothing to enter?
(b) Is it worth the stamp required to mail the lottery form?

*SOLUTION*

(a) Because

$$P(\$25,000) = \frac{1}{950,000} \qquad P(\$5000) = \frac{2}{950,000} \qquad P(\$1000) = \frac{4}{950,000}$$

the expected value is

$$(\$25,000)\frac{1}{950,000} + (\$5000)\frac{2}{950,000} + (\$1000)\frac{4}{950,000} = \$0.04$$

(b) Hardly!

Notice again that expected value is not something to be expected, in the ordinary sense of the word. It is a long-run average of repeated experimentation.

**PRACTICE PROBLEM**

Alfa Car Insurance Company insures 200,000 cars each year. Records indicate that during the year the company faces the likelihood of making the liability payments shown in Table 9-5 for accidents.
What amount should the company expect to pay per car insured?

*ANSWER* | $910

**TABLE 9-4**

| x | P(x) | xP(x) |
|---|------|-------|
| 0 | $\frac{1}{8}$ | 0 |
| 1 | $\frac{3}{8}$ | $\frac{3}{8}$ |
| 2 | $\frac{3}{8}$ | $\frac{6}{8}$ |
| 3 | $\frac{1}{8}$ | $\frac{3}{8}$ |
| Total | Expected value $= \frac{12}{8} = \frac{3}{2}$ | |

**TABLE 9-5**

| Liability | Corresponding Probabilities |
|-----------|------------------------------|
| $500,000 | 0.0001 |
| $100,000 | 0.001 |
| $50,000 | 0.004 |
| $30,000 | 0.01 |
| $5,000 | 0.04 |
| $1,000 | 0.06 |
| $0 | 0.8849 |

# SIMULATIONS

One of the most powerful tools available to both scientists and business analysts is the tool of simulation. Using simulations, rocket engines, marketing schemes, and weapons systems are tested before they are even built. We will certainly not test any rocket engines, but we can experience the flavor of simulations. For example, because the probability of having a boy child is roughly the same as the probability of having a girl child, we could use the flip of a coin to simulate the birth of a child. We could simulate possible outcomes for a family with 3 children by flipping a coin 3 times. By repeating this simulation 100 times, we could compute empirical probabilities for the experiment without ever having undergone the considerable difficulty of having 300 children.

A very useful tool in performing simulation is a table of random digits such as the one in Table 9-6. The numbers in Table 9-6 are collections of digits that are randomly generated by a computer. We will use this table in the following simulation.

**TABLE 9-6**

| | | | | | | | |
|---|---|---|---|---|---|---|---|
| 12135 | 65186 | 86886 | 72976 | 79885 | 07369 | 49031 | 45451 |
| 10724 | 95051 | 70387 | 53186 | 97116 | 32093 | 95612 | 93451 |
| 53493 | 56442 | 67121 | 70257 | 74077 | 66687 | 45394 | 33414 |
| 15685 | 73627 | 54287 | 42596 | 05544 | 76826 | 51353 | 56404 |
| 74106 | 66185 | 23145 | 46426 | 12855 | 48497 | 05532 | 36299 |
| 57126 | 99010 | 29015 | 65778 | 93911 | 37997 | 89034 | 79788 |
| 94676 | 32307 | 41283 | 42498 | 73173 | 21938 | 22024 | 76374 |
| 68251 | 71593 | *93397 | 26245 | 51668 | 47244 | 13732 | 48369 |
| 60907 | 17698 | 32865 | 24490 | 56983 | 81152 | 12448 | 00902 |
| 07263 | 16764 | 71261 | 52515 | 93269 | 61210 | 55526 | 71912 |
| 43501 | 10248 | 34219 | 83416 | 91239 | 45279 | 19382 | 82151 |
| 57365 | 84915 | 11437 | 98102 | 58168 | 61534 | 69495 | 85183 |
| 38161 | 22848 | 06673 | →35293 | 27893 | 58461 | 10404 | 17385 |
| 26760 | 51437 | 87751 | 41523 | 10816 | 54858 | 35715 | 47947 |
| 65592 | 93388 | 36555 | 21136 | 43900 | 89837 | 78093 | 28870 |
| 48651 | 16719 | 99032 | 86292 | 40668 | 72821 | 59266 | 44970 |
| 71495 | 84760 | 35193 | 06961 | 41211 | 33548 | 40026 | 63873 |
| 81242 | 06154 | 69109 | 60926 | 62177 | 72065 | 70225 | 86018 |
| 26574 | 84854 | 38915 | 83783 | 46780 | 08735 | 38781 | 94657 |
| 07736 | 70130 | 46808 | 18940 | 14795 | 34231 | 23671 | 05856 |
| 26533 | 06561 | 09049 | 67618 | 12560 | 59539 | 41937 | 18490 |
| 36335 | 84039 | 05960 | 38850 | 62976 | 65958 | 99682 | 64250 |
| 92074 | 87770 | 31924 | 99481 | 15505 | 55099 | 42072 | 57637 |
| 00243 | 48272 | 45390 | 24171 | 96173 | 98887 | 03335 | 45965 |
| 68900 | 91374 | 18868 | 45389 | 57567 | 89557 | 56764 | 59362 |
| 57663 | 88219 | 88929 | 03419 | 28838 | 89659 | 64710 | 60768 |
| 27715 | 05262 | 06208 | 96357 | 65700 | 82054 | 28590 | 95933 |
| 91798 | 54270 | 85403 | 30110 | 00426 | 19915 | 38883 | 43423 |
| 64221 | 42325 | 55273 | 68399 | 91856 | 76729 | 25130 | 64615 |
| 10852 | 21817 | 08641 | 82759 | 75389 | 96295 | 05934 | 53697 |

**EXAMPLE 6**

Simulate the experiment of tossing a pair of coins 50 times.

*SOLUTION*

We need to use the table to determine the result of tossing each coin. Because each digit is equally likely to be even or odd, we will use the first two digits in each entry to represent the outcomes of the coins. If the first digit is even, the first coin shows a head; if the first digit is odd, the first coin shows a tail. Similarly, the second digit will determine whether the second coin shows a head or tail. We need to randomly choose a place in the table to start; close your eyes and mark a starting point with your pencil. Suppose the arrow marks the spot where to start. Because the first two digits of 35293 are odd, the outcome of our first toss is two tails. Proceeding to the next number down the column, the 4 and 1 in 41523 indicate that the next pair of tosses results in a head and a tail. Continuing, we get the results from our 50 trials as shown in Table 9-7. How does this compare to the theoretical probabilities?

**TABLE 9-7**

|    | Frequency | Empirical Probability |
|----|-----------|----------------------|
| HH | 12        | $\frac{12}{50} = .24$ |
| HT | 12        | $\frac{12}{50} = .24$ |
| TH | 13        | $\frac{13}{50} = .26$ |
| TT | 13        | $\frac{13}{50} = .26$ |

**EXAMPLE 7**

A basketball player hits 40% of her shots from three-point range. Use simulation to determine the probability that in a game in which she takes 5 long shots (from beyond the three-point line) she will hit at least 3 of them.

*SOLUTION*

We will let a single digit represent each attempted long shot. Because she hits 40% of her shots from this distance, we will let the four digits 0, 1, 2, 3 represent successful shots and the digits 4, 5, 6, 7, 8, 9 represent missed attempts. (Note that 0, 1, 2, 3 comprise 40% of the digits 0 through 9.) A five-digit entry from the random number table will represent a game in which she attempts five shots. We randomly pick a place to begin. Suppose that it is the location marked with the asterisk in Table 9-6. The first entry is 93397. Because the 3's represent successes, but 9 and 7 represent misses, in the first game she made only 2 of her 5 shots. If we simulate this 50 times, we learn that in 13 of 50 games she hit 3 or more of her 3-point shots. Thus, from the simulation we would assign a probability of .26 to this event.

# Just for Fun

The "Birthday Problem" gives a result that defies intuition. Suppose that 35 people are in a room. What is the probability that at least 2 of them have their birthdays on the same day of the year? Because there are 365 days in the year and only 35 people, it seems that the probability should be fairly small. In fact, however, it is quite large—namely, 0.814.

**The Probability $P$ That Among $n$ People at Least 2 Will Have the Same Birthday**

| $n$ | $P$ |
|-----|------|
| 5 | .027 |
| 10 | .117 |
| 15 | .253 |
| 20 | .411 |
| 23 | .506 |
| 25 | .569 |
| 30 | .706 |
| 35 | .814 |
| 40 | .891 |
| 45 | .941 |
| 50 | .970 |
| 55 | .986 |
| 60 | .994 |
| 65 | .998 |
| 70 | .999 |
| 75 | .999 + |

# EXERCISE SET 3

$R$ 1. A basket contains 4 yellow balls and 5 green balls. One ball is drawn and laid aside, and a second ball is drawn. A tree diagram showing the possible outcomes and the probabilities is shown below.

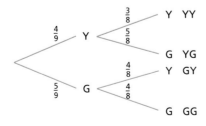

(a) What is the probability of drawing a yellow ball and then a green ball?

(b) What is the probability of drawing 2 green balls?

2. Consider the problem in Exercise 1 but in this case return the first ball to the basket after it is drawn and its color noted.

(a) Redraw the tree diagram from Exercise 1 with the correct probabilities.

(b) What is the probability of drawing 2 green balls?

3. In a lottery, 200 tickets are sold for $1 each. There are 4 prizes, worth $50, $25, $10, and $5. What is the expected value of a single ticket?

4. A fair die is rolled. What is the expected value of the number of dots?

5. The 6 letters of the word *LITTLE* are in a box. We draw 4 letters one at a time and place them in a row. We are interested in the probability that the letters spell the word *TELL*. We do not have to draw the whole tree. Indeed, the branch of the tree that spells *TELL* is our primary interest. Label this branch with the probabilities that are required and compute the probability of *TELL*.

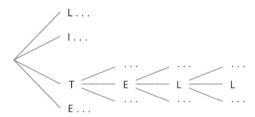

*T* 6. We spin the three spinners shown below. What is the probability that the resulting letters spell *NOT*?

7. From a box containing 5 red balls and 3 white balls, 2 balls are drawn successively at random, without replacement. What is the probability that the first is white and the second is red?

8. John must take a 4-question true–false quiz but has failed to study. What is the probability that he will score 100% if he guesses on each question?

9. Suppose the quiz (see Exercise 8) that John must take consists of 4 multiple-choice questions with answers (a) through (d). What is the probability that John will make 100%?

10. A candy jar contains 6 pieces of peppermint, 4 pieces of chocolate, and 12 pieces of butterscotch candy. A small boy reaches into the jar, snatches a piece, and eats it rapidly. He repeats this act quickly.

    (a) What is the probability that he eats a peppermint and then a chocolate?
    (b) What is the probability that he eats 2 chocolates?
    (c) What is the probability that he eats a chocolate and then a butterscotch?
    (d) What is the probability that he eats a chocolate and a butterscotch?

11. Suppose that the small boy of Exercise 10 is caught by his mother immediately after he snatches his first piece. She makes him return the candy to the jar. He waits an appropriate length of time and then again snatches a piece.

    (a) What is the probability that the frustrated thief snatches first a peppermint and then a chocolate?
    (b) What is the probability that he gets chocolate on both tries?

12. Assume that two cards are drawn from a standard deck of playing cards. What is the probability that a jack is drawn, followed by a queen,

    (a) if the first card is replaced before the second is drawn?
    (b) if the first card is not replaced before the second is drawn?

13. Use Table 9-6 to simulate the experiment of flipping a coin 3 times. Use the first 3 digits of each entry in the table to determine respectively whether the first, second, and third flips show a head or a tail. Simulate the experiment 50 times and compute the appropriate empirical probabilities.

14. Tiny Trinkles cereal is placing 1 of 3 tiny dolls in each box of cereal. Each of the dolls is equally likely to occur in a given box of cereal. We are interested in the probability that if we open 5 boxes of cereal, we will have a collection of all 3 dolls.

    (a) How could we use simulation with rolls of a die to answer this question?
    (b) Use simulation to answer this question by rolling 5 dice 20 times.

15. How could we use a die or dice to simulate

    (a) flipping a coin?       (b) flipping a coin 5 times?
    (c) 3 times at bat by a batter bating .333?
    (d) a family of 4 children, each classified as boy or girl?

16. How would we use a box containing 4 cards numbered 1 through 4 to simulate

    (a) flipping a coin?
    (b) flipping a coin 4 times?
    (c) a 3-game series by a team that wins 75% of its games?
    (d) 4 times at bat by a hitter hitting .250?

*C* 17. Ed and Jack play a match of tennis each Sunday afternoon. They play until 1 player has won 2 sets, then retire for the day. The probability that Ed will win a given set is .6 and the probability is .4 that Jack will win.

    (a) Draw a tree diagram showing the possible outcomes of a Sunday afternoon's play.

(b) What is the probability that the match will be over after 2 sets?

(c) What is the probability that Jack will win 2 sets?

(d) What is the probability that Ed will win 2 sets?

18. A candy bowl contains 4 pieces of candy in identical wrappers. Two of the pieces of candy are butterscotch and two are licorice. Joyce sits down to read, absentmindedly takes a piece of candy from the bowl, and begins to eat it. Suppose she will continue to eat the candy until she pops a licorice candy in her mouth.

(a) Draw a tree diagram that shows the possible outcomes of this eating session.

(b) What is the probability that Joyce will place only 1 piece of candy in her mouth?

(c) What is the probability that Joyce will place 3 pieces of candy in her mouth?

19. A baseball player is batting .300. In a typical game he will have 3 official at bats.

(a) Draw a 3-stage tree diagram describing the possible outcomes.

(b) What is the probability of 3 hits?

(c) What is the probability of at least 2 hits?

(d) What is the probability of no hits?

20. Answer the questions in Exercise 19 by simulation. Let the first 3 digits of an entry in Table 9-6 represent 3 times at bat. If the entry is 0, 1, or 2, record a hit; otherwise, record a failure. Simulate a game with 3 times at bat 50 times and compute the probabilities requested in Exercise 19.

21. A manufacturer receives a shipment of 20 articles. Unknown to him, 6 are defective. He selects 2 articles at random and inspects them. What is the probability that the first is defective and the second is satisfactory?

22. Tom and Savoy decide to play the following game for points. A single die is rolled. If it shows a non-prime, Tom receives points equal to twice the number of dots showing. If it shows a prime, Tom loses points equal to 3 times the number of dots showing. What is the expected value of the game?

23. A man who rides a bus to work each day determines that the probability that the bus will be on time is 7/16. The probability that the bus will be 5 minutes late is 3/16; 10 minutes late, 1/4; and 15 minutes late, 1/8. What is his expected waiting time if the man arrives at the bus stop at the scheduled time?

24. Two coins, each biased (or weighted) $\frac{1}{3}$ heads and $\frac{2}{3}$ tails, are tossed. The payoff is $5 for matching heads, $3 for matching tails, and −$2 if they don't match. What is the expected value of the game?

25. The probability that a test will be positive for a person with AIDS is .95, and the probability that the test will be positive for a person who does not have AIDS is .01. Of the people in a given area of a city, 2% have AIDS. If a person is selected at random from that area, what is the probability that the test will be positive? (HINT: The answer will come from two branches of the accompanying tree diagram.)

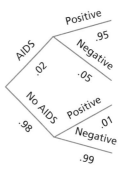

26. Use random digit simulation (using Table 9-6) to estimate the probability that two cards drawn from a standard deck with replacement will be of different suits. (HINT: One model could be as follows. Let 1 and 2 represent hearts, 3 and 4 spades, 5 and 6 diamonds, and 7 and 8 clubs. If either a 9 or 0 occurs, ignore that random number and move to the next one. Start at any place in Table 9-6, and consider sets of 2 digits. Record the number of times you get different suits and the number of times you get the same suit.)

27. In Exercise 1

(a) what is the probability of drawing at least 1 yellow ball?

(b) What is the probability of drawing 2 balls of different color?

28. In Exercise 2

(a) What is the probability of drawing at least 1 green ball?

(b) What is the probability of drawing 2 balls of the same color?

## REVIEW EXERCISES
. . . . . . . . . . . . . . . . . . . . . . . . . . . . .

29. A bag contains 6 red balls, 4 black balls, and 3 green balls. A ball is drawn from the bag. What is the probability that the ball is

(a) red or black?                    (b) blue?

(c) red or black or green

(d) not red and not green?

(e) not black?                       (f) green?

(g) not red or not black?            (h) not green?

30. A couple wishes to have either 3 children or 4 children. Further, they want exactly 2 girls. Is it more likely that they will get their wish with 3 children or with 4? Answer this question by using Table 9-6 to do the necessary simulations.

(a) Choose a starting point in Table 9-6 and simulate the experiment of having 3 children 50 times. You may use the first 3 digits of each entry in the table to represent the 3 children.

(b) Choose a starting point in Table 9-6 and simulate the experiment of having 4 children 50 times. You may use the last 4 digits in each number to represent the 4 children.

(c) On the basis of your simulations, answer the young couple's question.

# SECTION 4
# THE FUNDAMENTAL PRINCIPLE OF COUNTING AND PERMUTATIONS

**PROBLEM**

The Hardy College Bulldogs are purchasing uniforms. Members can purchase red or white shorts. They can choose red, white, or striped shirts. How many possible choices are there for the uniforms?

**OVERVIEW**

In a uniform sample space the probability of an event is computed by counting the outcomes in the event and the outcomes in the sample space and then dividing. In some cases it is easy to list all the outcomes and then count them. In other cases, it is difficult to list all the outcomes. Hence, we need to learn some procedures that will allow us to count sets without listing all of their elements.

**GOALS**

Illustrate the Data Analysis and Probability Standard, page 392.
Illustrate the Problem Solving Standard, page 15.

As we have seen earlier, tree diagrams are useful in listing all the possible outcomes of many experiments. Observe carefully the tree diagrams in the next two examples, and you may discover a hint about a powerful counting technique.

**EXAMPLE 1**

The college chorale is planning a concert tour with performances in Dallas, St. Louis, and New Orleans. In how many ways can its itinerary be arranged?

*SOLUTION*

If there is no restriction on the order of the performances, any one of the three cities can be chosen as the first stop. After the first city is selected, either of the other cities can be second, and the remaining city is then the last stop. (See Figure 9-15.)

Did you notice that there were three choices for the first stop, two choices for the second stop, and one choice for the third stop for a total of $3 \cdot 2 \cdot 1 = 6$ possible itineraries?

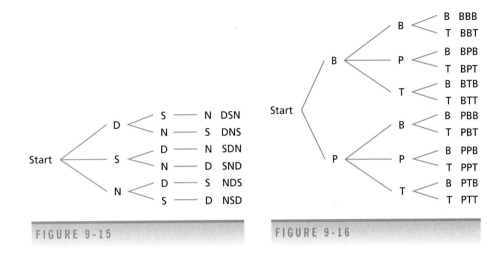

FIGURE 9-15                    FIGURE 9-16

**EXAMPLE 2**

The members of the chorale in Example 1 decided to sing first in New Orleans, next in Dallas, and finally in St. Louis. Now, they must decide on their modes of transportation. They can travel from the campus to New Orleans by bus or plane; from New Orleans to Dallas by bus, plane, or train; and from Dallas to St. Louis by bus or train. The tree diagram in Figure 9-16 indicates the chorale's options. The first part of the trip can be made in 2 ways, the second part in 3 ways, and the last part in 2 ways. Notice that the total number of ways the transportation can be chosen is $2 \cdot 3 \cdot 2 = 12$ ways.

These examples suggest the following principle, called the *Fundamental Principle of Counting.*

**Fundamental Principle of Counting**

1.  If an experiment consists of two steps, performed in order, with $n_1$ possible outcomes of the first step and $n_2$ possible outcomes of the second step, then there are

    $$n_1 \cdot n_2$$

    possible outcomes of the experiment.

2.  In general, if $k$ steps are performed in order, with possible number of outcomes $n_1, n_2, n_3, \ldots, n_k$, respectively, then there are

    $$n_1 \cdot n_2 \cdot n_3 \cdot \ldots \cdot n_k$$

    possible outcomes of the experiment.

**EXAMPLE 3**

A coin is tossed 5 times. If we classify each outcome as either a head or a tail, how many outcomes are in the sample space?

*SOLUTION*

Because there are 5 steps, each with 2 possible outcomes, there are $2 \cdot 2 \cdot 2 \cdot 2 \cdot 2 = 32$ different outcomes.

**PRACTICE PROBLEM**

A die is rolled 3 times. How many different outcomes are there?

*ANSWER*

216

The Fundamental Principle of Counting is helpful in solving problems such as the following.

**EXAMPLE 4**

In the state of Georgia, automobile license plates contain an arrangement of 3 letters followed by 3 digits. If all letters and digits may be used repeatedly, how many different arrangements are available?

*SOLUTION*

There are 26 letters to choose from for each of the 3 letter places, and there are 10 digits to choose from for the digit places. By the Fundamental Principle of Counting, the number of arrangements is

$$26 \cdot 26 \cdot 26 \cdot 10 \cdot 10 \cdot 10 = 17,576,000$$

**EXAMPLE 5**

If the letters and numbers on a license plate in Georgia (see Example 4) are assigned randomly, what is the probability that you will receive a license plate on which the letters read *DOG?*

*SOLUTION*

In Example 4 we learned that the sample space contains 17,576,000 outcomes, all of which are equally likely. By the counting principle there are

$$1 \cdot 1 \cdot 1 \cdot 10 \cdot 10 \cdot 10 = 1000$$

possible license plates that begin with *DOG* (because we have only 1 choice for each of the letters). Thus the probability of a license plate with the word *DOG* is

$$\frac{1000}{17,576,000} = \frac{1}{17,576}$$

**EXAMPLE 6**

An urn contains 5 red balls and 7 white balls. A ball is drawn, its color is noted, but the ball is not replaced. A second ball is drawn. What is the probability of drawing a red ball followed by a white ball?

*SOLUTION 1*

By the counting principle, there are 12 ways of drawing the first ball and 11 ways of drawing the second. Therefore, there are

$$12 \cdot 11 = 132$$

ways of drawing the 2 balls. All of these ways are equally likely. At the same time, there are 5 ways of drawing the red ball on the first draw and 7 ways of drawing the white ball on the second draw. Therefore, the number of ways of drawing a red ball and then a white ball is

$$5 \cdot 7 = 35$$

Thus, $$P(R \text{ followed by } W) = \frac{35}{132}$$

*SOLUTION 2*

We, of course, recognize that this problem could have been solved with a tree diagram as discussed in the previous section. Consider the tree diagram in Figure 9-17 and then observe that

$$P(R \text{ followed by } W) = \frac{5}{12} \cdot \frac{7}{11} = \frac{35}{132}$$

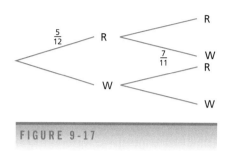

FIGURE 9-17

**EXAMPLE 7**

In how many ways can 6 students line up outside Wheeler's office to complain about their grades?

*SOLUTION*

As a first step, we can choose a person to be first in line; there are 6 ways to do this. Then we can choose a second person; only 5 persons are available after the first person is chosen. Similarly, the third place must be filled by one of 4 persons and so on.

| First | Second | Third | Fourth | Fifth | Sixth |
|-------|--------|-------|--------|-------|-------|
| 6 | 5 | 4 | 3 | 2 | 1 |

By the Fundamental Principle of Counting, there are $6 \cdot 5 \cdot 4 \cdot 3 \cdot 2 \cdot 1$ or 720 ways to accomplish this task.

# PERMUTATIONS

Notice in Example 7 that each of the 720 possible lineups is a different ordered arrangement of the set of 6 persons. Ordered arrangements of sets of objects are called **permutations** of the sets.

**EXAMPLE 8**

Count the number of different permutations of the letters A, B, and C, and then verify the count with a tree diagram.

*SOLUTION*

Because there are 3 ways to choose a first element, then only 2 ways to choose a second element, and finally only 1 way to choose a third, there are

$$3 \cdot 2 \cdot 1 = 6$$

different permutations. This fact is demonstrated with the tree diagram in Figure 9-18.

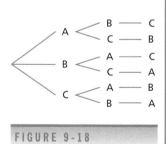

The 6 permutations are listed below:

| ABC | BAC | CAB |
|-----|-----|-----|
| ACB | BCA | CBA |

FIGURE 9-18

In the previous 2 examples, we observed that a set of 6 distinct elements has $6 \cdot 5 \cdot 4 \cdot 3 \cdot 2 \cdot 1$ permutations, whereas a set of 3 distinct elements has $3 \cdot 2 \cdot 1$ permutations. In general,

| **Number of Permutations of *n* Objects** | The number of permutations of *n* distinct objects is denoted $P(n, n)$ and $$P(n, n) = n(n - 1)(n - 2) \cdot \cdots \cdot 3 \cdot 2 \cdot 1$$ which can be written as *n*! (read as *n* factorial). |
|---|---|

The preceding discussion suggests that *n*! is the product of positive integers 1 to *n*, inclusive. The product $6 \cdot 5 \cdot 4 \cdot 3 \cdot 2 \cdot 1$ may be denoted by 6!, called *6 factorial*. We define both 1! and 0! to be 1. The statement that 0! = 1 may seem surprising, but you will learn later in your work with factorials that this definition is reasonable and consistent with the idea of factorials for positive integers.

## CALCULATOR HINT

On many calculators, there is a key $\boxed{!}$, $\boxed{x!}$, or $\boxed{n!}$ that can be used to compute factorials. We use the $\boxed{\text{PRB}}$ key to locate a menu on which is located a *!* entry. For example 6! can be found as follows:

6 $\boxed{\text{PRB}}$ (underline *!*) $\boxed{=}$ $\boxed{=}$ to get 720. Verify that 10! = 3,628,800.

**EXAMPLE 9**

A first-year class is to elect a president, a vice-president, a secretary, and a treasurer from among 6 class members who qualify. In how many ways can the class officers be selected?

*SOLUTION*

If we consider the order as president, vice-president, secretary, and treasurer, then (Maria, Tom, Jim, Tomoko) is certainly different from (Tomoko, Tom, Maria, Jim). Thus, each selection of officers is a permutation, not of the whole set, but of a subset of 4 chosen from the whole set. Using the Fundamental Principle of Counting, you can see that the position of president can be filled in 6 ways. After this occurs, the position of vice-president can be filled in 5 ways, so the 2 positions can be filled in $6 \cdot 5$ ways. Then the secretary can be selected in 4 ways, so the 3 positions in $6 \cdot 5 \cdot 4$ ways. Finally, only 3 people remain as candidates for treasurer. Hence, the number of ways that all 4 positions can be filled is $6 \cdot 5 \cdot 4 \cdot 3$.

An ordered arrangement of 4 things chosen from 6 things is called a *permutation of 6 things taken 4 at a time*. The number of permutations of 6 things taken 4 at a time is denoted $P(6, 4)$. In the previous example, we determined that

$$P(6, 4) = 6 \cdot 5 \cdot 4 \cdot 3$$

If we wished to express $P(6, 4)$ using factorials, we could observe that

$$P(6, 4) = 6 \cdot 5 \cdot 4 \cdot 3 = \frac{6 \cdot 5 \cdot 4 \cdot 3 \cdot 2 \cdot 1}{2 \cdot 1} = \frac{6!}{2!}$$

Thus, $$P(6, 4) = \frac{6!}{2!} = \frac{6!}{(6 - 4)!}$$

By reasoning in the same way, we find the following.

| Number of Permutations | The *number of permutations* of $n$ things taken $r$ at a time is given by $$P(n, r) = \frac{n!}{(n - r)!} \qquad 1 \le r \le n$$ |
| --- | --- |

## CALCULATOR HINT

The number of permutations, $P(n, r)$ can be computed using the factorial key of a calculator and the formula for $P(n, r)$. On some calculators there are specific keys for computing $P(n, r)$; sometimes these keys are named *nPr*. We compute the number of permutations by going to the ⎡PRB⎤ menu and selecting *nPr*. First we compute $P(6, 4)$ by using the definition involving factorials, and then we compute $P(6,4)$ by selecting the *nPr* entry.

6 ⎡PRB⎤ (underline *!*) ⎡=⎤ ⎡÷⎤ 2 ⎡PRB⎤ (underline *!*) ⎡=⎤ ⎡=⎤ , and

6 ⎡PRB⎤ (underline *nPr*) ⎡=⎤ 4 ⎡=⎤. Did you get 360 in both cases?

**EXAMPLE 10**

Professor Wheeler is asked to judge the Homecoming Pageant. Overcome by the sheer beauty and talent of the 10 contestants, he decides to randomly assign his rankings of first, second, and third. What is the probability that he will award Lori first prize, Jodi Lyn second prize, and Joy third prize?

*SOLUTION*

The sample space would consist of all permutations of 3 chosen from the 10 contestants. Hence,

$$n(S) = P(10, 3) = \frac{10!}{7!} = 10 \cdot 9 \cdot 8 = 720$$

Only one of the 720 outcomes ranks Lori first, Jodi Lyn second, and Joy third. Hence, the probability of this result is $\frac{7}{120}$.

# Just for Fun

Move only one glass so that empty glasses alternate with full glasses.

# EXERCISE SET 4

R 1. A student plans a trip from Atlanta to Boston to London. From Atlanta to Boston, he can travel by bus, train, or airplane. However, from Boston to London, he can travel only by ship or airplane.

(a) In how many ways can the trip be made?
(b) Verify your answer by drawing an appropriate tree diagram and counting the routes.

2. A sociology quiz contains a true–false question and 2 multiple-choice questions with possible answers (a), (b), (c), and (d).

(a) In how many possible ways can the test be answered?
(b) Draw a tree diagram and count the options to check your answer to (a).
(c) If Jodi guesses on each problem, what is the probability that she will get all 3 correct?

3. There are 6 roads from $A$ to $B$ and 4 roads between $B$ and $C$.

(a) In how many ways can Joy drive from $A$ to $C$ by way of $B$?
(b) In how many ways can Joy drive round trip from $A$ to $C$ through $B$?

4. Kate wants to buy an automobile. She has a choice of 2 body styles (standard or sports model) and 4 colors (green, red, black, or blue). In how many ways can she select the automobile?

5. (a) In how many ways can 2 speakers be arranged on a program?
(b) In how many ways can 3 speakers be arranged on a program?
(c) In how many ways can 4 speakers be arranged on a program?

🖩 6. Compute the following without a calculator and then check your answers with a calculator

(a) 4!    (b) 7!    (c) 0!    (d) 1!

7. (a) How many permutations are there of the set $P, Q, R$?
(b) Write down all of the permutations of $P, Q, R$.

8. (a) How many permutations taken 2 at a time are there of the set $W, X, Y, Z$?
(b) List all of the permutations in (a).

9. A die is tossed and a chip is drawn from a box containing 3 chips numbered 1, 2, and 3. How many possible outcomes can be obtained from this experiment? Verify your answer with a tree diagram.

🖩 10. Evaluate each of the following without a calculator and then check your answers with a calculator:

(a) $P(5, 3)$    (b) $P(6, 5)$    (c) $P(8, 1)$
(d) $P(9, 2)$    (e) $P(7, 2)$    (f) $P(8, 7)$

T 11. Write a simple expression for each of the following:

(a) $P(r, 1)$
(b) $P(k, 2)$
(c) $P(r, r - 1)$
(d) $P(k, k - 2)$
(e) $P(k, 3)$
(f) $P(k, k - 3)$

12. The license plates for a certain state display 3 letters followed by 3 numbers (examples: MFT-986, APT-098). How many different license plates can be manufactured if no repetitions of letters or digits are allowed?

13. Employee ID numbers at a large factory consist of 4-digit numbers such as 0133, 4499, and 0000.

(a) How many possible ID numbers are there?
(b) How many possible ID numbers are there in which all 4 digits are different?

C 14. (a) How many 3-digit numbers are there? (Remember, a 3-digit number cannot begin with 0.)
(b) How many 3-digit numbers are there that end in 3, 6, or 9?
(c) If you are randomly assigned a 3-digit number as an ID number, what is the probability that it will end in 3, 6, or 9?

15. An ice chest contains 5 cans of cola, 7 cans of ginger ale, and 3 cans of root beer. Al randomly selects a can, and then Sheila takes one. Compute each of the following probabilities in two ways. Use a uniform sample space with $15 \cdot 14$ elements. Use portions of a tree diagram as in Section 3.

(a) Al gets a cola and Sheila a root beer.
(b) Al gets a ginger ale and Sheila a cola.
(c) Both get root beers.
(d) Neither gets a cola.

16. Employee ID numbers at a large factory consist of 4-digit numbers possibly beginning with 0. What is the probability that, if a number is chosen at random from the list of ID numbers, all 4 of its digits will be different?

17. A typical social security number is 413-22-9802. If a social security number is chosen at random, what is the probability that all the digits will be the same? (Social security numbers may begin with 0.)

18. Consider the license plates in Exercise 12. What is the probability that a citizen will receive a license plate whose first 3 letters read *WHY*?

## REVIEW EXERCISES

19. A single ball is drawn from a basket that contains balls numbered 1 through 10. Find the probability that the number of the ball that is chosen is:

    (a) even.      (b) greater than 7.

    (c) even and greater than 7.

    (d) even or greater than 7.

    (e) What property of probability can be illustrated with the facts in (a) through (d)?

20. Two cards are drawn from a standard deck of playing cards. What is the probability that a king is drawn, followed by an ace,

    (a) if the first card is replaced before the second is drawn?

    (b) if the first card is not replaced before the second is drawn?

21. A desperate student offers to play the following game with her professor. A single die is rolled. If the roll is odd, the student receives points equal to twice the number of dots showing. If the roll is even, the student loses points equal to 3 times the number of dots showing. What is the expected value of this game for the student?

22. Two cards are drawn without replacement from a standard deck. What is the probability that

    (a) the first is red and the second is black?

    (b) both are red?          (c) both are black?

---

### ❖ PCR Excursion ❖

23. In this section we have been concerned with counting the ordered subsets of a set (called *permutations*). In the next section we will be interested in counting the ordinary unordered subsets of a set, which sometimes are called *combinations*. In this excursion, we do some preparatory work for the next section and discover interesting patterns that have fascinated mathematicians for centuries.

A. The set {a, b} has 4 subsets. { }, {a}, {b}, and {a, b}. (Remember that the null set, { }, is a subset of all sets.) Count the number of subsets for each of sets {a, b, c}, {a, b, c, d}, {a, b, c, d, e} by completing the table:

| Set | Number of Elements | Subsets | Number of Subsets |
|---|---|---|---|
| { } | 0 | { } | 1 |
| {a} | 1 | { }, {a} | 2 |
| {a, b} | 2 | { }, {a}, {b}, {a, b} | 4 |
| {a, b, c} | | | |
| {a, b, c, d} | | | |
| {a, b, c, d, e} | | | |

B. Do you see a pattern in the cases you examined in A? How many subsets does a set with 6 elements have? A set with 10 elements? A set with $n$ elements?

C. Let us look more carefully at the subsets of each set. If we classify the subsets of {a, b} by the number of elements in each subset, we observe that there are 1 subset with no elements, 2 subsets with 1 element, and 1 subset with 2 elements. Notice how this fact is recorded in row 3 of the following triangular chart. Similarly, the set {a, b, c} has 1 subset with no elements, 3 subsets with 1 element, 3 subsets with 2 elements, and 1 subset with 3 elements (row 4 of the following chart). By looking at your work from A, complete the rows in the following chart for sets with 4 elements (fifth row) and sets with 5 elements (sixth row).

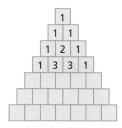

D. To analyze the subsets of a set with 6 elements, we would like to have the entries in row 7 of our triangular chart. Look at the pattern formed by the previous rows and complete the last row. Describe what the entries in row 7 tell us about the subsets of a set with 6 elements.

E. The triangular table you created is called *Pascal's Triangle*. It contains a wealth of information, but, among other things, it counts for us the unordered subsets (combinations) of various sizes in a set. Complete 2 more rows of Pascal's Triangle and then look for as many patterns as can be found. (HINT: What is the sum of the entries in a row? What can be said about the second entry in each row?)

# SECTION 5 COUNTING AND COMBINATIONS

| | |
|---|---|
| **PROBLEM** | Eight students each submit an essay for competition. In how many ways can 3 essays be chosen to receive a certificate of merit? |
| **OVERVIEW** | In the previous section we developed the ability to count the number of ordered subsets or permutations of a set. However, there are many circumstances in which we need to count the number of unordered subsets of some set. In the context of counting, unordered subsets of a sample space are called **combinations**. |
| **GOALS** | Illustrate the Data Analysis and Probability Standard, page 392.<br>Illustrate the Representation Standard, page 50.<br>Illustrate the Connections Standard, page 147.<br>Illustrate the Algebra Standard, page 4. |

Write all subsets with 3 elements that can be chosen from the set $\{P, Q, R, S\}$. Did you get

$$\{P, Q, R\}, \{P, Q, S\}, \{P, R, S\}, \{Q, R, S\}?$$

Notice that there are 4 subsets with 3 elements. In the context of counting, we say that there are 4 combinations of 3 objects that can be chosen from a set of 4 objects. In symbols, we write $C(4, 3) = 4$.

| | |
|---|---|
| **Combinations** | In general, a subset of $r$ elements chosen from a set $S$ with $n$ elements is called *an r combination of* S. The number of $r$ combinations that can be chosen from a set of $n$ elements is denoted $C(n, r)$. $C(n, r)$ is sometimes called *the number of combinations of* n *things taken* r *at a time*. |

Notice that each of the 3 combinations of $\{P, Q, R, S\}$ can be ordered in $3! = 6$ ways. For instance,

$$PQR, PRQ, QPR, QRP, RPQ, RQP$$

are all the different permutations of the combination (subset) $\{P, Q, R\}$. Thus, the number of permutations taken 3 at a time from a set with 4 elements is $3! \cdot C(4, 3)$; that is,

$$3! \cdot C(4, 3) = P(4, 3) \quad \text{or}$$

$$C(4, 3) = \frac{P(4, 3)}{3!} \quad \text{or because } P(4, 3) = \frac{4!}{(4 - 3)!}$$

$$C(4, 3) = \frac{4!}{3!1!}$$

This reasoning generalizes as follows:

| | |
|---|---|
| **Combination of *n* Things Taken *r* at a Time** | The number of ways of selecting $r$ objects from $n$ objects without regard to order (the number of combinations of $n$ things taken $r$ at a time) is $$C(n, r) = \frac{n!}{r!(n - r)!} \qquad 1 \le r \le n$$ |

## CALCULATOR HINT

Some books use $_nC_r$ or $\binom{n}{r}$ instead of $C(n, r)$, and some calculators have a $\boxed{nC_r}$ key. We use a key found under the $\boxed{\text{PRB}}$ menu, where we underline *nCr*. For the preceding example $C(4, 3)$ is found by 4 $\boxed{\text{PRB}}$ (underline *nCr*) $\boxed{=}$ 3 $\boxed{=}$. Did you get 4?

**EXAMPLE 1**

Compute the number of unordered subsets with 4 elements that can be chosen from a set with 9 elements.

*SOLUTION*

$$C(9, 4) = \frac{9!}{4!(9 - 4)!} = \frac{9 \cdot 8 \cdot 7 \cdot 6 \cdot 5 \cdot 4 \cdot 3 \cdot 2 \cdot 1}{4 \cdot 3 \cdot 2 \cdot 1 \cdot 5 \cdot 4 \cdot 3 \cdot 2 \cdot 1} = \frac{9 \cdot 8 \cdot 7 \cdot 6}{4 \cdot 3 \cdot 2 \cdot 1} = 126$$

Notice that to compute $C(n, r)$ easily we first write the answer in factorial notation, expand, divide out common factors, and multiply.

**PRACTICE PROBLEM**

Compute $C(7, 4)$ without a calculator. Then check your answer with a calculator.

*ANSWER*

35

It is important to learn to distinguish between counting problems in which you are counting permutations and counting problems in which you are counting combinations. Consider the problem with which we began this section.

**EXAMPLE 2**

Eight students each submit an essay for competition. In how many ways can 3 essays be chosen to receive a certificate of merit?

*SOLUTION*

In this problem we are counting the number of unordered subsets with 3 elements that can be chosen from a set of 8 elements. Hence, there are

$$C(8, 3) = \frac{8!}{3!5!} = \frac{8 \cdot 7 \cdot 6 \cdot 5 \cdot 4 \cdot 3 \cdot 2 \cdot 1}{3 \cdot 2 \cdot 1 \cdot 5 \cdot 4 \cdot 3 \cdot 2 \cdot 1} = 56$$

possible outcomes.

**EXAMPLE 3**

Eight students each submit an essay for competition. In how many ways can first, second, and third prizes be awarded?

*SOLUTION*

Because we are not only choosing 3 essays but also placing them in order by awarding first, second, and third prizes, we are counting permutations.

Thus there are

$$P(8, 3) = \frac{8!}{(8-3)!} = \frac{8 \cdot 7 \cdot 6 \cdot \cancel{5} \cdot \cancel{4} \cdot \cancel{3} \cdot \cancel{2} \cdot \cancel{1}}{\cancel{5} \cdot \cancel{4} \cdot \cancel{3} \cdot \cancel{2} \cdot \cancel{1}} = 8 \cdot 7 \cdot 6 = 336$$

possible outcomes.

**EXAMPLE 4**

The president of the Student Government Association wishes to appoint a committee of senators consisting of 3 men and 4 women. Currently there are 12 male senators and 8 female senators. In how many ways can this committee be formed?

*SOLUTION*

There are $C(12, 3)$ ways of choosing the male members of the committee and $C(8, 4)$ ways of choosing the female members of the committee. By the Fundamental Principle of Counting, the number of ways of appointing this committee is

$$C(12, 3) \cdot C(8, 4) = 220 \cdot 70 = 15,400$$

**PRACTICE PROBLEM**

A 5-card hand is chosen from a 52-card deck. How many hands contain 3 aces and 2 kings?

*ANSWER*

$C(4, 3) \cdot C(4, 2) = 24$

The ability to count combinations allows us to solve many interesting probability problems.

**EXAMPLE 5**

A basket contains 5 red marbles and 6 black marbles. A handful of 4 marbles is chosen from the basket.
(a) How many outcomes are in the sample space for this experiment?
(b) How many outcomes are in the event "2 red, 2 black"?
(c) What is the probability of this event?

*SOLUTION*

(a) The outcomes are all possible combinations of 4 chosen from the 11 marbles so the sample space contains $C(11, 4) = 330$ outcomes.
(b) There are $C(5, 2)$ ways to choose the red marbles and $C(6, 2)$ ways to choose the black marbles, so the event contains $C(5, 2) \cdot C(6, 2) = 150$ outcomes.
(c) The probability is $\dfrac{150}{330} = \dfrac{5}{11}$

**EXAMPLE 6**

Find the number of 5-card hands that can be drawn from an ordinary deck of cards. What is the probability of getting all hearts in a given hand?

*SOLUTION*

There are 52 cards in a deck of cards. We will be selecting 5 cards at a time, without regard to the order in which the cards are drawn. The outcome is a combination of 52 things taken 5 at a time:

$$C(52, 5) = \frac{52!}{(52-5)!5!} = \frac{52!}{47! \cdot 5!}$$

$$= \frac{52 \cdot 51 \cdot 50 \cdot 49 \cdot 48}{1 \cdot 2 \cdot 3 \cdot 4 \cdot 5}$$

$$= 2,598,960$$

There are 13 hearts. The number of ways in which 5 cards can be drawn from these 13 is $C(13, 5) = 1287$. Therefore,

$$P(\text{all hearts}) = \frac{1287}{2{,}598{,}960} \approx .0005$$

## CALCULATOR HINT

To check the computation in Example 6 with a calculator try

52 [PRB] (underline $nCr$) 5 [=] 2,598,960.

# EXERCISE SET 5

*R* 1. Consider the set of 4 objects {$W, X, Y, Z$}.

    (a) How many combinations of 2 objects can be chosen from this set? List them.

    (b) How many permutations of 2 objects can be chosen from this set? List them.

    (c) Verify that $P(4, 2) = (2!)C(4, 2)$.

    (d) How many combinations of 3 objects can be chosen from this set? List them.

    (e) How many permutations of 3 objects can be chosen from this set? List them.

    (f) Verify that $P(4, 3) = 3!C(4, 3)$.

*In Exercises 2 through 7, determine whether the problem is counting permutations or combinations and compute the correct answer.*

  2. In how many ways can a student select 3 books from a reading list of 10 books?

  3. In how many ways can 4 books be chosen from 12 books and arranged on a shelf?

  4. In how many ways can a starting 5 be chosen from a team of 12 basketball players if one disregards position?

  5. In how many ways can a starting 5 be chosen from a team of 12 basketball players if each of the 5 players is assigned a position?

  6. In how many ways can a chair and vice-chair be chosen from a committee of 6 members?

  7. In how many ways can a coach select 3 cocaptains from a team of 22 baseball players?

  8. Three students who are seniors in mathematics and 3 mathematics faculty members are to be placed on a student activities committee for the department. If there are 6 mathematics faculty members and 8 senior mathematics majors, in how many ways can this committee be formed?

  9. How many different hands consisting of 7 cards can be drawn from an ordinary deck of cards?

 10. A special committee of 3 persons must be selected from a 12-person board of directors. In how many ways can the committee be selected? Check your answer with a calculator.

 11. From a standard deck of cards, how many different 7-card hands can be drawn consisting of

    (a) 7 spades?

    (b) 5 clubs and 2 hearts?

    (c) 4 clubs, 1 spade, and 2 hearts?

    (d) 3 clubs, 2 hearts, and 2 diamonds?

*T* 12. Ten cities are competing to be selected as the site of a new Jupiter automobile assembly plant. In how many ways can the CEO of Jupiter Inc. select 4 cities as finalists for this new industrial development.

 13. A firm buys material from 3 local companies and 5 out-of-state companies. If 4 orders are submitted at once, in how many ways can 2 orders be submitted to a local firm and 2 to an out-of-state firm?

 14. A bowl contains 8 red marbles and 14 black marbles. At random, 3 marbles are selected from the bowl without replacement.

    (a) What is the probability that all 3 are black?

    (b) What is the probability that 1 is red and 2 are black?

 15. A hat contains 20 slips of paper numbered 1 to 20. If 3 are drawn without replacement, what is the probability that all are numbered less than 10?

*C* 16. (a) How many 5-card hands contain 2 aces and 3 kings?

    (b) How many 5-card hands can be drawn from a standard deck?

    (c) What is the probability that a 5-card hand chosen at random contains exactly 2 aces and 3 kings?

17. A hospital ward contains 12 patients, of whom 6 have heart disease. If an intern randomly selects 4 of the patients to examine, what is the probability that all of them will have heart disease?

18. If a 5-card hand is selected from a standard deck, what is the probability of

    (a) 4 aces?
    (b) exactly 1 ace?
    (c) no aces?
    (d) at least 1 ace?

## REVIEW EXERCISES

• • • • • • • • • • • • • • • • • • • • • • • • • • • • • • • •

19. A bag contains 6 red balls, 4 black balls, and 3 green balls. Two balls are drawn in succession from the bag without replacement. What is the probability of getting

    (a) 2 red balls?
    (b) a red ball followed by a green ball?
    (c) 2 black balls?

20. Two dice are tossed. What is the probability that the sum of the 2 numbers is

    (a) greater than 10?
    (b) equal to 9?
    (c) equal to 7?

21. In a certain college, 30% of the students failed mathematics, 20% failed English, and 15% failed both mathematics and English. What is the probability that a student failed mathematics or English?

# SOLUTION TO INTRODUCTORY PROBLEM

UNDERSTAND THE PROBLEM.  A couple wants to have either 3 or 4 children, including exactly 2 girls. We are to find the probability of exactly 2 girls with 3 children and the probability of exactly 2 girls with 4 children.

DEVISE A PLAN.  Use a tree diagram to describe all possible arrangements of children in order of birth for a family with 3 children and a family with 4 children. Let B represent boy and G represent girl. Count the number of possible arrangements and the number of arrangements with 2 girls.

CARRY OUT THE PLAN.  The resulting sample spaces are tabulated in the table. Because the probabilities of a girl and a boy are approximately the same, this is a uniform sample space. Hence,

$$P(\text{exactly 2 girls from 3 children}) = \frac{3}{8}$$

$$P(\text{exactly 2 girls from 4 children}) = \frac{6}{16} = \frac{3}{8}$$

Are you surprised?

LOOK BACK.  The probability of exactly 2 girls is the same whether the family has 3 children or 4 children. We can verify this result with simulation. For help with this verification, see the PCR Excursion at the end of Section 3 of this chapter.

| Three Children | | Four Children | | |
|---|---|---|---|---|
| BBB | (BGG) | BBBB | (BGGB) | (GBBG) |
| BBG | (GBG) | BBBG | (BBGG) | GBBB |
| BGB | (GGB) | BBGB | (BGBG) | GBGG |
| GBB | GGG | BGBB | (GBGB) | GGBG |
| | | BGGG | (GGBB) | GGGB |
| | | | | GGGG |

# SUMMARY AND REVIEW

## SAMPLE SPACE

(a) Any result of an experiment is called an *outcome*.

(b) If each outcome has the same chance of occurring as any other, the outcomes are equally likely.

(c) The outcomes of a sample space satisfy the following criteria:
   (i) The categories do not overlap.
   (ii) No result is classified more than once.
   (iii) The list exhausts all possibilities.

(d) An *event* is a subset of sample space.

(e) If each outcome of a sample space is equally likely to occur, the sample space is a *uniform* sample space.

(f) A tree diagram is useful in identifying the outcomes in a multistep experiment.

## COUNTING TECHNIQUES

(a) *Fundamental Principle of Counting:* If two steps are performed in order, with $n_1$ possible outcomes of the first step and $n_2$ possible outcomes of the second step, then there are $n_1 \cdot n_2$ combined possible outcomes of the first step followed by the second.

(b) $n! = n(n - 1) \cdot (n - 2) \cdots \cdot 1$

(c) The ordered arrangements of $r$ objects selected from a set of $n$ different objects ($r \leq n$) are called *permutations* of $n$ things taken $r$ at a time. The number of permutations of $n$ things taken $r$ at a time is
$$P(n, r) = \frac{n!}{(n - r)!}$$

(d) A $r$ combination is a subset (unordered) with $r$ elements. The number of $r$ combinations from a set with $n$ elements is
$$C(n, r) = \frac{n!}{r!(n - r)!}$$
where $1 \leq r \leq n$.

## PROBABILITY

(a) In a uniform sample space with $m$ outcomes, each outcome has a probability of $1/m$.

(b) In a uniform sample space the probability of event $A$ is given by
$$P(A) = \frac{n(A)}{n(S)}$$

(c) If an experiment is performed $N$ times, where $N$ is a large number, then
$$P(A) = \frac{\text{number of times } A \text{ occurs}}{N}$$

(d) If $A$ and $B$ are mutually exclusive (have no outcomes in common), then the probability of $A$ or $B$ is given by $P(A \cup B) = P(A) + P(B)$.

(e) For any two events $A$ and $B$,
$$P(A \cup B) = P(A) + P(B) - P(A \cap B)$$

(f) The complement of event $A$ is everything in the sample space except $A$, denoted by $\overline{A}$.
$$P(\overline{A}) = 1 - P(A)$$

(g) In a multistep experiment the probability of an outcome described by a path in a tree diagram can be computed by multiplying the probabilities along that path.

## ODDS

(a) The odds in favor of event $E$ are equal to the ratio $P(E)/P(\overline{E})$.

(b) The odds against event $E$ are equal to $P(\overline{E})/P(E)$.

## EXPECTED VALUE

Suppose that there are $n$ payoff values in an experiment: $A_1, A_2, A_3, \ldots A_n$. Then the expected value of the experiment is
$$E = A_1 P(A_1) + A_2 P(A_2) + \cdots + A_n P(A_n)$$
where $P(A_1) + P(A_2) + \cdots + P(A_n) = 1$.

# CHAPTER TEST

1. What is the probability of getting heads when a 2-headed coin is tossed?

2. In how many ways can 6 books be arranged on a shelf?

3. A jar contains 3 red balls, 2 green balls, and 1 yellow ball. Tabulate a sample space for the following experiments.
   (a) A single ball is drawn (a sample space with 3 outcomes).
   (b) A ball is drawn and pocketed. A second ball is drawn. (A tree diagram might be helpful.)
   (c) A ball is drawn, its color is recorded, and it is replaced in the jar. A second ball is drawn. (Try a tree diagram.)

4. Suppose that $P(A) = .35$, $P(B) = .51$, and $P(A \cap B) = .17$. Compute the following.
   (a) $P(\overline{A})$         (b) $P(A \text{ or } B)$

5. A box contains 3 red balls and 4 white balls. What is the probability of drawing 2 white balls
   (a) if the first ball is replaced before the second one is drawn?
   (b) if the first ball is not replaced?

6. From a group of 7 people, how many committees of 4 can be selected?

7. In a certain college, 30% of the students failed mathematics, 20% failed English, and 75% passed both mathematics and English. What is the probability that a student passed English or passed mathematics?

8. A file contains 20 good sales contracts and 5 canceled contracts. In how many ways can 4 good contracts and 2 canceled contracts be selected?

9. A card is drawn from a standard deck. What is the probability that it is a king or a spade?

10. The city jail of Rocky Hill has 5 cells, numbered 1 to 5. One evening, 4 drunks were arrested for disturbing the peace. In how many ways can they be assigned to separate cells?

11. A red die and a green die are tossed.
    (a) Describe a uniform sample space for this experiment.
    (b) What is the probability that the red die shows a 6 while the green die shows a 1?
    (c) What is the probability that the sum of the 2 numbers is 7?
    (d) What is the probability that the red die shows an even number?

12. A box contains 6 red and 4 black balls. Three balls are drawn at random. What is the probability of getting 2 red balls and 1 black ball?

13. In how many ways can a chairman, a treasurer, and a secretary be selected from a board of 12 persons?

14. Consider a family of 3 children. Find the probability that all 3 children are of the same sex.

15. A new pod is being formed for a middle school. The principal has available 3 mathematics specialists, 4 social studies specialists, 2 science specialists, and 4 language arts specialists. In how many ways can the principal choose 1 teacher from each area to form a team to teach in the pod?

# THE USES AND MISUSES OF STATISTICS

# 10

The management of Acme Manufacturing published an average annual salary for 50 employees of $29,280. The Trainsters Union, trying to organize the Acme plant, published an average annual salary of $24,000. Some of the employees made an informal survey and published an average annual salary of $26,000. The interesting point is that all were correct. How can this be?

The Incredible Shrinking Salary

When Aunt Jane asserts that smoking is not harmful to a person's health because Uncle Joe lived to be 88 years of age and smoked 2 packs of cigarettes every day of his adult life, Aunt Jane is using statistical thinking; that is, she has organized the data of her experience (Uncle Joe) and then made a statement on the basis of her data. However, she lacks an understanding of how many data are needed, how the data should be organized, and what conclusions are appropriate or inappropriate relative to the data. This chapter should help you avoid making the types of mistakes that Aunt Jane made.

Statistics can be divided into two subdivisions: in descriptive statistics, techniques are used to summarize and describe the characteristics of a set of data, and inferential statistics, generalizations or conclusions are made about the data in a large group (called a population) from a small portion (called a sample).

In this chapter, we focus on descriptive statistics: understanding a set of data by forming frequency distributions, drawing associated graphs, finding measures of central tendency (mean, median, and mode), and finding measures of the scattering of data.

We use the following problem-solving strategies:  ▪ Substitute in a formula. (throughout chapter) ▪ Make a chart or table. (throughout chapter)  ▪ Use a variable. (456)  ▪ Use an equation. (456)  ▪ Use reasoning. (456, 466)  ▪ Make a list. (throughout chapter)  ▪ Draw a picture. (throughout chapter)

SECTION 1

# FREQUENCY DISTRIBUTIONS AND GRAPHICAL REPRESENTATION

**PROBLEM**

Line graphs are often used to make predictions about the future. Based on the trend for the years 1980 to 1985, estimate the world population in the year 2000.

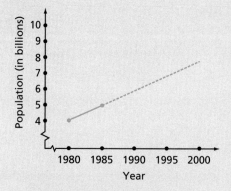

**OVERVIEW**

We are immersed daily in a torrent of numerical data flowing in bubbling splendor from our televisions, our radios, our newspapers, our hair stylist, and our favorite Uncle Al. Although such information is part of our daily lifestyle, we are often unsure of how to organize, interpret, or understand the messages it conveys. In this section, we consider how to organize and summarize data for a better understanding of statistics.

First, we discuss some basic techniques for classifying and summarizing a set of observed measurements (data). Then we represent the data with graphs.

**GOALS**

Illustrate the Data Analysis and Probability Standard, page 392.*
Illustrate the Representation Standard, page 50.*
Illustrate the Algebra Standard, page 4.*
Illustrate the Connections Standard, page 147.*
Illustrate the Problem Solving Standard, page 15.*
Illustrate the Reasoning and Proof Standard, page 23.*

* The complete statement of the standard is given on this page of the book.

# Organizing Data

The first objective of a statistician is to develop a plan to collect data for study. After the data have been collected, the second objective is to make sense of this large mass of information. Suppose that you have collected the numbers shown in Table 10-1. The table contains your tabulation of the number of colds experienced during one winter by each of a group of 30 elementary-school children.

| TABLE 10-1 | | | | | | | | | | | | | | |
|---|---|---|---|---|---|---|---|---|---|---|---|---|---|---|
| 7 | 1 | 1 | 0 | 3 | 4 | 5 | 5 | 3 | 2 | 3 | 3 | 6 | 6 | 2 |
| 4 | 2 | 1 | 0 | 0 | 3 | 4 | 5 | 6 | 3 | 1 | 4 | 1 | 3 | 4 |

A quick glance at Table 10-1 reveals little about what the numbers imply about the group of children represented in the data. Closer observation indicates that the largest number of colds experienced was 7 and the smallest number experienced was 0. The difference between the largest and smallest entries in the data is called the **range** of the data. In this case, the range is $7 - 0 = 7$.

Table 10-1 allows us to make only general observations because the numerical values have not been organized. Using a **frequency distribution table** to help organize the information enables us to uncover more meaning. To summarize data with a frequency distribution, we record the number of students who reported each number of colds. By the **frequency** of a number, we mean the number of times a number occurs in a set of data.

**EXAMPLE 1**    Make a frequency distribution for the data in Table 10-1.

*SOLUTION*    From the summary in Table 10-2, it is obvious that 3 colds was the number most often reported. The summary also shows how the number of colds was distributed among the 30 students.

| TABLE 10-2 | | |
| --- | --- | --- |
| **Number of Colds** | **Tally** | **Frequency** |
| 0 | ||| | 3 |
| 1 | 卌 | 5 |
| 2 | ||| | 3 |
| 3 | 卌 || | 7 |
| 4 | 卌 | 5 |
| 5 | ||| | 3 |
| 6 | ||| | 3 |
| 7 | | | 1 |
| | Total | 30 |

When many numerical values are involved, a frequency distribution may become cumbersome. In this case we choose **intervals** (or **classes**) of data and construct a grouped frequency distribution table for them.

In a **grouped frequency distribution,** we cover the range of data by intervals of equal length and record the number of values that fall into each interval. Constructing a grouped frequency distribution involves four main steps.

**Steps in Making a Grouped Frequency Distribution**

1. Find the range.
2. Choose the number and size of the classes into which the information should be grouped.
3. Find class limits.
4. Count the values in each class. This number is the frequency of the class.

**EXAMPLE 2**

Table 10-3 gives the lengths of engagements (in months) of 30 newly married students. Construct a grouped frequency distribution for these values.

*SOLUTION*

The range of the data is $18 - 1 = 17$. We arbitrarily select six classes for our grouping. Since $17 \div 6$ is 2.833, the length of the classes (if the classes are to be of equal length) must be more than 2.833 in order to include all the data in six classes. Whenever feasible, classes should be of equal integral length. Thus, we arbitrarily select the following class limits: 1–3, 4–6, 7–9, 10–12, and so on. The grouped frequency distribution is found in Table 10-4.

| TABLE 10-3 | | | | |
| --- | --- | --- | --- | --- |
| **Lengths of Engagements** | | | | |
| 10 | 2 | 9 | 6 | 11 |
| 17 | 4 | 10 | 7 | 3 |
| 1 | 4 | 11 | 6 | 3 |
| 8 | 15 | 12 | 9 | 12 |
| 8 | 18 | 12 | 6 | 10 |
| 8 | 18 | 12 | 6 | 9 |

| TABLE 10-4 | | |
| --- | --- | --- |
| **Class Intervals** | **Tallies** | **Frequency** |
| 1–3 | |||| | 4 |
| 4–6 | 卌 | | 6 |
| 7–9 | 卌 || | 7 |
| 10–12 | 卌 |||| | 9 |
| 13–15 | | | 1 |
| 16–18 | ||| | 3 |

In Table 10-4, the class interval 7–9 includes all measurements between 6.5 and 9.5. Thus, an engagement of 6.7 months would be placed in the class 7–9, but an engagement of 9.6 months would go in the class 10–12. The numbers that fall halfway between class limits (.5, 3.5, 6.5, 9.5, 12.5, 15.5, and 18.5) are called **class boundaries.**

The **length** of a class interval can be found by taking the difference between the boundaries of a class. In Table 10-4 the length of the class intervals is $3.5 - .5 = 3$. The middle value of a class interval is called a **class mark.** It can be computed as the average of the class boundaries of the class interval. Thus the class marks for the lengths of engagements in Table 10-4 are 2, 5, 8, 11, 14, and 17.

# STEM AND LEAF PLOTS

A grouped frequency distribution allows us to summarize the information in a set of data by replacing many values by a few representative values. The **stem and leaf plot** provides a way of summarizing data that does not lose the individual values in the data set. To illustrate the stem and leaf plot, let's consider the grades obtained by two classes who were tested on the material in Chapter 9. These grades are given in Table 10-5.

| TABLE 10-5 | |
|---|---|
| *Class I* | *Class II* |
| 56, 64, 72, 73, 84, | 99, 81, 50, 64, 76, |
| 98, 80, 86, 75, 68, | 63, 71, 78, 81, 92, |
| 46, 78, 75, 91, 63, | 87, 79, 74, 60, 68, |
| 84, 79, 69, 76, 58 | 92, 84, 86, 65, 78 |

In these stem and leaf plots, the first digit serves as the stem, and the second digit as the leaf. For example, the stem of the 46 in Class I is 4, and the leaf is 6. Likewise, 56 and 58 have stems of 5 and leaves of 6 and 8, respectively. The data for Class I are listed with six stems—4, 5, 6, 7, 8, and 9—and with appropriate leaves in Table 10-6. In Table 10-7, the same leaves are arranged in increasing order.

| TABLE 10-6 | |
|---|---|
| **Class I** | |
| *Stems* | *Leaves* |
| 4 | 6 |
| 5 | 6, 8 |
| 6 | 4, 8, 3, 9 |
| 7 | 2, 3, 5, 8, 5, 9, 6 |
| 8 | 4, 0, 6, 4 |
| 9 | 8, 1 |

| TABLE 10-7 | |
|---|---|
| **Class I** | |
| *Stems* | *Leaves* |
| 4 | 6 |
| 5 | 6, 8 |
| 6 | 3, 4, 8, 9 |
| 7 | 2, 3, 5, 5, 6, 8, 9 |
| 8 | 0, 4, 4, 6 |
| 9 | 1, 8 |

| TABLE 10-8 | | | |
|---|---|---|---|
| **Class II** | | | **Class I** |
| *Leaves* | *Stems* | *Leaves* | |
| | 4 | 6 | |
| 0 | 5 | 6, 8 | |
| 8, 5, 4, 3, 0 | 6 | 3, 4, 8, 9 | |
| 9, 8, 8, 6, 4, 1 | 7 | 2, 3, 5, 5, 6, 8, 9 | |
| 7, 6, 4, 1, 1 | 8 | 0, 4, 4, 6 | |
| 9, 2, 2 | 9 | 1, 8 | |

Now let's compare Class I and Class II, using the same stems. In Table 10-8, the leaves of Class I increase from left to right, and the leaves of Class II increase from right to left.

A quick inspection of how the leaves increase in Class I and in Class II suggests that the students in Class II did better on this test.

| **Steps in Making Stem and Leaf Plot** | |
|---|---|

1. Decide on the number of digits in the data to be listed under stems (1-digit, 2-digit, or 3-digit numbers).
2. List the stems in a column, from least to greatest.
3. List the remaining digits in each data entry as leaves. (You may wish to order these values from smallest to largest.)

| TABLE 10-9 | |
|---|---|
| **Weights of Members of the Basketball Team** | |
| *Stems* | *Leaves* |
| 15 | 8 |
| 16 | 4, 7 |
| 17 | 5, 6, 8 |
| 18 | 3, 4, 6, 8 |
| 19 | 4, 5 |
| 20 | 3 |

17 | 6 represents 176 lb.

It is always a good idea to describe or explain a stem and leaf plot with a heading. Likewise, it is a good idea to explain the notation. For example, 17 6 in Table 10-9 represents a weight of 176.

 BAR GRAPHS

We have seen throughout this text that we can greatly improve our understanding and our problem-solving ability if we can draw a graph, picture, or diagram. There are several ways to represent graphically a conglomeration of data. One such representation is a **bar graph.** To construct a bar graph, first construct a frequency distribution or a grouped frequency distribution, whichever is appropriate. Then plot the frequencies on the vertical axis and the data values or intervals on the horizontal axis. Finally, draw a bar to show the relationship between the values and the frequencies.

**EXAMPLE 3**

Draw a bar graph that represents the number of graduates each May as shown in Table 10-10.

*SOLUTION*

Figure 10-1 is a bar graph of the data in Table 10-10. Notice that the number of graduates is measured on the vertical axis and that the years are given on the horizontal axis. The break in the vertical axis, denoted by ⸸, indicates that the scale is not accurate from 0 to 150. The height of each bar represents the number of students who graduated in a given year. To determine from the bar graph the number of students who graduated in 1983, locate the bar labeled 1983 and draw a horizontal line from the top of the bar to the vertical axis. The point where this horizontal line meets the vertical axis identifies the number of graduating students. Thus, about 185 students graduated in 1983.

| TABLE 10-10 | | | | | | | | | | | | |
| --- | --- | --- | --- | --- | --- | --- | --- | --- | --- | --- | --- | --- |
| *May of Year* | 1980 | 1981 | 1982 | 1983 | 1984 | 1985 | 1986 | 1987 | 1988 | 1989 | 1990 | 1991 |
| *Number of Graduates* | 152 | 163 | 197 | 185 | 201 | 196 | 210 | 189 | 195 | 205 | 200 | 180 |

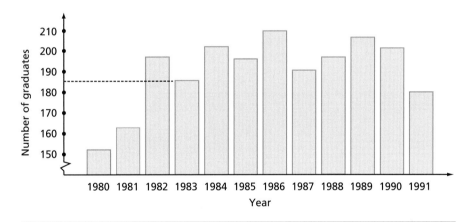

FIGURE 10-1

A **line graph** does a better job of showing fluctuations and emphasizing changes in the data than does a bar graph. The line graph in Figure 10-2 represents the distance (in meters) run in 6 min. by a group of freshmen in a physical education class. Looking at this graph, you can readily see the variations in the numbers of students who ran given distances in 6 min.

 HISTOGRAM

A bar graph representing a grouped frequency distribution is called a **histogram.** To construct a histogram, we first construct a grouped frequency distribution. Then we represent each interval with a bar or a rectangle. The height of the rectangle indicates the frequency of the interval. To label each

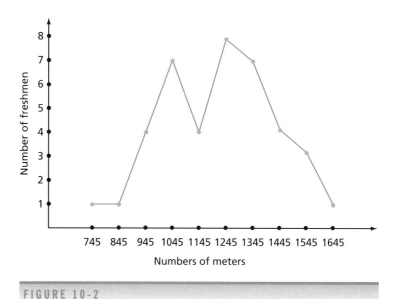

FIGURE 10-2

rectangle, we use the interval that the rectangle represents or the midpoint of the interval, called the *class mark*. Usually, the rectangles touch at a point halfway between each pair of class limits so that there are no gaps in the histogram. In Figure 10-3(c) the rectangles touch at 19.5, 24.5, 29.5, and 34.5. These are termed the *class boundaries*.

| EXAMPLE 4 |

Draw a histogram of the data in Table 10-11.

| **TABLE 10-11** | | |
|---|---|---|
| *Class Intervals* | *Tallies* | *Frequency* |
| 15–19 | \|\|\|\| | 4 |
| 20–24 | \|\|\|\| \|\| | 7 |
| 25–29 | \|\|\|\| | 5 |
| 30–34 | \|\| | 2 |
| 35–39 | \|\| | 2 |

*SOLUTION*

Figures 10-3(a), (b), and (c) are identical representations of the histogram for these data except that in (a) each rectangle is labeled according to its class interval, in (b) each rectangle is labeled according to the midpoint of its class interval (the class mark), and in (c) each rectangle is labeled by class boundaries.

# FREQUENCY POLYGON

When a line graph is used to represent a grouped frequency distribution, it is sometimes called a **frequency polygon.** To draw a frequency polygon, we plot the midpoints (class marks) of the intervals versus the frequency of the

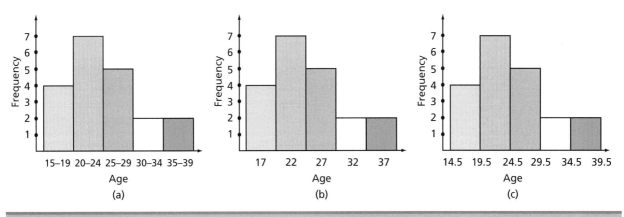

FIGURE 10-3

intervals, and then we connect the resulting points with straight-line segments. Finally, we connect the first and last class marks to points on the horizontal axis that are located one interval beyond these marks.

**EXAMPLE 5**

Draw a frequency polygon for the data in Table 10-11.

*SOLUTION*

Figure 10-4 presents a frequency polygon for the grouped frequency distribution of Table 10-11.

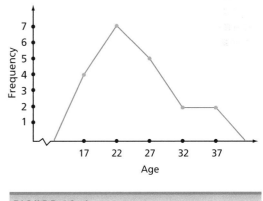

FIGURE 10-4

**PRACTICE PROBLEM**

Draw a frequency polygon for the data in Table 10-12.

*ANSWER*

Figure 10-5 shows the required polygon.

| TABLE 10-12 | |
|---|---|
| *Class* | *Frequency* |
| 4–8 | 4 |
| 9–13 | 3 |
| 14–18 | 7 |
| 19–23 | 2 |

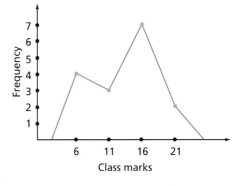

FIGURE 10-5

# CIRCLE GRAPH

One of the simplest types of graphs is the **circle graph,** sometimes called a **pie chart.** It consists of a circle partitioned into sectors, each of which represents a percentage of the whole.

**EXAMPLE 6**

Table 10-13 records examination grades in a class. Represent the data with a circle graph.

*SOLUTION*

The circle graph in Figure 10-6 is a visual representation of the data. The largest percentage of the class made C's; in fact, more than half of the class made C's.

**TABLE 10-13**

| Final Examination Grades | Frequency |
|---|---|
| A | 4 |
| B | 15 |
| C | 36 |
| D | 3 |
| F | 2 |
| Total | 60 |

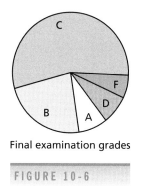

Final examination grades

FIGURE 10-6

In constructing the circle graph shown in Figure 10-6, we use a protractor to obtain angle measurements. Because 36 of 60 or $\frac{36}{60}$ or 60% of the students made C's, the sector representing C's has an angle of $0.60(360°) = 216°$. Similarly, because $\frac{1}{15}$ of the grades were A's, the section representing A's encompasses $(\frac{1}{15})(360°) = 24°$. The remaining sections are constructed in the same way.

# MISUSES OF STATISTICS

Throughout this chapter, we shall discuss the misuses of statistics. In fact, it is at least as important for you to be able to recognize the incorrect use of statistical concepts as it is to be able to use them correctly yourself. Suppose that Senator Cloghorn has announced that 70% of the people in his state oppose a 10¢ per gallon increase in the tax on gasoline. Here are some of the questions we should ask before accepting the validity of the senator's statistic:

1. How many people were surveyed?
2. How were those who participated in the survey selected?
3. Did Senator Cloghorn do a random survey? That is, did each adult in the state have an equal opportunity to be selected as part of the survey?

4. What proportion of people asked to participate actually responded to the survey?
5. How was the question stated?

Suppose that those surveyed were limited to individuals who had contributed to Senator Cloghorn's campaign. The thinking of this group would tend to be similar to that of the senator and in any case would not necessarily reflect the thinking of the population as a whole. Suppose that only 30% responded to the survey. Then you might be getting a sample of only those who felt strongly about the issue. Before accepting a statistic as completely accurate, ask some questions about the procedure used to obtain the statistic.

Many statistical ideas are communicated with graphs and charts. You need to be able to recognize when these visual representations have been constructed to misrepresent and to confuse. One way that erroneous conclusions can be suggested by grouped frequency distributions is through the use of unequal intervals. Table 10-14 is a retabulation of the data from Table 10-3 in Example 2. This table suggests that the most common length of engagement is in the interval 6–9. This is misleading. Why?

| TABLE 10-14 | | |
|---|---|---|
| **Class** | **Tallies** | **Frequency** |
| 1–3 | IIII | 4 |
| 4–5 | II | 2 |
| 6–9 | JHI JHI I | 11 |
| 10–12 | JHI IIII | 9 |
| 13–15 | I | 1 |
| 16–18 | III | 3 |

Another misuse of statistics involves graphs. By changing scales on graphs, it is possible to create false impressions. For instance, although Figures 10-7(a) and (b) represent exactly the same data, (b) has been drawn to conceal the fact that the second interval has at least twice the frequency of the other intervals.

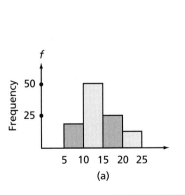
(a)

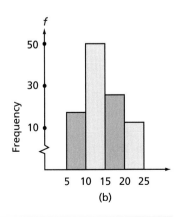
(b)

FIGURE 10-7

There are several ways to manipulate a line graph to produce a misleading impression. Compare the graph shown in Figure 10-8(a) with the graphs shown in (b) and (c). In (b), the vertical axis has been stretched, making the graph appear steeper. In (c), the horizontal axis has been stretched so that the graph appears flatter. The effect on the graph is especially dramatic if one of the axes is stretched while the other is compressed.

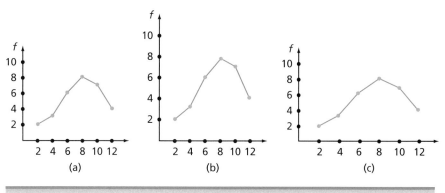

FIGURE 10-8

# Just for Fun

Guess which digit in this calendar appears most often. Then list the digits 0 through 9, and tabulate the frequencies. Did you correctly guess the answer?

**November**

| S | M | T | W | T | F | S |
|---|---|---|---|---|---|---|
|   |   |   |   | 1 | 2 | 3 |
| 4 | 5 | 6 | 7 | 8 | 9 | 10 |
| 11 | 12 | 13 | 14 | 15 | 16 | 17 |
| 18 | 19 | 20 | 21 | 22 | 23 | 24 |
| 25 | 26 | 27 | 28 | 29 | 30 |   |

# EXERCISE SET 1

**R** 1. In the given pie chart or circle graph,

  (a) what percent are professionals?
  (b) what percent are crafts workers?
  (c) what percent are managers or clerical workers?
  (d) what percent are neither managers nor professionals?

2. In a transportation survey, bus riders on the Friday evening run were asked how many times they had ridden the bus that week. Summarize the data in a frequency distribution.

| | | | |
|---|---|---|---|
| 4 | 8 | 6 | 4 |
| 7 | 2 | 2 | 8 |
| 2 | 5 | 8 | 1 |
| 7 | 9 | 8 | 3 |
| 8 | 2 | 4 | 8 |
| 10 | 3 | 3 | 9 |

3. The accompanying bar graph shows changes in population for the four largest cities in the United States.

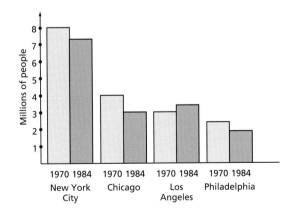

  (a) Which city increased in population between 1970 and 1984?
  (b) In which city did the population decrease the most?
  (c) What was the city with the second largest population in 1984?

4. Following is a tabulation of the ages of mothers of the first babies born in Morningside Hospital in 2001:

| Class | Tally | Frequency |
|---|---|---|
| 15–19 | IIII | 4 |
| 20–24 | IIII II | 7 |
| 25–29 | IIII | 5 |
| 30–34 | II | 2 |
| 35–39 | II | 2 |

  (a) What is the number of mothers in the tabulation?
  (b) What is the number of mothers who were younger than 30?
  (c) What is the number of mothers who were at least 20 years of age?

5. The following table lists how many students had a specific number of absences in a given semester.

| Number of Absences | Frequency |
|---|---|
| 0 | 25 |
| 1 | 18 |
| 2 | 20 |
| 3 | 31 |
| 4 | 34 |
| 5 | 14 |
| 6 | 13 |
| 7 | 12 |
| 8 | 8 |
| 9 | 3 |
| 10 | 1 |

  (a) Display these data with a bar graph.
  (b) Represent these data with a line graph.

6. The following stem and leaf plot records the distances in feet that 22 children in a recreation program could throw a softball.

  (a) Write the distances represented in the stem and leaf plot.
  (b) What is the shortest throw?
  (c) How many of the throws traveled more than 60 ft?

Lengths of Softball Throws

| | |
|---|---|
| 3 | 1, 7 |
| 4 | 1, 2, 6 |
| 5 | 1, 2, 3, 3, 7, 8, 9 |
| 6 | 2, 8, 8, 9 |
| 7 | 1, 3, 6 |
| 8 | 2, 2 |
| 9 | 6 |

7. Alabaster College has 1426 students. The accompanying table classifies them by age.

   (a) Find the class marks.
   (b) Present the data as a frequency polygon.
   (c) Present the data as a histogram.

   | Age | Number of Students |
   |-----|--------------------|
   | 15–19 | 562 |
   | 20–24 | 450 |
   | 25–29 | 350 |
   | 30–34 | 58 |
   | 35–39 | 6 |

8. The accompanying circle graphs record the contributions to federal candidates for office in the late 1970s.

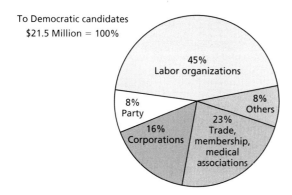

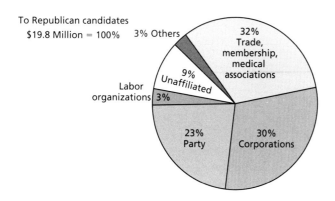

   (a) Compare the percent of contributions to Democratic and Republican candidates that came from labor organizations; repeat the comparison for contributions from corporations.
   (b) What dollar amount of support for Democratic candidates came from the party?
   (c) What dollar amount of support for Republican candidates came from corporations?

**T** 9. Comment on the following misuses of statistics.

   (a) More people die in hospitals than at home. There are severe deficiencies in our medical care system.
   (b) More people died in accidents in the United States last year than died in fighting in World War II. Therefore, it is more dangerous to live in the United States today than it was to fight in World War II.
   (c) Professor Jones gave more A's last semester than Professor Smith. Therefore, you should enroll next semester in Professor Jones's class.
   (d) Most automobile accidents occur near home. Therefore, you are much safer on a long trip.

10. (a) Make a bar graph comparing the number of students in a freshman mathematics course majoring in various academic fields, as recorded in the accompanying table.

   | Academic Fields | Number of Students |
   |-----------------|--------------------|
   | Business administration | 110 |
   | Social sciences | 100 |
   | Life sciences | 60 |
   | Humanities | 30 |
   | Physical sciences | 60 |
   | Elementary education | 140 |

   (b) Display the data given in part (a), using a circle graph.
   (c) Draw a line graph for these data.

*Exercises 11 through 14 refer to the following line graph.*

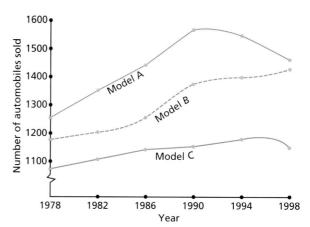

11. In 1998, how many more model A cars were sold than model B cars?

12. In which 4-year period did model C have the greatest decrease in sales? What was the decrease?

13. In which 4-year period did model B have the greatest increase in sales? What was the increase?

14. Compare the increase in sales of the 3 models from 1982 to 1998.

15. Park officials sought to analyze public use of a municipal park. One evening, 36 people were interviewed and their ages recorded.

| | | | | | |
|---|---|---|---|---|---|
| 7 | 18 | 35 | 73 | 18 | 28 |
| 15 | 19 | 41 | 61 | 16 | 24 |
| 51 | 65 | 12 | 65 | 61 | 26 |
| 16 | 62 | 14 | 73 | 72 | 48 |
| 17 | 59 | 16 | 62 | 43 | 68 |
| 21 | 16 | 17 | 19 | 32 | 72 |

Summarize the resulting data in a stem and leaf plot. Arrange the numbers in increasing order on each stem. What general observations can you make about the persons who use the park?

16. Summarize the data from Exercise 15 in a grouped frequency distribution with seven intervals. Let the first interval be 5 to 14.

17. Use a protractor to construct a pie chart showing the percent of women who work in various occupations. The percents are as follows:

| | |
|---|---|
| Professional | 16% |
| Managers | 4% |
| Clerical | 35% |
| Crafts | 2% |
| Operative | 14% |
| Service | 17% |
| All other | 12% |

18. A small country exports 20 main products each year, ranging from iron ore to toy medical kits to surgical instruments. The value of each export (in millions of dollars) is given in the accompanying table. Group the values into 6 intervals of minimum whole-number length, starting the first class at 60.

| | | | | |
|---|---|---|---|---|
| 86 | 62 | 239 | 290 | 207 |
| 285 | 232 | 214 | 131 | 195 |
| 424 | 343 | 476 | 140 | 398 |
| 363 | 348 | 156 | 222 | 370 |

19. The amounts (rounded to the nearest $1) that a sample of 50 freshmen spent on textbooks per class during a fall semester are listed in the accompanying table.

| | | | | | | | | | |
|---|---|---|---|---|---|---|---|---|---|
| 43 | 51 | 45 | 63 | 52 | 57 | 51 | 41 | 48 | 47 |
| 40 | 48 | 47 | 43 | 51 | 45 | 52 | 60 | 51 | 48 |
| 49 | 52 | 51 | 50 | 50 | 48 | 47 | 51 | 55 | 58 |
| 45 | 46 | 45 | 48 | 43 | 59 | 50 | 50 | 57 | 48 |
| 47 | 48 | 47 | 44 | 45 | 54 | 54 | 56 | 50 | 49 |

(a) Find the range.

(b) Create a grouped frequency distribution for this information using 6 intervals of minimum whole-number length, with the first class beginning at 40.

(c) Find the class marks.

(d) Draw a frequency polygon.

20. Make a stem and leaf plot for the data in Exercise 19.

*C* 21. Given the following histogram, tabulate a frequency distribution and find the class marks.

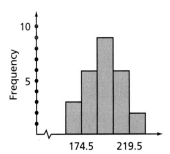

22. Given the following frequency polygon, tabulate a grouped frequency distribution.

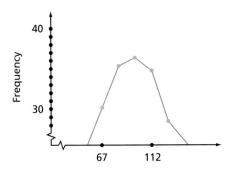

23. The new vice-president of the Seashore Oil Company claims that production has doubled during the first 12 months of his tenure. To present this fact to the board of directors, he has the following graph prepared in which both the height and width of the barrel are doubled. What is misleading about this graph? (HINT: Think in terms of volume.)

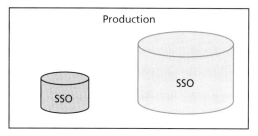

24. Two treatments are given for a disease, both to men and to women. The fraction cured, calculated as (number cured)/(number treated), is written as a decimal in the accompanying table.

|  | Treatment X | Treatment Y |
|---|---|---|
| **Men** | $\dfrac{20}{100} = 0.20$ | $\dfrac{50}{210} = 0.24$ |
| **Women** | $\dfrac{40}{60} = 0.67$ | $\dfrac{15}{20} = 0.75$ |

Glancing at the table, would you say that Treatment $Y$ is better than Treatment $X$? Be careful! Find the total number cured by each treatment divided by the number treated. Now which treatment is better? Explain!

# WHAT IS AVERAGE?

**PROBLEM**

Sandra receives grades of 69, 71, 78, 82, and 73 on her 5 tests in Math 102. She gives her average grade as 74.6, but her friend Sam claims that Sandra's average is 73. Which average is correct?

**OVERVIEW**

In this section, we learn that both averages are correct. Sandra found the mean, and Sam the median.

In the preceding section, we used stem and leaf plots, histograms, frequency polygons, circle graphs, and other graphs to summarize and explain sets of data. Sometimes, however, we need a more concise procedure for characterizing a set of data. Specifically, we want a simple number that estimates the location of the center of the set of data. Therefore, in this section we introduce averages, which are measures of central tendency. Three measures are in general use: the arithmetic mean, the median, and the mode. The fact that these three (as well as others) exist often invites misuses of statistics. One measure may be quoted, but the reader may automatically assume another. When a measure of central tendency is quoted, immediately ask the question, "Which one?"

**GOALS**

Illustrate the Data Analysis and Probability Standard, page 392.
Illustrate the Problem Solving Standard, page 15.

 FINDING MEANS

The most widely used measure of central tendency is the *arithmetic mean* (sometimes called the *arithmetic average*). The arithmetic mean of a set of $n$ measurements is the sum of the measurements divided by $n$.

<table>
<tr><td rowspan="2">DEFINITION</td><td>**Arithmetic Mean**</td><td>Consider $n$ measurements $x_1, x_2, x_3, \ldots, x_n$. The formula for the *arithmetic mean*, denoted by $\bar{x}$, is

$$\bar{x} = \frac{x_1 + x_2 + x_3 + \cdots + x_n}{n}$$</td></tr>
</table>

Mathematicians have developed a very useful notation to use in expressing complicated sums. When the Greek letter sigma, $\Sigma$, occurs in a mathematical expression, it means "add the indicated terms." For instance, the sum $x_1 + x_2 + x_3 + \ldots + x_n$ can be represented as $\sum_{i=1}^{n} x_i$. The index $i = 1$ at the bottom of $\Sigma$ and the $n$ at the top indicate that the $x$ terms should be added starting with the first one and stopping at the $n$th one. Using this notation, the formula for the mean can be expressed as follows:

$$\bar{x} = \frac{\sum_{i=1}^{n} x_i}{n} \text{ or } \bar{x} = \frac{\Sigma x}{n}$$

**EXAMPLE 1**

Find the arithmetic mean of 8, 16, 4, 12, and 10.

*SOLUTION*

$$\bar{x} = \frac{8 + 16 + 4 + 12 + 10}{5} = 10$$

 **CALCULATOR HINT**

We often use a calculator in computing the mean for a set of data. For instance, to find the mean of the three numbers {14 ,17, 22} we can use the keystrokes 14 $\boxed{+}$ 17 $\boxed{+}$ 22 $\boxed{=}$ to get 53. Then 53 $\boxed{\div}$ 3 $\boxed{=}$ 17.67. However, because we often work with large amounts of data, we are fortunate that many calculators have built-in programs that not only compute the mean of a data set, but other important statistics also. We describe how this works with a $\boxed{\text{STAT}}$ menu, a $\boxed{\text{DATA}}$ key, and a $\boxed{\text{STATVAR}}$ menu. For the data in Example 1, under the $\boxed{\text{STAT}}$ menu underline *1-VAR*. Using $\boxed{\text{DATA}}$ enter the data of your problem at $x_1 =, x_2 =, x_3 =$ etc. when prompted by the calculator. After you enter each value, press $\boxed{\blacktriangledown}$. Since each data value occurs only one time, use 1 for each frequency.

$\boxed{\text{STAT}}$ (underline *1-VAR*) $\boxed{=}$ $\boxed{\text{DATA}}$ (at $x_1$ enter 8) $\boxed{\blacktriangledown}$ (at $f$ enter 1) $\boxed{\blacktriangledown}$ (at $x_2$ enter 16) $\boxed{\blacktriangledown}$ (at $f$ enter 1) $\boxed{\blacktriangledown}$ 4 $\boxed{\blacktriangledown}$ 1 $\boxed{\blacktriangledown}$ 12 $\boxed{\blacktriangledown}$ 1 $\boxed{\blacktriangledown}$ 10 $\boxed{\blacktriangledown}$ 1 $\boxed{\text{STATVAR}}$ (underline $n$ to get an answer of 5) (underline $\bar{x}$ to get an answer of 10)

Do these agree with the answers in Example 1?

**EXAMPLE 2**

Find the arithmetic mean of 25, 25, 25, 25, 30, 30, 30, 40, 40, 40, 40, and 50.

*SOLUTION*

$$\bar{x} = \frac{25 + 25 + 25 + 25 + 30 + 30 + 30 + 40 + 40 + 40 + 40 + 50}{12}$$

$$= \frac{25(4) + 30(3) + 40(4) + 50(1)}{4 + 3 + 4 + 1} = \frac{400}{12} = 33\frac{1}{3}$$

In the preceding example, observe that the 4, 3, 4, and 1 are the frequencies of 25, 30, 40, and 50, respectively. The mean is obtained by multiplying each value by its frequency of occurrence and then dividing the sum of these products by the sum of the frequencies. Let us now generalize the formula for finding the arithmetic mean to include the frequencies of the observations.

---

**DEFINITION**

**Arithmetic Mean, Frequency Distribution**

Let $x_1, x_2, \ldots, x_m$ be different measurements. Then the formula for the arithmetic mean $\bar{x}$ is

$$\bar{x} = \frac{x_1 f_1 + x_2 f_2 + x_3 f_3 + \cdots + x_m f_m}{f_1 + f_2 + f_3 + \cdots + f_m}$$

where $f_i$ is the frequency of $x_i$ for $i = 1, 2, 3, \ldots, m$.

---

Using the Greek letter $\Sigma$, the formula for the arithmetic mean of a grouped frequency distribution can be written as

$$\bar{x} = \frac{\sum_{i=1}^{m} x_i f_i}{\sum_{i=1}^{m} f_i} \quad \text{or } \bar{x} = \frac{\Sigma xf}{n}$$

**EXAMPLE 3**

Find the arithmetic mean of the following set of data:

| x | 4 | 14 | 24 | 34 |
|---|---|----|----|----|
| f | 2 | 8 | 20 | 10 |

*SOLUTION*

$$\bar{x} = \frac{4(2) + 14(8) + 24(20) + 34(10)}{2 + 8 + 20 + 10} = \frac{940}{40} = 23.5$$

If data are presented in a grouped frequency table, we may have no way of knowing the distribution of the data within a class. We therefore must assume either that the data are uniformly distributed within a class interval around the class mark or that all the data within a class interval are located at the class mark. Thus, to compute the mean we use the formula for $\bar{x}$, where $x$ represents the class mark, $f$ the frequency of each class, and $m$ the number of class intervals.

**EXAMPLE 4**

Find the arithmetic mean of the data in Table 10-15.

*SOLUTION*

From Table 10-15,

$$\bar{x} = \frac{495}{20} = 24.75$$

| TABLE 10-15 | | |
|---|---|---|
| **x (Classmark)** | **f** | **xf** |
| 17 | 4 | 68 |
| 22 | 7 | 154 |
| 27 | 5 | 135 |
| 32 | 2 | 64 |
| 37 | 2 | 74 |
| Total | 20 | 495 |

## CALCULATOR HINT

Calculators often have built-in functions to compute the mean of data presented in a grouped frequency distribution. In our illustration we again use a calculator with a [STAT] menu, a [DATA] key, and a [STATVAR] menu. In entering the data we will use the down arrow [▼] affer entering each value. To compute the mean of the data in the frequency distribution in Example 4, we will proceed as follows.

[STAT] (underline *1-Var*) [=] [DATA] 17 [▼] 4 [▼] 22 [▼] 7 [▼] 27 [▼] 5 [▼] 32 [▼] 2 [▼] 37 [▼] 2 [=] [DATAVAR]

Underline $n$. Did you learn that there are 20 data values? Underline $\bar{x}$. Did you find the mean to be 24.75?

A small company has four employees with annual salaries of $35,500, $36,053, $37,144, and $37,553. The president of the company has an annual salary of $65,000. The mean of the 5 salaries is $42,250. This is a true average, as given by the arithmetic mean, but most people would not accept it as a meaningful measure of central tendency. The median salary is more representative, because the one large salary tends to weight the mean upward to a misleading extent.

# FINDING MEDIANS

The median of a set of observations is the middle number when the observations are ranked according to size.

| DEFINITION | **Median** | If $x_1, x_2, x_3, \ldots, x_n$ is a set of data placed in increasing or decreasing order, the *median* is the middle entry, if $n$ is odd. If $n$ is even, the median is the mean of the two middle entries. |
|---|---|---|

**EXAMPLE 5**

Consider the set of five measurements 7, 1, 2, 1, and 3. Arranged in increasing order, they may be written as

$$1, 1, \underset{\text{median}}{2}, 3, 7$$

Hence, the median is 2.

**EXAMPLE 6**

The array          25, 2, 5, 6, 5, 23, 7, 10, 22, 15, 21, 23

can be arranged in decreasing order as

$$25, 23, 23, 22, 21, \underset{\text{median}}{15, 10}, 7, 6, 5, 5, 2$$

So the median is      $\dfrac{15 + 10}{2} = 12.5$

#  FINDING MODES

The third measure of central tendency is called the *mode*. It refers to the measurement that appears most often in a given set of data.

| DEFINITION | **Mode** | The *mode* of a set of measurements is the observation that occurs most often. If every measurement occurs only once, then there is no mode. If the two most common measurements occur with the same frequency, the set of data is *bimodal*. It may be the case that there are three or more modes. |
| --- | --- | --- |

**EXAMPLE 7**

Baseball caps with the following head sizes were sold in a week by the Glo-Slo Sporting Goods Store: 7, $7\frac{1}{2}$, 8, 6, $7\frac{1}{2}$, 7, $6\frac{1}{2}$, $8\frac{1}{2}$, $7\frac{1}{2}$, 8, and $7\frac{1}{2}$. Find the mode of the head sizes.

*SOLUTION*

The mode is $7\frac{1}{2}$; it occurs 4 times, more times than any other size.

**PRACTICE PROBLEM**

Find the mode of 21, 23, 24, 22, 24, 20, 22, 24, 25, 20, 22, and 21.

*ANSWER*

There are two modes: 22 and 24.

The decision about which measure of central tendency to use in a given situation is not always easy. The mean is a good average of magnitudes, such as weights, test scores, and prices, provided that no extreme values are present to distort the data. When extraordinarily large or small values are included in the data set, the median is usually better than the mean. However, the mean is the average most often used because it gives equal weight to the value of each measurement. The median is a positional average. The mode is used when the "most common" measurement is desired. The most appropriate measure for the price of pizzas in town would be the arithmetic mean. However, to select the best-tasting pizza in town, you could make a survey and use the mode (if a large number of persons were involved). Unfortunately, people with a stake in the outcome tend to use the average that best suits the objectives they hope to accomplish, and they quote the result as an accomplished (and exclusive) fact. This, of course, leads to a widespread mistrust of statistics.

**EXAMPLE 8**

In one group of games against the Dodgers, the Reds won 6 of 7 games by the scores given in Table 10-16. Find the mean, median, and mode of these scores.

| TABLE 10-16 | | | | | | | |
|---|---|---|---|---|---|---|---|
| *Dodgers* | 2 | 6 | 1 | 15 | 4 | 2 | 2 |
| *Reds* | 4 | 7 | 2 | 1 | 5 | 3 | 3 |

*SOLUTION*

When the mean scores are computed, the following results are obtained:

$$\text{Dodgers' mean score} = 4.57$$
$$\text{Reds' mean score} = 3.57$$

Although the Reds dominated the series, the Dodgers' mean score was substantially higher. In this case, the mean is not a good average to use because the Dodgers' extraordinarily high score in one game biased the mean. In such cases, it is often better to use the median. Thus,

Reds' scores (placed in order):     1   2   3   ③   4   5   7

median

Dodgers' scores (placed in order):     1   2   2   ②   4   6   15

In this case, the median offers a better measure for comparing the scores. Coincidentally, the mode score for the Reds is 3, and the mode score for the Dodgers is 2.

# PERCENTILES

One way of reporting a person's relative performance on a test is to identify the percentage of people taking the test who scored lower than the person under consideration. For example, someone who scores higher than 70% of those who take a test is said to be in the 70th percentile. Conversely, a percentile score of 85 means that the person scored higher than 85% of those in the sample. In the next several examples, we will practice using percentiles.

DEFINITION

**Percentile**

Let $x_1, x_2, x_3, \ldots, x_n$ be a set of $n$ measurements arranged in order of magnitude. The *Pth percentile* is the value of $x$ such that $p\%$ of the measurements are less than the value of $x$ and $(100 - p)\%$ are greater than $x$. For small data sets, some percentiles cannot be computed.

**EXAMPLE 9**

Monica scored in the 68th percentile on the mathematics portion of the Iowa Test of Basic Skills. Can she conclude that she got 68% of the questions correct on this test?

*SOLUTION*

No, Monica has no information about how many problems she got correct. She does know that she scored higher than 68% of those who took the mathematics portion of the test. We say that the *percentile rank* of Monica's score is 68.

Because percentiles are of practical value only for very large data sets, we are concerned with how to interpret them, not how to compute them. However, there are three percentiles to which we wish to give closer attention. The 25th percentile is also called the **first quartile,** the 50th percentile is the **second or middle quartile,** and the 75th percentile is the **third or upper quartile.** The median is approximately equal to the second quartile (the 50th percentile).

For small data sets we can get an easy approximation of the first quartile by computing the median of the scores below the median. Similarly, we can approximate the upper quartile by computing the median of the scores above the median.

**EXAMPLE 10**

Find the median and approximate the first and third quartiles of the following data.

8, 14, 12, 64, 7, 9, 42, 84, 76, 92, 41, 15, 17, 26, 47, 16, 21, 22, 23, 24

*SOLUTION*

Arrange the data in order of increasing magnitude:

7, 8, 9, 12, 14, 15, 16, 17, 21, 22, 23, 24, 26, 41, 42, 47, 64, 76, 84, 92

There are 20 observations, so the median is the average of the 10th and 11th scores. The median is 22.5, the average of 22 and 23. There are 10 scores below the median, and their median is 14.5. This score of 14.5 is the first quartile. Likewise there are 10 scores above the median and their median is 44.5. This score of 44.5 is the third quartile.

# Just for Fun

The average salary of 6 office workers in the XYZ Corporation is $24,000. John remembers five of the six salaries: $20,000, $23,000, $29,000, $26,000, and $22,000. If the average is the median, describe the missing salary for John. Do the same if the average salary is the mean.

# EXERCISE SET 2

*R* 1. Compute the mean, the median, and the mode (or modes, if appropriate) for the given sets of data.

   (a) 3, 4, 5, 8, 10
   (b) 4, 6, 6, 8, 9, 12
   (c) 3, 6, 2, 6, 5, 6, 4, 1, 1
   (d) 7, 1, 3, 1, 4, 6, 5, 2
   (e) 21, 13, 12, 6, 23, 23, 20, 19
   (f) 18, 13, 12, 14, 12, 11, 16, 15, 21

2. An elevator has a maximum capacity of 15 people and a load limit of 2250 lb. What is the mean weight of the passengers if the elevator is loaded to capacity with people and weight?

3. Find the mean of the given distribution.

| x | Frequency |
|---|-----------|
| 10 | 2 |
| 20 | 6 |
| 30 | 8 |
| 40 | 4 |

4. Make up a set of data with 4 or more measurements, not all of which are equal, with each of the following characteristics.

   (a) The mean and median are equal.
   (b) The mean and mode are equal.
   (c) The mean, median, and mode are equal.
   (d) The mean and median have values of 8.
   (e) The mean and mode have values of 6.
   (f) The mean, median, and mode have values of 10.

5. The mean of a set of 8 scores is 65. What is the sum of the 8 scores?

6. The mean of 9 of 10 scores is 81. The tenth score is 100. What is the mean of the 10 scores?

7. Which of the three averages should be used for the following data?

   (a) The average salary of four salesmen and the owner of a small store
   (b) The average height of all male students in W. R. Berry High School
   (c) The average dress size sold at Acme Apparel

8. At the initial meeting of an athletic club, the weights of the members were found to be 220, 275, 199, 246, 302, 333, 401, 190, 286, 254, 302, 323, and 221.

   (a) Compute the mean, median, and mode of the data.
   (b) Check the value obtained for the mean by using the statistics menu on your calculator.
   (c) Which measure is most representative of the data?

9. The weights (in kg) of the members of the Laramy High School football squad are as follows: 75, 60, 62, 94, 78, 80, 72, 74, 76, 89, 95, 98, 97, 80, 98, 91, 96, 90, 84, 73, 80, 92, 94, 96, 99, 84, 60, 68, 74, 80, 92, 96, 88, 74, 84, 94, 72, 76, 64, and 80.

   (a) What is the mean weight of the football squad?
   (b) What is the median weight?
   (c) What is the modal weight?
   (d) If you were a sportswriter assigned to do a story on this squad, how would you describe the (average) weight?

10. The following grades were recorded for a test on Chapter 9. Find the arithmetic mean. Then check using the statistics menu on your calculator.

| Score | Frequency |
|-------|-----------|
| 100 | 3 |
| 90 | 5 |
| 80 | 7 |
| 70 | 15 |
| 60 | 14 |
| 50 | 3 |
| 40 | 3 |

11. For each of the given sets of observations, approximate the 3 quartiles.

   (a) 16, 14, 12, 13, 15, 18, 24, 8, 10, 4
   (b) 18, 47, 64, 32, 41, 92, 84, 27, 14, 12

12. In a class of 80 students, Jodi scored in the 80th percentile. How many students scored lower than Jodi?

13. Aaron scored in the 98th percentile of the verbal portion of the SAT. Interpret this result.

14. For the following data, find the median. 17, 26, 34, 41, 52, 14, 13, 18, 27, 31, 39, 43, 44, 47, 49.

*For the histograms shown in Exercises 15 and 16, find the mean of the distributions.*

*T* 15.

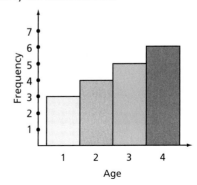

16.

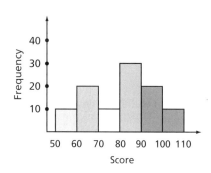

17. The president of Wargo Furniture Factory draws a salary of $110,000 per year. Four supervisors have salaries of $30,000 each. Twenty workers have salaries of $20,000 each. Discuss each of the following.

(a) The president says the average salary is $25,200.

(b) The union says the average salary is $20,000.

(c) Which average is more representative of the factory salaries?

18. If 99 people have a mean income of $18,000, by how much does the mean income increase when an employee is added who has an income of $160,000?

19. A student has a mean average grade of 89 on 9 tests. What must she make on the tenth test to have an average of 90?

20. An interesting property of the mean is that the sum of the differences in the mean and each observation (deviations of each score from the mean, considered as signed numbers) is 0. Show that this statement is true for the following data: 5, 8, 10, 12, 15.

 21. The accompanying table shows the distribution of scores on a test administered to freshmen at Laneville College. Find the arithmetic mean. Then check your answer using the statistics menu on your calculator.

| Score | Frequency |
|---|---|
| 140–149 | 3 |
| 130–139 | 4 |
| 120–129 | 8 |
| 110–119 | 13 |
| 100–109 | 4 |
| 90–99 | 2 |
| 80–89 | 0 |
| 70–79 | 1 |

HINT: The mid-point of each interval (class mark) is the data value for the interval.

22. Find the mean price–earnings ratio of 100 common stocks listed on the New York Stock Exchange, where the distribution is as follows. Then check using the statistics menu on your calculator.

| Interval | Frequency |
|---|---|
| 0–4 | 6 |
| 5–9 | 46 |
| 10–14 | 30 |
| 15–19 | 10 |
| 20–24 | 4 |
| 25–29 | 2 |
| 30–34 | 2 |

HINT: The mid-point of each interval (class mark) is the data value for the interval.

23. The following frequency distribution gives the weekly salaries, by title, of the employees of the Glasgow Light Bulb Factory.

| Title | Number | Weekly Salary |
|---|---|---|
| Manager | 1 | $550 |
| Supervisors | 3 | 450 |
| Inspectors | 3 | 350 |
| Line workers | 21 | 250 |
| Clerks | 5 | 150 |

Frank examined this list and concluded that the mean salary is

$$\frac{550 + 450 + 350 + 250 + 150}{5} = \$350$$

(a) Is he correct?

(b) If he is not correct, what should the mean be?

24. The data in the following table have been collected on the expenses (excluding travel and lodging) for 6 trips made by teachers in the mathematics department at Snelling College.

| Number of Days on Trip | Total Expenses | Expenses Per Day |
|---|---|---|
| 0.5 | $18.00 | $36.00 |
| 2.5 | 75.00 | 30.00 |
| 3 | 60.00 | 20.00 |
| 1 | 19.50 | 19.50 |
| 8 | 132.00 | 16.50 |
| 5 | 160,00 | 32.00 |
| 20 | $464.50 | $154.00 |

Let $\bar{x} = \dfrac{\$154.00}{20}$

= \$7.70 average expense per day

or $\bar{x} = \dfrac{\$464.50}{20}$

= \$23.23 average expense per day

or $\bar{x} = \dfrac{\$154.00}{6}$

= \$25.67 average expense per day

Which average is realistic?

25. A study of the number of oil spills into the nation's waterways in recent years gives the following number of spills of various sizes. Find the arithmetic mean. Then check your answer using the statistics menu on your calculator.

| Millions of Gallons of Oil | Number of Spills |
|---|---|
| 1–3 | 6 |
| 4–6 | 9 |
| 7–9 | 13 |
| 10–12 | 10 |
| 13–15 | 7 |
| 16–18 | 3 |

HINT: The mid-point of each interval (class mark) is the data value for the interval.

26. A company employs 50 men and 50 women. The table below summarizes the salary information of the company.

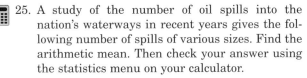

| Years of Service | Men | | Women | |
|---|---|---|---|---|
| | Number | Average Salary | Number | Average Salary |
| Less than 5 years | 10 | \$20,000 | 40 | \$22,500 |
| More than 5 years | 40 | \$27,400 | 10 | \$30,000 |

(a) What is the average (mean) salary for men in the company?

(b) What is the average (mean) salary for women in the company?

(c) Does the information from this table indicate that the company discriminates on the basis of gender? Explain.

## REVIEW EXERCISES

27. Group the following test scores, which were received by 24 students, into 10 classes (95–99, 90–94, 85–89, 80–84, 75–79, 70–74, 65–69, 60–64, 55–59, and 50–54):

63, 71, 85, 96, 94, 90, 75, 72, 77, 71, 62, 84,

81, 76, 61, 54, 87, 94, 62, 81, 94, 77, 63, 60

Construct (a) a histogram and (b) a frequency polygon.

# SECTION 3   HOW TO MEASURE SCATTERING

**PROBLEM**

The fact that the mean salary of management of the Doran Company exceeds the mean salary of management of the Wargo Company does not imply that the salaries of the Doran Company are superior to those of the Wargo Company. Compare four monthly salaries: \$4000, \$4100, \$4100, and \$21,000, with a mean of \$8,300; and four salaries: \$6,800, \$6,900, \$7,000, and \$7,200, with a mean of \$6,975. The mean salary of the first company exceeds the mean salary of the second company, but the lowest salary of the second company is better than all but the largest salary of the first company.

**OVERVIEW**

The preceding example indicates the need for a measure of dispersion or scattering of data. In other words, we need a measure that indicates whether or not the entries in a set of data are close to the mean. Thus, in this section we consider the

**GOALS**

Illustrate the Data Analysis and Probability Standard, page 392.
Illustrate the Representation Standard, page 50.
Illustrate the Connections Standard, page 147.

# MEASURES OF DISPERSION (SCATTERING)

There are several ways to measure dispersion of data. The one that is easiest to calculate is the range, introduced in Section 1.

**EXAMPLE 1**

For the set of data, 7, 3, 1, 15, 41, 74, and 35, the range is 73 because $74 - 1 = 73$.

Although the range is easy to obtain, it is not always a good measure of dispersion because it can be so radically affected by a single extreme value. For example, suppose that the 74 in the set of observations listed previously was miscopied and listed as 24 instead. As a result, the range would change from 73 to 40.

Because the range is significantly affected by extreme values, other measures of scattering or dispersion are preferable. In this section, we consider variance, denoted by $s^2$, and standard deviation, $s$.

---

**DEFINITION**

**Variance**

*Variance* for a set of data can be obtained in four steps:
1. Compute the mean $\bar{x}$.
2. Compute the difference between each observation and the arithmetic mean, $x - \bar{x}$.
3. Square each difference $(x - \bar{x})^2$.
4. Divide the sum of the differences squared by $n$, where $n$ is the number of observations.

---

In other words, the variance is the mean of the squares of the differences of the data from the mean, or the average of the squared deviations.

**EXAMPLE 2**

Find the variance for the data 5, 7, 1, 2, 3, and 6.

*SOLUTION*

1. Compute the mean of the data. (See Table 10-17.) In this case, we have

$$\bar{x} = \frac{24}{6} = 4$$

2. Determine the difference between each $x$ and $\bar{x} = 4$. (See the second column of Table 10-17.)
3. Compute the square of each of these differences; that is, compute $(x - \bar{x})^2$ (the third column of Table 10-17).

4. Sum the squares of differences (that is, sum the third column of Table 10-17), and divide by $n = 6$.

$$s^2 = \frac{28}{6} = 4.67$$

The variance is 4.67.

**TABLE 10-17**

| $x$ | $x - \bar{x}$ | $(x - \bar{x})^2$ |
|-----|-----|-----|
| 5 | 1 | 1 |
| 7 | 3 | 9 |
| 1 | −3 | 9 |
| 2 | −2 | 4 |
| 3 | −1 | 1 |
| 6 | 2 | 4 |
| Total  24 | | 28 |

The preceding four-step calculation is equivalent to the following formula for variance.

**Formula for Variance**

Variance, denoted by $s^2$, is calculated as

$$s^2 = \frac{(x_1 - \bar{x})^2 + (x_2 - \bar{x})^2 + (x_3 - \bar{x})^2 + \cdots + (x_n - \bar{x})^2}{n}$$

where $\bar{x}$ is the mean of the observations.

This formula is sometimes written in summation notation as

$$s^2 = \frac{\sum\limits_{i=1}^{n}(x_i - \bar{x})^2}{n} \qquad \text{or} \qquad \frac{\sum(x - \bar{x})^2}{n}$$

Because the variance is the average of the squared deviations from the mean, it is clear that it measures dispersion. However, the variance represents different units than the original data. If the original data was in inches, the variance is measured in square inches. Hence, the most often used measure of dispersion is the **standard deviation.** Standard deviation has the advantage of being expressed in the same units as the original data.

To compute standard deviation, remember that standard deviation is simply the square root of the variance.

$$s = \sqrt{s^2} = \sqrt{\frac{\sum\limits_{i=1}^{n}(x_i - \bar{x})^2}{n}}$$

**EXAMPLE 3**

Find the standard deviation of the data given in Example 2.

*SOLUTION*

$$s = \sqrt{s^2} = \sqrt{4.67} \qquad \text{\small $s^2 = 4.67$ in Example 2}$$

$$= 2.16$$

**PRACTICE PROBLEM**

Compute the standard deviation of 112, 108, 114, 100, and 116.

*ANSWER*

5.66

These formulas for variance and standard deviation are useful because they make it clear that they measure variation of the data from the arithmetic mean. However, a different formula is more accessible for calculator computations.

| Computational Formula for Variance | $s^2 = \dfrac{x_1^2 + x_2^2 + \cdots + x_n^2 - n\bar{x}^2}{n} = \dfrac{\sum\limits_{i=1}^{n} x_i^2 - n\bar{x}^2}{n}$ |
| --- | --- |

**EXAMPLE 4**

Use the computational formula for variance to find the variance of the data given in Example 2. Also find the standard deviation.

*SOLUTION*

The first column of Table 10-18 displays the values of the variable $x$, and the second column displays the values of $x^2$. Therefore,

$$\bar{x} = \frac{24}{6} = 4 \qquad s^2 = \frac{124 - 6(4)^2}{6} \qquad \sum_{i=1}^{n} x_i^2 = 124$$

$$n = 6$$

$$= \frac{28}{6} = 4.67 \qquad s = \sqrt{4.67} = 2.16$$

This is, of course, the same answer obtained in Example 2.

**TABLE 10-18**

| x | x² |
| --- | --- |
| 5 | 25 |
| 7 | 49 |
| 1 | 1 |
| 2 | 4 |
| 3 | 9 |
| 6 | 36 |
| 24 | 124 |

## CALCULATOR HINT

Since we often work with large sets of data, we are fortunate that many calculators have built-in programs that not only compute the mean of a data set, but also the standard deviation. In the Calculator Hint on page 451, we described how to enter data into a calculator with a [ STAT ] menu, a [ DATA ] key, and a [ STATVAR ] menu. If we follow the instructions in that Calculator Hint for entering the data, we can use the following keystrokes to find the standard deviation:

[ STATVAR ] (underline $\sigma_x$) Test this strategy using the data in Table 10-18. Remember to record a frequency of 1 for each data value you enter. Did you get 2.16? The $\sigma_x$ in the calculator is the $s$ in the textbook.

Most calculators provide two standard deviations. $\sigma_x$ is the standard deviation of a population (the standard deviation, s, we obtain in this section). The standard deviation of a sample is denoted by $s_x$. What is the difference in these two standard deviations? In inferential statistics, when computing the standard deviation of a sample in order to estimate the standard deviation of the population, we divide by $n-1$ instead of $n$ to get the variance. Because we are concerned primarily with descriptive statistics in this book, we will not need to be concerned with this subtlety.

When the data are presented in a frequency distribution, the formula for variance can be written as follows. As before, to find the standard deviation one computes the square root of the variance.

| **Variance of Frequency Distributions** | Suppose that $x_1, x_2, \ldots, x_m$ have respective frequencies $f_1, f_2, \ldots, f_m$. Then $$s^2 = \frac{x_1^2 f_1 + x_2^2 f_2 + \cdots + x_m^2 f_m - n\bar{x}^2}{n} = \frac{\sum\limits_{i=1}^{m} x_i^2 f_i - n\bar{x}^2}{n}$$ where $n = f_1 + f_2 + \ldots + f_m$. |
| --- | --- |

**EXAMPLE 5**

Find the variance and the standard deviation of the data tabulated in the first two columns of Table 10-19.

*SOLUTION*

$$\bar{x} = \frac{120}{20} = 6$$

$$s^2 = \frac{848 - (20)6^2}{20}$$

$$\sum_{i=1}^{n} x_i^2 f_i = 848$$
$$n = 20$$

$$= 6.4$$

$$s = \sqrt{6.4} = 2.53$$

**TABLE 10-19**

| x | f | xf | x² | x²f |
| --- | --- | --- | --- | --- |
| 2 | 3 | 6 | 4 | 12 |
| 4 | 4 | 16 | 16 | 64 |
| 6 | 6 | 36 | 36 | 216 |
| 8 | 4 | 32 | 64 | 256 |
| 10 | 3 | 30 | 100 | 300 |
| Total | 20 | 120 | | 848 |

## CALCULATOR HINT

STAT (underline *1-VAR*) = DATA 2 ▼ 3 ▼ 4 ▼ 4 ▼ 6 ▼ 6 ▼ 8 ▼ 4 ▼ 10 ▼ 3 STATVAR

$n = 20$, $\bar{x} = 6$, and $\sigma_x = 2.5298$. The $\sigma_x$ in the calculator is the $s$ in the textbook.

An important property of standard deviation is the fact that in a set of data most of the data are located within 2 standard deviations of the mean of the data. At least 75% of data lie within 2 standard deviations of the mean for any data set, and we will learn in the next section that, for many collections of data, 95% of the data lie within 2 standard deviations of the mean.

**EXAMPLE 6**

Babe Ruth won 12 American League home run championships in his career. The number of home runs he hit to win those championships were:

$$11 \quad 29 \quad 54 \quad 59 \quad 41 \quad 46$$
$$47 \quad 60 \quad 54 \quad 46 \quad 49 \quad 46$$

The mean of this data is 45.2 and the standard deviation is 13.0. What percentage of the data lie within 2 standard deviations of the mean?

*SOLUTION*

Mean + 2 standard deviations = 45.2 + 2(13.0) = 71.2
Mean − 2 standard deviations = 45.2 − 2(13.0) = 19.2
Eleven of the 12 values lie within 2 standard deviations of the mean, or 92% of the data lie in this interval.

Undoubtedly you are wondering, "How am I going to use this new idea of standard deviation?" There are many applications, one of which follows. One of the most common misuses of statistics consists of making inappropriate comparisons. Knowing both the mean and the standard deviation helps to improve comparisons.

**EXAMPLE 7**

Juan made scores of 90, 82, 70, 61, and 94 on 5 tests. Of course, Juan did his best work relative to the rest of the class on the last test.

Or did he? What do we know about how the other students scored? Maybe everyone in the class made a higher score than 90 on the last test.

Example 7 illustrates a need for standard scores by which to make comparisons. Such scores are called *z scores.*

**DEFINITION**

**z-Scores**

A score or measurement, denoted by $x$, from a distribution with mean $\overline{x}$ and standard deviation $s$ has a corresponding *z score* given by

$$z = \frac{x - \overline{x}}{s}$$

representing the number of standard deviations from the mean.

**EXAMPLE 8**

The average weight of bags of potato chips is 10 oz., with a standard deviation of 1 oz. One bag of potato chips is weighed and has a weight of 10.2 oz. Convert this measurement into standard units.

*SOLUTION*

We know that $x$ = 10.2 oz., $\overline{x}$ = 10 oz., and $s$ = 1 oz. Therefore,

$$z = \frac{x - \overline{x}}{s} = \frac{10.2 - 10}{1} = 0.2$$

The $z$ score is a measurement expressed in standard units or without units. For example, if $x$ and $\overline{x}$ are in feet, then $s$ is in feet, and the division eliminates the units. Consequently, $z$ scores are useful for comparing two sets of data with different units.

**EXAMPLE 9**

Teresa scores a 76 on the entrance test at school X and an 82 at school Y. At which school did she have the better score?

SOLUTION | To answer this question, we need to know that the mean score at school X was 70 with a standard deviation of 12 while the mean score at school Y was 76 with a standard deviation of 16. The $z$ scores are then as follows:

$$\text{School X:} \qquad z = \frac{76 - 70}{12} = 0.5$$

$$\text{School Y:} \qquad z = \frac{82 - 76}{16} = 0.375$$

Because 0.5 is greater than 0.375, Teresa's score at school X was better than her score at school Y in comparison to the scores of others who took the test.

# INTERQUARTILE RANGE

Remember that the first quartile $Q_1$ of a set of data is the score below which 25% of the data lies while the third quartile $Q_3$ is the score above which 25% of the data lies. Therefore the interval between the first quartile $Q_1$ and the third quartile $Q_3$ contains the middle 50% or half of the data. The length of this interval provides a useful measure of the variation of the central portion of data. The **interquartile range** (IQR) is computed by subtracting the first quartile from the third quartile.

| DEFINITION | Interquartile Range | The *interquartile range (IQR)* equals <br><br> $$Q_3 - Q_1$$ |
|---|---|---|

EXAMPLE 10 | The following are 16 grades received on Professor Wheeler's probability test, arranged in increasing order. Find the median, $Q_1$, $Q_3$, and the interquartile range.

SOLUTION | Remember that we can easily approximate $Q_1$ by computing the median of the scores below the median. Similarly, we can approximate $Q_3$ by computing the median of the scores above the median.

64  66  68  71  73  75  76  78  82  83  84  90  92  95  96  99
$\uparrow$             $\uparrow$             $\uparrow$
$Q_1 = 72$       median = 80       $Q_3 = 91$

$$IQR = 91 - 72 = 19$$

Therefore, a span of 19 points includes the middle 50% of the grades.

# BOX AND WHISKER PLOTS

Sometimes the median, $Q_1$, $Q_3$, and the range are displayed in a diagram called a **box and whisker plot.** Figure 10-9 shows such a plot for the data in Example 10. The interquartile range is represented as the box. The lines extending from $Q_1$ and $Q_3$ to the lowest and highest scores are the whiskers. To characterize data points widely separated from the rest of the data, we define an outlier.

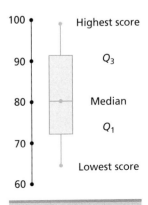

FIGURE 10-9

DEFINITION

**Outlier**

An *outlier* is any data point farther than 1.5 *IQRs* above $Q_3$ or farther than 1.5 *IQRs* below $Q_1$.

For the data in Example 10 we compute

$$Q_3 + 1.5(IQR) = 91 + 1.5(19) = 119.5$$
$$Q_1 - 1.5(IQR) = 72 - 1.5(19) = 43.5$$

Thus, for this set of data, there are no outliers.

Sometimes it is convenient to obtain the box and whisker plot from what we called in Section 1 a *stem and leaf plot*.

**EXAMPLE 11**

The heights in feet of the 14 tallest buildings in Minneapolis are:

| 960 | 775 | 668 | 579 | 561 | 447 | 440 |
| 416 | 403 | 366 | 356 | 355 | 340 | 337 |

Form a stem and leaf plot, and then make a box and whisker plot. Are there any outliers?

SOLUTION

### TABLE 10-20

| Stems | Leaves |
|-------|--------|
| 9 | 60 |
| 8 | |
| 7 | 75 |
| 6 | 68 |
| 5 | (79) 61 |
| 4 | 47, 40, 16, 03 |
| 3 | 66, (56) 55, 40, 37 |

$Q_3 = 579$

Median = 428

$Q_1 = 356$

Hence, $$IQR = 579 - 356 = 223$$

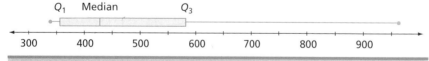

FIGURE 10-10

$$Q_3 + 1.5(IQR) = 913.5$$
$$Q_1 - 1.5(IQR) = 21.5$$

The value 960 is an outlier in this set of values.

# Just for Fun

In a class of 6 students, each guesses the number of pennies in a jar. The 6 guesses are 52, 59, 62, 65, 49, and 42. One guess missed by –12, and the others by 5, 8, 11, –2, and –5. How can you use a statistic to help you find the number of pennies in the jar?

# EXERCISE SET 3

*For the given sets of observations in Exercises 1 through 6, find the range, the variance, and the standard deviation.*

*R* 1. 6, 8, 8, 14

2. 10, 12, 13, 14, 16

3. 16, 14, 12, 13, 15, 18, 24, 8, 10, 4

4. 18, 47, 64, 32, 41, 92, 84, 27, 14, 12

5. 9, 7, 16, 14, 12, 13, 14, 18, 24, 8, 10, 4

6. 10, 17, 18, 47, 64, 32, 41, 92, 84, 27, 14, 12

7. For Exercise 3, construct a box and whisker plot. List all outliers.

8. For Exercise 4, construct a box and whisker plot. List all outliers.

9. For Exercise 5, construct a stem and leaf plot. From this chart, find $Q_3 + 1.5(IQR)$ and $Q_1 - 1.5(IQR)$. Are there any outliers?

10. For Exercise 6, construct a stem and leaf plot. From this chart, find $Q_3 + 1.5(IQR)$ and $Q_1 - 1.5(IQR)$. Are there any outliers?

11. Compute $s^2$ for the following frequency distribution.

| $x$ | 10 | 14 | 18 | 22 |
|---|---|---|---|---|
| $f$ | 4 | 6 | 8 | 2 |

12. The mean of a population is 100, with a standard deviation of 10. Convert the following to $z$ scores.

(a) 110     (b) 80      (c) 71
(d) 120     (e) 140     (f) 40

13. Find the standard deviation of the following data. Then check using the statistics menu on your calculator.

| Class | Frequency |
|---|---|
| 0–2 | 10 |
| 3–5 | 20 |
| 6–8 | 30 |
| 9–11 | 40 |

HINT: The mid-point of each interval is the data value for the interval.

14. Find the standard deviation of the given data. Then check using the statistics menu on your calculator.

| $x$ | Frequency |
|---|---|
| 1 | 10 |
| 4 | 20 |
| 7 | 30 |
| 10 | 40 |

15. Two instructors gave the same test to their classes. Both classes had a mean score of 72, but the scores of class A showed a standard deviation of 4.5, and those of class B showed a standard deviation of 9. Discuss the difference in the two classes' scores.

16. Joan decides to join the New Army to seek her fortune. She takes a battery of tests to determine placement into the appropriate corps. She makes 75 on the office work test and 80 on the outdoor activity test. The office work test has a mean of 60 and a standard deviation of 20, whereas the outdoor activity test has a mean 75 and a standard deviation of 10. Into which group should Joan be placed?

17. When Tran entered his profession in 1990, the average salary in the profession was $28,500, with a standard deviation of $1000. In 2000, the average salary in the profession was $38,000, with a standard deviation of $3000. Tran made $28,000 in 1990 and $37,000 in 2000. In which year did he do better in comparison with the rest of the profession?

*T* 18. The salaries of the 10 supervisors at the Aleo Automobile Works are $40,000, $43,000, $42,000, $41,000, $46,500, $41,500, $45,000, $47,000, $46,000, and $41,000. How many of the 10 supervisors have salaries within 1 standard deviation of the mean? How many have salaries within 2 standard deviations of the mean?

19. (Multiple choice) A student received a grade of 80 on a math final where the mean grade was 72 and the standard deviation was $s$. On the statistics final, he received a 90, where the mean grade was 80 and the standard deviation was 15. If the standardized scores (scores adjusted to a mean of 0 and a standard deviation of 1) were the same in each case, then $s =$

(a) 10     (b) 12     (c) 16     (d) 18     (e) 20

*C* 20. A teacher has just given an examination to his students. What does he know about his students' performance on the test if the distribution of scores has

(a) a large range but a small standard deviation?
(b) a mean higher than the median?

(c) a mean lower than the median?

(d) a small range but a large standard deviation?

(e) a standard deviation of 0?

21. (a) What effect does adding the same amount to each observation have on the standard deviation? Test your conjecture by adding 5 to each member of the set 10, 12, 13, 14, 16.

(b) What effect does dividing each entry by the same number have on the standard deviation? Check your conjecture by dividing 80, 85, 80, 70, and 80 by 5.

(c) What effect does subtracting a number from each observation and then dividing each result by a number have on the standard deviation? Check your conjecture by subtracting 75 from each entry and then dividing by 5 for 80, 75, 80, 70, and 80.

22. A pollster tabulated the ages of 30 users of a vitamin pill designed to make the person who takes it feel young. The results are shown in the accompanying table. Find the standard deviation. Then check using the statistic menu on your calculator.

| Age | Frequency |
|---|---|
| 20–29 | 1 |
| 30–39 | 2 |
| 40–49 | 4 |
| 50–59 | 5 |
| 60–69 | 9 |
| 70–79 | 6 |
| 80–89 | 3 |

HINT: The mid-point of each interval is the data value for the interval.

23. In Exercise 14, Exercise Set 2 find *IQR*.

24. In Exercise 11(a), Exercise Set 2 find *IQR*.

25. It can be shown that, for any set of data, most of the values lie within 2 standard deviations on either side of the mean. Examine the data in the following table to verify this fact. First find the mean and standard deviation of each class.

**Height in Centimeters**

| Class I | | | Class II | | |
|---|---|---|---|---|---|
| 156 | 158 | 182 | 168 | 180 | 183 |
| 178 | 159 | 176 | 180 | 187 | 190 |
| 160 | 176 | 174 | 176 | 176 | 178 |
| 166 | 160 | 172 | 188 | 186 | 174 |
| 189 | 187 | 154 | 179 | 192 | 188 |
| 153 | 180 | 198 | 176 | 179 | 181 |
| 159 | 162 | 176 | 173 | 174 | 180 |
| 180 | 166 | 192 | 178 | 176 | 175 |

(a) How many of the values from Class I lie within 2 standard deviations on either side of the mean? What percent of Class I is this?

(b) How many of the values from Class II lie within 2 standard deviations on either side of the mean? What percent of Class II is this?

26. The following table of data shows the miles per gallon reported by owners of 5 makes of 8-cylinder automobiles from different manufacturers.

**Manufacturer**

| A | B | C | D | E |
|---|---|---|---|---|
| 18 | 18 | 24 | 21 | 18 |
| 19 | 18 | 16 | 18 | 18 |
| 20 | 20 | 18 | 19 | 19 |
| 21 | 21 | 20 | 18 | 27 |
| 22 | 24 | 22 | 20 | 18 |
| 22 | 19 | 24 | 21 | 18 |

(a) Which sample suggests the best gasoline mileage?

(b) Which sample has the lowest standard deviation?

27. Starting with the original formula for variance on p. 461, show how this simplifies to the computational formula.

## REVIEW EXERCISES

28. Consider the set of scores 1, 8, 16, 18, 20, 20, 21, 23, 24, 29. Find the following.

(a) mean     (b) median     (c) mode     (d) range

29. The mean salary of all employees at the Brown Corporation is $30,000. Make up an example to show how this statistic may be misleading.

30. Find the mean for the following frequency distribution. Then check using the statistic menu on your calculator.

| x | 10 | 20 | 30 | 40 | 50 |
|---|---|---|---|---|---|
| f | 4 | 6 | 8 | 4 | 3 |

# SECTION 4 THE NORMAL DISTRIBUTION

| PROBLEM | The grades on a certain standardized test are normally distributed, with a mean of 70 and a standard deviation of 10. What is the probability that a randomly selected student who took the test scored between 70 and 85? |
|---|---|
| OVERVIEW | The key to answering this question is understanding the term **normally distributed.** One of the fortunate surprises in statistics is that many line graphs and bar graphs are approximately bell-shaped. In fact, if we modify some line graphs to indicate probability rather than frequency, the resulting graphs closely approximate a smooth, bell-shaped curve called the **normal probability curve. If this is true, the data involved are said to be normally distributed.** |
| GOALS | Illustrate the Data Analysis and Probability Standard, page 392. Illustrate the Representation Standard, page 50. |

## NORMAL PROBABILITY CURVE

To prepare to appreciate the characteristics of the normal probability curve, consider the line graph and histogram in Figure 10-11, which are based on the data in Table 10-21. Notice that the frequencies have been converted into relative frequencies or probabilities.

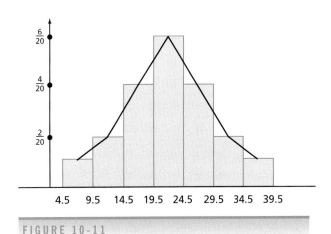

FIGURE 10-11

The height of each rectangle represents the probability that the variable falls in the associated interval. For example, the probability that the variable falls between 19.5 and 24.5 is 0.3. For a probability curve, however, we want the rectangle's area, rather than its height, to represent the probability. Because the width of each rectangle is 5, the area of each rectangle is 5

**TABLE 10-21**

| Class | Frequency | Relative Frequency |
|-------|-----------|--------------------|
| 5–9   | 1 | $\frac{1}{20} = 0.05$ |
| 10–14 | 2 | $\frac{2}{20} = 0.10$ |
| 15–19 | 4 | $\frac{4}{20} = 0.20$ |
| 20–24 | 6 | $\frac{6}{20} = 0.30$ |
| 25–29 | 4 | $\frac{4}{20} = 0.20$ |
| 30–34 | 2 | $\frac{2}{20} = 0.10$ |
| 35–39 | $\dfrac{1}{20}$ | $\frac{1}{20} = 0.05$ |

times the probability. Therefore, we can revise the graph so that the area of each rectangle is equal to the probability by dividing the height of each rectangle by 5. The resulting graph is shown in Figure 10-12.

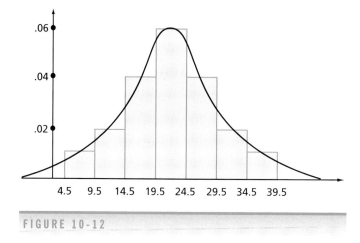

**FIGURE 10-12**

In Figure 10-12 the area of the rectangle over the interval from 9.5 to 14.5 is 0.10 and represents the probability that a randomly selected data value from the data set in Table 10-21 falls between 9.5 and 14.5. Said another way, the area in this rectangle is 10% of the total area of the rectangles in Figure 10-12 and represents 10% of the data in Table 10-21.

Overlaying the bar graph in Figure 10-12 is sketched a smooth curve. This smooth curve is an example of a normal curve. You can observe some interesting properties of the normal probability curve (often called a normal distribution) by studying the approximating bar graph.

The sum of the areas of all the rectangles is equal to 1. Therefore, it is reasonable to assume that the area under the smooth, approximating curve is equal to 1. The mean of the data is 22, and the curve is symmetric about a vertical line through the mean. The standard deviation of the data is approximately 7.5. The interval about the mean that extends for 1 standard deviation on either side of the mean is the interval from $(22 - 7.5)$ to $(22 + 7.5)$. This interval from 14.5 to 29.5 contains $\frac{4}{20} + \frac{6}{20} + \frac{4}{20} = \frac{14}{20}$, or 70%, of the data. (See Figure 10-13.) The interval about the mean that extends for 2 standard deviations on either side of the mean is the interval from (22 − 15) to (22 + 15), or from 7 to 37. Assume that half of the frequency in the first and last class intervals belongs in this range from 7 to 37. Then the interval (from mean − 2 standard deviations to mean + 2 standard deviations) contains 19/20, or 95%, of the data. (See Figure 10-14.)

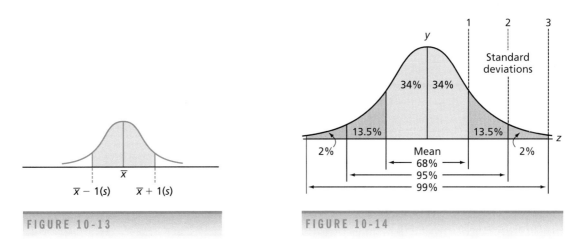

FIGURE 10-13

FIGURE 10-14

In the previous section, we discussed the process of converting data into standard units. Recall that to express $x$ in standard units, we subtract the mean and then divide by the standard deviation:

$$z = \frac{x - \bar{x}}{s}$$

When a data value from a normal distribution is standardized, the resulting data value lies in a special normal distribution called the *standard normal distribution*. The standard normal distribution has a mean of 0 and a standard deviation of 1.

The curve in Figure 10-15 is the standard normal distribution. We use $z$ to represent the standard normal variable and $y$ to represent the probability. The maximum value of the curve is attained at the mean, $z = 0$. The standard normal curve has perfect symmetry. Because of this characteristic, the mean and median of the distribution have the same value—namely, 0. The domain is not bounded because values occur as far out as you wish to go; that is, the curve never intersects the horizontal axis.

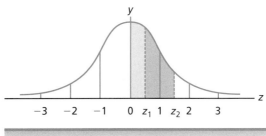

FIGURE 10-15

The area under the standard normal curve is equal to 1. To find the probability that $z$ is between $z_1$ and $z_2$, $P(z_1 \leq z \leq z_2)$, we obtain the area under the curve between $z_1$ and $z_2$ (the shaded region in Figure 10-15). Table 10-22 gives the area under the normal curve less than or equal to $z = z_1$, and greater than or equal to $z = 0$; that is, the area indicated by the light shading in Figure 10-15 is given in Table 10-22 at $z = z_1$. The area from $z = 0$ to $z = z_1$ is the same as the probability that $z$ is less than or equal to $z_1$ and greater than or equal to 0, $P(0 \leq z \leq z_1)$. The table stops at $z = 3.09$, because the area under the curve beyond $z = 3.09$ is negligible.

The fact that the standard normal curve is symmetric about $z = 0$ means that the area under the curve on either side of 0 is 0.5. This symmetry allows us to compute probabilities that do not specifically occur in the table below.

## TABLE 10-22

| z | .00 | .01 | .02 | .03 | .04 | .05 | .06 | .07 | .08 | .09 |
|---|-----|-----|-----|-----|-----|-----|-----|-----|-----|-----|
| 0.0 | .0000 | .0040 | .0080 | .0120 | .0160 | .0199 | .0239 | .0279 | .0319 | .0359 |
| 0.1 | .0398 | .0438 | .0478 | .0517 | .0557 | .0596 | .0636 | .0675 | .0714 | .0753 |
| 0.2 | .0793 | .0832 | .0871 | .0910 | .0948 | .0987 | .1026 | .1064 | .1103 | .1141 |
| 0.3 | .1179 | .1217 | .1255 | .1293 | .1331 | .1368 | .1406 | .1443 | .1480 | .1517 |
| 0.4 | .1554 | .1591 | .1628 | .1664 | .1700 | .1736 | .1772 | .1808 | .1844 | .1879 |
| 0.5 | .1915 | .1950 | .1985 | .2019 | .2054 | .2088 | .2123 | .2157 | .2190 | .2224 |
| 0.6 | .2257 | .2291 | .2324 | .2357 | .2389 | .2422 | .2454 | .2486 | .2517 | .2549 |
| 0.7 | .2580 | .2611 | .2642 | .2673 | .2704 | .2734 | .2764 | .2794 | .2823 | .2852 |
| 0.8 | .2881 | .2910 | .2939 | .2967 | .2995 | .3023 | .3051 | .3078 | .3106 | .3133 |
| 0.9 | .3159 | .3186 | .3212 | .3238 | .3264 | .3289 | .3315 | .3340 | .3365 | .3389 |
| 1.0 | .3413 | .3438 | .3461 | .3485 | .3508 | .3531 | .3554 | .3577 | .3599 | .3621 |
| 1.1 | .3643 | .3665 | .3686 | .3708 | .3729 | .3749 | .3770 | .3790 | .3810 | .3830 |
| 1.2 | .3849 | .3869 | .3888 | .3907 | .3925 | .3944 | .3962 | .3980 | .3997 | .4015 |
| 1.3 | .4032 | .4049 | .4066 | .4082 | .4099 | .4115 | .4131 | .4147 | .4162 | .4177 |
| 1.4 | .4192 | .4207 | .4222 | .4236 | .4251 | .4265 | .4279 | .4292 | .4306 | .4319 |
| 1.5 | .4332 | .4345 | .4357 | .4370 | .4382 | .4394 | .4406 | .4418 | .4429 | .4441 |
| 1.6 | .4452 | .4463 | .4474 | .4484 | .4495 | .4505 | .4515 | .4525 | .4535 | .4545 |
| 1.7 | .4554 | .4564 | .4573 | .4582 | .4591 | .4599 | .4608 | .4616 | .4625 | .4633 |
| 1.8 | .4641 | .4649 | .4656 | .4664 | .4671 | .4678 | .4686 | .4693 | .4699 | .4706 |
| 1.9 | .4713 | .4719 | .4726 | .4732 | .4738 | .4744 | .4750 | .4756 | .4761 | .4767 |
| 2.0 | .4772 | .4778 | .4783 | .4788 | .4793 | .4798 | .4803 | .4808 | .4812 | .4817 |
| 2.1 | .4821 | .4826 | .4830 | .4834 | .4838 | .4842 | .4846 | .4850 | .4854 | .4857 |
| 2.2 | .4861 | .4864 | .4868 | .4871 | .4875 | .4878 | .4881 | .4884 | .4887 | .4890 |
| 2.3 | .4893 | .4896 | .4898 | .4901 | .4904 | .4906 | .4909 | .4911 | .4913 | .4916 |
| 2.4 | .4918 | .4920 | .4922 | .4925 | .4927 | .4929 | .4931 | .4932 | .4934 | .4936 |
| 2.5 | .4938 | .4940 | .4941 | .4943 | .4945 | .4946 | .4948 | .4949 | .4951 | .4952 |
| 2.6 | .4953 | .4955 | .4956 | .4957 | .4959 | .4960 | .4961 | .4962 | .4963 | .4964 |
| 2.7 | .4965 | .4966 | .4967 | .4968 | .4969 | .4970 | .4971 | .4972 | .4973 | .4974 |
| 2.8 | .4974 | .4975 | .4976 | .4977 | .4977 | .4978 | .4979 | .4979 | .4980 | .4981 |
| 2.9 | .4981 | .4982 | .4982 | .4983 | .4984 | .4984 | .4985 | .4985 | .4986 | .4986 |
| 3.0 | .4987 | .4987 | .4987 | .4988 | .4988 | .4989 | .4989 | .4989 | .4990 | .4990 |

**EXAMPLE 1**   (a) Find $P(0 \leq z \leq 1.84)$        (b) Find $P(z \leq 1.84)$

*SOLUTION*

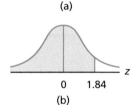

1.84
(a)

0   1.84
(b)

FIGURE 10-16

(a) To solve such problems it is important to sketch the area in question. Give careful attention to the shaded area in Figure 10-16(a). We find $P(0 \leq z \leq 1.84)$ by finding $z = 1.84$ in Table 10-22 and observing that the probability is .4671 in that table.

(b) Give careful attention to the shaded area in Figure 16(b). Since the area under either half of the curve is 0.5000,

$$P(z \leq 1.84) = .5000 + P(0 \leq z \leq 1.84)$$
$$= .5000 + .4671 = .9671$$

For the normal curve, $P(z \leq a) = P(z < a)$, and $P(z \geq a) = P(z > a)$, for all $a$. In general, for distributions (called *continuous distributions*) like the standard normal curve, the probability that $z$ is less than or equal to a number is the same as the probability that $z$ is less than the number.

**EXAMPLE 2**   Find $P(z < 1.2)$.

*SOLUTION*

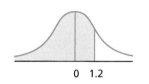

0   1.2

FIGURE 10-17

In Figure 10-17 we see that the area in question can be broken into two parts. The area to the left of 0 is .5 while the shaded area to the right of 0 can be found in Table 10-22 to be .3849. Thus,

$$P(z < 1.2) = P(z \leq 1.2) = .3849 + .5000 = .8849$$

If it is known that a set of data values is normally distributed, but that they are not described by the standard normal distribution, then we can compute their probabilities using the $z$ scores discussed in Section 3.

**EXAMPLE 3**   Find the probability that the normal variable $x$, with mean 175 and standard deviation 20, is less than or equal to 215.

*SOLUTION*   The $z$ value of 215 is

$$z = \frac{215 - 175}{20} = 2$$

Therefore,        $P(x \leq 215) = P(z \leq 2) = .9772$

Can you sketch the picture that shows that

$$P(z \leq 2) = .5 + .4772 = .9772?$$

**PRACTICE PROBLEM**   Find $P(z \leq 0.8)$.

*ANSWER*   .7881

Recall that the standard normal curve is symmetrical about $z = 0$. This fact is important to our discussion of areas under the curve. The fact that the standard normal curve is symmetric about the origin means that the areas

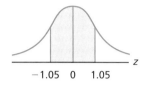

**FIGURE 10-18**

under the curve extending an equal distance on either side of 0 are the same. In Figure 10-18 we see that

$$P(-1.05 \leq z \leq 1.05) = 2P(0 \leq z \leq 1.05)$$
$$= 2(.3531) = .7062$$

Because the total area under the curve is 1, the area to the right of $z = 1.66$ is 1 minus the area to the left of 1.66. (See Figure 10-19.) In symbolic notation,

**FIGURE 10-19**

$$P(z \geq 1.66) = 1 - P(z < 1.66)$$
$$= 1 - [.5000 + P(0 < z < 1.66)]$$
$$= 1 - [.5000 + .4515]$$
$$= 1 - .9515 = .0485$$

**EXAMPLE 4**

The mileage data on a new model Pluto automobile is normally distributed. The mean in-city mileage is 32 miles per gallon, and the standard deviation is 1.5. What is the probability that the Pluto you purchase will average less than 30 miles per gallon?

*SOLUTION*

To solve this problem, we find the associated $z$ score for 30 miles per gallon.

$$z = \frac{30 - 32}{1.5} = -1.33$$
$$P(x < 30) = P(z < -1.33)$$

Careful attention to Figure 10-20 reveals that

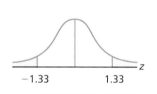

**FIGURE 10-20**

$$P(z < -1.33) = P(z > 1.33)$$
$$= 1 - P(z \leq 1.33)$$
$$= 1 - .9082 = .0918$$

Sometimes we need to compute the probability that $z$ is in a certain range—say, between 0.4 and 1.4. (See Figure 10-21.) This probability is indicated by $P(0.4 < z < 1.4)$. It can be obtained by finding the probability that $z$ is less than 1.4 and subtracting from this the probability that $z$ is less than 0.4.

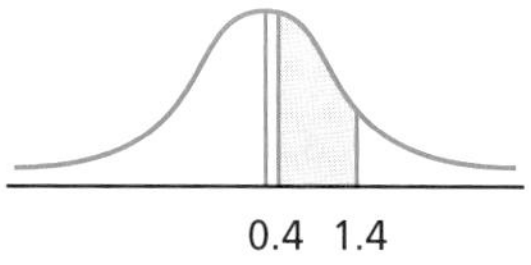

$$P(0.4 < z < 1.4) = P(z < 1.4) - P(z < 0.4)$$
$$= .9192 - .6554$$
$$= .2638$$

**FIGURE 10-21**

**PRACTICE PROBLEM**

Find $P(-0.51 \leq z \leq 1.98)$.

*ANSWER*

.6711

**EXAMPLE 5**

The Iron Fist Security Agency has uniforms to fit men ranging in height from 68 to 74 inches. The heights of adult males are normally distributed, with a mean of 70 inches and a standard deviation of 2.5 inches. What percent of male applicants to Iron Fist can be fitted in their existing uniforms?

*SOLUTION*   The $z$ values that correspond to 68 and 74 are

$$z = \frac{68 - 70}{2.5} = -0.8 \quad \text{and} \quad z = \frac{74 - 70}{2.5} = 1.6$$

Therefore,
$$P(68 \le x \le 74) = P(-0.8 \le z \le 1.6)$$
$$= P(-0.8 \le z \le 0) + P(0 \le z \le 1.6)$$
$$= P(0 \le z \le 0.8) + .4452$$
$$= .2881 + .4452 = .7333$$

Hence, the probability that a given applicant can be fitted in a uniform is .73, so 73% of the applicants can be fitted.

To this point we have been computing the probability that a data value from a normal distribution is less than some specific value. We can turn the question around and ask, "What is a $z$ score, $z_o$, with the property that $P(z \le z_0)$ is a specific number?"

**EXAMPLE 6**   Find the third quartile, $Q_3$, for the standard normal distribution.

*SOLUTION*   We need to find a $z$ score, $z_0$, so that $P(z < z_0) = .75$. Looking at Figure 10-22, we see that $P(0 \le z \le z_0) = .25$. Looking in the body of Table 10-22 for probabilities near .25, we see that $z_0$ is about .67. This is the third quartile for the standard normal curve.

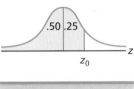

FIGURE 10-22

**EXAMPLE 7**   Acme Lights Inc. produces lightbulbs. The lifetimes of these bulbs are normally distributed with a mean of 800 h. and a standard deviation of 25 hours. Acme wishes to guarantee a lifetime for their bulbs so that 90% of the bulbs will burn longer than the guaranteed lifetime. What lifetime should they advertise?

*SOLUTION*   We must first find a $z$ score, $z_0$, above which 90% of the area lies and below which 10% of the area lies. Then we must find the corresponding score in a normal distribution with mean of 800 and standard deviation of 25.

Looking at Figure 10-23, we see that the desired $z$ score is a negative number, but that its opposite is the score that appears in Table 10-22 with an area of .4. Hence, $z$ is approximately −1.28.

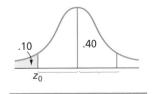

FIGURE 10-23

If we let $x$ be the lightbulb life associated with the $z$ score of −1.28, then

$$-1.28 = \frac{x - 800}{25}$$

Solving for $x$, we learn that the advertised lifetime should be 768 hours.

# Just for Fun

The combination of a lock is given by the following distribution of numbers: 14, 5, 6, 11, 13, 8, 3, and 4. Turn left the mean of the distribution, turn right the value of the median, turn left the largest positive integer in the standard deviation, and turn right $\frac{1}{2}$ plus half the range. Can you open the lock?

# EXERCISE SET 4

*R* 1. Find the area under the standard normal curve that lies between the following pairs of values of $z$.

  (a) $z = 0$ to $z = 2.40$
  (b) $z = 0$ to $z = 0.41$
  (c) $z = 0$ to $z = 1.67$
  (d) $z = -0.36$ to $z = 0.36$

2. Assuming that the following sketches represent the standard normal curve, compute the shaded areas.

  (a)
  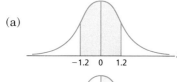

  (b)
  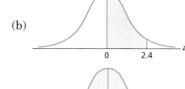

  (c)

  (d)
  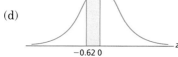

3. Sketch the area that is equal to each of the following probabilities and then find the probabilities by using Table 10-22.

  (a) $P(z \le 1.7)$
  (b) $P(z < .6)$

4. If $x$ is a variable having a normal distribution, with $\bar{x} = 12$ and $s = 4$, find the probability that $x$ assumes the following values.

  (a) $x \le 16$
  (b) $x \le 14$

5. The scores on a national achievement test are normally distributed, with a mean of 50 and a standard deviation of 10. What fraction of the students taking the test make below 65 on the test?

*T* 6. Sketch the areas equal to each of the following probabilities and then find the probabilities by using Table 10-22.

  (a) $P(z \le -2.1)$  (b) $P(z \ge -1.4)$
  (c) $P(z \le 0.1)$  (d) $P(z < -1.6)$
  (e) $P(z > 1.5)$  (f) $P(z > 2.4)$
  (g) $P(z > -2.1)$  (h) $P(z > -1.8)$
  (i) $P(1.3 \le z \le 2.4)$  (j) $P(2.1 < z < 2.8)$
  (k) $P(-1.2 \le z \le 0.3)$  (l) $P(-2.6 \le z \le 1.4)$

7. Given that $x$ is normally distributed, with mean 50 and standard deviation 10, find the following probabilities.

  (a) $P(x \ge 50)$  (b) $P(x \le 50)$
  (c) $P(x \ge 60)$  (d) $P(x \le 70)$

(e) $P(x \le 40)$     (f) $P(x \ge 46)$

(g) $P(38 \le x \le 54)$     (h) $P(32 \le x \le 61)$

*In Exercises 8 through 18, assume that the data are normally distributed.*

8. Radar is used to check the speed of traffic on Interstate 75 through Atlanta. If the mean speed of the traffic is 60 mph, with a standard deviation of 4 mph, what percentage of the cars are exceeding the legal speed of 55 mph?

9. Young rabbits placed on a certain high-protein diet for a month show a weight gain with a mean of 120 g and a standard deviation of 12 g.

   (a) What is the probability that a given rabbit will gain at least 100 g in weight?
   (b) If 15,000 rabbits are placed on this diet, how many can be expected to gain at least 140 g?

10. The heights of men in a certain army regiment have a mean of 177 cm and a standard deviation of 4 cm.

    (a) What percent of the men are between 173 and 181 cm in height?
    (b) What percent of the men are between 169 and 185 cm in height?

11. The scores on the entrance exam for the Kentucky Police Academy have a mean of 60 and a standard deviation of 6. What is the probability that a randomly selected test score will lie between 60 and 75?

12. The Goodbond Tire Company manufactures a superior-quality tire (sold at a superior price) that it guarantees for 50,000 miles. The average life of one of these tires is 55,000 miles (with a standard deviation of 4000 miles), but occasionally a tire fails in less than 50,000 miles. What percent of the tires will fail before 50,000 miles?

13. A test given to kindergarten students involves putting together a jigsaw puzzle. The mean time of completion of the puzzle is 150 seconds with a standard deviation of 20 seconds.

    (a) If the test were given to 1000 children, about how many would finish the puzzle in less than 200 seconds?
    (b) What percent of the children would finish the puzzle in 120 seconds or less?

14. The pulse rate per minute of American males aged 18 to 25 has a mean of 72 and a standard deviation of 9.7. If the requirements for employment by a law enforcement agency require that the applicant have a pulse rate of less than 95 beats per minute, what percent of males aged 18 to 25 would meet the requirements?

*C* 15. The life of a certain brand of batteries has a mean of 1200 days and a standard deviation of 100 days. If the manufacturer does not want to replace more than 12% of the batteries, for how long should the batteries be guaranteed?

16. The average time needed to complete an aptitude test is 80 min, with a standard deviation of 10 minutes. When should the test giver stop the test to make certain that 90% of the people taking it have had time to complete it?

17. The first decile is the 10th percentile, $P_{10}$; the second decile is the 20th percentile, $P_{20}$; and so on. The deciles divide a data set into 10 sections, each containing 10% of the data.

    (a) Find the first decile of the standard normal distribution.
    (b) Find the ninth decile of the standard normal distribution.
    (c) Find the first decile of the distribution in Exercise 13.
    (d) Find the ninth decile in Exercise 13.

18. The grades in a certain class have a mean of 76 and a standard deviation of 6. The lowest D is 61, the lowest C is 70, the lowest B is 82, and the lowest A is 91. What percentage of the class will make A's? What percentage will make B's? C's? D's? Fs?

## REVIEW EXERCISES

19. The president of a university quoted that the school's enrollment was up 4%. However, the registrar said it was down 1%. Give one possible reason for this discrepancy.

20. A city survey was made to determine the average family earnings of residents. This information was obtained from all the employers' records in the city. Would the result be misleading? Why?

21. One-third of the women students at Lonesome University became pregnant last year. Can you think of reasons why this statement might be misleading?

22. The following test scores were received by 24 students.

    63, 71, 85, 96, 94, 90, 75, 72, 77, 71, 62, 84,
    61, 54, 87, 94, 32, 81, 94, 77, 63, 60, 81, 76

    Find the following.

    (a) mean     (b) median
    (c) mode     (d) first quartile
    (e) range     (f) standard deviation
    (g) *IQR*     (h) outliers

23. For the data in Exercise 22, construct a stem and leaf plot.

24. For the data in Exercise 22, construct a box plot without whiskers.

25. Subtract 30 from each data value $x$, and divide the result by 10 to get a new data value $y$.

| $x$ | 10 | 20 | 30 | 40 | 50 |
|---|---|---|---|---|---|
| $f$ | 4 | 6 | 8 | 4 | 3 |

(a) Find the mean of the $x$ values and the mean of the $y$ values.
(b) Find the standard deviation of the $x$ values and the standard deviation of the $y$ values.
(c) Compare the means and standard deviations of the two sets of data.
(d) Then check using the statistics menu on your calculator.

26. The following IQ scores were made by 30 first-grade students in one classroom.

| | | | | | |
|---|---|---|---|---|---|
| 128 | 133 | 100 | 115 | 82 | 99 |
| 107 | 142 | 98 | 112 | 152 | 100 |
| 105 | 78 | 114 | 84 | 86 | 110 |
| 96 | 93 | 101 | 94 | 86 | 124 |
| 120 | 100 | 102 | 107 | 94 | 128 |

(a) Find the median.
(b) Form a stem and leaf chart.
(c) Construct a box and whisker plot.
(d) Find all outliers.

# SOLUTION TO INTRODUCTORY PROBLEM

UNDERSTANDING THE PROBLEM.  Of course, the key word in understanding the problem is *average*. What do you mean by average? Which average?

DEVISING A PLAN.  To investigate the claims made, we need to analyze the underlying salaries. Then we can identify the nature of each "average."

CARRYING OUT THE PLAN.   A tabulation of the salaries at Acme Manufacturing is given in the table at right.

The management used the arithmetic mean as average, and you can verify this to be $29,280. The union used the median of $24,000. The employees talked to those with the most common salary, $26,000, so they used the mode. All were correct; they simply used different averages.

| Salary | Frequency |
|---|---|
| $160,000 | 1 |
| 90,000 | 1 |
| 70,000 | 1 |
| 26,000 | 20 |
| 24,000 | 10 |
| 23,000 | 10 |
| 22,000 | 7 |
| Total | 50 |

# SUMMARY AND REVIEW

## ORGANIZING DATA

As you organize data, make certain you understand the meaning of the following terms:

(a) Frequency
(b) Frequency distribution
(c) Grouped frequency distribution
(d) Class boundaries
(e) Class marks
(f) Stem and leaf plot

## GRAPHING DATA

Make certain you understand the techniques for visually presenting statistical data:

(a) Bar graph        (b) Histogram
(c) Line graph       (d) Frequency polygon
(e) Circle graph

## MEASURES OF CENTRAL TENDENCY

(a) Formulas for the arithmetic mean:

$$\bar{x} = \frac{x_1 + x_2 + x_3 + \ldots + x_n}{n}$$

$$\bar{x} = \frac{x_1 f_1 + x_2 f_2 + \ldots + x_m f_m}{f_1 + f_2 + \ldots + f_m}$$

(b) The median is the middle measurement.

(c) The mode is the most common measurement.

## MEASURES OF SCATTERING

(a) Variance is given by the following formulas.

$$s^2 = \frac{(x_1 - \bar{x})^2 + (x_2 - \bar{x})^2 + \cdots + (x_n - \bar{x})^2}{n}$$

$$s^2 = \frac{(x_1 - \bar{x})^2 f_1 + (x_2 - \bar{x})^2 f_2 + \cdots + (x_m - \bar{x})^2 f_m}{n}$$

(b) Standard deviation is defined by

$$s = \sqrt{\text{variance}}$$

## INTERQUARTILE RANGE (IQR)

(a) Percentiles are values below which a given percent of the data is found.

(b) $Q_1$ represents the value below which 25% of the data are located.

(c) $Q_3$ represents the value below which 75% of the data are located.

(d) $IQR = Q_3 - Q_1$

(e) Box and whisker plot: Make sure you can draw a box and whisker plot and locate outliers (data above $Q_3 + 1.5\,(IQR)$ or below $Q_1 - 1.5(IQR)$).

## STANDARD VARIABLE

To change from the variable $x$ to the standard variable $z$, use

$$z = \frac{x - \bar{x}}{s}$$

## NORMAL DISTRIBUTION

The normal distribution has a bell-shaped graph with the following characteristics:

(a) 68% of data lie within 1 standard deviation of the mean.

(b) 95% of data lie within 2 standard deviations of the mean.

(c) 99% of data lie within 3 standard deviations of the mean.

# CHAPTER TEST

*For questions 1 through 5, consider the following set of values.*

$$x: 3,\ 7,\ 11,\ 15,\ 19,\ 23$$

1. Find the mean.
2. Find the median.
3. Find the variance of $x$.
4. Find the interquartile range.
5. Draw a box and whisker plot.

*Questions 6 through 9 refer to the following frequency distribution:*

| Interval | Frequency |
|----------|-----------|
| 1–4 | 5 |
| 5–8 | 7 |
| 9–12 | 10 |
| 13–16 | 5 |
| 17–20 | 3 |

6. Draw a histogram.
7. Draw a frequency polygon.
8. Find the mean.

9. Find the standard deviation.

*Questions 10 through 18 refer to the following scores on a mathematics test.*

33, 37, 42, 48, 52, 53, 54, 55, 57, 59, 62, 62,
64, 64, 64, 68, 69, 71, 72, 72, 73, 73, 74, 74,
78, 79, 82, 83, 85, 87, 88, 89, 93, 96, 98

10. Approximate the interquartile range.
11. Find the mean.
12. Find the median.
13. Find the mode.
14. Make a stem and leaf chart.
15. Make a box and whisker plot.
16. Are there any outliers? Name them.
17. Make a grouped frequency distribution with class intervals 31–40, 41–50, . . . .
18. Find the standard deviation of your grouped frequency distribution in Exercise 17.
19. A variable is normally distributed with mean 25 and standard deviation 5. Find $P(25 < x < 30)$.
20. A set of measurements is approximately normally distributed, with a mean of 100 and a standard deviation of 10. Find the percentage of measurements larger than 110.

# INFORMAL GEOMETRY

Cutting a circular cake produces interesting problems in geometry. If $n$ concurrent vertical cuts (that is, $n$ vertical cuts through the same point on top) are made, $2n$ pieces of the cake are produced. But suppose that the vertical cuts are not concurrent. The figures on the right show the maximum number of pieces. When $n = 1$, 2, 3, and 4, the maximum number of pieces are 2, 4, 7, and 11, respectively. Use your problem-solving techniques to conjecture the maximum number of pieces with 5 vertical cuts. Then draw a figure and verify your conjecture. What is the maximum number of pieces produced by 10 cuts?

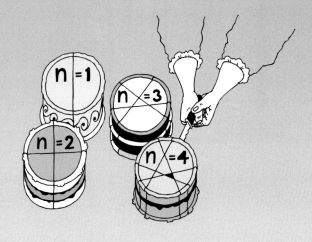

(M. Gardner, *Aha!* New York: W. H. Freeman, 1978)

The next three chapters of this book are devoted to the study of geometry. The name geometry, meaning "earth measurement," reveals that this branch of mathematics involves practical applications through measurement as well as a study of geometric concepts.

This chapter takes an informal approach to geometry. The approach includes a description of undefined terms and a discussion of definitions and axioms, with many examples, as well as statements and discussions of important theorems. You will sketch figures, look for models, visualize two-dimensional and three-dimensional figures, and understand relationships.

Later you can use what you have learned in this chapter to explore and experiment with new ideas.

This chapter calls for use of the following problem-solving strategies and suggestions for understanding problems: ▪ Make a table. (526) ▪ Try a simpler version of the problem. (509, 510) ▪ Look for a pattern. (509, 510, 526) ▪ Guess, test, and remove impossible situations. (489) ▪ Use reasoning. (522) ▪ Work backward. (526) ▪ Draw a picture. (throughout the chapter)

# SECTION 1  AN INTRODUCTION TO GEOMETRY

**PROBLEM**

Fold a piece of paper. What geometric figure is represented by the fold? Fold the piece of paper again so that the creases are not parallel. Do the two creases intersect? In how many points? What geometric truth have you demonstrated? What geometric figure does a flashlight beam represent?

**OVERVIEW**

If your answers to these introductory questions were "a line (or line segment), yes, one, two intersecting lines intersect in a unique point, and a ray," then you probably already understand much of this section.

**GOALS**

Use the Geometry Standard: ● analyze characteristics and properties of two- and three-dimensional geometric shapes and develop mathematical arguments about geometric relationships; ● specify locations and describe spatial relationships using coordinate geometry and other representational systems; ● apply transformations and use symmetry to analyze mathematical situations; ● use visualization, spatial reasoning, and geometric modeling to solve problems.

Can you imagine the challenges facing the mathematicians of the ancient civilizations of Egypt and Babylon (circa 3000 B.C.)? Not only did they need to develop a numeration system so that commerce could prosper and government could function (see Chapter 4), but also they needed to develop geometric tools to solve a number of pressing problems. In Egypt the Nile River overflowed its banks each year and erased evidence of boundaries in the fertile farms along the Nile. The ability to resurvey those fields accurately and efficiently was important. Significant amounts of geometry were needed to build the pyramids and monuments of the governing dynasties. In Babylon, geometry was needed not only for agriculture and architecture but also for astronomy. The Babylonians were avid astronomers and needed geometry to map the skies. As the histories of these two great civilizations drew to a close, their mathematicians possessed a number of practical, "rule of thumb" geometric procedures that solved a wide variety of problems.

The ancient Greeks had a different point of departure for their study of geometry. The tone of their investigations was set by the Pythagorean mystery cult who believed that "all is number" and regarded the search for mathematical truths as a sacred vocation. Thus the Greeks used deductive reasoning to establish geometric principles that were true in all cases. As their studies matured, they produced a formal geometrical system in which fundamental geometric definitions and postulates were established and all other geometric principles were deduced from them. This systematic mode of inquiry has served as an important model of mathematical investigation to this day.

Geometry is even more important today than it was in the ancient world. Many geometric concepts are used in navigation, surveying, engineering, architecture, art, and scientific research on the largest scale possible, the study of the basic structure of the universe. One of the most fertile new areas of inquiry in mathematics, chaos theory, is very geometric in character.

# POINTS AND LINES

In the study of geometry, we are concerned with sets of elements called **points.** All geometric figures are sets of points. A point is a unique, exact location in space. It has no dimensions. Physical representations of points in nature include a star seen with the naked eye or the tip of a pencil. However, we shall treat *point* as an undefined term. In a diagram, a point is usually named by a capital letter, such as *A, B, C,* or *D.* (See Figure 11-1.)

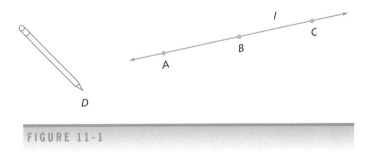

Now think of **space** as the set of all points. Thus, the physical universe may be visualized as being filled with points. The geometric figures we shall study consist of subsets of these points in space.

(a)

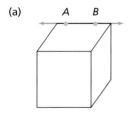

(b)

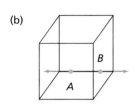

(c)

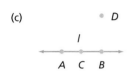

FIGURE 11-2

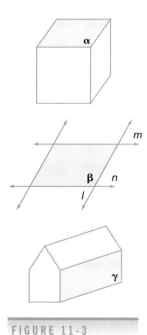

FIGURE 11-3

A **line** is a particular subset of points in space. (We use the word *line* to mean *straight line* in this book.) The edge of a desk, a sideline on a football field, and the edge of a box are physical models of parts of lines (called *line segments*). (See Figure 11-2.) Yet these are only physical models of lines; the idea of a line is an abstraction that we use without definition. The set of connected points in Figure 11-2(c) represents a line. The arrow on each end indicates that the line extends infinitely far in each direction. A line can be named either by a lowercase letter (line *l*) or by a pair of capital letters representing any two points on the line, such as line $\overleftrightarrow{AB}$.

A **plane** is another subset of points in space. Intuitively, we describe a plane as being "flat," like a sheet of paper, and as extending infinitely far in all directions; therefore, a plane contains infinitely many points and infinitely many lines.

When visualizing a plane, we usually think of a set of points lying on some flat surface. Tops of tables and the surfaces of floors and walls are common models of parts of planes.

Figure 11-3 shows several representations of a plane. Greek letters such as α (alpha), β (beta), and γ (gamma) are commonly used to name planes.

In statements relating points, lines, and planes, we use other undefined terms, such as *is on, is between,* and *separates.* If the members of a set of points are on a line, they are said to be **collinear,** and the line is said to **pass through** them or to contain them. In Figure 11-2(c), *A, B,* and *C* are collinear; however, *A, B,* and *D* are not. If three points are collinear, then one of the points is **between** the other two. Which one of the points in Figure 11-2(c) is between the other two? In Figure 11-2(c), point *A* is on line *l,* and line *l* is said to **contain** point *A.* In set terminology, *A* is an **element** of *l,* or $A \in l$.

When we fold a piece of paper (waxed paper is best), the resulting crease represents a piece of a line. Fold the paper again so that the creases are not parallel. The two creases intersect in how many points? Your answer demonstrates the following property: **If two distinct lines touch or cross (intersect), then the intersection is exactly one point.** (See Property 1 in Table 11-1.)

Consider a point in space. How many lines can contain or pass through this point? It should be easy to visualize that infinitely many lines can contain a single point. (See Property 2 in Table 11-1.)

**TABLE 11-1**

**Properties of Lines**

| *Property* | *Illustration* |
|---|---|
| 1. If two distinct lines intersect, they intersect in exactly one point. | |
| 2. Two or more lines are said to be **concurrent** if there is exactly one point common to all of them. (These lines may be in the same plane, or they may be lines in space.) | |
| 3. Exactly one line contains any two distinct points. | |

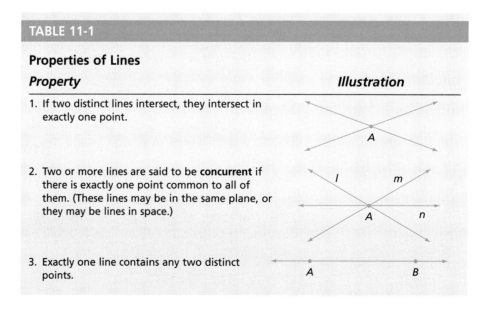

How many lines can contain the same pair of distinct points? Mark two points on a piece of paper. Draw a line containing the two points. Can you draw another line through them that is distinct from the first line? Take any two points in a room and stretch a string between them. Any other string stretched between these two points would occupy the same position as the first string. These physical models illustrate the fact that lines are uniquely determined by two fixed points. In general, **exactly one line contains two distinct points.** (See Property 3 in Table 11-1.)

If two distinct lines in a plane do not intersect, the lines are said to be *parallel*.

| DEFINITION | **Parallel Lines** | Two lines in a plane are *parallel* if and only if there is no point common to both lines. |
|---|---|---|

Straight railroad tracks suggest the idea of parallel lines. In Figure 11-4(a), lines *r* and *s* are parallel. If *l* and *m* are parallel, we write *l* ∥ *m*, where ∥ is read "is parallel to."

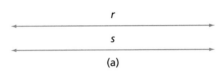

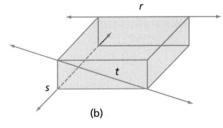

FIGURE 11-4

Can you imagine two lines that do not meet anywhere but are still not parallel? This situation can happen if the two lines are not in the same plane. Notice in Figure 11-4(b) that *r* and *s* will not intersect no matter how far they are extended. Similarly, lines *t* and *s* and lines *t* and *r* do not intersect. Moreover, they are not parallel because they do not lie in the same plane. Such lines are called **skew lines.**

Suppose that you have a line *m* and a point *P* not on *m*. How many lines containing *P* can be drawn parallel to the given line *m*? The answer is "only one line can be drawn through a point parallel to a given line where the given point is not on the line." (See Figure 11-5, where *n* ∥ *m*.)

FIGURE 11-5

**EXAMPLE 1**

In Figure 11-6, draw a line through point *P* parallel to $\overleftrightarrow{AB}$.

*SOLUTION*

Line $\overleftrightarrow{CD}$ in Figure 11-6 contains *P* and is parallel to $\overleftrightarrow{AB}$. Note that the ratio of vertical change to horizontal change from *A* to *B* is the same as from *C* to *D*.

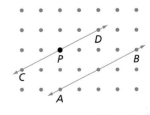

FIGURE 11-6

Certain subsets of a line are important and are given names. Let $A$ be a point on a line, as in Figure 11-7. All the points on one side of $A$ (excluding $A$) constitute a *half line.* (See Table 11-2.)

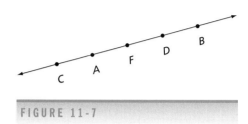

FIGURE 11-7

| TABLE 11-2 | |
|---|---|
| **Subsets of Lines** | |
| *Definition* | *Illustration* |
| A **half line** is a subset of a line, containing all the points of the line on one side of a given point (excluding the point). | ←---------○————•<br>⠀⠀⠀⠀⠀⠀⠀⠀⠀$A$⠀⠀⠀⠀$B$ |
| The **ray** $\overrightarrow{AB}$ is the union of point $A$ and the half line containing all points on line $\overleftrightarrow{AB}$ that lie on the same side of $A$ as point $B$. | ○————•————→<br>$A$⠀⠀⠀⠀$B$<br>⠀⠀$\overrightarrow{AB}$ |
| Two given points on a line and the portion of the line containing all the points between these two points constitute a line **segment**. | •————•<br>$A$⠀⠀⠀⠀$B$<br>$\overline{AB}$ or $\overline{BA}$ |

If the point that separates a line into two half lines is included with one of the half lines, that subset of a line is called a *ray.* (See Table 11-2.) Thus, in Figure 11-7, $A$ and all the points to the right of $A$ constitute a ray. **The first letter used to denote a ray names the endpoint; the second letter names some point on the half line and indicates direction.** $A$ is the endpoint of ray $\overrightarrow{AB}$. If $D$ is a point on $\overrightarrow{AB}$ other than $A$, then $\overrightarrow{AB}$ and $\overrightarrow{AD}$ name the same ray. In Figure 11-7, point $A$ determines two rays, $\overrightarrow{AB}$ and $\overrightarrow{AC}$, which are called **opposite rays.** Notice that

$$\overrightarrow{AB} \cap \overrightarrow{AC} = \{A\}$$

Consider the points $A$ and $B$ in Figure 11-7. These two endpoints and the portion of the line between them are called *segment* $\overline{AB}$ and are denoted by either $\overline{AB}$ or $\overline{BA}$. (See Table 11-2.) Thus, $\overline{AB}$ consists of the points that $\overrightarrow{AB}$ and $\overrightarrow{BA}$ have in common, or $\overline{AB} = \overrightarrow{AB} \cap \overrightarrow{BA}$.

| **EXAMPLE 2** | Find the following in Figure 11-7.<br><br>(a) $\overline{CD} \cap \overline{AB}$⠀⠀⠀(b) $\overrightarrow{CA} \cap \overrightarrow{AC}$⠀⠀⠀(c) $\overline{CA} \cap \overline{DB}$⠀⠀⠀(d) $\overline{AF} \cap \overline{DF}$ |
|---|---|

*SOLUTION*

(a) $\overline{AD}$ (because all the points in $\overline{AD}$ are common to both $\overline{CD}$ and $\overline{AB}$)
(b) $\overline{AC}$
(c) $\varnothing$
(d) $\{F\}$

**PRACTICE PROBLEM**

Using Figure 11-8, find each of the following.

(a) $\overrightarrow{DI} \cap \overline{AF}$      (b) $\overrightarrow{VI} \cup \overline{AI}$      (c) $\overrightarrow{ID} \cap \overline{AF}$

*ANSWER*

(a) $\overline{AI}$      (b) $\overrightarrow{AI}$      (c) $\overline{AI}$

**EXAMPLE 3**

Using Figure 11-9, name the following.

(a) $\overleftrightarrow{AB} \cap \overleftrightarrow{AD}$      (b) $\overleftrightarrow{AB} \cap \overleftrightarrow{CD}$      (c) $\overrightarrow{DC} \cap \overleftrightarrow{CD}$

*SOLUTION*

(a) $\{A\}$      (b) $\varnothing$      (c) $\overrightarrow{DC}$

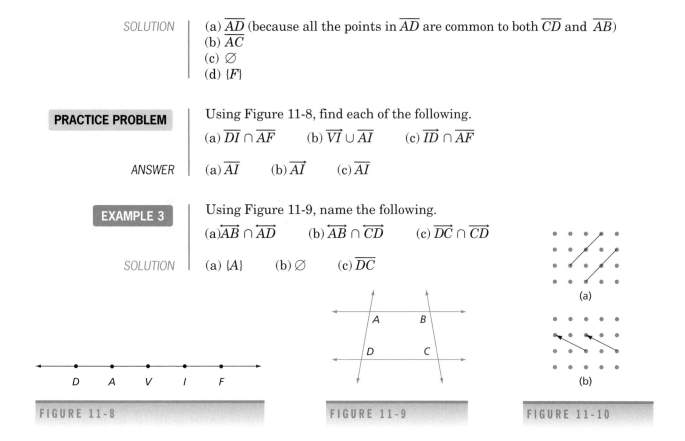

FIGURE 11-8

FIGURE 11-9

FIGURE 11-10

When we say that two segments—as in Figure 11-10(a)—or two rays—as in Figure 11-10(b)—are parallel, we mean that the lines containing these segments or rays are parallel.

 # PLANES

Let's shift our thinking now to planes. How many points are needed to determine a plane? If you attempt to balance a piece of cardboard on the tip of a pencil, does the cardboard tend to tilt in some direction? How many planes can contain one point? Now hold the tips of two pencils under the cardboard. Does the cardboard tilt? How many planes can contain two points? Finally, hold the tips of three pencils under the cardboard, and try to tilt the cardboard. What do you find? In like manner, if you sit on a three-legged stool whose legs are not quite the same length, all three legs will touch the floor. If you try sitting in a four-legged chair that has one leg slightly shorter than the other three, how many legs will touch the floor? Did you answer three? This discussion suggests the following property of planes. If $A$, $B$, and $C$ are three distinct noncollinear points, then one and only one plane contains $A$, $B$, and $C$. (See Property 1 in Table 11-3.)

Take a pencil representing a line, a stiff sheet of paper representing a plane, and a point in space. How many ways can you arrange the paper to contain the pencil and the point? Once the placement of the point and of the

| TABLE 11-3 | |
|---|---|
| **Properties of Planes** | |
| *Property* | *Illustration* |
| A plane can be uniquely determined by four combinations of points and/or lines.<br>1. Three noncollinear points:<br>    Points *A*, *B*, and *C* determine plane α. | |
| 2. A line and a point not on that line:<br>    $\overleftrightarrow{AB}$ and *C* determine plane α. | |
| 3. Two intersecting lines:<br>    $\overleftrightarrow{AB}$ and $\overleftrightarrow{CB}$ determine plane α. | |
| 4. Two parallel lines:<br>    $\overleftrightarrow{AB}$ and $\overleftrightarrow{CD}$, with $\overleftrightarrow{AB}$ parallel to $\overleftrightarrow{CD}$,<br>    determine plane α. | |

line is decided, the arrangement is unique. This suggests that **a line and a point not on the line uniquely determine a plane.** (See Property 2 in Table 11-3.)

Take two pencils and arrange them to represent intersecting lines. How many possible ways can you arrange your piece of paper to contain both pencils? This experiment suggests that **two intersecting lines uniquely determine a plane.** (See Property 3 in Table 11-3.)

Now arrange your pencils so that they seem to be parallel. Again try to arrange your sheet of paper in different ways to contain the pencils. This leads to the conclusion that **two parallel lines uniquely determine a plane.** (See Property 4 in Table 11-3.)

Just as a point separates a line, a line separates a plane into three parts: the line and two half planes. The points on the line do not belong to either half plane. However, the line is called the **edge** or boundary of both half planes. For example, in Figure 11-11, the shaded region represents one half plane and the lightly shaded region represents the other half plane. The por-

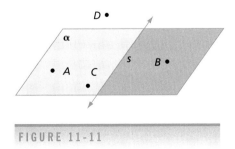

FIGURE 11-11

tion of plane α that contains point *A* is called the *A* side of line *s*, and the other half plane is called the *B* side of line *s*. If *A* and *B* were on the same side of line *s*, then segment $\overline{AB}$ would lie entirely in one half plane. Does $\overline{AB}$ in Figure 11-11 lie in one half plane? Why or why not? How about $\overline{AC}$? What is the relationship between the segment and the line when two points lie in opposite half planes?

Throughout this section, we have discussed points, lines, and planes in space, where space is the set of all points in three dimensions. Recall that a point separates a line into three disjoint sets: the point and two half lines on each side of the point. A line separates a plane into three disjoint sets: the line and the two half planes determined by the line. Similarly, a plane separates space into three disjoint sets: the plane and the two half spaces determined by the plane.

# Just for Fun

Can you form four equilateral triangles with six matches of the same length?

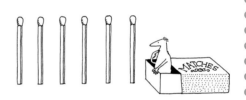

# EXERCISE SET 1

*R* 1. Which geometric concepts are suggested by these physical situations?

   (a) A piece of rubber band stretched between two points in space
   (b) A pair of opened scissors
   (c) The wall of your bedroom
   (d) A straightened paper clip
   (e) The tip of a needle
   (f) The blackboard
   (g) The spokes of a bicycle wheel

2. Draw sketches showing the following intersections.

   (a) A line and a ray that intersect in one point
   (b) A line and a line segment that intersect in the line segment itself
   (c) A plane and a line segment that intersect in one point
   (d) Two lines that neither intersect nor are parallel
   (e) A line segment and a ray that intersect in an endpoint

3. Imagine four points *A*, *B*, *C*, and *D*, no three of which are collinear. How many lines do they determine if they are coplanar—that is, if they lie in the same plane?

4. Consider the points and lines shown in the following figure.

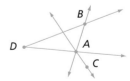

Find each of the following.

   (a) A segment containing point *B*
   (b) A ray with point *A* as the endpoint
   (c) Two rays with point *D* as the endpoint
   (d) A ray with point *B* as the endpoint
   (e) Two lines
   (f) Three segments containing point *A*

5. Using the following figure, classify the named pairs of lines as parallel, skew, or intersecting.

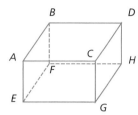

(a) $\overleftrightarrow{AB}$ and $\overleftrightarrow{CD}$     (b) $\overleftrightarrow{FH}$ and $\overleftrightarrow{AB}$
(c) $\overleftrightarrow{AB}$ and $\overleftrightarrow{FB}$     (d) $\overleftrightarrow{AE}$ and $\overleftrightarrow{DH}$
(e) $\overleftrightarrow{AE}$ and $\overleftrightarrow{BD}$     (f) $\overleftrightarrow{EG}$ and $\overleftrightarrow{HG}$

6. Label each statement as true or as false. Rewrite the false statements as true statements.

   (a) A line has an infinite number of different line segments that are subsets of it.
   (b) Space is the set of all points.
   (c) A ray has two endpoints.
   (d) A line segment contains infinitely many points.
   (e) $\overrightarrow{AB}$ is the same set of points as $\overrightarrow{BA}$.
   (f) If $A$, $B$, and $C$ are collinear with $B$ between $A$ and $C$, then $\overrightarrow{AB} \cap \overrightarrow{BC} = \overline{BC}$.
   (g) In part (f), $\overrightarrow{AB} \subset \overrightarrow{BC}$.
   (h) A ray separates a plane.
   (i) Two distinct lines either cross in a point or are parallel.
   (j) A half line separates a plane.
   (k) The endpoints of $\overline{BA}$ are $B$ and $A$.
   (l) $\overline{AB}$ is the same line segment as $\overline{BA}$.
   (m) A line has one endpoint.
   (n) If $A$, $B$, and $C$ are collinear points, where $\overline{AB} \cap C = \varnothing$, then $\overline{AB} \subset \overrightarrow{CA}$.
   (o) For any points $A$ and $B$, $\overrightarrow{AB} = \overrightarrow{BA}$.
   (p) If two lines are parallel to a third line, they are parallel to each other.

7. Are the following lines in space sometimes, always, or never coplanar?

   (a) Two lines
   (b) Two intersecting lines
   (c) Two parallel lines
   (d) Two skew lines

8. Classify the following pairs of segments as parallel or not parallel.

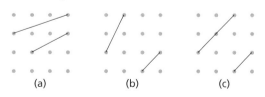

9. Consider three collinear points $D$, $E$, and $F$, with $E$ between $D$ and $F$.

   (a) Do $\overrightarrow{DE}$ and $\overrightarrow{ED}$ name the same ray?
   (b) Do $\overline{DE}$ and $\overline{ED}$ name the same line segment?
   (c) Do $\overrightarrow{DE}$ and $\overrightarrow{EF}$ have points in common?
   (d) Is each point on $\overrightarrow{EF}$ also on $\overrightarrow{ED}$?
   (e) Do $\overleftrightarrow{DE}$ and $\overleftrightarrow{EF}$ name the same line?

10. Consider $l$, $m$, and $n$ as three distinct lines in a plane. These lines may be related to each other, as shown in the accompanying figure.

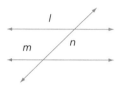

By sketching a figure, show other ways they may be related.

11. Using the accompanying line, determine each of the following, and tell which are empty.

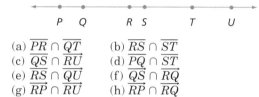

(a) $\overline{PR} \cap \overline{QT}$     (b) $\overline{RS} \cap \overline{ST}$
(c) $\overline{QS} \cap \overline{RU}$     (d) $\overrightarrow{PQ} \cap \overline{ST}$
(e) $\overrightarrow{RS} \cap \overrightarrow{QU}$     (f) $\overrightarrow{QS} \cap \overrightarrow{RQ}$
(g) $\overrightarrow{RP} \cap \overrightarrow{RU}$     (h) $\overrightarrow{RP} \cap \overrightarrow{RQ}$
(i) $\overline{RS} \cup \overline{QT}$

12. (a) Draw two segments $\overline{UV}$ and $\overline{XW}$ so that $\overline{UV} \cap \overline{XW}$ is empty but $\overleftrightarrow{UV} \cup \overleftrightarrow{XW}$ is one line.
    (b) Draw two segments $\overline{AB}$ and $\overline{CD}$ for which $\overline{AB} \cap \overline{CD}$ is not empty but for which $\overleftrightarrow{AB}$ is not the same as $\overleftrightarrow{CD}$.

13. In a plane, draw three lines and then four lines with the following number of intersections. Do you find that any of the drawings are impossible?

    (a) 0     (b) 1
    (c) 2     (d) 3
    (e) 4     (f) 5

*T* 14. Line $l$ contains the four points $A$, $B$, $C$, and $D$.

Name two rays on $l$ that have the following characteristics.

    (a) Their intersection is a ray.
    (b) Their intersection is a point.
    (c) Their intersection is a segment.
    (d) Their intersection is empty.

15. In Exercise 14, name two line segments such that

    (a) their intersection is one of the segments.
    (b) their intersection is $B$.
    (c) their intersection is $\overline{BC}$.
    (d) their intersection is empty.

16. Draw a figure to demonstrate each of your answers to the following.

   (a) Is it possible for a ray to be a proper subset of another ray?

   (b) Is it possible for the union of two rays to be a segment?

17. Do a point and a line necessarily determine a plane, regardless of where the point is located?

18. Classify the following pairs of rays as parallel or not parallel.

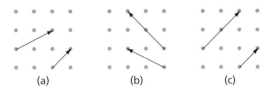

      (a)         (b)         (c)

19. The figure below represents a square pyramid with base on plane *ACB*.

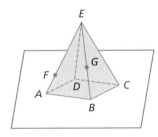

   (a) Name all the line segments that form the edges of the pyramid.

   (b) Which pairs of edges are part of parallel lines?

   (c) Which pairs of edges are part of skew lines?

20. Answer the following questions about the figure from Exercise 19.

   (a) Are the points *A*, *F*, and *D* collinear?

   (b) Are the points *B*, *C*, *D*, and *G* collinear?

   (c) Name five lines skew to a line drawn through *F* and *G*.

*C* 21. If all points are in one plane and no three points are collinear, how many lines are determined by the following?

   (a) 2 points     (b) 3 points

   (c) 4 points     (d) 5 points

   (e) 6 points     (f) *n* points

22. How many rays are determined by

   (a) three collinear points?

   (b) four collinear points?

   (c) five collinear points?

   (d) *n* collinear points?

23. If all the lines are in the same plane, each line intersects each of the other lines, and no point is common to three lines, how many points of intersection are determined by the following?

   (a) 2 lines     (b) 3 lines

   (c) 4 lines     (d) 5 lines

   (e) 6 lines     (f) *n* lines

24. Let *P* be a point not on line *m*. How many lines may be drawn through *P* intersecting *m*? How many planes contain *P* and *m*?

25. Two intersecting lines partition a plane into at most four disjoint regions, while three lines (no two of which are parallel) partition the plane into at most seven regions. Determine a pattern and complete the table.

| *Number of Lines* | 1 | 2 | 3 | 4 | 5 | ... | 10 |
|---|---|---|---|---|---|---|---|
| *Number of Regions* | 2 | 4 | 7 | | | | |

26. Using terms such as *segments* and *half lines,* name the following sets, where *x* represents points on a number line. ( $\leq$ means less than or equal to.)

   (a) $\{x \mid x > 1\}$

   (b) $\{x \mid x \leq -2\}$

   (c) $\{x \mid x \geq 3\} \cap \{x \mid x \leq 5\}$

   (d) $\{x \mid x \leq 3\} \cup \{x \mid x \geq 2\}$

27. Draw the intersection of the following figures so that the intersection is a line segment.

   (a) Two line segments

   (b) A ray and a line segment

   (c) Two rays

*For Exercises 28 through 32, determine all the different possible arrangements of the geometric objects, draw a picture of one such arrangement, and describe it using the terms introduced in this section.*

28. Three points in a plane

29. Four points in a plane

30. Three lines in a plane

31. Four lines in a plane

32. Three lines in space

# SECTION 2 LINES, PLANES, AND ANGLES

| | |
|---|---|
| **PROBLEM** | Complete each statement using *always, sometimes,* or *never.*<br>1. The sides of an angle are _____ rays.<br>2. A line and a plane that do not intersect are _____ parallel.<br>3. Adjacent angles _____ have a common side.<br>4. If the intersection of two planes is empty, then the two planes are _____ parallel.<br>5. A line perpendicular to a plane _____ intersects the plane in one point. |
| **OVERVIEW** | If your answer to each of the preceding questions was *always,* you already understand some of the concepts of this section. An intuitive understanding of the relationship between lines and planes, the relationship between planes and planes, and the characteristics of angles will be helpful as you pursue your study of geometry. |
| **GOALS** | Illustrate the Geometry Standard, page 482.<br>Illustrate the Representation Standard, page 50.<br>Illustrate the Problem Solving Standard, page 15.<br>Illustrate the Reasoning and Proof Standard, page 23. |

## RELATIONS AMONG LINES AND PLANES

Examine the walls of a room. Let the walls represent parts of planes. The intersection of two walls (if they intersect) is clearly a line segment. We generalize this idea by observing that if the intersection of two distinct planes is not empty, then the intersection is a line. (See Figure 11-12(a).)

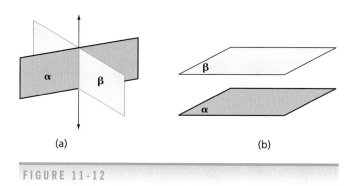

(a)                                                    (b)

FIGURE 11-12

Planes do not always intersect. Consider the opposite walls of a room as parts of planes. If these walls are accurately constructed, they will not intersect, no matter how far they are extended. Such planes are said to be *parallel.* (See Figure 11-12(b).) The top and bottom of a box represent parts of parallel planes.

**Parallel Planes**    If the intersection of two distinct planes is empty, then the planes are said to be *parallel*.

Planes α and β and planes α and γ in Figure 11-13 intersect along the dotted lines. However, plane β does not intersect plane γ because the two planes are parallel.

In the previous section, we considered how two lines might be related to each other. Let's consider now how a line and a plane might be related. There are three possibilities for the relationship between a line and a plane in space. (See Table 11-4.)

Place two tacks in a bulletin board and stretch a rubber band between them. Notice how the stretched rubber band seems to cling to the bulletin board. Intuitively, this result suggests that if a line contains two different points of a plane, the line lies in the plane. (See Figure 11-14(a).)

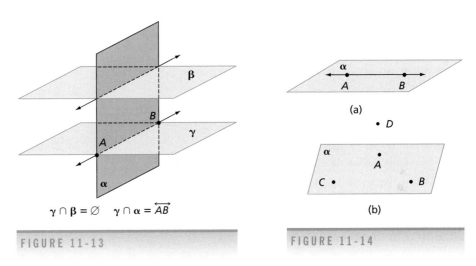

$\gamma \cap \beta = \varnothing \quad \gamma \cap \alpha = \overleftrightarrow{AB}$

FIGURE 11-13                                    FIGURE 11-14

A line can also intersect a plane or be parallel to a plane.

Points that lie in the same plane are said to be *coplanar*. Points *A*, *B*, and *C* in Figure 11-14(b) are coplanar. However, points *A*, *B*, *C*, and *D* are not coplanar. Lines and rays that lie in the same plane are also termed *coplanar*.

 ANGLES

We consider now two special rays that are coplanar. If these rays have a common endpoint, they define an *angle*.

**Angle**    An *angle* is the union of two rays that have the same endpoint.

**TABLE 11-4**

**Properties of a Line and a Plane**

| *Property* | *Illustration* |
|---|---|
| 1. A line and a plane may intersect in exactly one point. | Line *l* intersects plane α in point *A*. Figure 11-15 (a) |
| 2. The entire line may lie in the plane. | Plane γ contains line *m*. Figure 11-15 (b) |
| 3. The line may be parallel to the plane. | Line *n* is parallel to plane α. Figure 11-15 (c) |

(a)

(b)

(c)

**FIGURE 11-15**

In Figure 11-16(a), the rays $\overrightarrow{AB}$ and $\overrightarrow{AC}$ are called the **sides** of the angle; the common endpoint *A* is called the **vertex.** Although an angle is an abstract idea and thus cannot actually be drawn, a figure like the one in Figure 11-16(a) can be used as a visual representation of an angle. An angle may be named in at least three ways:

1. By using the symbol ∠ with the vertex, such as ∠*A*. (Use this notation only if point *A* is the vertex of exactly one angle in the figure.)
2. By using the symbol ∠ and three points *A*, *B*, *C* with *B* and *C* on different sides, such as ∠*BAC* or ∠*CAB*. The vertex point is indicated by the middle letter.
3. By using the symbol ∠ and a letter of the alphabet (such as *x* or *y*) or a Greek letter (such as ∠θ) or a numeral (such as ∠1). In a drawing, the letter is placed inside the angle near the vertex, as shown in Figure 11-16(a).

In Figure 11-16(a), if $\overrightarrow{AB}$ were rotated about point *A* until $\overrightarrow{AB}$ coincided with $\overrightarrow{AC}$, we would say that $\overrightarrow{AB}$ had been rotated through angle θ. We now need a unit of measure for such an angle.

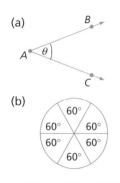

**FIGURE 11-16**

## MEASURING ANGLES

Over 2500 years ago, the Babylonians used a number system (and a measuring system) based on the number 60. The Babylonians divided a circle into 6 equal parts, as in Figure 11-16(b), and each of these 6 equal parts into 60 equally spaced units. Thus they divided the circle into 6(60) = 360 equally spaced units, now called **degrees.**

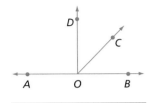

FIGURE 11-17

To further understand the idea of angle measure, let's consider $\overrightarrow{OB}$ in Figure 11-17 as rotating counterclockwise until it coincides with $\overrightarrow{OA}$. $\overrightarrow{OB}$ has rotated through a half plane to coincide with $\overrightarrow{OA}$. ($\overleftrightarrow{AB}$ divides the plane into two half planes.) Therefore, the measure of the interior of ∠BOA, (which we abbreviate as the measure of ∠BOA) is $\frac{1}{2} \cdot 360° = 180°$. Consequently, the other angles such as ∠$BOC$ possess measures of between 0° and 180°. For example, in Figure 11-17, ∠$BOD$ (which seems to have half of the measure of ∠$BOA$) has a measure of 90° and is called a **right angle.**

A protractor is an instrument commonly used for measuring the interior of angles. In Figure 11-18, each hash mark indicates an angle of 5°. Find the arrow or dot in the middle of the bottom of the protractor (which is usually indicated by an *O*). Place that dot on the vertex of the angle to be measured with the 0° line of the protractor on one side of the angle. The other side crosses the protractor at some line, indicating the measure of the interior of the given angle. In Figure 11-18, ∠$AOB$ has a measure of 20°; ∠$AOC$ has a measure of 120°; and ∠$COD$ has a measure of 45°. The fact that ∠$COD$ has a measure of 45° is sometimes denoted by "$m(\angle COD) = 45°$."

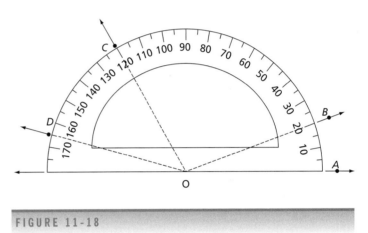

FIGURE 11-18

Not only are there 360 degrees in a full rotation, each degree is subdivided into 60 minutes and each minute into 60 seconds. Thus 1 degree (1°) = 60 minutes (60′) and 1 minute = 60 seconds (60″). To add or subtract degree measures we use a regrouping process similar to those we used in our algorithms for whole numbers. In this case, however, we regroup groups of 60 instead of groups of 10.

**EXAMPLE 1**

(a) Add $20°34'46'' + 10°40'32''$.

(b) Subtract $48°21'36'' - 22°36'40''$

*SOLUTION*

$20°34'46''$
$10°40'32''$
$\overline{30°74'78''} = 30°75'18''$    1′ = 60″
          $= 31°15'18''$    1° = 60′

Since 1° = 60′ and 1′ = 60″
$48°21'36'' = 47°81'36'' = 47°80'96''$
$47°80'96''$
$\underline{- 22°36'40''}$
$25°44'56''$

Angles are commonly classified according to the properties of their measures (see Table 11-5).

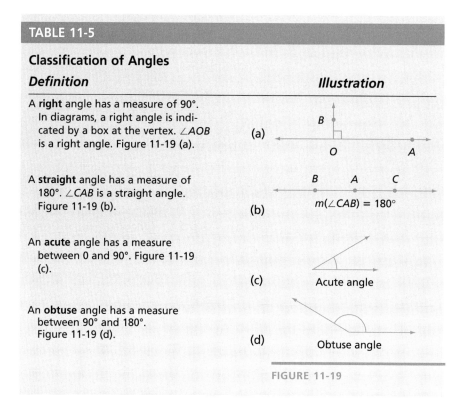

**TABLE 11-5**

**Classification of Angles**

*Definition*                                                                                                 *Illustration*

A **right** angle has a measure of 90°. In diagrams, a right angle is indicated by a box at the vertex. ∠AOB is a right angle. Figure 11-19 (a).

(a)

A **straight** angle has a measure of 180°. ∠CAB is a straight angle. Figure 11-19 (b).

(b)   $m(\angle CAB) = 180°$

An **acute** angle has a measure between 0 and 90°. Figure 11-19 (c).

(c)   Acute angle

An **obtuse** angle has a measure between 90° and 180°. Figure 11-19 (d).

(d)   Obtuse angle

**FIGURE 11-19**

# PERPENDICULAR LINES

Lines, rays, and line segments that form right angles are said to be **perpendicular**. In the illustration of a right angle in Figure 11-19(a), $\overrightarrow{OB}$ is perpendicular to $\overleftrightarrow{AO}$. Each long side of this page is perpendicular to each short side. If lines $r$ and $s$ are perpendicular, this relationship is denoted by $r \perp s$.

If a line and a plane intersect, it is possible for them to be perpendicular to each other. This occurs if and only if the line is perpendicular to every line in the plane that passes through the point of intersection.

The walls of a typical room that join in a corner are parts of planes that are perpendicular, and each wall of a room is perpendicular to the floor. Mathematically we say that two planes are perpendicular if and only if one plane contains a line perpendicular to the other plane.

# OTHER RELATIONS AMONG ANGLES

For some definitions, we need the concept of the interior of an angle. The interior of ∠BAC shaded in Figure 11-20(c) is the intersection of the half plane on the C side of $\overleftrightarrow{AB}$ in (a) and the half plane on the B side of $\overleftrightarrow{AC}$ in (b). $D$ is in the **interior** of ∠BAC, and $E$ is in the **exterior.**

Relationships between two angles are described in Table 11-6.

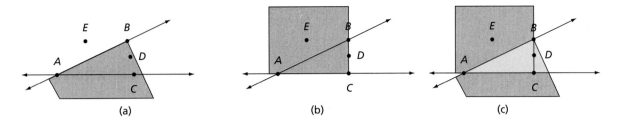

FIGURE 11-20

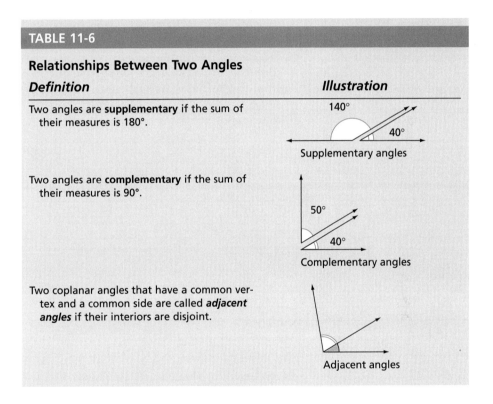

**TABLE 11-6**

**Relationships Between Two Angles**

| *Definition* | *Illustration* |
|---|---|
| Two angles are **supplementary** if the sum of their measures is 180°. | Supplementary angles |
| Two angles are **complementary** if the sum of their measures is 90°. | Complementary angles |
| Two coplanar angles that have a common vertex and a common side are called *adjacent angles* if their interiors are disjoint. | Adjacent angles |

Figure 11-21 shows the intersection of two lines. Observe that four angles—$\angle AOB$, $\angle AOC$, $\angle COD$, and $\angle DOB$—are formed by the intersection of these two lines. Because the sum of the measures of $\angle AOB$ and $\angle AOC$ is 180°, these angles are supplementary. Further, since $\angle AOB$ and $\angle AOC$ have a common side $\overrightarrow{OA}$, share a common vertex $O$, and have disjoint interiors, they are supplementary adjacent angles.

**DEFINITION**

**Vertical Angles**   The pairs of nonadjacent angles formed by the intersection of two lines are called *vertical angles*.

In Figure 11-21, $\angle COA$ and $\angle DOB$ are not adjacent angles. The non-adjacent angles formed by the intersection of two lines are called **vertical angles.** Angles $COD$ and $AOB$ are vertical angles.

In general, there are always two pairs of angles with the same measure formed when two lines intersect. In Figure 11-22, let m($\angle CBA$) = $x°$. Since $\angle CBA$ and $\angle ABE$ are adjacent and together form a straight line, m($\angle ABE$) = 180° − $x°$.

$$\text{m}(\angle DBE) + \text{m}(\angle ABE) = 180°$$

$$\text{m}(\angle DBE) + 180° − x° = 180°$$

$$\text{m}(\angle DBE) = x° = \text{m}(\angle CBA)$$

Vertical angles have the same measure, and angles having the same measure are called **congruent angles.** As we shall discuss later, line segments having the same measure are also said to be congruent.

Suppose that two lines intersect and form a right angle, as shown in Figure 11-23. Now $\angle PQR$ and $\angle PQS$ are supplementary angles; so m($\angle PQS$) = 90°. Likewise, m($\angle RQT$) and m($\angle TQS$) equal 90°. Therefore, all four angles are right angles.

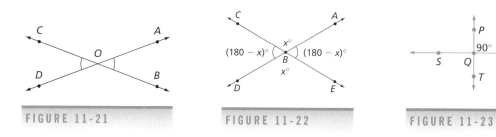

FIGURE 11-21          FIGURE 11-22          FIGURE 11-23

## TRANSVERSALS

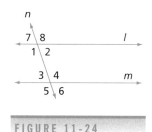

FIGURE 11-24

Any line that intersects a pair of lines in exactly two points is called a **transversal** for those lines. In Figure 11-24, line $n$ is a transversal of lines $l$ and $m$. For this transversal, angles 1, 2, 3, and 4 are called *interior* angles, whereas angles 5, 6, 7, and 8 are called *exterior* angles. $\angle 1$ and $\angle 4$ and $\angle 2$ and $\angle 3$ are pairs of **alternate interior angles.** $\angle 5$ and $\angle 8$ and $\angle 6$ and $\angle 7$ are pairs of **alternate exterior angles.** $\angle 6$ and $\angle 2$, $\angle 4$ and $\angle 8$, $\angle 5$ and $\angle 1$, and $\angle 3$ and $\angle 7$ are called **corresponding angles.** Recall that $\angle 3$ and $\angle 6$, $\angle 4$ and $\angle 5$, $\angle 1$ and $\angle 8$, and $\angle 7$ and $\angle 2$ are called *vertical angles.*

The following two properties state important relationships between parallel lines and the angles formed by their transversals.

**Properties of Angles and Transversals**

1. If lines $l$ and $m$ in a plane are intersected by a transversal in such a way that alternate interior angles are of equal measure, then $l$ is parallel to $m$.
2. Alternate interior angles formed by two parallel lines and a transversal are of equal measure.

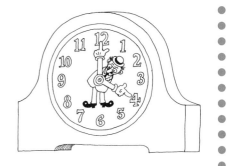

**FIGURE 11-25**

Using the fact that pairs of vertical angles are of equal measure, we can determine that several of the angles formed by a transversal of parallel lines are of equal measure. In Figure 11-24 if $l$ is parallel to $m$, $\angle 2$ and $\angle 3$ have the same measure because they are alternate interior angles. $\angle 2$ and $\angle 7$ have equal measures because they are vertical angles. Therefore, from the transitive property of equality, $\angle 7$ and $\angle 3$ have the same measure. In fact, corresponding angles formed by a transversal between parallel lines always have the same measure.

**PRACTICE PROBLEM**

Suppose that in Figure 11-25, $l \parallel n$ and $\mathrm{m}(\angle 1) = 70°$. Find each of the following angle measures.
(a) $\mathrm{m}(\angle 2)$      (b) $\mathrm{m}(\angle 3)$      (c) $\mathrm{m}(\angle 8)$

*ANSWER*      (a) $110°$      (b) $70°$      (c) $110°$

# Just for Fun

At four o'clock, what is the measure of the angle formed by the hands of the clock? What is the measure at three o'clock?

# EXERCISE SET 2

$R$ 1. Carefully fold a piece of paper in half. Hold the paper so that the fold is on a tabletop but neither side of the paper lies on the tabletop. Consider the three planes: the plane of the tabletop and the two planes of the two parts of the folded paper. What is the intersection of these three planes? Now stand the folded paper on its end. What is the intersection of the three planes?

2. Classify the following statements as always true, sometimes true, or never true.

(a) Two distinct planes intersect in a line.

(b) The intersection of two distinct lines determines a unique plane.

(c) If two parallel planes are cut by a third plane, the lines of intersection are parallel.

(d) If a plane contains one and only one of two parallel lines, it is parallel to the other line.

(e) Given a plane and a point not on the plane, there is exactly one plane through the given point parallel to the given plane.

(f) A line parallel to each of two intersecting planes is parallel to the line of intersection of the planes.

3. Give three names other than ∠A for the given angle. Use a protractor to find the measure of this angle.

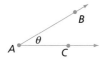

4. Classify each of the following statements as either true or false.

    (a) Adjacent angles are complementary.
    (b) Vertical angles have equal measures.
    (c) Adjacent angles must have a common vertex.
    (d) Any two intersecting lines form adjacent angles.
    (e) Any two right angles have equal measures.
    (f) Two obtuse angles cannot be adjacent.

5. Are angles *x* and *y* adjacent in the following figures? If not, explain.

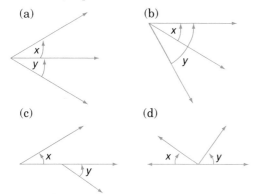

6. Find the following from the accompanying figure. (To label a plane, use three points.)

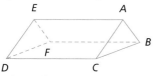

    (a) A pair of parallel planes
    (b) Two intersecting planes
    (c) Three planes that intersect in a point
    (d) A line parallel to a plane
    (e) A line that intersects a plane in one point
    (f) Two skew lines.

7. Each of two intersecting lines is parallel to a given plane. Is the plane determined by these lines parallel to the given plane?

8. Planes α and β are parallel and planes β and γ are parallel. Is α necessarily parallel to γ? Illustrate with a drawing.

9. Suppose that *l* is a line and *A* and *B* are points not on *l*.

    (a) How many planes contain *A* and *B* and at least one point of *l* if $\overrightarrow{AB}$ is parallel to *l*?
    (b) In general, how many planes contain *A* and *B* and at least one point of *l*?

10. If *l* is a line on plane α and *m* is a line on plane β, and if α and β are parallel, are *l* and *m* necessarily parallel?

11. Suppose that four points do not lie in the same plane. How many planes are determined by these points?

12. For each of the following sets of angles, tell which pairs of marked angles are adjacent and which are vertical.

    (a)             (b)

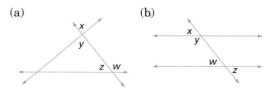

13. Make sketches that show the intersection of two angles as

    (a) a segment.    (b) empty.
    (c) a ray.        (d) exactly two points.
    (e) one point.    (f) exactly four points.

14. Find two adjacent angles and a vertical angle for each of the following angles in the accompanying diagram.

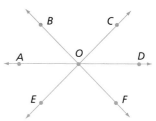

    (a) ∠*AOC*    (b) ∠*BOE*    (c) ∠*COD*

15. (a) If an angle has measure *x*, then its complements have measure _____ .
    (b) If an angle has measure *y*, then its supplements have measure _____ .
    (c) All supplements of an angle have the same _____ .
    (d) If two angles have the same measure and are supplementary, then the measure of each angle is _____ .

16. Given the following figure consisting of three parallel lines cut by a transversal, list the angles that have the same measure as ∠9, and explain.

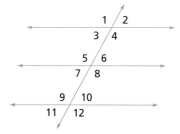

17. Identify each angle in the accompanying figure as acute, obtuse, or right.

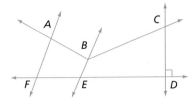

(a) $\angle FAB$     (b) $\angle ABC$     (c) $\angle CDE$

(d) $\angle EBC$     (e) $\angle FEB$     (f) $\angle AFD$

18. If $m(\angle COB) = 40°$ in the accompanying figure, find the following.

(a) $m(\angle AOB)$     (b) $m(\angle AOD)$     (c) $m(\angle COD)$

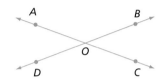

19. In the accompanying sketch, $m(\angle COB) = 35°$, and $\overrightarrow{EF}$ is perpendicular to $\overrightarrow{AC}$. Find the measure of each of the following angles.

(a) $\angle AOB$     (b) $\angle AOE$     (c) $\angle COD$

(d) $\angle DOF$     (e) $\angle AOD$     (f) $\angle FOC$

(g) $\angle EOB$     (h) $\angle FOB$

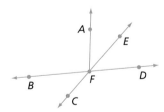

20. Complete the following statements about the given figure.

(a) Two angles adjacent to $\angle AFE$ are _____ and _____ .

(b) _____ and _____ are a pair of vertical angles.

(c) $\angle BFC$ and $\angle CFD$ are _____ and _____ angles.

(d) If $m(\angle BFC) = 40°$, then $m(\angle CFD) = $ _____ .

(e) If $m(\angle BFC) = 40°$, then $m(\angle EFD) = $ _____ .

(f) If $m(\angle CFD) = 120°$, then $m(\angle BFE) = $ _____ .

(g) Suppose that $\overleftrightarrow{AF} \perp \overleftrightarrow{BD}$. Then $m(\angle AFB) = $ _____ .

(h) Suppose that $\overleftrightarrow{AF} \perp \overleftrightarrow{BD}$ and $m(\angle AFE) = 30°$. Then $m(\angle EFD) = $ _____ .

21. Perform the following operations.

(a) $14°56'41'' + 8°24'33''$    (b) $18°14'8'' - 6°30'19''$

𝑇 22. (a) Planes $\alpha$ and $\beta$ have points $A$, $B$, and $C$ in common. Suppose that $A$, $B$, and $C$ are non-collinear. What conclusion can you draw about $\alpha$ and $\beta$?

(b) If $\alpha$ and $\beta$ are distinct planes having points $A$, $B$, and $C$ in common, what conclusion can you draw about points $A$, $B$, and $C$?

23. Complete the blank with *no, exactly one,* or *at least one.*

(a) Given any line $l$ and any point $P$ not on $l$, there is _____ plane parallel to $l$ and containing $P$.

(b) Given any plane $\alpha$ and any point $P$ not on $\alpha$, there is _____ plane containing $P$ and intersecting $\alpha$.

(c) Line $l$ is parallel to plane $\alpha$. There is _____ line in $\alpha$ skew to $l$.

(d) Given any plane $\alpha$ and any line $l$ parallel to $\alpha$, there is _____ plane containing $l$ and intersecting $\alpha$.

(e) Given any plane $\alpha$ and any point $P$ not on $\alpha$, there is _____ line parallel to $\alpha$ and containing $P$.

24. Each of two planes intersects a third plane in parallel lines. Are the two planes necessarily parallel? Illustrate with a drawing.

25. If you use point $A$ as the vertex, how many right angles can you draw at $A$ so that the sides contain at least one point of the lattice other than $A$?

```
·   ·   ·   ·   ·

·   ·   ·   ·   ·

·   ·   ·   ·   ·
        A
·   ·   ·   ·   ·
```

26. Two parallel planes intersect a third plane in lines $l$ and $m$. Are $l$ and $m$ necessarily parallel? Show with a drawing.

27. $\alpha$, $\beta$, and $\gamma$ are three distinct planes. Show with drawings the various ways the three planes can be related.

28. Is it possible for

(a) a line to be $\perp$ to a line in the plane but not be $\perp$ to the plane?

(b) a line not in the plane to be $\parallel$ to a line in the plane but not be $\parallel$ to the plane?

*C* 29. Consider the intersection of three lines in space.

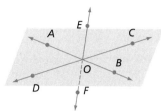

Two of these lines, $\overrightarrow{AB}$ and $\overrightarrow{CD}$, lie in the same plane; the third line, $\overrightarrow{EF}$, does not lie in their plane.

(a) How many angles are formed? Name them.

(b) Name six pairs of adjacent angles.

(c) How many pairs of angles are vertical? Name them.

30. In the accompanying figure, assume that m($\angle 4$) = m($\angle 8$). Prove that all the other pairs of corresponding angles are of the same measure.

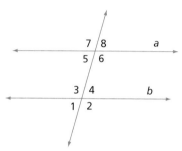

31. Suppose that you are given four points in space.

(a) What is the greatest number of lines determined by them? The greatest number of angles?

(b) Let the four points be in one plane. How many lines can be determined by them? Illustrate.

32. (a) Draw three noncollinear points on your paper, and label them *A*, *B*, and *C*. Connect the three with line segments. How many segments do you have?

(b) Measure $\angle ABC$, $\angle BCA$, and $\angle CAB$. What is the sum of the measure of these three angles?

(c) Repeat steps (a) and (b) for three other points.

(d) Make a conjecture as to the sum of the measures of the angles in the interior of a triangle.

## REVIEW EXERCISES

33. List three relationships between a line and a plane.

34. Can you take the wobble out of a table with four legs by shortening or lengthening one leg only? Explain.

35. List four ways to determine a plane.

36. How many lines are determined by four points, no three of which are collinear? How many line segments are determined by these four points?

37. Find the following.

(a) $\overrightarrow{AB} \cap \overrightarrow{CD}$    (b) $\overrightarrow{AB} \cap \overrightarrow{DC}$    (c) $\overrightarrow{CD} \cap \overrightarrow{DC}$

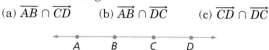

## LABORATORY ACTIVITIES

38. Let's look at a procedure for illustrating perpendicular lines. Using the accompanying figure as a model, fold the corner *A* until it assumes the marked position $A'$, and crease it to form line *l*. Fold the crease onto itself, *C* onto $C'$, to form line *m*. Now unfold. These two crease lines form a _____ angle.

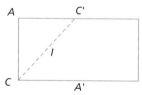

39. Draw a line $\overrightarrow{AB}$ containing point *C*. Use paper folding to show how to obtain a line perpendicular to $\overrightarrow{AB}$ at *C*. Describe your activity.

40. Draw a line $\overrightarrow{AB}$ and a point *C* not on $\overrightarrow{AB}$. Use paper folding to find a perpendicular line to $\overrightarrow{AB}$ through *C*. Describe your activity.

• C

41. A Mira is made of Plexiglas and reflects as a mirror, as well as being transparent. Study the accompanying figure to determine how to draw $\angle ABC'$ with the same measure as $\angle ABC$. Then use the Mira to find the bisector of $\angle BAC$. Describe your activity.

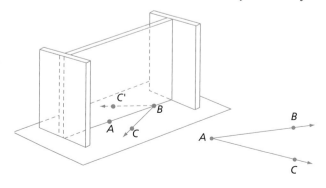

42. In this investigation you will use the *Geometer's Sketchpad* to discover relationships among the angles formed when parallel lines are intersected by a third line called a **transversal.**

A. Sketch

*Step 1:* Construct $\overleftrightarrow{AB}$ and point *C*, not on $\overleftrightarrow{AB}$.

*Step 2:* Construct a line parallel to $\overleftrightarrow{AB}$, through *C*.

*Step 3:* Construct $\overleftrightarrow{CA}$ and points *D*, *E*, *F*, *G*, and *H* as shown.

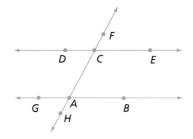

B. Measure the eight angles in your figure.

C. In the chart below, one example of each type of angle pair is given. Fill in the chart under pair 1 and under pair 2 with other angle pairs of that type, then state what relationship, if any, you observe between the angles in a pair. Note: there are more than two pairs of two of these types. Can you identify which types have more than two pairs?

| Angle Type | Pair 1 | Pair 2 |
|---|---|---|
| Corresponding | ∠FCE and | |
| Alternate Interior | ∠ECA and | |
| Alternate Exterior | ∠FCE and | |
| Vertical | ∠FCE and | |

D. Investigate

Drag point *F*, which moves $\overleftrightarrow{AC}$, and watch which angles stay equal. (Be careful not to change the point order on your lines—that will change the angles Sketchpad measures!) Do the relationships in the chart above still hold? Write your conjectures from your investigation.

E. Now drag point *E* or *B* so that the lines are no longer parallel. What do you discover about your conjectures in *D*?

F. Now explore the converse problems

(a) Set two corresponding angles equal. What about $\overleftrightarrow{AB}$ and $\overleftrightarrow{CD}$?

(b) Set two alternate interior angles equal. What about the two lines?

(c) Set two alternate exterior angles equal. What about the two lines?

# SIMPLE CLOSED CURVES

**PROBLEM**

Many interesting figures can be displayed on a geoboard. A geoboard is a board with rows of evenly spaced nails, as shown in the following figure.

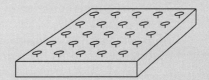

By stretching a rubber band around nails on the geoboard, see if you can form a rhombus; try an isosceles triangle.

**OVERVIEW**

Have you forgotten the meanings of *rhombus* and *isosceles triangle*? These terms will be defined in this section. It will be instructive for you to see how many figures introduced in this section can be illustrated by stretching a rubber band around nails in a geoboard. We shall discuss a number of familiar plane figures, such as the triangle and the square, and we shall discuss the angles associated with each. When we use phrases like "angle of a triangle" or "angle of a square," we mean, of course, the portion of the angle formed by the vertex and the lines containing the two sides of the figure. As you study these figures, observe that they bound regions in the plane.

**GOALS**

Illustrate the Geometry Standard, page 482.
Illustrate the Representation Standard, page 50.
Illustrate the Problem Solving Standard, page 15.
Illustrate the Connections Standard, page 147.

# SIMPLE CLOSED CURVES

Think of a piece of thread lying on a tabletop and of the many configurations into which it can be arranged. This image should serve as a model for a discussion of plane curves. A **plane curve** can be considered as a set of points that can be drawn without lifting the pencil. Thus lines, rays, line segments, and angles are all plane curves, as are the drawings in Figure 11-26.

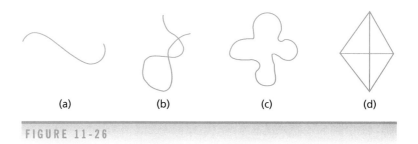

(a)　　　　(b)　　　　(c)　　　　(d)

FIGURE 11-26

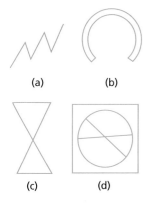

(a)　　　(b)

(c)　　　(d)

FIGURE 11-27

A curve is **simple** if, without lifting the pencil, we can draw it without retracing any of its points (with the possible exception of its endpoints). Drawings (a) and (c) in Figure 11-26 are simple; drawings (b) and (d) are not. A **closed curve** is drawn by starting and ending at the same point.

In Figure 11-27(a), the curve is simple but not closed; the diagram in (b) represents a simple closed curve; the diagram in (c) is closed but not simple; and the one in (d) is not a simple closed curve. Why not?

Simple closed curves have a common property. Can you discover it? No, it has nothing to do with size or shape. In fact, the property is so fundamental that you may have overlooked it. Every simple closed curve has an inside and an outside. Actually, a simple closed curve separates the points of the plane into three sets: the set of points on the curve, and two sets called the *interior* and the *exterior* of the curve. Although this property may seem obvious, a well-known theorem that states this result is quite difficult to prove.

| Jordan Curve Theorem | Any simple closed curve $c$ separates a plane into three disjoint sets of points called the *interior,* the *exterior,* and the *curve.* |
|---|---|

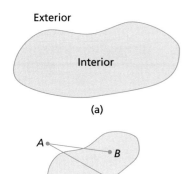

Exterior

Interior

(a)

(b)

FIGURE 11-28

Thus, for any simple closed curve $c$ in the plane, the plane is the union of three sets, no two of which intersect. The three sets are the set $c$, the interior of $c$, and the exterior of $c$. The union of a simple closed curve with its interior is called a **region.** (See Figure 11-28(a).)

Sometimes, when dealing with complicated simple closed curves, you may have difficulty determining whether a point is in the interior or the exterior of the curve. Using Figure 11-28(b) and a ruler, draw a line segment from $A$ to $B$. This segment cuts the closed curve in one point so $A$ (in exterior) and $B$ (in interior) are on opposite sides of the closed curve. Now draw segment $\overline{AC}$. This segment cuts the curve in two points. (Notice that $A$ and $C$ are both in the exterior.) Thus we may conjecture that, for two points $A$ and $B$ (one in the interior and one in the exterior), $\overline{AB}$ intersects the curve (excluding points of tangency) in an odd number of points. If $A$ and $B$ are both either in the interior or in the exterior, then $\overline{AB}$ intersects the curve (again, excluding points of tangency) in none or an even number of points. Check this conjecture on $A$ (in the exterior) and $B$, $C$, and $D$ for Figures 11-29(a) and (b).

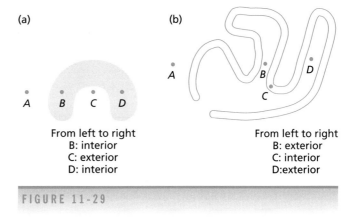

(a)

(b)

A    B    C    D

From left to right
B: interior
C: exterior
D: interior

A    B    C    D

From left to right
B: exterior
C: interior
D:exterior

FIGURE 11-29

Regions of a plane are classified as *convex* or *nonconvex.*

| DEFINITION | **Convex Region** | A region is *convex* if, for any two points in the region, the line segment joining them lies completely in the region. |
|---|---|---|

A region is **nonconvex** (sometimes called **concave**) if at least one segment that joins points of the region is not completely in the region. The region bounded by the closed curve in Figure 11-30(a) is convex, whereas the region in Figure 11-30(b) is nonconvex. Curves bounding convex regions are

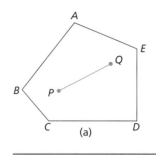

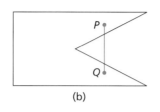

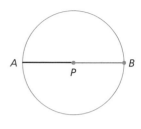

FIGURE 11-30

FIGURE 11-31

called **convex curves.** In this section we speak of convex polygons and concave polygons.

 CIRCLES

An example of a simple closed curve whose interior is a convex region is a circle. We define a circle using our intuitive concept of distance.

| DEFINITION | **Circle** | The set of all points in a plane at a given distance from a fixed point in the plane is a *circle*. |
|---|---|---|

In this definition of a circle, the fixed point is called the **center** of the circle, and any line segment from the fixed point to a point on the circle is called a **radius** of the circle. In Figure 11-31, a line is drawn through the center of the circle, intersecting the circle in points $A$ and $B$. The segment $\overline{AB}$ is called a diameter of the circle, and $\overline{PB}$ is called a radius ($\overline{PA}$ is also a radius).

 POLYGONS

Whereas all circles are convex curves, polygons may be either convex or concave.

| DEFINITION | **Polygon** | A *polygon* is a simple closed curve that is the union of three or more line segments, $\overline{AB}, \overline{BC}, \overline{CD}, \dots, \overline{PQ}$, such that the points $A, B, C, D, \dots, P, Q$ are coplanar and distinct, and no three consecutively named points are collinear. |
|---|---|---|

Angle,
vertex angle, or
interior angle
(a)

Central angle
of a
regular polygon
(b)

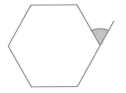

Exterior angle
(c)

**FIGURE 11-33**

If the segments described in the definition intersect in a point or points in addition to their endpoints, the polygon is not a simple polygon. (See Figure 11-32(b).) In this book, we are interested in simple polygons. Figure 11-32(a) is a convex simple polygon, and Figure 11-32(c) is a concave simple polygon.

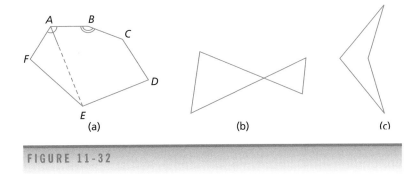

(a)   (b)   (c)

**FIGURE 11-32**

For the polygon given in Figure 11-32(a), $\overline{AB}$, $\overline{BC}$, $\overline{CD}$, $\overline{DE}$, $\overline{EF}$, and $\overline{FA}$ are called the **sides** or **edges** of the polygon $ABCDEF$. The points $A$, $B$, $C$, $D$, $E$, and $F$ are called the **vertices** of the polygon. **Adjacent vertices** are endpoints of the same side, and the **diagonals** of a polygon bounding a convex region are the line segments joining nonadjacent vertices such as $\overline{AE}$ in Figure 11-32(a).

Adjacent pairs of sides of a polygon are parts of angles of the polygon. **When we refer to the *angles* of a convex polygon, we mean the interior angles.** For example, $\angle FAB$ and $\angle ABC$ are two interior angles of the polygon in Figure 11-32(a). A polygon has the same number of interior angles as it has sides.

Three angles of interest in the discussion of polygons are shown in Figure 11-33.

| DEFINITION | **Regular Polygon** | A simple convex polygon with all sides of equal length and all angles of equal measure is called a *regular polygon*. Such a polygon is said to be **equilateral** and **equiangular**. |
|---|---|---|

A polygon is named according to the number of sides it has. Table 11-7 lists several types of polygons and provides illustrations of regular polygons.

 **TRIANGLES**

Several types of triangles are given special names because of the lengths of their sides. (See Figure 11-34.)

| TABLE 11-7 | | | |
|---|---|---|---|
| **Regular Polygons or *n*-gons (*n* sides)** | | | |
| *Name* | *Number of Sides* | *Number of Angles* | *Illustration* |
| Triangle | 3 | 3 | |
| Quadrilateral | 4 | 4 | |
| Pentagon | 5 | 5 | |
| Hexagon | 6 | 6 | |
| Heptagon or 7-gon | 7 | 7 | |
| Octagon | 8 | 8 | |

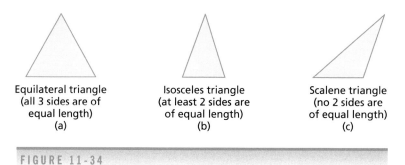

Equilateral triangle
(all 3 sides are of
equal length)
(a)

Isosceles triangle
(at least 2 sides are
of equal length)
(b)

Scalene triangle
(no 2 sides are
of equal length)
(c)

FIGURE 11-34

# QUADRILATERALS

Special types of quadrilaterals are listed in Table 11-8. A *trapezoid* can be defined as a quadrilateral with at least one pair of opposite sides parallel. Instead, we choose to use the definition of *trapezoid* that is found in most elementary school textbooks.

# ANGLES OF POLYGONS

We use the property of alternate interior angles to prove a very important property of angles associated with a triangle.

| TABLE 11-8 | |
|---|---|
| **Quadrilaterals** | |
| *Definition* | *Illustration* |
| A **parallelogram** is a quadrilateral with both pairs of opposite sides parallel. | |
| A **trapezoid** is a quadrilateral with exactly one pair of opposite sides parallel. | |
| A **rectangle** is a parallelogram with a right angle. (One right angle implies that all four angles are right angles.) | |
| A **square** is a rectangle with all sides of equal length. | |
| A **rhombus** is a parallelogram with all sides of equal length. | |
| A **kite** is a quadrilateral with two pairs of adjacent sides of equal length. | |

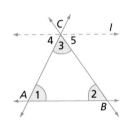

FIGURE 11-35

**EXAMPLE 1**

Show that the sum of the measures of the interior angles of a triangle is 180°.

*SOLUTION*

In triangle $ABC$ shown in Figure 11-35, draw line $l$ parallel to $\overrightarrow{AB}$. Mark $\angle 4$ and $\angle 5$. Parallel lines $l$ and $\overrightarrow{AB}$ are cut by transversal $\overleftrightarrow{CB}$, so alternate interior angles have equal measure; thus, $m(\angle 2) = m(\angle 5)$. Likewise, $m(\angle 1) = m(\angle 4)$. But $m(\angle 4) + m(\angle 3) + m(\angle 5) = 180°$. Why? Therefore, $m(\angle 1) + m(\angle 2) + m(\angle 3) = 180°$.

If the triangle in the preceding example is an equilateral (regular) triangle, then the measure of each angle is 60°.

**EXAMPLE 2**

Use the problem-solving strategy of considering simpler cases to look for a pattern that describes the number of degrees in each interior angle of a regular $n$-gon (polygon).

*SOLUTION*

We have already seen that the number of degrees in each angle of an equilateral triangle is

$$\frac{1 \cdot 180°}{3} = 60°$$

We can ascertain the sum of the measures of the interior angles of a square by drawing one of the two diagonals of the square. (See Figure 11-36(a).) Do you see that the sum of the angle measures of the square is equal to the sum of the angle measures of two triangles? Thus, the sum of the angle measures of a square is $2 \cdot 180°$ and, because there are four equal angles in the square, the measure of each angle is

$$\frac{2 \cdot 180°}{4} = 90°$$

(a)

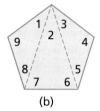

(b)

FIGURE 11-36

Next we consider the regular pentagon in Figure 11-36(b). By drawing two diagonals from the same vertex, we can divide the pentagon into three triangles, and we can label the resulting angles as $\angle 1$ through $\angle 9$. The sum of the measures of $\angle 1$, $\angle 8$, and $\angle 9$ is 180°, because these are the angles of a triangle. Likewise, the sum of the measures of $\angle 2$, $\angle 6$, and $\angle 7$ is 180°; and the sum of the measures of $\angle 3$, $\angle 4$, and $\angle 5$ is 180°. Thus, the sum of the measures of $\angle 1$ through $\angle 9$ is

$$3 \cdot 180° = 540°$$

But this is also the sum of the five interior angles of the pentagon. Moreover, because there are five interior angles, each interior angle is of measure

$$\frac{3 \cdot 180°}{5} = 108°$$

Now, by dividing a hexagon into four triangles, show that each interior angle is of measure

$$\frac{4 \cdot 180°}{6} = 120°$$

Now do you see the pattern?

Table 11-9 identifies the angle sums and vertex angle measures of regular $n$-gons.

When you connect the center of a regular polygon with a pair of adjacent vertices, you form a *central angle*. (See Figure 11-33(b).) Use the problem-solving strategy of considering simpler cases to show that the measure of the central angle of any regular $n$-gon is 360°/$n$. Start with a triangle. It has three central angles, so each is 360°/3 = 120°. Next consider a square; then a pentagon. Then find a pattern and generalize.

(a)

(b)

(c)

(d)

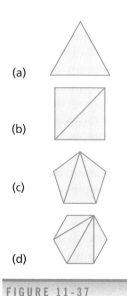

FIGURE 11-37

### TABLE 11-9

**Angle Measures of Regular *n*-gons**

| Number of Sides | Angle Sum | Measure of Vertex Angle | Illustration |
|---|---|---|---|
| 3 | $1 \cdot 180°$ | $\frac{(1 \cdot 180°)}{3} = 60°$ | Figure 11-37(a) |
| 4 | $2 \cdot 180°$ | $\frac{(2 \cdot 180°)}{4} = 90°$ | Figure 11-37(b) |
| 5 | $3 \cdot 180°$ | $\frac{(3 \cdot 180°)}{5} = 108°$ | Figure 11-37(c) |
| 6 | $4 \cdot 180°$ | $\frac{(4 \cdot 180°)}{6} = 120°$ | Figure 11-37(d) |
| $\vdots$ | $\vdots$ | $\vdots$ | |
| $n$ | $(n-2)180°$ | $\frac{(n-2)180°}{n}$ | |

Just for Fun

The unshaded set is a curve. Is it a simple curve?

# EXERCISE SET 3

*R* 1. Identify the figures that are plane curves, and classify these as simple or not simple.

   (a)      (b)      (c)

   (d)      (e)      (f)

   (g)   (h)      (i)

2. In Exercise 1, classify the simple plane curves as closed or not closed.

3. Which of the following figures are simple polygons?

   (a)      (b)      (c)

   (d)      (e)      (f)

4. Which of the regions in Exercise 3 are convex?

5. Draw if possible a polygonal region with the following characteristics.

   (a) Convex with four sides
   (b) Nonconvex with four sides
   (c) Convex with three sides
   (d) Nonconvex with three sides
   (e) Convex with two sides

6. Sketch a curve that is the union of line segments such that the curve is

(a) simple but not closed.
(b) simple and closed.
(c) a four-sided polygon bounding a convex region.
(d) a five-sided polygon bounding a nonconvex region.

7. Sketch a line that contains (a) exactly one point of some simple closed curve, (b) two points, (c) three points, (d) four points.

8. Draw two triangles whose intersection is

(a) one point.      (b) two points.
(c) three points.   (d) four points.
(e) five points.    (f) six points.

Are there other possible nonempty intersections?

9. If possible draw a circle that intersects a triangle in

(a) one point.      (b) two points.
(c) three points.   (d) four points.
(e) five points.    (f) six points.
(g) seven points.

10. An isosceles trapezoid is a trapezoid with two sides of equal length. On the following lattice diagram, draw an isosceles trapezoid, where one side (not one of the equal sides) is given.

11. Classify each of the following as true or as false.

(a) Every closed curve is simple.
(b) A square is a rhombus.
(c) The union of two segments cannot be a simple closed curve.
(d) A circle always bounds a convex region.
(e) A parallelogram is a rectangle with a right angle.
(f) An equilateral triangle has exactly two sides of equal measure.
(g) A square is a rectangle.
(h) An isosceles triangle can have a right angle.
(i) Every polygon has more than two sides.
(j) An angle is a simple closed curve.
(k) A rectangle is a regular polygon.

12. Choose from the given characteristics those that apply to each of the following figures.

(i)   Convex polygon
(ii)  Nonconvex or concave polygon
(iii) Regular polygon
(iv)  Plane curve
(v)   Simple curve
(vi)  Closed curve

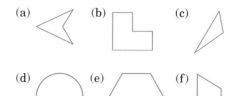

(a)    (b)    (c)

(d)    (e)    (f)

13. Consider the following polygons.

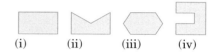

(i)    (ii)    (iii)    (iv)

(a) Which are convex?
(b) Which have a diagonal that intersects the exterior?
(c) Can we say, "A concave polygon is a polygon in which at least one diagonal intersects its exterior"?

14. For which of the quadrilaterals—A, rectangle; B, parallelogram; C, rhombus; and D, trapezoid—are the following properties always true?

(a) At least one pair of opposite sides parallel
(b) Opposite sides parallel
(c) Opposite sides of equal measure
(d) Opposite angles of equal measure
(e) All four sides of equal measure
(f) All four angles of equal measure
(g) All four sides and all four angles of equal measure

T15. (a) Can a right triangle also be an equilateral triangle?
(b) Can a right triangle also be an isosceles triangle?

16. Draw an equilateral hexagon that is not equiangular.

17. Draw an equiangular hexagon that is not equilateral.

18. Find the number of triangles in the following figure.

19. Name the regular $n$-gon that has this central angle?

(a) 10°     (b) 20°
(c) 30°     (d) 60°

*C* 20. What regular *n*-gons have each of the following angles as interior vertex angles?

(a) 162°      (b) 160°      (c) 150°      (d) 144°

21. How many diagonals do each of the following polygons have?

(a) Triangle
(b) Parallelogram
(c) Regular pentagon
(d) Regular hexagon
(e) Regular heptagon
(f) Regular *n*-gon

*In Exercises 22 and 23, let △XYZ denote the triangle with vertices X, Y, and Z.*

22. Let *T* be the union of *ABCD* with its interior (the shaded part of the figure). Which of the following are true?

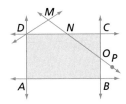

(a) △*NOC* ∩ *ABCD* = {*N, O*}
(b) $\overleftrightarrow{MP}$ ∩ *T* = $\overline{NO}$
(c) $\overline{MO}$ ⊂ *T*
(d) △*NOC* ⊂ *T*
(e) ∠*DAB* ⊂ *T*
(f) *D* is in the interior of ∠*ABC*.
(g) *M* ∈ $\overline{NO}$ ∪ $\overrightarrow{AD}$
(h) △*MDN* ∩ *T* = ∅

23. For the accompanying figure, describe each of these sets of points.

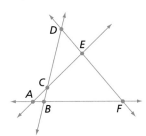

(a) △*DBF* ∩ $\overline{AC}$
(b) (Exterior of △*DBF*) ∩ $\overline{CE}$
(c) (Interior of △*DBF*) ∩ $\overleftrightarrow{CE}$
(d) (Interior of △*DBF*) ∩ (interior of △*CDE*)
(e) (Interior of △*DBF*) ∪ (interior of △*CDE*)
(f) (Interior of ∠*EAF*) ∩ (interior of △*CDE*)

24. *A* is in the exterior of this simple closed curve. Is *B* in the exterior or interior? *C*?

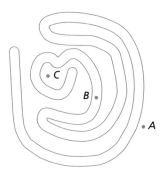

*Note: In Chapter 7 you studied the Pythagorean Theorem: The square of the length of the longest side of a right triangle is equal to the sum of the squares of the other two sides. The converse of this statement is also true. You will find these facts useful in Exercises 25 through 27.*

25. If one connects a lattice point with the endpoints of $\overline{CD}$, one forms a triangle.

(a) Find all right triangles that can be formed in this way.
(b) Find all isosceles triangles that can formed in this way.

26. Why is the following figure an isosceles triangle?

27. Why is the following figure a rectangle?

28. Use the fact that the sum of a vertex angle and the corresponding exterior angle is 180° to develop an expression for the measure of an exterior angle of a regular *n*-gon.

## REVIEW EXERCISES

29. Classify each statement as true or as false. If it is false, explain why.

(a) A line separates space.

(b) A point separates a line.

(c) For any two points $A$ and $B$, $\overrightarrow{AB} = \overrightarrow{BA}$.

(d) For any two points $A$ and $B$, $\overleftrightarrow{AB} = \overleftrightarrow{BA}$.

(e) If $\overrightarrow{AB} = \overrightarrow{AD}$, then $B = D$.

(f) Every hexagon is a polygon.

(g) Every square is a parallelogram.

(h) The sides of any regular polygon are of equal length.

(i) For a triangle to be isosceles, three angles must be of equal measure.

(j) If the sides of a polygon are of equal length, the polygon is a regular polygon.

(k) If 4 points are collinear, then they must be coplanar.

(l) If a line segment has the center of a circle for one endpoint and a point on the circle for the other endpoint, it is called a *diameter*.

(m) A circle is a polygon.

(n) A line could be called an infinite set of collinear points.

(o) The union of two half lines is a line.

(p) A segment separates a plane.

(q) Closed curves are simple curves.

(r) All straight lines are curves.

(s) A polygon has more than three sides.

(t) A plane region is a simple closed curve.

(u) Three points, if not all on the same line, determine a plane.

30. If possible draw an angle and a circle that intersect in exactly

(a) no points.

(b) one point.

(c) two points.

(d) three points.

(e) four points.

(f) five points.

(g) more than five points.

31. One acute angle of a triangle has a measure of $40°$. The difference in the other two angles is $20°$. Find the three angles.

## LABORATORY ACTIVITY

32. Using Geometer's Sketchpad place three points that could be the vertices of a triangle. Label them $A$, $B$, and $C$. Construct the sides from $A$ to $B$, from $B$ to $C$, and from $A$ to $C$. Measure the lengths of the three sides and the sizes of the three angles. Drag the vertices around and notice how the measures change. Make some scalene, isosceles, and equilateral triangles and note characteristics of the angle and side measurements.

## ⁘ PCR Excursion ⁘

33. In this excursion we will classify quadrilaterals by the properties of the diagonals. (A *diagonal* is a segment connecting two nonadjacent vertices of a quadrilateral.) The following is a list of properties that may or may not be true about the diagonals of a quadrilateral. Using *Geometer's Sketchpad*, construct several of each of the types of quadrilaterals listed and their diagonals. Tell which of the following properties each satisfies.

Property (a): The diagonals are equal in length.

Property (b): The angle between the diagonals is a right angle.

Property (c): The diagonals bisect each other.

Property (d): Only one diagonal is bisected.

| Quadrilaterals | Properties | | | |
| --- | --- | --- | --- | --- |
| | (a) | (b) | (c) | (d) |
| Rectangle | | | | |
| Parallelogram | | | | |
| Square | | | | |
| Rhombus | | | | |
| Trapezoid | | | | |
| Kite | | | | |

# SIMPLE CLOSED SURFACES

**PROBLEM**

The top view, the side view, and the end view of a three-dimensional figure formed by stacks of blocks are given as follows.

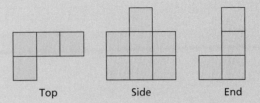

Top          Side          End

Can you visualize and draw the three-dimensional figure?

**OVERVIEW**

Simple closed surfaces are analogous to simple closed curves in a plane. Although you may find three-dimensional figures no more difficult to understand than plane figures, you may encounter some difficulty in drawing and visualizing them. In this section, you will learn to recognize common three-dimensional figures such as spheres, cubes, pyramids, and cylinders. Actually, a knowledge of these figures preceded most of modern mathematics. Over 5000 years ago, the civilization that occupied what we now know as Iran and Iraq used solid figures made from clay as parts of contracts in business transactions. These figures were called by different names, but archaeology has validated their existence.

**GOALS**

Illustrate the Geometry Standard, page 482.
Illustrate the Representation Standard, page 50.
Illustrate the Problem Solving Standard, page 15.
Illustrate the Reasoning and Proof Standard, page 23.
Illustrate the Communication Standard, page 78.

## VISUALIZING THREE-DIMENSIONAL FIGURES

FIGURE 11-38

One of the purposes of this section is to help you to visualize and draw three-dimensional figures. Let's piece together the three views shown in the introductory problem to make Figure 11-38. How did you know there was one block in front? (See the top and end views of the opening illustration.) How did you know there was one extra block in the middle of the back row, on top of the others? (See the side view.)

The world around us is three-dimensional, so it is important to have some understanding of three-dimensional geometry. We begin by examining and defining some well-known spatial figures.

The concept of a **simple closed surface** in space is analogous to the concept of a simple closed curve in a plane. Although we shall not attempt to define it rigorously, a simple closed surface cannot have holes through it and must separate the points of space into three disjoint sets of points: the set of points on the surface, the set of points interior to the surface, and the set of points exterior to the surface. Hence the surface in Figure 11-39(a) is a simple closed surface, and that in Figure 11-39(b) is not. We shall consider the union of the points on the surface and the points interior to the surface to be the **solid** bounded by the closed surface. We may also term this union of points a **space region.**

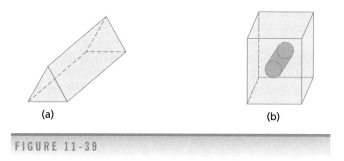

(a)          (b)

FIGURE 11-39

 POLYHEDRA

Let's begin with the types of space surfaces formed by polygonal regions. The general term for these space figures is *polyhedron.*

| | |
|---|---|
| **DEFINITION** | **Polyhedron** |

A *polyhedron* is a simple closed surface in space whose boundary is composed of polygonal regions.

Notice the different types of polyhedrons in Figure 11-40. In (a), a square bounds each polygonal region. In (b), the polygonal regions are bounded by triangles. In (c), the regions are bounded by pentagons. The polygonal regions that form the polyhedron are called *faces* of the polyhedron.

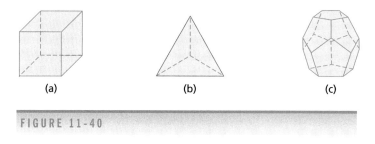

(a)          (b)          (c)

FIGURE 11-40

Now we consider a special kind of polyhedron, called a *prism.*

| DEFINITION | **Prism** | Polygons with the same size and shape are said to be congruent. A *prism* is a polyhedron formed by two congruent polygonal regions in parallel planes, along with three or more regions bounded by parallelograms joining the two polygonal regions so as to form a closed surface. |

The polygonal regions in parallel planes are called **bases** of the prism. The parallel edges joining the two bases are called **lateral edges,** and the regions bounded by parallelograms are called **lateral faces.** The bases and the lateral faces are called the **faces** of the prism. If the lateral faces are rectangular regions, and the planes containing the lateral faces are perpendicular to the planes containing the bases of the prism, then the prism is called a **right prism.**

A prism is named in terms of its bases. The **triangular prism** in Figure 11-41(a) resembles a wedge or a trough. The **quadrilateral prism** in (b) could be a closed shoe box or a room. Sometimes chalk or a pencil is shaped in the form of the **hexagonal prism** in (d).

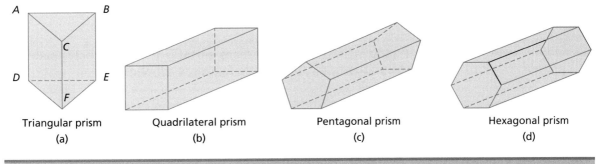

Triangular prism
(a)

Quadrilateral prism
(b)

Pentagonal prism
(c)

Hexagonal prism
(d)

FIGURE 11-41

Let us examine the triangular prism in Figure 11-41(a) in order to identify the important components of prisms.

1. The bases are the two triangular regions $\triangle ABC$ and $\triangle DEF$.
2. The lateral faces are $ACFD$, $ABED$, and $BCFE$.
3. The faces include both the lateral faces and the bases.
4. $\overline{BE}$, $\overline{CF}$, and $\overline{AD}$ are the lateral edges.
5. $\overline{AC}$, $\overline{BC}$, $\overline{AB}$, $\overline{DF}$, $\overline{FE}$, and $\overline{DE}$ are the sides of the bases.
6. Points $A$, $B$, $C$, $D$, $E$, and $F$ are called the **vertices** of the prism.

To get a better idea of the quadrilateral and pentagonal prisms, look at the patterns in Figure 11-42. These patterns can be cut out of cardboard and then folded and glued together to form the polyhedra.

**PRACTICE PROBLEM**

How many faces does a hexagonal prism have? How many edges? How many vertices?

*ANSWER*

8 faces, 18 edges, and 12 vertices

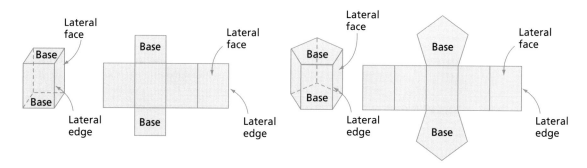

FIGURE 11-42

Another polyhedron of interest is the pyramid.

DEFINITION

**Pyramid**

A *pyramid* is a polyhedron formed by a simple closed polygonal region (called the **base**), a point (called the **vertex**) not in the plane of the region, and the triangular regions joining the point and the sides of the polygonal region.

In a **regular pyramid,** the base is a regular polygon, and the lateral faces are all congruent triangles. As a consequence the line joining the vertex and the center of the base is perpendicular to the base. A pyramid is classified according to the polygonal region that forms the base, as in Figure 11-43. When we think of pyramids, we usually think of the monumental quadrilateral pyramids of Egypt.

Again, a picture or a pattern is sometimes helpful in visualizing a pyramid. The surface pattern for a quadrilateral pyramid is shown in Figure 11-44.

(a)

Triangular pyramid

(b)

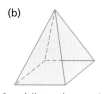

Quadrilateral pyramid

(c)

Pentagonal pyramid

FIGURE 11-43

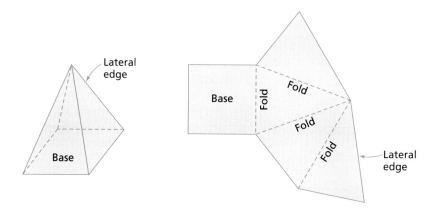

FIGURE 11-44

## EULER'S FORMULA

An unexpected relationship exists among the numbers of vertices, edges, and faces of polyhedra.

| | |
|---|---|
| **Euler's Formula** | Let $V$ represent the number of vertices, $E$ the number of edges, and $F$ the number of faces of a polyhedron. Then $$V + F - E = 2$$ |

Consider the triangular prism in Figure 11-41(a). Count the number of vertices, edges, and faces. Did you get 6 vertices, 9 edges, and 5 faces? Because $6 + 5 - 9 = 2$, Euler's formula is satisfied. Count the number of faces, edges, and vertices in the quadrilateral prism, and verify that the formula is satisfied here also.

**PRACTICE PROBLEM**    Verify Euler's formula for a pentagonal pyramid.

*ANSWER*    A pentagonal pyramid has 6 vertices, 6 faces, and 10 edges. By Euler's formula, $6 + 6 - 10 = 2$.

## REGULAR POLYHEDRA

Just as polygons are named according to their number of sides, so regular polyhedra are named according to their number of faces.

| | |
|---|---|
| **Regular Polyhedron** | If the faces of a convex polyhedron are congruent regular polygonal regions, and if each vertex is the intersection of the same number of edges, then the polyhedron is called a *regular polyhedron*. |

A regular **tetrahedron** is formed by 4 congruent triangular regions; a cube is formed by 6 congruent quadrilateral regions; an octahedron by 8 congruent triangular regions; a regular **dodecahedron** by 12 congruent pentagonal regions; and a regular **icosahedron** by 20 congruent triangular regions. Greek mathematicians (such as Theaetetus) discovered that these are the only possible regular polyhedra. Because of the prominence of Plato in the Greek world, they are sometimes called the *Platonic solids*. (See Table 11-10.)

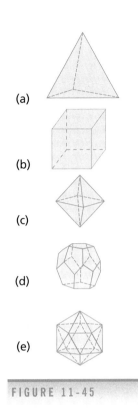

(a)

(b)

(c)

(d)

(e)

FIGURE 11-45

| TABLE 11-10 | | |
|---|---|---|
| **Regular Polyhedra** | | |
| *Name* | *Number of Faces* | *Illustrations* |
| Tetrahedron | 4 | Figure 11-45(a) |
| Hexahedron (cube) | 6 | Figure 11-45(b) |
| Octahedron | 8 | Figure 11-45(c) |
| Dodecahedron | 12 | Figure 11-45(d) |
| Icosahedron | 20 | Figure 11-45(e) |

## DIHEDRAL ANGLES

You may have observed that some of the angles in polyhedra are different from those studied previously. Whereas plane angles are formed by intersecting lines, dihedral angles are formed by intersecting planes. Thus two intersecting planes form four dihedral angles in the same manner that two intersecting lines form four plane angles.

| DEFINITION | **Dihedral Angle** | If two planes intersect, a *dihedral angle* is the union of two noncoplanar half planes and the line of intersection. |
|---|---|---|

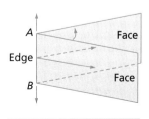

FIGURE 11-46

The two half planes are called the **faces** of the dihedral angle, and the common line is called the **edge** of the dihedral angle. Figure 11-46 illustrates a dihedral angle.

A plane angle formed by two rays, one in each face of a dihedral angle, with each ray having its endpoint on the edge of the dihedral angle and each ray perpendicular to the edge—is called a **plane angle of the dihedral angle.** The measure of a dihedral angle is the measure of the plane angle of the dihedral angle. Thus a right dihedral angle has a right plane angle.

# SPHERES, CYLINDERS, AND CONES

Can you visualize a set of points in space equidistant from a fixed point? This set of points can be considered as the union of infinitely many congruent circles, all having the same center. Such a three-dimensional surface is called a **sphere.** (See Figure 11-47.)

| | | |
|---|---|---|
| **DEFINITION** | **Sphere** | A set of all points in space at a given distance from a fixed point is called a *sphere*. |

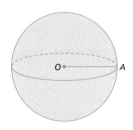

FIGURE 11-47

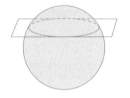

FIGURE 11-48

The fixed point of the sphere is called the **center** of the sphere, and any line segment from the center to a point on the sphere is called a **radius** of the sphere.

The **interior** of a sphere is the set of all points whose distance from the center of the sphere is less than the measure of the radius. The set of points whose distance from the center of the sphere is greater than the measure of the radius of the sphere is called the **exterior** of the sphere.

Consider the intersection of a sphere and a plane. If the intersection is not empty, it will be either a single point or a circle. (See Figure 11-48.) If the plane is **tangent** to the sphere, the intersection is a single point. Such a situation is intuitively understood by visualizing a ball resting on a flat surface.

Consider a circle and a fixed line that is not in the plane of the circle and is not parallel to the plane of the circle. The union of all lines through the circle that are parallel to the fixed line is called a *cylinder of infinite extent* (in some texts, simply a *cylinder*). A more general definition of a cylinder of infinite extent replaces the circle by a simple closed curve.

In this text we will use the term circular cylinder for the simple closed surface formed when we consider a second plane, parallel to the one mentioned in the previous paragraph. Each of the two planes intersect the cylinder of infinite extent in a circle. The union of the two resulting circular regions with the line segments from the cylinder of infinite extent that connect the circles forms a simple closed surface that we will call a **circular cylinder.** (See Figure 11-49)

A circular cylinder is analogous to a prism: It consists of bases that are congruent circular regions in parallel planes joined by parallel line segments.

In Figure 11-49(a), the cylinder is called a **right circular cylinder.** In such a cylinder, the line segment joining the centers of the bases is perpendicular to the bases. The circular cylinders in (b) and (c) are not right cylinders.

Another useful simple closed surface is the cone. A **circular cone** consists of a circular base and segments joining a fixed point (vertex) to points on the base. (See Figure 11-50(a).) In general, a cone can have any simple closed curve bounding the base. By this definition, pyramids are cones with polygonal bases. Circular cones have circular bases, and it is circular cones

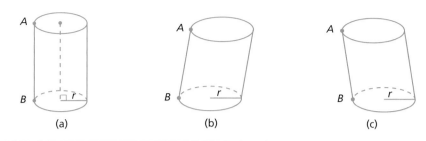

FIGURE 11-49

that we usually call *cones*. This is the terminology that will be used in this and succeeding chapters. The pattern in Figure 11-50(b) should help you to visualize the composition of a cone.

A circular cone in which the line segment from the vertex of the cone to the center of the circular base is perpendicular to the base is called a *right circular cone*. This is the type of circular cone shown in Figure 11-50(a).

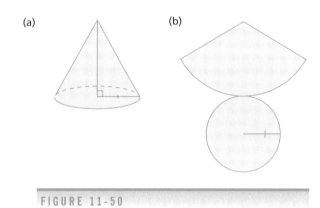

FIGURE 11-50

# Just for Fun

These are pictures of the same ABC block or letter block. What letter is opposite C? Is opposite E? Is opposite A?

# EXERCISE SET 4

*R* 1. Identify the following types of polyhedra. (Answers may not be unique.)

(a)    (b)

(c)    (d)

2. Verify Euler's formula for the figures in Exercise 1.

3. Answer true or false.

(a) A pyramid is a polyhedron.
(b) A pyramid is a prism.
(c) Every triangular prism is a right prism.
(d) Some polyhedra are prisms.
(e) The bases of a prism lie in perpendicular planes.
(f) The faces of a pyramid are parallelograms.
(g) A pyramid with seven faces has a hexagon as the boundary of its base.
(h) Euler's formula does not hold for pyramids.

4. Give the name for each spatial figure shown. List the segments that are lateral edges and the points that are vertices.

(a) 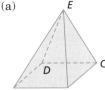   (b)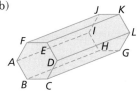

5. Draw a pyramid and a prism with each of the following as a base.

(a) Triangle
(b) Quadrilateral
(c) Pentagon
(d) Hexagon

6. For each of the following, what is the smallest number of faces possible?

(a) Prism
(b) Pyramid

7. Name the polyhedron that can be constructed by using each of the following patterns.

(a)    (b)

(c)    (d)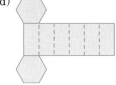

8. Sketch the simple closed curve you get when you intersect

(a) a plane and a sphere.

(b) a plane and a right circular cylinder with a plane parallel to the base of the cylinder.

(c) a plane and a right circular cone with a plane parallel to the base of the cone.

(d) a plane and a right circular cylinder with a plane not parallel to the base of the cylinder.

9. Indicate whether the following are true for a pyramid, a prism, or both.

   (a) Its lateral faces are parallelograms.
   (b) It has the same number of faces as vertices.
   (c) It has one base.
   (d) It always has an even number of vertices.
   (e) It always has an even number of lateral faces.
   (f) It can have as few as four faces.
   (g) It has two bases.
   (h) Its lateral faces are triangles.

10. Consider a prism with each of the following bases. How many lateral faces does it have? How many vertices? How many edges?

    (a) Triangle
    (b) Quadrilateral
    (c) Pentagon
    (d) Hexagon

11. Do Exercise 10 for a pyramid.

*T* 12. Verify Euler's formula for the number of faces, vertices, and edges of each spatial figure named.

    (a) A pentagonal pyramid
    (b) A hexagonal prism
    (c) A triangular prism
    (d) A quadrilateral pyramid
    (e) A hexagonal pyramid
    (f) A pentagonal prism
    (g) A quadrilateral prism
    (h) A triangular pyramid

13. Name the polyhedron that could be formed from each pattern below.

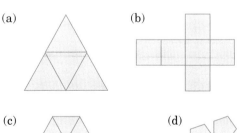

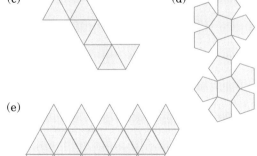

14. Complete the table, indicating the number of vertices, edges, and faces and confirming Euler's formula for each of the following regular solids.

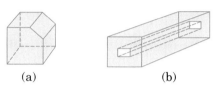

| Solid | V | E | F | V + F − E |
|-------|---|---|---|-----------|
| Tetrahedron | | | | |
| Cube | | | | |
| Octahedron | | | | |
| Dodecahedron | | | | |
| Icosahedron | | | | |

15. Does Euler's formula hold for each of the following?

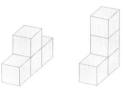

    (a)                    (b)

16. Four cubes glued together are called *Soma cubes*. Construct all possible Soma cubes, such as the two given. Draw pictures of your Soma cubes. How many did you construct?

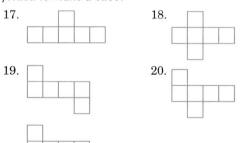

*In Exercises 17 through 21, which patterns can be folded to make a cube?*

17.                          18.

19.                          20.

21.

*C* 22. Draw the top, side, and end views of the following stacks of cubes.

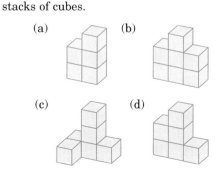

    (a)        (b)


    (c)        (d)

23. Count the number of vertices, edges, and faces on what is left of a quadrilateral prism after one corner is cut off. Does Euler's formula hold?

24. A quadrilateral prism has a tunnel cut through it in the form of a pentagonal prism. Find $V$, $E$, $F$, and $V + F - E$.

25. Generalize the answers for Exercise 10 for an $n$-sided base. Use your work to verify Euler's formula for all prisms.

26. Generalize the answers for Exercise 11 for an $n$-sided base. Use your work to verify Euler's formula for all pyramids.

27. The following are top (T), side (S), and end (E) drawings of stacks of cubes. Construct the spatial configurations.

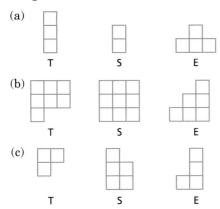

## REVIEW EXERCISES

28. Sketch
    (a) skew lines $r$ and $s$.
    (b) a plane that is parallel to line $r$ and intersects line $s$.
    (c) a plane containing line $r$, where $r$ and $s$ are skew lines.

29. Sketch
    (a) a convex hexagon.
    (b) a nonconvex quadrilateral.
    (c) a nonconvex triangle.
    (d) a nonconvex hexagon.
    (e) a nonconvex circle.

30. Consider the following four points.

Sketch a closed figure beginning and ending at $X$ and containing $A$, $B$, and $C$ such that
(a) the figure is convex.
(b) the figure is not a polygon.
(c) the figure is a convex polygon.

31. Determine whether $X$ is in the interior or the exterior of the closed curve.

32. Can $\overline{AC} \cap \overline{BD} = \varnothing$ and $\overleftrightarrow{AC} \cap \overrightarrow{BD}$ be a point? A line? A segment? Describe, in general, the various possible relationships that can exist between $\overline{AC} \cap \overline{BD}$ and $\overleftrightarrow{AC} \cap \overline{BD}$.

## LABORATORY ACTIVITIES

33. Enlarge the patterns in Exercise 13 and form the associated polyhedra.

34. Make a different pattern from that in the main text for
    (a) a quadrilateral pyramid.
    (b) an octahedron.
    (c) a tetrahedron.

35. Isometric dot paper allows us to represent a three-dimensional view. Draw two views of a $2 \times 2 \times 2$ cube.

### ❖ PCR Excursion ❖

36. A cube formed from white wood is painted black. Then it is cut into 27 smaller cubes, all of the same size. If a face of one of the small cubes was on the surface of the original cube, it is black. All other faces of the small cubes are white. What can we learn about the 27 small cubes?

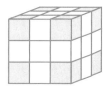

(a) What is the largest number of black sides on a cube?

(b) Where were these located in the original cube?

(c) What is the number of such cubes?

(d) Where in the original cube were the small cubes with two sides painted black?

(e) How many exist?

(f) Where in the original cube were the small cubes with one side black?

(g) How many exist?

(h) Where in the original cube were the small cubes with no black sides?

(i) How many exist?

(j) What is your sum of (c), (e), (g), and (i)?

(k) Explain how you would construct the given cube by using the pieces we have been discussing.

# SOLUTION TO INTRODUCTORY PROBLEM

**UNDERSTAND THE PROBLEM.** First, let's restate the problem. Cutting the cake is equivalent to using line segments (called *chords*) to divide a circle into regions. One chord divides the circle into 2 regions; 2 chords into a maximum of 4 regions; 3 chords into a maximum of 7; and 4 chords into a maximum of 11.

| Chords | Maximum Number of Regions | Increase |
|--------|---------------------------|----------|
| 1 | 2 | |
| 2 | 4 | 2 |
| 3 | 7 | 3 |
| 4 | 11 | 4 |

**DEVISE A PLAN.** From the accompanying table, we attempt to determine a pattern. Notice that the number of regions formed by 4 chords is the number formed by 3 plus 4. The number formed by 3 is the number formed by 2 plus 3. The number for 2 is the number formed by 1 plus 2. Let's work backward.

**CARRY OUT THE PLAN.**
$$n(4) = n(3) + 4$$
$$n(4) = [n(2) + 3] + 4$$
$$n(4) = n(2) + (3 + 4)$$
$$n(4) = [n(1) + 2] + (3 + 4)$$
$$n(4) = 2 + 2 + 3 + 4$$

At this point we may see a pattern. Do you see that

$$n(5) = 2 + 2 + 3 + 4 + 5$$

and that $n(k) = 2 + 2 + 3 + 4 + \cdots + k$?

Therefore, $n(10) =$

$$2 + 2 + 3 + 4 + 5 + 6 + 7 + 8 + 9 + 10 = 56$$

**LOOK BACK.** To obtain this solution, we looked at several simple cases and observed a pattern. We could also have obtained the solution by thinking carefully about how the subdivisions occurred. Suppose that $k$ line segments have been drawn, creating $n(k)$ regions. Let us start drawing the $(k + 1)$st segment at point $A$. Immediately, we divide the region adjacent to $A$, creating a new region. Then each time we intersect one of the first $k$ line segments, we enter a new region and divide it. Hence, with the $(k + 1)$st segment we create $k + 1$ new regions

$$n(k + 1) = n(k) + k + 1$$

Because with 1 segment we have 2 regions, with 2 segments we have $2 + 2$ regions. Because with 2 segments we have $2 + 2$ regions, with 3 segments we have $2 + 2 + 3$ regions, and so on. Thus, with $k$ segments we have $2 + 2 + 3 + 4 + \cdots + k$ regions.

# SUMMARY AND REVIEW

## POINTS, LINES, PLANES, AND SPACE

1. *Point, line, plane,* and *between* are undefined terms.

   (a) Points on a line are said to be *collinear.*
   (b) Lines containing a common point are *concurrent.*
   (c) If three distinct points are collinear, then one of the points is between the other two.
   (d) Points and/or lines that lie in the same plane are said to be *coplanar.*

2. A *plane* is determined by any of the following:

   (a) Three noncollinear points
   (b) Two intersecting lines
   (c) A line and a point not on the line
   (d) Two parallel lines

3. A unique line is determined by two distinct points.

4. Two lines in a plane are *parallel* if and only if there is no point on both lines.

5. If lines $r$ and $s$ are not in the same plane, they are said to be *skew* lines.

6. A *ray* is the union of a point on a line and one of the half lines determined by the point.

7. A line *segment* is a subset of a line consisting of two points on the line and the set of points between them.

8. Characteristics of planes:

   (a) If the intersection of two distinct planes is not empty, then the intersection is a line.
   (b) If a line contains two distinct points of a plane, then the line lies in the plane.
   (c) Planes are parallel if their intersection is empty.

9. An *angle* is the union of two rays that have the same endpoint.

   (a) A *right* angle has a measure of $90°$.
   (b) A *straight* angle has a measure of $180°$.
   (c) An *acute* angle has a measure between $0°$ and $90°$.
   (d) An *obtuse* angle has a measure between $90°$ and $180°$.
   (e) Two angles are *supplementary* if the sum of their measures is $180°$, and *complementary* if the sum of their measures is $90°$.
   (f) Two coplanar angles that have a common vertex and a common side are called *adjacent* angles if the interiors of the two angles are disjoint.

10. Plane curves

   (a) A curve is *simple* if, without lifting the pencil, you can draw the curve without retracing any of its points (with the possible exception of its endpoints).
   (b) A *closed* curve is drawn by starting and ending at the same point.
   (c) A simple closed curve separates a plane into three disjoint sets of points: the *interior,* the *exterior,* and the curve.
   (d) A region is *convex* if, for any two points in the region, the line segment joining them lies completely in the region.
   (e) A *polygon* is a closed curve that is the union of three or more line segments $\overline{AB}$, $\overline{BC}$, $\overline{CD}$, $\ldots$, $\overline{PQ}$ such that $A$, $B$, $C$, $D$, $\ldots$, $P$, $Q$ are coplanar and distinct, and no three consecutively named points are collinear.
   (f) Polygons can be classified as triangles (3 sides), quadrilaterals (4), pentagons (5), hexagons (6), heptagons (7), octagons (8), $\ldots$, $n$-gons ($n$).
   (g) In general, a triangle can be classified as acute, right, or obtuse; scalene, isosceles, or equilateral.
   (h) In general, a *quadrilateral* can be classified as a parallelogram, a trapezoid, a rectangle, a square, a rhombus, or a kite.

## SURFACES

1. A polyhedron is a simple closed surface in space whose boundary is composed of polygonal regions.

   (a) A prism is a polyhedron formed by two congruent polygonal regions in parallel planes, along with three or more regions bounded by parallelograms joining the two polygonal regions so as to form a closed surface.
   (b) A pyramid is a polyhedron formed by a simple closed polygonal region and a point not in the plane of the region, together with the triangular regions that join the point and the edges of the polygonal region. The triangular regions are congruent in a regular pyramid.

2. The Platonic solids consist of the following:

   (a) Regular tetrahedron
   (b) Regular hexahedron
   (c) Regular octahedron
   (d) Regular dodecahedron
   (e) Regular icosahedron

3. Let $V$ represent the number of vertices, $E$ the number of edges, and $F$ the number of faces of a polyhedron. Then Euler's formula states that $V + F - E = 2$.

4. Consider a circle and a fixed line that is not in the plane of the circle and is not parallel to the plane of the circle. The union of all lines through the circle that are parallel to the fixed line is called a cylinder of infinite extent. Consider a plane parallel to the one mentioned above. Each of the two planes intersect the cylinder of infinite extent in a circle. The union of the two resulting circular regions with the line segments from the cylinder of infinite extent that connect those circles is a circular cylinder.

5. All points in space equidistant from a fixed point make up a sphere.

6. A cone consists of segments joining a fixed point not in a given plane to points on a closed curve in the given plane. Usually the closed curve is a circle.

The following table summarizes classifications and relationships for many of the geometric concepts of this chapter. The relationships of congruence and similarity are covered in subsequent chapters.

| *Geometric Entity* | *Classifications* | *Relationships* |
|---|---|---|
| Lines | Plane, space | Intersecting, skew, perpendicular, parallel, concurrent |
| Curves | Plane, simple, closed, simple closed, polygon, circle | Intersecting, parallel |
| Angles | Acute, right, obtuse, straight | Vertical, adjacent, corresponding, alternate interior, alternate exterior, supplementary, complementary |
| Polygons | Concave, convex, triangle, quadrilateral, *n*-gon | Congruent, similar |
| Triangles | Sides: equilateral, isosceles, scalene angles: acute, right, obtuse | Congruent, similar |
| Quadrilaterals | Square, rectangle, parallelogram, kite, rhombus, trapezoid | Congruent, similar |
| Solids | Sphere, cylinder, cone, polyhedron | Congruent, similar |
| Polyhedrons | Prism, pyramid, octahedron, dodecahedron, icosahedron | |

# CHAPTER TEST

1. For each of the following, sketch two parallelograms, if possible, that satisfy the given condition.

   (a) Their intersection is a single point.
   (b) Their intersection is exactly two points.
   (c) Their intersection is exactly three points.
   (d) Their intersection is exactly one line segment.

2. Verify Euler's formula for the following.

   (a)　　　　　(b)

3. Fill in the blanks.

   (a) _____noncollinear points determine a plane, and _____ intersecting lines determine a plane.
   (b) An _____ is the union of two rays that have the same endpoint (called the _____).
   (c) A _____ is a polygon with five sides, while a _____ is a polygon with four sides.
   (d) The set of all points in a plane at an equal distance from a fixed point in the plane is a _____ .

4. Classify as true or false.

   (a) Skew lines are lines in a plane that do not intersect.
   (b) It is impossible for a right triangle to be isosceles.
   (c) In the figure, 1 and 2 are called vertical angles.

   (d) A ray is the union of a point and a line.
   (e) If a plane contains one and only one of two parallel lines, it is parallel to the other line.

(f) One line segment may be a subset of another line segment.

(g) Adjacent angles are complementary.

5. Make sketches to show the following pairs intersecting in two points.

   (a) Two rectangles   (b) Two angles

6. Can two planes have one and only one point in common?

7. In the following figure, you see a pyramid with its top cut off viewed from above. Classify the following statements as true or false.

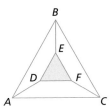

   (a) Edges $\overline{AD}$ and $\overline{CF}$ intersect.
   (b) $\overleftrightarrow{DF}$ and $\overleftrightarrow{BE}$ intersect.
   (c) $\overline{BE}$ and $\overline{CF}$ intersect.

8. How many lines are determined by the following points, no three of which are collinear?

   (a) 2   (b) 3   (c) 4   (d) $n$

9. Given the line shown in the accompanying drawing, find the following.

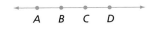

   (a) $\overrightarrow{AB} \cap \overrightarrow{DC}$   (b) $\overrightarrow{AB} \cap \overline{CD}$   (c) $\overrightarrow{AB} \cup \overrightarrow{BD}$

10. (a) List three different names for line $m$.

   (b) Name two different rays on $m$ with endpoint $B$.
   (c) Find a simpler name for $\overrightarrow{AB} \cap \overrightarrow{BA}$.
   (d) Find a simpler name for $\overrightarrow{AB} \cap \overrightarrow{BC}$.
   (e) Find a simpler name for $\overrightarrow{BA} \cap \overrightarrow{AC}$.

11. In the accompanying figure, $\overrightarrow{PQ}$ is perpendicular to $\alpha$.

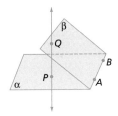

(a) Name a pair of skew lines.

(b) Using only the letters in the figure, name as many planes as possible, each perpendicular to $\alpha$.

(c) What is the intersection of the planes $APQ$ and $\beta$?

(d) Is there a single plane containing $A$, $B$, $P$, and $Q$? Explain your answer.

12. In the figure, $l$ is parallel to $m$, and m($\angle 1$) = 60°. Find each of the following.

   (a) m($\angle 3$)   (b) m($\angle 6$)   (c) m($\angle 8$)

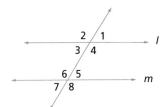

13. For the given figure, find two angles adjacent to $\angle DOC$ and one vertical angle for $\angle DOC$.

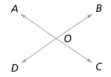

14. Classify as true or false.

   (a) A prism with a quadrilateral base has four faces.
   (b) A pyramid with a triangular base has four faces.
   (c) There are five Platonic solids.
   (d) Each of the Platonic solids has a triangular face.
   (e) A sphere is an example of a polyhedron.

15. Sketch each of the following:

   (a) A triangular prism
   (b) A square pyramid
   (c) A circular cone

16. (a) Sketch a convex hexagon.
   (b) What is the measure of each interior angle of a regular hexagon?
   (c) What is the measure of each central angle of a regular hexagon?

# MEASUREMENT AND THE METRIC SYSTEM

# 12

**P**lace a square on a lattice similar to the one shown in the introductory picture. Count the number of points on the square and the number of points interior to the square; then calculate the area of the square. Record the number of points on the square, the number of points interior to the square, and the area of the square in a table, using squares with sides of 1 unit, 2 units, 3 units, 4 units, 5 units, and so on. Can you discover a relationship among these three quantities recorded? Does this relationship hold for lattice figures other than squares? Does it hold for every polygonal region represented on a lattice?

The metric system is important to you, a resident of the United States, not only because of its growing adoption worldwide, but also because of its increasing use in this country. This increased use is evident in highway signs given in kilometers as well as in miles, temperatures given in Celsius and Fahrenheit, canned goods labeled in metric and English units, and soft drinks packaged in liter bottles.

In this chapter, you will be introduced to linear, two-dimensional, and three-dimensional measurements. Most of the units we use for these measurements will be metric. When you complete this chapter, you should understand the formulas for obtaining measurements and also be more proficient in using metric units in everyday life.

In this chapter, we use the following problem-solving strategies. ▪ Guess, test, and revise. (541, 543, 544, 559, 578) ▪ Try simpler versions of the problem. (578) ▪ Look for a pattern. (578) ▪ Use cases or break into parts. (578) ▪ Make a chart or a table. (578) ▪ Restate the problem. (576) ▪ Use direct reasoning. (541, 556, 576) ▪ Draw a diagram. (539, 547, 573, 576) ▪ Draw a picture. (throughout the chapter) ▪ Use dimensional analysis. (throughout the chapter)

# SECTION 1 CONGRUENCE, LINEAR MEASURE, AND MEASUREMENT

| **PROBLEM** | Albert, whose home is in Detroit, and Sherman, who lives in Germany, are working together on the design of an electric motor for an automobile. Albert suggests the length of a part should be 24, but Sherman has been using 61. How can these two competent design specialists be so far apart in their preliminary designs of this part? |
| --- | --- |
| **OVERVIEW** | Of course, the two design specialists are using approximately the same length part. Albert is expressing the length in inches, whereas Sherman is using centimeters. |
| **GOALS** | Use the Measurement Standard: ● understand measurable attributes of objects and the units, systems, and processes of measurement; ● apply appropriate techniques, tools, and formulas to determine measurements.<br>Illustrate the Geometry Standard, page 482.*<br>Illustrate the Representation Standard, page 50.*<br>Illustrate the Communication Standard, page 147.*<br><br>* The complete statement of the standard is given on this page of the book. |

## THE MEASUREMENT PROCESS

The measurement process is the process by which numerical values are assigned to geometric entities. What is the length of a football field? What should be the size of the cover for our boat? How much Gatorade does the container hold? The answers to all of these questions involve the measurement process. It is this process that provides the everyday use of geometry both as a language and in application problems. The measurement process can be described as a three-step procedure.

| **Steps in the Measurement Process** | 1. Select the attribute of the entity to be measured that represents the most appropriate characteristic of the entity. <br> 2. Select the unit to be used in the measurement process. For linear measure, for example, | you might select an inch or a centimeter. <br> 3. Use a measurement device (such as a marked ruler) to determine the number of units required to measure the entity. |
| --- | --- | --- |

# A SHORT HISTORY OF MEASUREMENT

The Bible and early Babylonian and Egyptian records indicate that measurements in ancient times used units associated with parts of the body. For example, the hand, the span, the cubit, and the foot were all used as units of measurement. (See Figure 12-1.)

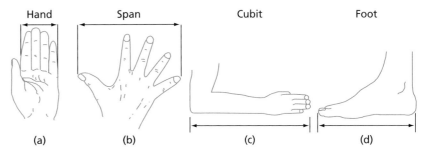

Hand (a)   Span (b)   Cubit (c)   Foot (d)

FIGURE 12-1

It is interesting to follow the evolution of terms associated with measurement. The Romans were responsible for introducing the cubit (approximately 26.6 in.) throughout the empire they ruled. Later King Henry I of England decreed that a yard (36 in.) was the distance from the tip of the nose to the end of the thumb of an outstretched arm. (See Figure 12-2(a).) The Roman mile was 2000 paces (about 2.5 ft. each). (See Figure 12-2(b).) Later the English standardized the mile to be 5280 ft.

FIGURE 12-2

(a)    Yard

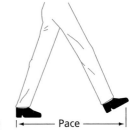

(b)    Pace

In early English measurements, an inch had a length of 3 barleycorns (grains of barley). A rod was the total length of the left feet of 16 men (later standardized to $16\frac{1}{2}$ ft). Of course, the English system was eventually refined and standardized into a system that has been used in many countries and is still in use in the United States today. You are familiar with today's standard English system of measurement, illustrated in part by linear measurements in Table 12-1.

### TABLE 12-1

12 inches (in.) = 1 foot (ft.)
3 feet (36 in.) = 1 yard
$5\frac{1}{2}$ yards ($16\frac{1}{2}$ ft.) = 1 rod
320 rods (1760 yards) = 1 mile

One big disadvantage of the English system is that there is no simple scheme for remembering the relationship between units. Without constant use, the average person forgets the number of feet in a rod, the number of square yards in an acre, and the number of gallons in a barrel. In addition, computations are often messy.

The rapid growth of science demanded a more convenient and consistent system of measurement. The French Academy of Sciences called a meeting of scientists from many countries at the end of the 18th century in an effort to establish an international standard of measures. This group developed the metric system, with the meter as the basic unit of length.

It was intended that the meter be one ten-millionth of the distance from the Equator to the North Pole. Unfortunately, this distance was originally measured incorrectly, and it cannot be measured accurately anyway. As a result, several redefinitions of the meter have been necessary. A distance between two marks on a platinum bar (when held at a temperature of 0° Celsius), on deposit in Washington, D.C., represents a copy of international standard for a meter. All other lengths used in the United States are defined in terms of the standard meter. For example, a yard is exactly 0.9144 m.

Computations in the metric system are much easier than computations in the English system, because all conversions are based on powers of 10. The entire SI (metric) system is based on common prefixes attached to root units. These prefixes and their meanings are listed in Table 12-2. They are used in expressing SI units for length, area, volume, mass, and capacity.

### TABLE 12-2

| Prefix | Meaning | Decimal Meaning |
|--------|---------|-----------------|
| mega | one million | 1,000,000 or $10^6$ |
| kilo | one thousand | 1000 or $10^3$ |
| hecto | one hundred | 100 or $10^2$ |
| deka | one ten | 10 |
| deci | one-tenth | 1/10 or 0.1 |
| centi | one-hundredth | 1/100 or 0.01 |
| milli | one-thousandth | 1/1000 or 0.001 |
| micro | one-millionth | 1/1,000,000 or 0.000001 |

## CONGRUENT FIGURES

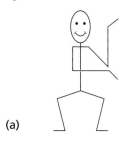

(a)

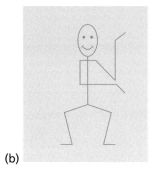

(b)

FIGURE 12-3

Before discussing measurement using SI units in more detail, we need to define what we mean by linear measure. The mathematical basis for linear measure involves congruent line segments. Hence we must discuss briefly the notion of congruence. To stimulate your intuition about congruence, think about the tracing of a given figure. A **tracing** is made by placing a thin sheet of paper over a figure and moving a pencil over every line of the figure so that an exact copy of the original figure is made on the thin paper.

Briefly study the stick figure in Figure 12-3(a). Take a thin sheet of paper and place it on top of the figure. Trace the figure on the new sheet of paper by going over every line. Notice that, as the figure is traced, the copy of the figure is exactly the same size and shape as the original. Every line and every point match. The stick figure in Figure 12-3(b) can be thought of as a trace of the given figure.

Tracing figures provides a basis for studying congruent figures. We consider two figures or two collections of points to be **congruent** if they have the same size and shape. If one of the congruent figures is superimposed on the other, the two coincide point by point; that is, one figure is the trace of the other. You can see that the two figures in the preceding example—the original and its trace—are congruent. The symbol "$\cong$" is a shorthand notation for "is congruent to."

---

**EXAMPLE 1**

Consider the three line segments $\overline{AB}$, $\overline{CD}$, and $\overline{EF}$ in Figure 12-4. Which pairs, if any, of these segments are congruent? One way of testing to see if two figures are congruent is by tracing. Trace $\overline{AB}$ to see if it is congruent to $\overline{CD}$ or $\overline{EF}$. Is the trace of $\overline{AB}$ the same size and shape as $\overline{CD}$ or $\overline{EF}$? Is $\overline{AB} \cong \overline{CD}$? Why is $\overline{AB}$ not congruent to $\overline{EF}$?

A •————————• B

C •——————• D

E •————————• F

FIGURE 12-4

## LINEAR MEASURE

Congruent line segments are the mathematical basis for linear measure. To measure length, we must first select a line segment to serve as a unit segment. In Chapter 7 we considered number lines such as that shown in Figure 12-5 and asserted that a one-to-one correspondence exists between the set of real numbers and points on the number line. We choose the interval from 0 to 1 to use as a unit interval.

FIGURE 12-5

**Properties of Linear Measure**

To every point on the number line, there corresponds exactly one real number, called its **coordinate.** Likewise, for every real number, there corresponds exactly one point, the **graph** of the number.

If we calculate distance by the process of counting the number of units between 2 points, then the distance between $E$ and $G$ in Figure 12-6 is 2. This is $3 - 1 = 2$, where 3 is the coordinate of $G$ and 1 is the coordinate of $E$. This suggests the following definition, using Figure 12-7.

FIGURE 12-6                    FIGURE 12-7

**DEFINITION**

**Distance**

If $A$ and $B$ are points on a number line with coordinates (real numbers) $a$ and $b$, respectively, then the *distance* between $A$ and $B$ is given by $|b - a|$. The distance between $A$ and $B$ is sometimes called the **measure** of $\overline{AB}$ and is denoted by $m(\overline{AB}) = m(\overline{BA}) = |b - a|$.

**EXAMPLE 2**

Using the preceding definition, we can obtain the following measures from Figure 12-6.

$$m(\overline{EG}) = 3 - 1 = 2 \qquad m(\overline{DG}) = 3 - 0 = 3$$
$$m(\overline{FC}) = |-1 - 2| = 3 \qquad m(\overline{BD}) = 0 - (-2) = 2$$

From the preceding discussion we have the following property.

**Congruent Segments**

When two segments are congruent, their measures are equal; conversely, if their measures are equal, then the segments are congruent.

When $\overline{AB}$ and $\overline{CD}$ are congruent, we use the notation $m(\overline{AB}) = m(\overline{CD})$ or the notation $AB = CD$. Thus, "$m(\overline{AB}) = m(\overline{CD})$" and "$AB = CD$" and "$\overline{AB} \cong \overline{CD}$" all mean the same thing for line segments $\overline{AB}$ and $\overline{CD}$.

We can identify five important properties of distance (or measure).

**Properties of Measure**

1. Distance is never negative; that is, $m(\overline{AB}) \geq 0$.
2. The distance between two points is zero only when the points are identical.
3. The distance between points is unaffected by the order in which they are considered; that is, $m(\overline{AB}) = m(\overline{BA})$. Henceforth, **$m(\overline{AB})$ will be written as $AB$ in this book.**
4. For any three noncollinear points $A$, $B$, and $C$, the distance between $A$ and $B$ plus the distance between $B$ and $C$ is greater than the distance between $A$ and $C$; that is, $AB + BC > AC$. (See Figure 12-8(a).)
5. If $A$, $B$, and $C$ are three collinear points, with $B$ between $A$ and $C$, then $AB + BC = AC$. (See Figure 12-8(b).)

Property 4, as illustrated in Figure 12-8(a), is known as the **Triangle Inequality.** In Figure 12-8(b), notice that $AB + BC = AC$ if and only if $A$, $B$, and $C$ are collinear and $B$ is between $A$ and $C$.

# Measuring Length

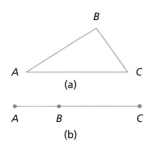

(a)

(b)

FIGURE 12-8

The SI units for measuring length are based on the **meter.** Using the information in Table 12-2, verify the results in Table 12-3, which gives the SI length measurements based on the meter. Notice the abbreviations or symbols for the measurements; these symbols are generally used with numerals.

| TABLE 12-3 | |
|---|---|
| 1 megameter (Mm) = $10^6$ meters (m) | 1 meter = 0.000001 or $10^{-6}$ megameter |
| 1 kilometer (km) = 1000 or $10^3$ meters | 1 meter = 0.001 or $10^{-3}$ kilometer |
| 1 hectometer (hm) = 100 or $10^2$ meters | 1 meter = 0.01 or $10^{-2}$ hectometer |
| 1 dekameter (dam) = 10 meters | 1 meter = 0.1 or $10^{-1}$ dekameter |
| 1 decimeter (dm) = 0.1 or $10^{-1}$ meter | 1 meter = 10 decimeters |
| 1 centimeter (cm) = 0.01 or $10^{-2}$ meter | 1 meter = $10^2$ centimeters |
| 1 millimeter (mm) = 0.001 or $10^{-3}$ meter | 1 meter = $10^3$ millimeters |
| 1 micrometer ($\mu$m) = 0.000001 or $10^{-6}$ meter | 1 meter = $10^6$ micrometers |

A kilometer is about the length of nine football fields (end zones included) laid end to end. A meter is approximately the distance from a doorknob to the floor, so most doorways are a little over 2 m high. A dime is about 1 mm thick, a nickel is about 2 cm across, a piece of chalk is about 1 cm in diameter, and the stitches on a baseball are about 1 mm apart. Figure 12-9 shows a 10-cm ruler on which a segment $\overline{AB}$ that is 1 cm long has been marked.

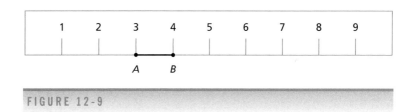

FIGURE 12-9

The kilometer, meter, centimeter, and millimeter are the most commonly used metric units for linear measurements. *Mega* and *micro* are used only when one needs very large or very small units, respectively. The prefixes *hecto, deka,* and *deci* are seldom used.

It is important in working with metric units to develop the ability to estimate the approximate length of objects expressed in metric units. Practice on the problems in the following example.

**EXAMPLE 3**

Select the answer that best estimates the length in each problem.
(a) Length of a library table:          30 m          3 m          3 km
(b) Length of a paper clip:          100 mm          3 cm          0.3 m

(c) Distance walked in 20 min:              80 m        5000 cm       1.5 km
(d) Length of a dinner knife:               20 cm        2 m          0.002 km
(e) Height of a church steeple:             50 m         50 km        500 m
(f) Distance from Maine to California:   48,000 m     480,000 km    4800 km

*SOLUTION*

(a) 3 m          (b) 3 cm          (c) 1.5 km
(d) 20 cm        (e) 50 m          (f) 4800 km

To convert from one metric unit to another, multiply or divide by a power of 10. To change from a larger unit to a smaller unit, multiply. There will be more smaller units. To change from a smaller unit to a larger unit, divide. There will be fewer larger units.

**EXAMPLE 4**

Express 13 km in meters.

*SOLUTION*

$$1 \text{ km} = 1000 \text{ m}$$
$$13 \text{ km} = 13{,}000 \text{ m}$$

**EXAMPLE 5**

Convert to centimeters.
(a) 3 m          (b) 43 mm          (c) 41 km          (d) $2 \cdot 10^8$ km

*SOLUTION*

(a) 1 m = 100 cm          (b) 10 mm = 1 cm
    3 m = 300 cm              1 mm = $\frac{1}{10}$ cm
                              43 mm = $\frac{43}{10}$ = 4.3 cm

(c)  1 km = 100,000 cm          (d)     1 km = $10^5$ cm
     41 km = 4,100,000 cm          $2(10)^8$ km = $2(10)^8(10)^5 = 2(10)^{13}$ cm

Some people like to perform conversions in metric units by moving the decimal point. (See Figure 12-10.) In going from a smaller unit to a larger unit, move the decimal point 1 place to the left for each unit crossed. For example, to change from meters to kilometers, move the decimal point 3 places to the left. Conversely, to change from kilometers to meters, move the decimal point 3 places to the right, adding appropriate zeros.

For example, to convert from hectometers to centimeters, we would move 4 units to the right on the scale. (See Figure 12-10.) Hence we would multiply by $10^4$ or move the decimal point 4 places to the right. To convert from decimeters to hectometers, we would move 3 units to the left on the scale. (See Figure 12-10.) Hence we would multiply by $10^{-3}$ or move the decimal point 3 places to the left.

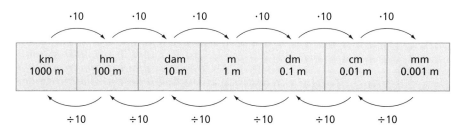

FIGURE 12-10

**EXAMPLE 6**

Convert the following by moving the decimal point.
(a) 55 mm to m        (b) 170 km to cm        (c) 860 cm to km

*SOLUTION*

(a) 55 mm = .055. m

(b) 170 km = 170.00000. cm

(c) 860 cm = .00860. km

## DIMENSIONAL ANALYSIS

You may sometimes need to convert from metric units to English units or from English units to metric units. This type of conversion provides an opportunity to introduce a new problem-solving technique, called **dimensional analysis.** In addition, dimensional analysis is a good technique for making conversions within a given measurement system.

In dimensional analysis, we use ratios that are equal to 1 and treat these ratios as fractions. The measurement units not needed simply divide out. This procedure eliminates the concern of whether to multiply or divide. For example, because 100 cm = 1 m and 2.54 cm = 1 in., the unit ratios can be written as

$$\frac{100 \text{ cm}}{1 \text{ m}} = \frac{1 \text{ m}}{100 \text{ cm}} = 1 \quad \text{and} \quad \frac{1 \text{ in.}}{2.54 \text{ cm}} = \frac{2.54 \text{ cm}}{1 \text{ in.}} = 1$$

**EXAMPLE 7**

(a) 12 cm is how many meters?        (b) Convert 4 mm to km.

*SOLUTION*

(a) $12 \text{ cm} = 12 \text{ cm} \cdot \dfrac{1 \text{ m}}{100 \text{ cm}} = .12 \text{ m}$

(b) $4 \text{ mm} = 4 \text{ mm} \cdot \dfrac{1 \text{ m}}{1000 \text{ mm}} \cdot \dfrac{1 \text{ km}}{1000 \text{ m}} = 4(10)^{-6} \text{ km}$

**EXAMPLE 8**

Convert 6 in. to centimeters.

*SOLUTION*

$6 \text{ in.} = 6 \text{ in.} \left( \dfrac{2.54 \text{ cm}}{1 \text{ in.}} \right) = 15.24 \text{ cm}$

**EXAMPLE 9**

Change 2 yards to centimeters.

*SOLUTION*

$2 \text{ yards} = 2 \text{ yards} \left( \dfrac{36 \text{ in.}}{1 \text{ yd.}} \right) \left( \dfrac{2.54 \text{ cm}}{1 \text{ in.}} \right) = 182.88 \text{ cm}$

**EXAMPLE 10**

Convert 16 miles to kilometers.

*SOLUTION*

$16 \text{ miles} = 16 \text{ miles} \left( \dfrac{5280 \text{ ft.}}{1 \text{ miles}} \right) \left( \dfrac{12 \text{ in.}}{1 \text{ ft.}} \right) \left( \dfrac{2.54 \text{ cm}}{1 \text{ in.}} \right) \left( \dfrac{1 \text{ m}}{100 \text{ cm}} \right) \left( \dfrac{1 \text{ km}}{1000 \text{ m}} \right)$

$\approx 25.7495 \text{ km}$

# PERIMETER

A useful measure associated with simple closed curves is the perimeter. For example, this distance is used to find the amount of weatherstripping needed to go around a window or the length of fence needed to enclose your backyard. The **perimeter** $P$ of a simple closed curve is the distance around the curve. The perimeter of a polygon is the sum of the measures of the sides of the polygon. Thus, the perimeter of the polygon in Figure 12-11 is

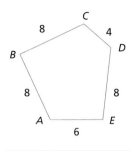

FIGURE 12-11

$$P = 8 + 8 + 8 + 6 + 4 = 34$$

**EXAMPLE 11**

Find the perimeter of the simple closed curve in Figure 12-12. The horizontal or vertical distance between two dots constitutes 1 unit.

*SOLUTION*

By the Pythagorean Theorem (explained in Chapter 7), the diagonal of a unit square (the hypotenuse of a right triangle) can be obtained from

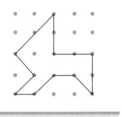

$$c^2 = a^2 + b^2$$
$$= 1^2 + 1^2 = 2$$
$$c = \sqrt{2}$$

The simple closed curve in Figure 12-12 consists of 6 segments of length $\sqrt{2}$ and 8 segments of length 1. Thus the perimeter of the figure is

FIGURE 12-12

$$P = 8 + 6\sqrt{2}$$

# PRECISION

In using any standard of measurement (such as a ruler) to measure the length of an object, it is important to recognize that the answer you obtain is an approximation. Errors are always involved in measurement. When you purchase 3.5 m of material, the actual size of the material you get is not exactly 3.5 m because the measuring process is not precise. Measurements are approximate for two reasons. First, they are estimated to a determined degree of precision, depending on the accuracy of the instrument. Second, the process of reading a scale or ruler introduces uncertainty and hence approximation. People vary their degree of precision from one measurement to another because the fallibility of their senses places limits on how precise they can be.

Because measurements are approximations, when you are working with measurements you should use the rules for adding (or subtracting) and multiplying (or dividing) approximate numbers, as discussed in Chapter 7.

**EXAMPLE 12**

A room is measured to be 5.3 m wide. Find the maximum error in this measurement.

*SOLUTION* | Because 5.3 is an approximation, or rounded-off number, the true width of the room is between 5.25 and 5.35 m. The maximum error or **greatest possible error,** abbreviated g.p.e., is 0.05 m.

Sometimes in working with measurements we use the term **precision.** The precision in the preceding example is 0.1; that is, the g.p.e. is one-half of the precision. Precision is sometimes inadequate to compare errors. To illustrate, 687.2 and 3.3 have the same precision; that is, both have been rounded to tenths. Yet, the error in the first number is not as large relative to its value as the error in the second number is to its value. Thus, we need an additional measure of accuracy, one that is provided by the concept of *relative error*. As the relative error becomes smaller, the measurement becomes more accurate.

---

**DEFINITION**

**Relative Error** | *Relative error* is the greatest possible error of a number $N$ divided by the number $N$.

$$\text{Relative error of } N = \frac{\text{g.p.e. of } N}{N}$$

---

**EXAMPLE 13** | Find the relative errors of the approximate numbers 0.05 and 2.1.

*SOLUTION* |
$$\text{Relative error of } 0.05 = \frac{0.005}{0.05} = 0.1$$

$$\text{Relative error of } 2.1 = \frac{0.05}{2.1} \approx 0.02$$

# Just for Fun

The chef baked a cake that measured 4 dm (decimeters) on each side. He wanted to cut the cake so that he would have 8 pieces of full thickness, each with each side measuring 2 dm. How did he cut the cake?

# EXERCISE SET 1

R 1. Convert the following measurements to meters.
   (a) 37,000 cm     (b) 389,000 mm
   (c) 12,000 mm     (d) $21 \cdot 10^8$ cm
   (e) 35,470 km     (f) 38,902 km

2. Convert the following measurements to millimeters.
   (a) 3789 cm    (b) 389,850 km    (c) 1284 m
   (d) $21 \cdot 10^8$ cm    (e) 35,478 km    (f) 38,902 m
   (g) 16 cm    (h) $412 \cdot 10^4$ m

3. Convert the following measurements to kilometers.
   (a) 1700 cm     (b) 120,000 mm     (c) 1600 m

4. Without using a ruler, draw segments that you think represent the lengths given; then measure them with a ruler to check your estimates.
   (a) 12 cm     (b) 5 cm     (c) 7 cm
   (d) 12 mm     (e) 24 mm     (f) 4 mm

5. Fill in the following blanks.
   (a) 32 mm = _____ m
   (b) 160 cm = _____ km
   (c) 3 km = _____ mm
   (d) 216 mm = _____ km
   (e) 1,340,000 μm = _____ cm
   (f) 1689 km = _____ cm
   (g) 48 m = _____ μm

6. Arrange in order, from shortest to longest.
   (a) 10 m, 100 cm, 100,000 mm
   (b) 10 cm, 1000 mm, 0.01 m
   (c) 0.1 cm, 0.01 m, 0.001 km
   (d) 0.1 cm, 0.001 km, 100 mm

7. Complete the following table.

| | Shortest Possible Length | Measurement | Longest Possible Length |
|---|---|---|---|
| (a) | _____ | 21.6 cm | _____ |
| (b) | _____ | 14.16 km | _____ |
| (c) | _____ | 2 mm | _____ |

8. Find the greatest possible error and the relative error for each part of Exercise 7.

9. The perimeter of a rectangle is 50 m. Its length is 12 m. What is its width?

10. Find the perimeter of a regular hexagon when each side is 3.2 cm.

11. Classify the following statements either as true or as false.

(a) If $\overline{AB}$ is congruent to $\overline{AD}$, then $B$ and $D$ name the same point. _____
(b) If $\overline{AB} \cong \overline{BC}$ and $\overline{BC} \cong \overline{CD}$, then $\overline{AB} \cong \overline{CD}$.
(c) If $m(\overline{AC}) = m(\overline{AD})$, then $C$ and $D$ name the same point.
(d) The union of two different segments ($\overline{AB} \cup \overline{CD}$) may have the same measure as one of the segments.
(e) The union of two different segments may be congruent to one of the segments.
(f) Two congruent figures have the same shape.
(g) If $\overline{AB}$ and $\overline{BC}$ are congruent, then they are of equal measure.

12. Consider the line segments determined by the points $A$, $B$, $C$, $D$, $E$, $F$, and $G$. Find each of the following measures, where $m(\overline{BC}) = \frac{1}{2}$ and $m(\overline{DE}) = m(\overline{EF}) = \frac{3}{4}$.

(a) $m(\overline{AB})$     (b) $m(\overline{CE})$     (c) $m(\overline{DB})$
(d) $m(\overline{GE})$     (e) $m(\overline{GA})$     (f) $m(\overline{CD})$

13. Would you use millimeters (mm), centimeters (cm), meters (m), or kilometers (km) to measure the
   (a) length of a room?
   (b) thickness of a button?
   (c) distance from Miami to Louisville?
   (d) length of a car?
   (e) width of a book?
   (f) width of a hand?
   (g) width of a paper clip?
   (h) length of a football field?
   (i) length of a photograph?
   (j) diameter of a screw?
   (k) altitude of an airplane?

14. Add or subtract.
   (a) 3.4 m + 689 cm + 46 cm
   (b) 4932 mm − 2.78 m
   (c) 7 m − 243 cm
   (d) 6.2 m + 541 cm + 4628 mm

15. Select the most sensible measurement for each item.
   (a) Height of your classroom: (i) 4 mm, (ii) 4 cm, (iii) 3 m
   (b) Distance of a cross-country race: (i) 10 m, (ii) 10 km, (iii) 1000 km
   (c) Length of a postcard: (i) 12 cm, (ii) 1.2 m, (iii) 12 m

(d) The length of an average book: (i) 2 mm, (ii) 24 cm, (iii) 24 m

(e) The width of an average door: (i) 0.01 km, (ii) 100 cm, (iii) 100 mm

16. Formulate rules for changing the following units of measure.

(a) Meters to kilometers

(b) Centimeters to meters

17. Using metric prefixes, identify each of the following.

(a) centifoot

(b) kilofoot

(c) millifoot

(d) decifoot

18. Choose the most sensible answer.

(a) The length of a tennis racket:
    68 mm    68 cm    68 m    68 km

(b) The average height of an adult:
    165 cm    250 cm    315 cm    500 cm

(c) The length of a canoe:
    4 mm    4 cm    4 m    4 km

(d) The distance from Denver to New Orleans:
    2130 mm    2130 cm    2130 m    2130 km

(e) The width of a light cord:
    4 mm    30 mm    60 mm    6 cm

*T* 19. Use the Pythagorean Theorem to find the length of the following segments.

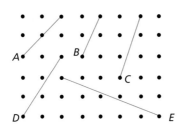

20. Find the perimeter of the following figures.

(a)              (b)

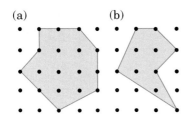

21. Arrange in increasing order.
    62 cm    6000 mm    0.6 m    61 dm

22. Arrange in decreasing order.
    2 km    1500 m    16,000 cm    160 dm

*C* 23. In the triplets of numbers given, find a coordinate $x$ so that $\overline{AB}$ is of the same length as $\overline{BC}$.

(a) $A$ at $\frac{1}{2}$, $B$ at 2, $C$ at $x$

(b) $A$ at $\frac{3}{4}$, $B$ at $x$, $C$ at 2

(c) $A$ at $-1$, $B$ at 3, $C$ at $x$

(d) $A$ at $x$, $B$ at 1, $C$ at 4

24. Find two possible coordinates for $B$ so that $\overline{AB}$ is a unit segment, where $A$ has the following coordinate.

(a) $\frac{1}{2}$        (b) $-3$        (c) $-4$

(d) $\frac{7}{8}$        (e) $-1$        (f) 3

25. (a) In the accompanying drawing, find the distance from $A$ to $B$.

(b) Add one-half of the answer in (a) to $A$ to get the midpoint of $\overline{AB}$.

(c) Add one-third of the answer in (a) to $A$ to get a point one-third of the way from $A$ to $B$.

(d) Add one-fourth of the answer in (a) to $A$ to get a point one-fourth of the way from $A$ to $B$.

26. Using Exercise 25, find

(a) the midpoint of $\overline{AC}$.

(b) a point one-third of the distance from $D$ to $E$ if $D$ is at 2 and $E$ is at 6.

27. A car averages 30 km per hour from $A$ to $B$. What is the average speed on the return trip from $B$ to $A$ if the average speed for the entire trip is 40 km per hour?

*In Exercises 28 through 32, estimate the answers based on your understanding of the sizes of various units of measurement in the English and metric systems. (1 km ≈ 0.6 miles)*

28. Convert the following measurements.

(a) 11 m to yards

(b) 150 ft. to centimeters

(c) 11 miles to kilometers

(d) 52 cm to inches

29. Rewrite the following highway signs (expressed in miles) in terms of SI units.

(a)                                    (b)

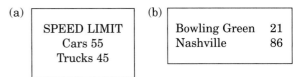

30. Some common footraces cover distances of 100 m and 500 m. How many yards are covered in each race? How many yards are covered in a 1500-m run? How many meters are covered in a 100-yard race?

31. Translate the measurements shown from English units into the appropriate metric units.

   (a) 24-in. waist
   (b) 15-in. neckline
   (c) 60 miles per hour
   (d) 5-ft., 11-in. date

32. Convert the following distances either to kilometers or to miles.

   (a) Louisville to Birmingham, 600 km
   (b) San Francisco to Los Angeles, 400 miles
   (c) Miami to New Orleans, 900 miles
   (d) Boston to Chicago, 1600 km

*In Exercises 33 through 37 use dimensional analysis to find the answer, knowing that 1 in. equals 2.54 cm.*

33. Work Exercise 31.     34. Work Exercise 28.

35. Work Exercise 29.     36. Work Exercise 30.

37. Work Exercise 32.

38. (a) In a precision machine shop, a platinum-alloy dowel is measured to be 0.00285 in. in radius. Express this measurement in yards.
   (b) In a machine shop in England, where a similar measurement was made, the measurement was found to be 0.0126 mm. Express this in meters.
   (c) Compare the difficulty of the two computations.

### ∴ PCR Excursion ∴

39. Using the questions below for background information, draw a polygon on the geoboard (vertices at points on the geoboard) that has the following properties:

   > It is a hexagon.
   > It is nonconvex (concave).
   > Two sides are parallel.
   > It has 1 right angle, 2 acute angles, 2 obtuse angles, and 1 angle greater than 180°.

It contains 7 points and encloses 4 points. It has a perimeter of $3\sqrt{2} + 2\sqrt{5} + 2$.

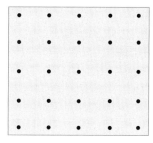

(a) When is a polygon a hexagon?
(b) What is the distinctive characteristic of a nonconvex polygon?
(c) Discuss the various ways line segments can be parallel and contain points of the geoboard.
(d) Discuss how angles with vertices at points of the geoboard and sides containing points of the geoboard can be right angles.
(e) As in (d), discuss how an angle can be an acute angle.
(f) As in (d), discuss how an angle can be an obtuse angle.
(g) As in (d), discuss how an interior angle can be greater than 180°.
(h) Form a hexagon that contains 7 points and encloses 4 points with the characteristics (a) through (g).
(i) Does your hexagon have the correct perimeter? If not, study ways that the length of a line segment can contain $\sqrt{2}$.
(j) How can the length of a line segment contain $\sqrt{5}$?
(k) Now use the problem-solving strategy of guess, test, and revise to find the hexagon with the correct perimeter that satisfies the given properties.

# SECTION 2  TWO-DIMENSIONAL MEASURE AND FORMULAS FOR AREA

**PROBLEM**

The famous Yin-Yang symbol shown below has been constructed from a circle of radius *r*. Compute the length of the arc that divides the colored and white regions.

**OVERVIEW**

Two-dimensional measure can best be understood by using the geoboard as a model. However, it is difficult (if not impossible) to find the area of a circle on a geoboard. Thus, in this section we develop formulas for computing the areas of parallelograms, triangles, and trapezoids; and then we enlarge our thinking to include the circumference and area of a circle. Using this knowledge, you should be able to solve the introductory problem.

 AREA

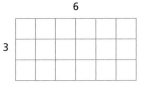

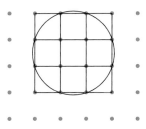

**Area** is a number that gives the measure of a region (or the interior of a closed curve). The measuring process for regions is similar to that for one-dimensional figures. To find a measure of a region we determine how many two dimensional units (usually squares) are required to completely cover the region without overlapping. Consider the region bounded by a rectangle with sides measuring 6 units and 3 units. (See Figure 12-13.)

Six congruent unit squares (the sides of which measure 1 unit), when placed in a contiguous but nonoverlapping position, cover the entire length of the rectangle. Three similarly placed unit squares cover the width of the rectangle. A total of 18 unit squares completely cover the region bounded by the rectangle. This discussion suggests that the unit square is quite reasonably designated as the unit for two-dimensional measure, called **area.**

Of course, unit squares cannot always be placed within many nonpolygonal regions to obtain the area visually. You note the difficulty in finding the area of a circle. (See Figure 12-14.) We shall learn later how to determine the area of a circle in square units.

Could other possibilities be chosen as the unit for two-dimensional area? How about a unit circle or a unit triangle? Can you see why the unit square is better than these?

The geoboard provides a grid of unit squares that can be used to demonstrate two-dimensional area.

**EXAMPLE 1**

Find the area of the regions bounded by the figures on the geoboards in Figure 12-15.

*SOLUTION*

(a) Count the 12 square units in the region bounded by the rectangle in Figure 12-15(a). Notice that the rectangle is 4 units long and 3 units wide and that $12 = 4 \cdot 3$.

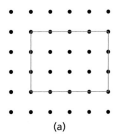

(a)

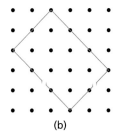

(b)

(b) For this rectangle we count the number of squares to be

$$7 \text{ full units} + 10 \text{ half units} = 12 \text{ full units}$$

For this rectangle, each of the longer sides is the hypotenuse of a right triangle, so

$$l^2 = 3^2 + 3^2 \quad \text{or} \quad l = \sqrt{18}$$

The width of the rectangle is also the hypotenuse of a right triangle, so

$$w^2 = 2^2 + 2^2 \quad \text{or} \quad w = \sqrt{8}$$

$$lw = \sqrt{18} \cdot \sqrt{8} = \sqrt{144} = 12$$

Thus it seems that (length) · (width) = area of a rectangular region.

The preceding examples suggest the following formula.

| | |
|---|---|
| **Area of a Rectangle** | If $l$ and $w$ (called *length* and *width*) are the linear measures of two consecutive sides of a rectangle forming the boundary of a rectangular region, then the *area* of the rectangular region is the product of the length and width, denoted by $A = lw$. |

Of course we can consider a square as a special case of a rectangle where $l = s$ and $w = s$ ($s$ being the side of the square). Thus, the area of the region bounded by a square is

$$A = s^2$$

 **PROPERTIES OF AREA**

Area, as discussed in the preceding examples, has the following properties.

| | |
|---|---|
| **Properties of Area** | 1. Area is additive; that is, the area of the whole is equal to the sum of the areas of the nonoverlapping parts. If a region is decomposed into a finite number of regions such that any two regions have in common only line segments, curves, or points, then the area of the whole region is the sum of the areas of the regions into which it is decomposed. |
| | 2. If figure $A \cong$ figure $B$, then the area bounded by $A$ equals the area bounded by $B$. |
| | 3. If a region is cut into parts and reassembled in a nonoverlapping manner to form another region, then the two regions have the same area. |

In Figure 12-16, the area of the rectangular region consisting of $A$ and $B$ in part (a) is the same as the area of the triangular region composed of $A$ and $B$ in part (b).

We use these properties of area to find the area of several figures on geoboards by counting the number of unit squares enclosed.

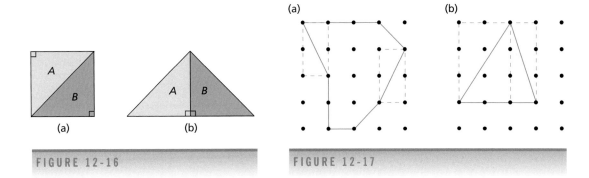

FIGURE 12-16

FIGURE 12-17

| EXAMPLE 2 | By counting the number of unit squares and the number of partial squares, find the area bounded by the closed figures in Figure 12-17. |

*SOLUTION*

(a) There are 7 complete unit squares in the figure. There are two half squares. Consider the left rectangle bounded by the broken lines. Half of the rectangle lies within the figure. Because the area of the rectangle is $1 \cdot 2$, the portion of interest is $\frac{1}{2}(1 \cdot 2) = 1$. The same is true for the rectangle on the right. Therefore, the area bounded by the figure is $7 + 2(\frac{1}{2}) + 2(1) = 10$ unit squares.

(b) Consider the triangle as a part of two rectangles (broken lines). The area of the rectangle on the left is 6 unit squares. Half of this rectangle lies within the triangle. Likewise, half of the rectangle on the right lies within the triangle. Therefore, the area of the triangle is $(\frac{1}{2})6 + (\frac{1}{2})3 = 4\frac{1}{2}$ unit squares.

## AREA MEASUREMENT

1 in.

1 square inch
(abbreviated 1 sq. in.
or 1 in.²)

1 cm

1 cm

1 square centimeter
(abbreviated 1 sq. cm
or 1 cm²)

FIGURE 12-18

Now let's extend our units of measurement to two dimensions. Suppose that you want to purchase carpet for your bedroom. The price of carpet is quoted in terms of a measurement of area called a *square yard*.

The basic units for area in the English system are the square inch (in.²), Figure 12-18, the square foot (ft.²), the square yard (yd.²), the acre (A), and the square mile (mi.²). Because 1 yd. = 3 ft., it follows that

$$1 \text{ yd.}^2 = 9 \text{ ft.}^2$$

$$1 \text{ ft.}^2 = 144 \text{ in.}^2$$

Table 12-4 summarizes units for area in the English system.

Conversion factors in the metric system are as follows:

$$1 \text{ m}^2 = 10^4 \text{ cm}^2 \qquad \text{or} \qquad 1 \text{ cm}^2 = 10^{-4} \text{ m}^2$$

$$1 \text{ km}^2 = 10^6 \text{ m}^2$$

Such conversion factors are found in Table 12-5. Other metric units for area include the **are** (a) (pronounced "air"), which is 100 m², and the **hectare** (ha), which is 100 a or 10,000 m².

| **TABLE 12-4** |
| --- |
| $1 \text{ ft.}^2 = 144 \text{ in.}^2$ |
| $1 \text{ yd.}^2 = 9 \text{ ft.}^2$ |
| $1 \text{ mi.}^2 = (1760)^2 \text{ yd.}^2$ |
| $1 \text{ A.} = 4840 \text{ yd.}^2$ |
| $1 \text{ mi.}^2 = 640 \text{ A.}$ |

| **TABLE 12-5** | | |
| --- | --- | --- |
| $1 \text{ km}^2 = 10^6 \text{ m}^2,$ | since | $1 \text{ km} = 10^3 \text{ m}$ |
| $1 \text{ hm}^2 = 10^4 \text{ m}^2,$ | since | $1 \text{ hm} = 10^2 \text{ m}$ |
| $1 \text{ dam}^2 = 10^2 \text{ m}^2,$ | since | $1 \text{ dam} = 10 \text{ m}$ |
| $1 \text{ dm}^2 = 10^{-2} \text{ m}^2,$ | since | $1 \text{ dm} = 10^{-1} \text{ m}$ |
| $1 \text{ cm}^2 = 10^{-4} \text{ m}^2,$ | since | $1 \text{ cm} = 10^{-2} \text{ m}$ |
| $1 \text{ mm}^2 = 10^{-6} \text{ m}^2,$ | since | $1 \text{ mm} = 10^{-3} \text{ m}$ |
| $1 \text{ }\mu\text{m}^2 = 10^{-12} \text{ m}^2,$ | since | $1 \text{ }\mu\text{m} = 10^{-6} \text{ m}$ |

Of course, Table 12-5 can be used for changing from one square metric unit to another; however, many people prefer the metric scale in Figure 12-19. Notice the two bars between units. This signifies that, if, for example, you were changing from square centimeters to square meters (this time moving to the left), you would move the decimal point four places to the left.

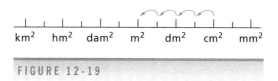

FIGURE 12-19

**EXAMPLE 3**

Change 30,000 cm² to square meters.

*SOLUTION*

$$1 \text{ cm}^2 = 0.0001 \text{ m}^2$$
$$30{,}000 \text{ cm}^2 = 30{,}000(0.0001) = 3 \text{ m}^2$$

**EXAMPLE 4**

Change 4 km² to square meters.

*SOLUTION*

$$1 \text{ km}^2 = 1{,}000{,}000 \text{ m}^2$$
$$4 \text{ km}^2 = 4(1{,}000{,}000) = 4{,}000{,}000 \text{ m}^2$$

To get a feel for the approximate size of areas expressed in SI units, consider the following areas:

Area of a postage stamp: 500 mm²
Area of a dollar bill: 100 cm²
Area of a wallet photo: 60 cm²

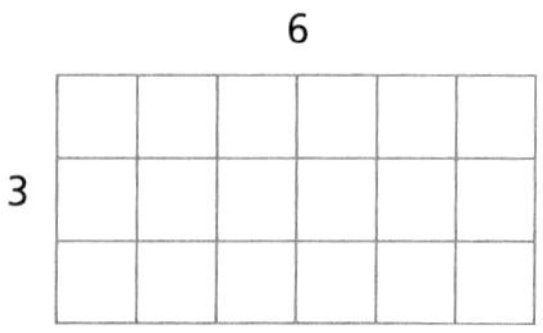

Suppose that the rectangle in Figure 12-20 is 6 cm long and 3 cm wide. When we multiply length times width to get area, we *can* think "6 cm times 3 cm equals 18 cm²," but this thinking is misleading. The multiplier tells the number of groups; the multiplicand tells how many in each group. We should think entirely in terms of area—a quantification of the squares needed to cover a surface. For the 6 cm × 3 cm rectangle, there are six 1 cm × 1 cm squares in each row (group) and three rows (groups); or there are six 1 cm × 1 cm squares per row, and there are three of these rows.

FIGURE 12-20

We multiply length times width for two reasons:

1. The number of segments in the length matches the number of squares in each row.
2. The number of segments in the width matches the number of rows.

By dimensional analysis,

$$\frac{6 \text{ cm}^2}{\text{row}} \cdot 3 \text{ rows} = 18 \text{ cm}^2$$

**EXAMPLE 5**

For a housing project, a residential property developer discovers a tract of land that is rectangular and measures 500 m by 1000 m. How many hectares are contained in this tract?

*SOLUTION*

$$\begin{aligned} A \text{ (of rectangle)} &= l \cdot w \\ &= 1000 \cdot 500 \text{ m}^2 \\ &= 500{,}000 \text{ m}^2 \\ &= \frac{500{,}000}{10{,}000} = 50 \text{ ha} \end{aligned}$$

**EXAMPLE 6**

Using dimensional analysis, convert 3 m² to square yards (2.54 cm = 1 in.).

*SOLUTION*

$$3 \text{ m}^2 \cdot \frac{100^2 \text{ cm}^2}{1 \text{ m}^2} \cdot \frac{1 \text{ in.}^2}{(2.54)^2 \text{ cm}^2} \cdot \frac{1 \text{ yard}^2}{(36)^2 \text{ in.}^2} \approx 3.59 \text{ yard}^2$$

# AREA OF A RIGHT TRIANGLE

Intuitively, we know that the diagonal of the rectangle in Figure 12-21(a) divides the rectangle into two right triangles of equal area. (We will prove later that these two right triangles are indeed congruent.) Thus the area of each right triangle is one-half the area of the rectangle, or

$$\text{Area bounded by a right triangle} = \frac{1}{2}lw$$

where $l$ and $w$ are the two legs of the right triangle.

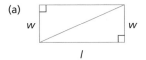
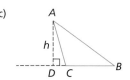

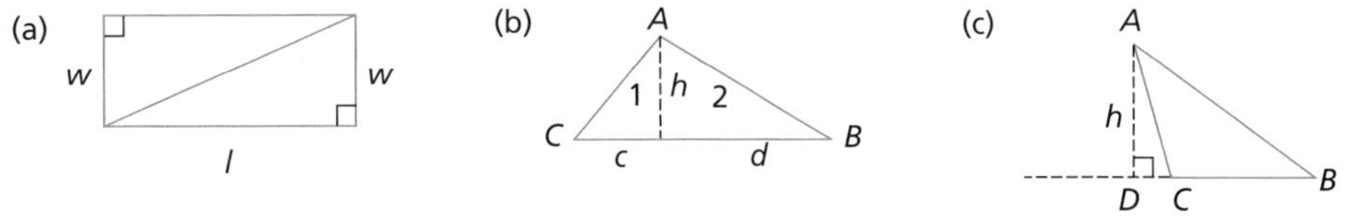

(a)

(b)

(c)

**FIGURE 12-21**

# AREA OF A TRIANGLE

We use the preceding formula for the area of a right triangle to find the area bounded by any triangle (called the *area of the triangle*). We consider two cases in Figure 12-21(b) and (c). In each case, we draw a perpendicular from the vertex $A$ to the opposite side. This perpendicular is called the **altitude,** and its measure is denoted by $h$. The opposite side to which the altitude is drawn is called the **base** of the triangle, and it is denoted by $b$. Notice that in Figure 12-21(b) the altitude lies inside the triangle, while in (c) it lies outside the triangle. In (b) the area of $\triangle ABC$ is computed as

$$\text{Area of } \triangle ABC = \text{area of right } \triangle 1 + \text{area of right } \triangle 2$$

$$= \frac{1}{2}ch + \frac{1}{2}dh$$

$$= \frac{1}{2}h(c + d)$$

$$= \frac{1}{2}hb \qquad c + d \text{ equals base } b.$$

In (c) the area of $\triangle ABC$ is computed as

$$\text{Area of } \triangle ABC = \text{area of right } \triangle ABD - \text{area of right } \triangle ACD$$

$$= \frac{1}{2}h(DB) - \frac{1}{2}h(DC)$$

$$= \frac{1}{2}h(DB - DC)$$

$$= \frac{1}{2}hb \qquad DB - DC = CB = b$$

# AREA OF A TRAPEZOID

We will use the formula for the area of a triangle to find the area bounded by a trapezoid. First, however, let's consider the area of the trapezoidal region on a geoboard. In Figure 12-22(a), the area of the region is

$$7 \text{ full units} + 2 \text{ half units} = 8 \text{ full units}$$

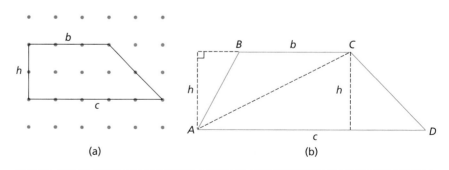

(a)  (b)

FIGURE 12-22

Now if we let $b$ and $c$ be the measures of the opposite parallel sides of the trapezoid (with $b = 3$ and $c = 5$) and if $h = 2$ is the perpendicular to each parallel side (called the *altitude*), then

$$\frac{1}{2}(b + c)h = \frac{1}{2}(3 + 5) \cdot 2 = 8$$

This is the same answer we obtain by counting.

In general, the area of the region bounded by a trapezoid may be considered as the sum of the areas of regions bounded by two triangles. For instance, consider trapezoid $ABCD$ in Figure 12-22(b). The area of the region bounded by $ABCD$ is

$$ABCD = \text{area of } \triangle ABC + \text{area of } \triangle ACD$$
$$= \frac{bh}{2} + \frac{ch}{2} = \left(\frac{b + c}{2}\right)h$$

Thus the area of the region bounded by a trapezoid is one-half the sum of the measures of the opposite parallel sides times the measure of the altitude—that is, $\frac{1}{2}(b + c)h$.

## AREA OF A PARALLELOGRAM

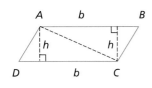

**FIGURE 12-23**

The area of a parallelogram can be treated as the sum of the areas of two triangles. In Figure 12-23, for example, the area of parallelogram $ABCD$ is

$$\text{Area of } ABCD = \text{area of } \triangle ABC + \text{area of } \triangle ACD$$
$$= \frac{1}{2}bh + \frac{1}{2}bh = bh$$

The formulas for the areas of several regions bounded by polygons are given next.

---

**Area Formulas for Polygonal Regions**

1. The area $A$ of the region bounded by a **parallelogram** is found by multiplying the measure of the base $b$ and the measure of the altitude or height $h$.
$$A = bh$$

2. The area $A$ of the region bounded by a **triangle** is one-half the product of the measure of the base $b$ and the measure of the altitude or height $h$.
$$A = \frac{1}{2}bh$$

3. The area $A$ of the region bounded by a **trapezoid** is the product of one-half of the sum of the measures of the opposite parallel sides $b$ and $c$ and the measure of the altitude or height $h$.
$$A = \frac{1}{2}(b + c)h$$

In this case, $(b + c)/2$ is sometimes called the **average base.**

**PRACTICE PROBLEM**

(a) What is the area of a region bounded by a parallelogram with a base of 5 cm and a height of 3 cm?

(b) Find the area of a triangular region with a base of 6 cm and a height of 3 cm.

(c) Find the area of a region bounded by a trapezoid whose parallel sides measure 6 cm and 10 cm and whose height measures 5 cm.

*ANSWER*    (a) 15 cm²    (b) 9 cm²    (c) 40 cm²

# AREA OF A REGULAR *N*-GON

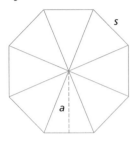

FIGURE 12-24

For a regular polygon with $n$ sides, called an ***n*-gon,** draw lines from the center of the polygon to each vertex. (See the octagon in Figure 12-24.) These lines from the center to each vertex, along with the sides of the polygon, form $n$ isosceles triangles. Because the polygon is regular, these $n$ triangles are congruent.

The area of the $n$-gon is the sum of the areas of the $n$ congruent isosceles triangles. Thus the area of the polygon is $n$ times the area of a single isosceles triangle. Let the length of each side of the polygon be $s$, and let the length of the perpendicular from the center to a side be $a$. (The distance $a$ from the center of a regular polygon to each of the sides is called the **apothem** of the polygon.) The area of each isosceles triangle is $\frac{1}{2}sa$. The area of the polygon is

$$A = n(\text{area of each triangle})$$

$$= n\frac{1}{2}sa$$

$$= \frac{1}{2}(ns)(a)$$

But $ns$ is the perimeter ($P$) of the $n$-gon. Thus, the area of the polygon is $\frac{1}{2}Pa$.

---

**Area of a Regular *n*-gon**

The area of a regular $n$-gon with sides $s$ is given by

$$A = \frac{Pa}{2}$$

where $P = ns$ is the perimeter of the $n$-gon and $a$ is the apothem.

---

**PRACTICE PROBLEM**

Find the area of a regular hexagon with side 2 and apothem $\sqrt{3}$.

*ANSWER*    $6\sqrt{3}$

# CIRCUMFERENCE OF A CIRCLE

The perimeter of a circle is called the **circumference** of the circle and is, of course, the total length of the simple closed curve. Intuitively, we can imagine stretching a string along the circle and then measuring the string. Thus the measurement of a circumference is given in linear units.

The ancient Greeks found by experimentation that, when the circumference of a circle was divided by the diameter, the same answer resulted, regardless of the size of the circle. At first they used 3 and then 3.14 for this number. Later, it was proved that the quotient of the circumference of the circle and the diameter is an irrational number called *pi* and denoted by the lowercase Greek letter $\pi$. Thus,

$$C = \pi d$$
$$C = 2\pi r$$

where $r$ is the radius.

## CALCULATOR HINT

Many calculators have a key for $\pi$ that gives the value of $\pi$ to eight or ten decimal places. Use your calculator to find the circumference of a circle with radius 2.71647541. $C = 17.06812$

**EXAMPLE 7**

If a bicycle wheel has a radius of 50 cm, how many times does it turn in going 1 km? Approximate $\pi$ by 3.14.

*SOLUTION*

$$\text{Circumference} = 2\pi(\text{radius})$$
$$= 100\pi \text{ cm}$$
$$\approx 100(3.14) \text{ cm}$$
$$\approx 314 \text{ cm}$$
$$1 \text{ km} = 100,000 \text{ cm}$$
$$\frac{100,000}{314} \approx 318 \text{ revolutions}$$

# AREAS OF CIRCULAR REGIONS

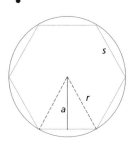

FIGURE 12-25

Suppose that a regular $n$-gon is inscribed in a circle, so that the circle contains each vertex of the $n$-gon. (See the hexagon in Figure 12-25.) Let the perimeter of the polygon be $P$ and the apothem $a$. We know that the area of the polygon is

$$\frac{aP}{2}$$

As the number of sides $n$ increases, $a$ approaches $r$ and $P$ approaches the circumference of the circle. Thus, for very large values of $n$, $aP/2$ is approximately

$$\frac{rC}{2} = \frac{r}{2}(2\pi r) = \pi r^2$$

Consequently, as the number of sides increases, the area of the polygon approaches the area of the circle and gets close to $\pi r^2$. It can be shown that $\pi r^2$ is actually the area of a circle.

| **Area of a Circle** | There is a real number $\pi$ in terms of which the area ($A$) of a circle is computed. $$A = \pi r^2$$ |
| --- | --- |

**EXAMPLE 8**

Find the area of the shaded region in Figure 12-26.

*SOLUTION*

To solve this problem, we need to employ some of our problem solving strategies. Because we have not developed a formula for regions like the shaded region, perhaps we should ask, "Is there a simpler problem that I can solve?" Certainly we can find the area of each of the circular regions.

$$\text{Area} = \pi r^2 = \pi(1)^2 = \pi$$

Likewise, we can find the area of the rectangle.

$$\text{Area} = lw = 4 \cdot 2 = 8$$

Now, can we put this information together to find the area of the shaded regions.

$$\text{Shaded region} = \text{area of rectangle} - \text{area of two circles} = 8 - 2\pi$$

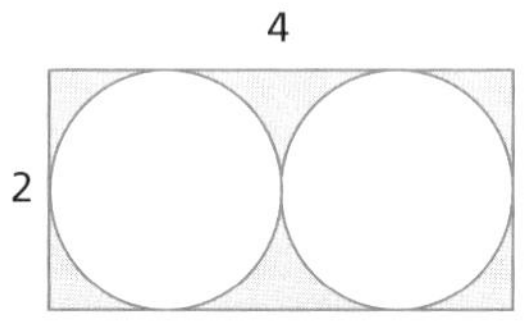

**FIGURE 12-26**

# RADIAN MEASURE

The ancient Babylonian civilization provided us with a method of measuring angles based on choosing the unit, the degree, so that 360 degrees make a full rotation. A more modern unit for measuring angles is called the **radian.** The radian has the property that $2\pi$ radians equals a full rotation, $\frac{\pi}{2}$ radians is equivalent to $90°$, and $\pi$ radians is equivalent to $180°$.

These relationships can be used to convert either from degree measure to radian measure or from radian measure to degree measure.

**EXAMPLE 9**

Write $1.5\pi$ radians in terms of degrees.

*SOLUTION*

$\pi$ radians $= 180°$. Therefore,

$$\frac{x°}{180°} = \frac{1.5\pi \text{ radians}}{\pi \text{ radians}}$$

$$x = 1.5(180°) = 270°$$

# ⋯ Arc and Sector of a Circle

The pie shaped portion of a circular region included between the sides of a central angle is called a **sector** of the circle. (See Figure 12-27.) The portion of the circle determined by the sides of the central angle is called an **arc** of the circle. The length of an arc of a circle divided by the radius of the circle gives the measure of the central angle in radians. If the radius of a circle and the arc length are of the same length, the central angle is said to have a measure of one radian.

**EXAMPLE 10**

If the radius of a circle is 4 cm, find the measure of the central angle in radians that gives an arc length of 2 cm.

*SOLUTION*

Measure of the central angle = 2 cm/4 cm = 0.5 radians

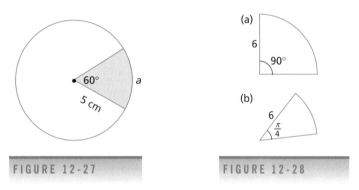

FIGURE 12-27                    FIGURE 12-28

**EXAMPLE 11**

Find the areas of the sectors of the circles shown in Figure 12-28 if each radius is 6.

*SOLUTION*

The area of a circle of radius 6 is $\pi 6^2 = 36\pi$. The area of the sector in Figure 12-28(a) is obtained from

$$\frac{90°}{360°} = \frac{A}{36\pi}$$

$$A = 9\pi$$

The area of the sector in Figure 12-28(b) is given by

$$\frac{\pi/4}{2\pi} = \frac{A}{36\pi}$$

$$\frac{1}{8} = \frac{A}{36\pi}$$

$$8A = 36\pi \text{ or } A = 4.5\pi$$

We now return to the introductory problem.

**EXAMPLE 12**

The famous Yin-Yang symbol in Figure 12-29 has been constructed from a circle of radius $r$. Compute the length of the arc that divides the colored and white regions.

*SOLUTION*

This arc consists of one-half the circumference of the two small circles, each with diameter $r$. Therefore, the length equals

$$\frac{1}{2}(\pi r) + \frac{1}{2}(\pi r) = \pi r$$

But $2\pi r$ is the circumference of the big circle. Therefore, the length of the dividing arc is one-half the circumference of the big circle.

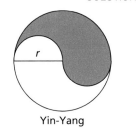

Yin-Yang

FIGURE 12-29

# Just for Fun

Al and Yute are racing around the track. If the center of Al's lane is 1 m outside the center of Yute's lane, in one lap, how much farther will Al run than Yute?

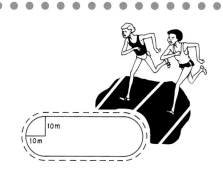

# EXERCISE SET 2

**R** 1. The following figures have been drawn on geoboards. For each figure, use the reasoning of Example 2 to count the square units bounded by the figure and thereby find the area. We will use this information in the next section to verify formulas.

(a)

(b)

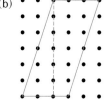

(c)

(d)

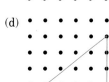

(e) (f)

(g) (h)

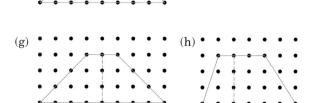

2. Find the areas bounded by each of the following polygons.

Rectangles:

(a) 3 m by 5 m

(b) 21.4 cm by 170 dm (express in dm²)

Triangles:
(c)  $b = 34$, $h = 52$
(d)  $b = 23$ cm, $h = 497$ mm (express in mm²)

Parallelograms:
(e)  $b = 24$, $h = 12$
(f)  $b = 53$ m, $h = 539$ dm (express in m²)

Trapezoids:
(g)  $h = 3$, $b = 3$, $c = 2$
(h)  $h = 3$ cm, $b = 124$ mm, $c = 5$ cm (express in cm²)

3. In (a) and (b) of Exercise 1, using the dashed line as an altitude, verify the formula for the area of a parallelogram.

4. In (c) and (d) of Exercise 1, verify the formula for the area of a right triangle.

5. In (e) and (f) of Exercise 1, using the dashed line as an altitude, verify the formula for the area of a triangle.

6. In (g) and (h) of Exercise 1, verify the formula for the area of a trapezoid.

7. Convert each measurement to square centimeters.
   (a) 34 m²      (b) 328 mm²      (c) 3100 mm²
   (d) 41 km²     (e) 42 km²       (f) 863 m²

8. Select the best approximation of each area.
   (a) The front of this book: (i) 4 m², (ii) 400 cm², (iii) 40 cm²
   (b) A football field: (i) 0.6 ha, (ii) 6 m², (iii) 600 m²

9. Complete the following conversions.
   (a) 6 yards² = _____ ft.²
   (b) 2 yards² = _____ in.²
   (c) 0.02 miles² = _____ yards².
   (d) 3600 ft.² = _____ yards².
   (e) 0.01 A. = _____ ft.²

10. Explain the difference between 4 m² and a 4-m square (one that is 4 m on each side).

11. Which of the two areas given in each pair is larger?
    (a) 34 km² or 129,000 m²
    (b) 1234 μm² or 3 cm²
    (c) 4 ha or 456,000 m²
    (d) 916 mm² or 1 cm²

12. A farm is square and measures 100 m on a side. What is the area of this farm in hectares?

13. Find the circumference and the area of a circular region if the radius is as follows. This time, approximate π by 3.1416. Assume that all measurements are accurate to 5 significant digits.
    (a) 4 ft      (b) 10 m
    (c) 12 cm     (d) 9 yards

14. The given figure is a semicircle; the area of the interior of this closed curve is $24.5\pi$. Find the distance around the closed curve.

15. Find the area of each figure.

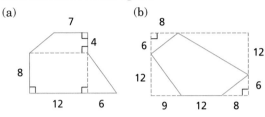

(a)                              (b)

16. Fill in the following blanks.
    (a) 60° = _____ radians
    (b) 240° = _____ radians
    (c) $\dfrac{\pi}{6}$ radians = _____ degrees
    (d) 4 radians = _____ degrees

17. For each of the following circles, find the missing arc length ($a$), angle measure ($\theta$), or radius ($r$). Then find the area of each sector shown.

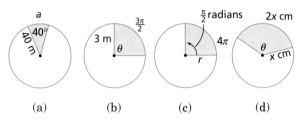

   (a)         (b)         (c)         (d)

18. For practice in metric conversions, perform the following additions and subtractions.
    (a)    2 m² + 80 dm² + 92 cm²
        + (10 m² + 46 dm² + 34 cm²)

    (b)    16 m² + 70 dm² + 56 cm²
        − (3 m² + 84 dm² + 72 cm²)

*In Exercises 19 through 23, find the area of each shaded region.*

19.

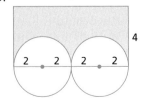

20.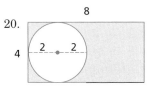

21.

22.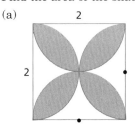

23.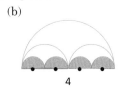

24. If the radii of two circles are in a ratio of 1 to 3, what is the ratio of their circumferences? Their areas?

*C* 25. *Estimate the answers based on your understanding of the sizes of various units of measurement in the English and metric systems.*

   (a) 1300 m² ≈ _____ acres
   (b) 15,793 mm² ≈ _____ square inches
   (c) 47 ft². ≈ _____ square meters
   (d) 3.77 km² ≈ _____ square miles
   (e) 21 km² ≈ _____ square yards
   (f) 29 A. ≈ _____ hectares
   (g) 36 in.² ≈ _____ square centimeters
   (h) 156 yards² ≈ _____ square meters

26. In Exercise 25, use dimensional analysis to find the answer, knowing that 1 in. equals 2.54 cm.

27. Find the area of the shaded regions.

   (a)           (b)

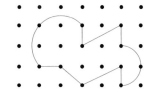

28. Find the area bounded by the following figure.

29. In trapezoid $ABCD$, $\overline{AB}$ is parallel to $\overline{CD}$, and $\overline{AD}$ is perpendicular to $\overline{CD}$. If $AB = 6$, $AD = 4$, and $DC = 8$, find the area of the trapezoidal region.

30. For a can containing three tennis balls, show that the circumference of the can is larger than the height of the can.

31. Find the area of an equilateral triangle with sides of length 10.

*For Exercises 32 through 34 find the circumference of the circle that is circumscribed (contains each vertex of the polygon) about the named polygon.*

32. A square with side 10.

33. A right triangle with legs of 6 and 8. (HINT: The center of the circle is at the middle of the hypotenuse.)

34. A regular hexagon with side 10.

35. If a car goes 80 km per hour, how many revolutions does a wheel 77 cm in diameter make in a half hour?

36. Find the ratio of the area of an inscribed square to the area of a circumscribed square for the same circle.

## REVIEW EXERCISES

37. Find the perimeter and the area of each of the following, where each unit is 1 cm.

   (a)          (b)

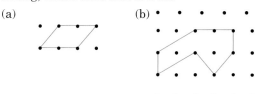

   (c)          (d)

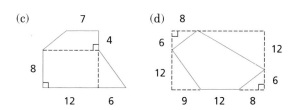

38. Perform these conversions.

   (a) 5001 cm = _____ km
   (b) 52 mm = _____ cm
   (c) 81 km = _____ m
   (d) 17 cm = _____ m
   (e) 2.61 km = _____ cm
   (f) 421 dm = _____ mm

39. Given the accompanying number line, find the following.

   (a) $AB$     (b) $CD$
   (c) $AC$     (d) $BD$

40. Which unit—the kilometer, the meter, the centimeter, or the millimeter—would you most likely use to measure the following?

    (a) Length of a pencil
    (b) Distance from Boston to New York
    (c) Thickness of glasses
    (d) Width of a postcard
    (e) Height of your classroom

41. To better understand the derivations in this chapter, consider the following investigations using *Geometer's Sketchpad*. (Suggested steps are given below.)

    1. Area of a parallelogram
    2. Area of a Triangle
    3. Area of a Trapezoid

    (a) First draw a rectangle *ABCD*, and measure the area of the rectangle. Now move *A* along $\overleftrightarrow{AB}$ forming parallelogram *ABCD*. Measure its area at different points for *A*. Make a conjecture.
    (b) Form triangle *ABC* and draw a line through *B* parallel to $\overleftrightarrow{AC}$. Move *B* along the line and compare the areas of the triangles. What do you find?
    (c) Now draw a line through *C* parallel to $\overleftrightarrow{AB}$ forming a parallelogram. Compare areas of △ABC and the parallelogram for each position of B. Make a conjecture. What do you find?
    (d) Now investigate the areas of trapezoids.

42. A metal band is stretched around the earth on a great circle. (Assume the earth is a perfect sphere and has a diameter of 40 million ft.)

    (a) Guess how much must the band be increased in length to be 1 ft above the earth all the way around? (i) The bands are the same length, (ii) the second band is just a bit longer (less than 10 ft), or (iii) the second band is much longer (at least greater than 10 ft).
    (b) Compute the length of the band stretched around the earth.
    (c) Now compute the length of the band that is 1 foot above the earth. Take the difference. Are you surprised at the answer?

    (d) Instead of the earth, place the same two bands around a beach ball of radius 4 ft. The second band is how much longer than the first band? Again, are you surprised at the answer?
    (e) Can you show algebraically, by using the formula for the circumference of a circle, why this is true?

43. Answer the questions below to help you draw a figure on the geoboard (vertices at points on the geoboard) that has the following attributes:

    > It is a pentagon.
    > It is nonconvex (concave).
    > Two sides are parallel.
    > It has one right angle, one 45° angle, one additional acute angle, one obtuse angle, and one interior angle > 180°.
    > It contains 6 points and encloses 2 points.
    > It bounds an area of 4 square units.

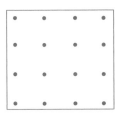

    (a) When is a polygon a pentagon?
    (b) What is the distinctive characteristic of a nonconvex polygon?
    (c) Discuss the various ways line segments can be parallel and contain points of the geoboard.
    (d) Discuss how angles with vertices at points of the geoboard and sides containing points of the geoboard can be right angles.
    (e) As in (d), discuss how an angle can be a 45° angle.
    (f) As in (d), discuss an obtuse angle on a geoboard.
    (g) Form a pentagon with the characteristics of (a) through (f) that contains six points and that encloses two points.
    (h) Does your pentagon bound a total area of 4 square units? If not, use the problem-solving strategy of guess, test, and revise to find the pentagon that bounds this area and encloses two points. Don't forget the characteristics you described in (a) through (f).

# VOLUME AND SURFACE AREA

**PROBLEM**

One gallon of paint covers 60 m² of surface area. If paint costs $12 per gallon, how much does it cost to paint the outside of the garage pictured below? Paint the garage door, but do not paint the roof.

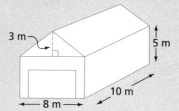

**OVERVIEW**

We now turn our attention to the volume of space regions. Just as a unit segment is used to measure length and a unit square is used to measure area, a unit cube is used to measure volume. Hence, our metric unit for volume will be the cubic meter—a cube each of whose six faces is a square measuring a meter on a side.

However, we shall not be concerned with volume alone in this section. The surface of a solid region must be measured, too; this measure is called *surface area*, and it is measured in terms of unit squares (for example, the square meter). You are interested in surface area when determining the amount of paint needed to paint a barn. You are interested in volume when computing the amount of space within the barn.

**GOALS**

Illustrate the Measurement Standard, page 532.
Illustrate the Geometry Standard, page 482.
Illustrate the Repesentation Standard, page 50.

## TOTAL SURFACE AREA OF A CYLINDER AND A PRISM

In the preceding section, we considered the concept of area for certain plane figures. In this section, we extend the concept of area to the surface area of space solids. The easiest way to introduce this concept is by looking at patterns that could be used to form these solids.

**EXAMPLE 1**

Consider the right prism in Figure 12-30. Imagine a pattern cut out of cardboard that could be glued to form this solid. The pattern is pictured in Figure 12-30(b). The area of the rectangle in the middle is called the **lateral area** of the solid. The combined area of the two bases and the lateral area is called the **total surface area** of the solid.

**EXAMPLE 2**

Look at the pattern for the right pentagonal prism in Figure 12-31. Again, the lateral surface area is given by the rectangle in the middle, and the total surface area is obtained by adding this to the area of the two bases.

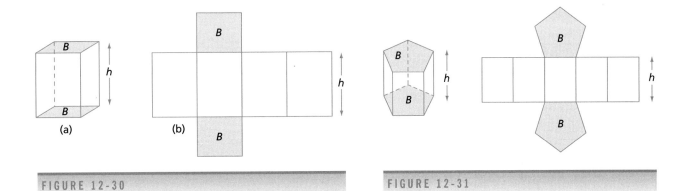

FIGURE 12-30

FIGURE 12-31

**EXAMPLE 3**

Consider the right cylinder in Figure 12-32. The lateral surface area is obtained from the rectangle in the middle, which is of length $2\pi r$ (circumference of the circle). To get the total surface area, we add this to the area of the two bases.

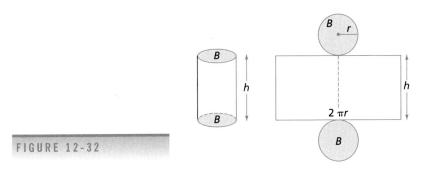

FIGURE 12-32

The preceding examples suggest the following definition, which can be used to determine the total surface area of many solids.

| Areas of a Surface | The *lateral surface area* of a surface is the sum of the areas of all the faces or sides (other than the bases) of the space figure. The *total surface area* is the lateral surface area added to the area of the bases of the space figure. |
|---|---|

**EXAMPLE 4**

Find the total surface area of a right rectangular prism, where $l = 2$, $w = 4$, and $h = 5$.

*SOLUTION*

$$\text{Total surface area} = 2(2 \cdot 4) + 2(2 \cdot 5) + 2(4 \cdot 5) = 76$$

**EXAMPLE 5**

Find the total surface area of a cylinder of height 6 cm and radius 4 cm.

*SOLUTION*

$$\text{Surface area} = \text{lateral surface area} + 2B$$
$$= 2\pi rh + 2\pi r^2$$
$$= 2\pi(24 + 16)$$
$$= 80\pi \text{ cm}^2$$

**PRACTICE PROBLEM**

Find the lateral area of the space region shown in Figure 12-33.

*ANSWER*    57

# ⋮VOLUME MEASURE

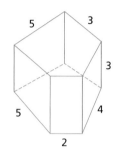

FIGURE 12-33

Just as length and area measures associate unique numbers, volume measure associates a unique number with each closed-space region. There is exactly one number for each closed-space region as measured with a given unit of measurement. The unit of measurement of volume is ordinarily the unit cube. In a unit cube, each edge of each square face measures 1 unit. As an illustration of the use of a unit cube, consider the volume bounded by a right rectangular prism (sometimes called a *rectangular parallelepiped*). A right rectangular prism is similar to a cube except that its faces are rectangles that are not necessarily squares. To find the volume of the region bounded by a rectangular prism, determine how many unit cubes can be fitted into it. Consider the rectangular prism in Figure 12-34(a). Here, 6 cubes can be placed on 1 edge (the length), 4 on a second edge (the width), and 3 on the third edge (the height). To find the total number of cubic units, we take the total number of cubes on the base ($4 \cdot 6 = 24$) and recognize that this number of cubes is used on each of the three levels ($24 \cdot 3 = 72$); so the total volume is 72 cubic units.

To obtain an understanding of volume quickly, students are often asked to place unit cubes into a space region or to count the number of unit cubes in a region.

**EXAMPLE 6**

Using the given unit cube, find the volume of the space region in Figure 12-34(b).

*SOLUTION*

Because there are only one row and two layers of cubes, the volume is obviously 8.

**EXAMPLE 7**

By counting, find the volume of the space regions in (a) and (b) of Figure 12-35.

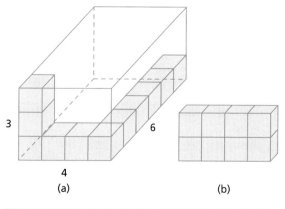

(a)                              (b)

FIGURE 12-34

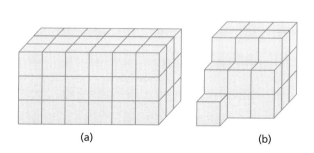

(a)                              (b)

FIGURE 12-35

*SOLUTION*

(a) In Figure 12-35(a), there are 3 layers of cubes. Each layer contains $6 \cdot 3$ cubes, so the volume is

$$(6 \cdot 3) \cdot 3 = 54 \text{ unit cubes}$$

(b) In Figure 12-35(b), the bottom layer has $3 \cdot 3 + 1 = 10$ cubes. The second layer has $3 \cdot 3 = 9$ cubes. The third layer has $2 \cdot 3 = 6$ cubes. Therefore, $V = 10 + 9 + 6 = 25$ cubes

To find the volume of a prism, we multiply base area times height. There are two reasons for this:

1. The number of squares in the base area matches the number of cubes or cubic units in the bottom layer.
2. The number of segments in the height matches the number of layers.

By dimensional analysis, a prism with a base of 3 cm $\times$ 4 cm and a height of 5 cm has 12 squares of area in the base, 12 cubic units per layer, and 5 layers.

$$\frac{12 \text{ cm}^3}{\text{layer}} \cdot 5 \text{ layers} = 60 \text{ cm}^3$$

The preceding discussion suggests the following formula for the volume of a right rectangular prism.

| | |
|---|---|
| **Volume of a Right Rectangular Prism** | The volume of a right rectangular prism is the product of the prism's length, width, and height.<br><br>$$V = lwh$$ |

We have already applied some properties of the volume of space regions in the preceding discussion. Can you recognize which ones?

| | |
|---|---|
| **Properties of Volume** | 1. Volume is additive; that is, the volume of the whole is equal to the sum of the volumes of the nonoverlapping parts. If $R$ and $S$ are nonoverlapping space regions (possibly with surfaces in common), then the volume of $R \cup S$ is equal to the volume of $R$ plus the volume of $S$.<br>2. If space region $R$ has the same size and shape as space region $S$, then the volume of space region $R$ equals the volume of space region $S$.<br>3. If a space region is cut into parts and reassembled without overlapping to form another space region, then the two space regions have the same volume. |

# UNITS OF MEASUREMENT FOR VOLUME

You are familiar with the basic units for volume in the English system. A representation of a cubic inch (not drawn to scale) is found in Figure 12-36.

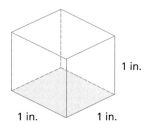

FIGURE 12-36

Two basic conversion equivalences in this system are

$$1728 \text{ in.}^3 \text{ (cubic inches)} = 1 \text{ ft}^3. \text{ (cubic feet)}$$

$$27 \text{ ft}^3. \text{ (cubic feet)} = 1 \text{ yard}^3 \text{ (cubic yards)}$$

Under the SI system, volume measure is based on the cubic meter ($m^3$). For small volumes, the cubic centimeter ($cm^3$) or the cubic millimeter ($mm^3$) is used as the unit of volume.

The volume of the figure depicted in Figure 12-37 illustrates a cubic centimeter ($cm^3$), which is about the size of a sugar cube. A cube is a special case of a right rectangular prism where $l = w = h = s$, so $V = s^3$. If the length of one edge of a cube is 4 m, then the volume of the cube is

$$V = s^3 = 4^3 = 64 \ m^3$$

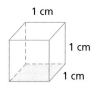

FIGURE 12-37

Table 12-6 summarizes the relationships among units of volume measure in the SI system. Notice that, because

$$1 \text{ km} = 10^3 \text{ m}$$

then

$$1 \text{ km}^3 = (10^3)(10^3)(10^3) \text{ m}^3$$

$$= 10^9 \text{ m}^3$$

| **TABLE 12-6** |
|---|
| $1 \text{ km}^3 = 10^9 \text{ m}^3$, since 1 km = $10^3$ m |
| $1 \text{ hm}^3 = 10^6 \text{ m}^3$, since 1 hm = $10^2$ m |
| $1 \text{ dam}^3 = 10^3 \text{ m}^3$, since 1 dam = 10 m |
| $1 \text{ dm}^3 = 10^{-3} \text{ m}^3$ (called a *liter*, symbolized by L) |
| $1 \text{ cm}^3 = 10^{-6} \text{ m}^3$ (called a *milliliter*, symbolized by mL) |
| $1 \text{ mm}^3 = 10^{-9} \text{ m}^3$, since 1 mm = $10^{-3}$ m |
| $1 \ \mu\text{m}^3 = 10^{-18} \text{ m}^3$, since 1 $\mu$m = $10^{-6}$ m |

The cubic meter, the cubic centimeter, and the cubic millimeter are the SI units of volume used in ordinary transactions.

You can use either Table 12-6 or a metric scale for cubic units to find equivalences, realizing that for each jump between consecutive cubic units you must move the decimal point 3 places.

**EXAMPLE 8**

Convert 321 $m^3$ to cubic centimeters.

*SOLUTION*

$$1 \text{ m}^3 = 1,000,000 \text{ cm}^3$$

$$321 \text{ m}^3 = 321(1,000,000) = 321,000,000 \text{ cm}^3$$

# CAVALIERI'S PRINCIPLE

We now discuss a principle first stated by Bonaventura Cavalieri.

In Figure 12-38, for example, if the shaded area of the circle in (a) is

| Cavalieri's Principle | Given: a plane and any two solids with bases in the plane. If every plane parallel to the given plane intersects the two solids in cross-sections that have the same area, then the solids have the same volume. |
| --- | --- |

equal to the shaded area of the triangle in (b), for planes $\alpha$ parallel to the two bases then the triangular prism and the cylinder have the same volume.

Cavalieri pictured a solid as being made up of infinitely many thin slices. (See Figure 12-39.)

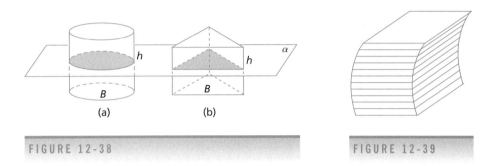

FIGURE 12-38                                    FIGURE 12-39

## VOLUME OF A PRISM AND A CYLINDER

Cavalieri's principle enables us to find the volume of an **oblique prism** (any prism that is not a right prism) and an oblique (not right) cylinder as easily as we could find the volume of a right prism and a right cylinder. Consider the right prism in Figure 12-40(a) and the oblique prism in Figure 12-40(b). Suppose it is known that the parallel planes $\alpha_1$, $\alpha_2$, and $\alpha_3$ intersect the prism in (a) and in (b) in cross-sections of equal area—that is, $A_1 = A'_1, A_2 = A'_2$, and $A_3 = A'_3$. Thus, by Cavalieri's principle, the volume of the prism in (a) equals the volume of the prism in (b), namely $A_1 h$. By similar arguments, oblique cylinders as in Figure 12-40(c) have the same volume as right cylinders with congruent bases and equal heights.

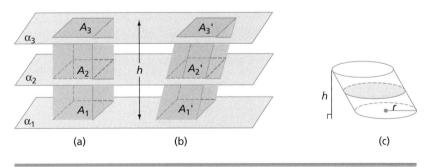

FIGURE 12-40

| Volume of Prism or Cylinder | The volume of any prism or cylinder is the area ($B$) of a cross-section cut by a plane parallel to the base multiplied by the altitude ($h$):<br><br>$$V = Bh$$ |
| --- | --- |

**EXAMPLE 9**

Consider a triangular prism with an altitude of 10. The base and altitude of the triangular bases are 6 and 8, respectively. Find the volume of the space region bounded by the prism.

*SOLUTION*

$$\text{Area of base} = \frac{1}{2}(6)(8) = 24$$

$$\text{Volume} = 24 \cdot 10 = 240$$

**EXAMPLE 10**

A tin can (circular cylinder) has a diameter of 10 and a height of 25. Find the volume and the total surface area of the can.

*SOLUTION*

$$\text{Volume} = \pi r^2 h = \pi(5)^2(25) = \pi(25)(25)$$

$$= 625\pi \approx 625\left(\frac{22}{7}\right) \approx 1964 \text{ cubic units}$$

$$\text{Total surface area} = (2\pi r)h + 2\pi r^2 = (2\pi \cdot 5) \cdot 25 + 2\pi(5)^2$$

$$= 250\pi + 50\pi = 300\pi \approx 943 \text{ square units}$$

## SURFACE AREA AND VOLUME OF A PYRAMID

If a pyramid is regular, its lateral faces are isosceles triangles. This is true regardless of the number of edges of the polygonal base. A perpendicular from the vertex of the pyramid to the base of one of the isosceles triangles is called the **slant height** of the pyramid. The area of each of these isosceles triangles is one-half the slant height times an edge of the polygonal base, or

$$\frac{ts}{2}$$

where $t$ represents the slant height and $s$ the side of the polygonal base. (See Figure 12-41.)

Now if the polygonal base has $n$ edges, the surface area of the $n$ triangles is

$$\frac{n(ts)}{2} = \frac{t(ns)}{2}$$

But $ns$ is the perimeter of the base. Therefore, the lateral surface area of a regular pyramid can be expressed as

$$\frac{tP}{2}$$

where $t$ is the slant height and $P$ is the perimeter of the base.

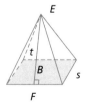

FIGURE 12-41

**Surface Area of a Pyramid**

For a regular pyramid with slant height $t$, perimeter of the base $P$, and area of the base $B$,

$$\text{Lateral surface area} = \frac{tP}{2}$$

$$\text{Total surface area} = \frac{tP}{2} + B$$

**EXAMPLE 11**

Find the total surface area of a regular pyramid with a square base that is 10 cm long on each side and has a slant height of 8 cm.

*SOLUTION*

Figure 12-42 presents a visual interpretation. The total area would be the area of 4 triangles plus the area of the square, or

$$4\left(\frac{8 \cdot 10}{2}\right) + 10^2 = 260$$

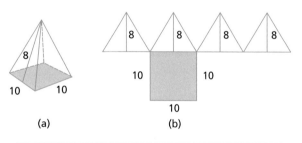

(a)    (b)

FIGURE 12-42

Consider in Figure 12-43 the triangular prism $PQCDEF$ with base of area $B$ and altitude $h$. It is possible to cut the prism into three pyramids that do not overlap, and it can be shown that these pyramids have equal volumes. Therefore, each has one-third the volume of the prism, which is $Bh$.

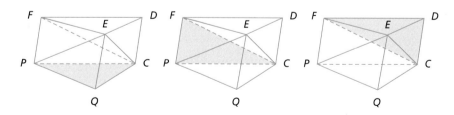

FIGURE 12-43

This result for triangular pyramids suggests the following that is true for all pyramids.

| **Volume of a Pyramid** | The volume of any pyramid is $$\frac{Bh}{3}$$ where $B$ is the area of the base and $h$ is the altitude. |

**PRACTICE PROBLEM**

Find the volume and total surface area of a regular pyramid that has a square base of side 3 m and an altitude of 5 m. (Remember that in a regular pyramid the line segment connecting the vertex to the center of the base is perpendicular to the base. Hence the slant height is $\sqrt{27.25}$. Why?)

*ANSWER*  Volume = 15 m³; total surface area = 40.32 m².

FIGURE 12-44

If we consider a pyramid with $n$ faces inscribed in a right circular cone and let $n$ increase without bound, then the surface area of the pyramid approaches the surface area of the cone. Similarly, the perimeter of the base of the pyramid approaches the circumference of the circular base of the cone. Thus, the lateral surface area of the cone is

$$\frac{tC}{2}$$

where $C$ is the circumference of the base and $t$ is the slant height. (See Figure 12-44.)

In Figure 12-44, as the number of faces of the pyramid increases, the volume of the pyramid approaches the volume of the cone. Thus the volume of the cone is

$$\frac{Bh}{3}$$

| **Volume and Surface Area of a Cone** | (a) The volume of any cone is $$\frac{Bh}{3}$$ where $B$ is the area of the base and $h$ is the height.<br>(b) Let $t$ be the slant height of a right cone, and let $C = 2\pi r$ be the circumference. Then $$\text{Lateral surface area} = \frac{Ct}{2} = \pi rt$$ $$\text{Total surface area} = \pi rt + \pi r^2$$ |

**EXAMPLE 12**

Find the total surface area of a right cone with slant height 10 cm and radius of base 4 cm.

SOLUTION

$$\text{Total surface area} = \pi r t + \pi r^2$$
$$= 40\pi + 16\pi = 56\pi \text{ cm}^2$$

**PRACTICE PROBLEM**

Find the lateral surface area of a regular triangular pyramid with base side of length 12 and slant height of length 8.

ANSWER

144

**EXAMPLE 13**

What is the volume of a water tank whose bottom is a cylinder of radius 3 m and height 4 m and whose top is a right cone of height 6 m?

SOLUTION

$$V = \pi r^2 \cdot (\text{height of cylinder}) + \frac{1}{3}\pi r^2 \cdot (\text{height of cone})$$
$$= \pi \cdot 3^2 \cdot 4 + \frac{1}{3}\pi \cdot 3^2 \cdot 6$$
$$= 54\pi \text{ m}^3$$

# VOLUME AND SURFACE AREA OF A SPHERE

In each formula we have considered in this chapter, we have used an intuitive approach to develop the formula. No such approach exists for developing the formulas for the surface area and volume of a sphere.

**Volume and Surface Area of a Sphere**

(a) The volume of a sphere of radius $r$ is

$$V = \frac{4}{3}\pi r^3.$$

(b) The surface area of a sphere with radius $r$ is:

$$\text{Total surface area} = 4\pi r^2.$$

**EXAMPLE 14**

How many cubic centimeters of air space are in a beach ball with a diameter of 20 cm?

SOLUTION

$$V = \frac{4}{3}\pi r^3$$

$$= \frac{4}{3}\pi 10^3 = \frac{4000}{3}\pi \text{ cm}^3$$

# Just for Fun

What is the ratio of the number of cubes that touch the floor to the number of cubes that do not touch the floor in the following stack of cubes?

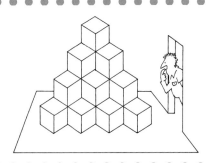

# EXERCISE SET 3

*R* 1. Find the volume and the total surface area of each figure.

(a) A cylinder with a diameter of 8 and an altitude of 10

(b) A right rectangular prism with $l = 4$, $w = 6$, and $h = 5$

(c) A right rectangular prism with $l = \frac{3}{4}$, $w = \frac{7}{8}$, and $h = 3\frac{1}{2}$

*In Exercises 2 through 10, find the lateral surface area.*

2.

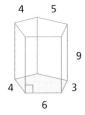

3. A right prism with a perimeter of 30 cm and a lateral edge of 10 cm

4. A right prism whose bases are equilateral triangles with sides of 6 cm and whose altitude is 10 cm

5. A right prism whose base is a right triangle with legs 3 ft. and 4 ft. and whose altitude is 8 ft.

6. A right hexagonal prism with altitude 10, base perimeter 50, and base area 35

7.

8.

*T* 9.

10. A right pentagonal prism with altitude 6, base area 40, and perimeter 60

*Find the total surface area and the volume of each solid in Exercises 11 through 14.*

11.

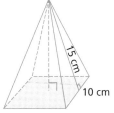

12.

$r = 5$ cm

13.

14. (a)

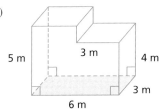

(b)

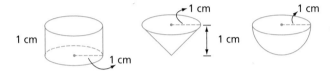

15. Find the volume of each container. What relation do you see among the three volumes? Replace $r = h = 1$ cm by $r = h = 2$ cm for these containers. Does the relation still hold?

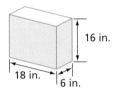

16. Find the volume of the prism in cubic feet.

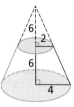

17. Find the total surface area in Exercise 16.

*C* 18. What is the length of the edge of a cube that has a total surface area of 288 in.²?

19. Convert each of the following to cubic centimeters.

    (a) 34 m³    (b) 328 km³    (c) 3100 mm³
    (d) 41 km³    (e) 42 mm³    (f) 863 km³

20. A cylindrical tank has a diameter of 10 ft. As the tank is being filled, how many cubic feet of water are needed to raise the water level in the tank by 5 in.?

21. The volume of a circular cone is $320\pi$ in.³, and the altitude is 20 in. What is the radius of the base?

22. If the surface area of a sphere is $400\pi$ units, find the volume.

23. Convert each of the following measures.

    (a) 6 cm³ = _____ mm³  (b) 6 cm³ = _____ dm³
    (c) 200 m³ = _____ km³  (d) 2 km³ = _____ cm³

24. Which of the following is larger if each has the given capacity?

    (a) One holding 13 m³ or one holding 132,000 cm³
    (b) One holding 4 km³ or one holding 40,000 m³
    (c) One holding 9 m³ or one holding $9(10)^6$ mm³
    (d) One holding 5 km³ or one holding 500,000 m³
    (e) One holding 123,456 mm³ or one holding 20 cm³

25. (a) A cone-shaped cup is filled with water to a depth equal to half the altitude. What part is this of a full cup of water?

(b) What should the water level be in order for the cup to be half full (that is, to contain one-half the volume of a full cup)?

26. Suppose that the radius of a sphere is tripled. What is the ratio of surface areas? Of volumes?

27. Find the volume and the surface area of the figure shown. The circular area of radius 2 is the top of the solid.

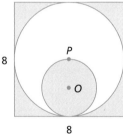

28. Find the volume of the walls of a pipe if the inner radius is 6 cm, the outer radius is 7 cm, and the length is 135 cm.

29. How many liters of paint are needed to paint the bottom and sides of a swimming pool that is 5 m by 10 m by 3 m? (The pool is 3 m deep all the way across.) Assume that 1 L of paint covers 24 m² and that the paint is purchased in 1-L cans.

30. How much does it cost to paint the garage described in the introduction to this section?

## REVIEW EXERCISES

31. Find the area of the shaded region.

32. The area of a circular region is equal in number to the circumference of the circle. Find the measure of the radius of the circle.

33. Find the area of the shaded region surrounded by the circles of radius 4 that touch as shown.

34. If an 80-cm-diameter wheel on a car traveling 50 km per hour makes 225 revolutions, how long has it been traveling?

# SECTION 4 OTHER SI UNITS OF MEASURE

**PROBLEM**

At an SPL fraternity party, Larry proclaimed that his 12-ounce can of soft drink did not weigh 12 ounces. Immediately there were disagreements and bets on both sides. Could Larry be right?

**OVERVIEW**

Larry's fraternity friends were surprised to learn that the 12 referred to fluid ounces, not weight. A fluid quart and a dry quart are different. You will find that the metric system is much less complicated.

A person, an apple, and a tree have *mass.* To understand how scientists approach the concept of mass, let us compare it to weight, a measure with which you are more familiar. How do you measure your weight? A scale could give your weight, but this weight would vary, depending on gravitational pull. For instance, a rock on the earth might show a weight of 30, whereas its weight might be only 5 on the moon.

**GOALS**

Illustrate the Measurement Standard, page 532.
Illustrate the Connectives Standard, page 147.

In addition to the ease with which we can convert among the units of length, the units of area, and the units of volume in the SI system, this system possesses useful relationships among its measures of volume, capacity, and mass. At certain specified conditions of temperature and atmosphere (4°C and 1 atm pressure), a cube of water 1 centimeter on a side occupies 1 milliliter of space (capacity) and has a mass of one gram. This may be summarized as follows:

Volume       1 cubic centimeter (cm³)
Capacity      1 milliliter (mL)
Mass   1 gram (g)

 CAPACITY

In the English system, the basic unit of capacity is the gallon. In the metric system, the basic unit is the liter (L), which is defined as the capacity of a box that is 10 cm long on each side. (See Figure 12-45.) One cubic centimeter is thus one-thousandth of a liter; this is the same quantity as a milliliter.

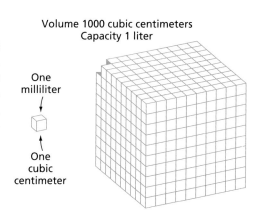

Volume 1000 cubic centimeters
Capacity 1 liter

One milliliter

One cubic centimeter

FIGURE 12-45

From the preceding discussion,

$$1 \text{ liter (L)} = 1000 \text{ mL} = 1000 \text{ cm}^3$$

$$= 10^3 \text{ cm}^3 = 1 \text{ dm}^3$$

Table 12-7 summarizes units of measurement for capacity. The most common units are the liter and the milliliter (often called a cc, for cm³).

Remember that a liter (L) is a little more than a quart and that a milliliter (mL) is approximately the amount of liquid in an eyedropper. (See Figure 12-46.) The metric scale in Figure 12-47 is useful for comparing metric units.

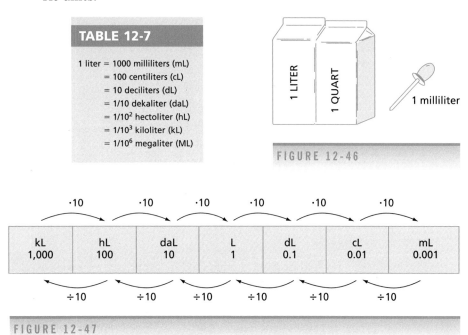

**TABLE 12-7**

1 liter = 1000 milliliters (mL)
= 100 centiliters (cL)
= 10 deciliters (dL)
= 1/10 dekaliter (daL)
= 1/10² hectoliter (hL)
= 1/10³ kiloliter (kL)
= 1/10⁶ megaliter (ML)

1 LITER   1 QUART   1 milliliter

FIGURE 12-46

| kL 1,000 | hL 100 | daL 10 | L 1 | dL 0.1 | cL 0.01 | mL 0.001 |

·10 ·10 ·10 ·10 ·10 ·10
÷10 ÷10 ÷10 ÷10 ÷10 ÷10

FIGURE 12-47

**EXAMPLE 1**

How many liters of liquid are there in 2000 milliliters?

*SOLUTION*

$$1 \text{ mL} = 0.001 \text{ L}$$

$$2000 \text{ mL} = 2000(0.001) = 2 \text{ L}$$

# GRAM

Although many people use *weight* and *mass* interchangeably, there is a difference between the two terms. Weight is affected by the force of gravity on an object; whereas mass is not. Because the force of gravity lessens as an object is moved further from the center of the earth, weight will diminish as the object is moved higher. However, because the changes in altitude on the earth's surface are minimal relative to the radius of the earth, measures of weight on the earth's surface fluctuate only miniscule amounts. Hence we can state with confidence that a mass of 1 kilogram weighs about 2.2 pounds. On the other hand, should we move elsewhere in our solar system,

**TABLE 12-8**

| | |
|---|---|
| 1 kilogram (kg) | = $10^3$ grams (g) |
| 1 hectogram (hg) | = $10^2$ grams |
| 1 dekagram (dag) | = 10 grams |
| 1 decigram (dg) | = 1/10 gram |
| 1 centigram (cg) | = 1/100 gram |
| 1 milligram (mg) | = 1/1000 gram |

this equivalence would not make sense. On the moon, mass in 1 kilogram would be about 0.4 pound.

In the metric system, the basic unit for mass is the gram. As noted earlier, 1 cm³ of water has a mass of 1 gram. Other units of mass are obtained by adding metric prefixes to *gram,* as in Table 12-8.

A widely used SI unit of mass is the **kilogram** (1000 g). A gram is a rather small mass (approximately that of a paper clip). A nickel has a mass of approximately 5 g. The kilogram, gram, and milligram are the units in practical use today. Figure 12-48 can be helpful for comparing metric units of mass.

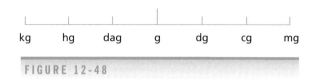

kg    hg    dag    g    dg    cg    mg

FIGURE 12-48

**EXAMPLE 2**

(a) Change 20 kg to grams.
(b) Change 36,000 g to kilograms.

*SOLUTION*

(a) 1 kg = 1000 g
   20 kg = 20(1000) = 20,000 g
(b) 1 g = 0.001 kg
   36,000 g = 36,000(0.001) = 36 kg

**EXAMPLE 3**

A detergent is packaged in super size at 2 kg for $2.10 and in regular size at 720 g for 60¢. Which box is more economical?

*SOLUTION*

The super size costs (2.10 ÷ 2000), or about $0.001 per gram. The regular size costs (0.60 ÷ 720), or about $0.0008 per gram. The regular size is the better buy.

Two units of mass used extensively in daily business transactions are the **metric ton** and what is often called the **kilo.** The metric ton (t) is equal to 1000 kg. Because 1 kg is approximately 2.20 lb., a metric ton is approximately 2200 lb., somewhat larger than a ton in the English system. A kilo is a kilogram. Saying "I ordered 4 kilos of potatoes" is the same as saying "I ordered 4 kilograms of potatoes." By ordering $\frac{1}{2}$ kilo of sugar, you would obtain about 1 lb. of sugar.

**EXAMPLE 4**

Use dimensional analysis to find the mass in kg of 4 m³ of water.

*SOLUTION*

We will use the fact that 1 cm³ of water has a mass of 1 g. It follows that

$$4 \text{ m}^3 \cdot \frac{10^6 \text{ cm}^3}{1 \text{ m}^3} \cdot \frac{1 \text{ g}}{1 \text{ cm}^3} \cdot \frac{1 \text{ kg}}{10^3 \text{ g}} = 4000 \text{ kg or 4 metric tons of water}$$

# TEMPERATURE

In the metric system, temperature is measured in degrees Celsius. The Celsius scale is named for the Swedish astronomer, Anders Celsius, who introduced it in 1712. In the English system, temperature is measured on the Fahrenheit scale. Both systems use the freezing and boiling points as reference points. These are defined on the Celsius scale as zero degrees Celsius (0°C) and 100 degrees Celsius (100°C) (see Figure 12-49). Because there are 100 divisions between freezing and boiling temperatures on the Celsius scale, this scale is sometimes called the *centigrade* scale. The Fahrenheit scale spans 180°F between freezing and boiling temperatures (32°F to 212°F). Because there are 180 divisions between freezing and boiling on the Fahrenheit scale, the relationship between the two scales is 100 to 180 or 5 to 9. That is, a change in temperature of 1°F represents a change in temperature of $\frac{5}{9}$°C. $\left(\frac{100}{180}\text{ reduces to }\frac{5}{9}.\right)$ Figure 12-49 gives comparisons between temperatures in degrees Celsius and in degrees Fahrenheit. The formulas relating the two are

$$F° = 32° + \frac{9}{5}C° \qquad \text{and} \qquad C° = \frac{5}{9}(F° - 32°)$$

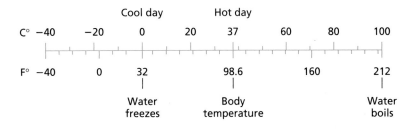

FIGURE 12-49

| | EXAMPLE 5 | |
|---|---|---|

Convert 41°F into degrees Celsius.

*SOLUTION*

$$C° = \frac{5}{9}(F° - 32°) = \frac{5}{9}(41° - 32°) = 5°C$$

# Just for Fun

A man with a 5-L and a 3-L bucket went to the well to get exactly 4 L of water. How was he able to get the 4 L of water by using only the two buckets?

Well, well, well.....

# EXERCISE SET 4

*R* 1. Convert into liters.

   (a) 125,000 mL      (b) 16 mL
   (c) 1230 kL         (d) 14 kL
   (e) 236 cL          (f) 135,000 cL

2. Perform the following conversions.

   (a) 1689 kg = _____ g     (b) 346 kg = _____ cg
   (c) 367 cg = _____ kg     (d) 6152 mg = _____ g
   (e) 14 cg = _____ mg      (f) 179 kg = _____ g

3. Choose the more reasonable measurement for mass.

   (a) Tennis ball (57 g or 57 kg)
   (b) Large dog (35 g or 35 kg)

4. Complete the table.

| Article | mg | g | kg |
|---------|-----|-----|-----|
| Vitamin C tablet | 250 | | |
| Adult | | | 80 |
| Chicken | | 610 | |

5. Choose the better measurement.

   (a) Full bathtub (210 mL or 210 L)
   (b) Eyedropper (1 mL or 1 L)
   (c) Bucket (8 mL or 8 L)
   (d) Gas tank in a car (80 mL or 80 L)

6. Choose the more sensible temperature.

   (a) A warm day (5°C or 30°C)

   (b) Cold milk (4°C or 40°C)
   (c) Hot soup (20°C or 80°C)
   (d) A snowy day (−8°C or 18°C)
   (e) A warm shower (50°C or 110°C)
   (f) Baseball weather (25°C or 75°C)

7. A right rectangular prism has edges that measure 16 cm, 24 cm, and 36 cm. How many liters will the container hold?

8. A container in the shape of a right prism is 110 cm high and contains 1 L. How many square centimeters are in the base?

9. A conical drinking cup is 8 cm high and has a diameter of 6 cm. How many of the cups can be filled from a 2-L bottle of cola?

10. One cubic centimeter of iron ore has a mass of about 0.008 kg. What is the mass of an iron bar that measures 50 cm by 20 cm by 5 cm?

11. Convert the following temperatures from degrees Fahrenheit to degrees Celsius.

    (a) 50°      (b) 68°      (c) 32°      (d) −2°

12. Convert the following temperatures from degrees Celsius to degrees Fahrenheit.

    (a) 20°      (b) 10°      (c) 30°      (d) 0°

13. A car is driven 180 km and uses 12 L of gasoline. What is the car's gasoline consumption in kilometers per liter?

14. Estimate the capacity of each item in liters or milliliters.

(a) A half-gallon carton of milk
(b) A cup of water
(c) A gallon can of gasoline
(d) A can of cola
(e) A quart jar of jelly
(f) A teaspoon of medicine

15. Estimate in grams or kilograms the mass of each of these items.

(a) A glass of water
(b) A quarter
(c) This math book
(d) A can of cola
(e) A ballpoint pen
(f) A regular-size automobile
(g) A sack of fertilizer      (h) A piece of chalk
(i) A basketball             (j) A tennis racket
(k) A compact automobile   (l) A bicycle

16. A distributor intends to sell a metric ton of sugar in 657-g containers at 22¢ each.

(a) How many containers will be obtained from the metric ton?
(b) For how much money will this metric ton sell?

17. Estimate these temperatures in degrees Celsius.

(a) Comfortable room temperature
(b) Too cool to go swimming
(c) Morning with light frost
(d) Boiling water

18. Formulate rules for performing the following conversions.

(a) Kilogram to gram
(b) Gram to milligram

C 19. Susan buys a kilogram of cocoa for $6. Larry buys 400 g for $2. Who made the better buy and why?

*Sometimes it is necessary to compare measurements in metric units and in English units. The following approximate conversion factors are useful.*

| | |
|---|---|
| 1 quart $\approx$ 0.946 L | 1 cup $\approx$ 236.7 mL |
| 1 fluid oz. $\approx$ 29.59 mL | 1 L $\approx$ 1.06 quart |
| 1 oz. $\approx$ 28.35 g | 1 mL $\approx$ 0.0338 fluid oz. |
| 1 lb. $\approx$ 0.454 kg | 1 g $\approx$ 0.0353 oz. |
| 1 tablespoon $\approx$ 14.79 mL | 1 kg $\approx$ 2.205 lb. |

20. Place a $<$, $=$, or $>$ in each blank.

(a) 6 quarts _____ 5L
(b) 32 L _____ 8 gal.
(c) 57 mL _____ 0.05 quart
(d) 24 L _____ 22 quarts
(e) 4 cups _____ $10^3$ mL
(f) 5 tablespoons _____ $10^2$ mL
(g) 6 fluid oz. _____ $10^2$ mL
(h) 20 mL _____ 1 fluid oz.

21. Select the larger

(a) 2000 g or 4 lb.       (b) 10 lb. or 4 kg
(c) 100 mg or 2 oz.      (d) 50 g or 2 oz.

22. Which of the following would you prefer?

(a) Sugar selling at 55¢/lb. or at 0.3¢/g
(b) Apples selling at 34¢/lb. or at 20¢/kg

23. Which would you prefer, a car that uses gasoline at a rate of 8 km/L or one that gets 30 miles/gal.

24. Insert a $<$, $=$, or $>$ in each blank.

(a) 3 lb. _____ $10^3$ g       (b) 3 L _____ 2 quarts
(c) 342 kg _____ $10^3$ lb.   (d) 2.3 oz. _____ 50 g
(e) 742 g _____ 10 oz.        (f) 1673 oz. _____ 20 kg

25. Convert each of the following measures into kilogram measure.

(a) A 110-lb. secretary
(b) A 6-lb. 7-oz. baby
(c) A 25-lb. watermelon
(d) A 2-ton truck
(e) A 10-lb. bag of potatoes

26. A bottle of shampoo contains 10 fluid oz. How many milliliters is this?

27. Cola is bottled in 1-L bottles. How many cups does each bottle contain?

28. Order each set from largest to smallest.

(a) 26 oz.     32.4 kg     36,734 g     4 lb.
(b) 5 quarts   6 L     2 gal.     7954 mL

29. Which price would you prefer as a purchaser?

(a) Flour at 15¢/lb. or at 35¢/kg
(b) Candy at $1.05/lb. or at 2¢/g
(c) Gasoline at $1.60/gal. or at 5¢/mL
(d) Milk at $1.70/gal. or at 29¢/L

## REVIEW EXERCISES

• • • • • • • • • • • • • • • • • • • • • • • • • • • •

30. Find the lateral surface area and the volume of each solid.

(a) 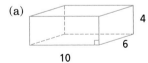     (b)

(c)

# SOLUTION TO INTRODUCTORY PROBLEM

UNDERSTANDING THE PROBLEM. We are to place a square on a geoboard and count the number of points ($P$) on the square and the number of points ($I$) interior to the square. Our task is to develop a relationship between the area ($A$) of the square and $P$ and $I$. We shall use a favorite strategy for solving a problem: making a table. (See Table 12-9.)

DEVISING A PLAN. The pattern is not obvious, so we rely on a trial-and-error inductive process to see if we can determine the answer.

CARRYING OUT THE PLAN. We shall first use a table to examine the values of $A$, $P$, and $I$ for several different squares with vertices on points of the geoboard. For instance, for the 2 by 2 square pictured below, $P = 8$, $I = 1$, and $A = 4$.

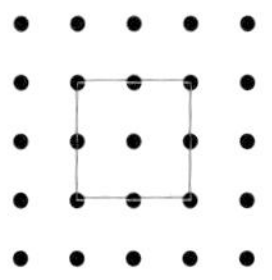

**TABLE 12-9**

| *Square* | *P* | *I* | *A* | *A − I* |
|---|---|---|---|---|
| 2 by 2 | 8 | 1 | 4 | 3 |
| 3 by 3 | 12 | 4 | 9 | 5 |
| 4 by 4 | 16 | 9 | 16 | 7 |
| 5 by 5 | 20 | 16 | 25 | 9 |
| 6 by 6 | 24 | 25 | 36 | 11 |

**TABLE 12-10**

| $\frac{1}{2}P + I$ | *A* |
|---|---|
| 2 | 1 |
| 5 | 4 |
| 10 | 9 |
| 17 | 16 |
| 26 | 25 |
| ⋮ | ⋮ |

Because $I$ represents points on the interior of the square, $I$ will always be less than $A$. When we focus on just how much $A$ exceeds $I$ (see the fifth column of our table), we notice that $A$ exceeds $I$ by roughly half of $P$. We look at this apparent relationship more closely in Table 12-10. Because $\frac{1}{2}P$ is smaller than each area, let's try adding a multiple of $I$. First we try $1 \cdot I$, or $I$ itself.

Wait a minute! Have we discovered a pattern? Is

$$A = \tfrac{1}{2}P + I - 1?$$

Check in Table 12-11.

**TABLE 12-11**

| *Number of Points P on a Square* | *Number of Points I in the Interior* | *Area A of the Square* |
|---|---|---|
| 4 | 0 | 1 |
| 8 | 1 | 4 |
| 12 | 4 | 9 |
| 16 | 9 | 16 |
| 20 | 16 | 25 |
| ⋮ | ⋮ | ⋮ |

LOOKING BACK. Are there other problems this formula will solve? Try this formula (called *Pick's formula*) not only on squares but also on other polygonal regions placed on a geoboard. You will receive a pleasant surprise.

# SUMMARY AND REVIEW

## METRIC UNITS

| *Prefix* | *Meaning* | *Decimal Meaning* |
|---|---|---|
| mega | one million | 1,000,000 *or* $10^6$ |
| kilo | one thousand | 1000 *or* $10^3$ |
| hecto | one hundred | 100 *or* $10^2$ |
| deka | ten | 10 |
| deci | one-tenth | 1/10 *or* 0.1 |
| centi | one-hundredth | 1/100 *or* 0.01 |
| milli | one-thousandth | 1/1000 *or* 0.001 |
| micro | one-millionth | 1/1,000,000 *or* 0.000001 |

Know the meaning of the following metric terms or formulas:

(a) hectare (b) metric ton
(c) gram (d) liter
(e) $C° = \frac{5}{9}(F° - 32°)$; $F° = \frac{9}{5}C° + 32$

## AREA

Some important formulas for area include:

(a) Parallelogram: $A = bh$
(b) Rectangle: $A = bh$
(c) Triangle: $A = \dfrac{bh}{2}$

(d) Trapezoid: $A = \frac{1}{2}(b + c)h$

(e) Regular $n$-gon: $A = \dfrac{Pa}{2}$

(f) Circle: $A = \pi r^2$

## SURFACE AREA

Some important formulas for total surface area include the following, where $B$ is the area of the base, $P$ is the perimeter of the base, $h$ is the height, $r$ is the radius, and $t$ is the slant height.

(a) $A = hP + 2B$      (b) $A = 2\pi rh + 2B$

(c) $A = tP/2 + B$      (d) $A = \pi rt + B$

(e) $A = 4\pi r^2$

Associate these with appropriate solids.

## VOLUME

Some formulas for volume include the following:

(a) $V = Bh$      (prism and cylinder)

(b) $V = \dfrac{Bh}{3}$      (pyramid and cone)

(c) $V = \dfrac{4\pi r^3}{3}$      (sphere)

## CAVALIERI'S PRINCIPLE

Given a plane and any two solids, if every plane parallel to the given plane intersects the two solids in cross-sections that have the same area, then the solids have the same volume.

# CHAPTER TEST

1. What is the area of the shaded region if the unit of measure is 1 cm?

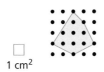

1 cm²

*Find the areas of the regions shown in Questions 2 through 4.*

2.          3.

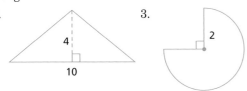

4.

5. (a) 328 km = _____ cm
   (b) 0.026 mm = _____ cm
   (c) 28 cm = _____ m

6. Insert a $<$, $>$, or $=$ between each pair of numbers.
   (a) 6.8 m    69.2 cm
   (b) 980 mm    0.0098 m
   (c) 0.0001 m    2 cm

7. Subtract     17 m + 4 cm + 6 mm
           − (5 m + 7 cm + 8 mm)

8. Choose the more reasonable measurement for
   (a) Pair of tennis shoes (1 g or 1 kg)
   (b) Nickel (5 g or 5 kg)

*In Questions 9 and 10, find the volume of the space regions shown.*

9. 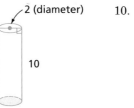     10.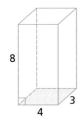

11. For Exercises 9 and 10, find the total surface area of each figure.

12. Complete the following.
   (a) 24 L = _____ cm³
   (b) 16 cm³ = _____ mm³
   (c) 50 L = _____ mL
   (d) This textbook has a mass of about 900 _____ .
   (e) 14,000 mg = _____ kg
   (f) 14 km³ = _____ cm³
   (g) 0.2 L has a volume of _____ dm³.
   (h) A cube measuring 1 m by 1 m by 1 m has a volume of _____ .

13. For each of the following, decide whether the situation is likely or unlikely.
   (a) Carrie's bath water has a temperature of 15°C.
   (b) Miranda found 26°C too warm and lowered the thermostat to 21°C.
   (c) Jim was drinking water whose temperature was −5°C.

(d) The water in the teakettle has a temperature of 90°C.

(e) The outside temperature dropped to 5°C, and ice appeared on the lake.

14. On the number line, find m($\overline{AB}$).

15. A circle has an area of $81\pi$ mm². What is its circumference?

16. If the perimeter of a square is equal to the circumference of a circle, compare the areas of the bounded regions.

17. In trapezoid $ABCD$, $\overline{AB}$ is parallel to $\overline{CD}$ and $\overline{AD}$ is perpendicular to $\overline{CD}$. If $AB = 6$, $AD = 4$, and $DC = 8$, find the area of the trapezoidal region.

18. Find the total surface area.

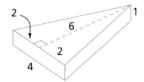

19. Fill in blanks with the word or phrase from the list below that makes a true statement.

| altitude | apothem |
|---|---|
| centi | milli |
| circumference | deci |
| gram | liter |
| faces | edges |
| lateral area | greatest possible error |
| slant height | |
| relative error | total surface area |
| vertices | |

(a) The distance around a circle is called its _____ .

(b) The prefix for one-thousandth in the SI system is _____ .

(c) 10 is the number of _____ of a right pentagonal pyramid.

(d) The basic unit for capacity in the SI system is the _____ .

(e) The _____ of a right rectangular prism consists of areas of six rectangles.

(f) If a measurement is made to the nearest meter, then the _____ is 0.5 m.

20. Find the ratio (one divided by the other) of the volume of a cube to the volume of an inscribed sphere.

# COORDINATE GEOMETRY AND TRANSFORMATIONS

# 13

Jack found a message inside a bottle that had washed onto the beach. Help him decipher the message. What techniques can you use?

One morning while meditating in bed (as was his custom), a Frenchman named René Descartes had a simple but profound idea. He observed that in order to describe the relationships between points in the plane it would be helpful to have a frame of reference on which to base the discussion. Out of this inspiration grew the rectangular coordinate system (or coordinate plane, sometimes called the Cartesian plane after Descartes) and the notion of coordinate graphing as a useful tool in understanding mathematical relationships.

In addition to studying coordinate geometry in this chapter, we consider geometric transformations. The idea of motion or movement is an important concept of geometry. We approach the subject intuitively through a study of flips, slides, and turns. These are mathematically defined concepts and are shown to be distance-preserving transformations. Transformation theory can be used to define congruence and similarity.

In this chapter, we use the following problem-solving strategies:  ▪ Guess, test, and revise. (598) ▪ Use reasoning. (throughout the chapter)  ▪ Draw a picture. (throughout the chapter)

# INTRODUCTION TO REFLECTIONS OR FLIPS AND COORDINATE GEOMETRY

**PROBLEM**

Write a large letter *T* on a piece of paper. Then fold the paper so that the *T* is folded under, and trace the *T* on the paper that remains on top. Unfold the paper. Your trace should be like the two "traced" *T*'s that are shown in color in the figure.

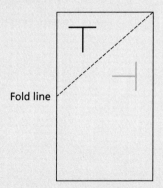
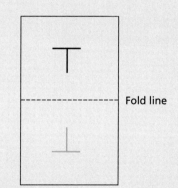

Fold line            Fold line

**OVERVIEW**

Reflections provide a basis for studying various properties of geometric figures. The idea of a reflection is very familiar to you. Every time you look in a mirror, you see a reflected image of yourself. A tracing is called the *image* of a reflection, and the line of the fold (shown as a dashed line above) is called the *line of reflection*.

**GOALS**

Illustrate the Geometry Standard, page 482.*
Illustrate the Algebra Standard, page 4.*
Illustrate the Representation Standard, page 50.*

* The complete statement of the standard is given on this page of the book.

In Chapter 7, we considered a real number line with one-dimensional coordinates. In this section, we use number lines in two directions to represent an ordered pair of numbers. These numbers are called **coordinates** (in two dimensions), and the coordinate system is called the *Cartesian coordinate system,* named for René Descartes.

**Coordinate System**

The $x$ axis is a horizontal number line that extends infinitely to the left and right; the $y$ axis is a vertical number line that extends infinitely up and down. The two axes cross at right angles, and their point of intersection is point 0 on the vertical number line and point 0 on the horizontal number line. This point is called the **origin.**

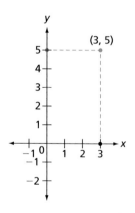

Because the two axes are number lines, units are marked off on each axis so that positive integers go up and to the right and negative integers go down and to the left. (See Figure 13-1.)

To locate a pair of coordinates in the Cartesian coordinate system, find the $x$ coordinate on the $x$ axis, and mentally draw a vertical line through this point. Then find the $y$ coordinate on the $y$ axis, and mentally draw a horizontal line through this point. The point of intersection of these two lines is the point that corresponds to the desired coordinates. Every pair of coordinates $(x, y)$ corresponds to a unique point in the coordinate plane, and every point on the plane has coordinates that can be written in the form $(x, y)$. Locating points on the plane is called **plotting** the points.

FIGURE 13-1

**EXAMPLE 1**   Plot the point that contains the coordinates $(3, 5)$.

SOLUTION   Move three places to the right of 0 on the $x$ axis. Mentally draw a vertical line through that point. Then move five places up on the $y$ axis. Mentally draw a horizontal line through that point. The intersection of these two lines, as shown in Figure 13-1, is the point $(3, 5)$.

**EXAMPLE 2**   Draw the triangle with vertices $A(-1, -2)$, $B(1, 3)$, and $C(4, 1)$.

SOLUTION   The triangle is shown in Figure 13-2.

Using the idea of the measure of a line segment, as described in Chapter 12, we can find the measure of a given line segment when one of the coordinates is fixed. For example, consider the line segment $\overline{AB}$ in Figure 13-3 (a) and the same line segment placed on coordinate axes in Figure 13-3 (b). The measure of this segment in Figure 13-3 (a) and (b) is

$$m(\overline{AB}) = 5 - 2 = 3$$

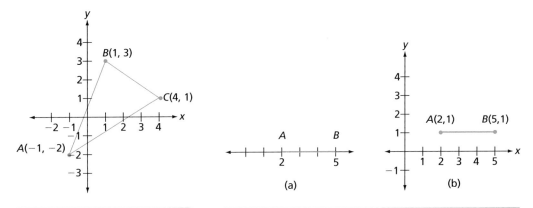

FIGURE 13-2        FIGURE 13-3

**EXAMPLE 3**

Find the length of the segment $\overline{AB}$, where
(a) $A$ is given by $(2, 3)$ and $B$ by $(2, -1)$.
(b) $A$ is given by $(2, 5)$ and $B$ by $(7, 5)$.

*SOLUTION*

(a) The change in $y$ is from 3 to –1, so $AB = 3 - (-1) = 4$
(b) The change in $x$ is from 2 to 7, so $AB = 7 - 2 = 5$

# DISTANCE FORMULA

The Pythagorean Theorem can be used to find the length of a line segment that is not parallel to an axis in the coordinate system. For example, suppose that you want to find the length of the line segment from $A(2, 3)$ to $B(5, 4)$ in Figure 13-4 (a). First draw a horizontal line through $(2, 3)$ and a vertical line through $(5, 4)$. These lines intersect at $C(5, 3)$. Then form right triangle $ABC$. $AC = 5 - 2 = 3$, and $CB = 4 - 3 = 1$. Therefore,

$$(AB)^2 = 3^2 + 1^2 = 10$$

$$AB = \sqrt{10}$$

In general, consider the segment from $A(x_1, y_1)$ to $B(x_2, y_2)$ in Figure 13-4 (b). The change in $y$ is given by $|y_2 - y_1|$, and the change in $x$ by $|x_2 - x_1|$. By the Pythagorean Theorem,

$$AB = \sqrt{(x_2 - x_1)^2 + (y_2 - y_1)^2}$$

## René Descartes
### 1596–1650

It is difficult to imagine a more varied career than that of René Descartes. One of three children, he was born in the town of La Haye (now La Haye–Descartes), Touraine, a former province in western France. He was educated from ages 8 to 16 at the Jesuit school of La Flèche, then took a law degree at the university in Poitiers (but never practiced law). He served in armies in Holland, Bavaria, and France but never became what could be called a professional soldier. He finally settled for a life of independent travel and study, residing most of the time in Holland.

Descartes' interest in mathematics began in 1618 when he sought a solution to a mathematical problem posted on a wall. Throughout his life he pursued mathematical ideas in a characteristically practical manner, even developing some of the mathematical notation that we use today. For example, he introduced the modern convention of using the beginning letters of the alphabet to represent known quantities (constants) and the last letters of the alphabet to represent unknown quantities (variables). He also conceived the notation of superscripts to represent repeated multiplication, as in $y^2$ to stand for $y$ multiplied by $y$.

But at heart, Descartes was a philosopher. He is known today as "the father of modern philosophy," solving the problem of his own existence with the phrase that made him famous, *Cogito, ergo sum* ("I think, therefore I am"). His most renowned philosophical work was Discourse on the *Method of Reasoning Well and Seeking Truth in the Sciences.*

In his work *La Géométrie,* he developed what we know today as analytic geometry. Analytic geometry allows the concepts of algebra to be brought to bear on the problems of geometry and vice versa. Descartes showed, for example, how ordinary multiplication, division, and extraction of square roots are easily performed by using geometry. Modern navigational charts, geological maps, and even common roadmaps follow Descartes' practice of finding locations by pairs of numbers or letters. We still honor this idea with such phrases as "the Cartesian product," which we studied in Chapter 3, and "Cartesian plane," which we study in this section.

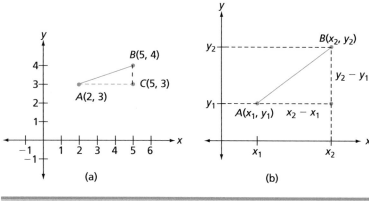

**FIGURE 13-4**

**Distance Formula**

The distance $AB$ between any two points $A$ and $B$ with coordinates $(x_1, y_1)$ and $(x_2, y_2)$ is given by

$$AB = \sqrt{(x_2 - x_1)^2 + (y_2 - y_1)^2}$$

| **EXAMPLE 4** | Find the length of the line segment $\overline{AB}$, where $A$ is $(2, -1)$ and $B$ is $(5, 3)$. |
|---|---|

*SOLUTION*

$$\text{m}(\overline{AB}) = \sqrt{(5 - 2)^2 + [3 - (-1)]^2}$$
$$= \sqrt{9 + 16}$$
$$= 5$$

 # REFLECTIONS

We will find the reflection or flip images of figures over a line by means of three techniques:

1. Folding a piece of paper along the line of reflection and marking the image
2. Viewing the reflection image with a mirror
3. Sketching the reflection image by using the definition of a reflection.

We begin by locating reflection images through paper folding.

| **EXAMPLE 5** | Find point $A'$ (called the **image** of $A$) corresponding to point $A$, by folding a piece of paper along line $m$ in (a), (b), and (c) of Figure 13-5. |
|---|---|

*SOLUTION*

Figure 13-5 shows the location of $A'$ in each case. In (c), where $A$ happens to be on line $m$, $A$ and $A'$ are the same point.

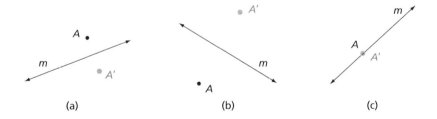

**FIGURE 13-5**  (a)  (b)  (c)

Now suppose that you have printed the word *MATH* on a piece of paper. Before the ink is dry, the paper is folded, covering the letters. When the paper is unfolded, you get an image as shown in Figure 13-6(a).

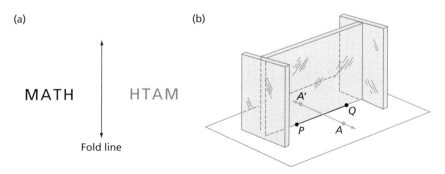

**FIGURE 13-6**

The preceding example is an illustration of a reflection about a line (sometimes called a *flip*).

A *mira,* made of Plexiglas, reflects like a mirror although it is transparent. Thus, in Figure 13-6 (b), $A'$ is the flip image of $A$, where $\overleftrightarrow{PQ}$ is the line of reflection.

Reflection images or flip images are easy to obtain by using paper folding, a mirror, a mira, or just common sense.

| | |
|---|---|
| **EXAMPLE 6** | Find the flip image of $\triangle ABC$ relative to line $m$ in Figure 13-7 (a). |
| *SOLUTION* | The flip image is shown in color in Figure 13-7 (a). |
| **PRACTICE PROBLEM** | Find $A'$ (image of $A$) and $B'$ (image of $B$) by folding along line $l$ in Figure 13-7 (b). |
| *ANSWER* | The images are as shown in Figure 13-7 (c). |
| **PRACTICE PROBLEM** | Find the flip image of the closed curve in Figure 13-8 (a) relative to line $m$. |
| *ANSWER* | The flip image is shown in Figure 13-8 (b). |

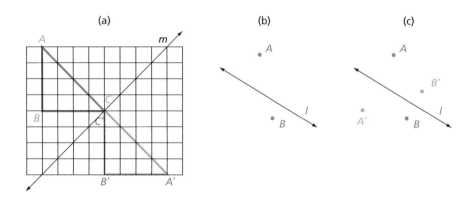

FIGURE 13-7

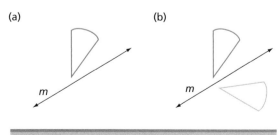

FIGURE 13-8

# TRANSFORMATIONS

The concept of a reflection is a special case of a transformation. Suppose that a rectangle $ABCD$ is drawn on a transparent sheet of rubber, which is stretched as shown in Figure 13-9. The vertices after the stretching are marked as $A'$, $B'$, $C'$, and $D'$. We associate a unique point with each given point:

$$A \leftrightarrow A' \qquad B \leftrightarrow B' \qquad C \leftrightarrow C' \qquad D \leftrightarrow D'$$

We call such a correspondence a *transformation*.

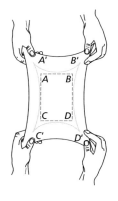

FIGURE 13-9

---

**DEFINITION**

**Transformation**

A *transformation* of a plane is a one-to-one correspondence between two sets of points in the plane.

---

A transformation is also called a **mapping.** In the preceding example, $A$ is mapped onto or into $A'$, $B$ onto $B'$, and so on. $A'$ is again called the *image* of $A$. A *reflection* is a transformation that has certain specific characteristics.

---

**DEFINITION**

**Reflection**

A *reflection* (or **flip**) about line $m$ (denoted by $R_m$) is a transformation that maps each point $P$ onto point $P'$ as follows:
(a) If $P$ is on $m$, then $P = P'$.
(b) If $P$ is not on $m$, then $m$ is the perpendicular bisector of $\overline{PP'}$.

---

In Figure 13-6 (b), a reflection maps $A$ onto $A'$ about line $\overleftrightarrow{PQ}$ because $\overleftrightarrow{PQ}$ is the perpendicular bisector of $\overline{AA'}$. In Figure 13-5 (a) and (b), if you placed a protractor along $\overline{AA'}$, you would see that line $m$ is perpendicular to $\overline{AA'}$. In fact, it is the perpendicular bisector of $\overline{AA'}$.

**EXAMPLE 7**

In Figure 13-10 (a), find the reflection image or flip image about line $m$ of
(a) $A$ and (b) $\overline{CD}$.

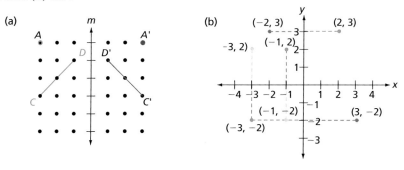

FIGURE 13-10

SOLUTION | The reflections are given in Figure 13-10 (a). To find the segment $\overline{C'D'}$, we could have found $C'$ (the image of $C$) and $D'$ (the image of $D$). Then we could have drawn $\overline{C'D'}$.

Reflect the point (2, 3) about the $y$ axis in Figure 13-10 (b). Do you see that the reflection image is (−2, 3)? Likewise, the reflection image of (−3, −2) is (3, −2).

| | |
|---|---|
| **Reflection about the $y$ Axis** | The *reflection* image of $(x, y)$ about the $y$ *axis* is the point $(-x, y)$. |

EXAMPLE 8 | Find the coordinates of $\triangle A'B'C'$, which is the reflection image of $\triangle ABC$ about the $y$ axis, where $\triangle ABC$ is defined by $A(2, 1)$, $B(3, -2)$, and $C(3, 4)$.

SOLUTION | The images of $A$, $B$, and $C$ are $A'(-2, 1)$, $B'(-3, -2)$, and $C'(-3, 4)$.

In Figure 13-10 (b), the reflection image of (−1, 2) about the $x$ axis is (−1, −2), and the reflection image of (−3, −2) about the $x$ axis is (−3, 2).

| | |
|---|---|
| **Reflection about the $x$ Axis** | The *reflection* image of $(x, y)$ about the $x$ *axis* is the point $(x, -y)$. |

EXAMPLE 9 | Find the coordinates of $\triangle A'B'C'$, which is the reflection image of $\triangle ABC$ about the $x$ axis, where $\triangle ABC$ is defined by $A(2, 1)$, $B(3, -2)$, and $C(3, 4)$.

SOLUTION | The images of $A$, $B$, and $C$ are $A'(2, -1)$, $B'(3, 2)$, and $C'(3, -4)$.

PRACTICE PROBLEM | Find the coordinates of the point obtained by reflecting (−1, 7) about the $y$ axis and then reflecting the resulting point about the $x$ axis.

ANSWER | (1, −7)

EXAMPLE 10 | Taking line segment $\overline{PQ}$, shown in Figure 13-11, reflect or flip it about line $l$ using paper folding.

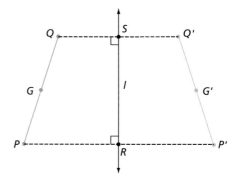

FIGURE 13-11

SOLUTION | Fold your paper along flip line *l*. Now trace the flip image, $\overline{P'Q'}$. Alternatively, fold your paper along flip line *l* and, with a pen, mark points $Q'$ and $P'$ corresponding to $Q$ and $P$. With a straightedge, draw $\overline{P'Q'}$, which is the flip image of $\overline{PQ}$.

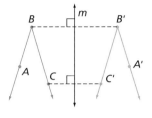

FIGURE 13-12

Note in Figure 13-11 that $PR = P'R$ and $QS = Q'S$, and that $\angle R$ and $\angle S$ seem to be right angles, satisfying the definition of a reflection. Likewise, $G'$, the image of $G$ (between $P$ and $Q$), is between $P'$ and $Q'$. Note also that $PQ = P'Q'$.

To see that reflection preserves angle measure, examine the reflection of $\angle ABC$ about line $m$ in Figure 13-12. $\angle ABC$ and $\angle A'B'C'$ seem to have the same measure.

This discussion suggests the following property.

| **Properties of Reflections** | Reflection is a transformation that preserves <br> (a) distance <br> (b) collinearity of points <br> (c) betweenness of points <br> (d) angle measure |
|---|---|

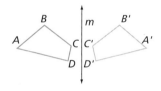

FIGURE 13-13

Figure 13-13 shows a quadrilateral *ABCD*, whose vertices are read in a clockwise direction. If we reflect the quadrilateral about line *m*, the vertices of $A'B'C'D'$ are read in a counterclockwise direction. Because the reflection reverses the order of the points or vertices, it appears that reflections do not preserve the orientation of points or vertices.

# Just for Fun

Is the distance around the unshaded region greater than, equal to, or less than the distance around the shaded area?

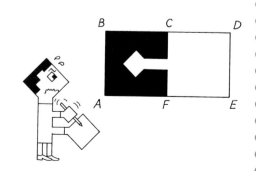

# EXERCISE SET 1

*In Exercises 1 through 4, trace each drawing onto thin paper. Then fold the paper and find the reflection image of each given point with respect to the given line m.*

*R* 1.

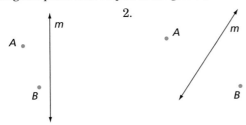

2.

3.

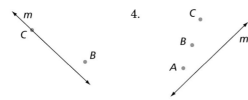

4.

5. (a) In Exercise 2, how are points *A* and *B* related?
   (b) In Exercise 3, what is the reflection image of *C*?
   (c) In Exercise 4, how are *A'*, *B'*, and *C'* related?

6. If *R* is the reflection image of point *S* over line *m*, how are *m* and $\overline{RS}$ related?

7. Find the images of the given figures or points after a reflection about line *m*.

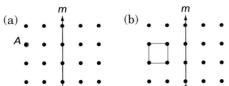

(a)    (b)

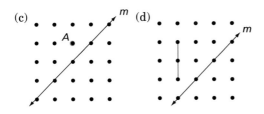

(c)    (d)

(e)    (f)

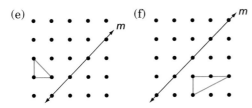

8. Trace the following figures on your paper. Then flip each figure about the given line, and draw the flip image.

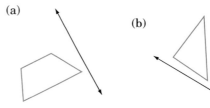

(a)    (b)

9. Classify each statement as true or false.
   (a) A segment and its reflection image have different lengths.
   (b) A point and its reflection image are the same distance from the reflecting line.
   (c) A reflection is a one-to-one correspondence.
   (d) A point and its image coincide if the point is on the reflecting line.
   (e) The reflection image of an acute angle is always acute.
   (f) The reflection image of a ray is always a ray.
   (g) When two nonintersecting lines are reflected over the same line, their images may intersect.

10. Could an angle coincide with its reflection image? Explain.

11. Name four properties preserved by reflections.

*In Exercises 12 through 15, plot the pairs of points and compute the distance between them.*

12. $A(-1, 3)$ and $B(7, -7)$   13. $A(2, 4)$ and $B(-2, -3)$

14. $A(-4, -1)$ and $B(3, -5)$   15. $A(-4, -3)$ and $B(-2, -7)$

*For Exercises 16 through 17, graph $\angle ABC$. Does the angle appear to be acute, right, or obtuse?*

16. $A(5, -1), B(2, -1), C(2, 3)$

17. $A(3, -2), B(-2, 4), C(7, 0)$

*In Exercises 18 and 19 use the converse of the Pythagorean Theorem.*

*T* 18. Show that $\triangle ABC$ is a right triangle: $A(5, -1)$, $B(2, -1), C(2, 3)$.

19. Show that $\triangle ABC$ is a right triangle: $A(6, 1)$, $B(2, -3), C(-4, 3)$.

20. Show that $\triangle ABC$ is an isosceles triangle: $A(1, 3)$, $B(5, 0), C(-2, -1)$.

21. Show that the diagonals of *ABCD* are of equal length if $A(3, 2), B(3, -1), C(7, -1)$, and $D(7, 2)$.

22. Given the accompanying triangle, find the following reflecting lines.
    (a) *l* so that *B* is the image of *A*
    (b) *m* so that *B* is the image of *C*
    (c) *n* so that *A* is the image of *C*

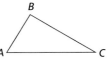

*Copy the drawings in Exercises 23 through 26. Then reflect ∠ABC over line m.*

23.
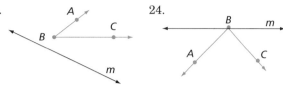

24.

25.
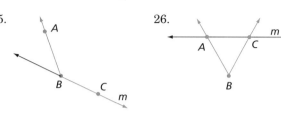

26.

27. Flip the following figures about line *l*, and show the flip image.

(a)    (b)

28. Find the image of the following figure after a reflection about line *m*.

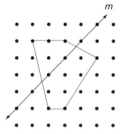

29. Draw a reflection of the figure about line *l* and label the corresponding points.

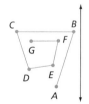

30. When do a point and its image coincide under a reflection?

31. Use the distance formula to determine if points *A*, *B*, and *C* are collinear.

(a) $A(1, 1)$, $B(3, 3)$, $C(5, 5)$
(b) $A(-1, 2)$, $B(1, 3)$, $C(3, 5)$

32. Sketch the image of the figure under the reflection about line *l*.

33. Describe the reflection—that is, find the reflection line—that maps

(a) $(x, y)$ onto $(2 - x, y)$.
(b) $(x, y)$ onto $(x, 4 - y)$.  (c) $(x, y)$ onto $(y, x)$.

*In Exercises 34 and 35, find the coordinates of the images reflected about the y axis and about the x axis. Draw these figures. Does your work look accurate?*

34. $A(1, 1)$, $B(-2, 3)$, $C(2, 4)$

35. $A(1, 1)$, $B(2, 3)$, $C(-1, 2)$, $D(-2, -1)$

36. Find the reflection of the triangle about line *m*.

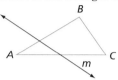

37. Describe in general terms the reflection that maps

onto rectangle $A'B'C'D'$.

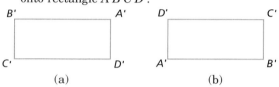

(a)  (b)

38. Let $P$ lie on line segment $\overline{AB}$, and let $r$ be the ratio $PA$ to $AB$. If $A = (x_1, y_1)$, $B = (x_2, y_2)$, and $P = (x_0, y_0)$, then $x_0 = x_1 + r(x_2 - x_1)$ and $y_0 = y_1 + r(y_2 - y_1)$. For example, the coordinates of the point 2/3 of the distance from $(1, 2)$ to $(4, 5)$ are:

$$x_0 = 1 + (2/3)(4 - 1) = 3$$
$$y_0 = 2 + (2/3)(5 - 2) = 4$$

Find the coordinates of the points

(a) 2/3 the distance from $(-3, 2)$ to $(4, 5)$.
(b) 1/4 the distance from $(-1, 2)$ to $(3, 6)$.

### ❖ PCR Excursion ❖

39. A. In this investigation we use *Geometer's Sketchpad* to discuss properties of reflections. First construct $\overleftrightarrow{AB}$. Then construct a triangle $CDE$. Find a reflection image of $\triangle CDE$ over $\overleftrightarrow{AB}$. Measure the length of some of the corresponding segments and form a conjecture. Measure some of the angles and make a conjecture. Do you think the two triangles are congruent? If you move around both triangles in a counterclockwise direction, are the vertices in the same order? Make a conjecture.

B. To find the bisector of an angle, find the line through the vertex of the angle so that one side of the angle is the reflection image of the other side over the line. The line obtained is the bisector of the angle.

C. Now show and explain how you would triple an angle.

# SLIDES OR TRANSLATIONS

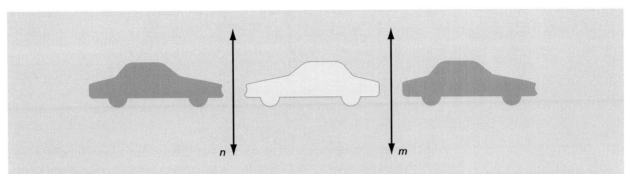

**PROBLEM**

Trace the leftmost figure above on thin paper. Fold on line *n* and show that the middle car is a reflection image of the car on the left. Fold on line *m* and note that the car on the right is a reflection image of the middle car about line *m*. Now mark corresponding points on the car on the left and the car on the right, and measure the distances between these points. Are all the distances the same? You have discovered a new transformation, called a *slide.* This new transformation is the result of two reflections about parallel lines.

**OVERVIEW**

In this section, we visualize the image of a figure after the figure has been translated to another position.

**GOALS**

Illustrate the Geometry Standard, page 482
Illustrate the Problem Solving Standard, page 15.
Illustrate the Representation Standard, page 50.
Illustrate the Communication Standard, page 78.
Illustrate the Algebra Standard, page 4.
Illustrate the Reasoning and Proof Standard, page 23.

## TRANSLATIONS

The only transformation we have defined so far is a reflection. Other transformations often result from successive reflections, as the following example illustrates.

**EXAMPLE 1**

In Figure 13-14, $\triangle A'B'C'$ is the reflection image of $\triangle ABC$ over line *m*. $\triangle A''B''C''$ is the reflection image of $\triangle A'B'C'$ over line *n*. Observe that *m* is parallel to *n*. Measure the distance from *A* to *A''*, from *B* to *B''*, and from *C* to *C''*, and make a conjecture about these distances.

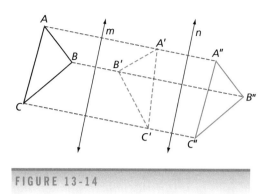

FIGURE 13-14

In the preceding example, it seems that $\triangle ABC$ has been moved a given distance in a given direction. This leads to the concept of a slide or, more formally, a translation.

**EXAMPLE 2**    Slide or translate the car in Figure 13-15 up the hill in the direction and for the distance indicated by the darkened arrow.

*SOLUTION*    Trace the car and the arrow. Slide your tracing paper along line $l$ until the tail of the arrow on the tracing paper is at the point of the arrow on the original paper. The trace of the figure in the new position coincides with the second figure in Figure 13-15, with $P$ and $Q$ corresponding to $P'$ and $Q'$.

The second car in Figure 13-15 is the **slide image** of the original figure. Keep this example in mind as we continue the discussion of a slide.

**EXAMPLE 3**    Slide a given line segment $\overline{MN}$, as in Figure 13-16, 5 cm in the direction of ray $l$.

*SOLUTION*    To begin, determine a point $M'$ 5 cm from point $M$ in the direction determined by ray $l$. In fact, slide each point on $\overline{MN}$ 5 cm from its original position, parallel to $l$. The point $N'$ is 5 cm from $N$, and $\overline{NN'}$ is parallel to $l$. Notice that segments like $\overline{MM'}$ and $\overline{NN'}$ are of length 5 cm and are parallel to $l$. The slide image of segment $\overline{MN}$ is segment $\overline{M'N'}$, and $M'N' = MN$.

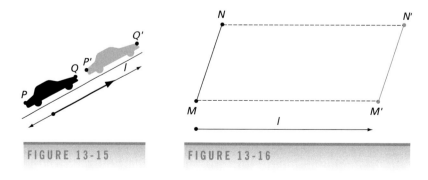

FIGURE 13-15    FIGURE 13-16

Example 2 showed how a figure was translated to a new position by sliding it a given distance in a given direction. In Example 3, individual points are translated for a given distance in a given direction, and the resulting slide image is a line segment. The intuitive understanding of slides obtained from these examples provides a valuable background for understanding the idea of translations.

| DEFINITION | **Translation** | A *translation* (or **slide**) determined by a ray *l* and a specified distance is a transformation that maps point *M* onto point *M'* so that $\overline{MM'}$ is parallel to and in the same direction as *l* and $\overline{MM'}$ equals the specified distance. |
| --- | --- | --- |

**EXAMPLE 4**

In Figure 13-17 (a), translate (or slide) the word *MATH* 2 blocks down. In (b), translate the triangle 3 blocks to the right.

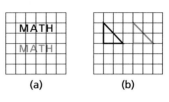

(a)          (b)

**FIGURE 13-17**

SOLUTION

The translation images are shown in color.

## TRANSLATION NOTATION

Let's return to the procedure of indicating the direction and magnitude of a translation by using an arrow. The translation defined by an arrow is often denoted as follows: $T_{AC}$ represents a translation indicated by the arrow from *A* to *C*. This notation is used in the following example.

**EXAMPLE 5**

Draw the image of the triangle under the translation $T_{AB}$ in Figure 13-18.

SOLUTION

The translation image is shown at the right in the figure.

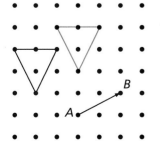

**FIGURE 13-18**

Sometimes a translation is denoted by subscripts on *T* that indicate the horizontal change in the translation by *x* and the vertical change by *y*. In this case, $T_{x,y}$ is a translation *x* units in a horizontal direction and *y* units in a vertical direction. For example, $T_{2,1}$ symbolizes that each point is translated 2 units to the right (horizontally) and 1 unit up (vertically).

## ⠿ TRANSLATION ON A COORDINATE SYSTEM

Suppose that $P(2, 3)$ is translated 3 units to the right and 1 unit up. Then $P'$ will have coordinates $(2 + 3, 3 + 1)$ or $(5, 4)$. This leads to the following characteristic.

| | |
|---|---|
| **Translation on a Coordinate System** | If $P(x, y)$ is translated $r$ units in an $x$ direction and $s$ units in a $y$ direction (denoted by $T_{r,s}$), the image of $P$ is $$P'(x + r, y + s)$$ |

**EXAMPLE 6**

Find the coordinates of $\triangle A'B'C'$, which is the image of $\triangle ABC$ translated 3 units to the right and 4 units down, where $\triangle ABC$ is defined by $A(2, 1)$, $B(3, -2)$, and $C(3, 4)$.

*SOLUTION*

In this example, $r = 3$ and $s = -4$. The images of $A$, $B$, and $C$ are $A'(5, -3)$, $B'(6, -6)$, and $C'(6, 0)$.

**PRACTICE PROBLEM**

Translate the triangle $\triangle ABC$ in Figure 13-19 in the direction and for the distance indicated by the arrow.

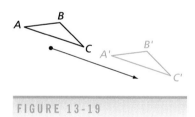

FIGURE 13-19

*ANSWER*

The translation image is shown in color.

## ⠿ TRANSLATION AS A COMPOSITE OF REFLECTIONS

The two reflections used to get a translation in the first example in this section are explained by the following definition.

<table>
<tr>
<td>D E F I N I T I O N</td>
<td>**Composite Reflections**</td>
<td>Let $R_m$ and $R_n$ represent two reflections about lines $m$ and $n$, respectively. Then the transformation "apply $R_m$ and then apply $R_n$ to the image obtained by $R_m$" is called the *composite* of $R_m$ and $R_n$ and is denoted by composite $R_n(R_m)$.</td>
</tr>
</table>

If the image of $P$ under $R_m$ is $P'$ and the image of $P'$ under $R_n$ is $P''$, then the image of $P$ under the composite $R_n(R_m)$ is $P''$.

| | |
|---|---|
| **Translation from Reflections** | A transformation is a *translation* ($T$), or slide, if and only if there exist parallel lines $m$ and $n$ such that $T =$ composite $R_n(R_m)$, where $R_m$ is a reflection over line $m$ and $R_n$ is a reflection over line $n$. |

**EXAMPLE 7**

In Figure 13-20, rectangle $WXYZ$ is reflected over line $m$, and the image is reflected over line $n$ to give rectangle $W'X'Y'Z'$. If $WXYZ$ is translated in the direction of $\overrightarrow{WW'}$ a distance of $WW'$, the same image $W'X'Y'Z'$ is obtained.

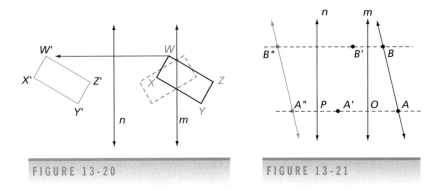

FIGURE 13-20          FIGURE 13-21

Because a translation can be considered a composite of two reflections, it is a mapping that preserves collinearity, betweenness, distance, and angle measure.

In Figure 13-21, $A'$ is the image of $A$ in a reflection over $m$, and $A''$ is a reflection of $A'$ over $n$. Notice that

$$AO = A'O \qquad \text{and} \qquad A'P = A''P$$

Now the distance between the parallel lines $m$ and $n$ is

$$A'O + A'P$$

while the distance of the translation from $A$ to $A''$ is

$$AO + A'O + A'P + A''P = 2A'O + 2A'P$$
$$= 2(A'O + A'P)$$

Thus, the distance of the translation for this example is twice the distance between the parallel lines.

By the definition of a reflection, $\overline{A''A}$ in Figure 13-21 is perpendicular to the parallel lines. This result is now stated as the following property.

| | |
|---|---|
| **Translation and Parallel Lines** | If a translation maps $A$ onto $A''$, then $\overline{AA''}$ is perpendicular to each of a set of parallel lines when the translation is considered in terms of reflections and $AA''$ is twice the distance between the parallel lines. |

See if you can establish from Figure 13-21 that $\overleftrightarrow{BA}$ is parallel to $\overleftrightarrow{B''A''}$, leading to the following property.

| Parallel Image | Under a translation of a line in which the direction of the translation is not parallel to the line, the line and its image are parallel. |
|---|---|

# Just for Fun

This Swiss miss has discovered a way of cutting the piece of wallpaper in her right hand into two pieces that will fit together to form the Swiss flag held in her left. The white cross in the center of the flag is actually a hole in the paper. The cutting must follow the lines ruled on the paper. Can you discover her secret?

# EXERCISE SET 2

R 1. Perform the translation $T_{AB}$ on point $K$ in (a), on the line segment $\overline{LM}$ in (b), and on $\triangle NOP$ in (c).

(a) *A*
  *B*   *K*

(b) *L* — • — *M*

(c) • • *N* — • — *O* •
              *P*

2. Perform the translation $T_{CD}$ on each of the geometric figures in Exercise 1.

*D*
  *C*

3. Perform $T_{1,2}$ on each of the geometric figures in Exercise 1.

4. Perform $T_{-1,1}$ on each of the geometric figures in Exercise 1.

5. Fill in the following blanks.

(a) $R_m$ represents a reflection about _____ .

(b) If $m\|n$, $R_m$ reflects about $m$, and $R_n$ reflects about $n$, then composite $R_n(R_m)$ is a transformation called a _____ .

(c) A translation preserves _____ .

(d) A translation is a composite of _____ .

(e) Under a translation, a line is parallel to _____ .

(f) If $S'$ and $T'$ are translation images of $S$ and $T$, then $S'T' =$ _____ .

(g) A point is its own image under a translation when _____ .

(h) A reflection acts like a flip; a translation acts like a _____ .

6. Trace each given figure on thin paper. Then slide each one as indicated, and draw the slide image.

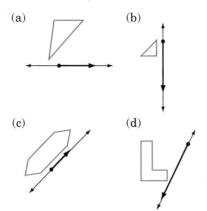

(a)  (b)  (c)  (d)

7. Slide each figure 2 cm in the direction of $\overrightarrow{PA}$.

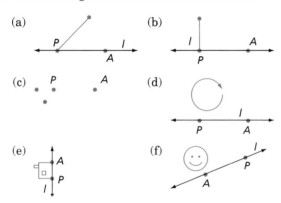

(a)  (b)  (c)  (d)  (e)  (f)

8. Trace each figure on your own paper, and then translate each traced figure 2 cm to the right.

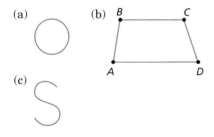

(a)  (b)  (c)

9. Draw an image for the letter $H$ with the following transformation. Translate $H$ until $P'$ is at the bottom of the original right-hand leg.

$T$ 10. Translate the following figures as indicated by the arrows.

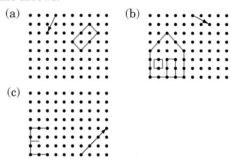

(a)  (b)  (c)

11. Find the coordinates of the image of $A(1, 1)$, $B(-2, 3)$, and $C(2, 4)$ by translating the figure 4 units to the right and 3 units down. Draw the new figure.

12. Work Exercise 11 for $A(1, 1)$, $B(2, 3)$, $C(-1, 2)$, and $D(-2, -1)$.

13. Line $l$ is parallel to $\overleftrightarrow{AB}$. Find the slide or translation image of point $C$ after a slide of distance $AB$ in a direction $\overrightarrow{AB}$.

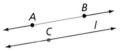

14. Draw 2 parallel lines in a vertical direction. Then perform the following transformation: Translate the parallel lines 4 cm in the direction of a ray that makes an angle of 45° counterclockwise from horizontal. Are the lines still parallel after the translation? Does parallelism appear to be preserved under translations?

15. In the accompanying figure, a translation maps $A$ onto $C$. Name the image of the following under the same translation.

   (a) $E$
   (b) $B$
   (c) $K$
   (d) $\overline{AE}$
   (e) $\angle EFK$
   (f) $BCGF$

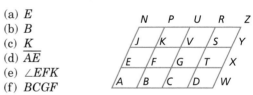

16. (a) In Exercise 15, find the images if $A$ is mapped onto $F$.
   (b) Find the images if $A$ is mapped onto $G$.

17. Draw a 30° angle with a protractor. Translate the angle 3 cm to the right. Now measure the angle. Do angles and their measures appear to be preserved under translations?

$C$ 18. Devise a method for determining the specified distance of the translation in the following figure. Does your method apply to any translation?

19. In the accompanying figure, a translation maps $A$ onto $A'$. Name the coordinates of the translation image of each of the following points if the same translation is applied.

    (a) $B$
    (b) $C$
    (c) $D$

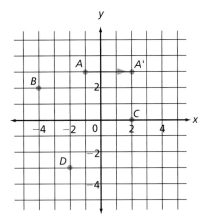

20. In the accompanying figure, a translation maps $A$ onto $A'$. Name the image of the following points under the same translation.

    (a) $C$
    (b) $B$
    (c) $D$
    (d) $E$

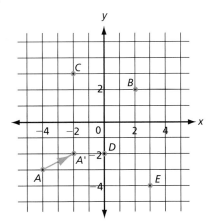

21. Draw two points ($A$ and $A'$) 4 cm apart. If $A'$ is the translation image of $A$, show two parallel lines ($m$ and $n$) so that the translation is composite $R_n(R_m)$.

22. Draw a pair of triangles, $\triangle ABC$ and a translation image $\triangle A'B'C'$. Find a pair of parallel lines ($m$ and $n$) such that the translation from $\triangle ABC$ to $\triangle A'B'C'$ is the composite $R_n(R_m)$.

23. (a) Under a translation, the image of the letter $L$ is 4 cm to the right of $L$. Find two parallel lines so that a composite of two reflections will give the same image.

    (b) Find a second set of parallel lines. (The answer in (a) is not unique.)

24. Describe the translation that maps

    (a) $(x, y)$ onto $(x + 2, y - 3)$.
    (b) $(x, y)$ onto $(x - 2, y + 5)$.

## REVIEW EXERCISES

• • • • • • • • • • • • • • • • • • • • • • • • • • • • • • • •

25. A reflection maps $K$ onto $Q$. Find the image of
    (a) $G$
    (b) $\overline{HI}$
    (c) $FGLK$

| A | B | C | D | E |
|---|---|---|---|---|
| F | G | H | I | J |
| K | L | M | N | O |
| P | Q | R | S | T |

26. In Exercise 25 find a reflection that maps $BCHG$ onto $TONS$.

27. Find the coordinates of the point obtained by reflecting $(-1, 7)$ about the $y$ axis and then reflecting the resulting point about the $x$ axis.

28. Find the vertices of the image of the triangle with vertices $(-1, 2)$, $(1, 4)$, and $(3, -1)$ under a reflection over the $y$ axis.

### ∵ PCR Excursion ∵

29. A. In this investigation we use *Geometer's Sketchpad* to investigate properties of reflections about parallel lines and about intersecting lines. Construct any irregular polygon. Then construct any two parallel lines. Reflect your figure over one line and then the other line. How does the distance from the original figure to the second image compare to the distance between your parallel lines? Place the parallel lines and the figure in different positions. (Make sure the distances change.) Again, reflect, reflect, and measure. Make a conjecture.

    B. Now translate a figure some fixed distance and direction. Can you find two parallel lines such that two reflections will yield the translation?

    C. As preparation for the next section, investigate reflecting figures across intersecting lines.

# ROTATIONS AND SUCCESSIVE MOTIONS

**PROBLEM**

Trace the figure and the lines *m* and *n*. Show that the dashed-line figure is the reflection of the figure about line *m*. Then show that the colored figure is the reflection of the dashed-line figure about line *n*. Now put a pen at point *P* and see if you can rotate the tracing paper until the first figure coincides with the second. This suggests that turns or rotations are composites of reflections. Thus, rotations preserve all the properties that reflections preserve.

**OVERVIEW**

It is usually not too difficult to visualize one motion or transformation. However, when two or more motions have been performed, it can be difficult to visualize what has happened. Do you get the same result if you perform a rotation followed by a reflection as you get by performing the same reflection followed by the rotation? You will learn the answer to this question and others in this section. In the process, you may stumble upon a bit of a surprise.

**GOALS**

Illustrate the Geometry Standard, page 482.
Illustrate the Representation Standard, page 50.
Illustrat the Problem Solving Standard, page 23.

## ROTATION

A *rotation* may be defined in terms of a center of rotation *P* and a turn angle θ. The center of rotation is often denoted by a dot, and the angle of turn is usually given by a curved arrow or by a specified number of degrees. The following example illustrates a rotation or a turn, using a turn arrow rather than a measured angle.

**EXAMPLE 1**

Cause the clock pendulum in Figure 13-22 (a) to swing in the direction and through the distance indicated by the turn arrow. Notice the center of rotation (point *P*) on which the pendulum swings.

SOLUTION

Trace the pendulum. Keeping the tracing paper fixed with a pin at point $P$, turn it until the tail of the arrow on the tracing paper is at point $V$. If possible, trace the pendulum in the new position. The second pendulum in Figure 13-22 (b) illustrates the resulting image.

**EXAMPLE 2**

Rotate the given segment $\overline{XY}$, shown in Figure 13-23, 45° in a counterclockwise direction about $X$.

SOLUTION

To perform the rotation, simply establish $X$ as the center of rotation, and draw a 45° angle with $\overrightarrow{XY}$ as one side of the angle. On the other side, mark point $Y'$ so that $XY = XY'$.

## NOTE

A rotation of 360° about any point will rotate a given figure onto itself. Such a rotation is called a **full-turn.** A rotation of 180° is called a **half-turn.**

**EXAMPLE 3**

Rotate $\overline{XY}$ 125° counterclockwise about $V$ in Figure 13-24.

SOLUTION

The rotation here is more complex than in Example 2 because $\overline{XV}$ and $\overline{YV}$ are on different lines. Rotate segment $\overline{XV}$ counterclockwise 125° about $V$; that is, with $V$ as center of rotation and $\overrightarrow{VX}$ as one ray, draw an angle of 125°. Mark $X'$ on the new ray so that $VX = VX'$. Similarly, rotate $\overrightarrow{VY}$ 125° about $V$, marking point $Y'$ such that $VY = VY'$. Now $X'Y'$ is equal to $XY$. Thus, $\overline{X'Y'}$ is the image of a 125° rotation of $\overline{XY}$ about $V$.

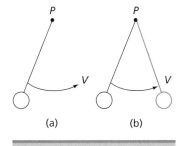

(a)          (b)

**FIGURE 13-22**

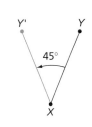

**FIGURE 13-23**

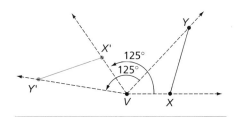

**FIGURE 3-24**

To determine a rotation, we need to know three things: the center of rotation, the size of the turn angle, and the direction of the turn (either clockwise or counterclockwise). This leads us to the following definition.

| DEFINITION | **Rotation** | A *rotation* (or turn), determined by a point $P$ and a given angle, maps point $R$ onto point $R'$ in such a way that $\angle RPR'$ is of the given measure and $RP = R'P$. The direction (clockwise or counterclockwise) of the turn must be specified. |

**PRACTICE PROBLEM**

Rotate the figure in Figure 13-25 (a) 90° in a clockwise direction about $A$.

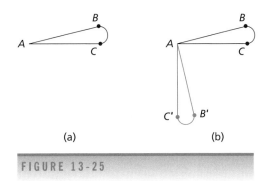

(a)                    (b)

FIGURE 13-25

*ANSWER*

The rotated figure is shown in Figure 13-25 (b).

**EXAMPLE 4**

(a) Find the image in Figure 13-26 (a) of point $A$ after a rotation of 90° about point $P$, as indicated.

(b) Find the image in Figure 13-26 (b) of point $B$ after a 180° rotation counterclockwise about point $P$, as indicated.

*SOLUTION*

The rotated images for (a) and (b) are shown as $A'$ and $B'$, respectively, in Figure 13-26.

**EXAMPLE 5**

Find the image in Figure 13-27 of the rotation of the triangle about point $P$. The 90° angle of rotation is given.

*SOLUTION*

The rotated image is shown in color in Figure 13-27.

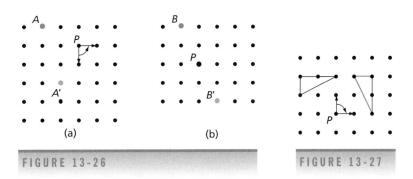

(a)                    (b)

FIGURE 13-26              FIGURE 13-27

# ROTATION AS A COMPOSITE OF REFLECTIONS

The composite of two reflections over parallel lines is a translation. Figure 13-28 shows a composite of two reflections over intersecting lines $m$ and $n$. Let $R_m$ reflect figure $ABC$ about line $m$ onto figure $A'B'C'$. Then let $R_n$ reflect figure $A'B'C'$ about line $n$ onto figure $A''B''C''$. In Figure 13-28, this could have been accomplished by rotating $\overline{OA}$ (with attached figure) through an angle $\theta$ until it coincided with $\overline{OA''}$. This example illustrates that, when two reflecting lines intersect, the composite of the two reflections results in a rotation. This characteristic is described as follows.

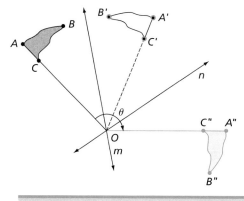

FIGURE 13-28

| **Rotation** | A transformation is a *rotation* (R) or *turn* about point P if and only if there exist two intersecting lines m and n, where R = composite $R_n(R_m)$; m and n intersect at P (called the *center of rotation*), and the angle of rotation is double the angle between m and n. |
|---|---|

Because a rotation can be considered as a composite of two reflections about intersecting lines, the properties that are preserved by reflections also are preserved by rotations. Consequently, the rotation transformation preserves distance, angle measure, betweenness, and collinearity.

## COMPOSITE TRANSFORMATIONS

One goal is to define congruence in terms of transformations. However, to accomplish this goal, we must allow for the possibility of successive transformations. As was noted earlier, images are sometimes produced as a result of several consecutive transformations.

**EXAMPLE 6**

Map the *E* on the left in Figure 13-29 onto the *E* on the right by successive transformations.

*SOLUTION*

First, we translate the figure 5 units to the right. Then we translate the figure 2 units up. Finally, we rotate the figure 90° counterclockwise about $P'$.

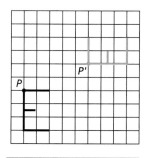

FIGURE 13-29

A translation of 5 cm to the right and a translation of 4 cm to the right can be accomplished by a translation of 9 cm to the right. Likewise, a rotation of 135° counterclockwise about some point P followed by a rotation of 60° clockwise about P can be accomplished by a rotation of 75° counterclockwise about P. This suggests that a translation followed by a translation is always a translation, and that a rotation followed by a rotation is always a rotation. Is the same thing true of reflections?

**EXAMPLE 7**

In Figure 13-30, a translation of 6 units up, followed by a rotation of 90° clockwise about the image of $B$, followed by a translation of 1 unit to the right produces image $\triangle A'B'C'$ from $\triangle ABC$. Notice that the same image $A'B'C'$ could have been obtained from $ABC$ by other sequences of transformations.

**PRACTICE PROBLEM**

Find the image onto which the figure on the left in Figure 13-31 is mapped by translating it 1 unit up, then translating it 2 units to the right, and finally rotating it 90° clockwise about the point $P$.

*ANSWER*

The image is shown in color in Figure 13-31.

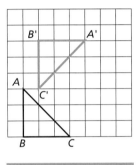

FIGURE 13-30

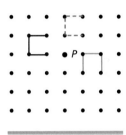

FIGURE 13-31

**EXAMPLE 8**

(a) In Figure 13-32 (a), first rotate $\angle ABC$ 90° clockwise about $P$, and then reflect it about line $m$.
(b) In Figure 13-32 (b), first reflect $\angle ABC$ about line $m$, and then rotate it 90° clockwise about $P$.

*SOLUTION*

The images for both (a) and (b) are shown in color in the two parts of Figure 13-32. Notice that the two final images are not the same.

**EXAMPLE 9**

In Figure 13-33, reflect the triangle about line $l$ so as to produce the first image. Then translate this image in the direction and distance indicated by arrow $m$ in order to get the colored image.

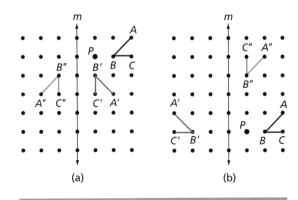

(a)                   (b)

FIGURE 13-32

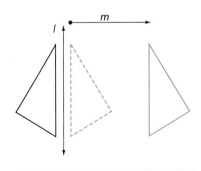

FIGURE 13-33

**EXAMPLE 10**

What motions have been performed in Figure 13-34 to produce the colored figure? (The intermediate figure is provided to help you determine the intermediate motion.)

SOLUTION

After studying the figure, you should realize that a rotation of 60° counterclockwise about the point $P$ and then a reflection about line $l$ have occurred.

**EXAMPLE 11**

Study the transformations that map the word *MATH* into the image shown in Figure 13-35. Do you see that this can be accomplished by a translation of 3 units to the right followed by a reflection about line $l$?

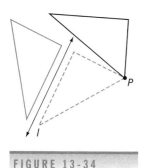

FIGURE 13-34

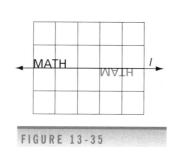

FIGURE 13-35

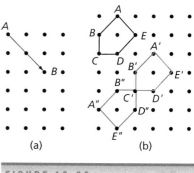

FIGURE 13-36

The preceding example introduces the following definition.

**Glide Reflection**

A translation followed by a reflection about a line parallel to the translation is called a *glide reflection*.

**EXAMPLE 12**

Use a glide reflection with $T_{AB}$ in Figure 13-36 (a) to find the image of the figure in Figure 13-36 (b). (Let the reflection line go through $C'$.)

SOLUTION

The image is shown in color in Figure 13-36 (b).

# CONGRUENT FIGURES

A transformation that preserves distances is called an **isometry.** In the preceding sections we have shown that reflections, translations, rotations, and glide reflections are isometries. Because these isometries also preserve angle measure, they give images that are the same size and shape as the original figures.

We have already intuitively investigated congruence in terms of tracing. Once the tracing has been made, the tracing can be moved to any position, yet it remains congruent to the given figure. The movement of the tracing can be explained in terms of a translation, a reflection, a rotation, or a combination of these.

<table>
<tr><td>DEFINITION</td><td>**Congruent Figures**</td><td>Two *figures A* and *B* are said to be *congruent* if and only if there is a composite of reflections, translations, and/or rotations that maps figure *A* onto figure *B* or figure *B* onto figure *A*.</td></tr>
</table>

Recall that rotations and translations are a composite of two reflections. A glide reflection can be considered as a composite of three reflections (two for the translation and one for the reflection). This leads to the following summary.

<table>
<tr><td>Congruence and Composites of Reflections</td><td>Given any two congruent figures in a plane, one can be obtained from the other by a composite of at most three reflections.</td></tr>
</table>

 **SIMILARITY**

Not all useful transformations preserve both size and shape. One useful transformation preserves only shape, while the size of the figure changes.

<table>
<tr><td>DEFINITION</td><td>**Similarity Transformation**</td><td>A *similarity* is a *transformation* of a plane that changes distances by a constant factor $k$, the **scale factor** of the similarity. If $A'$ and $B'$ are the images of $A$ and $B$, then<br><br>$$A'B' = k(AB)$$<br><br>where $k$ is a positive real number.</td></tr>
</table>

A **magnification** is a similarity transformation, with center $O$ and scale factor $k > 1$, that maps each point $P$ onto a point $P'$ such that $P'$ is on $\overrightarrow{OP}$ and $OP' = kOP$. (See Figure 13-37 (a).)

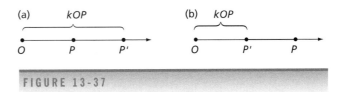

**FIGURE 13-37**

A **dilation** is a similarity transformation, with center $O$ and scale factor $k < 1$, that maps each point $P$ onto a point $P'$ such that $P'$ is on $\overrightarrow{OP}$ and

$OP' = kOP$. (See Figure 13-37 (b).) The following properties hold for similarity transformations.

| Properties of Similarity Transformations | A given similarity transformation with scale factor $k$<br>(a) maps each segment onto a segment $k$ times as long.<br>(b) maps each angle onto a congruent angle |
|---|---|

Similarity transformations enable us to define similar figures.

| Similar Figures | Geometric figures are similar (denoted by ∼ ) if and only if a similarity transformation maps one onto the other. |
|---|---|

**EXAMPLE 13**    Draw a magnification of $\overline{AB}$ in Figure 13-38, with a scale factor of 2, using center $O$.

*SOLUTION*    The required magnification is shown as $A'B'$. Each side of $\triangle OA'B'$ is twice the corresponding side of $\triangle OAB$.

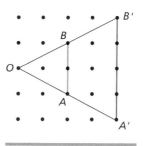

FIGURE 13-38

# Just for Fun

Can this figure be obtained by rotations of a triangle? Explain.

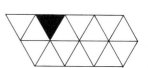

# EXERCISE SET 3

**R** 1. What is the image under a 90° clockwise rotation about $P$ of $K$ in (a), the line segment $\overline{LM}$ in (b), and $\triangle RST$ in (c)?

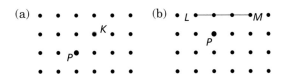

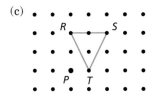

2. Find a 90° counterclockwise rotation of the geometric figures in Exercise 1.

3. Find a 180° clockwise rotation of the geometric figures in Exercise 1.

4. Find a 45° counterclockwise rotation of the geometric figures in Exercise 1.

5. Rotate each figure 90° clockwise about point $P$.

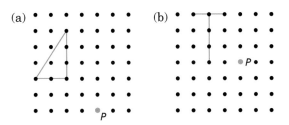

6. Rotate each figure 120° counterclockwise about $P$.

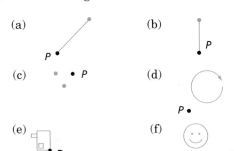

7. Sketch the image of the figure under the rotation shown.

8. Rotate each of the following figures 120° counterclockwise about point $A$.

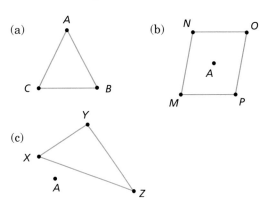

9. Draw an image for the letter $H$ with the following transformation. Rotate $H$ 180° clockwise about $P$.

10. Trace the following figures on thin paper. Then turn each figure about the given point through the rotation indicated by the turn angle.

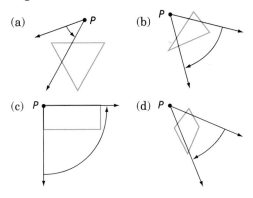

11. Translate $W$ by $T_{AB}$, and then rotate the image 90° counterclockwise about $P$.

12. In Exercise 11, replace $W$ with a horizontal segment 2 units long starting at $W$ and extending to the right.

13. In Exercise 11, replace $W$ with a unit square whose upper left corner is at $W$.

14. Perform a glide reflection on $X$, given $T_{AB}$, where the line of reflection is $\overleftrightarrow{AB}$.

15. In Exercise 14, replace $X$ with a horizontal segment 2 units long whose right end is at $X$.

16. In Exercise 14, replace $X$ with a unit square whose upper right corner is at $X$.

17. With $P$ as the center, draw a magnification of segment $\overline{AB}$, with a scale factor of 2.

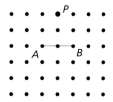

18. In Exercise 17, magnify with a scale factor of 3.

19. In Exercise 17, dilate with a scale factor of $\frac{1}{2}$.

20. In Exercise 17, dilate with a scale factor of $\frac{1}{3}$.

$T$ 21. Describe one composite of reflections, translations, or rotations that maps each of these designs onto itself.

(a) 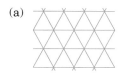    (b)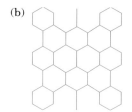

22. In each part, perform in succession the two slides in the direction and distance indicated by the arrow diagrams. Does the order of making a slide make a difference in the final result?

(a)    (b)

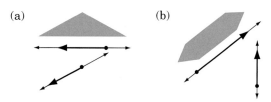

23. Perform in succession the two turns (order not important) for each figure shown. In each case, both turns occur about the same vertex $V$.

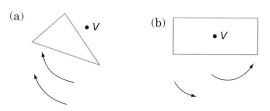

(a)    (b)

24. Perform two consecutive flips about the line of reflection given in each figure.

(a)    (b)

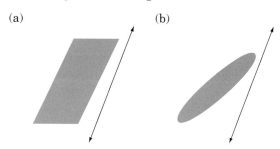

25. In each pair of figures shown, the shaded figure is the image of the other under a translation, reflection, or rotation. Identify the motion that has occurred.

(a)    (b)

(c)    (d)

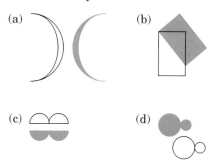

*Exercises 26 through 28 refer to the following figure.*

| A | B | C | D | E |
|---|---|---|---|---|
| F | G | H | I | J |
| K | L | M | N | O |
| P | Q | R | S | T |

26. What is the image of $\triangle GCH$ after a 90° clockwise rotation about $L$?

27. State the measure of the rotation about $K$ that maps $G$ onto $Q$.

28. Find a translation and a rotation about $N$ that maps $PQLK$ onto $IHMN$.

29. Describe the motion that will produce
    (a) the image on the right from the one on the left.
    (b) the image on the left from the one on the right.

C 30. Determine whether a translation, reflection, or rotation has been performed in the following figures.

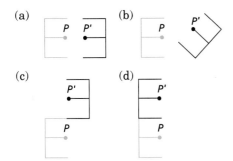

31. The quadrilaterals in the accompanying figure are images under two reflections about lines $m$ and $n$. Find a rotation that will accomplish the same result. What is the point of rotation?

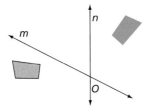

REVIEW EXERCISE

32. A translation maps $K$ onto $H$. Name the image of
    (a) $R$
    (b) $\overline{LM}$
    (c) $FGLK$

| A | B | C | D | E |
|---|---|---|---|---|
| F | G | H | I | J |
| K | L | M | N | O |
| P | Q | R | S | T |

⁘ PCR Excursion ⁘

33. The first cube is rotated to get the second cube, and the second cube is rotated with the same rotation to get the third cube. The fourth cube is obtained by rotating the third cube the same rotation. The same cube is used in (a), (b), and (c).

    (a) Verbally describe the rotation in each of the three parts. Draw a figure if necessary.
    (b) Label the sides of the fourth cube. (Place the numbers the way they would be seen on the cube.)
    (c) Verbally describe rotations in an opposite direction from the three rotations given.
    (d) Describe any other possible rotations of a cube.

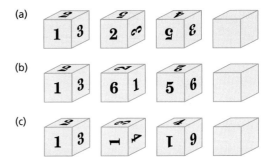

# GEOMETRIC PATTERNS IN NATURE AND ART

| **PROBLEM** | Find examples of natural objects that have both lines of symmetry and points of rotational symmetry. |
|---|---|
| **OVERVIEW** | Have you noticed geometric figures as they occur in nature? In this section, we shall call attention to a few of many examples. Triangles are found on a cross-section of the gem tourmaline. (See Figure 13-39 (a).) Regular polygons are seen in a cross-section of cadmium sulfide crystals. (See Figure 13-39 (b).) |

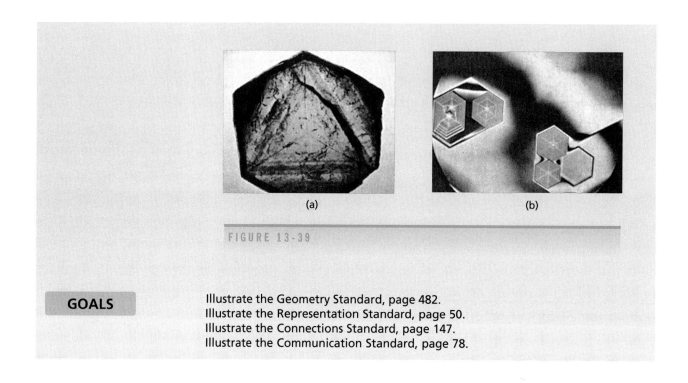

(a)                                                    (b)

FIGURE 13-39

**GOALS**

Illustrate the Geometry Standard, page 482.
Illustrate the Representation Standard, page 50.
Illustrate the Connections Standard, page 147.
Illustrate the Communication Standard, page 78.

# SYMMETRY AND PROPORTION

Consider the five rectangles in Figure 13-40. Which of these seems most pleasing to your eye? If you were an early Greek, you would choose rectangle (c) because the ratio of length to width is equal to the so-called **golden ratio** (approximately 1.6). Greek artists felt that artworks employing the golden ratio were pleasing to the eye and conveyed a sense of serenity and harmony. Among the Greek artworks that display the golden ratio are the Parthenon in Athens (Figure 13-41 (a)) and the famous statue of Apollo of Belvedere (Figure 13-41 (b)).

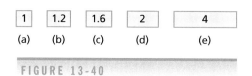

| 1 | 1.2 | 1.6 | 2 | 4 |
(a)    (b)    (c)    (d)         (e)

FIGURE 13-40

Evidently the golden ratio still appeals to the modern eye because it still occurs in many designs; check the ratio of length to width on the American flag.

Think back to your primary-school days of cutting paper hearts. Did you have difficulty making the two sides match? Finally you may have resorted to folding your paper, drawing one side of the heart, cutting both sides of paper along the design, and opening the fold to have a perfect heart. The fold

(a)

(b)

FIGURE 13-41

line is the **line of symmetry** of this heart. (See Figure 13-42.) Formally, this type of symmetry is called **reflectional symmetry.**

Some figures have more than one line of symmetry. To produce a figure with two lines of symmetry, fold a piece of paper, fold the crease back over itself, cut out the desired figure, and then unfold. The two perpendicular creases are lines of symmetry or reflection. (See Figure 13-43.)

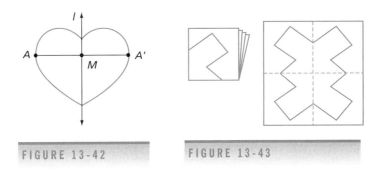

FIGURE 13-42     FIGURE 13-43

There are many examples of lines of symmetry in nature. Do you see a line of symmetry for the butterfly and the beetle in Figure 13-44?

**EXAMPLE 1**   Line $l$ is a line of symmetry for Figure 13-45. The figure on one side of the line is a reflection of the figure on the other side of the line.

Figures may also possess a second type of symmetry called **rotational symmetry.** Rotational symmetry involves a central point of symmetry, which might be called a *point of balance*. A figure has a rotational symmetry if there is a rotation (of less than 360°) about the point of symmetry that

FIGURE 13-44

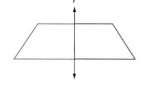

FIGURE 13-45

moves the figure onto itself. The following example uses the tracing method to help determine a point of symmetry.

**EXAMPLE 2**

Trace the rectangle in Figure 13-46 on another sheet. Determine the central point, and call it $P$. Test $P$ as a point of symmetry by rotating the tracing of the rectangle about the point $P$. If the tracing can be turned so that the original and the tracing coincide in less than a full rotation, then the point $P$ is a point of symmetry.

*SOLUTION*

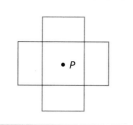

FIGURE 13-46

Notice that the original rectangle and the tracing do not coincide in Figure 13-46 when the tracing of the rectangle has been rotated 90°. However, the rectangle and its tracing do coincide after a rotation of 180°. Thus this rectangle has 180° rotational symmetry about point $P$.

There are many examples in nature of rotational symmetry. In *Snow Crystals* (Bentley & Humphrey, 1931) are found more than 2000 pictures of snowflakes that display examples of both rotational and reflectional symmetry. See if you can find some 60° rotational symmetries in the picture of snowflakes in Figure 13-47.

FIGURE 13-47

| EXAMPLE 3 | Determine whether the two figures in Figure 13-48 have rotational symmetry. If so, how much rotation is involved in each? |
|---|---|

*SOLUTION*   The square has 90°, 180°, and 270° rotational symmetry, and the pentagon has 72°, 144°, 216°, and 288° rotational symmetry.

# TESSELLATIONS

The geometric concepts of symmetry and proportion are always important in artistic design, but sometimes geometric patterns become the content of the art. For instance, works by artists such as M. C. Escher have intrigued mathematicians because of their emphasis on geometric form. Escher made use of patterns called **tessellations.** Note the repetition of the drawing of a man on a horse in Figure 13-49.

(a)           (b)

FIGURE 13-48

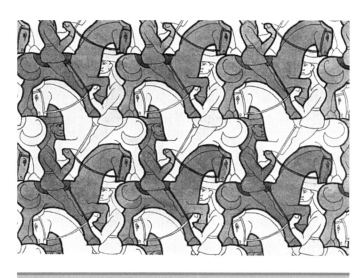

FIGURE 13-49

| DEFINITION | **Tessellation** | A *tessellation* is a pattern of shapes that covers the plane without overlapping the component parts and without leaving gaps. |
|---|---|---|

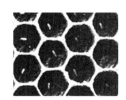

FIGURE 13-50

Another excellent example of a tessellation can be found in the geometry of a cross-section of a honeycomb. (See Figure 13-50.)

Tessellations have been used as patterns for rugs, paintings, fabrics, pottery, and architecture since ancient times, as illustrated in Figure 13-51. The Moors of Spain were quite proficient in decorating walls and floors with tessellations, using colored tiles.

**EXAMPLE 4**

The tessellation in Figure 13-52 (a) is a pattern from a floor tile. The basic element is a square tile.

The tessellation in Figure 13-52 (b) is another example of a repetition of a polygonal figure. In this figure, the repeated element is an equilateral triangle, one of which is shaded. The tessellation is formed by a 60° rotation of the triangle about one of its vertices.

FIGURE 13-51

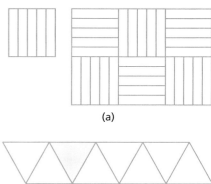

(a)

(b)

FIGURE 13-52

(a)

(b)

(c)

FIGURE 13-53

Any triangle can be used to form a tessellation. Figure 13-53 (a) shows a tessellation of isosceles triangles; (b) a tessellation of right triangles; and (c) a tessellation of scalene triangles.

Because the degree measure of a full rotation is 360°, it follows that the sum of the angles of the polygons that meet at a point in a tessellation must be 360 degrees. For instance, consider the tessellation by equilateral triangles in Figure 13-54 (a). Note that 6 equilateral triangles meet at each vertex and that 6(60°) = 360°. Similarly, in Figure 13-54 (b) four squares meet at a vertex and 4(90°) = 360°. Using this reasoning we can determine which of the regular n-gons tessellate the plane.

In Chapter 11 we learned that the sum of the interior angles of an n-gon is $(n - 2)180°$. For a hexagon this sum is 720° and hence in a regular hexagon each interior angle measures 120°. Because 3(120°) = 360 degrees, a regular hexagon will tessellate the plane with 3 hexagons meeting at each vertex. (See Figure 13-54 (c).) On the other hand, each of the interior angles of a regular pentagon measures 108° (Figure 13-55). If in a tessellation formed by regular pentagons, 3 pentagons meet at a vertex, the sum of the angles about this vertex would be 108° + 108° + 108° which is less than 360°. This cannot happen. Similarly, if 4 regular pentagons meet at a vertex, the sum of the angles about the vertex would be 4(108°) which is larger than 360°. This cannot happen. Can you see that no arrangement of regular pentagons can tessellate the plane?

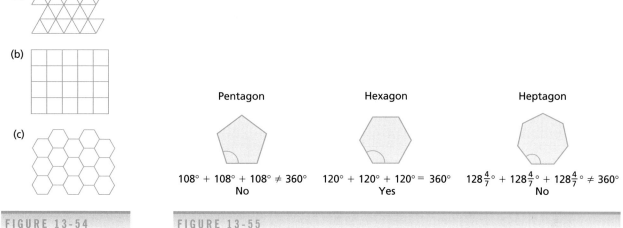

Pentagon

$108° + 108° + 108° \neq 360°$
No

Hexagon

$120° + 120° + 120° = 360°$
Yes

Heptagon

$128\frac{4}{7}° + 128\frac{4}{7}° + 128\frac{4}{7}° \neq 360°$
No

FIGURE 13-54

FIGURE 13-55

In a similar manner, we can show that no other regular $n$-gon beyond the hexagon (6-gon) will form a tessellation.

From the tessellations shown in Figure 13-56 it appears that any quadrilateral can form a tessellation.

We can form a tessellation based on quadrilaterals by rotating any quadrilateral 180° about the midpoint of one of the sides. (See Figure 13-57.)

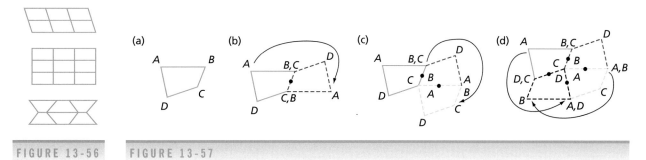

FIGURE 13-56   FIGURE 13-57

An interesting study involves finding non-regular n-gons that tessellate the plane. For example, in Exercise 16 you will show that the pentagon in Figure 13-58 will tessellate the plane. To date 13 different types of pentagons have been found that tessellate the plane.

FIGURE 13-58

# EXERCISE SET 4

*R* 1. In the given sketches, find the cases in which *l* appears to be a line of symmetry.

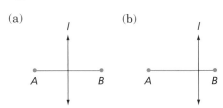

(a)  (b)

(c)  (d)

2. In the given sketches, decide which figures have a line of symmetry. Draw all lines of symmetry for these figures.

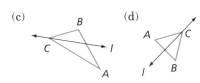

(a)  (b)

(c)  (d)

3. Determine which figures in Exercise 2 have a rotational symmetry about *P*. About *Q*.

4. How many lines of symmetry are there for each of the following regular polygons? How many degrees are there in the smallest rotational symmetry?

(a) Triangle
(b) Square
(c) Pentagon
(d) Hexagon
(e) Heptagon
(f) Octagon

5. (a) Which of the letters that follow have two lines of reflection?
(b) Which have rotational symmetry?
(c) Which have two lines of reflection but no rotational symmetry?

**A B H I M N O P T X Z**

*T* 6. With which of the following polygons can you tessellate the plane? Draw a portion of the plane to show the tessellations.

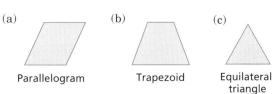

(a) Parallelogram  (b) Trapezoid  (c) Equilateral triangle

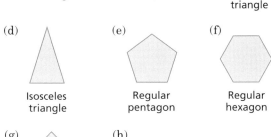

(d) Isosceles triangle  (e) Regular pentagon  (f) Regular hexagon

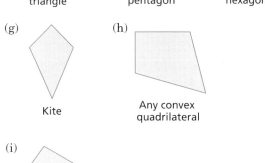

(g) Kite  (h) Any convex quadrilateral

(i) Scalene triangle

(j) Nonconvex quadrilateral  (k) Heptagon

7. Examine the following figures for both reflectional and rotational symmetry. What do you find?

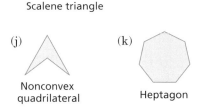

(a)  (b)

8. On a lattice, draw a tessellation based on each of the figures shown.

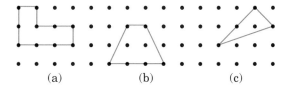

(a)  (b)  (c)

9. How many lines of symmetry are present in the following figures? What is the smallest rotational symmetry?

(a) 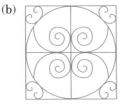   (b)

(c)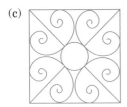

10. A tessellation is a *regular tessellation* if it is constructed from regular polygons of one size and shape such that each vertex figure (the polygon formed by connecting midpoints of edges emanating from a given vertex) is a regular polygon.

    (a) What is the vertex figure of a tessellation of squares?
    (b) Equilateral triangles?
    (c) Regular hexagons?

11. Referring to the tessellations in Exercise 10, draw vertex figures at each vertex, where possible. What do you discover?

12. From Exercise 11, the vertex figures of

    (a) triangle tessellations form a tessellation made up of _____ and _____ .
    (b) square tessellations form a _____ tessellation.
    (c) hexagon tessellations form a tessellation made up of _____ and _____ .

13. See if you can find an example of each of the following nonregular polygons that will form a tessellation.

    (a) Triangle   (b) Pentagon   (c) Hexagon

14. Sketch figures with the following qualities.

    (a) Exactly one line of symmetry
    (b) Exactly two lines of symmetry

(c) Exactly three lines of symmetry
(d) More than three lines of symmetry
(e) Rotational but not reflectional symmetry
(f) Reflectional but not rotational symmetry

15. If possible, sketch and describe a triangle that has

    (a) no lines of symmetry.
    (b) exactly one line of symmetry.
    (c) exactly two lines of symmetry.
    (d) exactly three lines of symmetry.

16. Show a tessellation using the irregular pentagon in Figure 13-58.

17. Form a tessellation made up of parallelograms. Then find three translations that will map the tessellation onto itself.

18. Show a tessellation made up of regular hexagons. Then find two rotations that will map the tessellation onto itself.

19. A tessellation is made up of equilateral triangles. Find some reflections that will map the tessellation onto itself.

20. Spatial figures have planes of symmetry for reflection and lines of symmetry for rotation. List the planes of symmetry and the lines of rotational symmetry for the following.

    (a) A right circular cylinder
    (b) A right triangular prism
    (c) A right circular cone

21. (a) From Exercise 20, find seven axes of rotational symmetry for a cube.
    (b) List six planes of symmetry for a cube.

22. Refer to Exercise 20. How many rotational axes of symmetry do the following (right) figures have?

    (a) Square pyramid
    (b) Pentagonal prism

23. Refer to Exercise 20. Discuss the planes of symmetry of the following nut.

# SECTION 5  SIMILAR GEOMETRIC FIGURES

**PROBLEM**

Take a pen flashlight and hold it in front of a rectangular piece of cardboard, with the light ray perpendicular to the center of the cardboard. Notice the shadow of the cardboard on the wall. The shadow is the same shape, but not the same size, as the cardboard. We intuitively describe the two figures as being similar.

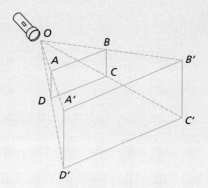

**OVERVIEW**

In Section 3 we learned that two geometric figures are similar provided that there is a similarity transformation that maps one onto the other. As a consequence similar figures have the same shape but not necessarily the same size. In this section we explore the connections between similar polygons and the measures of their sides and angles.

**GOALS**

Illustrate the Geometry Standard, page 482.
Illustrate the Measurement Standard, page 532.
Illustrate the Representation Standard, page 50.

## SIMILAR FIGURES

Congruent figures are exactly alike in size and shape. Sometimes, however, we are interested in figures that are alike in shape and yet are not congruent. In geometry, **similar** figures have the same shape but are not necessarily the same size.

Figure 13-59 illustrates 3 sets of 2 geometric figures. In each set, the top figure is similar to the one below it. In comparing the top figure to the bottom figure, notice that each segment or part of the bottom figure is increased in size by the same relative amount. In fact, there is a mathematical relationship between the corresponding parts of each figure. Such a relationship exists in the form of a ratio. The ratio of corresponding segments of similar figures is often called a **scale factor.** For example, if

$$\frac{A'B'}{AB} = k \qquad \text{then} \qquad A'B' = kAB$$

The length of $\overline{A'B'}$ is thus $k$ times the length of $\overline{AB}$. If $k > 1$, then $\overline{A'B'}$ is longer than $\overline{AB}$. If $k < 1$, then $\overline{A'B'}$ is shorter than $\overline{AB}$.

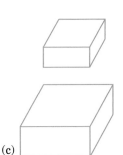

(a)          (b)          (c)

FIGURE 13-59

## SIMILAR TRIANGLES

If the two triangles $\triangle ABC$ and $\triangle A'B'C'$ in Figure 13-60 (a) are similar, there is a common ratio between their corresponding sides, as determined by points $A$ and $A'$, $B$ and $B'$, and $C$ and $C'$:

$$A'B' = kAB \qquad \text{or} \qquad \frac{A'B'}{AB} = k$$

$$B'C' = kBC \qquad \text{or} \qquad \frac{B'C'}{BC} = k$$

$$C'A' = kCA \qquad \text{or} \qquad \frac{C'A'}{CA} = k$$

Therefore,

$$\frac{A'B'}{AB} = \frac{B'C'}{BC} = \frac{C'A'}{CA}$$

**EXAMPLE 1**

In Figure 13-60 (b), the two triangles $\triangle ABC$ and $\triangle A'B'C'$ are similar with a scale factor of 2. This is shown by the fact that the length of each segment in $\triangle ABC$ is twice the length of the corresponding segment in $\triangle A'B'C'$.

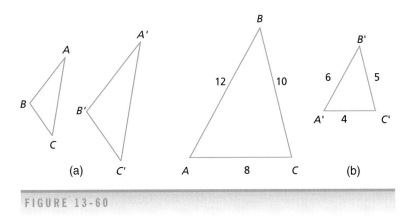

(a)          (b)

FIGURE 13-60

Special properties can be used as shortcuts to show that two triangles are similar (denoted by $\sim$ ).

| | |
|---|---|
| **Similar Triangles** | 1. Two triangles are *similar* if and only if corresponding angles are congruent. Thus, in Figure 13-61, $\triangle ABC \sim \triangle DEF$ if and only if<br><br>$$\angle A \cong \angle D, \qquad \angle B \cong \angle E, \qquad \angle C \cong \angle F$$<br><br>2. Two triangles are *similar* if and only if the ratios of the lengths of corresponding sides are equal. Thus, in Figure 13-60 (a),<br><br>$$\frac{AB}{A'B'} = \frac{BC}{B'C'} = \frac{CA}{C'A'}$$ |

**EXAMPLE 2**

Compare the similar triangles shown in Figure 13-62: $\overline{AC}$ corresponds to $\overline{DF}$, $\overline{AB}$ corresponds to $\overline{DE}$, and $\overline{CB}$ corresponds to $\overline{FE}$.
The following table may be helpful.

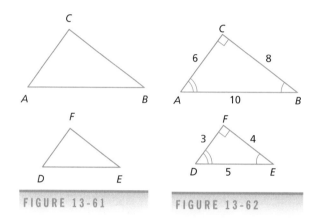

FIGURE 13-61          FIGURE 13-62

| | Side Opposite ⊿ | Side Opposite ⊿ | Side Opposite ∟ |
|---|---|---|---|
| Small △ | $DF = 3$ | $FE = 4$ | $DE = 5$ |
| Large △ | $AC = 6$ | $CB = 8$ | $AB = 10$ |

$$\frac{AC}{DF} = \frac{AB}{DE} = \frac{CB}{FE} \qquad \text{or} \qquad \frac{6}{3} = \frac{10}{5} = \frac{8}{4}$$

**EXAMPLE 3**

Suppose that $\triangle ABC \sim \triangle FGH$. (See Figure 13-63.)
(a) If $m(\angle A) = 40°$ and $m(\angle H) = 30°$, find $m(\angle B)$ and $m(\angle C)$.
(b) If $AB = 3$, $BC = 5$, $FH = 12$, and $GH = 10$, find $FG$ and $AC$.

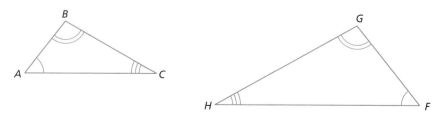

FIGURE 13-63

SOLUTION    (a) Because $\triangle ABC \sim \triangle FGH$, we know that $\angle C \cong \angle H$. Thus, $\text{m}(\angle C) = 30°$.

Therefore,
$$\text{m}(\angle B) + \text{m}(\angle C) + \text{m}(\angle A) = 180°$$
$$\text{m}(\angle B) + 30° + 40° = 180°$$
$$\text{m}(\angle B) = 110°$$

(b)

|  | Side Opposite ∡ | Side Opposite ∡ | Side Opposite ∡ |
|---|---|---|---|
| Small △ | 5 | AC? | 3 |
| Large △ | 10 | 12 | FG? |

Because $\triangle ABC \sim \triangle FGH$,

$$\frac{AB}{FG} = \frac{BC}{GH} \qquad \text{or} \qquad \frac{3}{FG} = \frac{5}{10} \qquad FG = 6$$

Similarly,

$$\frac{AC}{FH} = \frac{BC}{GH} \qquad \text{or} \qquad \frac{AC}{12} = \frac{5}{10} \qquad AC = 6$$

**EXAMPLE 4**    In Figure 13-64, assume that $\overline{DE}$ is parallel to $\overline{AB}$. In this circumstance $\triangle DCE$ is similar to $\triangle ACB$ because $\angle D$ and $\angle E$ are the corresponding angles to $\angle A$ and $\angle B$ between parallel lines. Find $y$.

SOLUTION    The side of length 6 in the smaller triangle corresponds to the side of length 8 in the larger triangle. The side of length $y$ in the smaller triangle corresponds to the side of length $y + 3$ in the larger triangle. Hence,

$$\frac{y + 3}{y} = \frac{8}{6} \qquad \text{or } y = 9$$

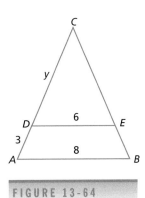

FIGURE 13-64

**EXAMPLE 5**    How high is a flagpole that casts a shadow of 24 m if a nearby post 3 m high casts a shadow of 6 m?

SOLUTION    Because the sun's rays make the same angles with the horizontal for the triangles involving the flagpole and the post, and because both the flagpole and the post are assumed to make right angles with the horizontal, the two triangles are similar, as indicated in Figure 13-65.

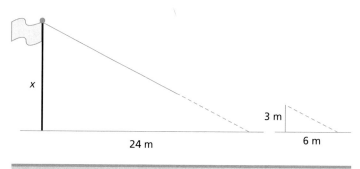

**FIGURE 13-65**

Thus,

$$\frac{x}{3} = \frac{24}{6}$$

$$6x = 72$$

$$x = 12$$

The flagpole is 12 m high.

## SIMILAR POLYGONS

Intuitively we know that two polygons are similar if they have the same shape. The following similarity properties of polygons come from the general definition of similar figures.

| Properties of Similar Polygons | Two *polygons* are *similar* if and only if there exists a one-to-one correspondence between the vertices of the polygons such that the following properties hold:<br>1. Corresponding angles (interior to interior and exterior to exterior) have equal measures.<br>2. Ratios of the lengths of corresponding segments are equal. |
|---|---|

Similar polygons provide the mathematical basis for most scale drawings. For instance, a floor plan of a house may be drawn on a small piece of paper and yet represent the shape of the house exactly. For such drawings, a scale is usually indicated; 1 cm on a scale drawing may represent 1 m in the actual measure of the house. In this case,

$$\frac{\text{measure of scale drawing of a segment}}{\text{measure of segment represented by scale drawing}} = \frac{1}{100}$$

The measure of the scale drawing of a room 10 m long is therefore

$$10 \cdot \frac{1}{100} = \frac{10}{100} = 0.1 \text{ m} = 10 \text{ cm}$$

### COMMON ERROR

Can you show that the following statement is incorrect? "Polygons are similar if corresponding angles are equal."

# Just for Fun

Because of optical illusions, the comparative size of different figures cannot always be visually determined. In part (a), can you determine which of the segments is longer by looking at them? In part (b), can you guess which set of parallel lines is longer? In part (c), which curve is longer?

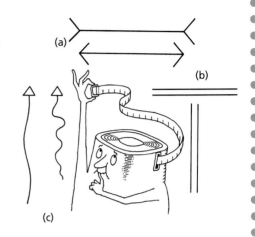

# EXERCISE SET 5

**R 1.** Assume that the following pairs of figures are similar, and find x.

(a)

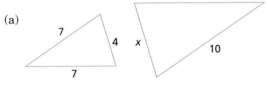

(b)

(c)

(d)

(e) In (d), find y.

**2.** Assume that the three triangles shown are similar. Find the measure of the unknown sides.

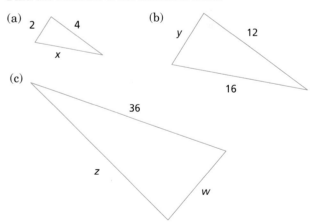

**3.** Form a triangle similar to the one shown, using a scale factor of 2. Then use a scale factor of 3.

4. Draw a triangle similar to the one shown, using a scale factor of $\frac{1}{2}$.

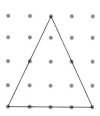

5. (a) Given △$ABC$ ~ △$DEF$. If m($\angle A$) = 70° and m($\angle E$) = 40°, find m($\angle C$) and m($\angle F$).
   (b) Given △$GHI$ ~ △$JKL$. If $GH$ = 5, $GI$ = 7, $JK$ = 8, and $KL = 9\frac{3}{5}$, find $HI$ and $JL$.
   (c) Given △$STU$ ~ △$VWX$, $TU$ = 8, $SU$ = 8, $VW$ = 4, and m($\angle TSU$) = 60°, find $ST$, $WX$, and m($\angle WVX$).

6. A snapshot is 4 cm wide and 6 cm long. It is enlarged so that it is 14 cm wide. How long is the enlarged picture? What is its perimeter?

7. At the same time that a yardstick held vertically casts a 4-ft. shadow, a vertical flagpole casts a 24-ft. shadow. How high is the flagpole?

8. On a certain map, 1.5 cm represents 60 km. If the distance between two cities on the map is 4 cm, what is the distance in kilometers between the two cities?

*T* 9. Find $x$ in the following figures, assuming in each part that the triangles are similar.

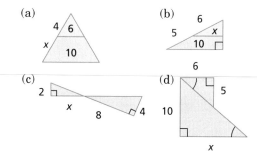

(a)
(b)
(c)
(d)

10. Prepare a scale drawing to represent a floor plan that is 8 m by 5 m, using a scale of 1 cm to 2 m.

11. Find the approximate measurements of the actual objects, using the scale drawings as given.

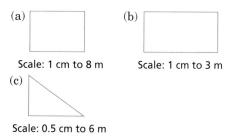

(a)
Scale: 1 cm to 8 m
(b)
Scale: 1 cm to 3 m
(c)
Scale: 0.5 cm to 6 m

12. If $CD = \frac{3}{2}$, find the length of $\overline{AB}$.

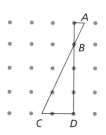

*C* 13. Two similar cylinders have radii of lengths 2 and 6, respectively. Find the ratio of their volumes. Find the ratio of their surface areas.

14. Suppose that △$ABC$ ~ △$DEF$, where $\angle A \cong \angle D$, $\angle B \cong \angle E$, and $\angle C \cong \angle F$. Let m($\overline{AC}$) = 7 m, m($\overline{DF}$) = 14 m, and m($\overline{DE}$) = 22 m. If the perimeter of △$ABC$ is 27 m, what is the perimeter of △$DEF$?

15. Judy has a picture of her boyfriend that is 15 cm wide and 25 cm long. She wants a smaller, wallet-size picture made and an enlargement made for her wall. Find the scale for the reduction and enlargement if the widths are to be 6 cm and 25 cm, respectively.

16. Merry Christmas is looking for a Christmas tree. The tree, however, can be no more than 4 m tall. Merry finds a tree that casts a shadow of 2 m, whereas Merry (120 cm tall) casts a shadow of 0.8 m. Will the tree fit in Merry's room?

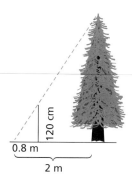

17. Prepare a scale drawing to represent the following floor plan, using a scale of 1 cm: 2 m.

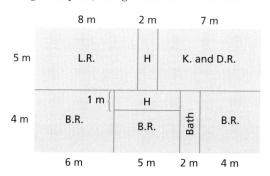

18. The idea of similar triangles can be expanded to include perimeters. The perimeters of two similar triangles are in proportion to any two corresponding sides.

    (a) Suppose that $\triangle A$ has a perimeter of 24. If one side measures 6 and it corresponds to a side of a similar $\triangle B$ measuring 2, what is the perimeter of $\triangle B$?

    (b) Suppose that $\triangle A$ has a perimeter of 12 and $\triangle B$ has a perimeter of 36. If two of the sides of $\triangle A$ measure 3 and 4, what is the measure of *each* side of $\triangle B$?

19. An observer on the shore saw a ship anchored off the coast. To find the distance to the ship, he made the measurements shown in the picture. How far is it from the shoreline to the ship?

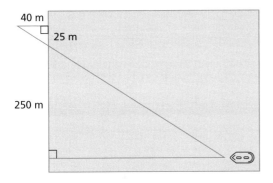

20. A farmer wishes to measure the distance across a river on his farm. Devise a plan to help the farmer find this distance.

21. At summer camp, a swimming course runs the length of a small lake. In order to determine the length of the swimming course, the counselors measure the two dry legs to a right triangle. What is the length of the swimming course? (HINT: Do not use similar triangles to make the indirect measurement.)

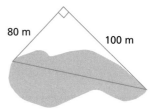

22. Two similar spheres have radii of lengths 4 ft. and 10 ft., respectively. Find the ratio of the volumes of the spheres.

# SOLUTION TO INTRODUCTORY PROBLEM

**UNDERSTANDING THE PROBLEM.** This printing is a mirror image or a reflection.

**DEVISING A PLAN.** The technique to use is another reflection. You can either hold the message up to a mirror upside down or hold it in front of a light upside down.

**CARRYING OUT THE PLAN.**

HAVEN'T
YOU GOT
ANYTHING
BETTER TO DO
THAN MESS WITH OLD BOTTLES?

# SUMMARY AND REVIEW

1. A transformation is a one-to-one correspondence between two sets of points on a plane.

2. A reflection $R_m$ over line $m$ is a transformation mapping each point $P$ onto point $P'$ as follows.

   (a) If $P$ is on $m$, then $P' = P$.

   (b) If $P$ is not on $m$, then $m$ is the perpendicular bisector of $\overline{PP'}$.

3. A composite of transformations results from successive applications of transformations. For two reflections, the composite $R_n(R_m)$ is obtained by applying $R_m$ and then $R_n$.

4. A transformation is a translation ($T$) if and only if $T$ = composite $R_n(R_m)$, where $R_m$ is a reflection over line $m$, $R_n$ is a reflection over line $n$, and $m$ is parallel to $n$.

5. A transformation is a rotation about point $P$ if and only if the rotation is a composite $R_n(R_m)$, where $m$ and $n$ intersect at $P$.

6. A similarity is a transformation of a plane that changes distances by a constant factor $k$, the scale factor of similarity. If $A'$ and $B'$ are the images of $A$ and $B$, then $A'B' = k(AB)$.

7. Two figures are *similar* if they have the same shape but not necessarily the same size.

8. Two triangles are similar if either of the following is true:

   (a) Corresponding angles are congruent.

   (b) Corresponding sides are proportional.

9. Two figures $A$ and $B$ are said to be congruent if and only if there is a composite of reflections, translations, or rotations such that the image of $A$ is $B$.

10. Congruent figures have the same size and shape.

# CHAPTER TEST

1. Identify each statement as true or false.

   (a) Different transformations may preserve different properties of figures.

   (b) The composite of two reflections is always a translation.

   (c) A segment and its reflection image may have different lengths.

   (d) There is a one-to-one correspondence between points and their images over a given reflection line.

   (e) A point $P$ and its reflection image coincide if $P$ is on the reflecting line.

   (f) Every composite of reflections gives an image congruent to the original figure.

   (g) Some transformations are not reflections.

   (h) A transformation is a rotation if and only if it is a composite of reflections about parallel lines.

   (i) Reflections, translations, and rotations preserve angle measure, betweenness, collinearity, and distance.

   (j) The rotation image of an obtuse angle may be an acute angle.

2. Reflect $L$ about the given line.

3. Translate $T$ the distance and direction given by the arrow.

4. Rotate $T$ in Question 3 by 60° counterclockwise about the given point $P$.

5. *A* is at (2, 1), and *B* is at (–2, 3). Find the image of each as reflected about the given line.

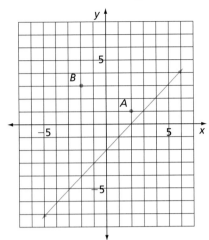

6. In Question 5, find the image of A translated 4 units to the left and 6 units down.

7. In Question 5, find the image of *B* rotated 90° counterclockwise about the origin 0.

8. Sketch the following diagram on your paper. Then reflect the triangle about $\overleftrightarrow{AB}$, and then rotate the image 90° clockwise about the image of *P*.

9. In Question 8, rotate the given triangle 90° clockwise about *P*, and then reflect about $\overrightarrow{AB}$. Are the answers the same?

10. In Question 8, reflect the triangle about $\overleftrightarrow{AB}$, and then translate it through the distance *AB* in the direction of $\overrightarrow{AB}$.

11. In Question 8, translate the triangle through a distance of *AB* in the direction of $\overrightarrow{AB}$, and then reflect it about $\overleftrightarrow{AB}$. Are the answers in Questions 10 and 11 the same?

12. The flag pole shown is 60 ft. high and casts a 100-ft. shadow. How tall is John, who casts a 9-ft. shadow?

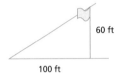

# ADDITIONAL TOPICS OF GEOMETRY AND GRAPH THEORY

# 14

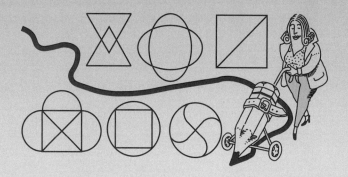

ry to trace the figures by beginning at one point, not raising your pencil until you have covered every point, and being certain not to trace any line or curve more than once. (You are allowed to cross a previously traced part.) Are there any that cannot be done? Make up more complicated figures and try to trace them. What is the secret?

In this chaper we continue our study of conguence, focusing on triangles. In our study of the Pythagorean Theorem we explore a proof of this theorem using the properties of congruent triangles. Then we use properties of congruent triangles to justify constructions, which provide intuitive understanding of many geometric concepts.

We conclude this chapter with two sections on graph theory, a branch of mathematics that can be applied to a multitude of diverse problems. We introduce the topic with problems encountered by a road inspector and a traveling salesman and conclude the study with shortest path problems, where weights are assigned to various edges of graphs.

# SECTION 1   CONGRUENCE AND TRIANGLES

**PROBLEM**

Leroy is helping his dad install siding on the end of their house. Leroy's dad says, "Cut the siding to fit snugly against the gable." Leroy is in luck because he has just studied congruent right triangles in geometry and knows how to apply them.

He needs to trim each piece of siding so that $\angle 1 \cong \angle 2$. This, of course, occurs when $\angle 3 \cong \angle 4$. Thus Leroy must cut from each piece of siding a congruent right triangle, with $\angle 3 \cong \angle 4$.

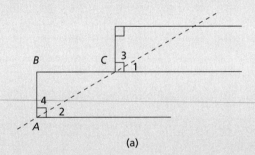

(a)

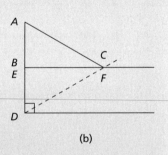

(b)

**OVERVIEW**

This is how he accomplishes the task: First he places a piece of siding against the gable, marks point $C$, and cuts off right triangle $ABC$. Then he places this triangle against the next piece of siding, marks point $F$, and cuts off right triangle $DEF$. The two right triangles $\triangle ABC$ and $\triangle DEF$ are congruent. Why? The answer to this question is one of the things you will learn in this section.

**GOALS**

Illustrate the Geometry Standard, page 482.*
Illustrate the Representation Standard, page 50.*
Illustrate the Reasoning and Proof Standard, page 23.*
Illustrate the Communication Standard, page 78.*

* The complete statement of the standard is given on this page of the book.

## ⠿ CONGRUENT POLYGONS

Although we are primarily interested in the congruence of triangles, we begin our discussion at a more general level with the congruence of polygons. As you recall, when two line segments $\overline{AB}$ and $\overline{CD}$ are of the same length, we say that they are congruent (denoted by $\overline{AB} \cong \overline{CD}$). Likewise, when two angles $\angle A$ and $\angle B$ have the same measure, we say that they are congruent (denoted by $\angle A \cong \angle B$). From the preceding chapter, we know that two geometric figures are congruent if they can be superimposed so as to coincide. But in order for two polygons to fit exactly together, all of their corresponding angles and sides must have the same measure. This leads to the following definition of *congruent polygons*.

| | |
|---|---|
| **Congruent Polygons** | Two *polygons* are *congruent* if and only if there is a correspondence between their vertices such that all of their corresponding sides and corresponding interior angles are of equal measure. |

## ⠿ CONGRUENT TRIANGLES

Consider the two congruent triangles $\triangle ABC$ and $\triangle DEF$ in Figure 14-1. In these triangles, we can choose this correspondence.

$$A \leftrightarrow D, B \leftrightarrow E, C \leftrightarrow F$$

| DEFINITION | **Congruent Triangles** | Two *triangles* $\triangle ABC$ and $\triangle DEF$ are *congruent* if and only if the vertices $A$, $B$, $C$ and $D$, $E$, $F$ can be paired so that corresponding angles and corresponding sides are congruent. If $A$ is paired with $D$, $B$ with $E$, and $C$ with $F$, then $\overline{AB} \cong \overline{DE}$, $\overline{BC} \cong \overline{EF}$, $\overline{CA} \cong \overline{FD}$, $\angle ABC \cong \angle DEF$, $\angle BCA \cong \angle EFD$, and $\angle CAB \cong \angle FDE$. |
|---|---|---|

Fortunately, it is unnecessary to check all six congruence relations in order to determine that two triangles are congruent. Let's consider intuitively how these requirements can be reduced. Suppose that we are given the two line segments in Figure 14-2(a) and the angle $A$ in Figure 14-2(b). Let's construct all the triangles that contain $\angle A$ included between sides congruent to the two given line segments. Using our compass we construct $\overline{AB}$ congruent to Segment 1 and $\overline{AC}$ congruent to Segment 2 as in Figure 14-2(b). Is it intuitively clear that there is only one way to complete our construction? The only triangle containing $\angle A$ included between two sides of the specified lengths is $\triangle ABC$. This suggests that two sides and an included angle are sufficient to fix uniquely the size and shape of a triangle; that is, any other triangle with sides congruent to Segment 1 and Segment 2 and included angle congruent to $\angle A$ will be congruent to $\triangle ABC$. This result that we have understood intuitively we will now state as a **postulate,** an assumed truth.

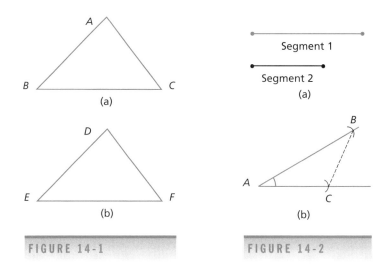

FIGURE 14-1

FIGURE 14-2

| **Side, Angle, Side:** **SAS** | If two sides and the included angle (the angle formed by the rays containing the two sides) of one triangle are congruent to two sides and the included angle of a second triangle, then the two triangles are congruent. This statement is abbreviated SAS, for *side, angle, side.* |
|---|---|

**EXAMPLE 1**

In Figure 14-3, $\angle RPS \cong \angle TPS$ and $RP = TP$. Is $\triangle RPS \cong \triangle TPS$?

*SOLUTION*

Because $\overline{PS}$ corresponds to and is congruent to $\overline{PS}$, and because $\overline{RP}$ corresponds to and is congruent to $\overline{TP}$ (given), and because $\angle RPS \cong \angle TPS$ (given) $\triangle RPS \cong \triangle TPS$ by SAS.

**EXAMPLE 2**

Is the pair of triangles in Figure 14-4 congruent? If so, why?

*SOLUTION*

In Figure 14-4, $\triangle MNP \cong \triangle OPN$ by SAS because the marks on the drawing indicate that $\angle MNP \cong \angle OPN$ *and* $\overline{MN} \cong \overline{OP}$. Clearly, $\overline{NP} \cong \overline{PN}$.

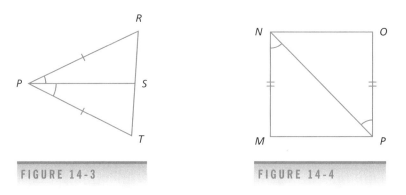

FIGURE 14-3

FIGURE 14-4

Suppose that you have three line segments, as shown in Figure 14-5, where $AB < EF$, $CD < EF$, and $AB + CD > EF$. Set your compass to span

$\overline{AB}$. With center at $E$, draw an arc using this compass setting in Figure 14-6. If $B'$ is any point on this arc, do you see that $\overline{EB'}$ is congruent to $\overline{AB}$? Similarly, set your compass to span $\overline{CD}$ and draw an arc centered at $F$ using this compass setting. If $D'$ is any point on this second arc, then $\overline{FD'}$ is congruent to $\overline{CD}$. Do you see that there is a single point above $\overline{EF}$ where segments congruent to $\overline{AB}$ and $\overline{CD}$ can meet to form a triangle? Evidently, three sides fix a triangle; therefore, two triangles with corresponding sides of equal length are congruent. In fact, the following theorem can be proved from the SAS postulate.

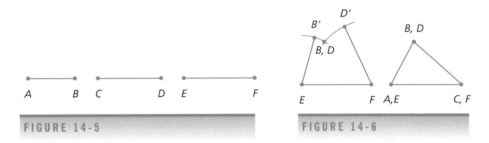

FIGURE 14-5                              FIGURE 14-6

| | |
|---|---|
| **Side, Side, Side: SSS** | If three sides of one triangle are congruent to three sides of a second triangle, then the two triangles are congruent. This statement is abbreviated SSS, for *side, side, side*. |

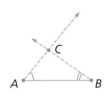

FIGURE 14-7

In a like manner, consider the two angles $\angle A$ and $\angle B$ of segment $\overline{AB}$ in Figure 14-7. Let the angles be such that the rays (not on $\overline{AB}$) of $\angle A$ and $\angle B$ intersect at $C$. Here again, the two angles and the included side seem to fix the triangle. Thus we see intuitively that, if two angles and the included side of one triangle are congruent to two angles and the included side of a second triangle, the two triangles are congruent.

This discussion suggests the following theorem, which can be proved from the SAS postulate.

| | |
|---|---|
| **Angle, Side, Angle: ASA** | If two angles and the included side of one triangle are congruent to two angles and the included side of another triangle, then the two triangles are congruent. This statement is abbreviated ASA, for *angle, side, angle*. |

**EXAMPLE 3**

In Figure 14-8, $\angle B \cong \angle E$ and $AB = DE$. Are the two triangles congruent? Why or why not?

*SOLUTION*

In solving this problem, we will use the problem-solving strategy of working backward. Observe that $\overline{AB}$ is congruent to $\overline{DE}$ and that $\angle B$ and $\angle E$ are congruent. Thus we will know that the two triangles are congruent by ASA if we can show that $\angle A$ and $\angle D$ are congruent.

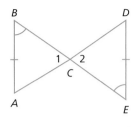

FIGURE 14-8

To see that $\angle A$ and $\angle D$ are congruent, we first observe that $\angle 1$ and $\angle 2$ are congruent because they are vertical angles. Because the sum of measures of angles of a triangle is 180°,

$$\text{m}(\angle A) = 180° - \text{m}(\angle B) - \text{m}(\angle 1)$$

$$= 180° - \text{m}(\angle E) - \text{m}(\angle 2) \quad \text{Because } \angle E \cong \angle B \text{ and } \angle 1 \cong \angle 2$$

$$= \text{m}(\angle D)$$

Thus, $\text{m}(\angle A) = \text{m}(\angle D)$ or $\angle A$ and $\angle D$ are congruent, and, of course, $\triangle ABC \cong \triangle DEC$ by ASA.

The discussion in the previous example demonstrates that, when two angles of one triangle are congruent to two angles of a second triangle, then the third angle of the first triangle must be congruent to the third angle of the second triangle, because the sum of the measures of the angles of any triangle is 180°. Thus, if any two angles and a corresponding side of one triangle are congruent to two angles and a corresponding side of another triangle, then the two angles and the included side of the first triangle are congruent to the two angles and the included side of the second triangle. Hence the two triangles are congruent, by ASA.

| **Angle, Angle, Side: AAS** | If two angles and a corresponding side of one triangle are congruent to two angles and a corresponding side of a second triangle, then the two triangles are congruent. The result is abbreviated AAS, for *angle, angle, side.* |
| --- | --- |

### COMMON ERROR

Two angles are of measure 65° and 45° and one side is of length 6 in triangles $A$ and $B$. Therefore, $\triangle A \cong \triangle B$ by AAS. This is not necessarily true. See Figure 14-9. The side must be a corresponding side; that is, it must be opposite the same angle in each triangle.

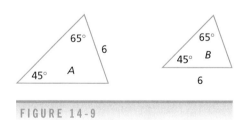

FIGURE 14-9

## CHARACTERISTICS OF RIGHT TRIANGLES

The conditions for the congruence of two right triangles are special cases of triangles in general. First recall that the sides of a right triangle have special names. The side opposite the right angle is called the **hypotenuse,** and the other two sides are called **legs.** (See Figure 14-10.)

The task of comparing two right triangles to determine congruence is often particularly easy. For instance, if two legs of one right triangle are congruent to two legs of another right triangle, we need only note that the angle between the legs for both right triangles is a right angle. Thus the triangles are congruent by SAS.

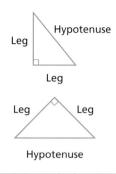

**FIGURE 14-10**

## ⠿ EQUILATERAL AND ISOSCELES TRIANGLES

Facts about the congruence of right triangles can be used to easily deduce the following facts about equilateral and isosceles triangles.

| **Properties of Isosceles and Equilateral Triangles** | 1. If a triangle is isosceles, the angles opposite congruent sides are congruent.<br>2. If two angles of a triangle are congruent, the triangle is an isosceles triangle, where the sides opposite congruent angles are congruent.<br>3. If a triangle is equilateral, all three angles are congruent.<br>4. If three angles of a triangle are congruent, the triangle is an equilateral triangle. |
| --- | --- |

## ⠿ THE PYTHAGOREAN THEOREM AND ITS CONVERSE

The Babylonians learned long ago that right triangles have a very special property: The square of the length of the side opposite the right angle is equal to the sum of the squares of the other two sides. For example, if the legs are 3 and 4, then the length of the hypotenuse must be 5, because $5^2 = 3^2 + 4^2$. This relationship can be demonstrated by counting the number of squares on the legs and then on the hypotenuse in Figure 14-11. Likewise, in Figure 14-12, $c^2 = a^2 + b^2$.

The converse of the Pythagorean Theorem is also true.

| **Converse of the Pythagorean Theorem** | If the square of the measure of one side of a triangle is equal to the sum of the squares of the measures of the other two sides, the triangle is a right triangle. |
| --- | --- |

## ⠿ PROOF OF THE PYTHAGOREAN THEOREM

Over the years, many proofs of the Pythagorean Theorem have been presented. The theorem was known to the Babylonians about a thousand years

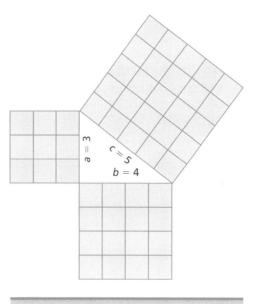

FIGURE 14-11

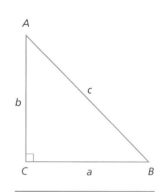

FIGURE 14-12

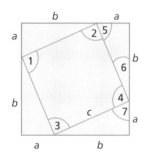

FIGURE 14-13

before Pythagoras. Nevertheless, the first general proof of the theorem is believed to have been given by Pythagoras in about 525 B.C. There have been many conjectures as to the type of proof Pythagoras might have given; it is generally believed today that it was a dissection type of proof, such as that shown in Figure 14-13. Notice that the properties of congruence figure prominently in the proof.

Figure 14-13 shows a square with sides that have measures of length $a + b$. In this square, we draw four right triangles with legs of measure $a$ and $b$. The four right triangles are congruent. (Why?) Thus, the four hypotenuses are congruent. Further,

$$m(\angle 6) + m(\angle 4) + m(\angle 7) = 180°$$
$$m(\angle 5) + m(\angle 6) = 90° \qquad \text{Why?}$$

Because the triangles are congruent,

$$m(\angle 7) = m(\angle 5) \qquad \text{Why?}$$

Hence, $m(\angle 7) + m(\angle 6) = 90°$ so $m(\angle 4) + 90° = 180°$ which means that $m(\angle 4) = 90°$. In like manner, it can be shown that

$$m(\angle 1) = m(\angle 2) = m(\angle 3) = 90°$$

Thus, the figure bounded by the four hypotenuses is a square because all sides are congruent and all angles have measures of 90°.

The area of the large square is equal to the area of the smaller square plus the area of the four congruent right triangles. The area of the large square is

$$(a + b)^2 = a^2 + 2ab + b^2$$

The area of the small square is $c^2$. Because the area of a right triangle is half the product of the legs, the total area of the four right triangles is

$$\frac{4(ab)}{2} = 2ab$$

Thus,

$$a^2 + 2ab + b^2 = c^2 + 2ab$$

Subtracting $2ab$ from each side of this equation yields $a^2 + b^2 = c^2$, thus completing the proof.

# Just for Fun

Henry Perigal, a London stockholder and amateur astronomer, discovered a paper-and-scissors "proof" of the Pythagorean Theorem. Can you?

(HINT: In the large square (2) on the leg of the right triangle, draw a line through the center of the square perpendicular to the hypotenuse. Then draw a line through the center of square (2) perpendicular to the first line. Cut out the pieces of square (2), and use them with (1) to form the square on the hypotenuse.)

# EXERCISE SET 1

*R* 1. Each pair of the following triangles appears to be congruent. Determine those cases in which the information given in the triangles is sufficient to ensure that the triangles are congruent. Explain your reasoning. Note: marks on line segments that are alike indicate congruent line segments. Congruent angles are indicated by ⊀ or ⊀.

(a)

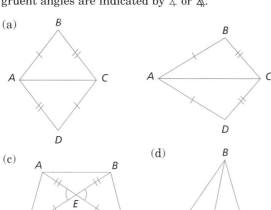

(c)

(d)

(e)

(f)

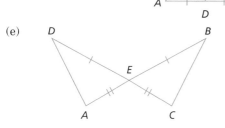

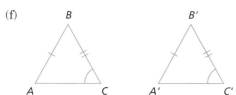

2. Given that △*ABC* ≅ △*DEF*, fill in the following blanks.

(a) ∠*ABC* ≅ _____

(b) ∠*FDE* ≅ _____

(c) ∠*BCA* ≅ _____

(d) *AB* = _____

(e) *EF* = _____

(f) *FD* = _____

3. The pairs of right triangles below appear to be congruent. Determine if enough information is given to be sure that they are congruent, and explain your reasoning.

(a)  (b)

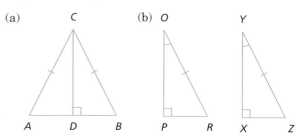

4. Find the length of the missing side on the following triangles.

(a)  (b)  (c)

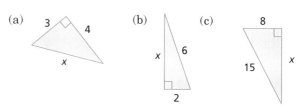

5. Which of the following triplets of numbers are the lengths of the sides of a right triangle?

(a) 5, 12, 13  (b) 20, 23, 31  (c) 2, 36, 42
(d) 4, 5, 6  (e) 4, $\sqrt{5}$, $\sqrt{21}$  (f) 18, 24, 30

6. Can an isosceles triangle have

(a) three acute angles?
(b) a right angle?  (c) an obtuse angle?

7. Show that *PQ* = *PR*.

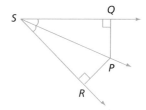

8. Verify that the following pairs of triangles are congruent.

(a)  (b)

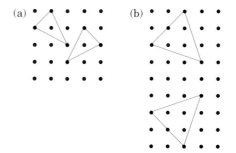

9. Prove that $\triangle ABC \cong \triangle DCB$.

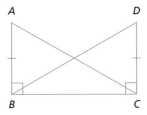

$T$ 10. Determine whether the conditions given are sufficient to prove that $\triangle ABC \cong \triangle DEF$.

(a) $\angle C \cong \angle F, \angle A \cong \angle D, BC = EF$
(b) $AB = DE, \angle E \cong \angle B, AC = DF$
(c) $\angle B \cong \angle E, AB = DE, \angle A \cong \angle D$
(d) $\angle A \cong \angle D, \angle C \cong \angle F, \angle B \cong \angle E$

11. (a) In triangle $ABC$, $\angle A$ and $\angle C$ are congruent. Discuss why this implies that $\overline{AB} \cong \overline{BC}$ and hence $\triangle ABC$ is isosceles.

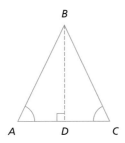

(b) Explain why the discussion from (a) proves Property 2, on page 637.

12. Show that $(BD)^2 + (AC)^2 = (AB)^2 + (DC)^2$.

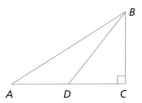

$C$ 13. Conditions for congruence specify that definite corresponding parts must be congruent. If possible, draw two triangles that are not congruent where the following conditions are satisfied.

(a) Three parts of one are congruent to three parts of the other.
(b) Four parts are congruent.
(c) Five parts are congruent.
(d) Six parts are congruent.

14. Show that $\triangle X \cong \triangle Y$ in the accompanying figures, and thus that the area of a parallelogram is $bh$. (HINT: First reason that $\angle D$ in the parallelogram is congruent to $\angle C$.)

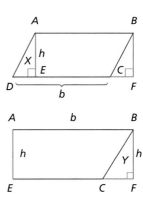

15. Show that $\triangle ACB \cong \triangle ADB$ in the accompanying figure, where $\overleftrightarrow{AD} \parallel \overleftrightarrow{CB}$ and $\overleftrightarrow{AC} \parallel \overleftrightarrow{BD}$. Then show that the area of $\triangle ACB = \frac{1}{2}hb$ given that the area of a parallelogram is $hb$.

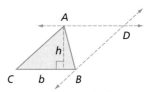

16. The minute hand of Big Ben is 8 ft. long, and the hour hand is 6 ft. long. What is the distance between the tips of the hands at 3 P.M.?

17. Prove that $(XJ)^2 + (HB)^2 = (XH)^2 + (JB)^2$.

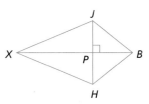

18. You want to take an 8-ft.-wide piece of plywood through a 36-in.-wide door. How tall must the door be in order to handle the plywood?

19. Reason why the following triangles are congruent.

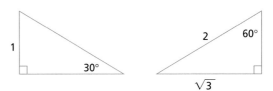

20. If $a = 2mn$, $b = n^2 - m^2$, and $c = n^2 + m^2$, show that

$$c^2 = a^2 + b^2$$

Thus, $a$, $b$, and $c$ satisfy the Pythagorean theorem. Values may be substituted for $m$ and $n$ to secure what are called *Pythagorean triplets*. Find 5 such triplets.

21. Using *Geometer's Sketchpad,* explore congruence of triangles using

    (a) SSS    (b) SAS    (c) ASA

22. In this section we verified the Pythagorean relationship between the length of the legs of a right triangle, $a$ and $b$, and the length of the hypotenuse, $c$, by constructing squares on each of the three sides of the triangle and comparing their areas. (See Figure 14-11.) Indeed, one geometric statement of the Pythagorean Theorem states that the sum of the areas of the squares constructed on the legs of the right triangle equals the area of the square constructed on the hypotenuse. In the explorations that follow, discover whether this statement can be made for figures other than squares.

    A. (a) On each of the legs of a right triangle with legs measuring 3 and 4 and hypotenuse measuring 5, construct an equilateral triangle. Compare the sum of the areas of the equilateral triangles on the legs with the area of the equilateral triangle on the hypotenuse. (Use the Pythagorean Theorem to find the height of the equilateral triangles.)

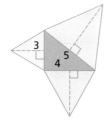

    (b) Construct a right triangle of your choice, and compare the sum of the areas of equilateral triangles on the legs with the area of an equilateral triangle on the hypotenuse.

    B. (a) Follow the Part (a) instructions of A, but construct semicircles on each side of the right triangle instead of equilateral triangles.

    (b) Follow the Part (b) instructions of A, but compare the areas of semicircles constructed on each side of the triangle.

    C. (a) Follow the Part (a) instructions of A, but construct rectangles on each side of the right triangle. Use rectangles whose width is half the length, and let the side of the triangle form the longest side of each rectangle.

    (b) Use rectangles as described in (a) on your right triangle from Part (b) of A. Compare the areas as before.

    Are you getting evidence for this geometric extension of the Pythagorean Theorem: "If figures with the same shape [similar figures] are constructed on the sides of a right triangle, then the sum of the areas of the figures on the legs is equal to the area of the figure on the hypotenuse"?

SECTION

# 2 ▷ JUSTIFICATIONS OF CONSTRUCTIONS

**PROBLEM**    Draw two triangles such that five parts of one triangle are congruent to five parts of the other but the two triangles are not congruent. Mark the congruent parts.

**OVERVIEW**    In this section we will learn several important geometric constructions and use our information about triangle congruence to justify the validity of those constructions.

# ◌ Geometric Constructions

Two tools used by ancient geometers to explore congruence were the straightedge (without units of measurement) and the compass. These tools serve today as physical devices for constructing congruent figures. Once we set a compass so that it spans a given line segment of a given length, we assume that the points of the compass subsequently produce two points the same distance apart as the given distance.

**CONSTRUCTION 1**

Construct a line segment that is congruent to line segment $\overline{CD}$ in Figure 14-14.

GIVEN:  Line segment $\overline{CD}$. With a straightedge, draw ray $\overrightarrow{AF}$.
(a) Set your compass points at $C$ and at $D$ in Figure 14-14 (a).
(b) Without changing the setting of your compass, place one end at $A$ and draw a portion of a circle (called an arc), cutting the ray at $B$ in part (b).
(c) $\overline{AB} \cong \overline{CD}$.

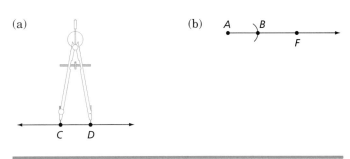

FIGURE 14-14

**CONSTRUCTION 2**

In Figure 14-15 (a), divide $\overline{AB}$ into two congruent segments.

GIVEN:  Line segment $\overline{AB}$.
(a) With $A$ as the center of a circle, draw an arc with radius greater than half of $AB$.
(b) Similarly, with $B$ as the center, draw an arc with the same radius as in (a).
(c) Use a straightedge to construct a line through the intersections of these two arcs.
(d) The point of intersection of this line and $\overline{AB}$, denoted by $C$, divides $\overline{AB}$ into the two congruent segments $\overline{AC}$ and $\overline{CB}$.

*JUSTIFICATION*

In steps (a) and (b), $\overline{BE}$ was constructed as congruent to $\overline{AE}$, and $\overline{BD} \cong \overline{AD}$. (See Figure 14-15 (b).) Moreover, $\overline{DE} \cong \overline{DE}$. Thus, $\triangle ADE \cong \triangle BDE$ (by SSS). It follows that $\angle BEC \cong \angle AEC$ (corresponding angles of congruent triangles), $\overline{BE} \cong \overline{AE}$ (by construction) and $\overline{CE} \cong \overline{CE}$. Therefore, $\triangle ACE \cong \triangle BCE$ (by SAS). Thus, $\overline{AC} \cong \overline{BC}$ (corresponding sides of congruent triangles).

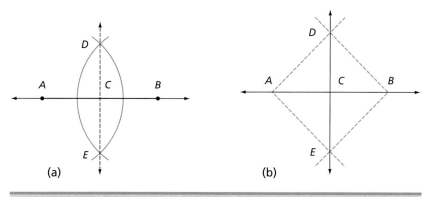

FIGURE 14-15

In a plane, one and only one line perpendicular to a given line can be drawn through a point not on the given line; this fact is illustrated by Construction 3.

**CONSTRUCTION 3**

Construct a perpendicular to a line from a given point not on the line.

GIVEN: Line $r$ and point $P$ not on $r$. See Figure 14-16 (a).

(a) With $P$ as a center, draw an arc that intersects $r$ in two points, $A$ and $B$. See Figure 14-16 (b).
(b) With $A$ and $B$ as centers, draw arcs with the same radii intersecting at $C$.
(c) Draw line $\overleftrightarrow{PC}$.
(d) $\overleftrightarrow{PC}$ is perpendicular to $\overleftrightarrow{AB}$.

*JUSTIFICATION*

We know that $\overline{PA} \cong \overline{PB}$ (by construction), $\overline{AC} \cong \overline{BC}$ (by construction), and $\overline{PC} \cong \overline{PC}$. Thus, $\triangle PBC \cong \triangle PAC$ (by SSS). It follows that $\angle BCP \cong \angle ACP$ (corresponding angles of congruent triangles), and, of course, $\overline{CO} \cong \overline{CO}$. Thus, $\triangle BCO \cong \triangle ACO$ (by SAS), and $\angle AOC \cong \angle BOC$ (corresponding parts of congruent triangles). But m($\angle AOC$) + m($\angle BOC$) = 180° (straight angle). Therefore,

$$\text{m}(\angle AOC) = \text{m}(\angle BOC) = 90°$$

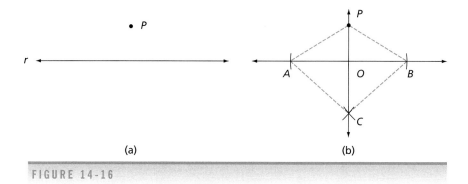

FIGURE 14-16

A similar construction exists for drawing a ray perpendicular to a given line at a given point on the line.

**CONSTRUCTION 4**

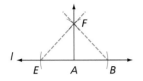

FIGURE 14-17

*JUSTIFICATION*

Construct a ray perpendicular to a given line at a given point on that line.

GIVEN:  Line *l* containing point *A*.
(a) Using point *A* as center, draw an arc in each direction. (See Figure 14-17.)
(b) Denote by *B* and *E* the intersections of these arcs and line *l*.
(c) Now with an arbitrary setting (except that the radius should be of length greater than *EA*), draw an arc in one half plane determined by *l*, with the fixed end of the compass at *E*.
(d) With the same compass setting, draw an arc with the fixed end of the compass at *B*. Denote by *F* the intersection of these two arcs.
(e) Next, using a straightedge, draw $\overrightarrow{AF}$.
(f) $\overrightarrow{AF}$ is perpendicular to line *l*.

We know that $\overline{EA} \cong \overline{BA}$ [by construction in (a) and (b)], $\overline{EF} \cong \overline{BF}$ [by construction in (c) and (d)], and $\overline{FA} \cong \overline{FA}$. Therefore, $\triangle EFA \cong \triangle BFA$ (by SSS). Thus, $\angle FAE \cong \angle FAB$. But $m(\angle FAE) + m(\angle FAB) = 180°$. Therefore,

$$m(\angle FAE) = m(\angle FAB) = 90°$$

**CONSTRUCTION 5**

Construct a perpendicular bisector of a segment $\overline{AB}$; that is, construct a perpendicular line to $\overline{AB}$ at *C* such that $AC = CB$.

GIVEN:  Line segment $\overline{AB}$.
Look again at Construction 2 and Figure 14-15. Note that Construction 2 produces a line $\overleftrightarrow{DE}$ that bisects segment $\overline{AB}$. Because $\triangle ACE \cong \triangle BCE$, $\angle ACE \cong \angle BCE$. As in the justifications of Constructions 3 and 4, this means that $\overleftrightarrow{DE}$ is perpendicular to $\overline{AB}$, hence $\overleftrightarrow{DE}$ is a perpendicular bisector of $\overline{AB}$; that is, Construction 2 produces the desired perpendicular bisector.

Congruent angles can be constructed by using a compass and straightedge, as demonstrated in Construction 6.

**CONSTRUCTION 6**

Construct an angle congruent to $\angle ABC$ in Figure 14-18 (a).

GIVEN:  $\angle ABC$.
(a) Draw $\overrightarrow{DE}$ as one side of the angle to be constructed. (See Figure 14-18 (b).)
(b) With *B* as the center, draw an arc of any radius, cutting $\overrightarrow{BA}$ at *H* and $\overrightarrow{BC}$ at *G*.
(c) With *D* as the center, draw an arc of the same radius as in part (b), cutting $\overrightarrow{DE}$ at *G'*.
(d) With *G'* as center, draw an arc of radius *HG*.
(e) Denote the intersection of these two arcs by *K*.
(f) Draw $\overrightarrow{DK}$.
(g) $\angle KDG' \cong \angle ABC$.

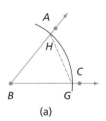

(a)

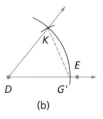

(b)

FIGURE 14-18

*JUSTIFICATION*

In steps (b) and (c), $\overline{DG'}$ was constructed congruent to $\overline{BG}$. In (d) and (e) $\overline{KG'} \cong \overline{HG}$ and $\overline{DK} \cong \overline{BH}$. Therefore, $\triangle HBG \cong \triangle KDG'$ (by SSS). Hence, $\angle KDG' \cong \angle HBG$ (corresponding angles of congruent triangles). Thus, $\angle KDG' \cong \angle ABC$ because $\angle HBG$ is another name for $\angle ABC$.

T
H
E
N

&

N
O
W

## Euclid
### 365? B.C.–275? B.C.

A student once asked Euclid why he should be laboring over geometry, so Euclid gave the student three coins, "since he must gain from what he learns." Euclid's own works in geometry, however, were clearly meant to display the beauty of geometric reasoning rather than to demonstrate the practical side of the subject.

In 323 B.C., Alexander the Great died and left behind a massive empire. Fortunately, the refined and insightful Ptolemy I ended up ruling the Egyptian portion of the empire. In Alexandria, Ptolemy founded a great university called the Museum to which he invited scholars from all over the world. Euclid, who was undoubtedly trained by students of Plato—he may even have studied at Plato's Academy—was among those summoned by Ptolemy to teach at the Museum.

Euclid's *Elements* on geometry comprise 13 books (we would probably call them chapters today): six books on plane geometry, three on number theory, one on irrational numbers, and three on solid geometry. The organization of the book into definitions, postulates, and theorems is the model for most of the mathematics that is presented today in textbook form.

Today, the geometry of Euclid has been extended and for some purposes superseded by other geometries. But we still use the basic concepts developed so beautifully by him. Euclidean geometry is used widely in modern engineering and technology. Euclid's ideas extend to other fields as well. In business and economics, the subject of linear programming is pervasive. In linear programming, polygons are basic, and polygons are an important topic in the *Elements*. The geometry needed for development of lenses, cameras, telescopes, and the like can all be found in Euclid's works.

Indeed, most of the mathematics you learned in Chapters 12 and 13 was included in Euclid's *Elements*. At a time when textbooks are rarely ever current for even 10 years, it is impressive to know that the *Elements* was used as a text for more than 2000 years. In addition to his books on geometry, Euclid composed treatises on such diverse subjects as astronomy, mechanics, music, and optics, but it is his writings about geometry that have forever secured his place in the history of mathematics.

Also of interest is the construction of the **bisector of an angle**—a ray that will form two angles of equal measure.

**CONSTRUCTION 7**

Construct the bisector of $\angle ABC$ in Figure 14-19.

GIVEN: $\angle ABC$.
(a) With $B$ as center, draw any arc intersecting $\overrightarrow{BA}$ and $\overrightarrow{BC}$. Call the points of intersection $D$ and $E$.
(b) Now with a compass setting greater than half of $DE$ and with the fixed end at $E$, draw an arc in the interior of $\angle ABC$.
(c) With the same compass setting as in (b) and with $D$ as the center, draw an arc intersecting the arc in (b) at $F$.
(d) Draw ray $\overrightarrow{BF}$.
(e) $\overrightarrow{BF}$ bisects $\angle ABC$.

JUSTIFICATION

We know that $BD = BE$ [from (a)], $EF = DF$ [from (b) and (c)], and $BF = BF$. Thus, $\triangle BEF \cong \triangle BDF$ (by SSS), so $\angle DBF \cong \angle EBF$.

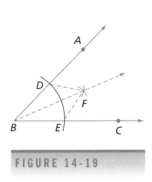

FIGURE 14-19

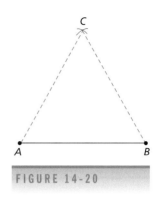

FIGURE 14-20

An easy construction that is useful in constructing certain $n$-gons is the construction of an equilateral triangle.

**CONSTRUCTION 8**

Construct an equilateral triangle with each side equal to line segment $\overline{AB}$. (See Figure 14-20.)

GIVEN: Line segment $\overline{AB}$.
(a) With a compass setting spanning $\overline{AB}$, draw an arc above $\overline{AB}$ with $A$ as the center.
(b) With the same compass setting as in (a), draw an arc with $B$ as the center intersecting the arc in (a) at $C$.
(c) Draw $\overline{AC}$ and $\overline{BC}$.
(d) $\triangle ABC$ is an equilateral triangle.

*JUSTIFICATION*

Because the compass setting used to get point $C$ is $AB$, it follows that

$$AB = AC = BC$$

Now let's consider a construction that allows us to obtain a circle that circumscribes a triangle.

**CONSTRUCTION 9**

Construct a circle circumscribing a triangle.

GIVEN: $\triangle ABC$ in Figure 14-21 (a).
(a) Construct a perpendicular bisector $l$ of segment $\overline{AB}$. (See Figure 14-21 (b).)
(b) Construct a perpendicular bisector $m$ of segment $\overline{AC}$.
(c) Mark the intersection $O$ of $l$ and $m$. This will be the center of the circumscribed circle.
(d) Set the fixed end of the compass at $O$ and the other end at $B$, and draw a circle with radius $OB$.
(e) The circle drawn in (d) contains $A$ and $C$ and thus is the circumscribed circle.

*JUSTIFICATION*

Because $O$ is on the perpendicular bisector of $\overline{AB}$, it follows that $OA = OB$. Because $O$ is on the perpendicular bisector of $\overline{CA}$, it follows that $OC = OA$. Thus,

$$OC = OA = OB = \text{radius of the circumscribed circle}$$

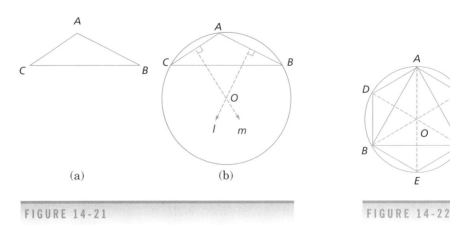

(a)                              (b)

FIGURE 14-21                              FIGURE 14-22

As an introduction to the construction of regular *n*-gons, we will use Constructions 8 and 9 to construct a regular hexagon.

**CONSTRUCTION 10**

Construct a regular hexagon. (See Figure 14-22.)
(a) Construct an equilateral triangle *ABC*.
(b) Construct a circumscribed circle about △*ABC*.
(c) Draw the bisectors of the three angles of the equilateral triangle, and indicate the intersections of the bisectors with the circle by *D*, *E*, and *F*.
(d) Draw $\overline{AD}$, $\overline{DB}$, $\overline{BE}$, $\overline{EC}$, $\overline{CF}$, and $\overline{FA}$, forming a hexagon.

*JUSTIFICATION*

In Exercise Set 2, you will prove that the bisectors of the angles of an equilateral triangle intersect in a common point. Knowing this, we can prove that the central angles are all equal; thus, each must equal 60°. Therefore, we have constructed 6 congruent equilateral triangles such as △*EOC*. Thus, $EC = FC = AF = DA = DB = BE$ , and the hexagon is regular.

In Exercise Set 2, you will construct additional *n*-gons.

# Just for Fun

A famous puzzle about a water lily was introduced by Henry Longfellow in his novel *Kavanagh.* When the stem of a water lily is vertical, the blossom is 22 cm above the surface of the water. When moved by a gentle breeze, keeping the stem straight, the blossom touches the water at a spot 45 cm from where the stem formerly cut the surface. How deep is the water?

# EXERCISE SET 2

*R* 1. Construct a triangle whose three sides are as given.

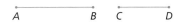

2. Draw a triangle with an acute angle at *A*. Construct each of the following.

   (a) The altitude from vertex *A*.
   (b) The median from vertex *A*. (A median is a segment drawn to the midpoint of the opposite side.)

3. Describe the points that are equidistant from the ends of segment $\overline{AB}$.

4. Construct a 45° angle.

5. Construct a 60° angle.

6. Construct a 30° angle.

7. Construct an isosceles triangle, using $\overline{AB}$ for two sides and $\overline{CD}$ for the third side.

*T* 8. Construct a triangle congruent to △*ABC* in the accompanying figure, justifying your construction by SSS.

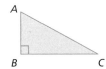

9. Construct a triangle congruent to △*ABC* in Exercise 8, justifying your construction by SAS.

10. Construct a triangle congruent to △*ABC* in Exercise 8, justifying your construction by ASA.

11. Construct a right triangle congruent to △*ABC* in the accompanying figure, justifying your construction by making leg $\overline{A'B'} \cong \overline{AB}$ and hypotenuse $\overline{A'C'} \cong \overline{AC}$ (SSS).

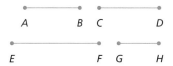

12. Construct a right triangle congruent to △*ABC* in Exercise 11, justifying your construction by making leg $\overline{A'B'} \cong \overline{AB}$ and leg $\overline{B'C'} \cong \overline{BC}$ (SAS).

13. Construct a right triangle congruent to △*ABC* in Exercise 11, justifying your construction by making leg $\overline{B'C'} \cong \overline{BC}$ and ∠*C'* ≅ ∠*C* (ASA).

14. (a) The point of intersection of the altitudes of a triangle with three acute angles lies where?
    (b) The point of intersection of the altitudes of a right triangle lies where?
    (c) The point of intersection of the altitudes of a triangle with an obtuse angle lies where?

15. Explain how you could obtain the following angles by constructions.

    (a) 75°      (b) 135°

16. Construct a line parallel to a given line through a specified point not on the line, and justify your construction.

17. Use the accompanying line segments to construct a quadrilateral. Is your answer unique?

18. Use the given angle and the following two sides to construct a triangle. Is your answer unique?

19. Construct an inscribed circle within an equilateral triangle *ABC*. (HINT: An inscribed circle is inside the triangle and touches but does not cross each side of the triangle.)

*C* 20. Construct an inscribed circle within some scalene triangle *ABC*, and justify your procedure.

21. Prove that the diagonals of a square intersect at the center of the square.

22. Construct a square.

23. Construct a circumscribed circle about the square in Exercise 22.

24. Using the construction in Exercise 23, draw the diagonals and bisect the central angles of the square to locate 8 points on the circle. Use these to form a regular 8-gon.

25. From Exercise 24, discuss how you would form a 16-gon; then generalize the family of *n*-gons you could form in this manner.

26. Discuss how you would construct a 12-gon from the hexagon in Construction 10.

27. From Construction 10 and Exercise 26, generalize the family of *n*-gons you could form in this manner.

28. Construct the three medians (see Exercise 2 (b).) of the triangle shown in the accompanying figure. Do the three medians intersect in a common point? Justify your answer.

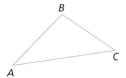

29. Construct the three altitudes of $\triangle ABC$ in Exercise 28. Do they intersect in a single point?

30. Prove that the bisectors of the angles of an equilateral triangle intersect in a common point.

31. For $\triangle ABC$, let $\angle B$ be $\angle$, let $\overline{BC}$ be, •────•, and let the altitude from $A$ be •──•. Construct the triangle.

### REVIEW EXERCISES
• • • • • • • • • • • • • • • • • • • • • • • • • • • • • •

32. Which of the following can be the sides of a right triangle?

   (a) $1, 5, \sqrt{26}$      (b) $15, 20, 25$
   (c) $2, 4\sqrt{2}, 6$      (d) $5, 12, 13$

33. Find $x$ in the following figure.

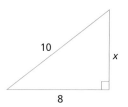

34. The outside ends of two 40-ft. wooden beams of a house are fixed. If each beam expands in length by 1 in., find how high the abutting ends are above the horizontal.

35. Two bugs are in a top corner of a room that is 12 ft. × 16 ft. × 8 ft. On the floor in the extreme opposite corner is a piece of candy. What is the shortest distance to the candy for a bug that

   (a) can fly?
   (b) must crawl?

---

# FROM THE HIGHWAY INSPECTOR TO GRAPH THEORY

**PROBLEM**

A highway inspector and a traveling salesman meet for breakfast and consider the map in Figure 14-23. The traveling salesman wishes to visit each town (marked by a dot) once during the day to ply his wares. However, for the sake of efficiency, he wants to design his trip in such a way that he never passes through the same town twice. The road inspector wishes to travel across every road on the map, but, for the sake of economy, she doesn't wish to travel on any segment of road more than once. Can they design trips that will accomplish their goals?

FIGURE 14-23

**OVERVIEW**

The preceding problem introduces a branch of mathematics known as graph theory, very useful in this century. With applications as diverse as mail routes, scheduling, computer design, and garbage collection, maps or graphs can be used to model a number of tasks and then devise ways to do those tasks more efficiently. You can easily construct a map. If you specify some points on a page and connect the points with line segments, you have a map.

**GOALS**

Illustrate the Representation Standard, page 50.
Illustrate the Problem Solving Standard, page 15.
Illustrate the Reasoning and Proof Standard, page 23.

Note in Figure 14-23 that there are six towns (represented by dots) and eight roads (represented by segments) connecting them.

**EXAMPLE 1**

The traveling salesman needs to visit every town exactly once, but he has no interest in driving on each stretch of road. The highway inspector needs a route that will allow her to travel over each section of road, but in the interest of economy, does not wish to travel on any section of road more than once. Find routes for the traveling salesman and the highway inspector in Figure 14-23.

*SOLUTION*

One solution for the salesman is shown in Figure 14-24(a) while a solution for the inspector is found in Figure 14-24(b). Has the inspector traveled on all roads exactly once? Has the salesman visited all six towns exactly once?

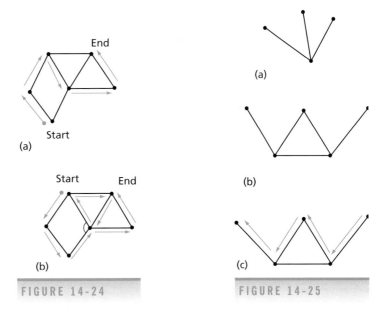

FIGURE 14-24        FIGURE 14-25

Now let us try to find routes for the inspector and the salesman for the simple maps in Figure 14-25.

With a little thought it is clear that neither the inspector nor the salesman can find a suitable route for the map in Figure 14-25(a). The map in Figure 14-25(b) can be traveled happily by the salesman (see Figure 14-25(c)), but there is no satisfactory route for the inspector.

# THE HIGHWAY INSPECTOR

Let us focus on whether there is a route for the highway inspector on maps we will consider. Clearly we could attempt to answer this question using the problem solving strategy, Guess, Test, and Revise. However, we would like to be able to identify some simple recognizable property of a map that would allow us to answer the question easily for each new map that we encounter. Consider the map given in Figure 14-26, and let us focus on what happens at Town A which is a junction for three sections of road.

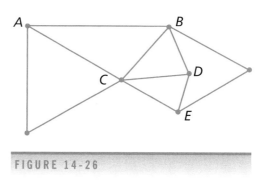

FIGURE 14-26

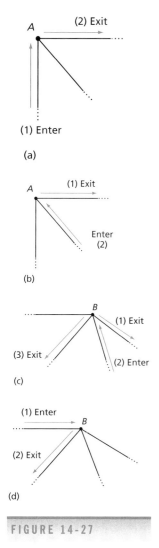

FIGURE 14-27

If Town $A$ is not where the inspector starts the trip, then on her first visit to Town $A$ she will enter on one of the three sections of road and exit on either of the two sections that remain. See Figure 14-27(a). It follows that in order to complete the inspection without driving on road sections already inspected, she must ultimately end her trip at Town $A$. Similarly, if the inspector begins at Town $A$, she inspects one of the sections of the road as she begins her trip. See Figure 14-27(b). When later she returns, she will use a second section and hence must leave to inspect the final section of road that connects Town $A$. Thus, if an economical (as discussed previously) path for the inspector can be found, she must either begin her trip at Town $A$ or end her trip at Town $A$. We can use similar reasoning to argue that since Town $C$ is a terminus for five sections of road, she must either begin or end her trip at Town $C$.

Now let us apply the same analysis at Town $B$, a connecting point for four sections of road. If the road inspector starts at $B$, then on her first trip out she will inspect one road, leaving three sections to be inspected. When she returns, she will inspect one road as she enters and another as she leaves, leaving one road uninspected. See Figure 14-27(c). Hence, she will need to end her trip at Town $B$ if she starts there. If the road inspector does not start at Town $B$, then after she enters and leaves $B$, she will have inspected two of the four sections. See Figure 14-27(d). The next time she passes through, she can both enter and leave on sections that have not yet been inspected. Further, she will then never need to return to Town $B$ because all sections connecting that town have been inspected.

It appears that the number of road sections terminating at each town is a very important number. The **degree** of the town is the number of sections of road that end at the town. By the reasoning we used above, we have the following theorems.

1. If the degree of a town is odd, then an economical trip by the inspector must either begin or end in that town.
2. If the degree of a town is even: (a) if the inspector begins at the town, then she must end at that town. (b) If she does not begin at the town, then each time she enters the town, she has an uninspected section on which to leave.

Looking back at Figure 14-26 we can make another observation. Since Towns $A$, $C$, $D$, and $E$ all have odd degree, it is impossible to find an economical trip (one where no road is traveled more than once) for the inspector. That is, since a town of odd degree must either be the beginning or the end of an economical trip for the inspector, a map that admits such a trip can have at most two towns of odd degree.

| | |
|---|---|
| **Map with Odd Degree Vertices** | If a map has more than two towns of odd degree, then the inspector has no economical route. |

Clearly a road system that has towns of odd degree is problematical. But what about a road system whose towns all have even degree? Before we answer this question we need a definition. A road system is **connected** if you can get from any town to any other town by traveling on roads in the system. The comments that follow will deal only with connected road systems. By capitalizing on the reasoning about when the inspector has uninspected sections of road on which to enter or leave a town, it is not hard to prove the following results: (See Exercise 28 for a brief discussion of the direction of the proof.)

| | |
|---|---|
| **Two Towns of Even Degree** | If each of the towns of a connected road system has even degree, the inspector has an economical tour. Further, the inspector ends the tour in the town where she starts. |
| **Two Towns of Odd Degree** | If a map of a connected road system has exactly two towns of odd degree, there is an economical route for the highway inspector. The inspector will begin at one town of odd degree and end at the other. |

This, of course, leaves only one open issue. What happens in maps with exactly one town of odd degree? This issue is closed remarkably easily with the following theorem.

| | |
|---|---|
| **Properties of Maps** | 1. The sum of the degrees of all towns on a map is an even number.<br>2. There are no maps with exactly one town of odd degree. |

Notice how nicely we have solved the highway inspector problem. To determine if an economical route exists, we need only compute the degree of each town on the map. (Note we have not done an equally nice job for the salesman. We will say more about this later.)

**EXAMPLE 2**

Find an economical trip for the highway inspector for the map in Figure 14-28(a).

*SOLUTION*

There are two odd vertices (*A* and *B*) and one even vertex. Thus the highway inspector has an economical route as illustrated in Figure 14-28(b). We will now demonstrate one of many applications of the inspector's problem.

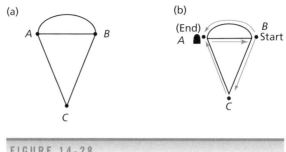

FIGURE 14-28

# GRAPH THEORY

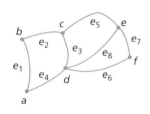

FIGURE 14-29

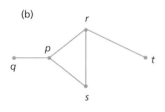

FIGURE 14-30

Not surprisingly, the work we have done with the problems of the highway inspector and the traveling salesman have application in many other areas of investigation. The same kinds of diagrams that we use to represent road maps can represent computer networks (with dots representing computers and connecting arcs representing communication links), telephone systems, and gas pipeline systems. Partially because of its many applications, the study of "road-map like" diagrams such as the one found in Figure 14-29 has been part of one of the most active and growing areas of mathematics for the last several decades. The diagrams are known to mathematicians as **graphs** and the branch of mathematics that includes their study is called **graph theory.** We will now spend a little time becoming familiar with terminology that is important in graph theory.

We could interpret the diagram in Figure 14-29 as a road map with towns represented by dots and sections of roads indicated by pieces of curves. In graph theory we call the dots **vertices** and the connecting pieces of curve we call **edges.** In Figure 14-29 the vertices are labeled $a$, $b$, $c$, $d$, $e$, and $f$. We might name the edges $e_1$, $e_2$, $e_3$, $e_4$, $e_5$, $e_6$, $e_7$, and $e_8$. Note that each edge could also be named by the pair of vertices it connects. Thus, $e_1$ might be named $(a,b)$, $e_2$ might be named $(b,c)$, and so on. Although we often think of graphs in terms of geometric representation, we have learned in mathematics it is both useful and necessary to represent objects in multiple ways. The formal definition of a graph is given by describing its vertices and edges using the language of sets.

| DEFINITION | **A Graph** | A graph $G = (V, E)$ consists of $V$, a nonempty finite set of vertices and $E$, a finite set of edges, with each edge associated with a pair of vertices. If vertices $u$ and $v$ are associated with a unique edge $e$, we may write $e = (u,v)$. |
|---|---|---|

In the graph in Figure 14-30, the edge $(p, q)$ connects vertices $p$ and $q$. In this case we say $p$ and $q$ are the endpoints of the edge, and the edge is incident with $p$ and $q$. In general if edge $e$ is associated with a vertex $u$, then we say $e$ **is incident with** $u$, and $u$ is an **endpoint** of $e$. If vertices $u$ and $v$ are distinct and there is an edge $e$ incident with both vertices $u$ and $v$, we say $u$ and $v$ are **adjacent vertices** and that $e$ **connects** $u$ and $v$.

**EXAMPLE 3**

Answer the following questions for the graph in Figure 14-31.

(a) Name the vertices that are endpoints of $e_1$.

(b) Name the edges that are both incident with vertex $x$ and incident with vertex $y$.

(c) Are vertices $z$ and $w$ adjacent? Why?

(d) Are vertices $x$ and $z$ adjacent? Why?

*SOLUTION*

(a) $x$ and $w$ are endpoints of $e_1$.

(b) $e_4$ and $e_5$ are incident with both vertices $x$ and $y$.

(c) Vertices $z$ and $w$ are adjacent because $e_2$ is incident with vertices $z$ and $w$.

(d) Vertices $x$ and $z$ are not adjacent because there is no edge that is incident to both vertices.

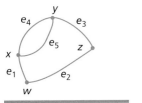

FIGURE 14-31

In Figure 14-31 we notice that we have two different edges incident with both vertex $x$ and vertex $y$. In this case it would not be appropriate to name $e_4$ or $e_5$ as $(x,y)$.

As we take a trip in a highway system, we alternate between travel along road sections and passing through towns connected by the road sections. By identifying the road sections traveled and the towns visited, we could completely describe such a trip to a friend. In graph theory we call such a trip a **path**, defined as follows.

| DEFINITION | | |
|---|---|---|
| | **Path** | Let $x$ and $y$ be vertices in a graph. A **path** from $x$ to $y$ of length $n$ is an alternating sequence of $n+1$ vertices and $n$ edges $(v_0, e_1, v_1, e_2, v_3 \ldots, v_n)$ where $x = v_0$, $y = v_n$, and edge $e_i$ is incident with vertices $v_{i-1}$ and $v_i$. |
| | **Circuit** | A path from $x$ to $x$ is called a **circuit.** A circuit begins and ends at the same vertex. |

In Figure 14-31 $(x, e_1, w, e_2, z, e_3, y)$ is a path from vertex $x$ to vertex $y$ of length 3. Similarly, $(x, e_5, y, e_4, x)$ is a circuit. Note that in a graph in which at most one edge connects each pair of vertices, there is no ambiguity if we omit reference to the edges. For instance, the path $(a, e_1, b, e_2, c, e_3, d)$ from $a$ to $d$ in Figure 14-29 could be named $(a,b,c,d)$.

**EXAMPLE 4**

In the graph of Figure 14-32,

(a) Find the path of length 3 from vertex $x$ to vertex $y$.

(b) Find the path of length 4 from vertex $x$ to vertex $y$..

(c) Find the path of length 5 from vertex $x$ to vertex $y$.

*SOLUTION*

(a) A path of length 3: $(x,a,b,y)$.

(b) A path of length 4: $(x,c,a,b,y)$.

(c) A path of length 5: We must use some of the edges more than once. Two possible answers are $(x,a,c,a,b,y)$ and $(x,a,x,a,b,y)$.

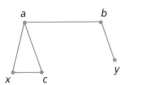

FIGURE 14-32

In the preceding example we saw that in some paths and circuits we repeat edges. In terms of the language of the highway inspector problem, a path in which an edge (a road segment) is repeated is not an economical trip since the highway inspector wishes to pass over each road segment exactly once. Hence we define a **simple path** to be a path in which no edge is repeated. In Example 4, the paths discussed in (a) and (b) are simple while the paths identified in (c) are not simple. Similarly, a **simple circuit** is a circuit in which no edge is repeated.

We now use the language of graphs to state the definitions and theorems that we discovered at the beginning of this section.

# EULER PATHS AND CIRCUITS

1. An **Euler path** in a graph is a simple path that includes each edge of the graph.
2. An **Euler circuit** is a circuit that is an Euler path.
3. The **degree of a vertex** *v* is the number of edges that are incident with *v*.
4. A graph is **connected** if for any two vertices *v* and *w* there is a path from *v* to *w*.

**EXAMPLE 5**

In the graph in Figure 14-33, determine whether the path
(a) (*a, c, d, a, c, b*) is an Euler path from vertex *a* to vertex *b*. Explain.
(b) (*c, b, a, c*) is an Euler circuit. Explain.
(c) Find the degree of vertex *a*. Explain.
(d) Find an Euler path in the graph in Figure 14-33.

*SOLUTION*

(a) This path fails to be an Euler path for two reasons; the edge (*a, c*) appears twice on the path, and the path does not include the edge (*a, b*).
(b) Since the path begins and ends with vertex *c,* it is a circuit. However, it fails to be an Euler circuit because it does not include the edges (*a, d*) and (*c, d*).
(c) Since three edges (*a, d*), (*a, c*), and (*a, b*) are incident with vertex *a,* the degree of *a* is 3.
(d) The path (*a, b, c, d, a, c*) is an Euler path that starts at *a* and ends at *c*.

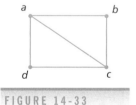

**FIGURE 14-33**

With this terminology available to us, we are able to restate the theorems of the Highway Inspector in the language of graph theory.

**Euler Paths and Circuits**

1. If a graph has more than two vertices of odd degree, then there is no Euler path for the graph.
2. If each of the vertices of a connected graph has even degree, then there is an Euler circuit for the graph.
3. If a connected graph has exactly two vertices *v* and *w* of odd degree, then there is an Euler path from *v* to *w*.

The salesman wished to travel a highway system passing through each town exactly once. For efficiency, he preferred to begin and end at the same town. In order to describe the problem of the salesman in the language of graphs, we need the following definitions.

1. A **Hamilton path** in a graph is a path that includes each vertex of the graph exactly once.
2. A **Hamilton circuit** is a circuit that includes each vertex of a graph exactly once except for the initial vertex and the final vertex which are the same.

The question of the salesman becomes, "Can I find a Hamilton path for the graph?" We have seen that it is relatively easy to determine whether Euler paths and circuits exist and to find those paths and circuits. For graphs with small numbers of vertices and edges, one can answer the Hamilton questions with trial and error techniques. For many large graphs the trial and error techniques require too much computation to be implemented even on high speed computers. Researchers in many different disciplines are actively trying to determine if there is a computationally efficient way to answer the Hamilton questions.

# EXERCISE SET 3

*R* 1. Working in pairs, have each person in the pair draw a map of a connected road system. Exchange maps and use the theorems of this section to determine whether there is an economical trip for the road inspector.

2. If possible find a route for the salesman in this map. Find a route for the road inspector.

3. Use the theorems of this section to complete the table for maps (a) through (g).

| Map | Number of Even Towns (Vertices) | Number Of Odd Towns (Vertices) | Economical Route for Inspector (Yes or No) |
|---|---|---|---|
| (a) | | | |
| (b) | | | |
| (c) | | | |
| (d) | | | |
| (e) | | | |
| (f) | | | |
| (g) | | | |

(a)   (b)   (c)

(d)   (e)

(f) 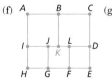  (g)

4. For each of the maps in Exercise 3, find the sum of the degrees of the towns. Does this agree with the theory of this section?

5. For each map in Exercise 3 for which it is possible, find an economical route for the road inspector.

6. For each map in Exercise 3 for which it is possible find an economical route for the salesman.

*In Exercises 7–10 describe each of the following graphs by determining the set of vertices* V *and the set of edges* E. *Since each pair of vertices is connected by at most one edge, describe each edge using a pair of endpoints.*

7.

8.

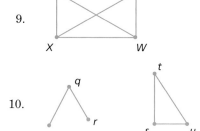

9.

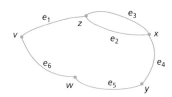

10.

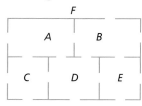

11. Sketch a geometric representation of these graphs.

(a) $V = \{a,b,c,d,e,f\}$; $E = \{(a,b), (a,c), (b,d), (b,e), (d,f), (e,f)\}$

(b) $V = \{x,y,z\}$; $E = \{(x,y), (x,z), (y,z)\}$

(c) $V = \{P,Q,R,T\}$; $E = \{\ \}$

12. In this graph find

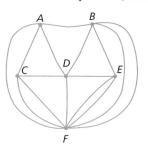

(a) Two edges incident with vertex $v$.

(b) Two vertices that are adjacent to vertex $x$.

(c) A simple path from vertex $v$ to vertex $x$ of length 2.

(d) A simple path from vertex $v$ to vertex $x$ of length 3.

(e) A path from vertex $v$ to vertex $x$ that is not simple.

(f) A circuit beginning and ending at vertex $y$.

(g) A vertex of degree 3.

13. Draw a graph with

(a) 4 vertices, each with degree 1.

(b) 4 vertices, each with degree 2.

(c) 4 vertices, each with degree 3.

𝑇 14. We wish to determine whether we can find a path through the house depicted below in such a way as to pass through each door exactly one time.

One strategy for solving this problem is to describe the house plan as a graph in which we identify each room as a vertex and each doorway as an edge. Consider the graph below in which we labeled the area outside the house with the vertex $F$. Use the theory of this section to determine whether it possible to find the required path though the house? If it is possible, find the path.

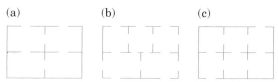

15. For each of the following floor plans, first draw a graph that corresponds to the floor plan and then determine whether it is possible to travel through the floor plan by passing through each door exactly once. Be sure to represent the exterior of the house as a vertex.

(a)    (b)    (c)

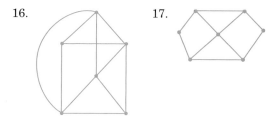

*For each of the graphs in Exercises 16–18, determine whether there is an Euler path or Euler circuit by looking at the degrees of the vertices. If the graph has Euler paths or Euler circuits, find one of them.*

16.    17.

18.

19. The highway inspector problem was first solved by Leonard Euler in 1735 when thinking about a puzzle problem that he described as follows: "In the town of Köenigsberg there is an island called Kneiphof, with two branches of the river Pregel

flowing around it. There are seven bridges crossing the two branches. The question is whether a person can plan a walk in such a way that he will cross each of these bridges once but not more than once."

Below is found a diagram of the seven bridges of Köenigsberg. Redraw this as a map with each island and the north and south shores represented as towns and the bridges represented as sections of road. Can you answer the Euler puzzle?

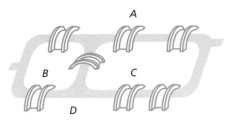

20. Look again at the Königsburg bridge problem. Could the citizens of this city find an acceptable walk by building one more bridge? Explain.

21. A city is built along both sides of a river and includes three islands and nine bridges as shown. Is it possible to walk around the city and cross each bridge exactly once?

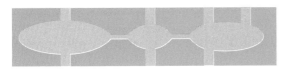

*For each of the graphs in Exercises 22–24 find a Hamilton path or circuit or determine that no such path or circuit exists. If no Hamilton path or circuit exists, explain why.*

22.      23.

24.

25. In 1857, the distinguished Irish mathematician and scientist, Rowan Hamilton, invented a game that was marketed under the name "A Voyage Round the World." The game consisted of a dodecahedron with a peg at each vertex. The object

was to use a string to mark a path from vertex to vertex (peg to peg) so that the path visited each vertex exactly once and then returned to the vertex from which the path started. Below is found a graph equivalent to Hamilton's game. Find a "Voyage Round the World," a Hamilton circuit, for this graph.

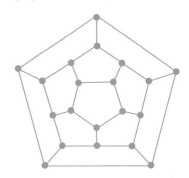

26. Give an example of a connected graph with

   (a) An Euler circuit and a Hamilton circuit.

   (b) Neither an Euler circuit nor a Hamilton circuit.

   (c) An Euler circuit but no Hamilton circuit.

   (d) A Hamilton circuit but no Euler circuit.

### ∵ PCR Excursions ∵

27. Suppose that Town $D$ is a town on a road map that has degree 6. Choose a partner and together write careful explanations of the following assertions. Use pictures to illustrate your reasoning.

   (a) If the road inspector begins an economical inspection trip at $D$, then the trip must end at $D$.

   (b) If the road inspector begins an economical inspection trip at some town other than $D$, then each time she enters Town $D$, she has an uninspected section on which to leave.

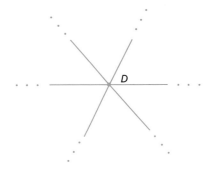

28. The theory of this section guarantees that when all towns (vertices) in a connected road map (in a connected graph) have even degree, we can always find an economical trip for the highway inspector (an Euler path). However, it does not tell us how to find that trip. Below you will find a sequence of steps that will always provide an economical path, even on a very complicated road map. Work with a partner to answer questions about each of the steps.

   (a) Start at any town on the map, call it *S*. Drive from town to town making random choices about which road sections to inspect, but not passing over any section of road twice. Eventually you must come to a town from which there are no new sections on which you may leave. Why? [HINT: How many sections of road are in the map?]

   (b) Because all towns are of even degree, this town at which we must stop is *S*. Why? [HINT: Think carefully about the reasoning you used in Exercise 27.]

   (c) If all sections of the map are inspected, we are done. If not, we observe that each town has an even number of uninspected sections since we use them two at a time (once to enter, once to leave). Pick a town *T* we have visited that is the endpoint of uninspected sections and start a side trip at *T*. If we are careful to not travel on any uninspected sections we must again stop. Why must our stopping point be *T*?

At this point you have covered the whole map, or there is some town with uninspected sections, and you can generate another side trip. When all road sections have been covered by side trips, we can design a single economical inspection tour by starting at *S* and traveling the original tour. However each time we come to a town that is the start of a side trip (like *T*) we will take the side trip beginning and ending at that town before continuing.

29. Work in pairs and use the strategy of Exercise 28 to find an economical route for the highway inspector in this graph. Have one student drive (trace the proposed path). Let the other student (while not looking at the graph) navigate by randomly choosing the section of the road taken at each town. [As the original trip and side trips are traced, the driver should number the unused sections at each town and ask the navigator to choose the number of the section on which to proceed.]

# SECTION 4 ▷ SHORTEST PATH PROBLEMS

**PROBLEM**

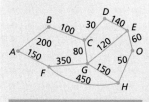

**FIGURE 14-34**

A truck driver studies the roadmap in Figure 14-34 hoping to find the route from Alpha (*A*) to Omega (*O*) that will be the shortest in terms of miles traveled. The number labeling each edge of the graph is the number of miles between the cities represented by the vertices.

**OVERVIEW**

In this section we explore the problem of finding a path through a graph that minimizes a quantity such as distance, time, or cost.

| **GOALS** | Use the Representation Standard, page 50. |
| | Use the Problem Solving Standard, page 15. |
| | Use the Reasoning and Proof Standard, page 23. |

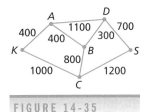

FIGURE 14-35

A controller studies the graph in Figure 14-35 hoping to find the least expensive route to move a shipment of widgets from Kappa $K$ to Sigma $S$. The number labeling each edge of the graph is the cost in dollars to move the shipment between the two terminals represented by the vertices.

The graphs in Figure 14-34 and 14-35 are examples of **weighted graphs,** graphs that have a number attached to each edge. The number attached to each edge of a weighted graph is called the **weight** of the edge. The **length** of a path in a weighted graph is the sum of weights of the edges in the path. In each of the situations described above, the problem is to find the path of least length between a specified pair of vertices.

One strategy for solving the "shortest path" problem is obvious; merely determine all simple paths between the two vertices, determine the length of each path, and choose the path of shortest length.

**EXAMPLE 1**

Find the shortest path from Kappa ($K$) to Sigma ($S$) for the graph in Figure 14-35 by identifying all simple paths from $K$ to $S$ and finding the length of each.

*SOLUTION*

As seen in Figure 14-35 there are seven paths from Kappa ($K$) to Sigma ($S$) that do not repeat edges.

$(K,A,D,S)$ with length $400 + 1100 + 700 = 2200$
$(K,A,D,B,C,S)$ with length 3800
$(K,A,B,D,S)$ with length 1800
$(K,A,B,C,S)$ with length 2800
$(K,C,B,A,D,S)$ with length 4000
$(K,C,B,D,S)$ with length 2800
$(K,C,S)$ with length 2200

Clearly the path with least length is $(K,A,B,D,S)$ with length 1800.

When using graphs to model and solve problems, it is often fairly easy to conceive of some solution to the problem, but much more difficult to find a solution that is efficient enough to actually be implemented for graphs with many vertices and edges. Hence, much of the creative work in graph theory involves designing efficient algorithms that accomplish the desired tasks.

Over the years there have been several algorithms developed for finding the shortest path through a weighted graph. We will examine an algorithm that can be described very nicely in terms of the geometric representation of the graph. It is related to an algorithm developed by the Dutch mathematician, E. Dijkstra, in 1959. This algorithm will find the shortest distance between two vertices. It can be described using four steps. In the algorithm note that Step 4 merely consists of performing Steps 2 and 3 iteratively until the process is complete.

**Geometric Algorithm for Shortest Path**

1. Suppose the vertex at which you wish to start is $A$, and the vertex at which you wish to end is $Z$. Circle $A$, then examine each vertex adjacent to $A$ and find the weights of the corresponding edges. Shade the edge with the smallest weight and circle the vertex that this edge shares with $A$.

2. Examine each of the uncircled vertices that are adjacent to circled vertices in the graph.

3. For each uncircled vertex $X$ being examined and each edge $e$ incident with that vertex and a circled vertex, find all simple paths from $A$ to $X$ that, in addition to $X$ and $e$, contain only circled vertices and those edges we have already shaded. Find the length of all these paths. Choose the vertex $X$ and edge $e$ that lie in the path of smallest length. Circle the vertex $X$ and shade the edge $e$.

4. Repeat Steps 2 and 3 until vertex $Z$ has been circled. The shaded path from $A$ to $Z$ is the shortest path from $A$ to $Z$. (NOTE: If in either Steps 1 or 3, there are two or more paths with the shortest length, break the tie arbitrarily.)

**EXAMPLE 2**

Use the Geometric Algorithm to find the shortest path from vertex $a$ to vertex $z$ in the weighted graph in Figure 14-36.

*SOLUTION*

We first circle the vertex from which we wish to start, vertex $a$. The two vertices adjacent to vertex $a$ are $b$ and $c$. The weight of $(a, b)$ is 4 while the weight of $(a, c)$ is 3. Hence we shade edge $(a,c)$ and circle vertex $c$. (Figure 14-37(a)).

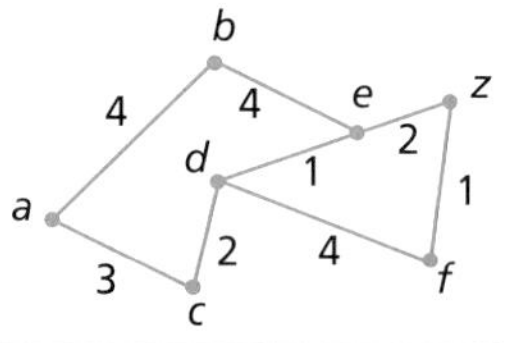

FIGURE 14-36

**FIRST ITERATION OF STEPS 2 AND 3.** The vertices adjacent to circled vertices are $b$ and $d$. The path of interest from $a$ to $b$ is the path $(a,b)$ of length 4. The path of interest from $a$ to $d$ is $(a,c,d)$ of length 5. Hence we shade the edge $(a,b)$ and circle vertex $b$. See Figure 14-37(b).

**SECOND ITERATION OF STEPS 2 AND 3.** The vertices adjacent to circled vertices are $d$ and $e$. The only simple path from $a$ to $e$ that (with the exception of $e$ and the edge incident to it) consists of shaded edges and circled vertices is $(a,b,e)$ of length 8. Similarly, the only interesting path from $a$ to $d$ is $(a,c,d)$ of length 5 Hence we select the edge $(c,d)$ and circle vertex $d$. (Figure 14-37(c)).

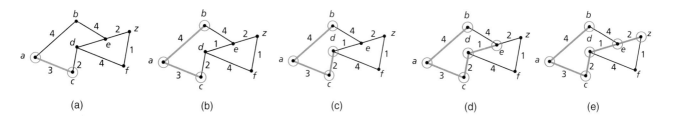

(a)  (b)  (c)  (d)  (e)

FIGURE 14-37

THIRD ITERATION OF STEPS 2 AND 3. The vertices adjacent to circled vertices are $e$ and $f$. The simple paths from $a$ to $e$ that (with the exception of $e$ and the edge incident to it) consist of shaded edges and circled vertices are: $(a,b,e)$ of length 8 and $(a,c,d,e)$ of length 6. Similarly, the only interesting path from $a$ to $f$ is $(a,c,d,f)$ of length 9. We shade the edge $(d,e)$ and circle vertex $e$. (Figure 14-37(d)).

FOURTH ITERATION OF STEPS 2 AND 3. The vertices adjacent to circled vertices are $f$ and $z$. The simple path from $a$ to $f$ that (with the exception of $f$ and the edge incident to it) consists of shaded edges and circled vertices is $(a,c,d,f)$ of length 9. Similarly the path of interest from $a$ to $z$ is $(a,c,d,e,z)$ of length 8. The path of shortest length is 8. We circle $z$ and shade $(e,z)$. (Figure 14-37(e).

We see that the path of shortest length has length 8 and that a path that has this length is $(a,c,d,e,z)$.

**EXAMPLE 3**

Suppose we are using the Geometric Algorithm to find the shortest path from $A$ to $F$ in the graph of Figure 14-38(a). In Figure 14-38(a) we have performed Steps 2 and 3 twice. What will be the next vertex circled and the next edge selected?

*SOLUTION*

The vertices adjacent to circled vertices are $C$ and $F$. The paths from $A$ to $C$ that (with the exception of $C$ and the edges incident to it) consist of shaded edges and circled vertices are $(A,B,C)$ of length 9 and $(A,B,E,C)$ of length 12. Similarly, the path of interest from $A$ to $F$ is $(A,B,E,F)$ of length 12. Hence we circle $C$ and shade $(B,C)$. See Figure 14-38(b). Now finish applying the Geometric Algorithm to find the shortest distance to $F$. Did you get a distance of 11?

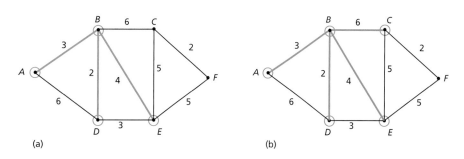

(a)   (b)

FIGURE 14-38

A cumbersome aspect of this algorithm is that each time we prepare to find a new vertex to circle, we must compute the length of paths all the way back to the starting vertex. For very large graphs, this is a severe short-coming because of the work involved. In Exercise 10 we show a procedure for keeping a running record of the weights or distances. In Exercise 13 we show how this information can be tabulated in a form useful for computers.

# EXERCISE SET 4

*In Exercise 1–3 find the length of the shortest path from a to z in the weighted graphs without using the Geometric Algorithm.* HINT: *Find all simple paths.*

*R* 1. 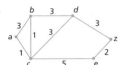  2.

3.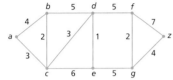

*In Exercises 4–6 find the length of the shortest paths in the weighted graphs using the Geometric Algorithm.*

4. Exercise 1.  5. Exercise 2  6. Exercise 3

7. For the problem in the introduction of this section find the shortest route for the truck driver.

*T* 8. Find the path that takes the least flying time to get from Beta to Gamma, where the numbers on edges are flying times (in hours and minutes) between cities.

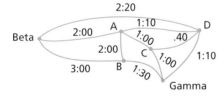

9. Use the Geometric Algorithm to find the shortest path from $A$ to $Z$ in this weighted graph.

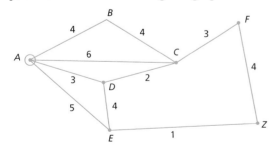

*C* 10. Suppose we are using the Geometric Algorithm to find the shortest path from $A$ to $Z$ in the weighted graph.

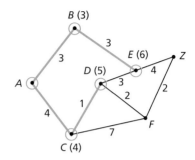

Circle $A$ and verify that the next vertex to circle is $B$ and the edge to shade is $(A,B)$. As you circle $B$, place the length of the path from $A$ beside $B$. Note the 3 by $B$ in the figure. Verify that we next circle $C$ and shade $(A,C)$. Note at $C$ that the length of the path is 4. Then verify that we next circle $D$ and shade $(C,D)$. The length of the path to $D$ is the 4 at $C$ plus 1, recorded at $D$. Next verify that we circle $E$ and shade $(B,E)$. The length of the path to $E$ is the 3 at $B$ plus 3, recorded at $E$. Complete the problem to find the shortest distance from $A$ to $Z$.

11. Follow the procedures of Exercise 10 to find the shortest path in Exercise 9. As each vertex is circled, label the vertex with the length of the path used to get there from $A$.

12. A company has branches in the cities of Alpha ($A$), Beta ($B$), Omega ($C$), Gamma ($G$), Kappa ($K$), and Sigma ($S$). The table below gives airline fares for direct flight between each of these cities (a hyphen indicates that there is no flight between the cities). Create a weighted graph summarizing this information and use the Geometric Algorithm to find the cheapest fare between each pair of cities.

|   | A | B | C | G | K | S |
|---|---|---|---|---|---|---|
| A | – | 120 | 250 | – | – | 50 |
| B | 120 | – | – | 240 | 90 | – |
| C | 250 | – | – | 100 | – | 270 |
| G | – | 240 | 100 | – | 130 | 150 |
| K | – | 90 | – | 130 | – | 290 |
| S | 50 | – | 270 | 150 | 290 | |

13. To implement a shortest path algorithm on a computer, we tabulate tables instead of drawing paths. For Exercise 10 we tabulate the following set of tables where the first column lists vertices circled, the second column is the distance $S$ from the beginning vertex to vertices that have been circled, and the third column is the preceding vertex $P$ in the shortest path to the circled vertex.

| First Iteration | | | Second Iteration | | |
|---|---|---|---|---|---|
| V | S | P | V | S | P |
| A | 0 | | A | 0 | |
| B | 3 | A | B | 3 | A |
| C | | | C | 4 | A |
| D | | | D | | |
| E | | | E | | |
| F | | | F | | |
| Z | | | Z | | |

| Third Iteration | | | Fourth Iteration | | |
|---|---|---|---|---|---|
| V | S | P | V | S | P |
| A | 0 | | A | 0 | |
| B | 3 | A | B | 3 | A |
| C | 4 | A | C | 4 | A |
| D | 5 | C | D | 5 | C |
| E | | | E | 6 | B |
| F | | | F | | |
| Z | | | Z | | |

Now complete the exercise. Did you get that the length of the shortest path from $A$ to $Z$ is 9? Tracing backwards through the table the path is $F$ to $Z$, $D$ to $F$, $C$ to $D$, $A$ to $C$. Thus the shortest path is $(A,C,D,F,Z)$.

*Find the shortest paths for the following exercises using only tables.*

14. Exercise 1      15. Exercise 2      16. Exercise 9

# SOLUTION TO INTRODUCTORY PROBLEM

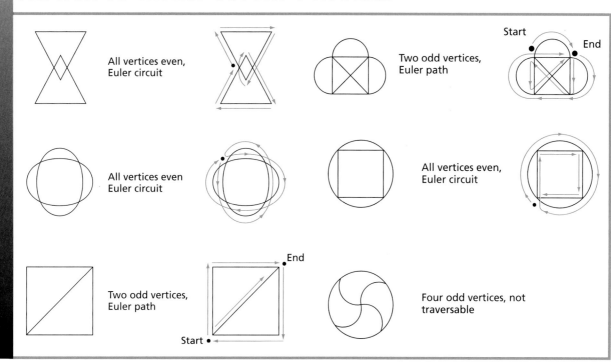

# SUMMARY AND REVIEW

1. Congruent figures have the same size and shape.

2. Two triangles are congruent if the following corresponding parts are congruent:
   (a) Three sides (SSS)
   (b) Two sides and the included angle (SAS)
   (c) Two angles and the included side (ASA)
   (d) Two angles and a corresponding side (AAS)

3. We can use our knowledge of congruent triangles to justify constructions.

4. We know how to complete the following constructions using straightedge and compass.
   (a) A line segment congruent to a given line segment.
   (b) A bisection of a line segment into two congruent line segments.
   (c) A perpendicular to a line from a point not on the line.
   (d) A ray that is perpendicular to a line at a given point on the line.
   (e) A perpendicular line that bisects a line segment.
   (f) An angle congruent to a given angle.
   (g) The bisector of an angle.
   (h) An equilateral triangle with each side congruent to a given segment.
   (i) A circle circumscribing a triangle.
   (j) A regular hexagon.

5. Graphs
   (a) A graph $G = (V, E)$ consists of $V$ a non-empty set of vertices, and $E$, a non-empty set of edges with each edge associated with a pair of vertices. Graphs are often represented with diagrams in which the vertices are displayed as dots and the edges as arcs that connect the vertices with which they are associated.
   (b) If edge $e$ is associated with vertices $u$ and $v$, then $e$ is said to be incident with $u$ and $v$; $u$ and $v$ are said to be endpoints of $e$; $e$ is said to connect $u$ and $v$; and $u$ and $v$ are said to be adjacent.

6. Paths:
   (a) A path in a graph from vertex $v$ to vertex $u$ is a list of vertices and edges in which $v$ is the first element of the list, pairs of adjacent vertices alternate with edges that connect them, and $u$ is the last item on the list.
   (b) A circuit is a path that begins and ends at the same vertex.
   (c) An Euler path in a graph is a simple path that includes each edge of the graph and an Euler circuit is a circuit that is an Euler path.
   (d) The degree of a vertex $v$ is the number of edges that are incident with $v$.
   (e) A graph is connected if for any two vertices $v$ and $w$ there is a path from $v$ to $w$.
   (f) If a graph has more than two vertices of odd degree, then there is no Euler path for the graph.
   (g) If a connected graph has exactly two vertices $v$ and $w$ of odd degree, then there is an Euler path from $v$ to $w$.
   (h) If each of the vertices of a connected graph has even degree, then there is an Euler circuit for the graph.
   (i) A Hamilton path in a graph is a path that includes each vertex of the graph exactly once.
   (j) A Hamilton circuit is a circuit that includes each vertex of a graph exactly once except for the initial vertex and the finl vertex.
   (k) The Geometric Algorithm gives the shortest path between two vertices for a weighted graph.

# CHAPTER TEST

1. If possible find a route for a road inspector in the following maps.

   (a) 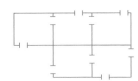  (b)

2. *ABCD* is a parallelogram. Prove the triangles are congruent.

   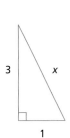

3. For the given floor plan, indicate whether you can go into each room and pass through each door only once. Can you start and stop in the same room? Show a possible path.

   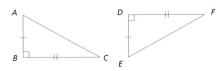

4. Find *x*.

   (a)   *x*   (b)

   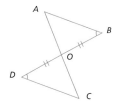

*Prove that the triangles in Questions 5 through 8 are congruent.*

5.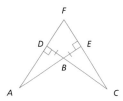

6.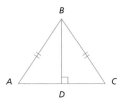

7.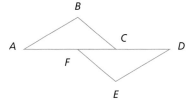

8. 

9. GIVEN: $\triangle ABC \cong \triangle DEF$
   PROVE: $CD = AF$

10. Construct a perpendicular to a line from a point not on the line, and justify your result.

11. Sketch a geometric representation of a graph with
    (a) $V = \{x,y,w,z\}$, $E = \{(x,y),\ (x,w),\ (x,z),\ (y,w),\ (w,z)\}$
    (b) $V = \{x,y,w,z\}$, $E = \{(x,y),\ (x,w),\ (y,z),\ (y,w)\ (w,z)\}$

12. Can you find a Hamilton circuit for the graphs in Exercise 1?

13. (a) Is there an Euler path or circuit for the given graph?

    (b) Is there a Hamilton path or circuit for this graph?

14. Find the shortest distance from *a* to *z* using the Geometric Algorithm.

    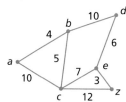

# APPENDIX: ANNUAL PERCENTAGE RATE TABLE

## TABLE A-1

### Annual Percentage Rate Table for Monthly Payment Plans
### (See instructions for use of table on page 338)
#### Annual Percentage Rate

| Number of Payments | 10.00 % | 10.25 % | 10.50 % | 10.75 % | 11.00 % | 11.25 % | 11.50 % | 11.75 % | 12.00 % | 12.25 % | 12.50 % | 12.75 % | 13.00 % | 13.25 % | 13.50 % | 13.75 % |
|---|---|---|---|---|---|---|---|---|---|---|---|---|---|---|---|---|
| | | | | | (Finance Charge per $100 of Amount Financed) | | | | | | | | | | | |
| 1 | 0.83 | 0.85 | 0.87 | 0.90 | 0.92 | 0.94 | 0.96 | 0.98 | 1.00 | 1.02 | 1.04 | 1.06 | 1.08 | 1.10 | 1.12 | 1.15 |
| 2 | 1.25 | 1.28 | 1.31 | 1.35 | 1.38 | 1.41 | 1.44 | 1.47 | 1.50 | 1.53 | 1.57 | 1.60 | 1.63 | 1.66 | 1.69 | 1.72 |
| 3 | 1.67 | 1.71 | 1.76 | 1.80 | 1.84 | 1.88 | 1.92 | 1.96 | 2.01 | 2.05 | 2.09 | 2.13 | 2.17 | 2.22 | 2.26 | 2.30 |
| 4 | 2.09 | 2.14 | 2.20 | 2.25 | 2.30 | 2.35 | 2.41 | 2.46 | 2.51 | 2.57 | 2.62 | 2.67 | 2.72 | 2.78 | 2.83 | 2.88 |
| 5 | 2.51 | 2.58 | 2.64 | 2.70 | 2.77 | 2.83 | 2.89 | 2.96 | 3.02 | 3.08 | 3.15 | 3.21 | 3.27 | 3.34 | 3.40 | 3.46 |
| 6 | 2.94 | 30.1 | 3.08 | 3.16 | 3.23 | 3.31 | 3.38 | 3.45 | 3.53 | 3.60 | 3.68 | 3.75 | 3.83 | 3.90 | 3.97 | 4.05 |
| 7 | 3.36 | 3.45 | 3.53 | 3.62 | 3.70 | 3.78 | 3.87 | 3.95 | 4.04 | 4.12 | 4.21 | 4.29 | 4.38 | 4.47 | 4.55 | 4.64 |
| 8 | 3.79 | 3.88 | 3.98 | 4.07 | 4.17 | 4.26 | 4.36 | 4.46 | 4.55 | 4.65 | 4.74 | 4.84 | 4.94 | 5.03 | 5.13 | 5.22 |
| 9 | 4.21 | 4.32 | 4.43 | 4.53 | 4.64 | 4.75 | 4.85 | 4.96 | 5.07 | 5.17 | 5.28 | 5.39 | 5.49 | 5.60 | 5.71 | 5.82 |
| 10 | 4.64 | 4.76 | 4.88 | 4.99 | 5.11 | 5.23 | 5.35 | 5.46 | 5.58 | 5.70 | 5.82 | 5.94 | 6.05 | 6.17 | 6.29 | 6.41 |
| 11 | 5.07 | 5.20 | 5.33 | 5.45 | 5.58 | 5.71 | 5.84 | 5.97 | 6.10 | 6.23 | 6.36 | 6.49 | 6.62 | 6.75 | 6.88 | 7.01 |
| 12 | 5.50 | 5.64 | 5.78 | 5.92 | 6.06 | 6.20 | 6.34 | 6.48 | 6.62 | 6.76 | 6.90 | 7.04 | 7.18 | 7.32 | 7.46 | 7.60 |
| 13 | 5.93 | 6.08 | 6.23 | 6.38 | 6.53 | 6.68 | 6.84 | 6.99 | 7.14 | 7.29 | 7.44 | 7.59 | 7.75 | 7.90 | 8.05 | 8.20 |
| 14 | 6.36 | 6.52 | 6.69 | 6.85 | 7.01 | 7.17 | 7.34 | 7.50 | 7.66 | 7.82 | 7.99 | 8.15 | 8.31 | 8.48 | 8.64 | 8.81 |
| 15 | 6.80 | 6.97 | 7.14 | 7.32 | 7.49 | 7.66 | 7.84 | 8.01 | 8.19 | 8.36 | 8.53 | 8.71 | 8.88 | 9.06 | 9.23 | 9.41 |
| 16 | 7.23 | 7.41 | 7.60 | 7.78 | 7.97 | 8.15 | 8.34 | 8.53 | 8.71 | 8.90 | 9.08 | 9.27 | 9.46 | 9.64 | 9.83 | 10.02 |
| 17 | 7.67 | 7.86 | 8.06 | 8.25 | 8.45 | 8.65 | 8.84 | 9.04 | 9.24 | 9.44 | 9.63 | 9.83 | 10.03 | 10.23 | 10.43 | 10.63 |
| 18 | 8.10 | 8.31 | 8.52 | 8.73 | 8.93 | 9.14 | 9.35 | 9.56 | 9.77 | 9.98 | 10.19 | 10.40 | 10.61 | 10.82 | 11.03 | 11.24 |
| 19 | 8.54 | 8.76 | 8.98 | 9.20 | 9.42 | 9.64 | 9.86 | 10.08 | 10.30 | 10.52 | 10.74 | 10.96 | 11.18 | 11.41 | 11.63 | 11.85 |
| 20 | 8.98 | 9.21 | 9.44 | 9.67 | 9.90 | 10.13 | 10.37 | 10.60 | 10.83 | 11.06 | 11.30 | 11.53 | 11.76 | 12.00 | 12.23 | 12.46 |
| 21 | 9.42 | 9.66 | 9.90 | 10.15 | 10.39 | .10.63 | 10.88 | 11.12 | 11.36 | 11.61 | 11.85 | 12.10 | 12.34 | 12.59 | 12.84 | 13.08 |
| 22 | 9.86 | 10.12 | 10.37 | 10.62 | 10.88 | 11.13 | 11.39 | 11.64 | 11.90 | 12.16 | 12.41 | 12.67 | 12.93 | 13.19 | 13.44 | 13.70 |
| 23 | 10.30 | 10.57 | 10.84 | 11.10 | 11.37 | 11.63 | 11.90 | 12.17 | 12.44 | 12.71 | 12.97 | 13.24 | 13.51 | 13.78 | 14.05 | 14.32 |
| 24 | 10.75 | 11.02 | 11.30 | 11.58 | 11.86 | 12.14 | 12.42 | 12.70 | 12.98 | 13.26 | 13.54 | 13.82 | 14.10 | 14.38 | 14.66 | 14.95 |
| 25 | 11.19 | 11.48 | 11.77 | 12.06 | 12.35 | 12.64 | 12.93 | 13.22 | 13.52 | 13.81 | 14.10 | 14.40 | 14.69 | 14.98 | 15.28 | 15.57 |
| 26 | 11.64 | 11.94 | 12.24 | 12.54 | 12.85 | 13.15 | 13.45 | 13.75 | 14.06 | 14.36 | 14.67 | 14.97 | 15.28 | 15.59 | 15.89 | 16.20 |
| 27 | 12.09 | 12.40 | 12.71 | 13.03 | 13.34 | 13.66 | 13.97 | 14.29 | 14.60 | 14.92 | 15.24 | 15.56 | 15.87 | 16.19 | 16.51 | 16.83 |
| 28 | 12.53 | 12.86 | 13.18 | 13.51 | 13.84 | 14.16 | 14.49 | 14.82 | 15.15 | 15.48 | 15.81 | 16.14 | 16.47 | 16.80 | 17.13 | 17.46 |
| 29 | 12.98 | 13.32 | 13.66 | 14.00 | 14.33 | 14.67 | 15.01 | 15.35 | 15.70 | 16.04 | 16.38 | 16.72 | 17.07 | 17.41 | 17.75 | 18.10 |
| 30 | 13.43 | 13.78 | 14.13 | 14.48 | 14.83 | 15.19 | 15.54 | 15.89 | 16.24 | 16.60 | 16.95 | 17.31 | 17.66 | 18.02 | 18.38 | 18.74 |

*(Continued)*

**TABLE A-1**

### Annual Percentage Rate Table for Monthly Payment Plans (Continued)
### (See instructions for use of table on page 338)
#### Annual Percentage Rate

| Number of Payments | 10.00 % | 10.25 % | 10.50 % | 10.75 % | 11.00 % | 11.25 % | 11.50 % | 11.75 % | 12.00 % | 12.25 % | 12.50 % | 12.75 % | 13.00 % | 13.25 % | 13.50 % | 13.75 % |
|---|---|---|---|---|---|---|---|---|---|---|---|---|---|---|---|---|
| | | | | | | (Finance Charge per $100 of Amount Financed) | | | | | | | | | | |
| 31 | 13.89 | 14.25 | 14.61 | 14.97 | 15.33 | 15.70 | 16.06 | 16.43 | 16.79 | 17.16 | 17.53 | 17.90 | 18.27 | 18.63 | 19.00 | 19.38 |
| 32 | 14.34 | 14.71 | 15.09 | 15.46 | 15.84 | 16.21 | 16.59 | 16.97 | 17.35 | 17.73 | 18.11 | 18.49 | 18.87 | 19.25 | 19.63 | 20.02 |
| 33 | 14.79 | 15.18 | 15.57 | 15.95 | 16.34 | 16.73 | 17.12 | 17.51 | 17.90 | 18.29 | 18.65 | 19.08 | 19.47 | 19.87 | 20.26 | 20.66 |
| 34 | 15.25 | 15.65 | 16.05 | 16.44 | 16.85 | 17.25 | 17.65 | 18.05 | 18.46 | 18.86 | 19.27 | 19.67 | 20.08 | 20.49 | 20.90 | 21.31 |
| 35 | 15.70 | 16.11 | 16.53 | 16.94 | 17.35 | 17.77 | 18.18 | 18.60 | 19.01 | 19.43 | 19.85 | 20.27 | 20.69 | 21.11 | 21.53 | 21.95 |
| 36 | 16.16 | 16.58 | 17.01 | 17.43 | 17.86 | 18.29 | 18.71 | 19.14 | 19.57 | 20.00 | 20.43 | 20.87 | 21.30 | 21.73 | 22.17 | 22.60 |
| 37 | 16.62 | 17.06 | 17.49 | 17.93 | 18.37 | 18.81 | 19.25 | 19.69 | 20.13 | 20.58 | 21.02 | 21.46 | 21.91 | 22.36 | 22.81 | 23.25 |
| 38 | 17.08 | 17.53 | 17.98 | 18.43 | 18.88 | 19.33 | 19.78 | 20.24 | 20.69 | 21.15 | 21.61 | 22.07 | 22.52 | 22.99 | 23.45 | 23.91 |
| 39 | 17.54 | 18.00 | 18.46 | 18.93 | 19.39 | 19.86 | 20.32 | 20.79 | 21.26 | 21.73 | 22.20 | 22.67 | 23.14 | 23.61 | 24.09 | 24.56 |
| 40 | 18.00 | 18.48 | 18.95 | 19.43 | 19.90 | 20.38 | 20.86 | 21.34 | 21.82 | 22.30 | 22.79 | 23.27 | 23.76 | 24.25 | 24.73 | 25.22 |
| 41 | 18.47 | 18.95 | 19.44 | 19.93 | 20.42 | 20.91 | 21.40 | 21.89 | 22.39 | 22.88 | 23.38 | 23.88 | 24.38 | 24.88 | 25.38 | 25.88 |
| 42 | 18.93 | 19.43 | 19.93 | 20.43 | 20.93 | 21.44 | 21.94 | 22.45 | 22.96 | 23.47 | 23.98 | 24.49 | 25.00 | 25.51 | 26.03 | 26.55 |
| 43 | 19.40 | 19.91 | 20.42 | 20.94 | 21.45 | 21.97 | 22.49 | 23.01 | 23.53 | 24.05 | 24.57 | 25.10 | 25.62 | 26.15 | 26.68 | 27.21 |
| 44 | 19.86 | 20.39 | 20.91 | 21.44 | 21.97 | 22.50 | 23.03 | 23.57 | 24.10 | 24.64 | 25.17 | 25.71 | 26.25 | 26.79 | 27.33 | 27.88 |
| 45 | 20.33 | 20.87 | 21.41 | 21.95 | 22.49 | 23.03 | 23.58 | 24.12 | 24.67 | 25.22 | 25.77 | 26.32 | 26.88 | 27.43 | 27.99 | 28.55 |
| 46 | 20.80 | 21.35 | 21.90 | 22.46 | 23.01 | 23.57 | 24.13 | 24.69 | 25.25 | 25.81 | 26.37 | 26.94 | 27.51 | 28.08 | 28.65 | 29.22 |
| 47 | 21.27 | 21.83 | 22.40 | 22.97 | 23.53 | 24.10 | 24.68 | 25.25 | 25.82 | 26.40 | 26.98 | 27.56 | 28.14 | 28.72 | 29.31 | 29.89 |
| 48 | 21.74 | 22.32 | 22.90 | 23.48 | 24.06 | 24.64 | 25.23 | 25.81 | 26.40 | 26.99 | 27.58 | 28.18 | 28.77 | 29.37 | 29.97 | 30.57 |
| 49 | 22.21 | 22.80 | 23.39 | 23.99 | 24.58 | 25.18 | 25.78 | 26.38 | 26.98 | 27.59 | 28.19 | 28.80 | 29.41 | 30.02 | 30.63 | 31.24 |
| 50 | 22.69 | 23.29 | 23.89 | 24.50 | 25.11 | 25.72 | 26.33 | 26.95 | 27.56 | 28.18 | 28.90 | 29.42 | 30.04 | 30.67 | 31.29 | 31.92 |
| 51 | 23.16 | 23.78 | 24.40 | 25.02 | 25.64 | 26.26 | 26.89 | 27.52 | 28.15 | 28.78 | 29.41 | 30.05 | 30.68 | 31.32 | 31.96 | 32.60 |
| 52 | 23.64 | 24.27 | 24.90 | 25.53 | 26.17 | 26.81 | 27.45 | 28.09 | 28.73 | 29.38 | 30.02 | 30.67 | 31.32 | 31.98 | 32.63 | 33.29 |
| 53 | 24.11 | 24.76 | 25.40 | 26.05 | 26.70 | 27.35 | 28.00 | 28.66 | 29.32 | 29.98 | 30.64 | 31.30 | 31.97 | 32.63 | 33.30 | 33.97 |
| 54 | 24.59 | 25.25 | 25.91 | 26.57 | 27.23 | 27.90 | 28.56 | 29.23 | 29.91 | 30.58 | 31.25 | 31.93 | 32.61 | 33.29 | 33.98 | 34.66 |
| 55 | 25.07 | 25.74 | 26.41 | 27.09 | 27.77 | 28.44 | 29.13 | 29.81 | 30.50 | 31.18 | 31.87 | 32.56 | 33.26 | 33.95 | 34.65 | 35.35 |
| 56 | 25.55 | 26.23 | 26.92 | 27.61 | 28.30 | 28.99 | 26.69 | 30.39 | 31.09 | 31.79 | 32.49 | 33.20 | 33.91 | 34.62 | 35.33 | 36.04 |
| 57 | 26.03 | 26.73 | 27.43 | 28.13 | 28.84 | 29.54 | 30.25 | 30.97 | 31.68 | 32.39 | 33.11 | 33.83 | 34.56 | 35.28 | 36.01 | 36.74 |
| 58 | 26.51 | 27.23 | 27.94 | 28.66 | 29.37 | 30.10 | 30.82 | 31.55 | 32.27 | 33.00 | 33.74 | 34.47 | 35.21 | 35.95 | 36.69 | 37.43 |
| 59 | 27.00 | 27.72 | 28.45 | 29.18 | 29.91 | 30.65 | 31.39 | 32.13 | 32.87 | 33.61 | 34.36 | 35.11 | 35.86 | 36.62 | 37.37 | 38.13 |
| 60 | 27.48 | 28.22 | 28.96 | 29.71 | 30.45 | 31.20 | 31.96 | 32.71 | 33.47 | 34.23 | 34.99 | 35.75 | 36.52 | 37.29 | 38.06 | 38.83 |

SOURCE: Courtesy of the Federal Reserve System, the Truth-in-Lending, Regulation Z, Annual Percentage Rate Tables.

# ANSWERS TO
# SELECTED EXERCISES

# CHAPTER 1

●●●●●●●●●●●●●●●●●●●●●●●●●●●●●●●●●●●●●●●●

EXERCISE SET 1, P. 12

**1.** The answer is (b). One possibility: (a), (c), (d), and (e) are modes of exercise without an aid; (b) would require something like a horse or a bicycle.

**2.** (a) $\triangle, \square$  (c) *b, a*  (e) *M, O*

**3.** (a) D, S, R  (b) B, B, J  (c) V, W, U

**4.** (a) 32, 64, 128

**5.** *HIJEFG, GHIJEF*  **6.** (a) 25

**7.** *bludo wasca*

**9.**

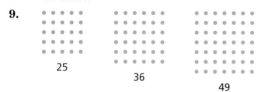

25    36    49

**11.** (a) 1234321  (b) 89991

**12.** (a) 2, 5, 10, 17, 26  (c) 3, 7, 11, 15, 19

**13.** (a) Starting with a first term of 7, add 5 to each term to get the next term.
(b) The first term is 5; to get the next term, add 3 to the previous term.
(c) Starting with a first term of 3, multiply each term by 3 to get the next term.
(d) Each term is the cube of the term number: $1^3 = 1$, $2^3 = 8$, $3^3 = 27$, etc.

**14.** (a) Starting with 2, each term is 3 times the preceding term: 486, 1458, 4374.
(c) Starting with 3, add successive even numbers (beginning with 2) to each term to obtain the following terms: 33, 45, 59.
(e) The odd-numbered terms start with 4 and increase each time by 1. The even-numbered terms start with 1 and increase each time by 1; 3, 7, 4.

**15.** (a) 17, 22  (b) 8  **16.** (a) 81, 243

**17.** (a)  (b)

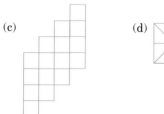

(c)  (d)

**19.** (a) 11  (b) 13  (c) 15  (d) 17

**21.** (a) 156  (b) 124

**23.** 8 pairs; 13 pairs; 21 pairs. It is a Fibonacci sequence.

**24.** (a) The number of ancestors in each generation is equal to the corresponding number in the Fibonacci sequence, where the first 1 represents the present generation.
(c) The number of female ancestors in each generation is equal to the corresponding number of the Fibonacci sequence.

**25.** (a) *One* rule is "the squares of the counting numbers"; *25*.
(b) Add to the previous term the sum of that term and its preceding term: 53.

**27.** (a) $7 + 5(n - 1)$ or $5n + 2$
(b) $5 + 3(n - 1)$ or $3n + 2$
(c) $3^n$  (d) $n^3$

EXERCISE SET 2, P. 21

**1.** (a) Find the number of children.
(b) 6 girls; each girl has 2 brothers. Do the girls share brothers?
(c) 8 children
(d) There are 8 children in the Brown family. (Summarize the information.)

**3.** (a) Find 8 coins.
(b) The coins must total 57 cents. Can you solve the problem with the coin of greatest value being a dime? Can you solve the problem with 1 quarter? Two quarters?
(c) Five dimes, 1 nickel, and 2 pennies; or 1 quarter, 1 dime, 4 nickels, and 2 pennies.
(d) Five dimes, 1 nickel, and 2 pennies total 57 cents. Also one quarter, 1 dime, 4 nickels, and 2 pennies total 57 cents. (Use guess, test, and revise.)

**5.** (a) Find the weight of each of 4 boxes.
(b) How many boxes are involved? What is the total weight of all boxes? What is the relationship between the lightest box and the next lightest? Between the second lightest and third lightest? Between the third lightest and the heaviest?
(c) 4, 8, 16, and 32 lb.
(d) The boxes weigh 4, 8, 16, and 32 lb. (Use guess, test, and revise or form an algebraic model.)

**7.** (a) The cost of a full tank of gasoline is unknown.
(b) What is the cost of gasoline per gallon? How many gallons of gasoline does the tank hold? Is the miles per gallon useful?
(c) Solution: $19.50
(d) A full tank of gasoline costs $19.50. (Summarize the information and strip the problem of irrelevant details.)

**9.** (a) The cost of each item
(b) The 28 miles and the 34 miles per gallon

**11.** (a) The number of motorcycles and the number of cars
(b) A motorcycle has 2 tires and a car has 4. How many tires and how many motors are in the parking lot?
(c) 28 cars and 17 motorcycles
(d) 17 motorcycles and 28 cars have $17(2) + 28(4) = 146$ tires. (Guess, test, and revise, or form an algebraic model.)

**15.** (a) How many nuts must be removed?
(b) Is the information about the numbers of cashews, pecans, and peanuts useful? What can happen on the first draw? What can happen on the second draw? If you do not have 2 of a kind after 3 draws, what has happened?
(c) 4 for 2 of a kind and 7 for 3 of a kind
(d) In drawing 4 nuts, you must have at least 2 of a kind; in drawing 7 nuts you must have at least 3 of a kind. (Strip the problem of irrelevant details.)

**17.** (a) All the ways of giving change for a quarter
(b) List the six possibilities for using pennies: 25, 20, 15, 10, 5, and 0. Then tabulate under each of these all possibilities for the other coins.
(c) 12 ways
(d) There are 12 ways to make change. (Make a table or chart for a quarter using nickels, dimes, and pennies.)

**19.** (a) How long will it take to cut it into five 2-foot pieces?
(b) A cut takes 2 minutes. How many cuts must be made?
(c) 8 minutes
(d) It takes 8 minutes to cut a 10-ft log into five 2-foot pieces. (Draw a picture.)

**21.** (a) Describe how to brown 3 pieces of toast in the shortest period of time.
(b) Each side must be browned for 30 seconds. Must we complete the first 2 pieces before starting the third?
(c) Brown side 1 of slices 1 and 2; side 2 of slice 1 and side 1 of slice 3; side 2 of slices 2 and 3. 90 seconds
(d) It takes 90 seconds to brown 3 pieces of toast in Sonya's kitchen. (Picture in your mind the order of browning the toast.)

**23.** There is the same amount of grape juice in the water as there is water in the grape juice. (Strip away the irrelevant details.)

**25.**

The given figure satisfies the requirements of the problem. (Do not impose conditions that are not there.)

**26.** (a) No

**27.** (a) T, I, R    (b) O, N, M
(c) 17, 21, 25    (d) 90, 93, 186

EXERCISE SET 3, P. 28

**1.** 18 min.    **3.** 900

**5.** Fill the 7-liter pail. Twice fill the 3-liter pail from the 7-liter pail. One liter remains.

**7.** 90 gifts. (Try a simpler version of the problem.)

**8.** (a)

```
      1
    6   5
  2   4   3
```

(c) Yes, 11 and 12

**9.** (a) 170    (b) 171. (Restate the problem.)

**11.**

```
   11 12  1
 10          2
9       •     3
 8          4
   7  6   5
```

(Guess, test, and revise.)

**13.** 11 coins. (Use a table, and then guess, test, and revise.)

**15.** 49 cubes. (Guess, test, and revise.)

**17.** Separate the coins into 2 stacks of 5 each, and weigh to determine the lighter stack. Weigh 4 coins of the lighter stack with 2 in each group.
(a) If they balance, the remaining coin is counterfeit.
(b) If they do not balance, take the lighter stack and weigh to determine the lighter coin.

**19.** To get 6 gallons in the 9-gallon container, we need to pour out 3 gallons. To pour 3 gallons in the 4-gallon container, it must contain 1 gallon. To get 1 gallon fill the 4-gallon container twice from the 9-gallon container, leaving 1 gallon. Pour this in the 4-gallon container. Then proceed as above. (Work backward.)

**21.** One cat catches 1 rat in 100 min. 100 cats catch 100 rats in 100 min. (Restate the problem.)

**23.** Mike (Make a table)

**25.** Cats vs Ants (5 to 1)
Dogs vs Ants (2 to 2)
Cats vs Dogs (2 to 0)

**27.**

**29.** No

**31.** (a) 8, 16, 32    (b) 7, 11, 16

**1.** (a) $4x + 5$    (b) $2x - 7$
    (c) $4x \div 6$    (d) $5x - 7$

**2.** (a) 34
    (c) 21

**3.** Guess the length is 10. The width is $10 - 3 = 7$. Then $10 + 10 + 7 + 7 = 34 \neq 26$. Let $x$ be the length. The width is $x - 3$. Then $x + x + (x - 3) + (x - 3) = 26$ or $4x - 6 = 26$.

**5.** Guess that 18 gal. of gasoline were used on interstates. Then $30 - 18$ gal. were used on city streets. $28 \cdot 18 + 20 \cdot (30 - 18) = 744 \neq 784$. Let $x$ be the number of gal. of gasoline used on interstates. Then $30 - x$ gal. were used on city streets. $28x + 20(30 - x) = 784$.

**7.** Guess he has 10 quarters. Then he has $30 - 10$ nickels. $25(10) + 5(30 - 10) = 350 \neq 390$. Let $x$ be the number of quarters; $30 - x$ is the number of nickels. $25x + 5(30 - x) = 390$.

**9.** (a) $x = 4$          Check    $2(4) + 4 = 12$
                                   $8 + 4 = 12$
                                   $12 = 12$
    (b) $x = 5$          Check    $3(5) - 5 = 10$
                                   $15 - 5 = 10$
                                   $10 = 10$
    (c) $x = 10$         Check    $5(10) = 50$
                                   $50 = 50$
    (d) $x = 9$          Check    $9 + 3 = 12$
                                   $12 = 12$
    ( e) $x = 5$         Check    $3(5) - 6 = 9$
                                   $15 - 6 = 9$
                                   $9 = 9$
    (f) $x = 1$          Check    $5(1) + 8 = 13$
                                   $5 + 8 = 13$
                                   $13 = 13$

**10.** (a) The length of the rectangle is 8.
    (c) 23 gallons of gasoline were used on interstate highways.
    (e) Marcellus has 12 quarters.

**11.** Let $x$ be the weight of Box A. $x + 2x + 4x + 8x = 60$. $x = 4$. Box A weighs 4 lb.; B weighs 8 lb.; C, 16 lb.; and D, 32 lb.

**13.** Let time $t$ be measured from when the first hose is turned on. $5t + 7(t - 30) = 1710$

$t = 160$ minutes.:The pool is full at 3:40 p.m.

**15.** Let $x$ be the number of miles traveled. $20x + 2500 = 5000$

$x = 125$. He can travel 125 miles.

**17.** Let $t$ be the time that grandpa drives. $40t + 80(t - 1) = 280$.

$t = 3$ They meet at 4:00 P.M.

**19.** 216

**1.**    1    5    10    10    5    1
        1    6    15    20    15    6    1

**2.** 17 chickens and 14 goats    **3.** 8, 14, 20, 26, 32

**4.** 24

**5.** Use 6 of the 9 coins to see if 3 will balance 3. If they balance, use 2 of the remaining 3 (not among the 6 selected) to see if they balance. If they balance, the remaining coin is the counterfeit. If they do not, select the lighter coin from the 2 that do not balance. If in the original weighing the 3 coins do not balance 3 coins, select 2 from the lighter 3. If they balance, the remaining coin is the counterfeit. If they do not balance, the lighter coin is counterfeit.

**6.** 122

**7.** (a) 22, 27, 32    (b) 1, 1, 1    (c) O, P, Q

**8.** 82 kg

**9.** $16 + 17 + 18 + 19 + 20 = 21 + 22 + 23 + 24$
    $25 + 26 + 27 + 28 + 29 + 30$
      $= 31 + 32 + 33 + 34 + 35$

In the first line, write 20 as $5 + 5 + 5 + 5$ and add to 16, 17, 18, and 19 to get the terms on the right.

**10.** Fill the 6-quart container and pour into the 10-quart container. Repeat the process. This time you can only pour 4 quarts into the 10-quart container. This leaves 2 quarts in the 6-quart container.

**11.** 4 days

**12.** 4 hours after Rolf starts

**13.** 2, 5, 8, 11, 14

**14.** (a) 162, 486, 1458    (b) $2(3)^9$    (c) $2(3)^{n-1}$

**15.** 140 adult tickets; 160 children tickets

## CHAPTER 2
••••••••••••••••••••••••••••••••••••••••

**1.** (a) False statement        (b) Not a statement
    (c) True statement
    (d) Not a statement unless the value of $x$ is known.
    (e) Not a statement        (f) False statement
    (g) Not a statement unless the value of $x$ is known.
    (h) Not a statement

**2.** (a) F        (c) F        (e) T

**3.** (a) F        (b) F        (c) F
(d) F        (e) T        (f) F

**4.** (a) F        (c) T        (e) T

**5.** (a) November has 30 days *and* Thanksgiving is always on November 25. *False.*
November has 30 days *or* Thanksgiving is always on November 25. *True.*

(b) The smallest counting number is 2 *or* 10 is not a multiple of 5. *False.*
The smallest counting number is 2 *and* 10 is not a multiple of 5. *False.*

(c) $2 + 3 = 4 + 1$ *and* $8 \cdot 6 = 4 \cdot 12$. *True.*
$2 + 3 = 4 + 1$ *or* $8 \cdot 6 = 4 \cdot 12$. *True.*

(d) Triangles are not squares *or* 3 is smaller than 5. *True.*
Triangles are not squares *and* 3 is smaller than 5. *True.*

**6.** (a) $A \wedge D$        (c) $C \wedge (\sim B)$

**7.** (a) It is snowing, and the roofs are not white.
(b) The roofs are not white, or the streets are slick.
(c) It is snowing, and the roofs are white or the streets are not slick.
(d) It is snowing or the streets are slick, and the trees are beautiful.
(e) It is not true that it is snowing, and the trees are not beautiful.
(f) It is not true that it is snowing, and the streets are slick.

**8.** (a) T        (c) F

**10.** (a) T        (c) F

**11.** (a)

| $p$ | $q$ | $\sim q$ | $p \vee \sim q$ |
|---|---|---|---|
| T | T | F | T |
| T | F | T | T |
| F | T | F | F |
| F | F | T | T |

(b)

| $p$ | $\sim p$ | $p \wedge \sim p$ |
|---|---|---|
| T | F | F |
| F | T | F |

**13.** (a) 2        (b) 4        (c) 8        (d) 16

**14.** (a) The stock market is not bullish, and the Dow average is increasing.

**15.** (a) $S \vee A$        (b) $\sim (S \vee A) \wedge G$

**16.** (a) $p \wedge q$

**17.** (a)

| $p$ | $q$ | $p \wedge q$ | $(p \wedge q) \vee q$ |
|---|---|---|---|
| T | T | T | T |
| T | F | F | F |
| F | T | F | T |
| F | F | F | F |

(b)

| $p$ | $q$ | $p \vee q$ | $(p \vee q) \wedge p$ |
|---|---|---|---|
| T | T | T | T |
| T | F | T | T |
| F | T | T | F |
| F | F | F | F |

**18.** (a) F

**19.** (a) T        (b) T        (c) T        (d) F

**20.** (a)

| $p$ | $q$ | $r$ | $\sim q$ | $p \vee \sim q$ | $(p \vee \sim q) \wedge r$ |
|---|---|---|---|---|---|
| T | T | T | F | T | T |
| T | T | F | F | T | F |
| T | F | T | T | T | T |
| T | F | F | T | T | F |
| F | T | T | F | F | F |
| F | T | F | F | F | F |
| F | F | T | T | T | T |
| F | F | F | T | T | F |

**EXERCISE SET 2, P. 57**

**1.** (a) $A \to D$        (b) $\sim A \to \sim B$
(c) $\sim C \to \sim A$        (d) $C \to \sim D$

**2.** (a) True        (c) False

**3.** (a) *Converse:* If one angle of a triangle has a measure of 90°, then the triangle is a right triangle.
*Inverse:* If a triangle is not a right triangle, then no angle of the triangle has a measure of 90°.
*Contrapositive:* If no angle of a triangle has a measure of 90°, then the triangle is not a right triangle.

(b) *Converse:* If a number is odd, then it is prime.
*Inverse:* If a number is not prime, then it is not odd.
*Contrapositive:* If a number is not odd, then it is not prime.

(c) *Converse:* If alternate interior angles are equal, then two lines are parallel.
*Inverse:* If two lines are not parallel, then alternate interior angles are not equal.

*Contrapositive:* If alternate interior angles are not equal, then two lines are not parallel.

(d) *Converse:* If Joyce is happy, then she is smiling.
*Inverse:* If Joyce is not smiling, then she is not happy.
*Contrapositive:* If Joyce is not happy, then she is not smiling.

**4.** (a) If a geometric figure is a triangle, then it is not a square.
(c) If she is an honest politician, then she does not accept bribes.

**5.** (a) *Converse:* If a geometric figure is not a square, then it is a triangle.
*Contrapositive:* If a geometric figure is a square, then it is not a triangle.

(b) *Converse:* If birds flock together, then they are birds of a feather.
*Contrapositive:* If birds do not flock together, then they are not birds of a feather.

(c) *Converse:* If a politician does not accept bribes, then he or she is an honest politician.
*Contrapositive:* If a politician accepts bribes, then he or she is not an honest politician.

**6.** (a) If it is not snowing, then the roofs are white.
(c) If the roofs are not white and the streets are not slick, then it is snowing.
(e) It is not true that, if it is snowing, then the roofs are not white.

**7.** (a) True   (b) True   (c) True   (d) True
(e) False   (f) True

**8.** (a) *Conditional:* If I teach third grade, then I do not teach in high school.
*Converse:* If I do not teach in high school, then I teach third grade.
*Inverse:* If I do not teach third grade, then I teach in high school.
*Contrapositive:* If I teach in high school, then I do not teach third grade.

**9.** In each part of this exercise, only the contrapositive has the same truth values as the original expression.

(a) *Converse:* If you have fewer cavities, then you brush your teeth with White-as-Snow.
*Inverse:* If you do not brush your teeth with White-as-Snow, then you do not have fewer cavities.
*Contrapositive:* If you do not have fewer cavities, then you do not brush your teeth with White-as-Snow.

(b) *Converse:* If you love mathematics, then you like this book.
*Inverse:* If you do not like this book, then you do not love mathematics.

*Contrapositive:* If you do not love mathematics, then you do not like this book.

(c) *Converse:* If you eat Barlies for breakfast, then you want to be strong.
*Inverse:* If you do not want to be strong, then you do not eat Barlies for breakfast.
*Contrapositive:* If you do not eat Barlies for breakfast, then you do not want to be strong.

(d) *Converse:* If your clothes are bright and colorful, then you use Wave.
*Inverse:* If you do not use Wave, then your clothes are not bright and colorful.
*Contrapositive:* If your clothes are not bright and colorful, then you do not use Wave.

**10.** (a) If one is careless, then one will have accidents.
(c) If two sides of a triangle are congruent, then the triangle is isosceles.

**11.** (a)

| $p$ | $q$ | $p \vee q$ | $\sim p$ | $\sim p \to q$ |
|---|---|---|---|---|
| T | T | T | F | T |
| T | F | T | F | T |
| F | T | T | T | T |
| F | F | F | T | F |

↑——— same ———↑

(b) If I do not pass English, then I go to summer school.
(c) I sleep tonight or I skip the luncheon.

**12.** (a)

| $p$ | $q$ | $\sim q$ | $p \wedge \sim q$ | $p \to q$ | $\sim (p \to q)$ |
|---|---|---|---|---|---|
| T | T | F | F | T | F |
| T | F | T | T | F | T |
| F | T | F | F | T | F |
| F | F | T | F | T | F |

↑——— same ———↑

**13.** (a)

| $p$ | $q$ | $p \vee q$ | $\sim (p \vee q)$ | $\sim p$ | $\sim q$ | $\sim p \wedge \sim q$ |
|---|---|---|---|---|---|---|
| T | T | T | F | F | F | F |
| T | F | T | F | F | T | F |
| F | T | T | F | T | F | F |
| F | F | F | T | T | T | T |

↑——— same ———↑

(b)

| $p$ | $q$ | $p \wedge q$ | $\sim (p \wedge q)$ | $\sim p$ | $\sim q$ | $\sim p \vee \sim q$ |
|---|---|---|---|---|---|---|
| T | T | T | F | F | F | F |
| T | F | F | T | F | T | T |
| F | T | F | T | T | F | T |
| F | F | F | T | T | T | T |

↑——— same ———↑

(c) I will not take history and I will not take physics.

(d) I am not on the baseball team or I am not on the basketball team.

**14.** (a) Equivalent

| $p$ | $q$ | $\sim p$ | $\sim p \vee q$ | $p \to q$ |
|---|---|---|---|---|
| T | T | F | T | T |
| T | F | F | F | F |
| F | T | T | T | T |
| F | F | T | T | T |

└── same ──┘

**15.** (a) If there is no inflation, then the interest rates will increase.

(b) It is not true that if Charles sings, Jane does not dance.

(c) It is not true that, if I carry my umbrella, then it will rain.

(d) If the moon is not made of cheese, then flight to Mars is impossible.

**16.** (a) If two lines are not parallel and not skew, then they intersect.

**17.** (a) If a geometric figure is a triangle, then it is not a square.

(c) If she is an honest politician, then she does not accept bribes.

**18.** (a)

| $p$ | $\sim p$ | $p \wedge \sim p$ |
|---|---|---|
| T | F | F |
| F | T | F |

Because the third column contains all F's, $p \wedge \sim p$ is a contradiction.

**19.** (a)

| $p$ | $p \to p$ |
|---|---|
| T | T |
| F | T |

Tautology

(b) Not a tautology

| $p$ | $q$ | $p \to q$ | $p \wedge (p \to q)$ | $\sim q$ | $[p \wedge (p \to q)] \to \sim q$ |
|---|---|---|---|---|---|
| T | T | T | T | F | F |
| T | F | F | F | T | T |
| F | T | T | F | F | T |
| F | F | T | F | T | T |

(c) Not a tautology

| $p$ | $q$ | $p \to q$ | $(p \to q) \to p$ |
|---|---|---|---|
| T | T | T | T |
| T | F | F | T |
| F | T | T | F |
| F | F | T | F |

(d) Tautology

| $p$ | $q$ | $\sim p$ | $\sim p \vee q$ | $(\sim p \vee q) \wedge p$ | $[(\sim p \vee q) \wedge p] \to q$ |
|---|---|---|---|---|---|
| T | T | F | T | T | T |
| T | F | F | F | F | T |
| F | T | T | T | F | T |
| F | F | T | T | F | T |

(e) Not a tautology

| $p$ | $q$ | $p \to q$ | $p \vee q$ | $(p \to q) \to (p \vee q)$ |
|---|---|---|---|---|
| T | T | T | T | T |
| T | F | F | T | T |
| F | T | T | T | T |
| F | F | T | F | F |

(f) Tautology

| $p$ | $q$ | $\sim p$ | $\sim p \vee q$ | $p \to q$ | $(\sim p \vee q) \to (p \to q)$ |
|---|---|---|---|---|---|
| T | T | F | T | T | T |
| T | F | F | F | F | T |
| F | T | T | T | T | T |
| F | F | T | T | T | T |

**21.** (a) If Joe does not play quarterback, then his team loses.

(b) If 3 does not divide 12, then 3 is not a factor.

(c) If you do not study regularly, then you will fail the course.

**23.** (a) Arithmetic is useful, or both logic is easy and algebra is hard.

(b) It is not true that logic is easy or algebra is hard.

(c) Arithmetic is not useful and algebra is not hard.

(d) Logic is easy, and either algebra is hard or arithmetic is not useful.

**EXERCISE SET 3, P. 64**

**1.** $s \to p$

$s$

$\therefore p$

**3.** $p \rightarrow c$
$c \rightarrow m$
$\therefore p \rightarrow m$

**4.** (a) Rule of detachment
(c) Chain rule

**5.** (a) Rule of contraposition
(b) Rule of detachment and chain rule
(c) Rule of detachment
(d) Rule of contraposition
(e) Chain rule

**6.** (a) If the sun shines, the garden flourishes (chain rule).
(c) Devron is sick (rule of contraposition).

**7.** (a) Susan takes mathematics (rule of detachment).
(b) You studied (rule of contraposition).
(c) You cry (rule of detachment).
(d) You do not cut the grass (rule of contraposition).

**9.** If Joan is tall, she wears contacts.
If she wears contacts, she has red hair.
If she has red hair, then she teaches seventh grade.
Therefore, if she is tall, she teaches seventh grade.

**11.** (a) Invalid    (b) Invalid    (c) Valid
(d) Valid    (e) Invalid    (f) Valid

**12.** (a) Invalid. Let $p$ represent "A man is a good speaker," and let $q$ represent "A man is a good teacher." $(p \rightarrow q) \wedge q$ implies $p$ is not always true as indicated by the circled truth values.

| $p$ | $q$ | $p \rightarrow q$ | $(p \rightarrow q) \wedge q$ |
|---|---|---|---|
| T | T | T | T |
| T | F | F | F |
| Ⓕ | T | T | Ⓣ |
| F | F | T | F |

**13.** (a) $G \rightarrow S$
$S \rightarrow B$
$\underline{\sim S}$
$\therefore \sim B$   Invalid

(b) $LM \rightarrow LC$
$\underline{\sim LM}$
$\therefore \sim LC$   Invalid

**15.** (a) This implies that you have been beating your wife, which may not be true.
(b) It is assumed that the customer will buy the suit because he looks good in it.

**17.** It could be merely a coincidence that you saw a black cat and lost your pocketbook.

**18.** (a) *Converse:* If we play tennis, then it will not rain.
*Contrapositive:* If we do not play tennis, then it will rain.

**19.** (a) F    (b) T    (c) F    (d) T

**1.** Diagrams will vary. One is shown for each.

(a)     (b)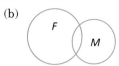

(c)     (d)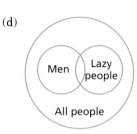

**2.** Diagrams will vary. One example is shown for each.

(a)     (c)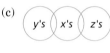

(e)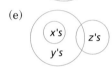

**3.** (a) $x = 3$    (b) $x = 11$
(c) $x = 4$ or $-4$    (d) Any value of $x$

**4.** (a) $x = 2$ (There are many possibilities.)
(c) $x = 3$ (There are many possibilities.)

**5.** (a) For some $x$, $x + 4 = 7$.
(b) For some $x$, $x - 7 = 4$.
(c) There exist numbers $x$ so that $x^2 = 16$.
(d) For all values of $x$, $x + 3 = 3 + x$.

**6.** (a) For all $x$, $x + 4 = 7$ or there is no $x$ such that $x + 4 = 7$.
(c) For all $x$, $x^2 = 16$ or there is no $x$ such that $x^2 = 16$.

**7.** (a) Some athletes over 7 ft. tall do not play basketball.
(b) No students work hard at their studies.
(c) Some men do not use Bob Bob hair oil.
(d) All professors are intelligent.
(e) Some men weigh more than 500 pounds.
(f) No rabbits have white tails.
(g) All bunnies are rabbits.

**8.** (a) There is no counting number greater than 50.
(c) There is no $x$ such that $x + 2 = 7$.
(e) For some $x$, $x(x + 2) \neq x^2 + 2x$.

**9.** (a) If an animal is a frog, then it is green.
(b) If an animal is a cat, then it is intelligent.
(c) If a person is a college professor, then he or she is brilliant.
(d) If a person has long legs, then he or she is not an athlete.

**11.**

2 Legs

Yes, no, no

**13.**

The diagram on the left satisfies the premises and hence shows it is possible that some beans are not edible. The same premises are satisfied by the diagram on the right. From these premises we cannot be certain that some beans are not edible.

**14.** (a) Diana is clever.
   (c) Kay is not a college woman.

**15.** (a) Some Wags are Tags.
   (b) All Blues are Greens.

**16.** (a) 3 is a counting number that is not even.

**17.** (a) All squares are rectangles.
   (b) All numbers that are divisible by 4 are divisible by 2.

**18.** (a) A geometric figure is a square, but it is not a rectangle.

**19.** $x$ is an irrational, real, and complex number.

**21.**                         **23.**

False                           False

**25.**

False

**27.** Consumption must come from production (chain rule).

**28.**                         (a) Not valid

**29.**                         (a) Not valid
                                (b) Valid
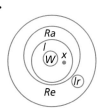            (c) Not valid
                                (d) Not valid

**30.**                         (a) Not valid

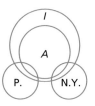

**31.** (a) Yes

   (b) No

**32.** (a)

**33.** It is raining and I will not carry an umbrella.

**35.** Lucy is late and she does not miss the bus.

**37.**

| $p$ | $q$ | $p \to q$ | $p \wedge q$ | $p \vee q$ |
|---|---|---|---|---|
| F | T | T | F | T |
| F | F | T | F | F |
| T | F | F | F | T |
| T | T | T | T | T |

**CHAPTER TEST, P. 76**

**1.** (a) Y      (b) N      (c) N      (d) Y

**2.** (a)                    (b)

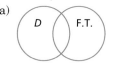          

**3.** (a) If something is a wolf, then it is a mammal.
   (b) If my watch is not fast, then Tom is late (or if Tom is not late, then my watch is fast).

**4.** (a)

| $p$ | $q$ | $r$ | $q \vee r$ | $p \wedge (q \vee r)$ |
|---|---|---|---|---|
| T | T | T | T | T |
| T | T | F | T | T |
| T | F | T | T | T |
| T | F | F | F | F |
| F | T | T | T | F |
| F | T | F | T | F |
| F | F | T | T | F |
| F | F | F | F | F |

(b)

| p | q | r | p ∧ q | p ∧ r | (p ∧ q) ∨ (p ∧ r) |
|---|---|---|---|---|---|
| T | T | T | T | T | T |
| T | T | F | T | F | T |
| T | F | T | F | T | T |
| T | F | F | F | F | F |
| F | T | T | F | F | F |
| F | T | F | F | F | F |
| F | F | T | F | F | F |
| F | F | F | F | F | F |

**5.** Yes, their truth values are identical.

**6.** (a)

| p | q | p → q | q → p |
|---|---|---|---|
| T | T | T | T |
| T | F | F | T |
| F | T | T | F |
| F | F | T | T |

(b) No; the truth values are not the same.

**7.** *Converse:* If a course is good, then it is hard.
*Contrapositive:* If a course is not good, then it is not hard.

**8.** (a) 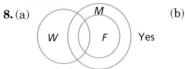 Yes (b) 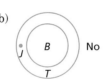 No

**9.** (a) Some birds are not black.
(b) There is no *x* such that *x* + 3 = 5.

**10.** (a) True (b) False

**11.** All triangles are plane figures (chain rule).

**12.** (a) If it is hot, I will miss supper (chain rule).
(b) Aaron did not sleep late (rule of contraposition).
(c) The stock market rises (rule of detachment).
(d) The sun does not shine (rule of contraposition).

**13.** (a)

| p | q | p → q | ~ (p → q) | ~ q | p ∧ ~ q |
|---|---|---|---|---|---|
| T | T | T | F | F | F |
| T | F | F | T | T | T |
| F | T | T | F | F | F |
| F | F | T | F | T | F |

↑————— same —————↑

(b) It is hot and I do not go to the beach.

# CHAPTER 3

●●●●●●●●●●●●●●●●●●●●●●●●●●●●●●●

**1.** (a) True
(b) False; 81 is not less than 16
(c) False; 5 ∈ A. {5} is a set and is not an element of this set.
(d) False; {1, 2, 3, . . . 15} is a subset.
(e) True
(f) False; 0 is not a counting number.

**2.** (a) {1, 2, 3, . . . , 16}  (c) { }

**3.** (a) {x | x is a counting number less than or equal to 16}
(b) {x | x is an even counting number}
(c) {x | x is a woman president of the United States}

**4.** (a) (i), (v)  (c) (i), (iii), (v)

**5.** (a) True  (b) False  (c) True
(d) False  (e) False

**6.** (a) No  (c) Yes

**7.** (a) ∈  (b) ⊆ or ⊂
(c) ⊆  (d) ∉

**8.** (a) R ∪ T = {5, 10, 15, 20}  R ∩ T = {15}
(c) A ∪ B = {0, 10, 100, 1000}  A ∩ B = {10, 100}
(e) A ∪ B = {x, y, z, r, s, t}  A ∩ B = {x, y}

**9.** (a) Set of all elements in A but not in B (A − B)
(b) Set of all elements in A and B (A ∩ B)
(c) Set of all elements in B but not in A (B − A)
(d) Set of all elements not in A and not in B ($\overline{A} \cap \overline{B}$)

**10.** (a) {a, c, e, g}  (c) {b, f, h}  (e) {c}

**11.** (a) A ∩ D  (b) B ∩ (C ∩ D)
(c) B ∩ ($\overline{C} \cap \overline{D}$)  (d) A ∩ (C ∪ $\overline{D}$)

**12.** (a) 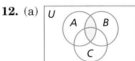  (c)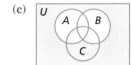

**13.** (a) 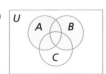  (b)
A ∪ B  $\overline{A} \cap \overline{C}$

(c)   (d)
A ∩ (B ∩ C)  A ∪ (B ∪ C)

**14.** (a) Yes; {∅} is a set with one element; ∅ is the empty set.    (c) No    (e) No

**15.** (a)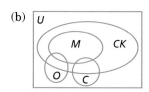

E: elementary teachers
H: high-school teachers
C: college teachers
U: all teachers in Ourtown

(b)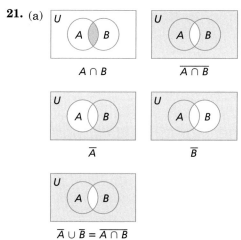

M: math teachers
O: teachers with overhead projector
C: teachers with cassette recorder
U: all teachers at Ourtown High
CK: teachers with chalkboards

**16.** (a) c, f, g, h

**17.** (a) {a}, {b}, {a, b}, ∅
(b) {a}, {b}, {c}, {a, b}, {b, c}, {a, c}, {a, b, c}, ∅
(c) {a}, {b}, {d}, {a, b}, {a, d}, {b, d}, ∅

**18.** (a) $B - A = \{7, 8\}$

**19.** (a) 8    (b) 16    (c) 32    (d) $2^n$

**20.** (a) $A \cap (\overline{B} \cap \overline{C})$ or $A \cap (\overline{B \cup C})$

**21.** (a)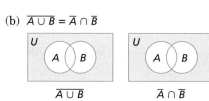

$A \cap B$    $\overline{A \cap B}$

$\overline{A}$    $\overline{B}$

$\overline{A} \cup \overline{B} = \overline{A \cap B}$

(b) $\overline{A \cup B} = \overline{A} \cap \overline{B}$

$\overline{A \cup B}$    $\overline{A} \cap \overline{B}$

**23.**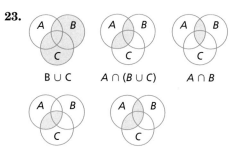

$B \cup C$    $A \cap (B \cup C)$    $A \cap B$

$A \cap C$    $(A \cap B) \cup (A \cap C)$

**24.** (a) $\{a, b\} \cup (\{b, c, d, e\} \cup \{d, e, f\})$
$= (\{a, b\} \cup \{b, c, d, e\}) \cup \{d, e, f\}$
as $\{a, b\} \cup \{b, c, d, e, f\}$
$= \{a, b, c, d, e\} \cup \{d, e, f\}$
as $\{a, b, c, d, e, f\} = \{a, b, c, d, e, f\}$

**25.** (a) Yes, $A \cap B \subseteq A \cup B$
(b) Not necessarily; $x$ could be an element in either $A$ or $B$ but not in $A \cap B$.

**27.** {A, B, C}, {A, B, D}, {A, B, E}, {A, C, D}, {A, C, E}, {A, D, E}, {B, C, D}, {B, C, E}, {B, D, E}, {C, D, E}, {A, B, C, D}, {A, B, C, E}, {A, B, D, E}, {A, C, D, E}, {B, C, D, E}, {A, B, C, D, E}

**28.** (a) {A, B} {A, C} {B, C} {A, B, C} {A, B, D} {A, B, E} {A, C, D} {A, C, E} {B, C, D} {B, C, E} {A, B, C, D} {A, B, C, E} {A, B, D, E} {A, C, D, E} {B, C, D, E} {A, B, C, D, E}

**29.** (a) $\overline{A} \cap \overline{B}$ or $\overline{A \cup B}$
(b) $B \cap \overline{A}$ or $B - A$    (c) $A \cap \overline{B}$ or $A - B$

EXERCISE SET 2, P. 97

**1.** (a) $x$ is a product of $y$ or $y$ is a manufacturer of $x$.
(b) $y$ is the number of wheels on $x$.

**2.** (a)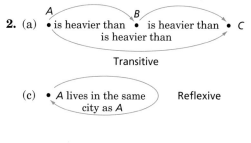

Transitive

(c)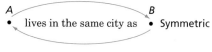

Reflexive

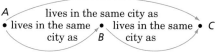

Symmetric

Transitive

(e)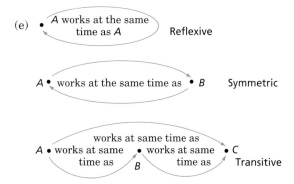

*A* works at the same time as *A*   Reflexive

*A* • works at the same time as • *B*   Symmetric

works at same time as
*A* • works at same   • works at same   • *C*
     time as        time as
              *B*              Transitive

(g) None

**3.** (a) Not reflexive, symmetric, not transitive
(b) Reflexive, not symmetric, transitive
(c) Not reflexive, not symmetric, transitive
(d) Equivalence relation
(e) Not reflexive, not symmetric, transitive
(f) Not reflexive, not symmetric, transitive
(g) Equivalence relation
(h) Reflexive, symmetric, not transitive
(i) Not reflexive, symmetric, not transitive
(j) Reflexive, not symmetric, transitive

**4.** (a) An equivalence relation
(c) Reflexive, transitive

**5.** (a) $A \times B = \{(a, r), (a, s), (a, t), (b, r), (b, s), (b, t), (c, r), (c, s), (c, t)\}$
(b) $A \times A = \{(a, a), (a, b), (a, c), (b, b), (b, a), (b, c), (c, c), (c, a), (c, b)\}$
(c) $B \times A = \{(r, a), (s, a), (t, a), (r, b), (s, b), (t, b), (r, c), (s, c), (t, c)\}$
(d) $B \times B = \{(r, r), (r, s), (r, t), (s, r), (s, t), (s, s), (t, t), (t, r), (t, s)\}$

**6.** (a) $A \times (B \cap C) = \{(a, c), (b, c), (c, c)\}$
(c) $(A \cap B) \times C = \{(c, a), (c, c), (c, x)\}$

**7.** (a) $\{(3, 0)\}$   (b) $\{(3, \varnothing)\}$

**8.** (a) $B = \{1, 4\}$, $C = \{1, 2, 3\}$
(c) $B = \{6\}$, $C = \{6, 7, 8\}$

**9.** $P \times S = \{$(Marie, Janet), (Marie, Jane), (Marie, Marion), (Boyd, Janet), (Boyd, Jane), (Boyd, Marion), (Kiefer, Janet), (Kiefer, Jane), (Kiefer, Marion)$\}$   **10.** (a) 9; 9

**11.** (a) $\{(1, 2), (1, 5), (2, 5), (3, 5), (4, 5)\}$
(b) $\{(1, 2), (1, 5), (2, 5), (3, 2), (3, 5), (4, 2), (4, 5)\}$
(c) $\{(2, 2)\}$
(d) $\{(2, 2), (3, 2), (4, 2)\}$   **12.** (a) Yes

**13.** (a) True   (b) False   (c) True
(d) False   (e) True

**14.** (a) {Asa, Jim, Ben, Bob, Joe, Amy}, {John, Jack, Alto, Bill, Jill}, {Allen, Betty, Alisa}

**15.** (a) $\{(w, w), (w, y), (w, z), (y, y), (y, z), (z, z), (x, x)\}$
(b) Reflexive, not symmetric, transitive

**16.** (a)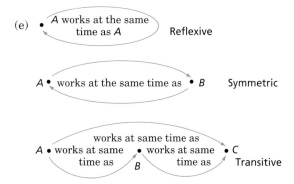

**17.** $\{(1, 1), (2, 2), (3, 3), (1, 2), (2, 1), (2, 3), (3, 2)\}$ on $\{1, 2, 3\}$

**18.** (a) $R = \{(1, 3), (3, 5), (1, 5)\}$

**19.** (a) $\{[1, (3, 4)], [1, (3, 5)], [1, (4, 4)], [1, (4, 5)], [1, (5, 4)], [1, (5, 5)], [2, (3, 4)], [2, (3, 5)], [2, (4, 4)], [2, (4, 5)], [2, (5, 4)], [2, (5, 5)]\}$
(b) $\{[(1, 3), 4], [(1, 4), 4], [(1, 5), 4], [(2, 3), 4], [(2, 4), 4], [(2, 5), 4], [(1, 3), 5], [(1, 4), 5], [(1, 5), 5], [(2, 3), 5], [(2, 4), 5], [(2, 5), 5]\}$

**20.** (a) $\{(1, 1), (1, 2), (1, 3), (1, 4), (2, 1), (2, 2), (2, 3), (2, 4), (3, 1), (3, 2), (3, 3), (3, 4), (4, 1), (4, 2), (4, 3), (4, 4), (3, 5), (4, 5), (5, 3), (5, 4), (5, 5)\}$

**21.** (a) $A \cap B = \{b, c\} = B \cap A$
(b) $B \cup C = C \cup B$; $\{b, c, d, e, g\} = \{b, c, d, e, g\}$
(c) $(A \cap B) \cap C = \{c\} = A \cap (B \cap C)$
(d) $(A \cup B) \cup C = A \cup (B \cup C)$
$\{a, b, c, d, e, g\} = \{a, b, c, d, e, g\}$

**22.** (a) $\{1, 2, 3, \ldots, 10\}$   **23.** No

**EXERCISE SET 3, P. 104**

**1.** (a) C   (b) O   (c) O   (d) O
(e) C   (f) O   (g) C   (h) O
(i) C   (j) C   (k) C   (l) O

**2.** (a) Yes   (c) Yes   (e) No

**3.** (a) $1 \leftrightarrow b$
(b) $1 \leftrightarrow \phi$, $2 \leftrightarrow \lambda$
(c) $1 \leftrightarrow a$, $2 \leftrightarrow b$, $3 \leftrightarrow c$, $4 \leftrightarrow d$, $5 \leftrightarrow e$
(d) $1 \leftrightarrow 4$, $2 \leftrightarrow 6$, $3 \leftrightarrow 8$
(e) $1 \leftrightarrow 10$, $2 \leftrightarrow 40$, $3 \leftrightarrow 30$, $4 \leftrightarrow 50$, $5 \leftrightarrow 70$
(f) $1 \leftrightarrow 2$

**4.** (a) 17   (c) 5   (e) 30   (g) 47   (i) 42

**5.** (a) True   (b) False   (c) True
(d) False   (e) False   (f) True
(g) True   (h) False   (i) True
(j) False   (k) True

**6.** (a) For $a \in A$, $a \notin R$. Therefore $A \neq R$.

**7.** (a) True   (b) False   (c) True
(d) True   (e) True   (f) True

**8.** (a) $A \cup B = \{2, 3, 4, \ldots, 10\}$, so $n(A \cup B) = 9$, $n(A) = 5$, and $n(B) = 5$.
$A \cap B = \{6\}$, so $n(A \cap B) = 1$.
Therefore, $n(A \cup B) = n(A) + n(B) - n(A \cap B)$ because $9 = 5 + 5 - 1$.

**9.** 890,000

**10.** (a) False: Let $A = \{a, b, c, d, e\}$ and $B = \{e, f, g, h, i\}$.
$n(A \cup B) = 9$, $n(A) = 5$, and $n(B) = 5$
$n(A \cup B) \neq n(A) + n(B)$ because $5 + 5 \neq 9$.

**11.** (a) 10    (b) 14    (c) 4    (d) 5

**12.** (a) 10

**13.** (a)

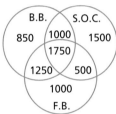

(b) 350
(c) 850
(d) 50
(e) 150
(f) 950

**14.** (a)

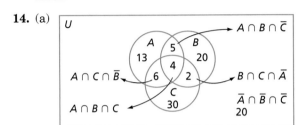

$A$ = car pools    $B$ = buses    $C$ = drove alone

**15.** No, only 49 took advantage of this sale. **16.** (a) 2

**17.** The poll was not accurate if all attended one game. This may be seen by the diagram drawn from the given information.

The total number of students is 7850, not 7900, as it should have been.

**19.** (a) $\varnothing$    (b) A    (c) A    (d) U
(e) $\varnothing$    (f) U    (g) $\varnothing$    (h) U

**20.** (a) {4, 5}    **21.** (a) $\overline{(A \cup B)} \cap C$    (b) $\overline{A \cap B}$

**EXERCISE SET 4, P. 112**

**1.** (a) Yes    (b) Yes    (c) Yes    (d) Yes
(e) Yes    (f) No    (g) Yes    (h) Yes
(i) No    (j) Yes

**2.** (a) No    (c) No    (e) Yes    (g) No

**3.** (a) Yes    (b) Yes    (c) No    (d) Yes

**4.** (a) $f(1) = 6; f(5) = 10$    (c) $f(1) = 4; f(5) = 16$
(e) $f(1) = 8; f(5) = 40$

**5.** (a) $f(2) = 1$    $f(3) = 2$    $f(4) = 3$
(b) $f(2) = 3$    $f(3) = 5$    $f(4) = 7$
(c) $f(2) = 4$    $f(3) = 9$    $f(4) = 16$
(d) $f(2) = 6$    $f(3) = 11$    $f(4) = 18$

**6.** (a) Yes    **7.** 4; 864

**9.** (a) {(0, 1), (1, 4), (2, 7), (3, 10)}

(b) 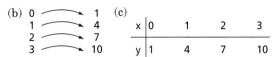    (c)

**(d)**

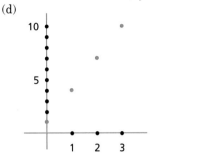

**10.** (a) $y = x + 2$

**11.** (a) Yes    (b) No    (c) Yes    (d) No

**12.** (a) $R(x) = (500 + 10x)(200 - x)$

**13.** $3.00; $4.50; $16.50; $37.50. Parking is $3 for the first hour and $1.50 for each additional hour.

**15.** 1000; 700; 400; 100

**17.**

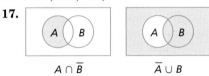

$A \cap \overline{B}$    $\overline{A} \cup B$

$\overline{A} \cup B$

**19.** (a) $\{e, f\}$    (b) $\varnothing$    (c) $\varnothing$    (d) A

**CHAPTER TEST, P. 115**

**1.** (a) $\{x \mid x \in A$ and $x \in B\}$    (b) $\{x \mid x \in A$ or $x \in B\}$
(c) $\{(x, y) \mid x \in A$ and $y \in B\}$

**2.** (a) F    (b) F    (c) T    (d) T    (e) T
(f) F    (g) F    (h) F    (i) F    (j) T
(k) T    (l) F    (m) F    (n) F

**3.** (a) {5, 11}    (b) {1, 5, 7, 17, 19}
(c) {1, 3, 5, 7, 11, 17}    (d) {1, 5, 7, 13, 17, 19, 23}
(e) {3, 5, 7, 11}

**4.** $A \cap \overline{(B \cup C)}$ or $A - (B \cup C)$

**5.** (a) False; $(b, c)$ and $(c, d)$ are elements of the relation, but $(b, d)$ is not.
(b) False; {(1, 1), (2, 2), (1, 2)} is reflexive on the set {1, 2} but not symmetric.
(c) False; the set has 1,000,000 elements.
(d) True
(e) False; $A \times B \neq B \times A$ and $(A \times B) \times C \neq A \times (B \times C)$
(f) False; {1, 2} and {3, 4} are equivalent but not equal.
(g) True

**6.** {3, 6, 8, 10}

**7.** (a) Symmetric; no     (b) Transitive; no
    (c) Reflexive, symmetric, and transitive; yes

**8.** (a) 15    (b) 4    (c) 40
    (d) 40    (e) They are equal.

**9.** (a)

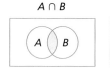 

(b)

(c) $\overline{A \cap B} = \overline{A} \cup \overline{B}$

**10.** (a) 8    (b) 2

**11.** {a}, {b}, {c}, {d}, {a, b}, {a, c}, {a, d}, {b, c}, {b, d},
{c, d}, {a, b, c}, {a, b, d}, {a, c, d}, {b, c, d}

**12.** (a) No    (b) Yes    (c) Yes

**13.**

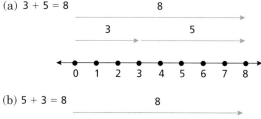

**14.** {(1, 3), (1, 6), (2, 3), (2, 6), (4, 3), (4, 6), (5, 3),
(5, 6)}

# CHAPTER 4

• • • • • • • • • • • • • • • • • • • • • • • • • • • • • • •

EXERCISE SET 1, P. 130

**1.** (a) True    (b) False    (c) False
    (d) True    (e) True

**2.** (a) Commutative (add.)    (c) Additive identity
    (e) Associative (add.)    (g) Commutative (add.)
    (i) Commutative (add.)

**3.** (a) 40 + (9 + 58) = 107
    (b) 50 + (6 + 78) = 134
    (c) 260 + (3 + 85) = 348

**4.** (a) 7    (c) 9

**5.** (a) 3 + 5 = 8

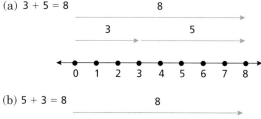

(b) 5 + 3 = 8

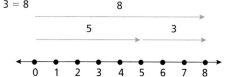

(c) The commutative property of addition

(d)

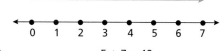

**6.** (a) Associative (add.)
    (c) Commutative (add.)

**7.** (a) 2    (b) No whole number answer
    (c) No whole number answer    (d) 5
    (e) 3    (f) No whole number answer

**8.** (a)

(c)

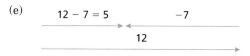

(e)

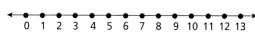

**9.** (a) True    (b) False    (c) False    (d) True
    (e) True    (f) True    (g) True    (h) False

**10.** (a) 21 − 16 = 5; 21 − 5 = 16
    (c) $f - a = c; f - c = a$

**11.** (a) 5 − 7 = ?    Whole numbers not closed for
                  subtraction
    (b) 5 − 3 ≠ 3 − 5    Subtraction is not commutative
    (c) 5 − (2 − 1) ≠ (5 − 2) − 1
         5 − 1 ≠ 3 − 1
             4 ≠ 2    Subtraction is not associative

**12.** (a) Yes    (c) No    (e) Yes    (g) Yes

**13.** (a) All whole numbers    (b) {5, 6, 7}
    (c) {8, 9, 10, . . .}        (d) {1, 2, 3, 4, . . .}
    (e) {2, 4}    (f) ∅

**14.** (a) {5}    (c) {3}    (e) ∅

**15.** (a) Set-subset model (girls are a subset of stu-
       dents)
    (c) Take-away model (3 pencils are taken from 11
       pencils)
    (e) Missing-addend model
       ($4 + x = 12$ or $x = 12 - 4$)
    (g) Comparison model (comparing Kendle's 11 let-
       ters to Kate's 3 letters)

**17.** $x + (x + 5) = 27$;  11 years of age

**19.** $x + 4 = 9$; 5 tickets

**21.** $x + (x + 5) + (x + 10) = 75$;  20, 25, 30

**23.** $c + 12 + 2 = 62$; $48

**24.** (a) $a \geq b$

**25.** (a) $8 + (5 + 2) = (8 + 5) + 2$    Associative (add.)
$\qquad\qquad\quad = 2 + (8 + 5)$    Commutative (add.)
$\qquad 8 + (5 + 2) = 2 + (8 + 5)$    Transitive (equality)
(b) $6 + (9 + 1) = 6 + (1 + 9)$    Commutative (add.)
$\qquad\qquad\quad = (6 + 1) + 9$    Associative (add.)
$\qquad\qquad\quad = 9 + (6 + 1)$    Commutative (add.)
$\qquad 6 + (9 + 1) = 9 + (6 + 1)$    Transitive (equality)
(c) $(a + b) + (c + d) = (b + a) + (d + c)$
$\qquad\qquad\qquad\qquad\qquad$    Commutative (add.)
$\qquad = [(b + a) + d] + c$    Associative (add.)
$\qquad = [b + (a + d)] + c$    Associative (add.)
$\qquad = [b + (d + a)] + c$    Commutative (add.)
$\qquad = [(b + d) + a] + c$    Associative (add.)
$\qquad = (b + d) + (a + c)$    Associative (add.)
$\qquad (a + b) + (c + d) = (b + d) + (a + c)$
$\qquad\qquad\qquad\qquad\qquad$    Transitive (equality)

**26.** (a) $a + (b + c) = a + (c + b)$    Commutative (add.)

**27.** (a) Let A $= \{l, m, n, o\}$, B $= \{p, q, r, s\}$ and
C $= \{t, u\}$.
$[A \cup (B \cup C)] = \{l, m, n, o\} \cup \{p, q, r, s, t, u\}$
$[A \cup (B \cup C)] = \{l, m, n, o, p, q, r, s, t, u\}$
$n[A \cup (B \cup C)] = 10$
$[(A \cup B) \cup C] = (\{l, m, n, o\} \cup \{p, q, r, s\}) \cup \{t, u\}$
$\qquad\qquad = \{l, m, n, o, p, q, r, s, t, u\}$
$n[(A \cup B) \cup C] = 10$
Thus, $n[(A \cup B)\cup C] = n[A \cup (B \cup C)]$ or
$n(A \cup B) + n(C) = n(A) + n(B \cup C)$; so
$(a + b) + c = a + (b + c)$.
(b) Let $A = \{l, m, n, o, p, q\}$ and $B = \{r, s, t, u\}$.
$A \cup B = \{l, m, n, o, p, q, r, s, t, u\} = B \cup A$.
Thus, $n(A \cup B) = n(B \cup A)$. Therefore,
$n(A) + n(B) = n(B) + n(A)$, or $6 + 4 = 4 + 6$.

**28.** (a) Yes

**29.** (a) No
(b) No
(c) No

**30.** (a) 6

EXERCISE SET 2, P. 144

**1.** (a) False    (b) True
(c) False    (d) True
(e) False    (f) False

**2.** (a) Commutative (mult.)
(c) Multiplicative property of 0
(e) Associative (mult.)

**3.** (a) $18 \cdot 5 = (9 \cdot 2) \cdot 5 = 9 \cdot (2 \cdot 5) = 9(10)$
(b) $24 \cdot 9 = (8 \cdot 3) \cdot 9 = 8 \cdot (3 \cdot 9) = 8(27)$
(c) $8 \cdot 36 = 8 \cdot (4 \cdot 9) = (8 \cdot 4) \cdot 9 = (32)9$
(d) $35 \cdot 14 = (7 \cdot 5) \cdot 14 = 7 \cdot (5 \cdot 14) = 7(70)$

**4.** (a) $(2 \cdot 3) + (2 \cdot 4)$
(c) $4(2 + 3)$
(e) $2x(2ay + z)$
(g) $(a + 5)x$
(i) $(2 + 3)(x + 4) = 5 (x + 4)$

**5.** (a) Commutative (add.)    (b) Associative (mult.)
(c) Commutative (mult.)
(d) Commutative (mult.)
(e) Additive identity   (f) Associative (mult.)
(g) Multiplicative identity and commutative (add.)
(h) Commutative (mult.)
(i) Distributive (mult. over add.)
(j) Distributive (mult. over add.)
(k) Commutative (add.)
(l) Commutative (mult.)
(m) Commutative (mult.)

**6.** (a) Measurement model (How many subsets of
24 each are in 480?)
(c) Missing factor model. ($10x = 100$ or
$x = 100/10$) or Measurement model
(How many subsets of 10 in 100?)

**7.** (a) $5(3 + 7 + 5 + 6) = 5(21) = 105$ or
$5 \cdot 3 + 5 \cdot 7 + 5 \cdot 5 + 5 \cdot 6 = 105$
(b) $11(5 + 8 + 10 + 4) = 11(27) = 297$ or
$11 \cdot 5 + 11 \cdot 8 + 11 \cdot 10 + 11 \cdot 4 = 297$

**8.** (a) 8

**9.** $24 \div 8 = 3$ (subtractions)

**10.** (a) $36 \div 9$

| 36 | 27 | 18 | 9 |
|----|----|----|----|
| $- 9$ | $- 9$ | $- 9$ | $-9$ |
| 27 | 18 | 9 | 0 |

$36 \div 9 = 4$ (subtractions)

**11.** (a) $6(8 - 5) = 6(3) = 18$
$6(8 - 5) = 6(8) - 6(5) = 18$
(b) $8(5 - 3) = 8(2) = 16$
$8(5 - 3) = 8(5) - 8(3) = 16$
(c) $17(10 - 4) = 17(6) = 102$
$17(10 - 4) = 17(10) - 17(4) = 102$

**12.** (a) False

**13.** (a) $4 - 3 \neq 3 - 4$. If $x = y$, this is true; $5 - 5 =$
$5 - 5$.
(b) $(5 - 3) - 1 \neq 5 - (3 - 1)$, as $1 \neq 3$. If $z = 0$,
$(5 - 3) - 0 = 5 - (3 - 0)$.
(c) $2 - 0 \neq 0 - 2$. If $x = 0$, $0 - 0 = 0 - 0 = 0$.
(d) $(2 + 3) \div 5 \neq 2 + 3 \div 5$. If $x = 0$, $(0 + 10) \div$
$2 = 0 + 10 \div 2$.

**14.** (a) $x = 2$

**15.** $x + 2x = 675$; $225, $450

**17.** $x + x + (x + 4) + (x + 4) = 48$
length is 14 cm; width is 10 cm.

**19.** $3x - x = 10$; $x = 5$      Perimeter $= 35$

**20.** (a) $(8 + 4) \div 2 = (8 \div 2) + (4 \div 2)$
$\qquad\qquad 12 \div 2 = 4 + 2$
$\qquad\qquad\qquad 6 = 6$

**21.** (a) This is true when $a = b$ and $a \neq 0$
   (b) 12   (c) 3
   (d) Division is not associative.

**22.** (a) 40   Check: $420 \boxed{\div} 3 \boxed{-} 50 \boxed{\times} 2 \boxed{=} 40$

**23.** (a) No, example $5 \cdot 4 = 20$.   Multiplicative identity is 1.
   (b) Yes.   Identity is 1.   (c) Yes.   No identity.
   (d) Yes.   Identity is 1.   (e) Yes.   No identity.
   (f) Yes.   Identity is 1.   (g) Yes.   Identity is 1.
   (h) Yes.   Identity is 1.

**24.** (a) No; $2 + (3 + 5) \neq (2 + 3) + (2 + 5)$, since $10 \neq 5 + 7$

**25.** (a) $4x + 12$   (b) $a^2 + 7a$
   (c) $x^2 + xa$  $(x \cdot x = x^2)$   (d) $4a + 12b + 8c$
   (e) $4y^2 + 4yx + 8yz$   (f) $xy + xz + 2x$
   (g) $12x - 6xy$   (h) $10abc + 20abd$

**26.** (a) $3 \cdot 10 + 3 \cdot 2 = 36$

**27.** (a) 248   (b) 441

**29.** (a)   (b)

   $2 \cdot 3 = 6$
   $4 \cdot 3 = 12$
   (c) No rectangle can be drawn for $2 \cdot 0 = 0$.

**31.** (a) $x(x + 2) + 3(x + 2) = x^2 + 5x + 6$
   (b) $a(a + 1) + 4(a + 1) = a^2 + 5a + 4$
   (c) $x(x + 3y) + 2y(x + 3y) = x^2 + 5xy + 6y^2$
   (d) $a(3a + b) + 2b(3a + b) = 3a^2 + 7ab + 2b^2$

**32.** (a) $8x + 16 = 8(x + 2)$   (one possibility)

**33.** (a) Yes   (b) Yes
   (c) Yes   (d) Yes.  It is 1.   **34.** (a) Yes

**35.** (a) $(2 \cdot 3) \cdot 4 = 4(2 \cdot 3)$   Commutative (mult.)
   $\qquad\quad = (4 \cdot 2) \cdot 3$   Associative (mult.)
   $(2 \cdot 3) \cdot 4 = (4 \cdot 2) \cdot 3$   Transitive (equality)
   (b) $(8 + 2) \cdot 7 = 7(8 + 2)$   Commutative (mult.)
   $\qquad\quad = 7(2 + 8)$   Commutative (add.)
   $(8 + 2) \cdot 7 = 7(2 + 8)$   Transitive (equality)
   (c) $(ab)c = (ba) \cdot c$   Commutative (mult.)
   $\qquad = b \cdot (ac)$   Associative (mult.)
   $\qquad = (ac) \cdot b$   Commutative (mult.)
   $\qquad = (ca) \cdot b$   Commutative (mult.)
   $(ab)c = (ca)b$   Transitive (equality)

**36.** (a) Associative property of addition

**EXERCISE SET 3, P. 155**

**1.** (a) $4(10,000) + 3(1,000) + 7(100) + 3(10) + 6$
   (b) $3(100,000) + 5(10,000) + 4(1,000) + 5(100)$
        $+ 5(10) + 5$
   (c) $1(10,000) + 4(1,000) + 1(100) + 3(10) + 6$
   (d) $3(100,000) + 1(10,000) + 5(1,000) + 1(100)$
        $+ 6(10) + 1$
   (e) $3(1,000,000) + 1(100,000) + 1(10,000)$
        $+ 1(1,000) + 0(100) + 0(10) + 5$
   (f) $2(10,000) + 0(1,000) + 0(100) + 0(10) + 4$

**2.** (a) 34,786   (c) 740

**3.** (a) Thousands   (b) Ones
   (c) Ones   (d) Millions

**4.** (a) Thousands   (c) Ten thousands

**5.** (a) 2   (b) 1

**6.** (a) LXXVI   (c) CLXXXIX   (e) CXLVIII

**7.** (a) 753   (b) 357

**9.** (a) 43,210   (b) 10,234

**10.** (a) 29   (c) 1776   (e) 10,000, 649

**11.** (a) 4   (b) 22   (c) 34
   (d) 1,001,020   (e) 200,320   (f) 23,000

**12.** (a) 2   (c) 11   (e) 4862

**13.** (a) 9   (b) 90   (c) 99
   (d) 900   (e) 990   (f) 999

**14.** (a) 6   (c) 7   (e) 9   (g) 843

**15.** (a) ∩|||| (b) ∩∩∩∩∩∩∩|||||
   (c) 𝟗∩∩∩∩∩∩∩∩|||| 
   (d) ⸙𝟗𝟗𝟗𝟗𝟗𝟗𝟗∩∩||||||||

**16.** (a) ⟨  ▼▼▼▼   (c) ▼▼▼   ▼▼▼▼

**17.** (a) ⁝⁝ (b) ⁝ (c) ⁝ (d) ⁝

**19.** (a) $200,320 > 36,601 > 2182$
   (b) $1092 > 81 > 34$

**20.** (a) MLXXXVIII; MXC

**21.** (a) Sum is 78   (b) Sum is 3752
   Difference is 50   Difference is 200
   (c) Sum is 874   (d) Sum is 365
   Difference is 490   Difference is 81

**23.** (a) This year  𝟗∩∩∩∩∩
   Next year  𝟗∩∩∩∩∩∩|||||
   (b) Barley  ▼▼  ⟨⟨⟨
   Corn  ▼  ⟨▼▼▼▼▼

**24.** (a) {2}   (c) {$x > 4$ and a whole number}
   (e) {0, 1, 2}

**25.** (a) Distributive property of multiplication over addition
   (b) Commutative property of multiplication
   (c) Multiplicative identity

**26.** (a) $(110)(9) - (9)(6) = 9(110) - 9(6)$
   $\qquad\qquad\qquad\qquad\quad$ Commutative (mult.)
   $\qquad\qquad = 9(110 - 6)$
   $\qquad\qquad\qquad\qquad\quad$ Distributive (mult. over sub.)
   $(110)(9) - (9)(6) = 9(110 - 6)$
   $\qquad\qquad\qquad\qquad\quad$ Transitive property of equality

**EXERCISE SET 4, P. 166**

**1.**

| | | |
|---|---|---|
| $10^9$ | 1,000,000,000 | billion |
| $10^8$ | 100,000,000 | hundred million |
| $10^7$ | 10,000,000 | ten million |
| $10^6$ | 1,000,000 | million |
| $10^5$ | 100,000 | hundred thousand |
| $10^4$ | 10,000 | ten thousand |
| $10^3$ | 1000 | thousand |

$10^2$          100     hundred
$10^1$           10     ten
$10^0$            1     one

**2.** (a) $2^5$     (c) $7^3 \cdot 5^2$     (e) $3^7$

**3.** (a) 81     (b) $x^{11}$     (c) 1     (d) 8

**4.** (a) $12 \cdot 10^7$

**5.** (a) 4,578    (b) 90,703    (c) 8,035
(d) 35,070    (e) 405    (f) 2
(g) 46,307

**6.** (a) 1, 2, 3, 4, 5, 6, 10, 11, 12, 13, 14, 15, 16, 20, 21, 22, 23, 24, 25, 26 (base seven)

**7.** (a)

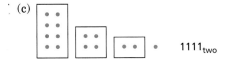

$15_{\text{ten}}$

(b)

$21_{\text{seven}}$

(c)

$1111_{\text{two}}$

**8.** (a) $15_{\text{seven}}, 20_{\text{seven}}$     (c) $110_{\text{two}}, 1000_{\text{two}}$

**9.** (a) 295    (b) 336    (c) 14    (d) 45

**10.** (a) $12_{\text{seven}}$     (c) $106_{\text{seven}}$     (e) $2626_{\text{seven}}$

**11.** (a) $1001_{\text{two}}$    (b) $100011_{\text{two}}$    (c) $110111_{\text{two}}$
(d) $100011101_{\text{two}}$     (e) $1111101000_{\text{two}}$
(f) $1010010100000_{\text{two}}$

**13.** (a) 0    (b) 4    (c) 5    (d) 0

**15.** 1, 2, 3, 4, 10, 11, 12, 13, 14, 20, 21, 22, 23, 24, 30, 31, 32, 33, 34, 40, 41, 42, 43, 44, 100, 101, 102, 103, 104, 110, 111, 112, 113, 114, 120, 121, 122, 123, 124, 130, 131, 132, 133, 134, 140, 141, 142, 143, 144, 200 (base five)

**16.** (a) $EET_{\text{twelve}}$ $1000_{\text{twelve}}$     (c) $221_{\text{three}}$ $1000_{\text{three}}$

**17.** (a) 133    (b) 138    (c) 184
(d) 132    (e) 280    (f) 17414

**18.** (a) $9_{\text{twelve}}$     (c) $47_{\text{twelve}}$     (e) $6E4_{\text{twelve}}$

**19.** (a) $14_{\text{five}}$    (b) $120_{\text{five}}$    (c) $210_{\text{five}}$
(d) $2120_{\text{five}}$    (e) $13000_{\text{five}}$    (f) $132110_{\text{five}}$

**20.** (a) $101_{\text{three}}$

**21.** (a) $2^{15}$     Check: 2 [∧] 3 [×] 8 [∧] 4 [=] 32,768
and 2 [∧] 15 [=] 32,768

(c) $2^{10}$     Check: 16 [$x^2$] [×] 2 [$x^2$] [=] 1024
and 2 [∧] 10 [=] 1024

**22.** (a) 4

**23.** (a) $4^0, 10^2, 7^3, 2^{10}, 3^7, 9^4, 4^9$
Check: 1, 100, 343, 1024, 2187, 6561, 262,144
(c) $21^2 + 22^2 + 23^2 + 24^2 = 25^2 + 26^2 + 27^2$
(e) 5; 6; 9     Check: 5 [∧] 9 ends in 5; 2 [∧] 8
ends in 6; 3 [∧] 6 ends in 9.

**24.** (a) Base eleven

**25.** 1241 erasers

**27.** (a) $333_{\text{four}}$; $888_{\text{nine}}$; $EEE_{\text{twelve}}$
(b) $1000_{\text{four}}$; $1000_{\text{nine}}$; $1000_{\text{twelve}}$;
(c) 63; 728; 1727; 64; 729; 1728

**28.** (a) $39_{\text{twelve}}$

**29.** Yes, commutative property of multiplication

**30.** (a) $5(10)^2 + 13(10)$
$= 5(10)^2 + (10 + 3)10$  Renaming 13
$= 5(10)^2 + [10(10) + 3(10)]$  Distributive (mult. over add.)
$= [5(10)^2 + 1(10)^2] + 3(10)$  Associative (add.) and renaming 100
$= (5 + 1)(10)^2 + 3(10)$  Distributive (mult. over add.)
$= 6(10)^2 + 3(10)$  Addition
$5(10)^2 + 13(10)$
$= 6(10)^2 + 3(10)$  Transitive (equality)

**31.** Yes; no; yes; no     **33.** $242_{\text{five}}$

**35.** (a) 𝍩||||||||
(b) ▼  ⟨⟨⟨⟨▼▼▼▼▼▼▼
(c) ⚏     (d) CIX

**36.** (a) 810     (c) 718

**EXERCISE SET 5, P. 178**

**1.** (a) 474    (b) 578    (c) 712    (d) 547

**2.** (a)   700 +  60 +  8
           500 +  70 +  4
          ―――――――――――――――
          1200 + 130 + 12 = 1342

**3.** 31,000; 31,471 rounds to 31,000.

**4.** (a) 300 + 70 + 10 = 380

**5.** (a) 23          (b) 276

**6.** (a)   8   6   4   2
            7   5   3   1
           ――――――――――――――
           16   1   7   3

**7.** (a) $8765 - 1234 = 7531$
(b) $5123 - 4876 = 247$

**8.** (a)    40 +  6
             70 +  5
           ―――――――――――――
            110 + 11 = 121

| $10^2$ | $10$ | $1$ | |
|---|---|---|---|
| | | | 1 |
| | 4 | 6 | 46 |
| + | 7 | 5 | + 75 |
| | | | ――― |
| | 11 | 1̶1̶ | 121 |
| | 1 | 1 | |
| 1 | 2 | 1 | |

(c)
$$60 + 0$$
$$-[30 + 7]$$
or
$$50 + 10$$
$$-[30 + \ \ 7]$$
$$20 + \ 3 = 23$$

| 10 | 1 |
|----|---|
| 6  | 0 |
| − 3 | 7 |

or

| 10 | 1 |
|----|---|
| 5  | 10 |
| − 3 | 7 |
| 2  | 3 |

$$\overset{5 \ 10}{\cancel{6}\cancel{0}}$$
$$- \ 37$$
$$\overline{23}$$

**9.** (a) $1001_{two}$    (b) $10010_{two}$
   (c) $223_{seven}$    (d) $306_{seven}$

**10.** (a) $1003_{seven}$    (b) $10100_{two}$

**11.** (a) 10 hrs., 32 min., 8 sec. and 2 hrs., 48 min., 54 sec.
   (b) 14 hrs., 10 min., 15 sec. and 1 hr., 49 min., 45 sec.

**12.** (a)

| + | 0 | 1 | 2 |
|---|---|---|---|
| 0 | 0 | 1 | 2 |
| 1 | 1 | 2 | 10 |
| 2 | 2 | 10 | 11 |

(c)

| + | 0 | 1 | 2 | 3 | 4 |
|---|---|---|---|---|---|
| 0 | 0 | 1 | 2 | 3 | 4 |
| 1 | 1 | 2 | 3 | 4 | 10 |
| 2 | 2 | 3 | 4 | 10 | 11 |
| 3 | 3 | 4 | 10 | 11 | 12 |
| 4 | 4 | 10 | 11 | 12 | 13 |

**13.** (a) $4T_{twelve}$    (b) $2430_{five}$    (c) $5_{twelve}$
   (d) $220_{three}$    (e) $E51_{twelve}$    (f) $123_{five}$
   (g) $214_{twelve}$    (h) $616_{twelve}$    (i) $222_{five}$
   (j) $7T6_{twelve}$

**14.** (a)

```
                  ┌ - - - - - → 1
   30  +  4  │  34      34
   20  +  7  │  27      27
   50  + 11 -┘→ 11 - ┐   61
   └ - - - - → 50      ↑
                 ──     │
                 61 - - ┘
```

**15.** (a) $(5000 + 700 + 6) - (3000 + 400 + 7)$
     $(5000 + 700 + 16) - (3000 + 400 + 10 + 7)$
     $(5000 + 700 + 100 + 16) - (3000 + 500$
         $+ \ 10 + 7)$
     $(2000 + 200 + 90 + 9) = 2299$

(b) $(300 + 20 + 9) - (100 + 40 + 6)$
     $(300 + 120 + 9) - (200 + 40 + 6)$
     $(100 + 80 + 3) = 183$

(c) $(1000 + 600 + 30 + 4) - (900 + 80 + 5)$
     $(1000 + 600 + 30 + 14) - (900 + 90 + 5)$
     $(1600 + 130 + 14) - (1000 + 90 + 5)$
     $(600 + 40 + 9) = 649$

**16.** (a)
$$6299$$
$$-4000$$
$$\overline{2299}$$

**17.** (a)
$$\overset{16}{5706}$$
$$- \ 3407$$
$$\overline{2\cancel{30}}$$
$$229$$
$$\overline{2299}$$

(b)
$$\overset{12}{329}$$
$$-146$$
$$\overline{\cancel{2}}$$
$$183$$

(c)
$$\overset{13\,14}{1634}$$
$$- \ 985$$
$$\overline{\cancel{7}}$$
$$6\cancel{5}$$
$$\overline{649}$$

**18.** (a) (i)
$$364 \qquad 300 + 60 + 4$$
$$+ \ 423 \ = \ 400 + 20 + 3$$
$$\overline{\phantom{xx} 700 + 80 + 7 = 787}$$

   (ii)

| $10^2$ | 10 | 1 |
|--------|----|---|
| 3 | 6 | 4 |
| + 4 | 2 | 3 |
| 7 | 8 | 7 |

   (iii)
$$364$$
$$+ \ 423$$
$$\overline{787}$$

**19.** (a) In base 2 there are no digits 8 or 9. The answer should read $10110011_{two}$.
   (b) It should read $127_{nine}$.
   (c) It should read $604_{twelve}$.
   (d) The answer is $1162_{seven}$.

**20.** (a) Base eight

**21.** Any base larger than 7

**22.** (a) $301_{seven}$

**23.** (a) 23    (b) 35    (c) 245

**24.** (a) $65_{seven}$; $101111_{two}$

**25.** (a) 276    (b) 13

**26.** (a) Units digit

**27.** (a) 30
   (b) 4 less
   (c) 100

**28.** (a) $k = 3$

EXERCISE SET 6, P. 190

**1.** (a)

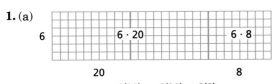

$$6(28) = 6(20) + 6(8)$$

(b)

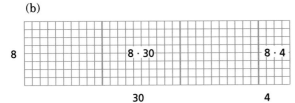

$$8(34) = 8(30) + 8(4)$$

(c)

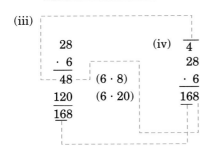

$$3(26) = 3(20) + 3(6)$$

**2-3.** (a) (i) $6 \cdot 28 = 6(20 + 8) = 120 + 48 = 168$

(ii)

| $10^2$ | $10$ | $1$ |
|---|---|---|
|  | 2 | 8 |
| + | 2 | 8 |
| + | 2 | 8 |
| + | 2 | 8 |
| + | 2 | 8 |
| + | 2 | 8 |
|  | 12 | 48 |
| 1 | 4 |  |
| 1 | 6 | 8 |

(iii)

$$
\begin{array}{r}
28 \\
\cdot\ 6 \\
\hline
48 \quad (6 \cdot 8) \\
120 \quad (6 \cdot 20) \\
\hline
168
\end{array}
$$

(iv)

$$
\begin{array}{r}
4 \\
28 \\
\cdot\ 6 \\
\hline
168
\end{array}
$$

(c) (i) $3 \cdot 26 = 3(20 + 6) = 60 + 18 = 78$

(ii)

| $10^2$ | $10$ | $1$ |
|---|---|---|
|  | 2 | 6 |
| + | 2 | 6 |
| + | 2 | 6 |
|  | 6 | 18 |
|  | 1 |  |
|  | 7 | 8 |

(iii)

$$
\begin{array}{r}
26 \\
\cdot\ 3 \\
\hline
18 \quad (3 \cdot 6) \\
60 \quad (3 \cdot 20) \\
\hline
78
\end{array}
$$

(iv)

$$
\begin{array}{r}
1 \\
26 \\
\cdot\ 3 \\
\hline
78
\end{array}
$$

**4.** (a) $q = 6, r = 6$    Check: 72 [INT÷] 11 [=] 6, 6

(c) $q = 0, r = 11$    Check: 11 [INT÷] 18 [=] 0, 11

**5.-6.** (a)

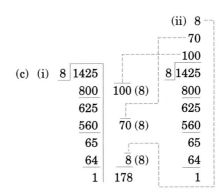

(i)
$$
\begin{array}{r}
6\ \overline{)166} \\
120 \\
\hline
46 \\
42 \\
\hline
4
\end{array}
\quad
\begin{array}{l}
20\ (6) \\
\\
7\ (6) \\
27
\end{array}
$$

(ii)
$$
\begin{array}{r}
7 \\
20 \\
6\ \overline{)166} \\
120 \\
\hline
46 \\
42 \\
\hline
4
\end{array}
$$

(iii)
$$
\begin{array}{r}
27 \\
6\ \overline{)166} \\
12 \\
\hline
46 \\
42 \\
\hline
4
\end{array}
$$

(c) (i)
$$
\begin{array}{r}
8\ \overline{)1425} \\
800 \\
\hline
625 \\
560 \\
\hline
65 \\
64 \\
\hline
1
\end{array}
\quad
\begin{array}{l}
100\ (8) \\
\\
70\ (8) \\
\\
8\ (8) \\
178
\end{array}
$$

(ii)
$$
\begin{array}{r}
8 \\
70 \\
100 \\
8\ \overline{)1425} \\
800 \\
\hline
625 \\
560 \\
\hline
65 \\
64 \\
\hline
1
\end{array}
$$

(iii)
$$
\begin{array}{r}
178 \\
8\ \overline{)1425} \\
8 \\
\hline
62 \\
56 \\
\hline
65 \\
64 \\
\hline
1
\end{array}
$$

**7.** (a)
$$
\begin{array}{r}
47 \\
\cdot\ 84 \\
\hline
188 \\
376 \\
\hline
3948
\end{array}
$$

(b)
$$
\begin{array}{r}
246 \\
\cdot\ 578 \\
\hline
1968 \\
1722 \\
1230 \\
\hline
142188
\end{array}
$$

**8-9.** (a) (i) $26(30 + 2) = 26 \cdot 30 + 26 \cdot 2$
$= (20 + 6)30 + (20 + 6)2$
$= 600 + 180 + 40 + 12 = 832$

(iii)
```
        32
       · 26
      ─────
        12  ┐ (6 · 2)
       180 ─┘ (6 · 30)
        40  ┐ (20 · 2)
       600 ─┘ (20 · 30)
      ─────
       832
```

(iv)
```
          1
        32
       · 26
      ─────
       192
        64
      ─────
       832
```

(c) (i) $74(90 + 2) = 74 \cdot 90 + 74 \cdot 2$
$= (70 + 4)90 + (70 + 4)2$
$= 6300 + 360 + 140 + 8 = 6808$

(iii)
```
        92
       · 74
      ─────
         8  ┐ (4 · 2)
       360 ─┘ (4 · 90)
       140  ┐ (70 · 2)
      6300 ─┘ (70 · 90)
      ─────
      6808
```

(iv)
```
        92
       · 74
      ─────
       368
       644
      ─────
      6808
```

**10.** (a) $8 \cdot 60 + 8 \cdot 3 = 480 + 24 = 504$

**11.** (a) $10 \cdot 84 = 840$   (b) $10 \cdot 138 = 1380$

**12.** (a) $(240 \div 3) + (9 \div 3) = 80 + 3 = 83$

**13.** Estimate: 64,000,000   Answer: 62,669,376

**15.** Estimate: 3000   Answer: 3184

**17.** (a) $100111_{two}$   (b) $20133_{seven}$
(c) $130366_{seven}$   (d) $1000001_{two}$

**18.** (a)

| · | 0 | 1 | 2 |
|---|---|---|---|
| 0 | 0 | 0 | 0 |
| 1 | 0 | 1 | 2 |
| 2 | 0 | 2 | 11 |

(c)

| · | 0 | 1 | 2 | 3 | 4 |
|---|---|---|---|---|---|
| 0 | 0 | 0 | 0 | 0 | 0 |
| 1 | 0 | 1 | 2 | 3 | 4 |
| 2 | 0 | 2 | 4 | 11 | 13 |
| 3 | 0 | 3 | 11 | 14 | 22 |
| 4 | 0 | 4 | 13 | 22 | 31 |

**19.** (a) $341_{five}$   (b) $TT_{twelve}$   (c) $31044_{five}$
(d) $13762_{twelve}$   (e) $120022_{three}$   (f) $212201_{three}$

**20.** (a) largest is $444_{five}$   (c) largest is $222_{three}$
smallest is $100_{five}$   smallest is $100_{three}$
difference is $344_{five}$   difference is $122_{three}$

**21.** (a) $1001_{two}$ R. $10_{two}$
(b) $20_{seven}$   (c) $16_{seven}$   R. $31_{seven}$

**22.** (a) 400

**23.**
```
    586
   · 88
  ─────
   4688
  4688
  ──────
 51,568
```

**25.** 121; 12,321; 1,234,321; 123,454,321

**26.** (a) (i)
```
 32 │1728
    │1600   50(32)
    │────
    │ 128
    │ 128   4(32)
    │────
    │  54
```

```
      4│54
     50│
```

(ii)
```
 32 │1728
    │1600
    │────
    │ 128
    │ 128
```

(iii)
```
      54
 32 │1728
    │ 160
    │────
    │ 128
    │ 128
```

**27.** (a)

34(176) = 5984

(b)

56(742) = 41,552

**28.** (a)

| Halving | Doubling |
|---------|----------|
| ~~14~~ | ~~36~~ |
| 7 | 72 |
| 3 | 144 |
| 1 | 288 |
|   | 504 |

**29.**
```
 23 │2473
    │−2300   100 · 23  (subtracting 23 one hundred
    │────                 times)
    │  173
    │ −161     7 · 23  (subtracting 23 seven times)
    │────     ─────
    │   12    107      107 rem 12; Yes. 107 · 23 +
    │                  12 = 2473
```

**30.** (a) Not exact; division is exact if there are no digits to the right of the decimal point. The remainder is 4.

**31.** (a) Multiplicative identity
(b) Commutative (mult.)
(c) Multiplicative property of zero
(d) Associative (mult.)
(e) Distributive (mult. over add.)

**32.** (a) (i)
$$400 + 30 + 6$$
$$\underline{200 + 40 + 3}$$
$$600 + 70 + 9 = 679$$

(ii)

| | $10^3$ | $10^2$ | $10$ | $1$ |
|---|---|---|---|---|
| | | 4 | 3 | 6 |
| | | 2 | 4 | 3 |
| + | | | | |
| | | 6 | 7 | 9 |

(iii)
$$436$$
$$\underline{+\ 243}$$
$$679$$

**CHAPTER TEST, P. 193**

**1.** (a) 1,000,212    (b) 4212

**2.** (a) $2^6$   (b) $2^{12}$   (c) $(xy)^5$   (d) $3^9$

**3.** (a) $2(10)^6 + 4(10)^3 + 6$   (b) 700   (c) 20,340

**4.** (a)
$$20 + \ 4$$
$$\underline{30 + \ 7}$$
$$50 + 11 = 61$$

(b)

| | $10$ | $1$ | |
|---|---|---|---|
| | | | 1 |
| 2 | | 4 | 24 |
| 3 | | 7 | 37 |
| + | | | 61 |
| 5 | | 11 | |
| 6 | | 1 | |

(c)
$$24$$
$$37$$
$$61$$

**5.**

24

|  | **20** | **4** |
|---|---|---|
| **10** | $10 \cdot 20$ | $10 \cdot 4$ |
| **16** **6** | $6 \cdot 20$ | $6 \cdot 4$ |

$$16 \cdot 24 = 10 \cdot 20 + 10 \cdot 4 + 6 \cdot 20 + 6 \cdot 4$$

**6.**
```
                    ┌----------------------- 2
  ┌ 24                                        24
  │ ·16                                       ·16
  └ 24 ┐      6(4)            ┌---------------144
120 ┘        6(20)           ┌----------------240
                                               ────
  ┌ 40       10(4)                             384
  └ 200      10(20)
     ────
     384
```

**7.** $101100_{\text{two}}$   **8.** (a) $100010100_{\text{two}}$   (b) $543_{\text{seven}}$

**9.** $q = 5, r = 2$

**10.** (a) Associative (add.)
(b) Distributive (mult. over add.)
(c) Multiplicative identity
(d) Distributive (mult. over add.)

**11.** (a) 0   (b) 0   (c) Not 0   (d) 0   (e) 0
(f) Not 0   (g) Not 0

**12.** (a) True   (b) False   (c) True   (d) False
(e) False   (f) True   (g) True   (h) False

**13.** $\{0, 1, 2, 3, 4\}$

**14.**
$$\begin{array}{ccccc} 20 & 16 & 12 & 8 & 4 \\ \underline{-\ 4} & \underline{-\ 4} & \underline{-\ 4} & \underline{-4} & \underline{-4} \\ 16 & 12 & 8 & 4 & 0 \end{array}$$

Subtraction of five 4's;
$20 \div 4 = 5$

**15.** Let $x$ be the distance that Lisa runs. John runs
$6 - 2x$. $x + (6 - 2x) = 4$

**16.** $x = 2; 6 - 2x = 2$

**17.** (a) $12 - 8 = 4$ because $4 + 8 = 12$
(b) $15 \div 3 = 5$ because $5 \cdot 3 = 15$

**18.** (a)

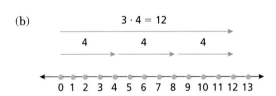

(b)

(c)

**19.** (a) $1000010_{\text{two}}$   (b) $221_{\text{seven}}$
(c) $1001101_{\text{two}}$   (d) $245_{\text{seven}}$

**20.** (a) 202   (b) 17

**21.** (a)
$$\begin{array}{l} \ \ \ 2000 + 400 + 30 + 6 \\ -\ \ \ \ \ \ \ \ [500 + 60 + 8] \\ \hline 1000 + 1300 + 120 + 16 \end{array}$$
or
$$\begin{array}{l} -\ \ \ \ \ \ \ [\ 500 + \ 60 + \ 8] \\ \hline 1000 + \ 800 + \ 60 + \ 8 = 1868 \end{array}$$

(b)

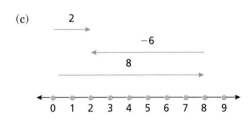

(c)
$$\begin{array}{r} {\scriptstyle 1\ 13\ 12\ 16} \\ \cancel{2}\ \cancel{4}\ \cancel{3}\ 6 \\ \underline{5\ 6\ 8} \\ 1\ 8\ 6\ 8 \end{array}$$

(b)

**22.** (a)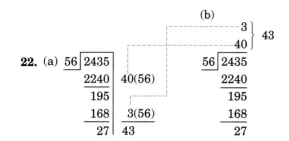

(c) $56\overline{\smash{)}2435}$
$\underline{224}$
$195$     Answer: 43 R.27
$\underline{168}$
$27$

**23.** (a) $21132_{six}$     (b) $5321_{twelve}$

**24.** (a) Base seven    (b) Base seven

**25.** $(6 + 2)5 = 6 \cdot 5 + 2 \cdot 5$   Distributive (mult. over add.)
             $= 2 \cdot 5 + 6 \cdot 5$   Commutative (add.)
             $= 2 \cdot 5 + 5 \cdot 6$   Commutative (mult.)
   $(6 + 2)5 = 2 \cdot 5 + 5 \cdot 6$   Transitive property of equality

## CHAPTER 5

• • • • • • • • • • • • • • • • • • • • • • • • • • • • • • • • •

EXERCISE SET 1, P. 208

**1.** (a) $^-5$     (b) 3     (c) 0
   (d) 8     (e) $^-a$     (f) $a$

**2.** (a)

(c)

**3.** (a)

(b)

(c)

**4.** (a) Yes     (c) No     (e) Yes     (g) No

**5.** (a) $^-3$    Check: $2\boxed{-}5\boxed{=}{}^-3$
   (e) $^-13$   Check: $\boxed{(-)}10\boxed{+}\boxed{(-)}3\boxed{=}{}^-13$
   (k) 4     Check: $\boxed{(}6\boxed{-}4\boxed{)}\boxed{-}\boxed{(-)}2\boxed{=}4$
   (o) $a$

**6.** (a)

(c)

(e)

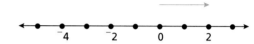

**7.** (a) No: one counterexample
      $(5 - 3) - 2 = 0$
      $5 - (3 - 2) = 4$
            $0 \neq 4$
   (b) No: one counterexample
      $5 - 3 \neq 3 - 5$
        $2 \neq {}^-2$

**8.** (a)   $x = 5$

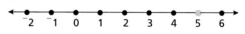

(c)   $x = 3$

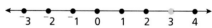

**9.** (a) $^-6$     Check: $\boxed{(-)}3\boxed{+}4\boxed{+}7\boxed{+}\boxed{(-)}$
                   $6\boxed{+}\boxed{(-)}8\boxed{=}{}^-6$

**11.** $^-8°C$

**13.** $188

**14.** (a) Yes, 5 yards more

**15.** (a) $^-10,500$ feet     (b) $^-16$ yards
   (c) 3500 feet

**16.** (a) False

**17.** (a) $^-150$ meters     (b) Aristotle
   (c) $^-12°C$         (d) (a)

**18.** (a) 7

**19.** (a) 26      (b) $a - b$
   (c) $c$       (d) $^-|x|$
   (e) 5       (f) $^-|x - y|$

**20.** (a) 237

**21.** (a) 0     (b) 1     (c) There is none.
   (d) There is none.     (e) $^-1$     (f) No

**22.** (a) ⁻10*x*

**23.** (a) ∅    (b) ∅
(c) The set of integers except 0
(d) I    (e) I⁺    (f) W

**24.** (a)

**25.** (a)    ⁻2 + ⁻4 = ⁻6

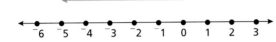

(b)    ⁻3 + 2 = ⁻1

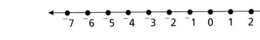

**27.** 6 + ⁻2 = (4 + 2) + ⁻2     Rewriting 6
= 4 + (2 + ⁻2)     Associative (add.)
= 4 + 0     Additive inverses
= 4     Additive identity
= 6 − 2     Rewrite 4
6 + ⁻2 = 6 − 2     Transitive (equality)

**29.** (a) (8 + 4) + ⁻2 = ⁻2 + (8 + 4)     Commutative (add.)
= (⁻2 + 8) + 4     Associative (add.)
= (8 + ⁻2) + 4     Commutative (add.)
(8 + 4) + ⁻2 = (8 + ⁻2) + 4     Transitive property of equality
(b) (6 + ⁻6) + 0 = 0 + (6 + ⁻6)     Commutative (add.)
= (0 + 6) + ⁻6     Associative (add.)
(6 + ⁻6) + 0 = (0 + 6) + ⁻6     Transitive property of equality
(c) (⁻4 + 6) + ⁻2 = ⁻4 + (6 + ⁻2)     Associative (add.)
= (6 + ⁻2) + ⁻4     Commutative (add.)
= (⁻2 + 6) + ⁻4     Commutative (add.)
(⁻4 + 6) + ⁻2 = (⁻2 + 6) + ⁻4     Transitive property of equality

### EXERCISE SET 2, P. 215

**1.** (a) ⁻45     (b) 28     (c) 0     (d) 8
(e) ⁻322     (f) ⁻210     (g) 30     (h) ⁻12
(i) 2     (j) 6     (k) ⁻135     (l) 24

**2.** (a) 72     Check: ⟦(−)⟧ 3 ⟦×⟧ ⟦(⟧ ⟦(−)⟧ 4 ⟦×⟧ 6 ⟦)⟧ ⟦=⟧ 72

(e) ⁻3     Check: ⟦(⟧ 23 ⟦−⟧ 5 ⟦)⟧ ⟦÷⟧ ⟦(−)⟧ 6 ⟦=⟧ ⁻3

(i) 24     Check: ⟦(⟧ ⟦(−)⟧ 4 ⟦+⟧ ⟦(−)⟧ 8 ⟦)⟧ ⟦×⟧ ⟦(⟧ 5 ⟦+⟧ ⟦(−)⟧ 7 ⟦)⟧ ⟦=⟧ 24

(k) ⁻*xyz*

**3.** (a) ⁻1(3 + ⁻5) = ⁻1(⁻2) = 2 or
⁻1(3 + ⁻5) = (⁻1)(3) + (⁻1)(⁻5) = ⁻3 + 5 = 2
(b) (3 + ⁻2)4 = 1(4) = 4 or
(3 + ⁻2)4 = (3)(4) + (⁻2)(4) = 12 + ⁻8 = 4

**4.** (a) ⁻5*a* + ⁻5*b* + ⁻5*c*     (c) ⁻3*y* + 2*yx* + ⁻4*yz*

**5.** (a) If both are positive or both negative
(b) If one or both of the factors is 0
(c) If one is positive and the other is negative

**6.** (a) |*a*| must be equal to |*b*|; *a* ≠ 0.     (c) 3

**7.** (a) *x* = 1     (b) *x* = 1     (c) *x* = ⁻4     (d) *x* = 3

**9.** Let *x* be the smallest integer.
*x* + (*x* + 1) + (*x* + 2) = ⁻75; *x* = ⁻26
The integers are ⁻26, ⁻25, and ⁻24.

**11.** Let *x* be the number of months Dalley Doughnut Shop has been in operation.
12,000 + 900*x* = 16,500; *x* = 5 months

**13.** Let *x* be the temperature 5 days ago.
*x* − 4 · 5 = ⁻45; *x* = ⁻25°C

**14.** (a) Let *x* be the temperature 2 days ago.
*x* − 4 · 2 = ⁻45; *x* = ⁻37°C

**15.** ⁻40° − 6 (d)°

**17.** Let *t* be the time in days until the temperature is 88°F.
94 − 2 · *t* = 88; *t* = 3 days

**19.** (a) ⁻29     (b) ⁻4     (c) 81     (d) ⁻512
(e) (a) and (b) are arithmetic; (c) and (d) are geometric.

**20.** (a) $\dfrac{^-1}{x^3}$

**21.** (a)

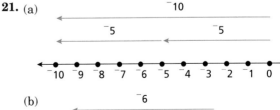

(b)

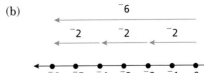

**22.** (a) All

**23.** *x* = ⁻*y* when *x* and *y* are additive inverses.

**24.** (a) Operation: addition

| Property | (Yes or No) |
|---|---|
| Closure | Yes |
| Commutativity | Yes |
| Associativity | Yes |
| Identity | Yes |
| Inverse | Yes |
| Distributive over addition | No |
| Distributive over subtraction | No |

**25.** (a) Positive  (b) Positive  (c) Negative
(d) Negative  (e) Positive  (f) Negative

**26.** (a) ⁻2  Check: [ [(−) 72 − [(−) 64 ]
÷ [ [(−) 56 + 60 ] = ⁻2

**27.** (a) (i)  Does not exist
(ii)  Does not exist
(iii) ⁻1
(b) (i)  Smallest whole number is 0
(ii)  Smallest integer does not exist
(iii) Smallest negative integer does not exist
(iv) Smallest positive integer is 1

**29.** (a) ⁻4  (b) 12

**30.** (a) 66

**31.** (a) $x^2 - 4$  (b) $9x^2 - 25$
(c) $4x^2 - y^2$  (d) $9x^2 - 16y^2$

**32.** (a) $(x - 3)(x + 3)$

**33.** (a) $\{x \mid x < {}^-3 \text{ and is an integer}\}$
(b) $\{x \mid x < {}^-3 \text{ and is an integer}\}$
(c) $\{x \mid x > {}^-1 \text{ and is an integer}\}$
(d) $\{x \mid x > 1 \text{ and is an integer}\}$

**34.** (a) ⁻3 < x < 4 and an integer

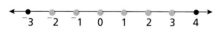

**35.** (a) 25  Check: 36 − [ [(−) 7 ] − 18 = 25

**36.** (a)

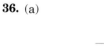

EXERCISE SET 3, P. 228

**1.** (a) 5, 7, 35  (b) 2, 3, 4
(c) 7, 11, 13  (d) 1, 13, 23

**2.** (a) False; $9 \cdot 5 = 45$  (c) False; $8 = 2 \cdot 4$
(e) False; $8 \mid 24$  (g) False; $6 \mid 12$  (i) True

**3.** (a) 4 yes, 8 no  (b) 4 yes, 8 yes
(c) 4 yes, 8 no

**4.** (a) Yes  (c) Yes

**5.** (a) Yes  (b) Yes  (c) Yes

**6.** (a) By 5 and 10  (c) By neither

**7.** (a) $2^4 \cdot 3^2$  (b) $2^4 \cdot 37$
(c) $2^2 \cdot 3^2 \cdot 17$  (d) $2 \cdot 3^4$

**8.** (a) Prime  (c) Not a prime
(e) Not a prime  (g) Not a prime

**9.** (a) 9  (b) 14  (c) 22
(d) 23  (e) 13  (f) 36

**10.** (a) $54 = 2 \cdot 3 \cdot 3 \cdot 3$  (c) $120 = 2 \cdot 2 \cdot 2 \cdot 3 \cdot 5$
(e) $141 = 3 \cdot 47$  (g) $144 = 2 \cdot 2 \cdot 2 \cdot 2 \cdot 3 \cdot 3$
(i) $256 = 2 \cdot 2 \cdot 2 \cdot 2 \cdot 2 \cdot 2 \cdot 2 \cdot 2$

**11.** (a) 15; 25; 35  (b) 33; 73; 113

**12.** (a) $1512 = 2 \cdot 2 \cdot 2 \cdot 3 \cdot 3 \cdot 3 \cdot 7$

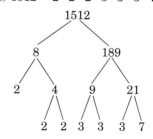

(c) $1836 = 2 \cdot 2 \cdot 3 \cdot 3 \cdot 3 \cdot 17$

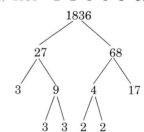

**13.** (a) 2)1512  (b) 2)810
2)756  5)405
2)378  3)81
3)189  3)27
3)63  3)9
3)21  3)3
7) 7  1
1
$1512 = 2^3 \cdot 3^3 \cdot 7$  $810 = 2 \cdot 5 \cdot 3^4$
(c) 2)1836
2)918
3)459
3)153  $1836 = 2^2 \cdot 3^3 \cdot 17$
3)51
17)17
1

**14.** (a) 23

**15.** (a) A  (b) D  (c) P  (d) A  (e) D
(f) A  (g) D  (h) A  (i) P

**17.** The answer you get is not an integer because it contains non-zero digits to the right of the decimal point.

**18.** (a) 15 rows

**19.** (a) Yes     (b) Yes     (c) Yes     (d) No

**21.** (a) 7, 8     Check: 6944 [INT÷] 7 [=] has a remainder of 0. 6944 [INT÷] 8 [=] has a remainder of 0. But each 6944 [INT÷] 5 [=]; 6944 [INT÷] 6 [=]; and 6944 [INT÷] 9 [=] provide a non-zero remainder

  (i) 8     Check: 482,144 [INT÷] 8 [=] has a remainder of 0, and 5, 6, 7, 9, and 11 provide non-zero remainders.

  (k) 11     Check: 23,606 [INT÷] 11 [=] has a remainder of 0 and 5, 6, 7, 8, and 9 provide non-zero remainders.

**22.** (a) Yes; $2^3 | 2^4$, $3^2 | 3^9$, and $5^7 | 5^7$

**23.** (a) Yes     (b) Yes     (c) Yes     (d) Yes
  (e) Yes     (f) No     (g) No     (h) Yes

**24.** (a) $4(2) + 1 = 9$, a composite number

**25.** One counterexample is given for each.
  (a) False   $3 | (2 + 4)$ but $3 \nmid 2$
  (b) False   $5 | 10$ and $5 | 25$ but $15 \nmid 25$
  (c) False   $6 \nmid 10$ and $6 \nmid 15$ but $6 | 150$     (d) True
  (e) False   $6 | (3 \cdot 10)$ but $6 \nmid 3$ and $6 \nmid 10$
  (f) True
  (g) False   $2 \nmid 5$ and $2 \nmid 7$ but $2 | (5 + 7)$     (h) True

**26.** (a) $a | a$   $a \neq 0$

**27.** No, 53 is not divisible by any number except 1.

**28.** (a) 18,225 is one of many possibilities.

**29.** $726,664 = 7(100,000) + 2(10,000) + 6(1000) + 664$
  Now $8 | 7(100,000)$, because $8 | 100,000$
  $8 | 2(10,000)$, because $8 | 10,000$
  $8 | 6(1000)$, because $8 | 1000$
  If $8 | 664$, then $8 | 726,664$.

**30.** (a) $c = 6 \cdot 5 \cdot 4 \cdot 3 \cdot 2 \cdot 1 = 720$

**31.** $11 | 143$, so no one will be left out.

**32.** (a) Always true

**33.** $1184 = 1 + 2 + 5 + 10 + 11 + 22 + 55 + 110 + 121 + 242 + 605$;
  $1210 = 1 + 2 + 4 + 8 + 16 + 32 + 37 + 74 + 148 + 296 + 592$

**34.** (a) {1, 4, 7, 10, 13, 16, 19, 22, 25, 28, 31, 34, 37, 40, 43, 46, 49, 52, 55, 58}

**35.** Yes, $k | (n! + k)$ for $k = (2, 3, \ldots, n)$

**36.** (a) 1 is the only proper divisor of a prime. Therefore, the sum of the proper divisors is less than $p$, and every prime is deficient.

**37.** (a) A prime is a number greater than 1 that has factors of only 1 and itself. In this set, the first 10 primes are 3, 5, 7, 11, 13, 17, 19, 23, 29, 31.
  (b) Yes

**38.** (a) A number in base seven is divisible by 2 if the sum of the digits is divisible by 2. Likewise, the number is divisible by 3 if the sum of the digits is divisible by 3.

**39.** (a) $^-41$   Check: [(−)] 6 [×] [(] 3 [−] [(−)] 5 [)] [−] [(] [(−)] 7 [)] [=] $^-41$
  (c) $xz - xy$

**40.** (a) $4(^-3) + {}^-4(2) = 4(^-3) + [(^-1)(4)](2)$   Rename $^-4$
  $= 4(^-3) + [4 \cdot {}^-1] \cdot (2)$   Commutative (mult.)
  $= 4(^-3) + 4 \cdot (^-1 \cdot 2)$   Associative (mult.)
  $= 4(^-3) + 4(^-2)$   Multiplication
  $= 4(^-3 + {}^-2)$   Distributive (mult. over add.)
  $4(^-3) + {}^-4(2) = 4(^-3 + {}^-2)$   Transitive property of equality

**EXERCISE SET 4, P. 240**

**1.** (a) g.c.d. = 15   Check: [MATH] (underline *g.c.d.*) 105 [,] 30 [=] 15
  *l.c.m.* = 210   Check: [MATH] (underline *l.c.m.*) 105 [,] 30 [=] 210
  (e) g.c.d. = 2   Check: [MATH] (underline *g.c.d.*) 16 [,] 42 [=] 2
  *l.c.m.* = 336   Check: [MATH] (underline *l.c.m.*) 16 [,] 42 [=] 336

**2.** (a) 6; 840     (c) 4; 7,341,600     (e) 1; 2730

**3.** (a) $3 \cdot 5 \cdot 2$     (b) $r \cdot s^2 \cdot t^2$
  (c) $3^3 \cdot 5^2 \cdot 2^3$     (d) $r^4 \cdot s^3 \cdot t^5$

**5.** $2^3 \cdot 3^3 \cdot 5^2 = 5400$

**6.** (a) When $a$ and $b$ are relatively prime
  (c) When $a$ and $b$ are both 1

**7.** (a) $44 = 2 \cdot 2 \cdot 11$   or   l.c.m. $= \dfrac{44 \cdot 92}{4} = 1012$
  $92 = 2 \cdot 2 \cdot 23$   (g.c.d. = 4)
  l.c.m. $= 2^2 \cdot 11 \cdot 23 = 1012$

  (b) $45 = 3^2 \cdot 5$   or   l.c.m. $= \dfrac{45 \cdot 72}{9} = 360$
  $72 = 3^2 \cdot 2^3$   (g.c.d. = 9)
  l.c.m. $= 2^3 \cdot 3^2 \cdot 5 = 360$

  (c) $146 = 2 \cdot 73$   or   l.c.m. $= \dfrac{146 \cdot 124}{2} = 9052$
  $124 = 2^2 \cdot 31$   (g.c.d. = 2)
  l.c.m. $= 2^2 \cdot 31 \cdot 73 = 9052$

  (d) $840 = 2^3 \cdot 5 \cdot 3 \cdot 7$   or   l.c.m. $= \dfrac{840 \cdot 1800}{2^3 \cdot 3 \cdot 5}$
  $1800 = 2^3 \cdot 3^2 \cdot 5^2$   $= 12,600$
  l.c.m. $= 2^3 \cdot 3^2 \cdot 5^2 \cdot 7$   (g.c.d. $= 2^3 \cdot 3 \cdot 5$)
  $= 12,600$

**8.** (a) 3   Check 1122 $\boxed{\text{INT}\div}$ 105 $\boxed{=}$ . The remainder is 72.

    105 $\boxed{\text{INT}\div}$ 72 $\boxed{=}$ . The remainder is 33.

    72 $\boxed{\text{INT}\div}$ 33 $\boxed{=}$ . The remainder is 6,

    33 $\boxed{\text{INT}\div}$ 6 $\boxed{=}$ . The remainder is 3.

    6 $\boxed{\text{INT}\div}$ 3 $\boxed{=}$ . The remainder is 0.

    3 is the g.c.d.

**9.** (a) g.c.d. = 1   l.c.m. = $ab$
(b) g.c.d. = $a$   l.c.m. = $b$
(c) g.c.d. = $a$   l.c.m. = $a$
(d) If g.c.d.$(a, b) = 1$   (e) $b\,|\,a$
(f) l.c.m. $(1, 4) = 4$     l.c.m. $(1, 101) = 101$
    g.c.d.$(1, 4) = 1$      g.c.d.$(1, 101) = 1$
    l.c.m. $(1, a) = a$
    g.c.d.$(1, a) = 1$
(g) l.c.m. = $ab$      (h) g.c.d. = $a$, l.c.m. = $a^2$

**11.** 20 min.

**13.** 7

**15.** 7:30 pm

**17.** 48

**19.** (a) 5     (b) 3 pears, 5 apples, 7 oranges

**21.** (a) Yes   (b) Yes   (c) Yes   (d) No   (e) No

**22.** (a) Yes

**23.** (a) Yes      (b) Yes

**24.** (a) Let $d$ = g.c.d. $(a, b)$. Then $xd\,|\,xa$ and $xd\,|\,xb$. So $xd$ is a divisor of $xa$ and $xb$. By definition, g.c.d. $(xa, xb)$ must contain $x$ as a factor. Let g.c.d. $(xa, xb)$ be some $xk > xd$. $k\,|\,a$ and $k\,|\,b$. Also $k > d$. This is a contradiction. Therefore, g.c.d. $(xa, xb) = x \cdot$ g.c.d. $(a, b)$.

**25.** Let $k = 1$: $2^2 \cdot 3 = 12$   (abundant)
Proper divisors of 12 are 1, 2, 3, 4, and 6.
$1 + 2 + 3 + 4 + 6 > 12$
Let $k = 2$: $2^3 \cdot 3 = 24$   (abundant)
Proper divisors of 24 are 1, 2, 3, 4, 6, 8, and 12.
$1 + 2 + 3 + 4 + 6 + 8 + 12 > 24$

**27.** (a) Not prime      (b) Prime
(c) Prime      (d) Not prime

**28.** (a) $2 \cdot 3 \cdot 3 \cdot 7$

**29.** No, the seventh number after a prime (not 2) will be even, not a prime.

**31.** (a) $x = 2$     (b) $x = 12$     (c) $x = {}^-1$
(d) $x = 3$     (e) $x < 2$ and $x$ an integer

**EXERCISE SET 5, P. 249**

**1.** (a) 1     (b) 8     (c) 8     (d) 3
(e) 0     (f) 9     (g) 6     (h) 4

**2.** (a) 1     (c) 3     (e) 2     (g) 3

**3.** (a) 4 P.M.     (b) Tomorrow at 10 P.M.
(c) 1 P.M. the next day

**4.** (a) 3     (c) 10     (e) 3

**5.** (a) 15     (b) 21     (c) 22
(d) 4     (e) 3     (f) 19

**7.** 2 years; 1 year

**9.** (a) 5, 9, 13     (b) 10, 17, 24

**11.** Tuesday

**12.** (a) Yes

**13.** (a) False   (b) True   (c) False   (d) False
(e) True   (f) False   (g) True   (h) False

**14.** (k) 8 (mod 9)   Check: 5 $\boxed{\wedge}$ 3 $\boxed{\text{INT}\div}$ 9 has a remainder of 8.

**15.** (a) 2     (b) 3     (c) 2
(d) 2     (e) 2     (f) 1 or 2

**17.** (a) Off     (b) Off     (c) Brightest
(d) Dim     (e) Brightest     (f) Off

**18.** (a) Tuesday

**19.** (a) Yes     (b) No     $1 < 2$
    $1 + 2 < 2 + 2$
    $3 < 0$ not true

**20.** (a) 21   Check: $\boxed{\text{MATH}}$ (underline *g.c.d.*)
126 $\boxed{,}$ 525 $\boxed{=}$ . The answer is 21.

**21.** g.c.d. = 2   Check: $\boxed{\text{MATH}}$ (underline *g.c.d.*)
5734 $\boxed{,}$ 12862 $\boxed{=}$ . The answer is 2.
l.c.m. = 36,875354 Check: $\boxed{\text{MATH}}$
(underline *l.c.m.*) 5734 $\boxed{,}$ 12862 $\boxed{=}$ .
The answer is 36,875,354.

**23.** (a) 24, 25, 26, 27, 28
(b) Has a factor of 2; has a factor of 3

**24.** (a) $n^3 - n = n(n - 1)(n + 1)$. So $n(n - 1)(n + 1)$ is a product of 3 consecutive natural numbers. Of any 3 consecutive natural numbers, one will be a multiple of 3. Thus, $3\,|\,(n^3 - n)$.

**25.** 60

**CHAPTER TEST, P. 253**

**1.**

| $\oplus$ | 0 | 1 | 2 | 3 |
|---|---|---|---|---|
| **0** | 0 | 1 | 2 | 3 |
| **1** | 1 | 2 | 3 | 0 |
| **2** | 2 | 3 | 0 | 1 |
| **3** | 3 | 0 | 1 | 2 |

**2.** (a) Prime     (b) Not prime     (c) Not prime

**3.** Divisible by 3     **4.** $252 = 2 \cdot 2 \cdot 3 \cdot 3 \cdot 7$

**5.** 6        **6.** 1710

**7.** (a) 3        (b) ⁻17        (c) ⁻20        (d) 3

**8.** (a)

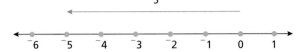

(b)

(c)

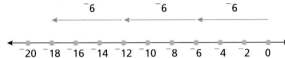

**9.** (a) False        (b) False        (c) False        (d) True
(e) True        (f) False

**10.** (a) ⁻11        (b) $2xw - xy$

**11.** Thursday; Friday        **12.** It is their product.

**13.**
$$7286 = 1684(4) + 550$$
$$1684 = 550(3) + 34$$
$$550 = 34(16) + 6$$
$$34 = 6(5) + 4$$
$$6 = 4(1) + 2$$
$$4 = 2(2) + 0$$
g.c.d. (7286, 1684) = 2

**14.** g.c.d. = 96; l.c.m. = 576        **15.** Thursday

**16.** Monday        **17.** 4 (mod 7)

**18.** (a) When $a = b$        (b) $a \,|\, b$        (c) $b \,|\, a$
(d) g.c.d. = $a$; l.c.m. = $b$
(e) g.c.d. = $b$; l.c.m. = $b^2$

# CHAPTER 6
●●●●●●●●●●●●●●●●●●●●●●●●●●●●●●●●

**EXERCISE SET 1, P. 265**

**1.** (a) $\frac{3}{8}$        (b) $\frac{1}{3}$        (c) $\frac{1}{2}$
(d) $\frac{2}{3}$        (e) $\frac{7}{8}$

**2.** (a) $\frac{6}{8}, \frac{12}{16}, \frac{18}{24}, \frac{24}{32}, \frac{36}{48}, \ldots$        (c) $\frac{0}{1}, \frac{0}{2}, \frac{0}{3}, \frac{0}{100}, \ldots$

**3.** (a) Yes        (b) No        (c) Yes
(d) Yes        (e) No        (f) Neither is defined.

**4.** (a) 3        (c) 3

**5.** (a) 3        (b) 3        (c) 0

**6.** (a) $\frac{1}{2}$        (c) $\frac{1}{4}$

**7.** (a) $\frac{81}{44}$   Check: ⎡FracMode⎤ (underline $d/e$) ⎡=⎤
⎡FracMode⎤ (underline $auto$) ⎡=⎤ 162 ⎡/⎤
88 ⎡=⎤ $\frac{81}{44}$

(c) $\frac{14}{19}$   Check: 308 ⎡/⎤ 418 ⎡=⎤ $\frac{14}{19}$

(e) $\frac{x^2}{z^2}$        (g) $\frac{x}{yz^3}$

**8.** (a) $\frac{49}{21}$   Check: 49 ⎡/⎤ 21 ⎡=⎤ $\frac{7}{3}$
(c) Impossible
(e) $\frac{30}{40}$   Check: 30 ⎡/⎤ 40 ⎡=⎤ $\frac{3}{4}$

**9.** (a) $\left\{\frac{0}{5}, \frac{1}{4}, \frac{2}{3}, \frac{3}{2}, \frac{4}{1}\right\}$        (b) $\left\{\frac{0}{4}, \frac{1}{4}, \frac{2}{3}, \frac{3}{2}, \frac{4}{1}\right\}$
(c) $\left\{\frac{2}{1}, \frac{3}{2}, \frac{4}{3}, \frac{4}{4}, \frac{5}{4}\right\}$        (d) $\left\{\frac{2}{1}, \frac{3}{1}, \frac{4}{1}, \frac{3}{2}, \frac{4}{2}\right\}$
(e) $\left\{\frac{5}{5}, \frac{4}{4}, \frac{3}{3}, \frac{2}{2}, \frac{6}{6}\right\}$

**10.** (a) $\left\{\frac{3k}{5k} \,\middle|\, k = 1, 2, 3, \ldots\right\}$
(c) $\left\{\frac{xk}{yk} \,\middle|\, k = 1, 2, 3, \ldots\right\}$        $y \neq 0$

**11.** (a) Yes   Check: Set ⎡FracMode⎤ on $d/e$ and $auto$.
9 ⎡/⎤ 17 ⎡=⎤ $\frac{9}{17}$
(c) No   Check: 7 ⎡/⎤ 14 ⎡=⎤ $\frac{1}{2}$
(e) Yes   Check: 3 ⎡/⎤ 17 ⎡=⎤ $\frac{3}{17}$

**13.** (a) >        (b) >        (c) >        (d) >        (e) <        (f) <

**14.** (a) $\frac{1}{2}$        (c) $\frac{1}{3}$

**15.** Answers will vary—one solution is given.
(a) $\frac{1}{3} < \frac{12}{30} < \frac{1}{2}$        (b) $\frac{3}{4} < \frac{39}{48} < \frac{5}{6}$

**16.** (a) Undefined        (c) Undefined        (e) 0

**17.** (a) $\frac{15}{16}$        (b) $\frac{1}{8}$        (c) $\frac{-17}{18}$

**18.** (a) $\frac{3}{8} = \frac{3}{8}$   Check: Set ⎡FracMode⎤ on $d/e$ and
$auto$. 750 ⎡/⎤ 2000 ⎡=⎤ $\frac{3}{8}$
Check: 39 ⎡/⎤ 104 ⎡=⎤ $\frac{3}{8}$

**19.** (a) $x = 15$        (b) $x = -8$        (c) $x = 45$        (d) $x = 12$

**20.** (a) $\frac{9750}{26000} = \frac{9750}{26000}$

**21.** (a) True        (b) True        (c) True
(d) False        (e) True        (f) False

**22.** (a) Yes; $ad = cd$

**23.** (a) $\dfrac{3 \cdot 5^2 \cdot 7}{2 \cdot 11} = \dfrac{525}{22}$　(b) $\dfrac{a^2 \cdot b^3}{2}$　(c) $\dfrac{x^2}{2\,y^2}$

(d) $\dfrac{c + 2d}{3e}$　(e) $\dfrac{a}{b^2}$　(f) $\dfrac{1}{2 + b}$

(g) $\dfrac{a}{2a + b}$　(h) $\dfrac{ab}{a + b}$　(i) $\dfrac{ab(a + b)}{a - b}$

**24.** (a) $a = b$　**25.** Tom

**26.** (a) The 4's are addends, not factors, and cannot be divided out.

**27.** This is true if $b > 0$ and $d > 0$.

Proof:　If $\frac{a}{b} < \frac{c}{d}$ and $b > 0$ and $d > 0$, then $ad < bc$. Thus $ab + ad < ab + bc$ or $a(b + d) < b(a + c)$.
$\frac{a}{b} < \frac{a + c}{b + d}$ because $b > 0$ and $b + d > 0$.
Similarly, if $\frac{a}{b} < \frac{c}{d}$, then $ad < bc$. Thus $ad + cd < bc + cd$ or $d(a + c) < c(b + d)$.
$\frac{a + c}{b + d} < \frac{c}{d}$ because $d > 0$ and $b + d > 0$.

**29.** (a) $9 \cdot 2000 \overset{?}{=} 15 \cdot 1200$
$18{,}000 = 18{,}000$; yes, the fractions are equal.
Check: Set ⎡FracMode⎤ on $d/e$ and *auto*.
$9$ ⎡/⎤ $15$ ⎡=⎤ $\frac{3}{5}$, Check: $1200$ ⎡/⎤ $2000$ ⎡=⎤ $\frac{3}{5}$, so the two rational numbers are equal.

(c) $(33)(189) \overset{?}{=} (43)(62)$
$6237 \neq 2666$; no, the fractions are not equal.
Check: $33$ ⎡/⎤ $43$ ⎡=⎤ $\frac{33}{43}$ and $62$ ⎡/⎤ $189$ ⎡=⎤ $\frac{62}{189}$, so the two simplified fractions are not equal.

**31.** (a) Reflexive:　$\frac{a}{b} = \frac{a}{b}$ because $ab = ba$.
Symmetric:　If $\frac{a}{b} = \frac{c}{d}$, then $\frac{c}{d} = \frac{a}{b}$. If $ad = bc$, then $cb = da$.
Transitive:　If $\frac{a}{b} = \frac{c}{d}$ and $\frac{c}{d} = \frac{e}{f}$, then $\frac{a}{b} = \frac{e}{f}$. Assume $\frac{a}{b} = \frac{c}{d}$ and $\frac{c}{d} = \frac{e}{f}$ with $b \neq 0$, $d \neq 0$, and $f \neq 0$. Then, $ad = bc$ and $cf = de$. Thus, $c = \frac{de}{f}$. Then $ad = bc = b \cdot \frac{de}{f}$. So, $adf = bde$ and $af = be$. Therefore, $\frac{a}{b} = \frac{e}{f}$.
Thus, "is equivalent to" is an equivalence relation on the set of fractions.

(b) One equivalence class is $\left\{ \cdots \dfrac{-4}{-6}, \dfrac{-2}{-3}, \dfrac{2}{3}, \dfrac{4}{6} \cdots \right\}$.

**32.** (a) $\dfrac{1}{2}$　Check: ⎡(⎤ $2$ ⎡^⎤ $10$ ⎡−⎤ $2$ ⎡^⎤ $9$ ⎡)⎤ ⎡÷⎤ ⎡(⎤ $2$ ⎡^⎤ $11$ ⎡−⎤ $2$ ⎡^⎤ $10$ ⎡)⎤ ⎡=⎤ ⎡▶F⎤ ⎡=⎤ $\frac{1}{2}$.

**EXERCISE SET 2, P. 279**

**1.** (a) $\dfrac{3}{4}$　(b) $\dfrac{37}{24}$　(c) $\dfrac{29}{36}$

(d) $\dfrac{19}{15}$　(e) $\dfrac{13}{2x}$　(f) $\dfrac{7 + 4x}{xy}$

**2.** (a) $\dfrac{3}{4}$　(c) $\dfrac{11}{36}$　(e) $3/2x$

**3.** (a) l.c.d. = 8; $\dfrac{19}{8}$　(b) l.c.d. = 32; $\dfrac{73}{32}$

(c) l.c.d. = 54; $\dfrac{10}{9}$　(d) l.c.d. = 60; $\dfrac{229}{60}$

(e) l.c.d. = 385; $\dfrac{1187}{385}$　(f) l.c.d. = 9; $\dfrac{28}{9}$

**4.** (a) $19\frac{1}{3}$　Check: $58$ ⎡/⎤ $3$ ⎡A b/c ◀ ▶ d/e⎤ ⎡=⎤ $19\frac{1}{3}$

**5.** (a) $\dfrac{23}{6}$　Check: $3$ ⎡Unit⎤ $5$ ⎡/⎤ $6$ ⎡A b/c ◀ ▶ d/e⎤ ⎡=⎤ $\frac{23}{6}$

(c) $\dfrac{49}{3}$　Check: $16$ ⎡Unit⎤ $1$ ⎡/⎤ $3$ ⎡A b/c ◀ ▶ d/e⎤ ⎡=⎤ $\frac{49}{3}$

**6.** (a) $\dfrac{1}{8} + \dfrac{3}{16}$　(c) $\dfrac{15}{4} + \dfrac{19}{4}$

**7.** (a) Sum $= 10\frac{13}{21}$　Check: ⎡FracMode⎤ (underline $A\ b/c$) ⎡=⎤ $8$ ⎡Unit⎤ $4$ ⎡/⎤ $21$ ⎡+⎤ $2$ ⎡Unit⎤ $3$ ⎡/⎤ $7$ ⎡=⎤ $10\frac{13}{23}$

Difference $= 5\frac{16}{21}$　Check: $8$ ⎡Unit⎤ $4$ ⎡/⎤ $21$ ⎡−⎤ $2$ ⎡Unit⎤ $3$ ⎡/⎤ $7$ ⎡=⎤ $5\frac{16}{21}$

**8.** (a) $2\frac{1}{24}$　(c) $\dfrac{17}{24}$
(e) Associative property of addition

**9.** (a) $11\frac{7}{12}$　Check: $5$ ⎡Unit⎤ $1$ ⎡/⎤ $2$ ⎡+⎤ $4$ ⎡Unit⎤ $3$ ⎡/⎤ $4$ ⎡+⎤ $1$ ⎡Unit⎤ $1$ ⎡/⎤ $3$ ⎡=⎤ $11\frac{7}{12}$

(e) $7\frac{49}{60}$　Check: $5$ ⎡/⎤ $12$ ⎡+⎤ $7$ ⎡Unit⎤ $2$ ⎡/⎤ $5$ ⎡=⎤ $7\frac{49}{60}$

**10.** (a) $\left(\dfrac{3}{4} + \dfrac{2}{5}\right) \overset{?}{=} \left(\dfrac{2}{5} + \dfrac{3}{4}\right)$

$\dfrac{15 + 8}{20} \overset{?}{=} \dfrac{8 + 15}{20}$

$\dfrac{23}{20} = \dfrac{23}{20}$

(c) $\dfrac{3}{4} + \dfrac{0}{1} \overset{?}{=} \dfrac{3}{4}$

$\dfrac{3 + 0(4)}{4} \overset{?}{=} \dfrac{3}{4}$

$\dfrac{0}{1} + \dfrac{3}{4} = \dfrac{3}{4}$

**11.** (a)

(b)

(c)

**12.** (a) $1\frac{1}{2}$ (commutative and associative)

   (c) 7 (commutative and associative)

   (e) $1\frac{3}{4}$ (commutative and associative)

**13.** (a) Between 13 and 16    (b) $18 < \text{sum} < 21$

**14.** (a) 15    **15.** (a) 15    (b) 19

**16.** (a) Yes

   (c) No, $\frac{2}{5} - \left(\frac{4}{5} - \frac{1}{5}\right) \neq \left(\frac{2}{5} - \frac{4}{5}\right) - \frac{1}{5}$

   $-\frac{1}{5} \neq -\frac{3}{5}$

   (e) No, $\frac{2}{5} - 0 = \frac{2}{5}$, but $0 - \frac{2}{5} \neq \frac{2}{5}$

**17.** (a) $\frac{11}{12}$    (b) $\frac{5}{6}$    (c) $\frac{10}{9}$    (d) $\frac{7}{3}$

**18.** (a) $\frac{-59}{40}$    (c) $\frac{-19}{45}$

**19.** (a) $\frac{5}{6}$    (b) $\frac{34}{100}$

**20.** (a) $S = \frac{55}{273}$    $D = \frac{43}{273}$

   Check: Set $\boxed{\text{FracMode}}$ on $d/e$ and $auto.$ 7 $\boxed{/}$ 39 $\boxed{+}$ 2 $\boxed{/}$ 91 $\boxed{=}$ $\frac{55}{273}$.

   Check: 7 $\boxed{/}$ 39 $\boxed{-}$ 2 $\boxed{/}$ 91 $\boxed{=}$ $\frac{43}{273}$

**21.** (a) Sum is $\dfrac{3 + 3y^2z}{xy^3z}$   Difference is $\dfrac{3 - 3y^2z}{xy^3z}$

   (b) Sum is $\dfrac{28a^4 + 5x^4}{14a^3x}$   Difference is $\dfrac{28a^4 - 5x^4}{14a^3x}$

   (c) Sum is $\dfrac{132 + n^3x}{12mn^2}$   Difference is $\dfrac{132 - n^3x}{12mn^2}$

   (d) Sum is $\dfrac{db^2 + a^3c}{a^3b\,^3c^2}$   Difference is $\dfrac{db^2 - a^3c}{a^3b^3c^2}$

**22.** (a) (i) Let $x = $ total weight.

$$x = \tfrac{1}{2} + 6\tfrac{1}{4} + 2\tfrac{3}{4} + 12\tfrac{1}{8} + 1$$

$$= 22\tfrac{5}{8} \text{ lb.} \qquad \text{(ii) } 19\tfrac{7}{8} \text{ lb.}$$

**23.** (a) $\dfrac{2 + 3}{2} = \dfrac{2}{2} + \dfrac{3}{2}$, not $\dfrac{2}{2} + 3$

   (b) $\dfrac{8}{4 + 4} \neq \dfrac{8}{4} + \dfrac{8}{4}$

   (c) $\dfrac{3}{4} + \dfrac{7}{9} = \dfrac{27 + 28}{36} = \dfrac{55}{36}$

**24.** (a) $\frac{26}{36} + \frac{29}{45} \approx \frac{27}{36} + \frac{30}{45} \approx \frac{3}{4} + \frac{2}{3} = \frac{17}{12}$

   (too large)

**25.** Yes; commutative property of addition

**26.** (a) $\dfrac{1 + x^2y^2}{x^3y^4}$

**27.** Let $a = \dfrac{1}{2}$   $b = \dfrac{1}{4}$   $c = \dfrac{3}{4}$   $d = 1$

$$a - (b - c - d) = \frac{1}{2} - \left(\frac{1}{4} - \frac{3}{4} - 1\right) = 2$$

$$(a - b - c) - d = \left(\frac{1}{2} - \frac{1}{4} - \frac{3}{4}\right) - 1 = -\frac{3}{2}$$

Thus, $a - (b - c - d) \neq (a - b - c) - d$

**28.** (a) $\dfrac{1}{n} + \dfrac{1}{n + 1} = \dfrac{2n + 1}{n(n + 1)}$   $n \geq 2$

**29.** (a) $\dfrac{1}{4} + \dfrac{1}{4}$   (b) $\dfrac{1}{22} + \dfrac{1}{22}$   (c) $\dfrac{1}{200} + \dfrac{1}{200}$

   (d) $\dfrac{1}{3} + \dfrac{1}{3}$   (e) $\dfrac{1}{5} + \dfrac{1}{5}$   (f) $\dfrac{1}{11} + \dfrac{1}{11}$

**30.** (a) $4 \cdot 5 + \dfrac{1}{4}$

**31.** $\dfrac{3}{8} = \dfrac{6}{16} = \dfrac{9}{24} = \dfrac{12}{32}$

**32.** (a) $\frac{2}{3}$  Check: Set $\boxed{\text{FracMode}}$ on $d/e$ and $auto.$

   24 $\boxed{/}$ 36 $\boxed{=}$ $\frac{2}{3}$

**EXERCISE SET 3, P. 290**

**1.** (a)

$\frac{1}{3} \cdot \frac{1}{3} = \frac{1}{9}$

(b)

$\frac{2}{3} \cdot \frac{1}{5} = \frac{2}{15}$

**2.** (a) $\frac{1}{7}$; $\frac{7}{1} \cdot \frac{1}{7} = \frac{7}{7} = 1$    (c) Impossible

   (e) $z/w$; $w/z \cdot z/w = wz/wz = 1$

**3.** (a) $\dfrac{1}{4}$  Check: Set $\boxed{\text{FracMode}}$ on $d/e$ and

   $auto.$ 3 $\boxed{/}$ 4 $\boxed{\times}$ 1 $\boxed{/}$ 2 $\boxed{\times}$ 2 $\boxed{/}$ 3 $\boxed{=}$ $\frac{1}{4}$

   (e) 36  Check: 4 $\boxed{\div}$ $\boxed{(}$ 1 $\boxed{/}$ 9 $\boxed{)}$ $\boxed{=}$ 36

   (i) $\dfrac{64}{63}$  Check: $\boxed{(}$ 4 $\boxed{/}$ 3 $\boxed{\div}$ 7 $\boxed{/}$ 8 $\boxed{)}$ $\boxed{\times}$ $\boxed{(}$ 1 $\boxed{/}$ 2 $\boxed{\div}$ 3 $\boxed{/}$ 4 $\boxed{)}$ $\boxed{=}$ $\frac{64}{63}$

**4.** (a) $a$    (c) $ab(ad + be)/cde$    (e) $x^2/y^2$

**5.** (a) $44\frac{17}{32}$  Check: Set $\boxed{\text{FracMode}}$ on $A\,b/c$ and $auto.$ 7 $\boxed{\text{Unit}}$ 1 $\boxed{/}$ 8 $\boxed{\times}$ 6 $\boxed{\text{Unit}}$ 1 $\boxed{/}$ 4 $\boxed{=}$ $44\frac{17}{32}$

   (c) $17\frac{1}{2}$  Check: 2 $\boxed{\text{Unit}}$ 1 $\boxed{/}$ 10 $\boxed{\times}$ 8 $\boxed{\text{Unit}}$ 1 $\boxed{/}$ 3 $\boxed{=}$ $17\frac{1}{2}$

   (e) $1\frac{19}{26}$  Check: 7 $\boxed{\text{Unit}}$ 1 $\boxed{/}$ 2 $\boxed{\div}$ 4 $\boxed{\text{Unit}}$ 1 $\boxed{/}$ 3 $\boxed{=}$ $1\frac{19}{26}$

   (g) $2\frac{11}{14}$  Check: 4 $\boxed{\text{Unit}}$ 7 $\boxed{/}$ 8 $\boxed{\div}$ 1 $\boxed{\text{Unit}}$ 3 $\boxed{/}$ 4 $\boxed{=}$ $2\frac{11}{14}$

**6.** (a) $1\frac{1}{2}$   Check: Set $\boxed{\text{FracMode}}$ on $A\ b/c$ and

*auto.* $3\ \boxed{/}\ 8\ \boxed{\div}\ 4\ \boxed{/}\ 16\ \boxed{=}\ 1\frac{1}{2}$

(c) $3\frac{1}{7}$   Check: $\boxed{(}\ 1\ \boxed{/}\ 3\ \boxed{+}\ 2\ \boxed{/}\ 5\ \boxed{)}\ \boxed{\div}$

$\boxed{(}\ 5\ \boxed{/}\ 6\ \boxed{-}\ 3\ \boxed{/}\ 5\ \boxed{)}\ \boxed{=}\ 3\frac{1}{7}$

(e) $3\frac{15}{91}$   Check: $\boxed{(}\ 2\ \boxed{/}\ 3\ \boxed{+}\ 6\ \boxed{/}\ 7\ \boxed{)}\ \boxed{\div}$

$13\ \boxed{/}\ 27\ \boxed{=}\ 3\frac{15}{91}$

**7.** (a) $x = \frac{5}{2}$   (b) $x = \frac{8}{5}$   (c) $x = \frac{5}{28}$   (d) $x = \frac{3}{4}$

(e) $x = 3$   (f) $x = \frac{1}{2}$   (g) $x = 2$   (h) $x = 1$

**8.** (a) Two examples

$$\frac{1}{2} \div \frac{1}{4} \overset{?}{=} \frac{1}{4} \div \frac{1}{2}$$

$$2 \neq \frac{1}{2} \quad \text{Division is not commutative.}$$

$$\frac{1}{2} \div \left(\frac{1}{4} \div \frac{1}{2}\right) \overset{?}{=} \left(\frac{1}{2} \div \frac{1}{4}\right) \div \frac{1}{2}$$

$$\frac{1}{2} \div \left(\frac{1}{2}\right) \overset{?}{=} (2) \div \frac{1}{2}$$

$$1 \neq 4 \quad \text{Division is not associative.}$$

**9.** These are estimated ranges.

(a) 24 to 35   (b) 48 to 63

(c) 2 to 3   (d) $2\frac{1}{4}$ to $3\frac{1}{3}$

**10.** (a) $7 \cdot 4 = 28$   (c) $8 \div 4 = 2$

**11.** (a) $6\frac{1}{2} \cdot 4 = 26$   (b) $8\frac{1}{2} \cdot 7 = 59\frac{1}{2}$

(c) $8\frac{1}{2} \div 4 = 2\frac{1}{8}$   (d) $9\frac{1}{2} \div 3 = 3\frac{1}{6}$

**12.** (a) Less than 2   (c) Less than 2

(e) Greater than 2

**13.** (a) $\left(4 \cdot \frac{1}{2}\right) \cdot \frac{1}{5} = 2 \cdot \frac{1}{5} = \frac{2}{5}$

(b) $\left(18 \cdot \frac{1}{6}\right) \cdot \frac{3}{4} = 3 \cdot \frac{3}{4} = \frac{9}{4}$

(c) $4(16) + \frac{7}{8}(16) = 64 + 14 = 78$

(d) $8(18) + \frac{5}{9}(18) = 144 + 10 = 154$

(e) $(24 \div 3) + \left(\frac{1}{2}\right) \div 3 = 8\frac{1}{6}$

(f) $(25 \div 5) + \left(\frac{1}{3}\right) \div 5 = 5\frac{1}{15}$

**15.** Let $x$ = number of inches

$x = 12(6\frac{1}{2} - 5\frac{3}{4})$   $x = 9$ in.

**17.** Let $x$ = number of suits

$(3\frac{1}{2})x = 60$   $x = 17\frac{1}{7}$ or 17 suits

**19.** Let $x$ = number of hours

$\frac{7}{6}x + \frac{6}{5}x = 1$   $x = \frac{30}{71}$ hr.

**21.** Let $x$ = original number of faculty members

$168 = \frac{4}{5} \cdot \frac{7}{8} \cdot x$   $x = 240$ faculty members

**23.** Let $x$ = fraction owned by family

$\frac{2}{3} \cdot \frac{3}{7} + \frac{3}{7} = x$   $x = \frac{5}{7}$ of the stock

**25.** Let $x$ = Linda's weight

$x + \frac{2}{3}x = 190$   Linda weights 114 lb; Shirley, 76.

**27.** (a) $2\frac{1}{2} \cdot 3\frac{1}{3} = d$   $d = 8\frac{1}{3}$ km

(b) $1724\frac{1}{2} = (20\frac{3}{4})r$   $r = 83\frac{9}{83}$ km/hr.

**29.** $x + (\frac{1}{10})x = 39{,}600;$   $x = \$36{,}000$

**30.** (a) $\frac{1}{4}$ lb.

**31.** (a) 1300 mL flour          12 eggs

24 mL salt             240 mL butter

180 mL sugar           1500 mL milk

32 mL baking powder

(b) $108\frac{1}{3}$ mL flour        1 egg

2 mL salt             20 mL butter

15 mL sugar           125 mL milk

$2\frac{2}{3}$ mL baking powder

**32.** (a) $\dfrac{1}{b^3}$

**33.** (a) \$2.04   (b) \$25,525.63

**34.** (a) \$53,240

**35.** $\dfrac{2}{2}$ is one example

$$\frac{2}{2} \text{ vs } \frac{2+2}{2} \text{ or } \frac{4}{2} \quad \text{This works for any fraction equal to 1.}$$

**37.** (a) $\dfrac{adf + cbf + ebd}{bdf}$   (b) $\dfrac{acfh + bdeg}{bdfh}$

(c) $\dfrac{fh(ad + bc)}{bd(eh + gf)}$   (d) $\dfrac{adfg}{bceh}$

**39.** (a) $\left(\dfrac{1}{11}\right)_{\text{seven}}$   (b) $\left[\dfrac{817\text{E}}{68}\right]_{\text{twelve}}$

**40.** (a) $\frac{34}{35}, \frac{6}{35}$   (c) $\frac{33}{20}, \frac{3}{20}$

**41.** (a) $\dfrac{6}{8} = \dfrac{3}{4}$   (b) $\dfrac{1}{2}$   (c) 0   (d) 4

**42.** (a) 5   (c) 1   (e) 3

**EXERCISE SET 4, P. 299**

**1.** (a) $\dfrac{5}{3}$   (b) $\dfrac{7}{11}$   (c) $\dfrac{2}{7}$

(d) $\dfrac{7}{24}$   (e) $\dfrac{1}{230}$

**2.** (a) 1:9   (c) 34:45

**3.** (a) 8   (b) 9   (c) $15\frac{6}{23}$

(d) $2\frac{2}{31}$   (e) 40   (f) $10\frac{11}{21}$

**4.** (a) 2:3   (c) 2:5

**5.** (a) $21\frac{1}{2}$ miles per gallon    (b) $\dfrac{7527}{215}\ \dfrac{\text{students}}{\text{teacher}}$

  (c) $\dfrac{3}{17}$ administrator per employee

  (d) $\dfrac{16,432}{9}\ \dfrac{\text{dollars profit}}{\text{Lot}}$

**6.** (a) $x = \frac{6}{5}$    (c) $x = 5\frac{1}{7}$

**7.** (a) 7:3    (b) 3:10    (c) 12 boys

**8.** (a)

|  | 900-g can | 850-g can |
|---|---|---|
| Price | 49¢ | 46¢ |
| Grams | 900 | 850 |
| Ratio | $\dfrac{49}{900}$ | $\dfrac{46}{850}$ |

  The 850-g can for 46¢ is a better buy.

**9.** (a) 29-oz. can for 98¢
  (b) $2.47 for 84 oz.
  (c) 3-lb. ham for $5.49
  (d) 16 oz. for 58¢
  (e) 8 oz. for 60¢

**10.** (a) vanilla $\frac{1}{2}$ teaspoon; flour $1\frac{1}{2}$ cups; sugar $\frac{7}{8}$ cup

**11.** 125 km

**13.** $1\frac{7}{8}$ g

**15.** 18 min.

**17.** 10 L of green; 10 L of yellow; 40 L of white

**19.** The number of boys is larger in the second class.

**21.** 54 gal.

**23.** $28\frac{4}{9}$ g

**25.** (a) $2\frac{2}{3}$    (b) $5\frac{5}{11}$    (c) 4    (d) 12

**27.** (a) 2:3    (b) $\frac{3}{5}$    (c) $\frac{2}{5}$

**29.** $\dfrac{a}{b} = \dfrac{c}{d}$    Given

  $ad = bc$    Exercise 28

  $bc = ad$    Equality Symmetric

  $\dfrac{b}{a} = \dfrac{d}{c}$    Exercise 28

**31.** $\dfrac{a}{b} = \dfrac{c}{d}$    Given

  $ad = bc$    Exercise 28

  $bc = ad = da$    Commutative (mult.) and symmetric (equality)

  $\dfrac{b}{d} = \dfrac{a}{c}$    Exericse 28

**32.** (a) $\dfrac{a}{b} = \dfrac{c}{d}$    Given

  $\dfrac{a}{b} + 1 = \dfrac{c}{d} + 1$    If $a = b$, then $a + k = b + k$.

  $\dfrac{a + b}{b} = \dfrac{c + d}{d}$    Addition of fractions

**33.** No.    $\dfrac{5}{6} = \dfrac{10}{12}; \dfrac{5-3}{6-3} = \dfrac{2}{3} \neq \dfrac{10}{12}$

**34.** (a) $\frac{8}{35}; \frac{10}{7}$    (c) $\frac{27}{40}; \frac{6}{5}$

**35.** (a) $x = -3$    (b) $y = 3$

**EXERCISE SET 5, P. 309**

**1.** (a) $\frac{9}{10} > \frac{7}{11}$    (b) $33/9 > 21/7$    (c) $\frac{5}{7} > \frac{2}{5}$
  (d) $\frac{29}{3} > \frac{16}{2}$    (e) $\frac{17}{51} < \frac{6}{7}$    (f) $\frac{22}{73} < \frac{25}{71}$

**2.** (a) $\frac{71}{100} < \frac{23}{30} < \frac{3}{2}$    (c) $\frac{16}{27} < \frac{11}{18} < \frac{2}{3} < \frac{67}{100}$
  (e) $\frac{156}{50} < \frac{22}{7} < \frac{10}{3} < \frac{7}{2}$

**3.** (a) $\{x \mid x$ is an integer $< 2\}$
  (b) $\{x \mid x$ is an integer $> 3\}$

**4.** (a) False    (c) False

**5.** (a) $Q, I, N$    (b) $Q$    (c) $Q, I$
  (d) $F, Q$ (excluding division by 0)
  (e) $W, F, Q, I, N$
  (f) $W, F, Q, I$
  (g) $Q, I$
  (h) $F, Q$ (excluding division by 0)

**6.** (a) If $\frac{a}{b} < \frac{c}{d}$, then $\frac{a}{b} + \frac{e}{f} < \frac{c}{d} + \frac{e}{f}$.
  (c) If $\frac{a}{b} < \frac{c}{d}$ and $\frac{e}{f} > 0$, then $\frac{a}{b} \cdot \frac{e}{f} < \frac{c}{d} \cdot \frac{e}{f}$.
  (e) If $\frac{a}{b} < \frac{c}{d}$ and $\frac{e}{f} < 0$, then $\frac{a}{b} \cdot \frac{e}{f} > \frac{c}{d} \cdot \frac{e}{f}$.
  (g) Commutative property of multiplication of rational numbers

**7.** (a) Yes    (b) No    (c) Yes
  (d) No    (e) No

**9.** Let $\frac{a}{b}$ and $\frac{c}{d}$ be fractions such that $b$ and $d$ are positive and $\frac{a}{b} < \frac{c}{d}$. Then $\frac{a+c}{b+d} < \frac{c}{d}$ if $ad + cd < bc + cd$ or $ad < bc$. But $ad < bc$, because $\frac{a}{b} < \frac{c}{d}$. Thus, $\frac{a+c}{b+d} < \frac{c}{d}$. Likewise, $\frac{a}{b} < \frac{a+c}{b+d}$ if $ab + ad < ab + bc$ or $ad < bc$. This is true, so $\frac{a}{b} < \frac{a+c}{b+d}$.

**10.** (a) $\{7\}$
  (c) $\{z \mid z$ is a rational number $\geq 4\}$
  (e) $\{2\}$
  (g) $\{x \mid x$ is a rational number $< -9\}$
  (i) $\{x \mid x$ is a rational number $\leq \frac{5}{8}\}$

**11.** (a) Not possible    (b) Not possible

**12.** (a) $(x + 8) + (-8)$
    $= 15 + (-8)$    If $\frac{a}{b} = \frac{c}{d}$, then $\frac{a}{b} + \frac{e}{f} = \frac{c}{d} + \frac{e}{f}$
  $(x + 8) + -8 = 7$    Addition
  $x + [8 + (-8)] = 7$    Associative (add.)
  $x + 0 = 7$    Additive inverse
  $x = 7$    Additive identity

**13.** (a) $16 \cdot \left(18 \cdot \frac{1}{3}\right) = 16 \cdot 6 = 96$    Commutative and associative (mult.)
  (b) $5\left(\frac{1}{3} \cdot 9\right) = 15$    Commutative and associative (mult.)
  (c) $\left(\frac{1}{3} + \frac{2}{3}\right) + \frac{3}{4} = 1\frac{3}{4}$    Associative and commutative (add.)

(d) $\frac{3}{8} + \left(\frac{5}{6} + \frac{1}{6}\right) = 1\frac{3}{8}$   Commutative and associative (add.)

(e) $\left(2 + \frac{5}{8}\right)16 = 2 \cdot 16 + \frac{5}{8} \cdot 16 = 42$   Distributive (mult. over add.)

(f) $4 \cdot 9 + \frac{1}{3} \cdot 9 = 39$   Distributive (mult. over add.)

**14.** (a) $(x + 2) + (-2) <$
$6 + (-2)$   If $\frac{a}{b} < \frac{c}{d}$ then $\frac{a}{b} + \frac{e}{f} < \frac{c}{d} + \frac{e}{f}$
$(x + 2) + (-2) < 4$   Addition
$x + (2 + -2) < 4$   Associative (add.)
$x + 0 < 4$   Additive inverses
$x < 4$   Additive identity
$\{x \mid x$ is a rational number $< 4\}$

**15.** (a) Closure for addition, additive identity, additive inverses
(b) Closure for addition and multiplication, multiplicative identity, multiplicative inverses
(c) Closure for addition and multiplication, additive and multiplicative inverses
(d) Closure for addition and multiplication, additive and multiplicative inverses

**17.** Let $x$ be Mrs. Jones' bill.
$6x + x = 168$
$x = \$24$ (Mrs. Jones); $6x = \$144$ (Mrs. Smith)

**19.** Let $x$ be Ralph's weight.
$3x + 54 = 300$
$x = 82$ kg (Ralph's weight)

**21.** Each positive rational number's additive inverse is not a positive rational number. Also, 0 (zero), additive identity, is not a positive rational number

**22.** (a) Let $p/q$ and $r/s$ be expressed with positive denominators and $v > 0$.
$p/q < r/s$   Given
$ps < qr$   Definition of less than
$psv^2 < qrv^2$   If $a < b$ and $c > 0$, $ac < bc$
$psv^2 + qvst < qrv^2 + qvst$   If $a < b$, then $a + c < b + c$
$(pv)(sv) + (qt)(sv) < (qv)(rv) + (qv)(st)$
        Commutative and associative (mult.)
$(pv + qt)(sv) < (qv)(rv + st)$   Distributive (mult. over add.)
$\frac{pv + qt}{qv} < \frac{rv + st}{sv}$   Definition of less than for rational numbers
$\frac{p}{q} + \frac{t}{v} < \frac{r}{s} + \frac{t}{v}$   Addition of rational numbers

**23.** (a) Yes (excluding division by 0)
(b) Yes (excluding division by 0)

**24.** (a) $\frac{p}{q} + \frac{t}{v} < \frac{r}{s} + \frac{t}{v}$   Given

$\left(\frac{p}{q} + \frac{t}{v}\right) + -\frac{t}{v} < \left(\frac{r}{s} + \frac{t}{v}\right) + -\frac{t}{v}$   Additive property of inequalities

$\frac{p}{q} + \left(\frac{t}{v} + -\frac{t}{v}\right) < \frac{r}{s} + \left(\frac{t}{v} + -\frac{t}{v}\right)$   Associative (add.)

$\frac{p}{q} + 0 < \frac{r}{s} + 0$   Additive inverses

$\frac{p}{q} < \frac{r}{s}$   Additive identity

**25.** 2 hr.
**26.** (a) $\frac{9}{32}$   **27.** $-10$   **28.** (a) $x = -1$

### CHAPTER TEST, P. 314

**1.** (a)

$\frac{3}{8}$

(b) $\frac{4}{6}, \frac{6}{9}, \frac{8}{12}$

**2.** (a) $\frac{56}{3}$   (b) $54\frac{1}{3}$   **3.** $1\frac{23}{24}$   **4.** $\frac{19}{36}$   **5.** $-\frac{3}{5}$

**6.** (a) $\frac{7}{5}$   (b) $\frac{1}{15}$   (c) Does not exist   (d) 12

**7.** $\frac{2}{3}$   **8.** $34\frac{3}{4}$   **9.** (a) 9   (b) $\frac{32}{5}$

**10.** $-\frac{26}{27}$   **11.** 1 quarter, 2 dimes, 2 nickels

**12.** (a) $\frac{2}{xy}$   (b) $a$   **13.** $4\frac{1}{7}$

**14.** 18   **15.** 3 hr.

**16.** (a) $x = \frac{19}{6}$   (b) $x = \frac{23}{3}$

**17.** (a) Multiplicative inverse
(b) Additive inverse
(c) Either (a) or (b)
(d) Additive identity, multiplicative identity, closure, etc.

**18.** (a) Yes   (b) Yes
(c) There is not an additive identity for the whole set.   (d) No

## CHAPTER 7

### EXERCISE SET 1, P. 327

**1.** (a) $(1 \cdot 10) + (4 \cdot 1) + \left(7 \cdot \frac{1}{10}\right) + \left(2 \cdot \frac{1}{100}\right)$

(b) $(4 \cdot 1) + \left(1 \cdot \frac{1}{1000}\right) + \left(6 \cdot \frac{1}{10000}\right)$

(c) $\left(1 \cdot \frac{1}{10}\right) + \left(4 \cdot \frac{1}{100}\right) + \left(6 \cdot \frac{1}{1000}\right)$

**2.** (a) 43.26   (c) 7.0042

**3.** (a) $\frac{17}{200}$   (b) $\frac{13}{4}$   (c) $\frac{637}{50}$

(d) $\frac{21}{25}$   (e) $\frac{11}{4}$   (f) $\frac{1}{4000}$

**4.** (a) 31.6    Check: 316 ÷ 10 = 31.6

   (e) 5.425    Check: 651 ÷ 120 = 5.425

   (g) 0.375    Check: 3 ÷ 8 = 0.375

**5.** 4100; 4070.79

**7.** 0.08; 0.08365

**9.** 453,000; 453,039

**11.** (a) Terminating       (b) Terminating
   (c) Not terminating    (d) Terminating
   (e) Not terminating    (f) Not terminating

**12.** (a) 2.4
   (c) Not terminating
   (e) Not terminating

**13.** (a) 850    (b) 32,500    (c) 127,400
   (d) 8400    (e) 27,500    (f) 2.5

**14.** (a) 0.000085    (c) 0.01274    (e) 0.00275

**15.** (a) Two and thirty six thousandths
   (b) Three and seventy one ten-thousandths
   (c) Sixteen ten-thousandths
   (d) Seven ten-thousandths

**16.** (a) $(2 + 3) + (.01 + .56) = 5.57$
   (c) $(14.35 + .02) - (8.98 + .02); 14.37 - 9 = 5.37$

**17.** (a) 91.48, 91.5, 91.5

   Check: For the nearest hundredth: MATH
   (underline *round*) 91.4833 , 2 = 91.48

   (c) 0.85, 0.9, 0.854

   Check: For the nearest hundredth: MATH
   (underline *round*) .8535 , 2 = .85

   (e) 273.87, 273.9, 274

   Check: For the nearest hundredth: MATH
   (underline *round*) 273.871 , 2 = 273.87

**18.** (a) The largest number (with two decimal places) is 4499.99; there is no largest real number; 3500.

   (c) The largest number with three decimal places is 2.149; there is no largest real number; 2.05

**19.** (a) $1.1 \cdot 10^7$    (b) $3.3 \cdot 10^{10}$    (c) $3.3 \cdot 10^{-6}$
   (d) $6.73 \cdot 10^{13}$    (e) $3.6754 \cdot 10^{-2}$    (f) $3.891 \cdot 10^3$

**20.** (a) 0.46    Check 4.6 EE (−) 1 = .46

   (c) 0.000006    Check: 6 EE (−) 6 = .000006

**21.** (a) 35.841    Check: 28.32 + 7.521 = 35.841

   (c) 5.30311    Check: .3 + 5.00311 = 5.30311

**22.** (a) $28\dfrac{32}{100} + 7\dfrac{521}{1000} = 35\dfrac{841}{1000} = 35.841$

   (c) $\dfrac{3}{10} + 5\dfrac{311}{100,000} = 5\dfrac{30311}{100,000} = 5.30311$

   (e) $354\dfrac{51}{100} - 38\dfrac{64}{100} = 315\dfrac{87}{100} = 315.87$

**23.** (a) $2 \cdot 10^1 + 8 \cdot 10^0 + 3 \cdot 10^{-1} + 2 \cdot 10^{-2}$
$$\dfrac{7 \cdot 10^0 + 5 \cdot 10^{-1} + 2 \cdot 10^{-2} + 1 \cdot 10^{-3}}{2 \cdot 10^1 + 15 \cdot 10^0 + 8 \cdot 10^{-1} + 4 \cdot 10^{-2} + 1 \cdot 10^{-3}}$$
$3 \cdot 10^1 + 5 \cdot 10^0 + 8 \cdot 10^{-1} + 4 \cdot 10^{-2} + 1 \cdot 10^{-3}$
$= 35.841$

   (b) $5 \cdot 10^{-1} + 6 \cdot 10^{-2}$
$$+ \dfrac{6 \cdot 10^{-3}}{5 \cdot 10^{-1} + 6 \cdot 10^{-2} + 6 \cdot 10^{-3} = 0.566}$$

   (c) $3 \cdot 10^{-1}$
$$\dfrac{5 \cdot 10^0 \qquad + 3 \cdot 10^{-3} + 1 \cdot 10^{-4} + 1 \cdot 10^{-5}}{5 \cdot 10^0 + 3 \cdot 10^{-1} + 3 \cdot 10^{-3} + 1 \cdot 10^{-4} + 1 \cdot 10^{-5}}$$
$= 5.30311$

   (d) $8 \cdot 10^1 + 4 \cdot 10^0 + 5 \cdot 10^{-1}$
$$- \dfrac{9 \cdot 10^0 + 7 \cdot 10^{-1} + 2 \cdot 10^{-2}}{7 \cdot 10^1 + 4 \cdot 10^0 + 7 \cdot 10^{-1} + 8 \cdot 10^{-2} = 74.78}$$

   (e) $3 \cdot 10^2 + 5 \cdot 10^1 + 4 \cdot 10^0 + 5 \cdot 10^{-1} + 1 \cdot 10^{-2}$
$$- \dfrac{[3 \cdot 10^1 + 8 \cdot 10^0 + 6 \cdot 10^{-1} + 4 \cdot 10^{-2}]}{3 \cdot 10^2 + 1 \cdot 10^1 + 5 \cdot 10^0 + 8 \cdot 10^{-1} + 7 \cdot 10^{-2}}$$
$= 315.87$

   (f) $3 \cdot 10^2 + 8 \cdot 10^1 + 9 \cdot 10^0 + 2 \cdot 10^{-1} + 7 \cdot 10^{-2}$
$$- \dfrac{[6 \cdot 10^1 + 3 \cdot 10^0 + 9 \cdot 10^{-1} + 9 \cdot 10^{-2}]}{3 \cdot 10^2 + 2 \cdot 10^1 + 5 \cdot 10^0 + 2 \cdot 10^{-1} + 8 \cdot 10^{-2}}$$
$= 325.28$

**24.** (a) $11 < (4.11 + 7.68) < 13$

**25.** (a) $4 + 8 = 12$    (b) $4 + 9 = 13$
   (c) $8 + 4 = 12$    (d) $12 + 8 = 20$

**26.** (a) 1400

**27.** (a) $4 <$ difference $< 6$    (b) $2 <$ difference $< 4$

**28.** (a) 4

**29.** (a) 4.24    (b) 72.2

**30.** (a) 152.43

**31.** Let $x$ be the number of miles for Sue.
$x + 16.2 + 13.5 = 40; x = 10.3$ miles
Check: $10.3 + 16.2 + 13.5 = 40.0$ miles

**33.** Let $x$ be the amount of gasoline in the tank.
$x + 14.6 = 20; x = 5.4$ gal.
Check: $5.4 + 14.6 = 20$ gal.

**35.** Let $x$ be the distance the two trains are apart.
$x = 53.6 + 47.3 - 61.4; x = 39.5$ km
Check: $39.5 + 61.4 = 53.6 + 47.3$
$100.9 = 100.9$

**36.** (a) 18.88979 (18.7)

**37.** (a) 10    (b) 16    (c) 1    (d) 9

**38.** (a) $100.111_{two}$

**39.** (a) $.48_{ten}$    (b) $1.56_{ten}$
   (c) $1.625_{ten}$    (d) $0.375_{ten}$

**EXERCISE SET 2, P. 338**

**1.** (a) 0.16002    Check: 7.62 × .021 = .16002
   (c) 2.64    Check: 32.736 ÷ 12.4 = 2.64

**2.** (a) $\dfrac{16002}{100,000}$    (c) $\dfrac{264}{100}$

**3.** (a) 16.3    Check: 66.83 $\boxed{\div}$ 4.1 $\boxed{=}$ 16.3
   (c) 17.1    Check: 5.301 $\boxed{\div}$ .31 $\boxed{=}$ 17.1

**4.** Part (a) is the same as parts (b), (e), and (f). We determine which quotients are identical by moving the decimal point.

**5.** (a) 0.0423    Check: 7.05 $\boxed{\times}$ .006 $\boxed{=}$ 0.0423
   (c) 6.63    Check: 2.04 $\boxed{\times}$ 3.25 $\boxed{=}$ 6.63

**6.** (a) 12    (c) 1

**7.** (a) 3160    (b) 0.00316    (c) 22
   (d) 2.125    (e) 90    (f) 78

**8.** (a) $48 < (8.576)(6.459) < 63$
   (c) $3 < (12.76 \div 3.41) < 5$

**9.** (a) $9 \cdot 6 = 54 \approx 50$    (b) $9 \cdot 8 = 72 \approx 70$
   (c) $10 \div 3 \approx 3$    (d) $10 \div 4 \approx 3$

**10.** (a) $>$    (c) $>$    (e) $<$

**11.** 3000; 3398.68

**13.** 0.0003; 0.000216 (approximately)

**15.** 3,000,000; 4,717,741.935

**17.** (a) Probably correct    (b) Not correct
   (c) Not correct    (d) Not correct
   (e) Not correct

**18.** (a) 4    (c) 3    (e) 3

**19.** (a) 28; 30.3366    (b) 10; 8.750406504
   (c) 350; 300    (d) 0.008; 0.00972

**20.** (a) 57.76    (c) 0.00595    (e) 943

**21.** (a) $1.2(10)^2$    (b) $2.8(10)^9$
   (c) $1(10)$    (d) 0.43

**22.** (a) $0.\overline{142857}$    Check: .142857142857 $\boxed{\blacktriangleright F}$ $\boxed{=}$ $\frac{1}{7}$

**23.** (a) $\dfrac{173}{500}$    Check: Set $\boxed{\text{FracMode}}$ on $d/e$ and
   *auto.* .346 $\boxed{\blacktriangleright F}$ $\boxed{=}$ $\frac{173}{500}$

   (c) $\dfrac{2}{11}$    Check: .1818181818 $\boxed{\blacktriangleright F}$ $\boxed{=}$ $\frac{2}{11}$

   (e) $\dfrac{823}{50}$    Check: 16.4599999999 $\boxed{\blacktriangleright F}$ $\boxed{=}$ $\frac{823}{50}$

   (g) $\dfrac{69}{55}$    Check: 1.25454545454 $\boxed{\blacktriangleright F}$ $\boxed{=}$ $\frac{69}{55}$

   (i) $\dfrac{1}{110}$    Check: .00909090909 $\boxed{\blacktriangleright F}$ $\boxed{=}$ $\frac{1}{110}$

**24.** (a) $\dfrac{6}{11}, \dfrac{5}{9}, \dfrac{19}{34}, \dfrac{26}{41}$

**25.** Nonterminating

**27.** Let $x$ be the weight of the satchel.
   $x + 2(1.03) + 3(.72) = 5.46$; $x = 1.24$ kg
   Check: $2.06 + 2.16 + 1.24 = 5.46$

**29.** Let $a$ represent average speed.
   $6(a) = 300$; $a = 50$ miles per hour
   Check: $50(6) = 300$

**30.** (a) $51.82

**31.** (a) 17.62 kg
   (b) Divinity = $12.50
   Fudge = $13.76
   Chocolate-covered peanuts = $3.24
   Mixed nuts = $13.70
   (c) $43.20

**33.** (a) 16, 16.259, 16.2591, 16.3
   (b) 0.013, 0.0136, 0.01365, 0.013654
   (c) 0.036, 0.063, 0.306, 0.603
   (d) 2.024, 2.204, 2.402, 2.420

**34.** (a) $x = 0.81$ (rounded)

**35.** (a) 2.925    Check: .175 $\boxed{+}$ 2 $\boxed{\text{Unit}}$ 3 $\boxed{/}$ 4 $\boxed{=}$ 2.925

   (c) 3.765    Check: 6 $\boxed{\text{Unit}}$ 3 $\boxed{/}$ 8 $\boxed{-}$ 2.61 $\boxed{\blacktriangleright \square}$ $\boxed{=}$ 3.765

   (e) 1.375    Check: 2 $\boxed{\text{Unit}}$ 1 $\boxed{/}$ 2 $\boxed{\times}$ .55 $\boxed{\blacktriangleright \square}$ $\boxed{=}$ 1.375

   (g) 0.315    Check: .21 $\boxed{\times}$ 2 $\boxed{\text{Unit}}$ 1 $\boxed{/}$ 4 $\boxed{\div}$ 1 $\boxed{\text{Unit}}$ 1 $\boxed{/}$ 2 $\boxed{\blacktriangleright \square}$ $\boxed{=}$ 0.315

**36.** (a) 6.7015, 6.705, 6.71, 6.715

**37.** (a) $7 \cdot 10^8$    Check: 3.4 EE 3 $\boxed{\times}$ 2 EE 5 $\boxed{=}$ $7 \times 10^8$, when rounded to the correct number of significant digits.

   (c) $-3.4(10)^{-2}$    Check: 1.61 EE $\boxed{(-)}$ 4 $\boxed{\times}$ $\boxed{(-)}$ 2.1 EE 2 $\boxed{=}$ $-3.4 \times 10^{-2}$, when rounded to the correct number of significant digits.

**38.** (a) $2(10)^2$    Check: 3.4 EE $\boxed{(-)}$ 3 $\boxed{\div}$ 2 EE $\boxed{(-)}$ 5 $\boxed{=}$ $2 \times 10^2$, when rounded to the correct number of significant digits.

**39.** (a) $\dfrac{1409}{99}$    Check: Set $\boxed{\text{FracMode}}$ on $d/e$ and
   *auto.* 14.232323232323 $\boxed{\blacktriangleright F}$ $\boxed{=}$ $\frac{1409}{99}$

   (c) $\dfrac{4952}{165}$    Check: 30.01212121212121 $\boxed{\blacktriangleright F}$ $\boxed{=}$ $\frac{4952}{165}$

**40.** (a) $0.\overline{6}$

**41.** (a) $4.1 \cdot 10^3$    Check: 3.6 EE 3 $\boxed{-}$ 2.1 EE 3 $\boxed{+}$ 2.61 EE 3 $\boxed{=}$ $4.1 \times 10^3$, when rounded to the correct number of significant digits.

   (c) $1.7 \cdot 10^4$    Check: 4.6 EE 3 $\boxed{+}$ 1.02 EE 2 $\boxed{+}$ 1.2 EE 4 $\boxed{=}$ $1.7 \times 10^4$, when rounded to the correct number of significant digits.

**43.** No; when you divide by something less than 1, the answer is greater than the dividend.

**45.** (a) $(24.02)_{\text{five}}$  (b) $(26.744)_{\text{twelve}}$
  (c) $(1654.134)_{\text{seven}}$  (d) $(0.111111)_{\text{two}}$
  (e) $(10.0011)_{\text{two}}$  (f) $(100.002)_{\text{three}}$

**46.** (a) 5.2

**47.** 3.655    Check: $7 \boxed{-} 3.285 \boxed{-} .06 \boxed{=} 3.655$

EXERCISE SET 3, P. 348

**1.** (a) 0.055    (b) 0.00012    (c) 0.0031
  (d) 4.26    (e) 0.436    (f) 0.184

**2.** (a) 14.7%  Check: $.147 \boxed{\blacktriangleright \%} \boxed{=} 14.7\%$
  (c) 0.5%  Check: $.005 \boxed{\blacktriangleright \%} \boxed{=} 0.5\,\%$
  (e) 861%  Check: $8.61 \boxed{\blacktriangleright \%} \boxed{=} 861\%$

**3.** (a) 120%  Check: $18 \boxed{/} 15 \boxed{\blacktriangleright \%} \boxed{=} 120\%$
  (c) 40%  Check: $6 \boxed{/} 15 \boxed{\blacktriangleright \%} \boxed{=} 40\%$
  (e) 27.5%  Check: $11 \boxed{/} 40 \boxed{\blacktriangleright \%} \boxed{=} 27.5\%$

**4.** (a) 34    (c) 123.45    (e) 0.66

**5.** (a) 34    (b) 276    (c) 123.45
  (d) 1,025    (e) 0.66    (f) 5.28

**7.**

| Paring knife | 0.36 | ($1.56) |
|---|---|---|
| Electric clock | 2.52 | ($10.92) |
| Can opener | 0.51 | ($2.21) |
| Wastebasket | 1.08 | ($4.68) |

**8.** (a) 25%    (c) 32    (e) 12.48    (g) 300%

**9.** (a) $1736    (b) 25%    (c) $70

**10.** (a) 19.35%    **11.** $16\frac{2}{3}\%$

**13.** $988; $\approx 20.24\%$    **15.** 20 hr.    **17.** $20.00

**19.** 31.82%    **21.** 200 g    **23.** $\approx 75.5\%$

**24.** (a) $318.08    **25.** $1578    **27.** $\approx 57.97\%$

**29.** $19.99    **30.** (a) $35.32    **31.** $21,767.10

**33.** 91%    **35.** 150 lb.

**37.** (a) 7.2036  Check: $2.001 \boxed{\times} 3.6 \boxed{=} 7.2036$

**38.** $\frac{3410}{999}$  Check: $3.413413413 \boxed{\blacktriangleright F} \boxed{=} \frac{3410}{999}$

EXERCISE SET 4, P. 360

**1.** (a) Irrational    (b) Irrational
  (c) Irrational    (d) Rational
  (e) Irrational    (f) Rational
  (g) Rational    (h) Irrational
  (i) Irrational    (j) Rational
  (k) Rational

**2.** (a) False    (c) False    (e) True    (g) False
  (i) False    (k) True    (m) True    (o) False
  (q) True

**3.** Let $A$ be a whole number, $B$ an integer, $C$ a rational number, $D$ an irrational number and $E$ a real number.

| Number | A | B | C | D | E |
|---|---|---|---|---|---|
| −5 | No | Yes | Yes | No | Yes |
| $2.\overline{71}$ | No | No | Yes | No | Yes |
| 0.717117111 . . . | No | No | No | Yes | Yes |
| 0 | Yes | Yes | Yes | No | Yes |
| $5\sqrt{2}$ | No | No | No | Yes | Yes |
| $3\pi$ | No | No | No | Yes | Yes |
| $0.\overline{45}$ | No | No | Yes | No | Yes |
| 4 | Yes | Yes | Yes | No | Yes |

**4.** (a) Yes, 7    **5.** No

**6.** (a) $\sqrt{2} \cdot \sqrt{3} = \sqrt{6}$, irrational; $\sqrt{2} \cdot \sqrt{2} = 2$, rational

**7.** $(3) + (\sqrt{5}) = 3 + \sqrt{5}$, which is irrational.

**9.** (a) 4
  (b) Infinitely many: $-3/2, 0, 1.2\overline{13}$
  (c) Infinitely many: $-2 + \sqrt{2}, \frac{\pi}{2}, 1.131131113 \ldots$

**10.** (a) 9

**11.** $\frac{3}{4}, 0.7510110111 \ldots, 0.75\overline{1}, 0.755, 0.\overline{75}$

**13.** (a) 4    (b) 3    (c) 2    (d) 15

**14.** (a) False

**15.** $\sqrt{0.2\overline{16}}$ and 0.466141141114 . . .

**17.** 4.5121121112 . . . , 4.5232232223 . . . , 4.5343343334 . . .

**18.** (a) 0.56521

**19.** (a) 7    (b) 11    (c) 3
  (d) 2    (e) −9    (f) −14

**20.** (a) $4\sqrt{2}$

**21.** (a) $3\sqrt{5}$    (b) $-3\sqrt{7}$    (c) $4\sqrt{2}$
  (d) $2\sqrt{6}$    (e) $20\sqrt{3}$    (f) $7\sqrt{5}$

**22.** (a) 6  Check: $36 \boxed{\wedge} \boxed{(} 1 \boxed{/} 2 \boxed{)} \boxed{=} 6$

**23.** (a) $5\sqrt{2}$  Check: $\boxed{\sqrt{}} 50 \boxed{=}$ approximately 7.07107 and $5 \boxed{\times} \boxed{\sqrt{}} 2 \boxed{=}$ 7.07107. This suggests that $\sqrt{50} = 5\sqrt{2}$.

  (c) $\frac{7}{6}\sqrt{3}$  Check: $\boxed{\sqrt{}} 147 \boxed{/} 36 \boxed{=}$ approximately 2.0207 and $\boxed{(} 7 \boxed{/} 6 \boxed{)} \boxed{\times} \boxed{\sqrt{}} 3 \boxed{=}$ approximately 2.0207. This suggests that $\sqrt{147/36} = (7/6)\sqrt{3}$.

**24.** (a) $\frac{8 - \sqrt{2}}{4}$

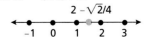

**25.** (a) $\frac{5}{2}\sqrt{2}$ (b) $\frac{7\sqrt{5}}{5}$ (c) $\frac{1}{2}\sqrt{14}$ (d) $\frac{1}{3}\sqrt{21}$

(e) $\frac{1}{7}\sqrt{105}$ (f) $\frac{1}{3}\sqrt{15}$ (g) $\frac{1}{4}\sqrt{6}$ (h) $\frac{1}{6}\sqrt{10}$

**26.** (a) $\frac{1}{160}$  Check: 4 ⌃ (−) 2 × 5 ⌃ (−) 1 ÷

2 ▶F = $\frac{1}{160}$

**27.** (a) (3rd) (b) (5th) (c) (7th) (d) (5th)

(e) (4th)

**29.** (a) $-3\sqrt[3]{3}$ (b) $2\sqrt{2}$ (c) $-2\sqrt[3]{3}$

(d) $-2$ (e) $2$ (f) $2\sqrt{2}$

**31.** $\frac{1}{4}+\frac{\sqrt{2}}{1000}, \frac{1}{4}+\frac{2\sqrt{2}}{1000}, \frac{1}{4}+\frac{3\sqrt{2}}{1000}, \ldots, \frac{1}{4}+\frac{10\sqrt{2}}{1000}$

**33.**

**35.**

**37.** (a) 16.3771 (b) 0.1516575

(c) 0.046637 (d) 68.425872

**38.** (a) $1.47 \cdot 10^2$ **39.** 0.405123 (rounded)

**41.** (a) $2 + \frac{1}{2} = 2\frac{1}{2}$ and $2\frac{1}{2} > 2$

(b) $0.5 + \frac{1}{0.5} = 2\frac{1}{2}$ $2\frac{1}{2} > 2$

(c) $4 + \frac{1}{4} = 4\frac{1}{4}$ and $4\frac{1}{4} > 2$

(d) $n + \frac{1}{n} = \frac{n^2 + 1}{n} \geq 2$

or $n^2 - 2n + 1 \geq 0$

$(n-1)^2 \geq 0$

$(n-1)^2 > 0$ for all $n > 1$

**43.** Let $a = 4$ and $b = 16$.  (counterexample)

$\sqrt{4 + 16} \neq \sqrt{4} + \sqrt{16}$

$\sqrt{20} \neq 2 + 4$ or $2\sqrt{5} \neq 6$

**44.** (a) Suppose $3\sqrt{2}$ is rational. Then there exist integers $a$ and $b$ (with $b \neq 0$) such that $3\sqrt{2} = \frac{a}{b}$. Then, $\sqrt{2} = \frac{a}{3b}$, which is rational. But, this contradicts $\sqrt{2}$ being irrational. Thus, $3\sqrt{2}$ is irrational.

**45.** $3.7\overline{\overline{5}}75 = \frac{124}{33}$

**47.** (a) False  4 is rational (b) True (c) True

(d) False  $\sqrt{2}$ is irrational and real

(e) False  $\sqrt{2} \cdot \sqrt{2} = 2$, which is rational

(f) True

(g) False  Irrationals do not repeat as decimals

(h) False  $\frac{a}{0}$ is not defined

(i) False  If $b = 0$, $a$ does not need to be 0

(j) False  Division by 0 is not possible

(k) False  There is a larger number $\frac{291}{292}$, etc. . . .

(l) False  There are positive rational numbers less than 1.

(m) True

(n) False  0 is neither positive nor negative.

(o) True

(p) False  If $a = 0$, $a^2 = 0$

(q) False  $\frac{-10}{11} > -1$

**48.** (a) $3.30

**49.** (a) 26.97% (b) $720 (c) $35

**CHAPTER TEST, P. 364**

**1.** 0.0234

**2.** (a) $8\sqrt{3}$ (b) 3

**3.** (a) $\frac{366}{100} = \frac{183}{50}$ (b) $\frac{25}{100} = \frac{1}{4}$

**4.** 2.43515515551 . . . ; 2.43616616661 . . .

**5.** (a) Rational (b) Irrational (c) Irrational

**6.** (a) 3.17 (b) 7.25 (c) 13.24

**7.** (a) $2.1(10)^{-4}$ (b) $4.67300(10)^2$

**8.** 6.6

**9.** 428.3

**10.** 50

**11.** 30.4

**12.** (a) 143.24872 (b) $\frac{14{,}324{,}872}{100{,}000}$ (c) Yes

**13.** $\frac{69}{55}$

**14.** $x < 24$

**15.** $2086.96

**16.** $x = \frac{3 - \sqrt{3}}{\pi - \sqrt{5}}$

# CHAPTER 8

● ● ● ● ● ● ● ● ● ● ● ● ● ● ● ● ● ● ● ● ● ● ● ● ● ● ● ● ●

**EXERCISE SET 1, P. 376**

**1.** (a)$80 (b) $36 (c) $100

**2.** (a) $580 (c) $600

**3.** $I = $360$, $A = $3360$

**4.** (a) $I = $624.32$, $A = $5624.32$

(c) $I = $978.09$, $A = $5978.09$

**5.** (a) $I = $339.72$, $A = $2339.72$

(b) $I = $343.32$, $A = $2343.32$

(c) $I = $345.16$, $A = $2345.16$

**6.** (a) $I = $1020.29$, $A = $4020.29$

(c) $I = $1042.05$, $A = $4042.05$

**7.** (a) $4400 (b) $3090 (c) $124

**8.** (a) 9% (c) 16% **9.** $3736.29

**11.** $5525.86

**13.** $r = \dfrac{I}{Pt}$

**15.** Between 18.25 and 18.5 years

**17.** $21.81

**19.** 22.5 or 23 days

**21.** (a) $134,391.64    (b) $2.02
(c) $6.72    (d) $19.69

**22.** (a) $179,084.77

**23.** (a) $215,892.50    (b) $3.24
(c) $10.79    (d) $31.63

**24.** (a) Between 23 and 24 years

**25.** 88,814 (truncated)

EXERCISE SET 2, P. 381

**1.** $1318.08. Amount of an annuity of $100 a year at 6% compounded annually for 10 years (one possibility)

**3.** $4842.98. Amount of an annuity of $200 a year at 8% compounded annually for 14 years (one possibility)

**5.** $1318.08    **7.** $2687.04

**9.** $106,029.77

**11.** (a) $45,761.96    (b) $13,288.67    (c) $6569.83

**13.** $5866.60    **15.** $265.81    **17.** $263.38

**19.** $3224.44; $7224.44

EXERCISE SET 3, P. 388

**1.** $368.00. Present value of an annuity of $50 a year at an interest rate of 6% compounded annually for 10 years (one possibility)

**3.** $1648.85. Present value of an annual annuity of $200 at an interest rate of 8% compounded annually for 14 years (one possibility)

**5.** $736.01

**7.** (a) 54.94    (b) 13.5%    **8.** (a) $50

**9.** (a) $9818.15    (b) $9909.41    (c) $4866.54

**10.** (a) $78.85

**11.** $44,160.52    **13.** $55.24    **15.** $267.17

**17.** $455.02    **19.** $180

CHAPTER TEST, P. 389

**1.** $790.31    **2.** $19,668.05    **3.** 40%

**4.** $2539.47    **5.** $811.09    **6.** $2000

**7.** $2000    **8.** $2090    **9.** $1213.28

**10.** 12.68%; No    **11.** Approximately 35 years.

**12.** $199,193.99

# CHAPTER 9

EXERCISE SET 1, P. 400

**1.** (a) 4    (b) {blue, black, yellow, green}
(c) $\dfrac{1}{4}$    (d) $\dfrac{1}{4}$    (e) Yes

**2.** (a) {H, O, P, E, F, U, L}    (c) 1/7

**3.** $\dfrac{1}{4}$

**5.** (a) 6    (b) {A, B, C, D, E, F}
(c) $\dfrac{11}{60}$    (d) $\dfrac{1}{6}$
(e) Empirical probability only approximates theoretical probability.
(f) Possibly. As the number of trials increases, empirical probability should more closely approximate theoretical probability.

**6.** (a) $\dfrac{19}{40}$
(c) The answer in (a) is an approximation. The accuracy of this answer will increase with additional tosses of the coin.

**7.** (a) {A, B, C}    (b) {R, G}
(c) {1, 2, 3, 4, 5, 6}    (d) {P, R} or {P, P, P, R}

**8.** (a) $\dfrac{1}{3}$    (c) $\dfrac{1}{6}$    **9.** $\dfrac{2}{2}$ or 1

**11.** (a) 5000
(b) No, because accuracy increases with the number of tosses

**13.** {1, 2, 3, 4, 5, 6}
$P(1) = .181, P(2) = .152,$
$P(3) = .144, P(4) = .138,$
$P(5) = .156, P(6) = .229$

**15.** {bass, bream, carp, catfish}
$P$(bass) = .235; $P$(bream) = .327
$P$(carp) = .144; $P$(catfish) = .294

**16.** (a) No, probability cannot be negative.    (c) Yes
(e) No, probability cannot be larger than 1.
(g) No, probability cannot be larger than 1.

**17.** (a) {HHH, HHT, HTH, HTT, THH, THT, TTH, TTT}    (b) 1/8

**18.** (a) {(1, 1), (1, 2), (1, 3), (1, 4), (2, 1), (2, 2), (2, 3), (2, 4), (3, 1), (3, 2), (3, 3), (3, 4), (4, 1), (4, 2), (4, 3), (4, 4)}

**19.** {H1, H2, H3, H4, H5, H6, T1, T2, T3, T4, T5, T6}

**20.** (a) {R, B}

**21.** (a) No; neglects 0 heads as an outcome
(b) No; neglects 3 heads as an outcome
(c) No, 4 is not a possible outcome.    (d) Yes

**22.** (a) Yes, but the odd number must be 1 or 3.

**23.** $\dfrac{1}{2}$

**1.** (a) $\dfrac{1}{5}$  (b) $\dfrac{2}{5}$  **2.** (a) $\dfrac{5}{11}$

**3.** (a) $\dfrac{1}{2}$  (b) $\dfrac{1}{3}$  (c) $\dfrac{2}{3}$

(d) 1  (e) $\dfrac{1}{3}$

**4.** (a) $\dfrac{1}{4}$  (c) $\dfrac{1}{52}$  (e) $\dfrac{1}{26}$  **5.** $\dfrac{1}{5}; \dfrac{4}{5}$

**6.** (a) Any 1 of the 35 students would be an equally likely outcome.  (c) $\dfrac{1}{35}$  (e) 1

**7.** (a) $\dfrac{2}{7}$  (b) $\dfrac{3}{14}$  (c) $\dfrac{1}{2}$

(d) $\dfrac{5}{7}$  (e) $\dfrac{11}{14}$  (f) 1

**8.** (a) $\dfrac{1}{2}$  (c) $\dfrac{1}{4}$

**9.** (a) Getting less than 2 heads (0 or 1)
(b) Getting 0, 1, 2, 6, or 7 tails
(c) Getting 0, 2, 3, 4, 5, 6, or 7 tails
(d) Getting at least 1 head

**11.** .35  **12.** (a) $\dfrac{1}{2}$  (c) $\dfrac{12}{13}$

**13.** (a) $\dfrac{1}{4}$  (b) $\dfrac{1}{4}$  **14.** (a) $\dfrac{3}{4}$

**15.** (a) .48  (b) .52  (c) .1  (d) .9

**16.** (a) $\dfrac{31}{125}$  **17.** 1 to 3  **19.** 12 to 1

**21.** (a) .4  (b) .8  (c) .7  (d) .9

**22.** (a) $\dfrac{1}{2}$

**23.** (a) $\dfrac{3}{4}$  (b) $\dfrac{3}{5}$  (c) $\dfrac{17}{20}$  (d) $\dfrac{7}{10}$

**25.** $P(A) = \dfrac{1}{4}$  $P(\overline{A}) = \dfrac{12}{16} = \dfrac{3}{4}$

$P(A) = 1 - P(\overline{A}) = 1 - \dfrac{12}{16} = \dfrac{1}{4}$

**26.** (a) .5

**27.** (a) $\dfrac{1}{16}$  (b) $\dfrac{6}{16} = \dfrac{3}{8}$

(c) $\dfrac{5}{16}$  (d) $\dfrac{4}{16} = \dfrac{1}{4}$

**28.** (a)

**1.** (a) $\dfrac{5}{18}$  (b) $\dfrac{5}{18}$

**2.** (a)

**3.** −$.55

**5.**

$$\underset{T}{\dfrac{2}{6}} \quad \underset{E}{\dfrac{1}{5}} \quad \underset{L}{\dfrac{2}{4}} \quad \underset{L}{\dfrac{1}{3}}$$

$P(\text{TELL}) = \dfrac{1}{90}$

**7.** $\dfrac{15}{56}$  **9.** $\dfrac{1}{256}$  **10.** (a) $\dfrac{4}{77}$

**11.** (a) $\dfrac{6}{121}$  (b) $\dfrac{4}{121}$  **12.** (a) $\dfrac{1}{169}$

**13.** Starting with the first entry in Table 9-6 and using 0–4 as a head and 5–9 as a tail:

|  | Frequency | Empirical Probability |
|---|---|---|
| HHH | 4 | .08 |
| HHT | 4 | .08 |
| HTH | 4 | .08 |
| HTT | 10 | .20 |
| THH | 9 | .18 |
| THT | 8 | .16 |
| TTH | 8 | .16 |
| TTT | 3 | .06 |

**14.** (a) One or 2 represent doll 1; 3 or 4 represent doll 2; 5 or 6 represent doll 3. Roll 5 times.

**15.** (a) If an even number is rolled on the die, the coin shows "heads." If an odd number is rolled, the coin shows "tails."
(b) Roll the die 5 times, letting an even number represent a head and an odd number a tail.

(c) Let 1 and 2 represent a hit and 3, 4, 5, and 6 a miss; roll the die 3 times for the 3 times at bat.

(d) Let an odd number represent a boy and an even number a girl; roll the die 4 times for 4 children.

**16.** (a) If an even-numbered card is drawn, the coin shows "heads"; if an odd-numbered card is drawn, the coin shows "tails."

**17.** (a)

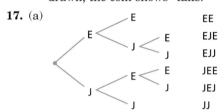

(b) .52    (c) .352    (d) .648

**18.** (a)

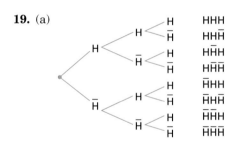

**19.** (a)

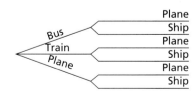

(b) .027    (c) .216    (d) .343

**21.** $\dfrac{21}{95}$    **23.** $5\dfrac{5}{16}$ min.    **25.** .0288

**27.** (a) $\dfrac{13}{18}$    (b) $\dfrac{5}{9}$    **28.** (a) $\dfrac{65}{81}$

**29.** (a) $\dfrac{10}{13}$    (b) 0    (c) 1    (d) $\dfrac{4}{13}$

(e) $\dfrac{9}{13}$    (f) $\dfrac{3}{13}$    (g) 1    (h) $\dfrac{10}{13}$

**EXERCISE SET 4, P. 426**

**1.** (a) 6 ways

(b) Atlanta    Boston    London

Bus
Train
Plane
Plane
Ship
Plane
Ship
Plane
Ship

**2.** (a) $2 \cdot 4 \cdot 4 = 32$ ways    (c) $\dfrac{1}{32}$

**3.** (a) 24    (b) 576

**5.** (a) 2    (b) 6    (c) 24

**6.** (a) 24    Check: 4 $\boxed{\text{PRB}}$ (underline *!*) $\boxed{=}$ $\boxed{=}$ 24

(c) 1    Check: 0 $\boxed{\text{PRB}}$ (underline *!*) $\boxed{=}$ $\boxed{=}$ 1

**7.** (a) 6    (b) *PQR; PRQ; QPR; QRP; RPQ; RQP*

**8.** (a) 12    **9.** 18

**10.** (a) 60    Check: 5 $\boxed{\text{PRB}}$ (underline $_nP_r$) 3 $\boxed{=}$ 60

(c) 8    Check: 8 $\boxed{\text{PRB}}$ (underline $_nP_r$) 1 $\boxed{=}$ 8

(e) 42    Check: 7 $\boxed{\text{PRB}}$ (underline $_nP_r$) 2 $\boxed{=}$ 42

**11.** (a) $r$    (b) $k^2 - k$    (c) $r!$

(d) $\dfrac{k!}{2}$    (e) $k(k - 1)(k - 2)$    (f) $\dfrac{k!}{6}$

**13.** (a) 10,000    (b) 5040    **14.** (a) 900

**15.** (a) $\dfrac{1}{14}$    (b) $\dfrac{1}{6}$    (c) $\dfrac{1}{35}$    (d) $\dfrac{3}{7}$

**17.** $\dfrac{1}{10^8}$

**19.** (a) $\dfrac{1}{2}$    (b) $\dfrac{3}{10}$    (c) $\dfrac{1}{5}$    (d) $\dfrac{3}{5}$

(e) $P(A \cup B) = P(A) + P(B) - P(A \cap B)$

**20.** (a) $\dfrac{1}{169}$    **21.** −3 points    **22.** (a) $\dfrac{13}{51}$

**EXERCISE SET 5, P. 431**

**1.** (a) 6: *WX, WY, WZ, XY, XZ, YZ*

(b) 12: *WX, WY, WZ, XW, XY, XZ, YW, YX, YZ, ZW, ZX, ZY*

(c) $12 = 2 \cdot 6$

(d) 4: *WXY, WXZ, WYZ, XYZ*

(e) 24: *WXY, WYX, WXZ, WZX, WYZ, WZY, XWY, XYW, XWZ, XZW, XYZ, XZY, YWX, YXW, YWZ, YZW, YXZ, YZX, ZWX, ZXW, ZWY, ZYW, ZXY, ZYX*

(f) $24 = 6 \cdot 4$

**3.** $P(12, 4) = 11,880$    **5.** $P(12, 5) = 95,040$

**7.** $C(22, 3) = 1540$    **9.** 133,784,560

**10.** (a) 220 Check: 12 $\boxed{\text{PRB}}$ (underline $_nC_r$)

3 $\boxed{=}$ 220

**11.** (a) 1716    (b) 100,386

(c) 725,010    (d) 1,740,024

**13.** 30    **14.** (a) $\dfrac{13}{55}$    **15.** $\dfrac{7}{95}$    **16.** (a) 24

**17.** $\dfrac{1}{33}$    **18.** (a) $\dfrac{1}{54,145} = .0000185$

**19.** (a) $\dfrac{5}{26}$    (b) $\dfrac{3}{26}$    (c) $\dfrac{1}{13}$

**20.** (a) $\dfrac{1}{12}$    **21.** $\dfrac{7}{20}$

**CHAPTER TEST, P. 434**

**1.** 1    **2.** 720

**3.** (a) {R, G, Y}
   (b) {RR, RG, RY, GR, GG, GY, YR, YG}
   (c) {RR, RG, RY, GR, GG, GY, YR, YG, YY}

**4.** (a) .65    (b) .69    **5.** (a) $\frac{16}{49}$    (b) $\frac{2}{7}$

**6.** 35    **7.** .75    **8.** 48,450    **9.** $\frac{4}{13}$

**10.** 120

**11.** (a) List all 36 possibilities to obtain a uniform sample space. {(1, 1) (1, 2) . . . (1, 6) (2, 1) . . .}

   (b) $\frac{1}{36}$    (c) $\frac{1}{6}$    (d) $\frac{1}{2}$

**12.** $\frac{1}{2}$    **13.** 1320    **14.** $\frac{2}{8} = \frac{1}{4}$    **15.** 96

# CHAPTER 10

• • • • • • • • • • • • • • • • • • • • • • • • • • • •

**EXERCISE SET 1, P. 447**

**1.** (a) 14%    (b) 21%    (c) 19%    (d) 75%

**3.** (a) Los Angeles    (b) Chicago    (c) Los Angeles

**4.** (a) 20    (c) 16

**5.** (a)

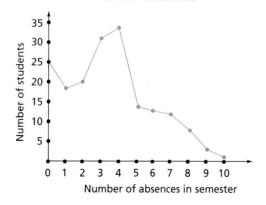

(b)    Number of absences

**6.** (a) 31, 37, 41, 42, 46, 51, 52, 53, 53, 57, 58, 59, 62, 68, 68, 69, 71, 73, 76, 82, 82, 96
   (c) 10

**7.** (a) 17, 22, 27, 32, 37

(b)

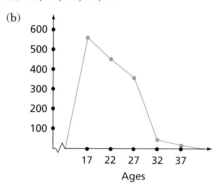

(c)

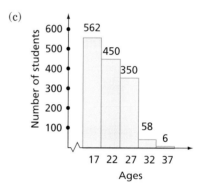

**8.** (a) 45% for Democratic candidates and 3% for Republican candidates; 16% and 30%.
   (c) 5.94 million dollars

**9.** Comments will vary.

   (a) False. People with serious illnesses are more prone to die, and these people are usually in hospitals.
   (b) Misleading because there were many more drivers last year than soldiers in World War II.
   (c) The characteristics of the students and classes are not given.
   (d) Misleading because there are many more miles driven near home than on long trips.

**10.** (a)

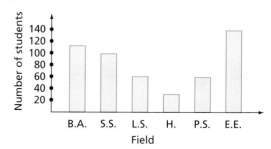

**11.** About 35

**13.** 1986–1990, approximately 110 cars

**15.**

```
0 | 7
1 | 2  4  5  6  6  6  6  7  7  8  8  9  9
2 | 1  4  6  8
3 | 2  5
4 | 1  3  8
5 | 1  9
6 | 1  1  2  2  5  5  8
7 | 2  2  3  3
```

The two groups who most heavily use the park are teenagers and people near retirement age.

**17.**

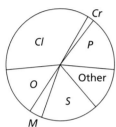

**19.** (a) 23

|  (b) Class | Frequency | (c) Class Marks |
|---|---|---|
| 40–43 | 5 | 41.5 |
| 44–47 | 12 | 45.5 |
| 48–51 | 20 | 49.5 |
| 52–55 | 6 | 53.5 |
| 56–59 | 5 | 57.5 |
| 60–63 | 2 | 61.5 |

(d)

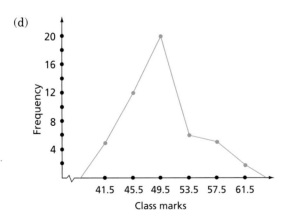

Class marks

**21.**

| Class marks | Class | Frequency |
|---|---|---|
| 167 | 160–174 | 3 |
| 182 | 175–189 | 6 |
| 197 | 190–204 | 9 |
| 212 | 205–219 | 6 |
| 227 | 220–234 | 2 |

**23.** Doubling the height and radius multiplies the *volume* by 8, not by 2. So, the visual presentation has 8 times the impact rather than 2, as it should be.

EXERCISE SET 2, P. 457

**1.** (a) Mean = 6　　　median = 5　　mode = none
　　(b) Mean = 7.5　　median = 7　　mode = 6
　　(c) Mean = $3.\overline{7}$　　median = 4　　mode = 6
　　(d) Mean = 3.625　median = 3.5　mode = 1
　　(e) Mean = 17.125　median = 19.5　mode = 23
　　(f) Mean = 14.67　median = 14　mode = 12

**3.** Mean = 27

**4.** (a) 5, 6, 7, 8　　(c) 3, 4, 4, 5　　(e) 3, 6, 6, 9

**5.** 520

**7.** (a) Median　　(b) Mean　　(c) Mode

**8.** (a) 273.23; 275; 302

　　(b) STAT (underline *1-VAR*) = DATA $(x_1 =)$ 220
　　　▼ (FRQ =) 1 ▼ $(x_2 =)$ 275 ▼ (FRQ =) 1
　　　▼ $(x_3 =)$ 199 ▼ (FRQ = ) 1 ▼ $(x_4 =)$ 246
　　　▼ (FRQ =) 1 ▼ $(x_5 =)$ 302 ▼ (FRQ =) 1
　　　▼ $(x_6 =)$ 333 ▼ (FRQ =) 1 ▼ $(x_7 =)$ 401
　　　▼ (FRQ =) 1 ▼ $(x_8 =)$ 190 ▼ (FRQ =) 1
　　　▼ $(x_9 =)$ 286 ▼ (FRQ =) 1 ▼ $(x_{10} =)$ 254
　　　▼ (FRQ =) 1 ▼ $(x_{11} =)$ 302 ▼ (FRQ =) 1
　　　▼ $(x_{12} =)$ 323 ▼ (FRQ =) 1 ▼ $(x_{13} =)$ 221
　　　▼ (FRQ =) 1 STATVAR (underline *n* and
　　　$\bar{x}$.) Read $n = 13$ and $\bar{x} = 273.23$

**9.** (a) Mean = 82.725 kg
　　(b) Median = 82 kg
　　(c) Mode = 80 kg
　　(d) The mean weight seems to be more useful. However, if you wish to downplay the extreme values you might use the median.

**11.** (a) 10, 13.5, 16
　　(b) 18, 36.5, 64

**13.** 98% of those taking the test made lower scores than Aaron

**15.** $2.\overline{7}$

**17.** (a) The president is using the mean, which is misleading because the distribution is heavily skewed on one end.
　　(b) The union is using the mode or median to show average.
　　(c) The median or mode of $20,000 is more representative.

**19.** 99

**21.** 118.214   Check: $\boxed{\text{STAT}}$ (underline *1-VAR*) $\boxed{=}$
$\boxed{\text{DATA}}$ ($x_1$ =) 144.50 $\boxed{\blacktriangledown}$ (FRQ =) 3 $\boxed{\blacktriangledown}$ ($x_2$ =)
134.5 $\boxed{\blacktriangledown}$ (FRQ =) 4 $\boxed{\blacktriangledown}$ ($x_3$ =) 124.5 $\boxed{\blacktriangledown}$
(FRQ =) 8 $\boxed{\blacktriangledown}$ ($x_4$ =) 114.5 $\boxed{\blacktriangledown}$ (FRQ =) 13 $\boxed{\blacktriangledown}$
($x_5$ =) 104.5 $\boxed{\blacktriangledown}$ (FRQ =) 4 $\boxed{\blacktriangledown}$ ($x_6$ =) 94.5 $\boxed{\blacktriangledown}$
(FRQ =) 2 $\boxed{\blacktriangledown}$ ($x_7$ =) 84.5 $\boxed{\blacktriangledown}$ (FRQ =) 0 $\boxed{\blacktriangledown}$
($x_8$ =) 74.5 $\boxed{\blacktriangledown}$ (FRQ =) 1 $\boxed{\text{STATVAR}}$ (underline
$\bar{x}$) Read $\bar{x}$ = 118.214.

**23.** (a) No. He should have multiplied each salary by
the frequency and divided by the sum of the
frequencies (number of employees).
(b) $271.21

**25.** 8.75 million gal.   Check: $\boxed{\text{STAT}}$ (underline
*1-VAR*) $\boxed{=}$ $\boxed{\text{DATA}}$ ($x_1$ =) 2 $\boxed{\blacktriangledown}$ (FRQ =) 6 $\boxed{\blacktriangledown}$
($x_2$ =) 5 $\boxed{\blacktriangledown}$ (FRQ =) 9 $\boxed{\blacktriangledown}$ ($x_3$ =) 8 $\boxed{\blacktriangledown}$ (FRQ =)
13 $\boxed{\blacktriangledown}$ ($x_4$ =) 11 $\boxed{\blacktriangledown}$ (FRQ =) 10 $\boxed{\blacktriangledown}$ ($x_5$ =) 14
$\boxed{\blacktriangledown}$ (FRQ=) 7 $\boxed{\blacktriangledown}$ ($x_6$ =) 17 $\boxed{\blacktriangledown}$ (FRQ) = 3
$\boxed{\text{STATVAR}}$ (underline $\bar{x}$) Read $\bar{x}$ = 8.75.

**26.** (a) $25,920

**27.** (a)

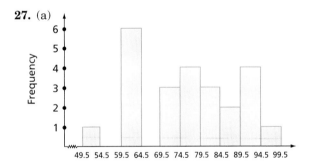

(b)

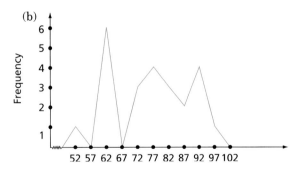

**EXERCISE SET 3, P. 467**

**1.** Range, 8; variance, 9; standard deviation, 3

**3.** Range, 20; variance, 27.44; standard deviation,
5.24

**5.** Range, 20; variance, 26.74; standard deviation
5.17

**7.**

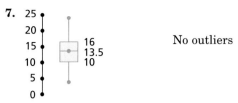

No outliers

**9.**

| Stems | Leaves |
|---|---|
| 0 | 4, 7, 8, 9 |
| 1 | 0, 2, 3, 4, 4, 6, 8 |
| 2 | 4 |

No outliers

**11.** 13.44   **12.** (a) 1   (c) −2.9   (e) 4

**13.** 3   Check: $\boxed{\text{STAT}}$ (underline *1-VAR*) $\boxed{=}$ $\boxed{\text{DATA}}$
($x_1$ =) 1 $\boxed{\blacktriangledown}$ (FRQ =) 10 $\boxed{\blacktriangledown}$ ($x_2$ =) 4 $\boxed{\blacktriangledown}$ (FRQ
=) 20 $\boxed{\blacktriangledown}$ ($x_3$ =) 7 $\boxed{\blacktriangledown}$ (FRQ =) 30 $\boxed{\blacktriangledown}$ ($x_4$ =) 10
$\boxed{\blacktriangledown}$ (FRQ =) 40 $\boxed{\text{STATVAR}}$ (underline $\sigma_x$) The
standard deviation is 3.

**15.** The scores in class B were much more scattered
from the mean than those in class A.

**17.** 2000   **19.** (b) 12

**20.** (a) Most students scored near the average, but a
small number had either extremely low or
extremely high scores.

**21.** (a) No effect
(b) Reduces to $\frac{1}{5}$ the original
(c) Reduces to $\frac{1}{5}$ the original

**23.** $IQR$ = 26

**25.** (a) Class I mean + 2s = 196.65 mean − 2s =
146.1. All but one value is in this interval;
95.8%
(b) Class II mean + 2s = 191.64 mean − 2s =
168.11. All but two values are in this interval;
91.7%

**26.** (a) *C*

**27.** $$s^2 = \frac{\sum\limits_{i=1}^{n}(x_i - \bar{x})^2}{n} = \frac{\sum\limits_{i=1}^{n}(x_i^2 - 2\bar{x}x_i + \bar{x}^2)}{n}$$

$$= \frac{\sum\limits_{i=1}^{n} x_i^2 - 2\bar{x}\sum\limits_{i=1}^{n} x_i + n\bar{x}^2}{n}$$

$$= \frac{\sum\limits_{i=1}^{n} x_i^2 - 2\bar{x}n\dfrac{\sum\limits_{i=1}^{n} x_i}{n} + n\bar{x}^2}{n}$$

$$= \frac{\sum\limits_{i=1}^{n} x_i^2 - 2n\bar{x}^2 + n\bar{x}^2}{n} = \frac{\sum\limits_{i=1}^{n} x_i^2 - n\bar{x}^2}{n}$$

**28.** (a) 18

**29.** Perhaps there are four employees with salaries of $20,000, $22,000, $24,000, and $19,000, and the president's salary is $65,000.

**EXERCISE SET 4, P. 476**

**1.** (a) .4918   (b) .1591   (c) .4525   (d) .2812

**2.** (a) .7698       (c) .9070

**3.** (a)

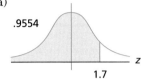

.9554

1.7

   (b)

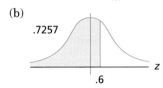

.7257

.6

**4.** (a) .8413

**5.** .9332 or 93.32%

**6.** (a)

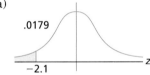

.0179

−2.1

**7.** (a) .5000   (b) .5000   (c) .1587
   (d) .9772   (e) .1587   (f) .6554
   (g) .5403   (h) .8284

**9.** (a) .9525   (b) Approximately 713

**10.** (a) 68.26%

**11.** 49.38%

**13.** (a) Approximately 994   (b) 6.68%

**15.** 1082 days

**17.** (a) −1.28       (b) 1.28
   (c) 124.4 seconds   (d) 175.6 seconds

**19.** Perhaps the president was speaking of only undergraduate enrollment, and the registrar of total enrollment.

**21.** Lonesome University is primarily a male institution. There are only 3 females at this school, one of whom (a married student) became pregnant.

**22.** (a) 75   (c) 94   (e) 64   (g) 23

**23.**

| Stems | Leaves |
|-------|--------|
| 3 | 2 |
| 4 | |
| 5 | 4 |
| 6 | 0, 1, 2, 3, 3 |
| 7 | 1, 1, 2, 5, 6, 7, 7 |
| 8 | 1, 1, 4, 5, 7 |
| 9 | 0, 4, 4, 4, 6 |

**25.** (a) Mean of $x = 28.4$; mean of $y = -.16$
   (b) $s$ of $x = 12.22$; $s$ of $y = 1.222$
   (c) (mean of $x - 30)/10 = $ (mean of $y$) or (mean of $x$) $= 10 \cdot$ (mean of $y$) $+ 30$
       ($s$ of $x$) $= 10$ ($s$ of $y$)
   (d) (For $x$'s) $\boxed{\text{STAT}}$ (underline $1$-VAR) $\boxed{=}$ $\boxed{\text{DATA}}$ $(x_1 =) 10$ $\boxed{\blacktriangledown}$ (FRQ $=) 4$ $\boxed{\blacktriangledown}$ $(x_2 =)$ $20$ $\boxed{\blacktriangledown}$ (FRQ $=) 6$ $\boxed{\blacktriangledown}$ $(x_3 =) 30$ $\boxed{\blacktriangledown}$ (FRQ $=) 8$ $\boxed{\blacktriangledown}$ $(x_4 =) 40$ $\boxed{\blacktriangledown}$ (FRQ $=) 4$ $\boxed{\blacktriangledown}$ $(x_5 =)$ $50$ $\boxed{\blacktriangledown}$ (FRQ $=) 3$ $\boxed{\text{STATVAR}}$ (underline $n$, then $\bar{x}$, then $\sigma_x$). The number of data values is 25; the mean is 28.4; the standard deviation is 12.22.

   (For $y$'s) $\boxed{\text{STAT}}$ (underline $1$-VAR) $\boxed{\text{DATA}}$ $(x_1 =)$ $\boxed{(-)}$ $2$ $\boxed{\blacktriangledown}$ (FRQ $=) 4$ $\boxed{\blacktriangledown}$ $(x_2 =)$ $\boxed{(-)}$ $1$ $\boxed{\blacktriangledown}$ (FRQ $=) 6$ $\boxed{\blacktriangledown}$ $(x_3 =) 0$ $\boxed{\blacktriangledown}$ (FRQ $=) 8$ $\boxed{\blacktriangledown}$ $(x_4 =) 1$ $\boxed{\blacktriangledown}$ (FRQ $=) 4$ $\boxed{\blacktriangledown}$ $(x_5 =) 2$ $\boxed{\blacktriangledown}$ (FRQ $=) 3$ $\boxed{\text{STATVAR}}$ (underline $n$, then $\bar{x}$, then $\sigma_x$). The number of data values is 25; the mean is $-0.16$; the standard deviation is 1.222.

**26.** (a) 101.5

**CHAPTER TEST, P. 479**

**1.** 13   **2.** 13   **3.** 46.67   **4.** 12

**5.**

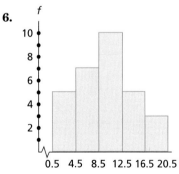

3   5   7   9   11   13   15   17   19   21   23

**6.**

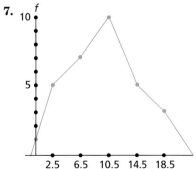

0.5   4.5   8.5   12.5   16.5   20.5

**7.**

2.5   6.5   10.5   14.5   18.5

**8.** 9.7    **9.** 4.78    **10.** 25    **11.** 68.857

**12.** 71    **13.** 64

**14.**

| Stems | Leaves |
|-------|--------|
| 3 | 3, 7 |
| 4 | 2, 8 |
| 5 | 2, 3, 4, 5, 7, 9 |
| 6 | 2, 2, 4, 4, 4, 8, 9 |
| 7 | 1, 2, 2, 3, 3, 4, 4, 8, 9 |
| 8 | 2, 3, 5, 7, 8, 9 |
| 9 | 3, 6, 8 |

**15.** 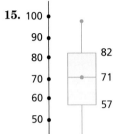    **16.** No outliers

**17.**

| Class | Frequency |
|-------|-----------|
| 31–40 | 2 |
| 41–50 | 2 |
| 51–60 | 6 |
| 61–70 | 7 |
| 71–80 | 9 |
| 81–90 | 6 |
| 91–100 | 3 |
|       | 35 |

**18.** 15.71    **19.** .3413    **20.** 15.87%

## CHAPTER 11

•••••••••••••••••••••••••••••••••••••

EXERCISE SET 1, P. 489

**1.** (a) Line segment    (b) Two intersecting lines
   (c) A plane    (d) A line segment
   (e) A point    (f) A plane
   (g) Intersecting segments or lines through a point.

**2.** (a)    (c)

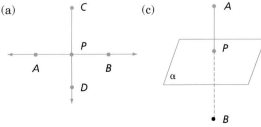

**(e)**

**3.** 6    **4.** (a) $\overrightarrow{AB}$ or $\overrightarrow{DB}$    (c) $\overrightarrow{DB}, \overrightarrow{DA}$    (e) $\overleftrightarrow{AB}, \overleftrightarrow{AC}$

**5.** (a) Parallel    (b) Skew    (c) Intersecting
   (d) Parallel    (e) Skew    (f) Intersecting

**6.** (a) True    (c) False; a ray has one endpoint.
   (e) False; $\overline{AB}$ is the same set of points as $\overline{BA}$. But $\overrightarrow{AB} \neq \overrightarrow{BA}$    (g) False; $\overline{AB} \subset \overleftrightarrow{AB}$
   (i)  False; two lines can be skew lines.
   (k)  False; a line has no endpoints.
   (m) False; a line has no endpoints.    (o) True

**7.** (a) Sometimes    (b) Always
   (c) Always    (d) Never

**8.** (a) Not parallel    (c) Parallel

**9.** (a) No    (b) Yes    (c) Yes
   (d) No    (e) Yes

**11.** (a) $\overline{QR}$    (b) {$S$}    (c) $\overrightarrow{RS}$
   (d) $\varnothing$    (e) $\overline{RS}$    (f) $\overrightarrow{QR}$
   (g) {$R$}    (h) $\overrightarrow{RP}$ or $\overrightarrow{RQ}$    (i) $\overrightarrow{QT}$

**12.** (a) 

**13.** (a) 

   (b) 

   (c)    Not possible with 4 lines

   (d) 

   (e) Not possible with 3 lines

   (f) Not possible with 3 lines

**14.** (a) $\overleftrightarrow{AB}, \overleftrightarrow{CD}; \overleftrightarrow{AC}, \overleftrightarrow{CD}$; and so on.

**15.** Many answers are possible.
   (a) $\overline{AC}$ and $\overline{BC}$    (b) $\overline{AB}$ and $\overline{BC}$
   (c) $\overline{AC}$ and $\overline{BD}$    (d) $\overline{AB}$ and $\overline{CD}$

**16.** (a) Yes, $\overline{BC} \subset \overline{AC}$

**17.** No, not if the point is located on the line

**18.** (a) Not parallel

**19.** (a) $\overline{AB}, \overline{BC}, \overline{CD}, \overline{DA}, \overline{AE}, \overline{BE}, \overline{CE}, \overline{DE}$

(b) $\overline{AB}$ and $\overline{CD}, \overline{AD}$ and $\overline{BC}$

(c) $\overline{AB}$ and $\overline{CE}, \overline{AB}$ and $\overline{DE}, \overline{BC}$ and $\overline{AE}, \overline{BC}$ and $\overline{DE}, \overline{DC}$ and $\overline{BE}, \overline{DC}$ and $\overline{AE}, \overline{AD}$ and $\overline{BE}, \overline{AD}$ and $\overline{CE}$

**20.** (a) No

**21.** (a) 1        (b) 3        (c) 6

(d) 10        (e) 15        (f) $\dfrac{n(n-1)}{2}$

**22.** (a) 4        **23.** (a) 1        (b) 3        (c) 6

(d) 10        (e) 15        (f) $\dfrac{n(n-1)}{2}$

**25.**

| Number of lines | 4 | 5 | ... | 10 |
|---|---|---|---|---|
| Number of regions | 11 | 16 | ... | 56 |

**26.** (a) Half line

**27.** One example for each is shown.

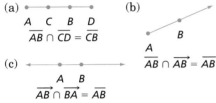

(a)
A  C  B  D
$\overline{AB} \cap \overline{CD} = \overline{CB}$

(b)
B
A
$\overline{AB} \cap \overrightarrow{AB} = \overline{AB}$

(c)
A  B
$\overrightarrow{AB} \cap \overrightarrow{BA} = \overline{AB}$

**29.** Four points in a plane can be collinear:

A  B  C  D

$B$ is between $A$ and $C$ and $A$ and $D$. $C$ is between $A$ and $D$ and $B$ and $D$. The 4 points can be non-collinear in two ways: with no 3 points collinear or with exactly 3 points collinear:

A    B            D

or

C    D         A  B  C

**31.** A partial list: four lines in a plane can be concurrent:

four lines can be nonconcurrent in several ways: three concurrent and one not. (It could be parallel to one of the lines or could intersect all three lines at points other than where three are concurrent.)

 or

no three concurrent, no two parallel:        all parallel:

three parallel and one not:        parallel in pairs:

two are parallel, two are not:

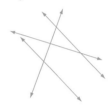

**EXERCISE SET 2, P. 499**

**1.** The intersection of the 3 planes with the fold on the tabletop is a line. The intersection of the 3 planes with the folded paper on its end is a point.

**2.** (a) Sometimes true: the planes may be parallel
(c) Always true        (e) Always true

**3.** $\angle BAC, \angle CAB, \angle\theta$; approximately $25° - 30°$

**4.** (a) False        (c) True        (e) True

**5.** (a) Yes
(b) No; the intersection of the interiors is not empty.
(c) No; no common vertex
(d) No; they do not have a common side.

**6.** Several answers possible in some cases.
(a) Planes $EDF$ and $ABC$
(c) Planes $ABC, ACDE$, and $BCDF$ intersect in point $C$.
(e) $\overleftrightarrow{AB}$ and plane $ACDE$        **7.** Yes

**9.** (a) Only one if all are in the same plane; otherwise an infinite number
(b) An infinite number

**11.** 4

**12.** (a) $\angle z$ and $\angle w$ are adjacent; $\angle x$ and $\angle y$ are vertical.

**13.** Drawings will vary.

(a) 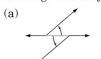 (b)

(c) 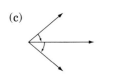 (d)

(e)  (f)

**14.** Several answers are possible for adjacent angles.

(a) ∠COD, ∠AOE adjacent
∠EOD vertical

(c) ∠COB, ∠DOF adjacent
∠AOE vertical

**15.** (a) $(90 - x)°$ (b) $(180 - y)°$
(c) measure (d) $90°$

**17.** (a) Acute (b) Obtuse (c) Right
(d) Obtuse (e) Obtuse (f) Acute

**18.** (a) $140°$ (c) $140°$

**19.** (a) $145°$ (b) $90°$ (c) $145°$ (d) $55°$
(e) $35°$ (f) $90°$ (g) $55°$ (h) $125°$

**20.** (a) ∠AFB, ∠EFD; or ∠AFC, ∠EFC
(c) Adjacent and supplementary
(e) $40°$ (g) $90°$

**21.** (a) $23°21'14''$ (b) $11°43'49''$

**22.** (a) They are the same plane.

**23.** (a) At least one (b) At least one
(c) At least one (d) At least one
(e) At least one

**25.** 10 angles

**27.**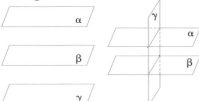

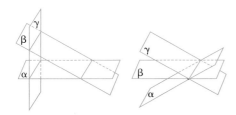

**28.** (a) Yes.

**29.** (a) 12 angles less than 180°; ∠EOC, ∠EOB, ∠EOD, ∠EOA, ∠FOC, ∠FOB, ∠FOD, ∠FOA, ∠COB, ∠COA, ∠BOD, ∠DOA
(b) 6 pairs: ∠EOC and ∠EOD, ∠EOC and ∠COF, ∠BOF and ∠EOB, ∠BOF and ∠AOF, ∠BOD and ∠BOC, ∠DOA and ∠AOC
(c) 6 pairs: ∠EOC and ∠DOF, ∠DOE and ∠FOC, ∠BOD and ∠AOC, ∠DOA and ∠BOC, ∠EOA and ∠FOB, ∠BOE and ∠FOA

**31.** (a) 6 lines; 48 angles (b) 6 lines

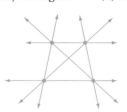

**32.** (a)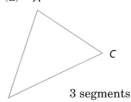

3 segments

**33.** (a) A line and a plane may intersect in exactly one point.
(b) The entire line may be in the plane.
(c) The line may be ∥ to the plane.

**35.** (a) Three noncollinear points
(b) A line and a point not on the line
(c) Two intersecting lines
(d) Two parallel lines

**37.** (a) No intersection or ∅ (b) $\overline{AD}$ (c) $\overline{CD}$

**EXERCISE SET 3, P. 511**

**1.** (a) Plane curve, simple (b) Plane curve, simple
(c) Plane curve, simple (d) Plane curve, simple
(e) Plane curve, simple
(f) Plane curve, not simple
(g) Not a plane curve
(h) Plane curve, not simple
(i) Not a plane curve

**2.** (a) Not closed (c) Closed (e) Closed

**3.** (a) Simple, but not (b) Simple polygon
a polygon
(c) Not simple (d) Simple polygon
(e) Simple polygon (f) Simple polygon

**4.** (a) Convex (c) Not convex (e) Not convex

**5.** Drawings will vary.

(a) 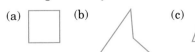 (b)     (c)

(d) Not possible    (e) Not possible

**6.** One example is shown.

(a) 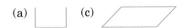 (c)

**7.** One example is shown.

(a) 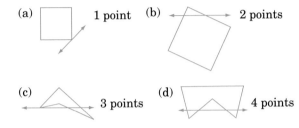 1 point    (b)     2 points

(c)     3 points    (d)     4 points

**8.** Many sketches are possible.

(a) 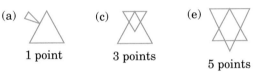 (c)     (e)

1 point    3 points    5 points

**9.** One example is shown.

(a) 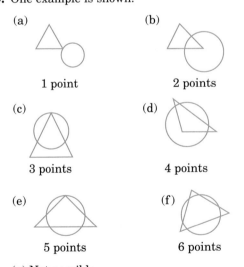 (b)

1 point    2 points

(c)     (d)

3 points    4 points

(e)     (f)

5 points    6 points

(g) Not possible

**11.** (a) False  (b) True  (c) True  (d) True
(e) False  (f) False  (g) True  (h) True
(i) True  (j) False  (k) False

**12.** (a) *ii, iv, v, vi*    (c) *i, iv, v, vi*
(e) *i, iii, iv, v, vi*

**13.** (a) *i, iii*    (b) *ii, iv*    (c) Yes

**14.**

|     | A | B | C | D |
|-----|-----|-----|-----|-----|
| (a) | Yes | Yes | Yes | Yes |
| (c) | Yes | Yes | Yes | No |
| (e) | No | No | Yes | No |
| (g) | No | No | No | No |

**15.** (a) No    (b) Yes    **17.**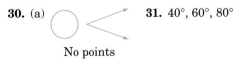

**19.** (a) 36-gon    (b) 18-gon
(c) 12-gon    (d) 6-gon

**20.** (a) 20-gon

**21.** (a) None    (b) 2    (c) 5
(d) 9    (e) 14
(f) $2(n - 3) + (n - 4) + (n - 5) + \ldots + 2 + 1$ or
$\dfrac{n(n - 3)}{2}$

**22.** (a) False

**23.** (a) $\{C\}$    (b) $\varnothing$
(c) $\overline{CE}$, excluding $C, E$    (d) Interior of $\triangle CDE$
(e) Interior of $\triangle DBF$    (f) $\varnothing$

**25.** (a) 4    (b) 4

**27.** Opposite sides are parallel and equal in length, and the angles are right angles.

**29.** (a) False—a line separates a plane or a plane separates space.
(b) True    (c) False—
There are two different rays.    $A \quad B$
(d) True    (e) False—
$A \quad B \quad D$
(f) True    (g) True    (h) True
(i) False—two are sufficient.
(j) False—angles must also be equal.
(k) True    (l) False—it is a radius.
(m) False—a circle is not composed of line segments.
(n) True    (o) False—
(p) False—a line separates a plane.
(q) False—not all closed curves are simple.
(r) True
(s) False—it could have 3 sides (triangle).
(t) False—the region includes the interior of a closed curve.
(u) True

**30.** (a)

No points

**31.** 40°, 60°, 80°

EXERCISE SET 4, P. 523

**1.** (a) Triangular prism    (b) Pentagonal pyramid
(c) Pentagonal prism
(d) Rectangular pyramid or quadrilateral pyramid

**2.** (a) $V = 6, F = 5, E = 9; V + F - E = 2$
(c) $V = 10, F = 7, E = 15; V + F - E = 2$

**3.** (a) True    (b) False    (c) False    (d) True
(e) False    (f) False    (g) True    (h) False

**4.** (a) Quadrilateral pyramid; lateral edges $\overline{AE}, \overline{DE},$
$\overline{CE}, \overline{BE}$; vertices $A, B, C, D, E$

**5.** Drawings will vary.

(a)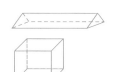

(b)

(c)

(d)

**6.** (a) 5

**7.** (a) Hexagonal pyramid    (b) Square pyramid
(c) Square prism    (d) Hexagonal prism

**8.** (a) ◯ or a point

(c) ◯ or a point

**9.** (a) Prism    (b) Pyramid    (c) Pyramid
(d) Prism    (e) Neither    (f) Pyramid
(g) Prism    (h) Pyramid

**10.** (a) 3, 6, 9    (c) 5, 10, 15

**11.** (a) 3, 4, 6    (b) 4, 5, 8
(c) 5, 6, 10    (d) 6, 7, 12

**12.** (a) $6 + 6 - 10 = 2$

**13.** (a) Tetrahedron (triangular pyramid)
(b) Cube (square prism)
(c) Octahedron    (d) Dodecahedron
(e) Icosahedron

**15.** (a) Yes  (b) Yes    **17.** No    **19.** Yes    **21.** Yes

**22.** (a)

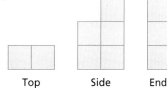

Top        Side        End

**23.** Yes; $9 + 7 - 14 = 2$ or $10 + 7 - 15 = 2$

**25.** Lateral faces: $n$; Total faces: $n + 2$
Vertices: $2n$; Edges: $3n$
$V + F - E = 2n + n + 2 - 3n = 2$

**27.** (a)     (b)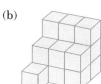

(c)

**28.** (a)

**29.** Sketches will vary.

(a)     (b)

(c) Not possible    (d)

(e) Not possible

**30.** (a)     **31.** Interior

CHAPTER TEST, P. 528

**1.** (a)     (b)

(c)    (d)

**2.** (a) $V + F - E = 12 + 8 - 18 = 2$
(b) $V + F - E = 9 + 7 - 14 = 2$

**3.** (a) Three; two    (b) Angle; vertex
(c) Pentagon; quadrilateral    (d) Circle

**4.** (a) False    (b) False    (c) False    (d) False
(e) True    (f) True    (g) False

**5.** (a)  (b)

**6.** No    **7.** (a) False    (b) False    (c) True

**8.** (a) 1    (b) 3    (c) 6    (d) $1 + 2 + \cdots + (n - 1)$
or $n(n - 1)/2$

**9.** (a) $\overrightarrow{DA}$    (b) $\varnothing$    (c) $\overleftrightarrow{AD}$

**10.** (a) $\overleftrightarrow{AB}; \overleftrightarrow{AC}; \overleftrightarrow{BC}$    (b) $\overrightarrow{BC}, \overrightarrow{BA}$    (c) $\overline{AB}$
(d) $\overline{AB}$    (e) $\overrightarrow{AB}$

**11.** (a) $\overrightarrow{PQ}$ and $\overleftrightarrow{AB}$
(b) Planes $APQ$ and $BPQ$    (c) $\overleftrightarrow{AQ}$
(d) No; two sets of three noncollinear points determine two distinct planes.

**12.** (a) 60°    (b) 120°    (c) 120°

**13.** $\angle BOC, \angle AOD$ (adjacent); $\angle AOB$ (vertical)

**14.** (a) False    (b) True    (c) True
(d) False    (e) False

**15.** (a) 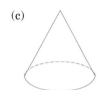 (b)

(c)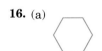

**16.** (a)    (b) 120°    (c) 60°

## CHAPTER 12

EXERCISE SET 1, P. 542

**1.** (a) 370 m    (b) 389 m
(c) 12 m    (d) $21 \cdot 10^6$ m
(e) 35,470,000 m    (f) 38,902,000 m

**2.** (a) 37,890 mm    (c) 1,284,000 mm
(e) $35.478 (10)^9$ mm    (g) 160 mm

**3.** (a) 0.017 km    (b) 0.12 km    (c) 1.6 km

**5.** (a) 0.032 m    (b) 0.0016 km    (c) $3 \cdot 10^6$ mm
(d) 0.000216 km    (e) 134 cm
(f) 168,900,000 cm    (g) $48 \cdot 10^6$ μm

**6.** (a) 100 cm, 10 m, 100,000 mm
(c) 0.1 cm, 0.01 m, 0.001 km

**7.** (a) 21.55; 21.65    (b) 14.155, 14.165    (c) 1.5; 2.5

**8.** (a) 0.05; 0.0023    (c) 0.5; 0.25    **9.** 13 m

**11.** (a) False    (b) True    (c) False    (d) True
(e) True    (f) True    (g) True

**12.** (a) m$(\overline{AB}) = 1\frac{1}{2}$    (c) m$(\overline{DB}) = 2$    (e) m$(\overline{GA}) = 6$

**13.** (a) m    (b) mm    (c) km    (d) m
(e) cm    (f) cm    (g) mm    (h) m
(i) cm    (j) mm    (k) km

**14.** (a) 1075 cm or 10.75 m    (c) 457 cm or 4.57 m

**15.** (a) 3 m    (b) 10 km    (c) 12 cm
(d) 24 cm    (e) 100 cm

**16.** (a) (m)/1000 = km  (divide by 1000)

**17.** (a) $\frac{1}{100}$ of a foot    (b) 1000 feet
(c) $\frac{1}{1000}$ of a foot    (d) $\frac{1}{10}$ of a foot

**18.** (a) 68 cm    (c) 4 m    (e) 4 mm

**19.** (a) $\sqrt{8}$ or $2\sqrt{2}$ units    (b) $\sqrt{5}$ units
(c) $\sqrt{10}$ units    (d) $\sqrt{13}$ units    (e) $\sqrt{29}$ units

**20.** (a) $5 + 4\sqrt{2} + \sqrt{5} \approx 12.89$ units

**21.** 0.6 m, 62 cm, 6000 mm, 61 dm

**23.** (a) $3\frac{1}{2}$ or $\frac{1}{2}$    (b) $1\frac{3}{8}$
(c) 7 or −1    (d) −2 or 4

**24.** (a) $B = 1\frac{1}{2}$ or $B = -\frac{1}{2}$

**25.** (a) 6    (b) 1    (c) 0    (d) $-\frac{1}{2}$

**26.** (a) $\frac{1}{2}$    **27.** 60 km/hr.

**28.** (a) Approximately 12 yards

**29.** (a) Cars, approximately 90 km/hr.
Trucks, approximately 70 km/hr.
(b) Bowling Green, approximately 35 km
Nashville, approximately 140 km

**31.** (a) Approximately 60 cm
(b) Approximately 40 cm
(c) Approximately 100 km/hr.
(d) Approximately 180 cm

**32.** (a) Approximately 350 miles

**33.** (a) 60.96 cm    (b) 38.1 cm
(c) $\approx$96.56 km/hr.    (d) 180.34 cm

**34.** (a) $\approx$12.0297 yd.

**35.** (a) Cars, $\approx$88.5139 km/hr.
Trucks, $\approx$72.4205 km/hr.
(b) $\approx$33.796 km; $\approx$138.404 km

**37.** (a) $\approx$372.8227 miles    (b) $\approx$643.74 km
(c) $\approx$1448.4096 km    (d) $\approx$994.194 miles

**38.** (a) 0.0000792 yd.

EXERCISE SET 2, P. 556

**1.** (a) 8    (b) 18    (c) 8    (d) 6
(e) 16    (f) 10    (g) 15    (h) $13\frac{1}{2}$

**2.** (a) 15m$^2$    (c) 884 square units
(e) 288 square units    (g) 7.5 square units

**3.** (a) 8; $4 \cdot 2 = 8$, A $= b \cdot h$
(b) 18; $3 \cdot 6 = 18$, A $= b \cdot h$

**4.** (c) 8 squares; $8 = \frac{1}{2} \cdot 4 \cdot 4$

**5.** (e) 16; $\frac{4 \cdot 8}{2} = 16$, $A = \frac{b \cdot h}{2}$

   (f) 10; $\frac{5 \cdot 4}{2} = 10$, $A = \frac{b \cdot h}{2}$

**6.** (g) 15 squares; $15 = \frac{1}{2}(8 + 2) \cdot 3$

**7.** (a) 340,000 cm²    (b) 3.28 cm²
   (c) 31 cm²    (d) $41 \cdot 10^{10}$ cm²
   (e) $42 \cdot 10^{10}$ cm²    (f) 8,630,000 cm²

**8.** (a) 400 cm²

**9.** (a) 54 ft.²    (b) 2592 in.²    (c) 61,952 yd.²
   (d) 400 yd.²    (e) 435.6 ft.²

**11.** (a) 34 km²    (b) 3 cm²
   (c) 456,000 m²    (d) 916 mm²

**13.** (a) C = 25.133 ft.    A = 50.266 ft.²
   (b) C = 62.832 m    A = 314.16 m²
   (c) C = 75.398 cm    A = 452.39 cm²
   (d) C = 56.549 yd.    A = 254.47 yd.²

**15.** (a) 158    (b) 294

**16.** (a) $\pi/3$ radians    (c) 30°

**17.** (a) Arc length $a = \frac{80}{9}\pi$ m;   Area $= \frac{1600}{9}\pi$ m²

   (b) $\theta = 90° = \pi/2$ radians;   Area $= (9/4)\pi$ m²
   (c) $r = 8$ units;   Area $= 16\pi$ square units
   (d) $\theta = 2$ radians;   Area $= x^2$ cm²

**18.** (a) 13 m² + 27 dm² + 26 cm²

**19.** $32 - 4\pi$    **21.** $12 + 4.5\pi$    **23.** $72\pi$

**25.** (a) 0.3 acre    (b) 25 in.²    (c) 4 m²
   (d) 1.5 mi.²    (e) $2(10)^7$ yd.²    (f) 10 ha
   (g) 230 cm²    (h) 130 m²

**26.** (a) 1300 m² $\cdot \frac{(10^2 \text{ cm})^2}{1\text{m}^2} \cdot \frac{1 \text{ in.}^2}{(2.54 \text{ cm})^2} \cdot \frac{1 \text{ ft.}^2}{(12 \text{ in.})^2} \cdot$

   $\frac{1 \text{ acre}}{43,560 \text{ ft.}^2} \approx 0.321$ acre

**27.** (a) $2\pi - 4$ square units    (b) $\pi/2$ square units

**29.** 28 square units

**31.** $25\sqrt{3}$ square units

**33.** $10\pi$    **35.** 16,536 revolutions

**37.** (a) Perimeter: $4 + 2\sqrt{2}$ cm; area: 2 cm²
   (b) Perimeter: $4 + 2\sqrt{2} + 2\sqrt{5}$ cm; area: 5 cm²
   (c) Perimeter: $47 + \sqrt{41}$; area: 158
   (e) Perimeter: $47 + 3\sqrt{65}$; area: 294

**38.** (a) 0.05001 km   (c) 81,000 m   (e) $2.61(10)^5$ cm

**39.** (a) 3    (b) 3    (c) 5    (d) 5

**40.** (a) cm    (c) mm    (e) m

### EXERCISE SET 3, P. 570

**1.** (a) $V = 160\pi$, $SA = 112\pi$
   (b) $V = 120$, $SA = 148$
   (c) $V = 2\frac{19}{64}$   $SA = 12\frac{11}{16}$

**3.** 300 cm²    **5.** 96 ft.²    **7.** 310 square units

**9.** 208 square units

**11.** $V = \frac{1000}{3}\sqrt{2}$ cm³; $SA = 400$ cm²

**13.** $V = 320\pi$ cubic units; $SA = 200\pi$ square units

**14.** (a) $V = 81$ m³; $SA = 120$ m²

**15.** The volume of the cone is $\frac{1}{3}$ the volume of the cylinder, and the volume of the half-sphere is $\frac{2}{3}$ the volume of the cylinder. Yes.

**17.** 984 in.²

**19.** (a) $34(10)^6$ cm³    (b) $328(10)^{15}$ cm³
   (c) 3.1 cm³    (d) $41(10)^{15}$ cm³
   (e) 0.042 cm³    (f) $863(10)^{15}$ cm³

**21.** $4\sqrt{3}$ in.

**23.** (a) 6000 mm³    (b) 0.006 dm³
   (c) $2(10)^{-7}$ km³   (d) $2 \cdot 10^{15}$ cm³   **24.** (a) 13 m³

**25.** (a) $\frac{1}{8}$

   (b) It is 79.37% of the altitude of the cone (cup).

**27.** $V = 56\pi$; $SA = 20\pi + 12\pi\sqrt{10}$

**29.** 6 L

**31.** $64 - 12\pi$ square units

**33.** $16\sqrt{3} - 8\pi$ square units

### EXERCISE SET 4, P. 576

**1.** (a) 125 L    (b) 0.016 L   (c) 1,230,000 L
   (d) 14,000 L    (e) 2.36 L    (f) 1,350 L

**2.** (a) 1,689,000 g    (c) 0.00367 kg    (e) 140 mg

**3.** (a) 57 g    (b) 35 kg

**5.** (a) 210 L    (b) 1 mL    (c) 8 L    (d) 80 L

**6.** (a) 30°C    (c) 80°C    (e) 50°C

**7.** 13.824 L    **9.** 26 full cups

**11.** (a) 10°C   (b) 20°C   (c) 0°C   (d) $(-18\frac{8}{9})$°C

**12.** (a) 68°F   **13.** 15 km/L   **14.** (a) 1.9 L

**15.** Estimates will vary.

   (a) $300 - 500$ g   (b) $8 - 12$ g    (c) $1 - 2$ kg
   (d) $350 - 500$ g   (e) $30 - 100$ g
   (f) $1400 - 1600$ kg
   (g) $20 - 40$ kg    (h) $5 - 12$ g    (i) $1 - 3$ kg
   (j) $600 - 900$ g   (k) $900 - 1000$ kg
   (l) $17 - 21$ kg

**16.** (a) 1522 containers

**17.** Estimates will vary

   (a) 23°C    (b) 20°C    (c) 0°C    (d) 100°C

**18.** (a) Multiply kilograms by 1000 to get grams.

**19.** Larry made the better buy because he got more grams per dollar.

**20.** (a) $>$

**21.** (a) 2000 g    (b) 10 lb.    (c) 2 oz.    (d) 2 oz.

**22.** (a) 55¢/lb.    **23.** 30 miles/gal.

**25.** (a) ≈49.9 kg    (b) ≈2.9 kg    (c) ≈11.4 kg
(d) ≈1816 kg    (e) ≈4.54 kg

**27.** 4 full cups

**28.** (a) 36,734 g, 32.4 kg, 4 lb., 26 oz.

**29.** (a) 15¢/pound    (b) $1.05/pound
(c) $1.60/gallon    (d) 29¢/liter

**30.** (a) $LSA = 128$; $V = 240$

CHAPTER TEST, P. 579

**1.** $8\frac{1}{2}$ cm²    **2.** 20 square units

**3.** $3\pi$ square units    **4.** 48 square units

**5.** (a) 32,800,000 cm    (b) 0.0026 cm    (c) 0.28 m

**6.** (a) >    (b) >    (c) <    **7.** 11 m 96 cm 8 mm

**8.** (a) 1 kg    (b) 5 g

**9.** $10\pi$ cubic units    **10.** 96 cubic units

**11.** $22\pi$ square units; 136 square units

**12.** (a) 24,000 cm³    (b) 16,000 mm³    (c) 50,000 mL
(d) grams    (e) $14(10)^{-3}$ kg    (f) $14(10)^{15}$ cm³
(g) 0.2 dm³    (h) 1 m³

**13.** (a) U    (b) L    (c) U    (d) L    (e) U

**14.** 3    **15.** $18\pi$ mm

**16.** The area of the circle is $4/\pi$ times the area of the square.

**17.** 28 square units

**18.** $28 + 4\sqrt{10}$ square units

**19.** (a) Circumference    (b) Milli    (c) Edges
(d) Liter    (e) Total surface area
(f) Greatest possible error

**20.** 6 to $\pi$

# CHAPTER 13

• • • • • • • • • • • • • • • • • • • • • • • • • • • • • •

EXERCISE SET 1, P. 591

**1.**

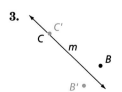

**3.**

**5.** (a) The image of $A$ is $B$, and the image of $B$ is $A$.
(b) The reflection of $C$ is $C$.    (c) They are
collinear.

**7.** (a)    (b)

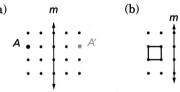

(c)    (d)

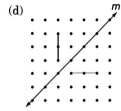

(e)    (f)

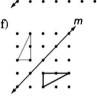

**8.** (a)

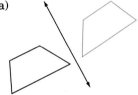

**9.** (a) False    (b) True    (c) True    (d) True
(e) True    (f) True    (g) False

**11.** Distance, collinearity of points, betweenness of points, and angle measure

**13.** $\sqrt{65}$

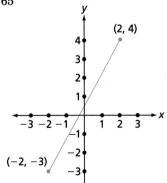

**15.** $2\sqrt{5}$

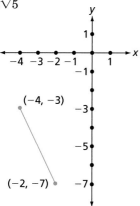

**17.** Acute

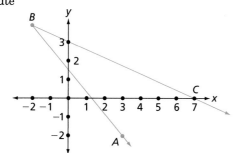

**19.** $AB = \sqrt{32}$; $BC = \sqrt{72}$; $AC = \sqrt{104}$. Thus, $(AB)^2 + (BC)^2 = (AC)^2$. **21.** $AC = 5$ and $BD = 5$.

**22.** (a) $l$ is the perpendicular bisector of $\overline{AB}$.

**23.**      **25.**

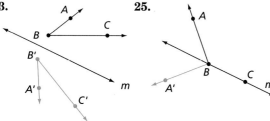

**27.** (a)      (b)

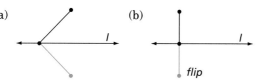

**29.**

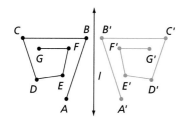

**31.** (a) Yes    (b) No

**33.** (a) Reflection line is $x = 1$.
    (b) Reflection line is $y = 2$.
    (c) Reflection line is $y = x$.

**35.** About the $y$ axis:
    $A'(-1, 1)$, $B'(-2, 3)$, $C'(1, 2)$, $D'(2, -1)$

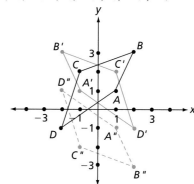

About the $x$-axis:
$A''(1, -1)$, $B''(2, -3)$, $C''(-1, -2)$, $D''(-2, 1)$

**37.** (a) Reflection line is the perpendicular bisector of $\overline{AB}$.
    (b) Reflection line is the perpendicular bisector of $\overline{AD}$.

**38.** (a) $(\frac{5}{3}, 4)$

**EXERCISE SET 2, P. 598**

**1.** (a)        (b)      (c)

**2.** (a)        (c)

**3.** (a)        (b)      (c)

**4.** (a) 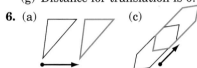 (c)

**5.** (a) Line *m*
(b) Slide or translation
(c) Distance, angle measure, betweenness of points, collinearity of points
(d) Two reflections about parallel lines
(e) Its image (f) ST
(g) Distance for translation is 0. (h) Slide

**6.** (a) 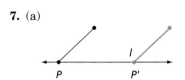 (c)

**7.** (a)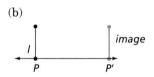

(b)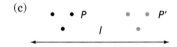

(c)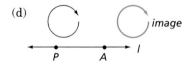

(d)

(e) 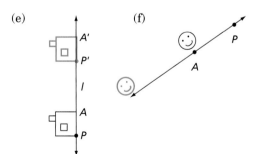 (f)

**8.** (a)  (c)

**9.**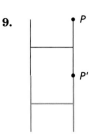

**10.** (a) 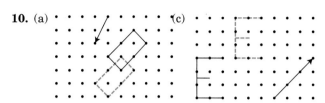 (c)

**11.**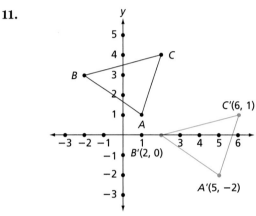

**13.**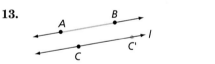

**15.** (a) $G$ (b) $D$ (c) $S$
(d) $\overline{CG}$ (e) $\angle GTS$ (f) $DWXT$

**16.** (a) $K; G; U; \overline{FK}; \angle KVU; GTSV$

**17.** Yes

**19.** (a) $(-1, 2)$ (b) $(5, 0)$ (c) $(1, -3)$

**20.** (a) $(0, 4)$ (c) $(2, -1)$

**21.**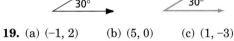

(one possibility)

**23.** (a)

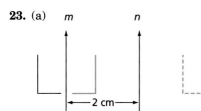

**24.** (a) Translates any point 2 units to the right and 3 units down.

**25.** (a) *M*    (b) $\overline{HC}$    (c) *RMLQ*
**27.** (1, −7)

EXERCISE SET 3, P. 609

**1.** (a)

(c)

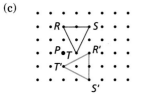

**2.** (a)    (c)

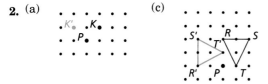

**3.** (a)    (b)

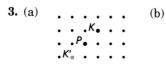

(c)

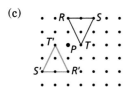

**4.** (a)    (c)

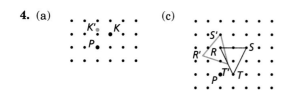

**5.** (a)

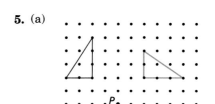

(b)

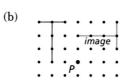

**6.** (a)

(c)    (e)

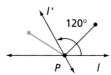

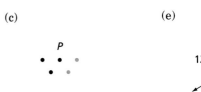

**7.**

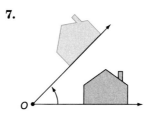

**8.** (a)

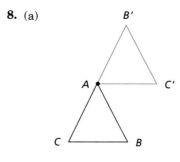

(c)

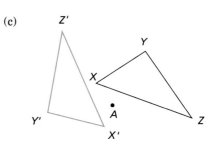

**9.**
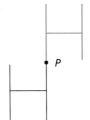

**10.** (a) 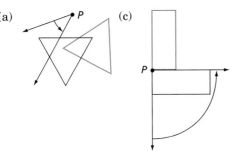 (c)

**11.** W     **13.** W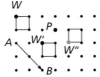

**15.**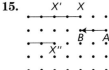

**17.** P     **19.** P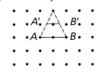

**21.** (a) The equilateral triangles can be obtained by reflections about sides or as 60° rotations about vertices, or as translations the length of sides.

(b) Reflections about sides or translations the distances between parallel sides.

**22.** (a)
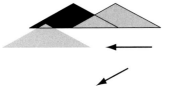

(Answers demonstrate that the order makes no difference.)

**23.** (a) 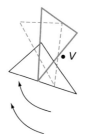    (b)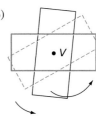

**24.** (a) (two flips)
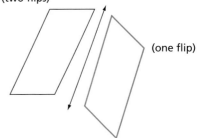
(one flip)

**25.** (a) Reflection    (b) Rotation
(c) Reflection    (d) Slide (translation)

**27.** 90° clockwise

**29.** (a) Rotate 90° clockwise about lower right corner; reflect about the vertical line through the tip.

(b) Reflect across a vertical line through the leftmost vertex, then rotate 90° counterclockwise about the bottom left vertex.

**30.** (a) Reflection

**31.**
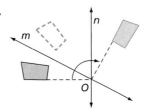

**32.** (a) O    (c) *CDIH*

**EXERCISE SET 4, P. 618**

**1.** (a) Yes    (b) No    (c) No    (d) Yes

**2.** (a)     (c)

**3.** (a) Rotational symmetry about $Q$ (multiples of 90°)

(b) No; no

(c) Rotational symmetry about $P$ at 60°, 120°, 180°, 240°, 300°

(d) No; no

**4.** (a) 3, 120°   (c) 5, 72°   (e) 7, 51.43°
**5.** (a) *H, I, O, X*   (b) *H, I, O, X, N, Z*   (c) None
**6.** (a) Tessellates the plane

**7.** (a) Reflectional and rotational   (b) Rotational symmetry

**8.** (a)

**9.** (a) No lines of symmetry; 45° rotational symmetry
   (b) 2 lines of symmetry, 180° rotational symmetry
   (c) 4 lines of symmetry; 90° rotational symmetry

**10.** (a) Squares

**11.** Not all vertex figures tessellate the plane. The hexagon does not tessellate the plane.

**12.** (a) Hexagons and equilateral triangles

**13.** (a) Yes, any triangle   (b)
   (c) Yes

**14.** (a)

**15.** (a)   (b)
   (c) Impossible   (d)

**17.**

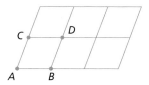

Translate *A* to *B*; *A* to *C*; and *A* to *D*.
**19.** Reflect across any side.
**20.** (a) Line of rotational symmetry is through the center of the cylinder. Planes of symmetry are any planes that go through the line of rotational symmetry. Another plane of symmetry is perpendicular to this line of symmetry at the center of the cylinder. Any line in this

plane through the center of the cylinder is a line for 180° rotational symmetry.

**21.** (a) Three through centers of opposite faces, and four along the diagonals of the cube.
   (b) 3 planes that intersect the midpoints of opposite edges parallel to faces; 6 planes that intersect opposite vertices (the diagonals of the faces)

**22.** (a) One

**23.** Three planes of symmetry are perpendicular bisectors of opposite lateral faces of the nut; one plane of symmetry is the perpendicular bisector of all lateral faces.

**EXERCISE SET 5, P. 625**

**1.** (a) $5\frac{5}{7}$   (b) 12   (c) 9   (d) $1\frac{1}{2}$   (e) 12
**2.** (a) $x = \frac{16}{3}$   (c) $z = 27, w = \frac{27}{2}$
**3.** (a), (b)

**5.** (a) m($\angle C$) = 70°   m($\angle F$) = 70°
   (b) $HI = 6, JL = 11.2$
   (c) m($\angle WVX$) = 60°, $WX = 4, ST = 8$

**7.** 18 ft.

**9.** (a) $2\frac{2}{3}$   (b) $5\frac{5}{11}$   (c) 4   (d) 12

**11.** (a) 8 m × 12 m   (b) 3 m × 6 m
   (c) 12 m, 18 m, 21.6 m

**13.** *V*: 1:27; *SA*: 1:9

**15.** For reduction, change 1 cm to $\frac{2}{5}$ cm.
   For enlargement, change 1 cm to $\frac{5}{3}$ cm.

**18.** (a) 8

**19.** 400 m from the shoreline to the ship.

**21.** 128.06 m

**CHAPTER TEST, P. 628**

**1.** (a) True   (b) False   (c) False   (d) True
   (e) True   (f) True   (g) True   (h) False
   (i) True   (j) False

**2.**   **3.**

**4.**

**5.** $A'(3, 0)$; $B'(5, -4)$    **6.** $(-2, -5)$    **7.** $(-3, -2)$

**8.** 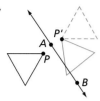    **9.** No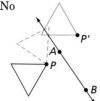

**10.** 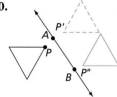    **11.** Yes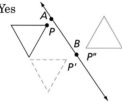

**12.** 5.4′

**13.**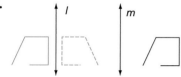

Yes, a translation perpendicular to $l$ and $m$ of magnitude twice the distance between $l$ and $m$.

**14.** (a) ⌐    (b) ♡

## Chapter 14

EXERCISE SET 1, P. 640

**1.** (a) Yes, SSS  (b) Yes, SSS    (c) Yes, SAS
    (d) No    (e) Yes, SAS    (f) No, not necessarily

**2.** (a) $\angle DEF$    (c) $\angle EFD$    (e) $BC$

**3.** (a) Yes, SSS    (b) Yes AAS

**4.** (a) 5    (c) $\sqrt{161}$

**5.** (a) Yes    (b) No    (c) No    (d) No
    (e) Yes    (f) Yes

**6.** (a) Yes    (c) Yes

**7.** $\triangle SQP \cong \triangle SRP$, by AAS. Therefore, $PQ = PR$.

**8.** (a) Yes SSS

**9.** $\triangle ABC \cong \triangle DCB$, by SAS    **10.** (a) Yes AAS

**11.** $\angle A \cong \angle C$; $\overline{BD} \cong \overline{BD}$
    $\triangle ABD \cong \triangle CBD$ by ASA so $AB = BC$
    Because $AB = BC$, $\triangle ABC$ is isosceles

**13.** (a)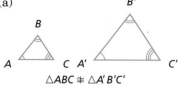
    $\triangle ABC \not\cong \triangle A'B'C'$

AAA holds but not congruent.

(b) Four parts of one are congruent to four of the other, but the triangles are not congruent.

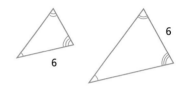

(c) Five parts of one are congruent to five of the other. Triangles are not congruent.

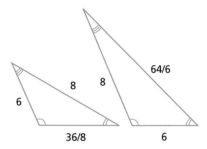

(d) Not possible

**15.** $\triangle ACB \cong \triangle ADB$ by ASA because
    $\angle CAB \cong \angle ABD$ (alternate interior angles)
    $\angle BAD \cong \angle ABC$ (alternate interior angles)
    $AB = AB$
    Area parallelogram $ACBD = b \cdot h$. So, area $\triangle ACB = \frac{1}{2} b \cdot h$.

**17.** $(XJ)^2 + (HB)^2$
    $= (XP)^2 + (JP)^2 + (PH)^2 + (PB)^2$
    $= [(XP)^2 + (PH)^2] + [(JP)^2 + (PB)^2]$
    $= (XH)^2 + (JB)^2$

**19.** By $c^2 = a^2 + b^2$, the unknown side of the $\triangle$ on the right is 1. The third angle of the $\triangle$ on the left is 60°. Therefore, by ASA, the two triangles are congruent.

**1.** Using one end of the longest side as the center, draw an arc with radius the length of the second side. Using the other end of the longest side as the center, draw an arc with radius the length of the shortest side. The intersection of these two arcs is the third vertex of the △.

**2.** (a) With *A* as the center, draw an arc cutting the line of the side opposite *A* in two points, *C* and *D*. With *C* and *D* as centers, draw two arcs with the same radii until they intersect. Connect *A* and the point of intersection. The portion of this segment from *A* to $\overline{CD}$ is the altitude.

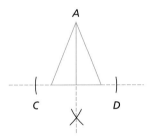

**3.** They lie on the perpendicular bisector of $\overline{AB}$.

**5.** Consider a line segment $\overline{AB}$. At the ends of $\overline{AB}$, draw two arcs of radius *BA*. Draw a ray from *A* to the intersection. The angle formed has a measure of 60°.

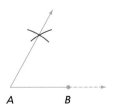

**7.** Draw arcs of length *AB* with centers at *C* and *D* until they intersect. Draw segments from *C* and *D* to the intersection.

**9.**

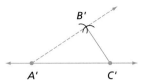

On a line mark two points *A'* and *C'* so that $\overline{A'C'} \cong \overline{AC}$. At *A'*, draw an arc of radius *AB*. At *C'*, draw an arc of radius *BC*. Through the intersection of these two arcs, draw a ray. ∠*A'* ≅ ∠*A*. By construction *A'B'* = *AB*. Then draw *B'C'*. △*A'B'C'* ≅ △*ABC*, by SAS.

**11.**

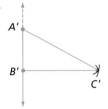

On a line mark points *A'* and *B'* so that $\overline{A'B'} \cong \overline{AB}$. At *B'*, draw an arc of radius *BC*. At *A'*, draw an arc of radius *AC*. Connect the intersection *C'* of these two arcs to *B'* and *A'*, respectively. By construction $\overline{C'A'} \cong \overline{CA}$ and $\overline{B'C'} \cong \overline{BC}$. Thus the two triangles are congruent by SSS.

**13.**

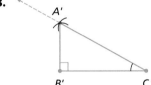

Construct $\overline{B'C'} \cong \overline{BC}$. Then at *C'*, construct ∠*C'* ≅ ∠*C*. At *B'* construct a perpendicular to $\overline{B'C'}$. Let the intersection of the ray from *C'* and the perpendicular at *B'* be *A'*: △*A'B'C'* ≅ △*ABC* by ASA.

**14.** (a) Inside the triangle

**15.** (a) Construct a 90° angle adjacent to one angle of an equilateral △. This produces a 150° angle. Bisect this angle.

(b) Construct perpendicular lines. Bisect one of the 90° angles. This produces a 45° angle. 45° + 90° = 135°. Copy the 135° angle using construction 6.

**17.**

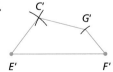

Construct $\overline{E'F'} \cong \overline{EF}$. Now at $E'$, draw an arc of radius $AB$; and at $F'$, draw an arc of radius $GH$. On the last arc, at any point $G'$, draw an arc of radius $CD$, getting intersection $C'$. Draw the quadrilateral. The quadrilateral is not unique, because we could take any point $G'$.

**19.**

Draw the perpendicular bisectors of the three sides of the equilateral triangle. You can prove that these intersect in a point equidistant from the three sides (or the center of the circle). Draw the circle.

**21.**

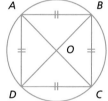

$AB = DC$, $\angle AOB \cong \angle COD$, and $\overrightarrow{AB} \parallel \overrightarrow{DC}$ (given) and $\angle OAB \cong \angle OCD$ and $\angle OBA \cong \angle ODC$ (alternate interior angles of a transversal cutting parallel lines). Therefore, $\triangle AOB \cong \triangle COD$ by ASA. In a like manner, $\triangle AOB \cong \triangle AOD \cong \triangle BOC$. Thus, $AO = BO = CO = DO$.

**23.** See Exercise 21.

**25.** Circumscribe a circle about the 8-gon. Bisect the 8 central angles of the 8-gon to locate $8 + 8 = 16$ points on the circle. Connect these to form a 16-gon. In this manner you can form a $2^n$-gon ($n \geq 2$).

**27.** You could construct a family of $3 \cdot 2^{n-1}$-gons. ($n \geq 1$)

**29.**

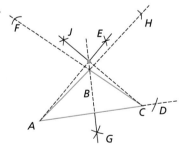

From each vertex construct a perpendicular to the opposite side. At $A$, draw an arc with radius $AC$ intersecting $\overrightarrow{CB}$ at $F$. With a fixed compass setting, draw arcs with centers $F$ and $C$, intersecting at $E$. Draw $\overline{AE}$, a part of which is the altitude

from $A$. With $B$ as a center, draw an arc with radius $AB$ intersecting $\overrightarrow{AC}$ at $D$. With $A$ and $D$ as centers, draw arcs intersecting at $G$. Draw $\overrightarrow{BG}$, obtaining the altitude from $B$. At $C$ draw an arc of radius $AC$ intersecting $\overrightarrow{AB}$ at $H$. With a fixed compass setting, draw arcs from $A$ and $H$ intersecting at $J$. Draw $\overline{CJ}$ to obtain the altitude from $C$. Are you surprised that they are concurrent?

**31.** Construct $\overline{BC}$ as given. Construct $\angle B$ as given. At $B$ construct a perpendicular to $\overline{BC}$. With your compass, mark point $F$ so that $BF$ is equal the length of the altitude. At $F$, construct a line parallel to $\overline{BC}$. The intersection of this line and the ray of $\angle B$ is $A$. Draw $\overline{AC}$.

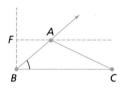

**32.** (a) Yes

**33.** $x = 6$

**35.** (a) $\sqrt{464} \approx 21.54$ ft.     (b) $28'$

EXERCISE SET 3, P. 657

**3.**

| Map | Number of Even Vertices | Number of Odd Vertices | Economical Route |
|---|---|---|---|
| (a) | 3 | 4 | No |
| (b) | 4 | 2 | Yes |
| (c) | 4 | 2 | Yes |
| (d) | 5 | 0 | Yes |
| (e) | 1 | 4 | No |
| (f) | 4 | 8 | No |
| (g) | 8 | 0 | Yes |

**4.** (a) 18, yes   (c) 12, yes   (e) 24, yes   (g) 20, yes

**5.**

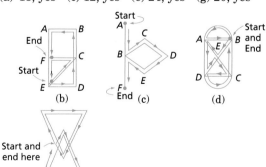

**6.** (a)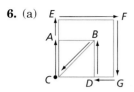

(c) and (g) are impossible

(e)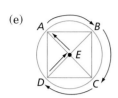

**7.** V = {P,Q}, E = {(P,Q)}

**9.** V = {X,Y,Z,W}, E = {(X,Y), X,Z), (X,W), (Y,W), (Y,Z), (Z,W)}

**11.** (a)     (b)

(c)

**12.** (a) $e_1$ and $e_6$    (c) $(v, e_1, z, e_2, x)$ or $(v, e_1, z, e_3, x)$
(e) $(v, e_1, z, e_2, x, e_3, z, e_2, x)$    (g) $x$ and $z$

**13.** (a)            (c)
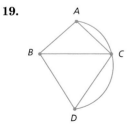

**15.** (a) Yes

(b) No    (c) No

**17.** Neither

**19.**

Because the Konigsberg graph has four odd vertices, there is not an economical path.

**21.** Yes

**23.** Hamilton Circuit

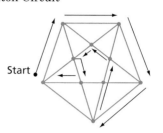

**25.** Hamilton Circuit

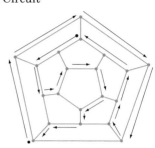

**EXERCISE SET 4, P. 664**

**1.** $(a,c,d,z)$ of length 7

**3.** $(a,c,d,e,g,z)$ of length 16

**5.** $(a,b)$ is length 2; circle $b$.
$(a,b,d)$ is of length 5; circle $d$.
$(a,b,d,c)$ is of length 6; circle $c$.
$(a,b,d,c,e)$ is of length 8; circle $e$.
$(a,b,d,c,e,z)$ is of length 11; circle $z$.

**7.** $(A,B,C,D,E,O)$ is of length 530.

**9.** $(A,D)$ is of length 3. Circle $D$.
$(A,B)$ is of length 4. Circle $B$.
$(A,D,C)$ is of length 5. Circle $C$.
$(A,E)$ is of length 5. Circle $E$.
$(A,E,Z)$ is of length 6; circle $Z$.

**11.**
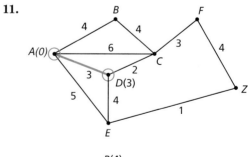

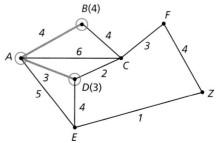

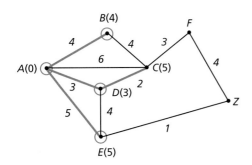

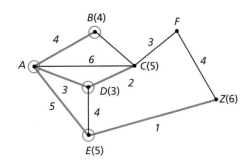

$(A,E,Z)$ is of length 6.

**13.** $(A,C,D,F,Z)$ is of length 9.

**15.** First Iteration

| V | S | P |
|---|---|---|
| a | 0 | |
| b | 2 | a |
| c | | |
| d | | |
| e | | |

Second Iteration

| V | S | P |
|---|---|---|
| a | 0 | |
| b | 2 | a |
| c | | |
| d | 5 | b |
| e | | |

Third Iteration

| V | S | P |
|---|---|---|
| a | 0 | |
| b | 2 | a |
| c | 6 | d |
| d | 5 | b |
| e | | |
| z | | |

Fourth Iteration

| V | S | P |
|---|---|---|
| a | 0 | |
| b | 2 | a |
| c | 6 | d |
| d | 5 | b |
| e | 8 | c |
| z | | |

Fifth Iteration

| V | S | P |
|---|---|---|
| a | 0 | |
| b | 2 | a |
| c | 6 | d |
| d | 5 | b |
| e | 8 | c |
| z | 11 | e |

The shortest path is $(a,b,d,c,e,z)$ of length 11.

**1.** (a) Yes

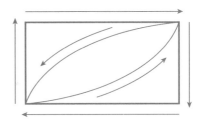

(b) No, 4 towns (vertices) of odd degree

**2.** $\triangle ABD \cong \triangle CBD$, by ASA ($\angle 1 \cong \angle 3$ and $\angle 2 \cong \angle 4$, alternate angles of a transversal cutting parallel lines; $BD = BD$)

**3.** Yes, no

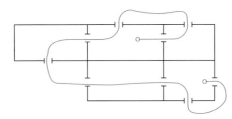

**4.** (a) 8    (b) $\sqrt{10}$

**5.** Since $\angle B \cong \angle D$ (right angles), $\triangle ABC \cong \triangle EDF$, by (SAS)

**6.** Since $\angle AOB$ and $\angle DOC$ are vertical angles and hence congruent, $\triangle AOB \cong \triangle COD$, by ASA

**7.** $\angle DBA \cong \angle EBC$ since they are vertical angles. Thus $\triangle ABD \cong \triangle CBE$, by (ASA)

$\angle A \cong \angle C$; $\overline{AB} \cong \overline{BC}$; $\overline{AE} \cong \overline{CD}$; since $\angle F$ is in both triangles, $\triangle AEF \cong \triangle CDF$, by (AAS)

**8.** By the Pythagorean Theorem $AD = DC$ so $\triangle ABD \cong \triangle CBD$, by (SSS)

**9.**     $AC = FD$ since the triangles are congruent
$AC - FC = FD - FC$
$AF = CD$

**10.** With $P$ as the center, draw an arc that intersects line $r$ in two points $A$ and $B$. With $A$ and $B$ as centers, draw arcs with the same radii, intersecting at $C$. Draw line $\overleftrightarrow{PC}$. $\overleftrightarrow{PC}$ is perpendicular to $\overleftrightarrow{AB}$.

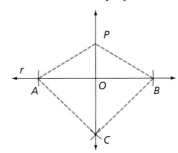

$\overline{PA} \cong \overline{PB}; \overline{AC} \cong \overline{BC}; \overline{PC} \cong \overline{PC}$. Therefore, $\triangle PBC \cong \triangle PAC$ (SSS), $\angle BCP \cong \angle ACP$, $\overline{CO} \cong \overline{CO}$. Thus, $\triangle BCO \cong \triangle ACO$, by SAS. $\angle AOC \cong \angle BOC$. m($\angle AOC$) + m($BOC$) = 180°, so m($\angle AOC$) = m($\angle BOC$) = 90°.

**11.** (a)
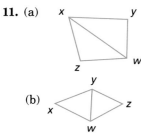

(b)

**12.** (a) Yes        (b) Yes

**13.** (a) Euler path        (b) Hamilton path

**14.** $(a,b)$ is of length 4. Circle $b$.
$(a,b,c)$ is of length 9. Circle $c$.
$(a,b,d)$ is of length 14. Circle $d$.
$(a,b,c,e)$ is of length 16. Circle $e$.
$(a,b,c,e,z)$ is of length 19.

# INDEX